TI-83 Plus Overv w

Key	Feature	Location in Text
– vs. (⁻)	Subtraction vs. Negation	Section 1.1 page 4
MATH abs	Absolute Value	Section 1.1 page 5
MATH Frac	Decimals to fraction	Section 1.2 page 16
^	Powers	Section 1.3 page 22
"E"	Scientific Notation	Section 1.3 page 28
Use of parentheses ( )	TI-83 Plus follows the order of operations	Section 2.1 page 50
MODE	GridOn vs. GridOff	Section 2.3 page 67
WINDOW	Set Xmin, Xmax, Ymin, Ymax	Section 2.3 page 67
ZOOM	Create special windows	Section 2.3 page 67
GRAPH	Display graph	Section 2.3 page 67
Arrow keys	Move free cursor	Section 2.3 page 68
Y =	Enter function	Section 2.4 page 76
2nd TblSet	Set up table	Section 2.4 page 76
2nd TABLE	Display table	Section 2.4 page 76
TRACE	Trace points on a graph	Section 2.4 page 77
Friendly Windows	Tables of Friendly Windows	Section 2.4 page 77 and pages 509–510 in appendix
2nd CALC Intersect	Find point of intersection	Section 4.3 page 150
2nd CALC Min and Max	Find turning points	Section 7.2 page 271
DRAW horizontal and vertical lines	Create a "squeegee" for domain and range; testing for a function	Section 8.2 page 311
DRAW Circle	Circle with center and radius	Section 8.2 page 312
VARS Y-VARS	Refer to functions Y1, Y2, etc.	Section 9.4 page 376
STAT	Enter data	Section 9.3 page 365
2nd StatPlot	Display data	Section 9.3 page 365
STAT CALC	Regression	Section 9.4 page 376
MODE	Dot vs. Connected	Section 11.2 page 445
Y = with shading	Graphing inequalities	Section 12.3 page 491

ELEMENTARY AND INTERMEDIATE ALGEBRA

A Practical Approach

ELEMENTARY AND INTERMEDIATE ALGEBRA

A Practical Approach

Timothy Craine
Jeffrey McGowan
Thomas Ruben

Houghton Mifflin Company Boston New York

Publisher: Jack Shira
Senior Sponsoring Editor: Lynn Cox
Development Editor: Laura Wheel
Assistant Editor: Melissa Parkin
Senior Project Editor: Kathryn Dinovo
Manufacturing Manager: Florence Cadran
Senior Marketing Manager: Ben Rivera
Marketing Assistant: Lisa Lawler

Cover image: 2003 © Russell Kaye / Getty Images

Printed in the U.S.A.

Library of Congress Control Number: 2002109424

Student Text ISBN: 0-618-10337-6
Instructor's Annotated Edition ISBN: 0-618-10339-2

123456789-QV-07 06 05 04 03

Contents

Preface

This text is a product of our experience teaching algebra to students at Central Connecticut State University. The idea for the project emerged when two of us were assigned the task of creating common final examinations for the elementary and intermediate algebra courses. This experience helped us think more clearly about what essential concepts need to be covered in the two courses. Not entirely satisfied with existing textbooks, we decided to write our own. Here are the principles we incorporated in our "practical approach" that we believe distinguish this text from others.

Brevity

Our book is brief and covers the concepts that should be mastered by all students in the algebra sequence, regardless of which mathematics courses they may subsequently take. A brief text conveys the message that students are responsible for learning all of the material and that instructors are responsible for completing the entire text. The text is designed for both semesters of a two-semester sequence and for the intensive course that combines the two courses into one semester.

Balanced Approach

We emphasize problem-solving and real-world applications of algebra without neglecting core mathematical concepts and essential symbol manipulation skills. We take full advantage of the technology of graphing calculators (the TI-83) while still requiring students to master pencil-and-paper techniques for tasks such as solving linear and quadratic equations and factoring polynomials.

Readability

We strive to make our text readable—and to set the expectation that students will read the text. In the earlier chapters the language is simple and direct. As the more complicated ideas are developed in later chapters, the prose becomes more complex. Yet throughout the text we write directly for the student, with a minimum of mathematical jargon. We introduce appropriate mathematical vocabulary only when we feel it is necessary to understand the underlying concept.

Visualization

Wherever possible we use visual images to convey mathematical ideas. In addition to using graphs extensively as a problem-solving tool, we also encourage students to model problem-solving situations through the use of pictures and diagrams.

Support for Instructors

This textbook may be adapted to a variety of teaching styles. Because we want the text to be readable, we expect teachers to incorporate reading the text into their homework assignments. We hope that our emphasis on visual images and encouragement of mental mathematics will prompt teachers to incorporate these strategies into their teaching.

We have written a detailed instructor's guide, containing teaching tips for each section of the text. In this way we can help instructors make more effective use of this book, regardless of their individual teaching styles. At the suggestion of reviewers we added eight short optional sections that may be included at the instructor's disretion.

Features of the Text

Chapter Opener and Exploration

Each chapter opens with an application of a concept introduced in the chapter. An exploration at the end of the chapter gives students the opportunity to apply the concept to answer a specific question related to the opener.

Exercises

At the end of each section are 15 exercises on the new material and 10 skill and review exercises. We typically assign all 25 exercises for homework. Students needing more practice may be assigned exercises at the end of the chapter or referred to additional exercises on the web site.

Key Concepts

Important definitions, facts, and procedures are summarized at the end of each chapter in a section titled *Key Concepts*.

Chapter Review and Test

Review questions at the end of each chapter, organized by section, provide additional practice on the objectives of each section. The chapter test includes items covering all of the essential sections of the chapter.

Calculator Integration

TI-83 graphing calculators are used throughout. As new features are required, their use is explained in the body of the text. A chart on the page facing the inside cover guides students to the place in the text where various features are first introduced.

A guide for students using TI-85, TI-86, or TI-89 calculators is included in Appendix B.

Transition from Elementary to Intermediate Algebra

Students who use the text only for intermediate algebra are advised to consult Appendix A, which shows how skills and concepts introduced in the first six chapters are applied in the second half of the book.

Exercise Answers

Answers to odd numbered exercises are found in the back of the student textbook.

Instructor Resources

Elementary and Intermediate Algebra: A Practical Approach has a complete set of teaching aids for the instructor.

Instructor's Annotated Edition This edition contains a replica of the student text and additional items just for the instructor. Answers to all exercises are provided.

Instructor's Resource Manual with Solutions The *Instructor's Resource Manual* includes a section-by-section overview, a list of all Examples, Facts and Definitions, Vocabulary, and Symbols covered in each section. Helpful Teaching Tips are also offered on a section-by-section basis. The *Solutions* contains worked-out solutions for all exercises in the text.

HM ClassPrep w/ HM Testing CD-ROM *HM ClassPrep* contains a multitude of text-specific resources for instructors to use to enhance the classroom experience. These resources can be easily accessed by chapter or resource type and can also link you to the text's web site. *HM Testing* is our computerized test generator and contains a database of algorithmic test items as well as providing **on-line testing** and **gradebook** functions.

Instructor Text-specific website The resources available on the *Class Prep CD* are also available on the instructor web site at math.college.hmco.com/instructors. Appropriate items are password protected. Instructors also have access to the student part of the text's web site.

Student Resources

Student Study Guide This manual contains study tips for students as well as complete solutions to all odd-numbered exercises in the text.

Math Study Skills Workbook ***by Paul D. Nolting*** This workbook is designed to reinforce skills and minimize frustration for students in any math class, lab, or study skills course. It offers a wealth of study tips and sound advice on note taking, time management, and reducing math anxiety. In addition, numerous opportunities for self-assessment enable students to track their own progress.

HM eduSpace® online learning environment *eduSpace®* is a text-specific online learning environment which combines an algorithmic tutorial program with homework capabilities. Specific content is available 24 hours a day to help you further understand your textbook.

SMARTHINKING live, on-line tutoring Houghton Mifflin has partnered with SMARTHINKING to provide an easy-to-use and effective on-line tutorial service. **Whiteboard Simulations** and **Practice Area** promote real-time visual interaction.

Three levels of service are offered.

- **Text-specific Tutoring** provides real-time, one-on-one instruction with a specially qualified 'instructor.'
- **Questions Any Time** allows students to submit questions to the tutor outside the scheduled hours and receive a reply within 24 hours.
- **Independent Study Resources** connect students with around-the-clock access to additional educational services, including interactive web sites, diagnostic tests and Frequently Asked Questions posed to SMARTHINKING e-structors.

Houghton Mifflin Instructional Videos and DVDs This text offers text-specific videos and DVDs, hosted by Dana Mosely, covering all sections of the text and providing a valuable resource for further instruction and review.

Next to every objective head, the  icon serves as a reminder that the objective is covered in a video/DVD lesson.

Student Text-specific web site On-line student resources can be found at this text's website at math.college.hmco.com/students.

The Great

John Nash

Acknowledgments

We would like to thank our colleagues who taught at least one section of elementary or intermediate algebra at Central Connecticut State University from 2000 through 2003, using a pilot edition of this text. We appreciate their many constructive suggestions.

Ali Antar
David Applegate
Richard Barton
Brian Benigni
Lee Braken
Timothy Brzezinski
Lisa Braverman
Isaac Brobbey
Nancy Castaneda
Marylou Chadziewicz
Richard Charette
Marvett Cobourne
Karen Collin
Marian Collins
Alan Dingfelder
John Driscoll
Marie Dubois
Mohamed El-Kafrawy
Nabawia El-Ramly
Samia Elsafty
Richard Feinn
William Gale
Ivan Gotchev
Tatiana Gotcheva
Jane Hurwitz
Brenda Janik
Ann Johnson
Jane Keleher
Denise Kelly
Richard Kramer
Kris Larsen
Sheryl Leone
Tara Leonetti
Sally Lesik
Elliston Little
Deborah Litwinko
Judyth Marzi
Adele Miller
Hendree Milward
Kunle Olumide
Carrie O'Neil
Lynn Page
Jane Piorkowski
Kurt Ragis
Eugene Rohr
William Sarmuk
Sylvia Schindelman
Angela Shaw
Robert Smellie
Hermine Smikle
Kathleen Stankewicz
Alison Stotz
Timothy Talmadge
Robert Vaden-Goad
Alvin Vaz
Robert Voytek
Danijela Vujic
Thomas Watson
Charles Waiveris
Grace Wright
Karlo Zvonarek

We also thank their students, who helped us find typos in the pilot edition.

We appreciate many colleagues from other institutions who reviewed the manuscript and offered advice.

Brenda Ammons, *Pellissippi State Technical Community College, TN*

Kenneth R. Anderson, *Chemeketa Community College, OR*

Ron Beeler, *Southern Illinois University Edwardsville, IL*

Alice Burstein, *Middlesex Community College, CT*

Shelia Chewning, *Germanna Community College, VA*

Robert Decker, *University of Hartford, CT*

Elaine Dinto, *Naugatuck Valley Community College, CT*

Jennifer M. Dollar, *Grand Rapids Community College, MI*

Maggie Williams Flint, *Northeast State Technical Community College, TN*

Barbara A. Gentry, *Parkland College, IL*

Meredith Anne S. Higgs, *Middle Tennessee State University, TN*

Ben Hill, *Lane Community College, OR*

William L. Hoard, *Front Range Community College Larimer Campus, CO*

Byron D. Hunter, *College of Lake County, IL*

Maureen Shields Kelley, *Northern Essex Community College, MA*

Doug Mace, *Kirtland Community College*

Elizabeth A. Mefford, *Walters State Community College, TN*

Kathryn S. Plum, *DeAnza College, CA*

Judith Pranger, *Binghamton University, NY*

Barbara Sausen, *Fresno City College, CA*

John Searcy

Lauri Semarne

April Strom, *Glendale Community College, AZ*

Katherine R. Struve, *Columbus State Community College, OH*

Jane H. Theiling, *Dyersburg State Community College, TN*

Dr. Bettie A. Truit, *Black Hawk College, IL*

Christi Verity

Cindy L. Wilson, *Henderson State University, AR*

We are grateful to Laurie Coulter of Houghton Mifflin Custom Publishing for publishing the pilot edition.

We thank Tim Brzezinski, Class of 2002, Central Connecticut State University, for writing the initial version of the student solution manual.

Michelle Lynes, administrative assistant, Department of Mathematical Sciences at CCSU provided invaluable assistance with the logistics of pilot testing.

We appreciate the careful attention given by Leslie Galen of Integre Technical Publishing Company to the design and page layout of the text.

We would also like to thank some of the many teachers who have inspired us over the years: Richard Powell, Ken Hoffman, David Kelly, Józef Dodziuk, Diane Guerette, and Eugene Smith.

Finally we must thank members of our respective families who supported us throughout the process of writing, rewriting, and editing the manuscript: Leslie Craine; Kim Sobel, Benny and Simcha McGowan; Amy and Benjamin Ruben.

TIM CRAINE
JEFF MCGOWAN
TOM RUBEN

Chapter Opening Features

CHAPTER

5

More Applications of Linear Equations

Coffee, Anyone?

The student center at a major university has just been renovated. There is now additional space to lease to commercial enterprises. A distributor of gourmet coffee is considering opening a franchise at the student center. However, the executives of this company want to know whether this is a good location for their business.

The coffee distributor hires a marketing firm to conduct a survey. The firm selects a random sample of students from the university and asks students in this sample whether or not they would patronize a gourmet coffee shop if one were opened on the campus. The results of this poll are used by the company to make its decision. In this chapter you learn how to apply proportions and percents to analyze information obtained from samples. (See Exploration on p. 206.)

Need help? For on-line resources, visit this web site: **math.college.hmco.c**

174

Chapter Opener

Each chapter opens with a list of the objectives covered in the chapter, an application linking chapter topics to a real world example, and a photo to enhance the application. The sections are listed clearly and serve as an organizational tool for the chapter. The same section titles appear throughout the chapter, along with objectives listed below them. Students are able to use these objectives as a guide for what they should learn in each section.

Exploration

Explorations at the end of the chapter give students the opportunity to work problems and answer questions related to the application introduced in the Chapter Opener. By using the strategies they were taught within the chapter, students are able to complete the Explorations and connect the mathematics to real world experiences.

206 ■ Chapter 5 More Applications of Linear Equations

Mixtures and Investments These problems often involve equations in standard form $Ax + By = C$. (Section 5.3)

Exploration

A sample of 500 students at a university are polled to see if they would patronize a proposed gourmet coffee shop. Of the students polled, 119 say yes. The total student population of the university is 18,500.

a. Set up and solve a proportion to estimate how many students would patronize the gourmet coffee shop.

b. What percent of the student population does the poll indicate would patronize the coffee shop?

c. All polls have a *margin of error*. The actual percentage in the population is likely to fall within a range determined by the sample percentage plus or minus the margin of error. For this poll the margin of error is 3.7%. Find the range of percents for this poll.

d. Given the margin of error, find the maximum and the minimum number of students (from the population) who are likely to patronize the coffee shop.

CHAPTER 5 • REVIEW EXERCISES

Section 5.1 Proportions

1. Use the cross-multiplication property to solve each proportion.

a. $\frac{10}{4} = \frac{x}{18}$ **b.** $\frac{9}{x} = \frac{2}{3}$

c. $\frac{x}{5} = 3$ **d.** $\frac{5}{2} = \frac{8}{x+1}$

2. In Example 3 we found the proportion

$$\frac{9 \text{ dollars}}{6 \text{ gallons}} = \frac{c \text{ dollars}}{20 \text{ gallons}}$$

Write two other proportions that are also correct. *Note:* See setting up proportions on page 180 for guidelines on how to write a proportion.

3. A formula for Celsius temperature in terms of Fahrenheit temperature is

$$C = \frac{F - 32}{1.8}$$

Notice that this equation is a proportion. Find the Fahrenheit temperature when the Celsius temperature is 20°.

4. The cost of heating oil is directly proportional to the number of gallons purchased. The Richards spent \$210 on 175 gallons of heating oil. How much does 1 gallon cost? Use this unit price to find the cost of 150 gallons of heating oil.

Use proportions to solve Exercises 5–8.

5. At a certain time of day a 5-foot person casts a 4-foot shadow. How tall is a building that casts a 17-foot shadow?

6. At a local gas station, 15 gallons of regular gasoline cost \$25. How many gallons of regular gasoline may be bought for \$18?

7. Sid can perform a job in 2 hours. Joanne isn't nearly as skillful, so the same job takes her 6 hours. Suppose they work together on this job.

Student Pedagogy

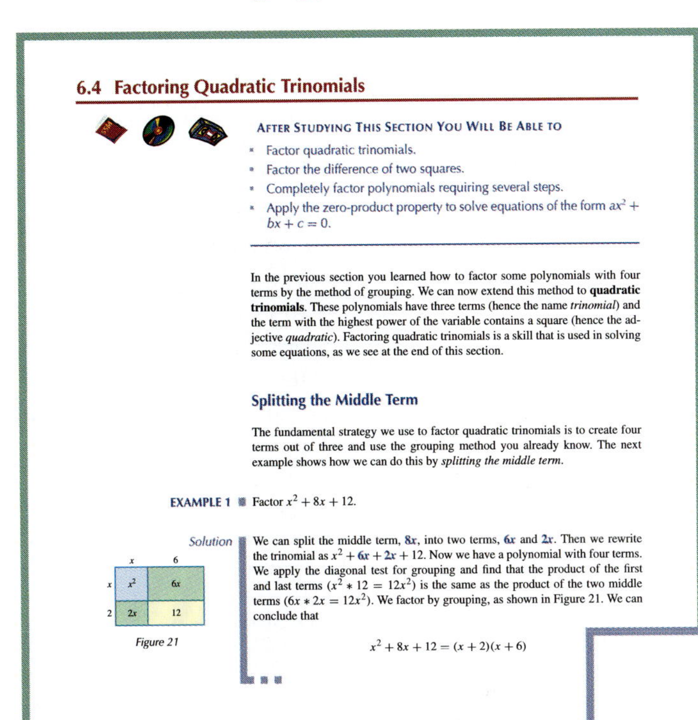

6.4 Factoring Quadratic Trinomials

After Studying This Section You Will Be Able To

- Factor quadratic trinomials.
- Factor the difference of two squares.
- Completely factor polynomials requiring several steps.
- Apply the zero-product property to solve equations of the form $ax^2 + bx + c = 0$.

In the previous section you learned how to factor some polynomials with four terms by the method of grouping. We can now extend this method to **quadratic trinomials**. These polynomials have three terms (hence the name *trinomial*) and the term with the highest power of the variable contains a square (hence the adjective *quadratic*). Factoring quadratic trinomials is a skill that is used in solving some equations, as we see at the end of this section.

Splitting the Middle Term

The fundamental strategy we use to factor quadratic trinomials is to create four terms out of three and use the grouping method you already know. The next example shows how we can do this by *splitting the middle term.*

EXAMPLE 1 Factor $x^2 + 8x + 12$.

Solution We can split the middle term, $8x$, into two terms, $6x$ and $2x$. Then we rewrite the trinomial as $x^2 + 6x + 2x + 12$. Now we have a polynomial with four terms. We apply the diagonal test for grouping and find that the product of the first and last terms ($x^2 * 12 = 12x^2$) is the same as the product of the two middle terms ($6x * 2x = 12x^2$). We factor by grouping, as shown in Figure 21. We can conclude that

$$x^2 + 8x + 12 = (x + 2)(x + 6)$$

Figure 21

Icons

for the Student Solutions Manual 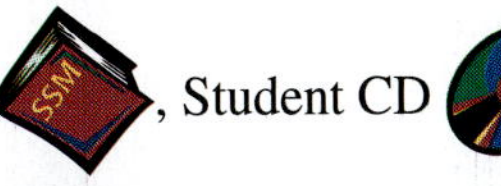, Student CD , and Videos appear at the beginning of every section. These make students aware of the additional study resources that are available for each section.

Key Terms

Bolded and italicized terms appear throughout the chapters to highlight key points. The bolded terms are also reviewed in the Key Concepts section at the end of each chapter.

7.3 Algebraic Techniques for Solving Quadratic Equations ■ 277

This is one version of the *quadratic formula.*

Definition The Quadratic Formula

Any solutions to a quadratic equation $ax^2 + bx + c = 0$ are given by the **quadratic formula**,

$$x = -\frac{b}{2a} \pm \frac{\sqrt{b^2 - 4ac}}{2a}$$

One important thing to notice is that by taking the square root in the quadratic formula, we have several different possibilities for how many solutions we get.

Definition Discriminant

The term $b^2 - 4ac$ found inside the radical in the quadratic formula is called the **discriminant**.

1. If the discriminant is positive, we get two real number solutions, or **roots**.
2. If the discriminant is zero, we get one real number solution, $x = \frac{-b}{2a}$.
3. If the discriminant is negative, we get *no* real number solution, because we are not able to take the square root of a negative number.

Note: There is no *real number* solution if the discriminant is negative; however, if we add $i = \sqrt{-1}$ to our number system, we will have imaginary, or complex, number solutions. Imaginary numbers are discussed in Section 7.5.

Definition and Property Boxes

are incorporated into each chapter to emphasize important terms and formulas.

Notes and Cautions

appear, as needed, to keep students focused and to make them aware of possible common errors.

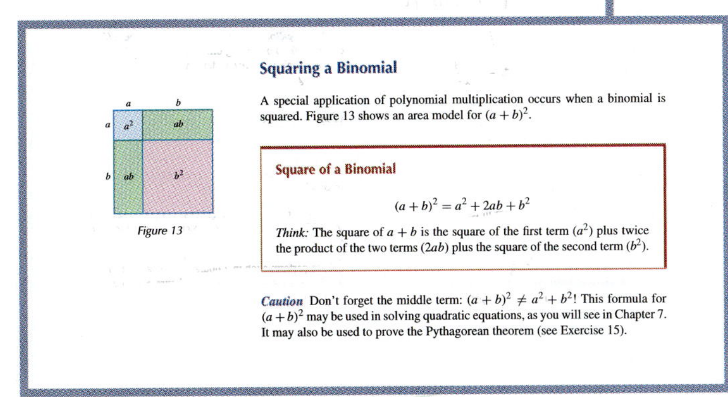

Squaring a Binomial

A special application of polynomial multiplication occurs when a binomial is squared. Figure 13 shows an area model for $(a + b)^2$.

Square of a Binomial

$$(a + b)^2 = a^2 + 2ab + b^2$$

Think: The square of $a + b$ is the square of the first term (a^2) plus twice the product of the two terms ($2ab$) plus the square of the second term (b^2).

Caution Don't forget the middle term: $(a + b)^2 \neq a^2 + b^2$! This formula for $(a + b)^2$ may be used in solving quadratic equations, as you will see in Chapter 7. It may also be used to prove the Pythagorean theorem (see Exercise 15).

Figure 13

Applications

Example and Solution

Example and Solution Applications are used throughout the material to enhance understanding of the mathematics. Examples of realistic situations are presented, followed by step-by-step solutions.

EXAMPLE 4 Many graphs we see in everyday life do not look like most of the graphs in this book. Statistical information is often presented in a graphical format, because it is easy to scan a graph quickly while you are reading the newspaper but much more difficult to read through a table of numbers. For example, *bar graphs* are often used to display data, as in Figure 14. Does the bar graph in Figure 14 represent a function?

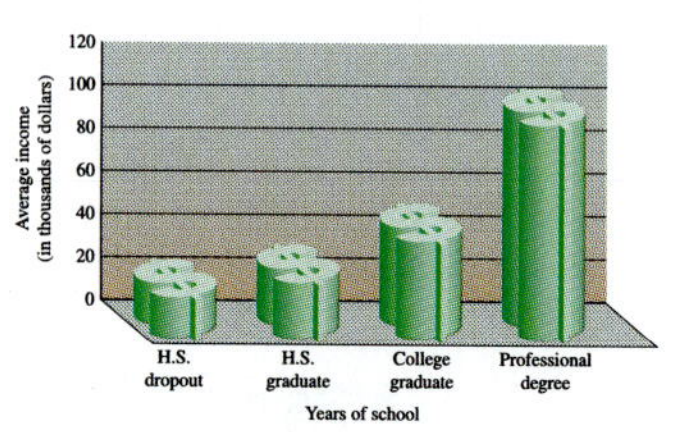

Figure 14 There are many graphical ways to represent a function. A bar graph is very useful when the input set is finite.

Solution The graph in Figure 14 shows how average annual income is related to educational level (data from the U.S. Census Bureau). To see that this is a function, let the domain be the set

{H.S. dropout, H.S. graduate, college graduate, professional degree}

The elements in this set indicate categories based on the level of education completed.

The range of the function is a finite set of numbers,

{\$18,900, \$25,900, \$45,400, \$99,300}

indicating income in dollars. We could certainly have put this information in a table, but it is immediately obvious from the graph that income rises with schooling, and the pattern in that rise is also obvious.

Exercises 7.4

1. Review Example 1 and explain why knowing the vertex is useful for picking a friendly window for viewing the graph.
2. Suppose the second astronaut in Example 1 throws the ball back to the first. Because her height and arm strength are different, the equation is different. The height of the ball is now given by the equation $h = {}^{-}2.7t^2 + 35t + 7$.
   a. Find the vertex.
   b. Use the vertex to pick a friendly window for viewing the graph.
3. In Example 2 the solutions to the equation $2x^2 + 10x - 600 = 0$ were $x = {}^{-}20$ and $x = 15$. Where do these solutions appear on the graph of $y = 2x^2 + 10x - 600$?
4. Outline the seven steps (*understand, visualize, assign* variable(s), *write* equation(s), *solve* equation(s), *answer* the question, *check* the answer) used to solve the problem in Example 3(a).
5. Why did we find the vertex in Example 3 and not the x-intercepts?
6. In Example 3 suppose we have 160 feet of fence and the gate is 4 feet wide. Use the seven-step process to find the dimensions of the dog run that maximizes the area.
7. Use the seven steps for problem solving to outline the work performed in Example 4.
8. Why did we find the x-intercepts in Example 4 and not the vertex?
9. Suppose the skid marks in Example 4 were 150 feet. How fast was the car going?
10. Assume a person can throw a baseball 25 m high on Earth. The same person's throw on the moon can be modeled by the equation $h = {}^{-}.82t^2 + 22t + 1.7$, where h is the height in meters above the moon after t seconds.
    a. How many times higher is the maximum height on the moon than on Earth?
    b. What does this indicate about the strength of gravity on the moon in relation to Earth?
11. Suppose an advanced civilization has a spaceship in orbit around our sun and they drop a probe into the sun to do research (their probe should withstand the heat of the sun). The equation for the height of the probe is $h = {}^{-}448t^2 + 1{,}056{,}000$, where h is the distance in feet from the sun after t seconds from release.
    a. How far is the spaceship from the sun at release?
    b. How many seconds would it take the probe to hit the sun?
    c. Suppose the probe is not as well designed as thought and would melt at 100 miles from the sun. How many seconds after release would the probe melt?

Section Exercises

Each section ends with two types of problems.

Exercises and Skills and Review

reinforce what students have learned. Exercises are comprehensive, manageable in length, and intended to be assigned in full.

Skills and Review 7.4

16. Find the exact vertex for the graph of $y = 2x^2 - 12x + 1$.
17. Use the quadratic formula or the discriminant to show that the graph of the equation $y = 4x^2 + x + 5$ has no x-intercepts.
18. Let $y = x^2 + 8x + 6$.
    a. Find the y-intercept.
    b. Approximate the x-intercepts.
    c. Find the exact vertex.
    d. Use the information from the preceding parts to sketch a graph by hand.
19. Use a graph to approximate the solutions to the equation $x^2 + (x + 6)^2 = 196$.
20. Solve exactly the equation $4x^2 - 25 = 0$.
21. Simplify the expression and write without negative exponents:
$$\left(\frac{x^5}{x^2y}\right)^{-3}.$$
22. A snowboard that normally sells for \$250 is on sale for \$210. Find the percent discount.
23. Solve the equation
$$\frac{x+5}{6} = \frac{1}{3}.$$
24. Solve the system of equations by either elimination or substitution
$$x + 2y = 52$$
$$3x - y = 23$$
25. Locate $\frac{1}{3}$, 2.7, ${}^{-}4\frac{2}{5}$, and $\frac{11}{4}$ on a number line.

Graphing Calculator

$$\text{slope} = \frac{307.2 - 172.8}{16 - 12}$$
$$= \frac{134.4}{4}$$
$$= 33.6 \qquad \textit{Slope from point B to point C}$$
$$\text{slope} = \frac{480 - 307.2}{20 - 16}$$
$$= \frac{172.8}{4}$$
$$= 43.2 \qquad \textit{Slope from point C to point D}$$

The slopes are increasing each time, and the points don't fit together to form a straight line. We can interpret this using rate of change; as the pizzas get bigger, the rate of which you need to add cheese *increases*.

EXAMPLE 3 Rachel and Mike think that if they use 250 g of cheese on a pizza, they could sell the pizza for \$10 and make a profit.

a. Use a graph to estimate how big a pizza they can make with 250 g of cheese.
b. Solve an equation to find how big a pizza they can make with 250 g of cheese.

Solution **a.** We can use our formula from Example 1(b). Substitute 250 into cheese $= 1.2d^2$.

$$250 = 1.2d^2$$

You may enter the formula cheese $= 1.2d^2$ into your calculator as Y1 = 1.2X ∧ 2, and the graph (Figure 3) should help you solve for x. You may use the window from Example 1(c): Xmin = 0, Xmax = 23.5, Ymin = 0, Ymax = 500.

Using the TRACE key to move along the graph until Y is about 250, you see that X should be between 14 and 15.

If y is 250
x is between 14 and 15

Figure 3 You can trace along the graph to find the approximate solution point.

b. We can get a more precise answer algebraically.

$$250 = 1.2d^2$$
$$\frac{250}{1.2} = d^2 \qquad \textit{Dividing by 1.2 on both sides}$$
$$208.3 \approx d^2 \qquad \textit{Rounding to the nearest .1}$$
$$\sqrt{208.3} \approx \sqrt{d^2} \qquad \textit{Taking a square root on both side}$$
$$14.4 \approx d$$

Rachel and Mike decide to make pizzas about 14.4 inches in diam

Graphing Calculator

examples are incorporated throughout the chapters to guide students through problems. Screens are shown directly on the page to provide a visual example. These also help students follow along step-by-step on their own graphing calculator.

5.

6.

7. Let $^-y^2 = x$.
   a. Complete a table of values for $y = {}^-2, {}^-1, 0, 1, 2$.
   b. Draw a graph by hand.
   c. Is this a function? Explain.
8. Let $|y| = x$.
   a. Complete a table of values for $y = {}^-2, {}^-1, 0, 1, 2$.
   b. Draw a graph by hand.
   c. Is this a function? Explain.
9. Here is a bar graph of per capita carbonated soft-drink consumption for the years 1996 through 2001. (http://www.bevnet.com)

Individual Soda Consumption by Year

Gallons; Year: 1996, 1997, 1998, 1999, 2000, 2001

   a. Is this a graph of a function?
   b. Estimate the soda consumption for each year and write ordered pairs in the form of (year, soda consumption).
   c. Express the domain and range using set notation.
10. The dot plot shows the ages of children from three families labeled A, B, and C. Does the graph represent a function? Explain.

Age of Children from Three Different Families

Age; Family: A, B, C

11. Match the domain and range with the appropriate graph.
   a. Domain $= \{x \mid {}^-\infty < x < \infty\}$; Range $= \{y \mid y > {}^-3\}$
   b. Domain = (number line, $^-3$); Range = (number line, $^-2$)
   c. Domain $= \{x \mid -\infty < x < \infty\}$; Range $= \{y \mid y = {}^-3\}$

1.

2.

3.

In Exercises 12–15, express the domain and range of each function on a number line and in set builder notation.

The graphing theme is carried throughout section and chapter-end material.

End of Chapter

Key Concepts

A chapter summary is provided at the end of every chapter. A list of all essential information, with definitions and page references, serves as a study and review tool for students before moving on to the Review Exercises and Chapter Test.

CHAPTER 11 ▪ KEY CONCEPTS

Rational Expression A rational expression is a fraction in which both the numerator and denominator are polynomials. (Page 426)

Equivalent Rational Expressions Two rational expressions are equivalent if one is found from the other by multiplying numerator and denominator by the same expression. (Page 426)

Combining Rational Expressions The sum or difference of two rational expressions may be expressed as a single rational expression. To add or subtract the expressions, a common denominator must be found. (Page 429)

Simplifying Rational Expressions A rational expression may be simplified by removing a common factor from both numerator and denominator. (Page 434)

Rational Function A rational function is a function of the form $f(x) =$ rational expression in x. (Page 438)

Hyperbola The graph of the function $y = f(x) = \frac{1}{x}$ is called a hyperbola. When the graph is stretched, shifted, or reflected across the x-axis, the curve remains a hyperbola. These hyperbolas have one horizontal asymptote and one vertical asymptote. (Page 440)

Asymptote If a curve approaches a line when it is traced, the line is called an asymptote to the curve. (Page 440)

Rational Equation An equation in which an unknown appears in one or more denominators is called a rational equation. (Page 447)

Applications of Rational Equations Work-rate and distance-rate-time problems are often solved using rational equations. (Pages 449, 452)

Indirect Variation If $b < 0$, then a function of the form $y = ax^b$ is called an indirect variation function. We can say that y varies indirectly as x. The number a is called the **variation constant**. Sometimes indirect variation is called **inverse variation.** (Page 459)

Characteristics of Indirect Variation In a real-world situation, indirect variation may be a good model if the following conditions apply:

1. For positive values of x, as x increases, y decreases.
2. As x gets close to zero, y becomes very large, that is y increases without bound.

The graph of an indirect variation function is a downward sloping curve in the first quadrant. Both the x-axis and the y-axis are asymptotes to this curve. (Page 459)

Solving Indirect Variation Problems (Page 460)

Step 1 Translate a verbal statement into a general variation equation.
Step 2 Use one pair of values to solve for the variation constant. Then write a specific variation equation with this constant.
Step 3 Use the specific equation to find an unknown value.

CHAPTER 11 ▪ REVIEW EXERCISES

Section 11.1 Rational Expressions

1. For each expression, find two equivalent rational expressions. *Note:* There are many possible answers.
   a. $\frac{5}{6}$ b. $\frac{x^3}{y^2}$
   c. $\frac{x-4}{x+5}$
2. Show that
$$\frac{x-4}{x+5} \text{ and } \frac{x^2-3x-4}{x^2+6x+5}$$
are equal when substituting the following values of x: $^-4, 3, 10$. Why are the two expressions not equal when $x = {}^-1$?
3. Find the missing numerators.
   a. $\frac{2}{5} = \frac{?}{15}$ b. $\frac{y}{x} = \frac{?}{x^2y}$
   c. $\frac{x-1}{x+3} = \frac{?}{x^2+3x}$
4. Express each sum or difference as a single rational expression.
   a. $\frac{2}{x} + \frac{3}{x^4}$
   b. $\frac{1}{x-2} - \frac{3x}{x^2-5x+6}$
   c. $\frac{4}{y} - \frac{1}{x}$
   d. $\frac{5}{x+1} + \frac{2}{x-4}$
   e. $\frac{3}{yz^2} + \frac{8}{y^2z^3}$
   f. $\frac{4}{x^2+2x} - \frac{1}{x^2+4x+4}$
5. Simplify each rational expression.
   a. $\frac{6xy^3}{9x^3y^2}$ b. $\frac{x^2-36}{x^2+6x}$
   c. $\frac{x^2-9}{x^2-6x+9}$ d. $\frac{x^2+7x}{x(x+7)^2}$
   e. $\frac{x-1}{1-x}$ (*Hint:* Factor out a $^-1$ from the denominator.)
6. Perform the indicated operation and simplify.
   a. $\frac{2x^3}{5y^3} \div \frac{x^4}{10y^2}$
   b. $\frac{15x^2}{x^2-x-12} * \frac{x-4}{3x}$
   c. $\frac{7x}{x+7} * \frac{x^2+15x+56}{2x^2+16x}$
   d. $\frac{x^2}{3x-24} \div \frac{6x^3}{x^2-64}$

Review Exercises

Review Exercises appear at the end of each chapter. These provide a complete review of all topics covered and allow students to link together sections and learning objectives.

Chapter Test

The Chapter Test exercises are designed to simulate a possible test of the material in the chapter.

CHAPTER 11 ▪ TEST

1. Express the sum as a single rational expression:
$$\frac{5}{x+2} + \frac{1}{x+3}.$$
2. Express the difference as a single rational expression:
$$\frac{1}{x^2-4} - \frac{3}{x^2-2x-8}.$$
3. Perform the multiplication and simplify:
$$\frac{x^2-x}{x^2} * \frac{(x+1)^2}{x^2-1}.$$
4. Perform the division and simplify:
$$\frac{x+5}{6-x} \div \frac{x^2+10x+25}{(x-6)^2}.$$

CHAPTER

1

Geometry and Numbers

The Ups and Downs of Wall Street

Investors pay close attention to how stock markets around the world are performing. In the United States the most closely watched market is the New York Stock Exchange, located on Wall Street in Manhattan. Every evening newscasters report on how the major stock indices such as the Dow Jones, the NASDAQ, and the S & P 500 performed that day. These numbers, which represent an average of many individual stocks, are usually reported in terms of change, that is, whether the stock price is up or down that day and by how many points. Thus investors are accustomed to thinking in terms of positive and negative numbers.

For many years individual shares traded on the New York Stock Exchange were sold in increments of $\frac{1}{8}$ of a dollar. A typical price for a share might be quoted as $42\frac{5}{8}$. At the beginning of the 21st century, however, the Exchange converted from fractions to decimals. Prices of stocks are now quoted to the nearest cent. (See Exploration on p. 42.)

Need help? For on-line resources, visit this web site: **math.college.hmco.com/students.**

1.1 Positive and Negative Numbers

After Studying This Section You Will Be Able To

- Add, subtract, multiply, and divide signed numbers.
- Use models to represent signed numbers.

If you have a checking account, you probably receive a statement from the bank at the end of each month showing the total amount of money you deposited and the amount you withdrew from the bank by writing checks. The deposits may be considered **positive numbers** and the checks, **negative numbers**. If the total of the checks is more than the total of the deposits, you may end up with a negative balance, which means at least one of your checks will bounce.

Positive and negative numbers are also called signed numbers. Positive numbers may be shown with a raised plus sign (e.g., $^{+}3$) and negative numbers, with a raised "minus" sign (e.g., $^{-}4$).

One application of signed numbers is in chemistry, where particles may carry a positive or negative charge. A single positive charge and a single negative charge neutralize each other.

EXAMPLE 1

a. An oxide ion consists of 8 protons, with positive charges, and 10 electrons, with negative charges. Find the charge on an oxide ion.

b. An aluminum ion consists of 13 protons and 10 electrons. Find the charge on an aluminum ion.

Solution

a. This situation is represented in Figure 1. Notice how pairs of positive and negative charges are lined up. There are two negative charges that are not paired. The net charge for the oxide ion is $^{-}\mathbf{2}$.

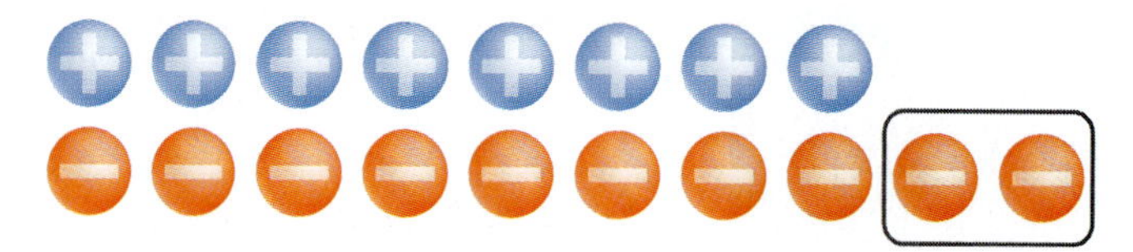

Figure 1

The combination of the charges may be written as a sum: $^{+}\mathbf{8} + {}^{-}\mathbf{10} = {}^{-}\mathbf{2}$. (Read "positive 8 plus negative 10 equals negative 2.")

b. This situation is represented in Figure 2. There are three positive charges that are not paired. The net charge for the aluminum ion is $^{+}\mathbf{3}$ because $^{+}\mathbf{13} + {}^{-}\mathbf{10} = {}^{+}\mathbf{3}$.

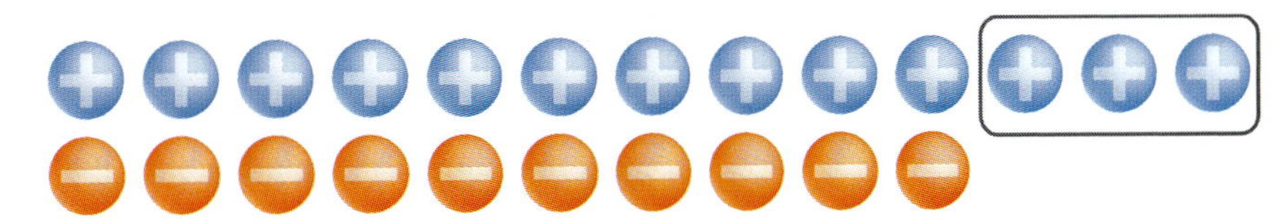

Figure 2

EXAMPLE 2 Find the net charge when you combine the ions.

a. A calcium ion (charge $^{+}2$) with a sodium ion (charge $^{+}1$)

b. A sulfate ion (charge $^{-}2$) with a phosphate ion (charge $^{-}3$)

c. A sodium ion (charge $^{+}1$) with a chloride ion (charge $^{-}1$)

Solution

a. This combination may be written $^{+}2 + {}^{+}1 = {}^{+}3$. The net charge is $^{+}3$, as shown in Figure 3.

Figure 3

b. This combination may be written $^{-}2 + {}^{-}3 = {}^{-}5$. The net charge is $^{-}5$, as shown in Figure 4.

Figure 4

c. $^{+}1 + {}^{-}1 = 0$. The net charge is zero. One sodium ion neutralizes one chlorine ion. See Figure 5. These two ions are found in pairs in the compound sodium chloride, which is ordinary table salt.

Figure 5

Absolute Value

Every number has an **absolute value**. In Examples 1 and 2 the absolute value of each group of like charges may be thought of as the number of particles. Thus, in Example 1(a) the absolute value of $^{+}8$ is 8 and the absolute value of $^{-}10$ is 10. When these two charges are combined, pairs of positives and negatives cancel each other out, leaving a net charge of $^{-}2$. Its absolute value is 2.

A common way of writing absolute value in algebra is to use two vertical lines: $| \, |$. You will often see these absolute value statements written as

$$|{}^{+}8| = 8 \qquad |{}^{-}10| = 10 \qquad |{}^{-}2| = 2$$

Adding Signed Numbers

Examples 1 and 2 illustrate the addition rules for positive and negative numbers.

Adding Signed Numbers

To add (or combine) two signed numbers:

- If the numbers have the *same* sign, *add* their absolute values and keep the sign.

(continued)

- If the numbers have *opposite* signs, *subtract* their absolute values and take the sign of the number with the *larger* absolute value.

Example 3 illustrates how to use the addition rules.

EXAMPLE 3 Use the addition rules for signed numbers to find these sums:

a. $^{-}23 + {}^{-}15$
b. $^{+}426 + {}^{-}782$
c. $^{+}\frac{1}{2} + {}^{+}\frac{1}{4}$
d. $^{+}91.2 + {}^{-}43.913$

Solution

a. The numbers have the same sign. Add their absolute values ($23 + 15 = 38$) and keep the sign. The answer is $^{-}38$ because both numbers are negative.

b. The numbers have opposite signs. Subtract their absolute values ($782 - 426 = 356$) and then take the sign of the number with larger absolute value ($^{-}782$). The answer is $^{-}356$.

c. The numbers have the same sign. Add their absolute values ($\frac{1}{2} + \frac{1}{4} = \frac{2}{4} + \frac{1}{4} = \frac{3}{4}$) and keep the sign. The answer is $^{+}\frac{3}{4}$ because both numbers are positive.

d. The numbers have opposite signs. Subtract their absolute values ($91.2 - 43.913 = 47.287$) and then take the sign of the number with larger absolute value. The answer is $^{+}47.287$.

Note: In parts (a) and (b) the numbers are **integers**. Integers include positive and negative whole numbers as well as zero. In parts (c) and (d) we observe that the rules for adding signed numbers apply to fractions and decimals, not just to integers.

Using Calculators

On most calculators the raised plus sign ($^{+}$) is omitted. Positive numbers are entered without signs. In this book we will usually omit the plus sign; that is, we will write 7 rather than $^{+}7$.

There is a special key, (−), for entering negative numbers.

Caution: Do not confuse the negative key with the subtraction key:

Although you may use a calculator to find the sum of two signed numbers, you should be able to do relatively simple problems such as $^{-}12 + 3$ using mental math and the rules for addition of signed numbers.

Calculators and computers often abbreviate absolute value as *abs*. On the TI-83, abs is found under the MATH NUM menu. If you enter ABS(8) the calculator will display 8. If you enter ABS($^-8$) it will also display 8.

The Number Line Model

As you saw in Examples 1 and 2, one model of signed numbers is charged particles. Another model for signed numbers is a number line. When a horizontal line is used, positive numbers are to the right of 0 and negative numbers are to the left of 0. See Figure 6.

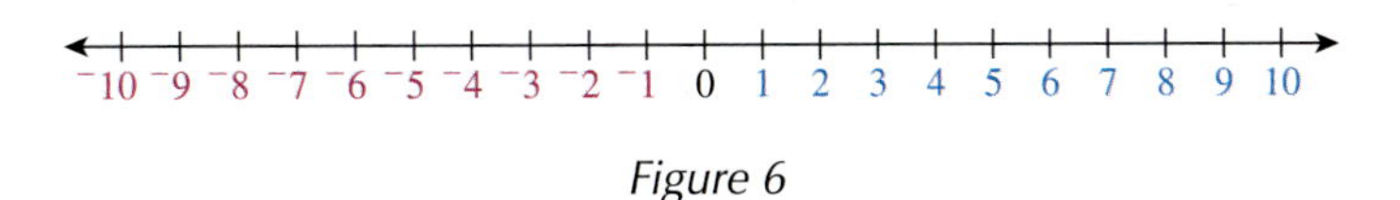

Figure 6

An example of an application that uses a vertical number line is temperature. Temperatures *above zero* are considered positive; temperatures *below zero* are negative. See Figure 7.

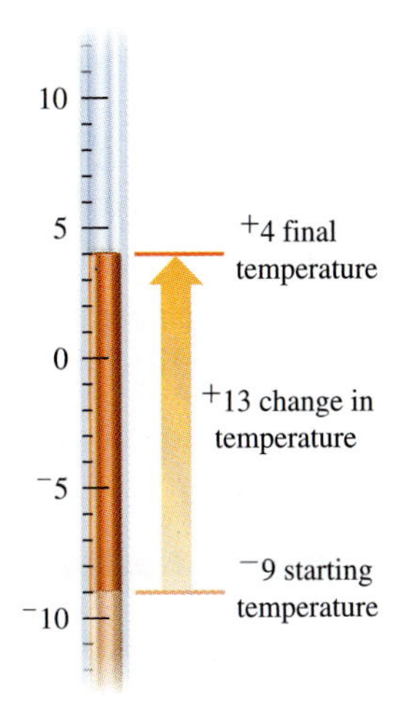

Figure 7

EXAMPLE 4 On a winter day the temperature at 6 A.M. was $^-9$°C. By noon the temperature had risen 13°. What was the temperature at noon?

Solution The thermometer in Figure 7 shows the starting point and the change in temperature. The final temperature, indicated on the thermometer, is $^+4$°C. This result may also be obtained by applying the addition rules for signed numbers: $^-9 + {}^+13 = {}^+4$.

The number line model for addition involves three elements: starting position, change in position, and final position. The relationship among these three elements is summarized as follows.

Number Line Model for Addition

starting position *plus* change in position = final position

start + change = final

The number line model for subtraction involves these same three elements. An application of subtraction is shown in Example 5.

Number Line Model for Subtraction

final position *minus* starting position = change in position

final − start = change

EXAMPLE 5 On the same winter day, at noon the temperature was $^{+}4°C$. By midnight that night it had fallen to $^{-}12°C$. By how much had the temperature changed?

Solution The temperature dropped 16°, so the change may be represented by $^{-}16$. You can verify this result by counting along the number line thermometer in Figure 8.

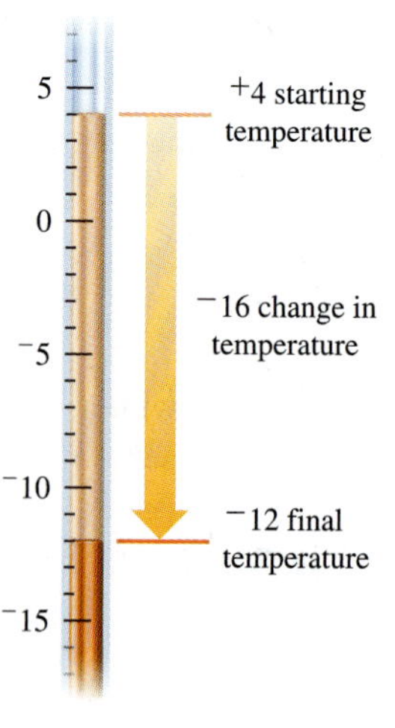

Figure 8

You can also use the number line model for subtraction to write

$$\text{final} - \text{start} = \text{change}$$
$$^{-}12 - {}^{+}4 = {}^{-}16$$

The change in temperature was $^{-}16$°C.

A Rule for Subtracting Signed Numbers

The answer to Example 5 is not found by combining the two temperatures. If we find the sum $^{-}12 + {}^{+}4 = {}^{-}8$, we clearly have an incorrect result, as a glance at the thermometer will show. Notice, however, that if the second number is changed to $^{-}4$, then we have $^{-}12 + {}^{-}4 = {}^{-}16$, which is the correct solution. In conclusion, the subtraction problem $^{-}12 - {}^{+}4$ gives the same result as the addition problem $^{-}12 + {}^{-}4$. The numbers $^{+}4$ and $^{-}4$ are called *opposites*, which suggests the following rule.

Subtracting Signed Numbers

To subtract two signed numbers, change the sign of the second number and change the operation to addition. Then follow the rules for the addition of signed numbers. *Think:* a minus $b = a$ plus the opposite of b; that is

$$a - b = a + {}^{-}b$$

Example 6 illustrates the use of the rule for subtraction.

EXAMPLE 6 Apply the rule for subtraction.

a. $^{+}7 - {}^{+}11$
b. $^{+}7 - {}^{-}11$
c. $^{-}7 - {}^{+}11$
d. $^{-}7 - {}^{-}11$

Solution

a. $^{+}7 - {}^{+}11 = {}^{+}7 + {}^{-}11$ *For opposite signs, subtract absolute values*
$= {}^{-}4$

b. $^{+}7 - {}^{-}11 = {}^{+}7 + {}^{+}11$ *For the same signs, add absolute values*
$= {}^{+}18$

c. $^{-}7 - {}^{+}11 = {}^{-}7 + {}^{-}11$ *For the same signs, add absolute values*

$= {}^{-}18$

d. $^{-}7 - {}^{-}11 = {}^{-}7 + {}^{+}11$ *For opposite signs, subtract absolute values*

$= {}^{+}4$

Multiplying Signed Numbers

To multiply two signed numbers, we cannot apply the rule for addition. The next two examples allow us to discover the correct rule.

Note: In the first few chapters of this book we use the symbol $*$ for multiplication. It is the same symbol that appears on the screen of a TI-83 graphing calculator.

EXAMPLE 7 Look for a pattern in the following equations. Notice that the first number is decreasing by 1, but the second number is always 5. What numbers should replace the question marks?

$$4 * 5 = 20$$
$$3 * 5 = 15$$
$$2 * 5 = 10$$
$$1 * 5 = 5$$
$$0 * 5 = 0$$
$$^{-}1 * 5 = ?$$
$$^{-}2 * 5 = ?$$
$$^{-}3 * 5 = ?$$

Solution The numbers on the right of the equal sign decrease by 5. Thus,

$$^{-}1 * 5 = {}^{-}5$$
$$^{-}2 * 5 = {}^{-}10$$
$$^{-}3 * 5 = {}^{-}15$$

From Example 7 it appears that when a positive number is multiplied by a positive number the product is positive, and when a negative number is multiplied by a positive number, the product is negative. What happens when a positive number is multiplied by a negative number, for example, $5 * {}^{-}3$? If we recognize that the order in which we multiply two numbers does not matter, we see that $5 * {}^{-}3 = {}^{-}3 * 5 = {}^{-}15$, as shown in Example 7. (The fact that $a * b = b * a$ is called the *commutative property of multiplication.*)

We may now ask what happens when two negative numbers are multiplied together. The next example addresses that question.

EXAMPLE 8 Look for a pattern in the following equations. Notice that the first number is decreasing by 1, whereas the second number is always $^-5$. What numbers should replace the question marks?

$$4 * {}^-5 = {}^-20$$
$$3 * {}^-5 = {}^-15$$
$$2 * {}^-5 = {}^-10$$
$$1 * {}^-5 = {}^-5$$
$$0 * {}^-5 = 0$$
$${}^-1 * {}^-5 = ?$$
$${}^-2 * {}^-5 = ?$$
$${}^-3 * {}^-5 = ?$$

Solution The numbers on the right of the equal sign are increasing by 5. Thus,

$${}^-1 * {}^-5 = 5$$
$${}^-2 * {}^-5 = 10$$
$${}^-3 * {}^-5 = 15$$

■ ■

From Example 8 we may conclude that when two negative numbers are multiplied together, their product is positive.

Examples 7 and 8 illustrate the rules for multiplying positive and negative numbers. Exercise 15 shows that the rules for division are similar.

Multiplying and Dividing Signed Numbers

If the signs are the same, the product or quotient is positive. If the signs are different, the product or quotient is negative.

positive ∗ positive = positive	positive ÷ positive = positive
positive ∗ negative = negative	positive ÷ negative = negative
negative ∗ positive = negative	negative ÷ positive = negative
negative ∗ negative = positive	negative ÷ negative = positive

There are special rules for multiplying and dividing with zero.

> **Multiplying and Dividing with Zero**
>
> - 0 multiplied by any number equals 0.
> - 0 divided by any positive or negative number equals 0.
> - Division by 0 is undefined.

Exercises 1.1

1. Write a sum for each of these charged-particle models.

   a.

   b.

   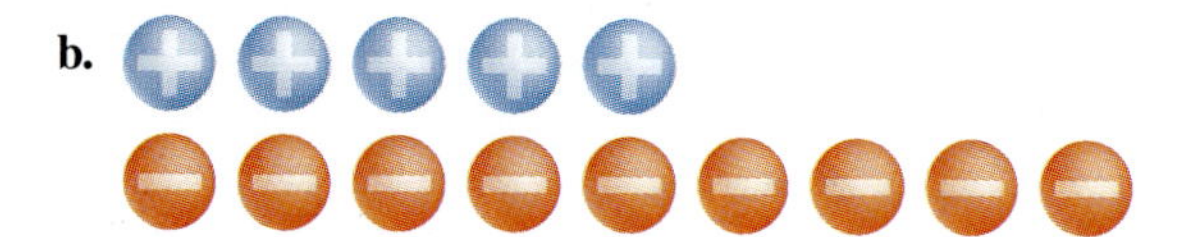

2. Find the absolute value for each of the resulting charges in Exercise 1.
3. Use a charged-particle model to find each sum.
   a. $^{-}6 + ^{+}7$ **b.** $^{-}8 + ^{+}2$ **c.** $^{-}6 + ^{-}2$
   d. $^{+}5 + ^{+}4$ **e.** $^{-}6 + ^{+}6$
4. Use the rule for addition to find each sum.
   a. $17 + ^{-}8$ **b.** $^{-}321 + ^{-}593$ **c.** $0 + ^{-}6$
   d. $^{-}34.7 + ^{+}71.8$ **e.** $^{-}62.3 + ^{-}41.75$ **f.** $^{-}\frac{1}{3} + \frac{1}{2}$
5. At 8:00 P.M. on a winter evening the temperature was $^{-}6$°C. By midnight the temperature had dropped 8°.
   a. What was the temperature at midnight?
   b. Represent this situation using the number line model for addition.
6. At 3:00 P.M. on a summer day, the temperature was 35°C. At 10:00 P.M. the temperature was 24°C.
   a. Find the change in temperature.
   b. Represent this situation with the number line model for subtraction.
7. Signed numbers may be applied to situations involving money. Think of deposits to your bank account as positive ($^{+}$) and checks or withdrawals as negative ($^{-}$). Alice has a beginning balance of $100 in her checking account. She writes a check for $250 and then deposits her paycheck, which is $475. She then withdraws $50 from an automatic teller machine (ATM).
   a. Find her balance after all these transactions.
   b. Represent this situation as a sum of all the transactions.
8. Signed numbers may be applied to elevations above ($^{+}$) or below ($^{-}$) sea level. Mt. Whitney is the highest spot in California, with an elevation of 14,494 feet above sea level. Death Valley has the lowest elevation in the state, 282 feet below sea level.
   a. Use positive or negative signs to represent the two elevations.
   b. Find the difference in elevations.
9. Apply the rule for subtraction to find each difference.
   a. $^{+}5 - ^{+}13$ **b.** $^{-}37 - ^{+}42$
   c. $^{+}11.8 - ^{+}5.7$ **d.** $^{-}\frac{2}{5} - \frac{^{-}1}{10}$
10. Explain the difference between these two keys on the calculator:

11. What happens when we use the negative key more than once? Experiment by entering

    etc. Describe what you observe.
12. Use the number line model for addition, start + change = final, to show that each pair of sums is the same. The first example is worked out for you.

Example: $^{-}3 + 9 = 9 + {^{-}3}$

Solution: $^{-}3$ is the starting position, 9 is the change, 6 is the final position.

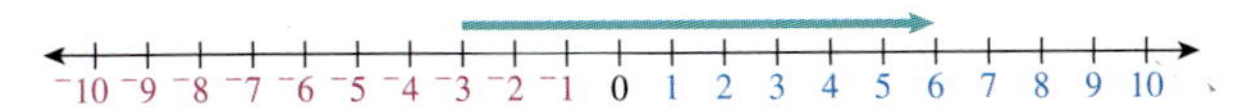

9 is the starting position, $^{-}3$ is the change, 6 is the final position.

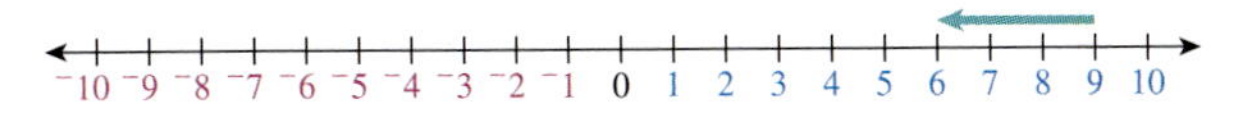

a. $^{-}6 + {^{-}2} = {^{-}2} + {^{-}6}$ **b.** $2 + {^{-}7} = {^{-}7} + 2$

c. $2 + 6 = 6 + 2$

13. Use the number line model for subtraction, final − start = change, to find each difference. Then check using the rule for subtraction. The first example is worked out for you.

Example: $^{-}7 - {^{-}2}$

Solution: $^{-}2$ is the starting position, $^{-}7$ is the final position, $^{-}5$ is the change.

$$^{-}7 - {^{-}2} = {^{-}7} + 2 = {^{-}5}$$

a. $^{-}3 - 5$ **b.** $5 - {^{-}3}$

c. $2 - 6$ **d.** $4 - {^{-}3}$

14. Find each product or quotient. Then complete each statement with the correct word, positive or negative.

a. $5 * 2$; positive * positive = ?

b. $5 * {^{-}2}$; positive * negative = ?

c. $^{-}5 * 2$; negative * positive = ?

d. $^{-}5 * {^{-}2}$; negative * negative = ?

e. $10 \div 2$; positive ÷ positive = ?

f. $10 \div {^{-}2}$; positive ÷ negative = ?

g. $^{-}10 \div 2$; negative ÷ positive = ?

h. $^{-}10 \div {^{-}2}$; negative ÷ negative = ?

15. Think carefully about each of these temperature situations. Answer the questions and show how each example illustrates the rules for dividing signed numbers on page 9.

a. The temperature at noon was 0°. Two hours later it was 6° above 0 ($^{+}6$). How fast did the temperature rise? (*Answer:* $^{+}6 \div {^{+}2} = ?$ so the temperature rose ____ degrees per hour.)

b. The temperature at noon was 0°. Two hours later it was 6° below 0 ($^{-}6$). How fast did the temperature rise? Think of a falling temperature as a negative rise. (*Answer:* $^{-}6 \div {^{+}2} = ?$ so the temperature rose ____ degrees per hour.)

c. The temperature at noon was 0°. Two hours before that (think negative 2) it was 6° below 0 ($^{-}6$). How fast did the temperature rise? (*Answer:* $^{-}6 \div {^{-}2} = ?$ so the temperature rose ____ degrees per hour.)

d. The temperature at noon was 0°. Two hours before that it was 6° above 0 ($^{+}6$). How fast did the temperature rise? (*Answer:* $^{+}6 \div {^{-}2} = ?$ so the temperature rose ____ degrees per hour.)

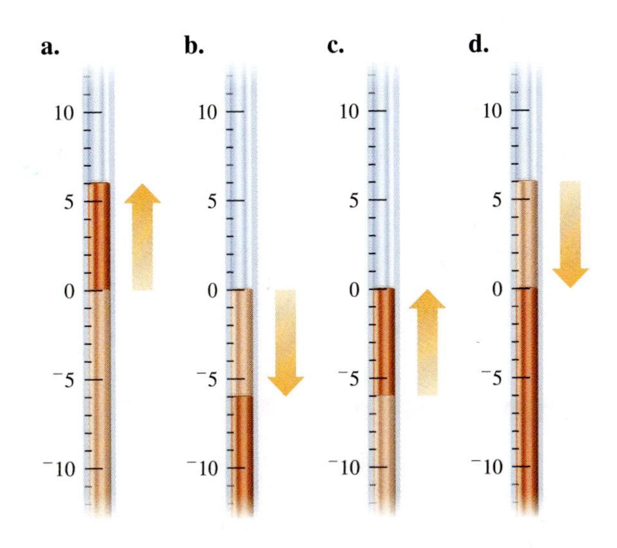

Skills and Review 1.1

16. Tisha eats $\frac{1}{2}$ of a whole pizza for lunch and $\frac{1}{3}$ of the same pizza for dinner. What fraction of the pizza did she eat in all? Use the diagrams to check your work.

Notice that $\frac{1}{2}$ and $\frac{1}{3}$ are different-size slices of pizza and cannot be added in fraction form until the slices are broken up into equal-size slices. The diagram shows the same pizza broken into six equal slices.

Equivalent fractions represent the same amount. Rewrite $\frac{1}{2}$ and $\frac{1}{3}$ with equivalent fractions using the new slice pattern in the diagram. Add the fractions.

17. **a.** Add: $\frac{1}{5} + \frac{1}{5} + \frac{1}{5} + \frac{1}{5} + \frac{1}{5}$

b. Explain why the answer to (a) is not $\frac{5}{25}$.

c. Multiply: $5 * \frac{1}{5} = \frac{5}{1} * \frac{1}{5}$.

d. Explain why the answers in (a) and (c) are the same.

18. Find each sum or difference without using a calculator.

a. $^{-}\frac{1}{3} + \frac{3}{4}$ **b.** $^{-}\frac{2}{3} - \frac{^{-}1}{2}$

c. $^{-}.72 + {}^{-}.08$ **d.** $1.43 - .23$

19. Often word problems have key words that indicate mathematical operations. Use the chart in the next column to convert each word phrase into a mathematical phrase.

a. 10 m per second

b. $\frac{1}{2}$ of 4 miles

Operation	Symbol	Words used to indicate operation
Multiplication	$\times$ or $*$	Product, of, factor, multiple (*Note:* $\times$ will not often be used to indicate multiplication in this book because of its confusion with the variable x.)
Division	/ or — or $\div$	Quotient, per, dividend, divisor
Addition	$+$	Sum, more, together, combine
Subtraction	$-$	Difference, less, left, remaining

c. 6 more than 68

d. 9 less than 81. (*Hint:* Think about what value is obtained; then write the mathematical phrase.)

20. One-half of one-half dollar is how much money? This question may be represented numerically as $\frac{1}{2} * \frac{1}{2}$. Perform the multiplication and describe your process for multiplying fractions.

21. Find each product or quotient without using a calculator.

a. $\frac{3}{5} * \frac{^{-}7}{2}$ **b.** $^{-}0.4 * {}^{-}5$ **c.** $^{-}1.8 \div 2$

22. Estimate 21.45 ounces + $^{-}$7.8 ounces − 5.32 ounces. Estimate by rounding each number to the nearest whole number.

23. How many seconds are in $1\frac{1}{2}$ hours? (*Hint:* 1 hour = 60 minutes and 1 minute = 60 seconds.)

24. How many ounces are in $\frac{3}{4}$ of a pint? (*Hint:* 1 pint = 16 ounces.)

25. Which number is greater, .1 or .07? Explain.

1.2 Rates and Ratios

After Studying This Section You Will Be Able To

- Apply the concept of rate to problems involving distance and work.
- Represent rational numbers as fractions, decimals, and percents.

Distance, Rate, and Time

When something is moving, its **speed** is given in units such as miles per hour or kilometers per second. The speed, or **rate**, of a moving object may be found by dividing the distance it travels by the time it takes. Thus, rate is the **ratio** of the distance to the time.

$$\text{rate} = \frac{\text{distance}}{\text{time}}$$

EXAMPLE 1 Sam travels by walking, riding her bicycle, and driving her car.

a. On Monday Sam walks 12 miles in 4 hours.
b. On Tuesday she rides her bicycle a distance of 5 miles in $\frac{1}{3}$ hour.
c. On Wednesday she drives a car 48 miles in $\frac{2}{3}$ hour.

Find her rate for each trip.

Solution A diagram can help you visualize what is happening in each case.

a. Distance = 12 miles and time = 4 hours. Assuming that her speed is constant, Sam is traveling 3 miles every hour, as shown in Figure 9.

$$\text{rate} = 12 \text{ miles} \div 4 \text{ hours} = 3 \text{ miles/hour}$$

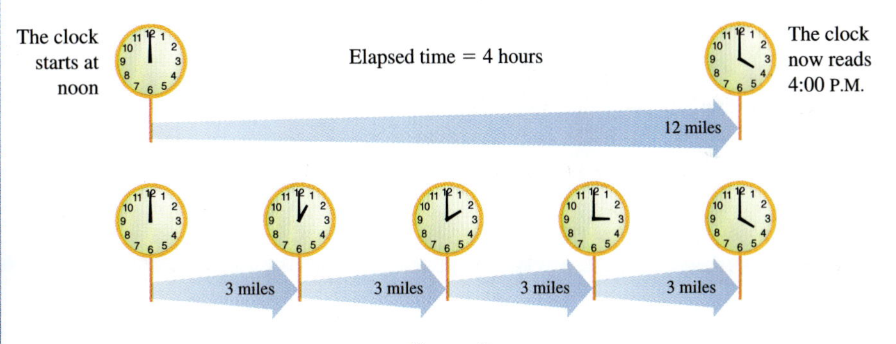

Figure 9

Notice that the word "per" may be written with a forward slash (/), which is also used as a symbol for division. When you see 3 miles/hour, read "3 miles per hour."

b. Distance = 5 miles and time = $\frac{1}{3}$ hour. Figure 10 shows that Sam travels 5 miles in $\frac{1}{3}$ hour. Figure 11 shows her traveling at the same rate for an entire hour.

$$\text{rate} = \frac{5 \text{ miles}}{\frac{1}{3}\text{hour}}$$

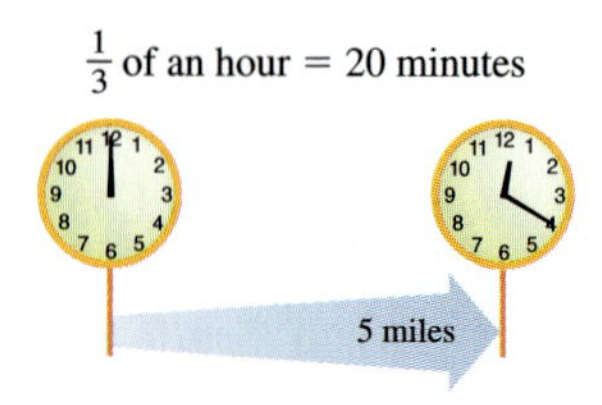

Figure 10

Remember, a fraction bar means division. Rewrite this as

$$\text{rate} = 5 \text{ miles } \div \tfrac{1}{3} \text{ hour}$$

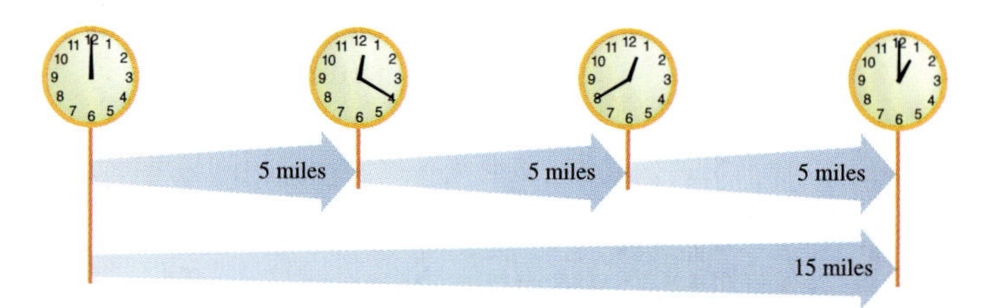

Figure 11 If Sam travels 5 miles every $\frac{1}{3}$ hour, she will travel 15 miles in 1 hour.

Now use the fact that dividing by a fraction (such as $\frac{1}{3}$) is the same as multiplying by its reciprocal ($\frac{3}{1}$).

$$\text{rate} = 5 \text{ miles} * \frac{3}{1 \text{ hour}} = 15 \text{ miles/hour}$$

c. Distance = 48 miles and time = $\frac{2}{3}$ hour. Figure 12 shows this, and Figure 13 shows that if Sam travels 48 miles in $\frac{2}{3}$ hour, she will travel 24 miles in $\frac{1}{3}$ hour and 72 miles in 1 hour.

$$\text{rate} = 48 \text{ miles } \div \tfrac{2}{3} \text{ hour} = 48 * \tfrac{3}{2} = 72 \text{ miles/hour}$$

Figure 12

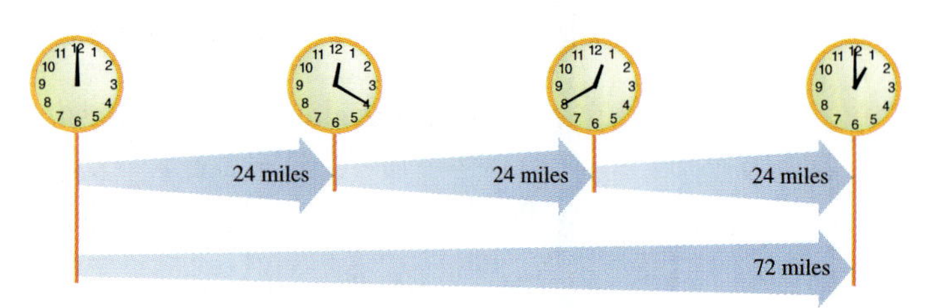

Figure 13 24 miles every $\frac{1}{3}$ hour means 48 miles in $\frac{2}{3}$ hour, which is the same as in Figure 12.

Work, Rate, and Time

Situations involving how quickly a job can be completed are similar to those involving distance, rate, and time. In fact, just as we define speed as

$$\text{rate} = \frac{\text{distance}}{\text{time}}$$

we can also define **work rate**:

$$\text{work rate} = \frac{\text{work}}{\text{time}}.$$

Example 2 shows how to apply this concept.

EXAMPLE 2 Joe earns money for his college expenses by running a lawn-care business. He uses the concept of work rate to estimate how much time he'll need for a particular job.

a. His best mower can mow a lawn with an area of 8000 square feet in 25 minutes. What is its rate in square feet per minute?

b. One of Joe's customers is Mr. Brown. Joe can do all of the work required on Mr. Brown's lawn in 2 hours. What is Joe's work rate in jobs per hour?

c. Sometimes Joe employs his friend Frank as a helper. Working alone, Frank can do all of the work required on Mr. Brown's lawn in 3 hours. What is Frank's work rate in jobs per hour?

d. Suppose Joe and Frank work together on Mr. Brown's lawn. What is their combined work rate? About how long will it take them to do the job?

Solution

a.
$$\text{rate} = \frac{\text{work}}{\text{time}} = \frac{8000 \text{ square feet}}{25 \text{ minutes}} = 320 \text{ square feet/minute}$$

b.
$$\text{rate} = \frac{\text{work}}{\text{time}} = \frac{1 \text{ job}}{2 \text{ hours}} = \frac{1}{2} \text{ job/hour}$$

c.
$$\text{rate} = \frac{\text{work}}{\text{time}} = \frac{1 \text{ job}}{3 \text{ hours}} = \frac{1}{3} \text{ job/hour}$$

d. If they are working together, it makes sense to *add* their individual work rates:

$$\text{Joe's rate} + \text{Frank's rate} = \frac{1 \text{ job}}{2 \text{ hours}} + \frac{1 \text{ job}}{3 \text{ hours}}$$

A common denominator for the two fractions is 6, so the combined rate is

$$\frac{3 \text{ jobs}}{6 \text{ hours}} + \frac{2 \text{ jobs}}{6 \text{ hours}} = \frac{5 \text{ jobs}}{6 \text{ hours}} = \frac{5}{6} \text{ jobs/hour}.$$

Because $\frac{5}{6}$ is a little less than 1, they can do almost the entire job in 1 hour. Together it will take them a little more than an hour to complete the job for Mr. Brown. Figure 14 shows this result.

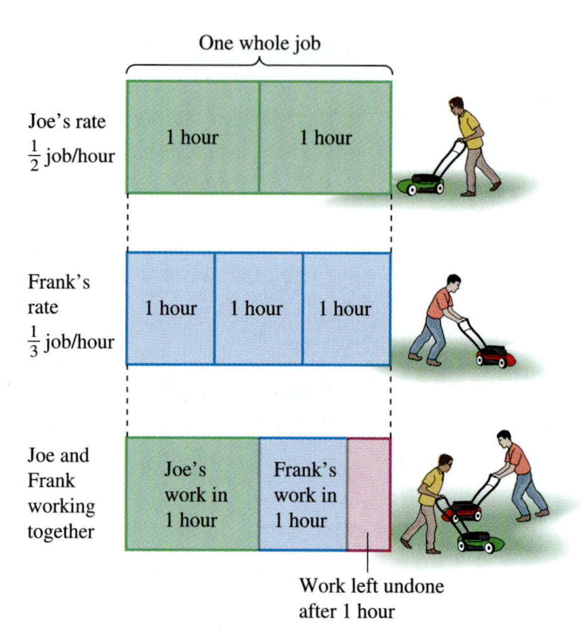

Figure 14

Symbols for Division and Fractions

As you can see from Examples 1 and 2, rates often involve fractions. You should be familiar with at least three different ways of symbolizing division and writing fractions. The symbol ÷ appears on calculator keys. The forward slash, /, appears on calculator and computer screen displays. Finally, the horizontal fraction bar, —, is commonly used in textbooks. The first of the two numbers being divided is called the **dividend** or **numerator**. The second of the two numbers is called the **divisor** or **denominator**. The dividend divided by the divisor gives the answer, or **quotient**.

$$\text{dividend} \div \text{divisor} = \text{dividend/divisor} = \frac{\text{dividend}}{\text{divisor}} = \text{quotient}$$

Division is not commutative. For instance, $6 \div 3 = 2$, but $3 \div 6 = .5$ (or $\frac{1}{2}$). Sometimes students confuse the order in which the dividend and divisor are written. This confusion arises from the fact that in performing a paper-and-pencil algorithm for long division, the divisor is written on the outside, to the left of the dividend, whereas in all other representations the dividend is written first.

$$\text{divisor}\overset{\text{quotient}}{\overline{\big)\text{dividend}}}$$

You may think of a common fraction as the quotient of the numerator and denominator. Thus, $\frac{2}{3}$ may be described as "two-thirds" or as "two divided by three."

Caution: When you use a calculator to multiply or divide fractions, you should enclose the fraction in parentheses. In part (c) of Example 1, for instance, $48 \div \frac{2}{3}$ appears on a calculator screen as 48/(2/3) and gives the correct quotient, 72. If the parentheses are omitted, as in 48/2/3, the calculator will first divide 48 by 2 to give 24 and then divide the result by 3 to give an incorrect answer of 8.

Rational Numbers

A **rational number** is one that may be expressed as the quotient, or ratio, of two integers. Notice that the word "ratio" appears in the word "rational" to help remember this definition. There are three immediate consequences of this definition.

Properties of Rational Numbers

1. Every integer is a rational number because every integer may be written as a ratio in which the divisor is 1: $4 = \frac{4}{1}$, $0 = \frac{0}{1}$, and $^{-}7 = \frac{^{-}7}{1}$.
2. Every fraction is a rational number because it is already expressed as the quotient of two integers: $\frac{5}{3}$, $\frac{2}{5}$, and $\frac{^{-}3}{4}$.
3. The one integer that never appears in the denominator of a rational number is 0, because division by 0 is undefined. Thus $\frac{0}{1} = 0$, but $\frac{1}{0}$ is not a number.

Rational numbers may also be expressed as **decimals** or **percents**. To express a common fraction as a decimal, divide the numerator by the denominator. In general, use a calculator for this purpose.

EXAMPLE 3 Express each of these common fractions as a decimal.

a. $\frac{13}{40}$

b. $\frac{19}{22}$

c. $\frac{7}{4}$

d. $\frac{52}{9}$

e. $\frac{3}{500}$

Solution

a. The calculator gives $\frac{13}{40} = .325$ (read "325 thousandths").

b. The calculator gives $\frac{19}{22} \approx .8636363636$. This is an approximation, rounded to 10 decimal places, the most displayed by the calculator. In fact, $\frac{19}{22}$ is an example of a *repeating decimal*. Actually,

$$\frac{19}{22} = .8636363636363636363636363636363\ldots$$

(and so on, forever).

You may have seen a repeating decimal written with a bar over the digits that repeat, as in $\frac{19}{22} = .8\overline{63}$. For most purposes you will encounter in this textbook, a three-digit approximation may be used, so we can say $\frac{19}{22} \approx .864$.

Note: Be sure to use the symbol $\approx$ to indicate that two numbers are approximately equal. Use $=$ only to indicate exact equality.

c. $\frac{7}{4} = 1.75$. As in (a) the calculator gives an exact answer. $\frac{7}{4}$ is called an **improper fraction** because the numerator is larger than the denominator. It may also be expressed as the **mixed fraction**, or mixed number, $1\frac{3}{4}$.

d. $\frac{52}{9}$ gives another repeating decimal, $5.\overline{7}$. Or you may write $\frac{52}{9} \approx 5.778$. As a mixed fraction, $\frac{52}{9} = 5\frac{7}{9}$.

e. $\frac{3}{500} = .006$

Percents

Fractions and decimals may be expressed as percents. To express a fraction as a percent, first write it as a decimal. To express a decimal as a percent, move the decimal point two places to the right and use the percent symbol (%).

EXAMPLE 4 ■ Express each of the fractions in Example 3 as a percent.

Solution

a. $\frac{13}{40} = .325 = 32.5\%$

b. $\frac{19}{22} \approx .864 = 86.4\%$

c. $\frac{7}{4} = 1.75 = 175\%$

d. $\frac{52}{9} \approx 5.778 = 577.8\%$

e. $\frac{3}{500} = .006 = .6\%$

Notice that a percent may be greater than 100% (c, d) or less than 1% (e).

Common Equivalents

Although a calculator may be used to find decimal and percent equivalents of common fractions, you should know (and have a good intuitive feel for) some frequently encountered fractions. Three families of fractions with small denominators are displayed in the table.

A table of fraction-decimal-percent equivalents.

Fourths and Eighths	Fifths and Tenths	Thirds and Sixths
	$\frac{1}{10} = .1 = 10\%$	
$\frac{1}{8} = .125 = 12.5\%$		
		$\frac{1}{6} \approx .167 = 16.7\%$
	$\frac{1}{5} = \frac{2}{10} = .2 = 20\%$	
$\frac{1}{4} = \frac{2}{8} = .25 = 25\%$		
	$\frac{3}{10} = .3 = 30\%$	
		$\frac{1}{3} = \frac{2}{6} \approx .333 = 33.3\%$
$\frac{3}{8} = .375 = 37.5\%$		
	$\frac{2}{5} = \frac{4}{10} = .4 = 40\%$	
$\frac{1}{2} = \frac{2}{4} = \frac{4}{8} = .5 = 50\%$	$\frac{1}{2} = \frac{5}{10} = .5 = 50\%$	$\frac{1}{2} = \frac{3}{6} = .5 = 50\%$
	$\frac{3}{5} = \frac{6}{10} = .6 = 60\%$	
$\frac{5}{8} = .625 = 62.5\%$		
		$\frac{2}{3} = \frac{4}{6} \approx .667 = 66.7\%$
	$\frac{7}{10} = .7 = 70\%$	
$\frac{3}{4} = \frac{6}{8} = .75 = 75\%$		
	$\frac{4}{5} = \frac{8}{10} = .8 = 80\%$	
		$\frac{5}{6} \approx .833 = 83.3\%$
$\frac{7}{8} = .875 = 87.5\%$		
	$\frac{9}{10} = .9 = 90\%$	
$\frac{2}{2} = \frac{4}{4} = \frac{8}{8} = 1.00 = 100\%$	$\frac{5}{5} = \frac{10}{10} = 1.0 = 100\%$	$\frac{3}{3} = \frac{6}{6} = 1.000 = 100\%$

Rational Numbers and the Number Line

Rational numbers help fill in the gaps between the integers on the number line. In fact, between any two integers there are *infinitely* many rational numbers. However, there are some points on the number line that are not represented by rational numbers. We encounter irrational numbers in the next section.

EXAMPLE 5 ■ Sketch the location of each number in Example 3 on a number line.

Solution

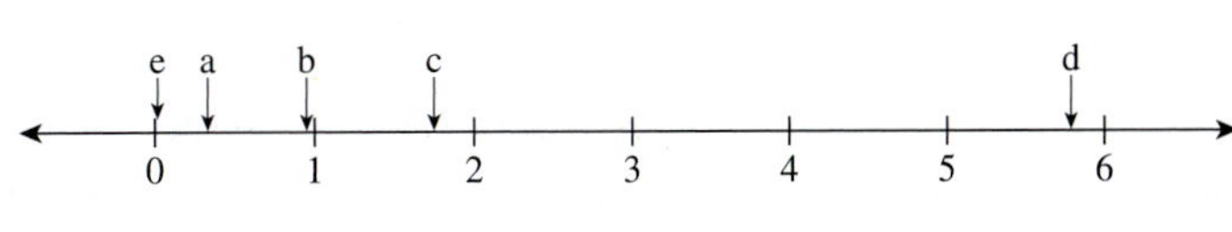

Figure 15

When comparing two numbers on a horizontal number line, the one to the left is smaller and the one to the right is greater. The symbols $<$ and $>$ may be used to express the relations less than and greater than. Thus $a < b$ means that a is to the left of b, and $a > b$ means that a is to the right of b on the number line.

EXAMPLE 6 Determine which symbol, $<$ or $>$, to insert between each pair of numbers to make true sentences.

a. $^{-}1$ ___ 0

b. 2 ___ $^{-}3$

c. $\frac{7}{4}$ ___ $\frac{19}{22}$

d. $\frac{3}{500}$ ___ $\frac{13}{40}$

Solution

a. $<$ because $^{-}1$ is to the left of 0

b. $>$ because 2 is to the right of $^{-}3$

c. $>$ because $\frac{7}{4}$ (c in the figure in Example 5) is to the right of $\frac{19}{22}$ (b in the figure in Example 5)

d. $<$ because $\frac{3}{500}$ (e in the figure in Example 5) is to the left of $\frac{13}{40}$ (a in the figure in Example 5)

Exercises 1.2

1. The Ramsey family drove from Hartford, Connecticut, to Washington, D.C., a distance of 342 miles, in 6 hours.
   - **a.** Draw a picture to illustrate this situation.
   - **b.** Find their average rate of speed.
2. The formula for finding rate when distance and time are known is

$$\text{rate} = \frac{\text{distance}}{\text{time}}$$

   - **a.** Consider the case where the rate and time are known. At a rate of 60 miles/hour, what distance will you travel in 1 hour, 2 hours, and 3 hours?
   - **b.** Find a formula for distance when rate and time are known.
   - **c.** Use your formula to find how far an airplane can travel in 2.5 hours if its speed is 350 miles/hour.
3. Use the diagram in Example 1(b) to explain why dividing by $\frac{1}{3}$ is the same as multiplying by 3.
4. In Example 1(c), Sam travels 48 miles in $\frac{2}{3}$ hour. There are 60 minutes in 1 hour, so $\frac{2}{3}$ hour is the same as 40 minutes.
   - **a.** Find Sam's speed in miles per minute.
   - **b.** Use your answer in (a) to find her speed in miles per hour. Compare the result with the solution to the example.
5. Determine the work rate in each of these situations.
   - **a.** Joseph can pick 250 apples in 2 hours. What is Joseph's work rate in apples per hour?
   - **b.** Rachel can paint 2 houses in 8 days. What is Rachel's work rate in houses per day?
   - **c.** Randy can repair 1 bicycle in 3 hours. What is Randy's work rate in bicycles per hour?
6. Study the table of common fraction-decimal-percent equivalents and describe any patterns you notice.
7. Name a common fraction that is equal or approximately equal to each of the percents.

a. 75% **b.** 40% **c.** 70%
d. 33.3% **e.** 62.5% **f.** 66.7%

8. Place the appropriate symbol ($=$ or $\approx$) in each blank space.

a. $\frac{7}{12}$ ____ .583 **b.** $\frac{13}{16}$ ____ .8125
c. $\frac{18}{13}$ ____ 1.385 **d.** $\frac{8}{25}$ ____ .32

9. What happens on your calculator when you enter $1 \div 0$? Use the properties of rational numbers on page 17 to explain why your calculator gave what it did.

10. There are two ways to decide which of two fractions is greater: (1) by finding equivalent fractions with a common denominator, and (2) by finding decimal equivalents. For instance, if $\frac{3}{4}$ and $\frac{2}{3}$ are compared by method 1, we compare $\frac{9}{12}$ with $\frac{8}{12}$; by method 2 we compare .75 with the approximate value .667. In both cases we determine that $\frac{3}{4} > \frac{2}{3}$. Use both methods to determine which symbol, $<$ or $>$, to insert between each pair of fractions.

a. $\frac{1}{4}$ ____ $\frac{2}{5}$ **b.** $\frac{1}{2}$ ____ $\frac{4}{9}$
c. $\frac{13}{16}$ ____ $\frac{4}{5}$ **d.** $\frac{3}{8}$ ____ $\frac{5}{12}$

11. Locate these fractions on the number line.

a. $\frac{1}{3}$ **b.** $\frac{1}{4}$ **c.** $\frac{3}{4}$
d. $\frac{3}{8}$ **e.** $\frac{3}{10}$ **f.** $\frac{3}{100}$

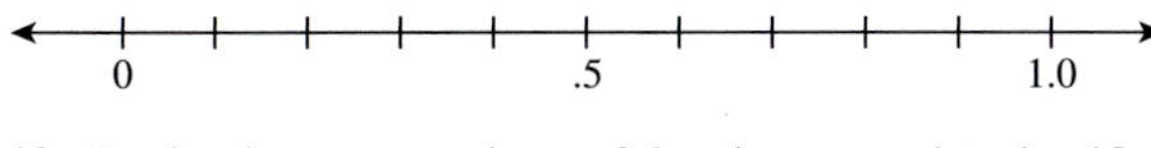

12. Decimal representations of fractions may be classified as repeating or terminating. For instance:

$\frac{1}{3} = .\overline{3} = .333333333333333\ldots$ *Repeating decimal*

$\frac{1}{2} = .5$ *Terminating decimal*

Classify each of these fractions. Then see if you can find a pattern that will help you predict which fractions are terminating and which are repeating.

$\frac{1}{4}$ $\frac{1}{5}$ $\frac{1}{6}$ $\frac{1}{7}$ $\frac{1}{8}$ $\frac{1}{9}$ $\frac{1}{10}$ $\frac{1}{12}$ $\frac{1}{15}$ $\frac{1}{20}$ $\frac{1}{24}$

13. A terminating decimal may be written as a fraction and then expressed in lowest terms. Here are two examples:

$$.45 = \frac{45}{100} = \frac{5 * 9}{5 * 20} = \frac{9}{20}$$

and

$$.642 = \frac{642}{1000} = \frac{2 * 321}{2 * 500} = \frac{321}{500}$$

Rewrite each decimal as a fraction in lowest terms.

a. .65 **b.** 1.24 **c.** .575
d. .400 **e.** .804

14. Many calculators have a feature that will write decimals as simplified common fractions. On the TI-82 and TI-83 this feature works for fractions with denominators less than or equal to 1000 and is found under the MATH menu. For example, if you type

the answer 9/20 will appear on the screen.

Use the fraction-conversion feature to check your work for Exercise 13(a)–(e).

15. Use the fraction-conversion feature of a calculator on each of the decimals. Then describe what you observe.

a. .3 **b.** .33
c. .333 **d.** .3333
e. .3333333333 **f.** .333333333333

g. A student concluded from part (f) that $\frac{1}{3}$ is *exactly equal* to .333333333333. Do you agree? Explain your reasoning.

Skills and Review 1.2

16. Suppose you drive 1480 miles in 29 hours. Which is the best estimate of your rate of speed in miles per hour? (*Hint:* Round each number before finding an answer).

a. 700 miles/hour **b.** 5 miles/hour
c. 70 miles/hour **d.** 50 miles/hour

17. Two signed numbers are multiplied. Give a rule for determining the resulting sign.

18. Find each product and simplify if possible.

a. $2 * \frac{^-1}{8}$ **b.** $\frac{^-3}{5} * 10$ **c.** $\frac{^-3}{2} * \frac{^-4}{9}$

19. Find each quotient and simplify if possible.
a. $\frac{^-1}{3} \div \frac{5}{6}$ **b.** $0 \div 4$ **c.** $\frac{2}{3} \div {^-4}$

20. Without using a calculator, find each value.
a. $5 + {^+2} = 5 + 2$ **b.** $5 - {^+2} = 5 - 2$
c. $5 - {^-2} = 5 + 2$ **d.** ${^-5} + 2$
e. ${^-5} - 2$ **f.** ${^-5} - {^-2}$

21. Use a number line to find each sum or difference.
a. $6 + {^-8}$ **b.** ${^-5} + 9$
c. ${^-2} - 4$ **d.** $3 - {^-1}$

22. Find each sum or difference without using a calculator.
a. $48 - 72$ **b.** ${^-.7} + {^-.25}$ **c.** ${^-\frac{3}{5}} - \frac{^-1}{4}$

23. How many inches are in 4 yards? (*Hint:* 1 yard = 3 feet and 1 foot = 12 inches.) Explain how you found your answer and show your work.

24. Which number is less, .23 or .217?

25. Is $\frac{2}{3}$ closer to 0, $\frac{1}{2}$, or 1? Explain.

1.3 Exponents and Radicals

AFTER STUDYING THIS SECTION YOU WILL BE ABLE TO

- Raise numbers to integer powers.
- Write numbers in scientific notation.
- Find square roots of positive numbers.

Positive Integer Exponents

Repeated addition may be represented as multiplication. For instance, $5 * 3$ is the same as $5 + 5 + 5$. Similarly, repeated multiplication may be represented with an operation called exponentiation: 5^3 means the same as $5 * 5 * 5$. Read 5^3 as "5 raised to the third power" or "5 to the third." The **base** is 5 and the **exponent** is 3.

Typically we write the exponent as a **superscript**; there is no operation symbol between the base and the exponent. However, you should think of exponentiation as an operation just like addition or multiplication. On calculators there is usually a special key for this operation. On the TI-83 it appears directly above the division key; the symbol is the **caret** ($\wedge$). Thus, on a calculator you write 5^3 as $5 \wedge 3$.

EXAMPLE 1 Find each value, without using a calculator.

a. 7^2
b. 2^5
c. $\left(\frac{1}{3}\right)^4$
d. 10^1
e. 1^{23}

Solution

a. $7^2 = 7 * 7 = 49$

b. $2^5 = 2 * 2 * 2 * 2 * 2 = 32$

c. $\left(\frac{1}{3}\right)^4 = \left(\frac{1}{3}\right) * \left(\frac{1}{3}\right) * \left(\frac{1}{3}\right) * \left(\frac{1}{3}\right) = \frac{1}{81}$

d. $10^1 = 10$

e. $1^{23} = 1$ (No matter how many times you multiply 1 by 1, the product is still 1.)

Negative Bases

When a negative number is raised to a power, the result is positive or negative, depending upon whether the exponent is even or odd.

EXAMPLE 2 Find each value without using a calculator.

a. $(^-2)^1$

b. $(^-2)^2$

c. $(^-2)^3$

d. $(^-2)^4$

e. $(^-2)^5$

Solution

a. $(^-2)^1 = {}^-2$

b. $(^-2)^2 = {}^-2 * {}^-2 = 4$

c. $(^-2)^3 = {}^-2 * {}^-2 * {}^-2 = {}^-8$

d. $(^-2)^4 = {}^-2 * {}^-2 * {}^-2 * {}^-2 = 16$

e. $(^-2)^5 = {}^-2 * {}^-2 * {}^-2 * {}^-2 * {}^-2 = {}^-32$

From the pattern shown in Example 2, we can make the following generalization.

Powers of Negative Numbers

When a negative number is raised to an odd power, the result is negative. When a negative number is raised to an even power, the result is positive.

Caution: When you wish to raise a negative number to a power, you must use parentheses. As we see in Example 2(d), $(^-2)^4 = 16$. If you enter $^-2 \wedge 4$ in your calculator, however, the result will be $^-16$ because $^-2^4$ is the opposite of 2^4, or $^-16$.

Zero as an Exponent

Exponentiation may be represented by repeated multiplication when the exponent is a positive integer. The definition of exponent can be extended to include other numbers. In this section we develop a definition for zero and negative integer exponents. In Chapter 9, we extend the meaning of exponentiation to include rational numbers as exponents.

What does it mean to raise a number to the zeroth power? For example, what is 2^0? You might be tempted to think that $2^0 = 0$, but that is not the definition mathematicians use. When any number (except 0) is raised to the zeroth power, the result is defined to be 1.

Definition Zero Exponent

Let a be any number except 0. Then $a^0 = 1$.

The justification for this definition of a zero exponent is illustrated by the following activity. Take a sheet of paper and fold it in half. Open it, and observe that the creases divide the paper into two regions. Then fold it back and make a second fold (in the other direction). Unfold the paper and you have four regions. Figure 16 shows how the number of folds is related to the number of regions. The information shown in Figure 16 may be organized into a chart:

Number of Folds	Number of Regions
0	1
1	2
2	4
3	8
4	16
5	32

Note that powers of 2 appear in the chart:

Number of Folds	Number of Regions
0	1
1	$2 = 2^1$
2	$4 = 2^2$
3	$8 = 2^3$
4	$16 = 2^4$
5	$32 = 2^5$

Thus, it appears that number of regions $=$ 2 raised to the power "number of folds," or $2^{\text{number of folds}}$. When the number of folds is 0, the number of regions is 1. So it makes sense to define $2^0 = 1$.

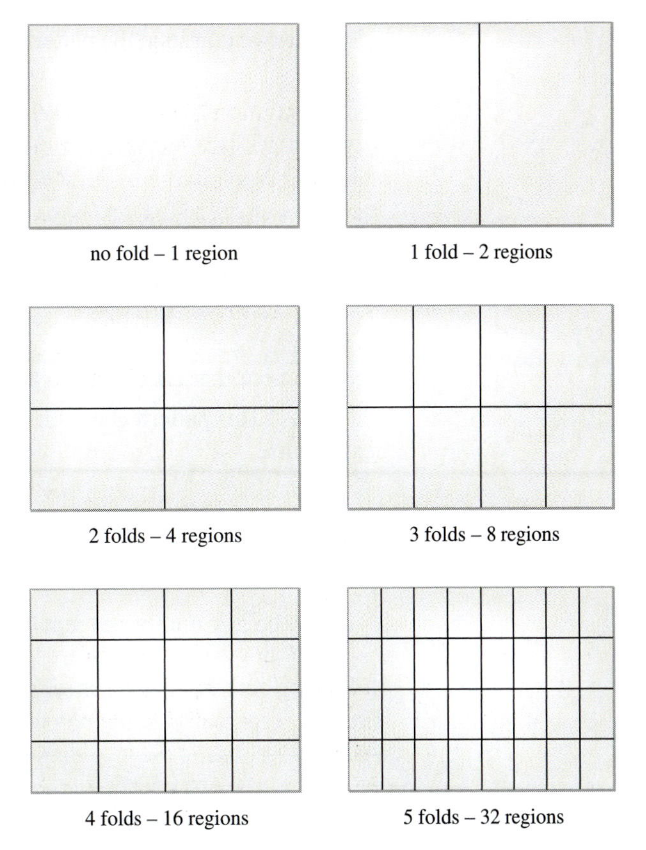

Figure 16

The definition of the zeroth power, like many others in mathematics, is based on looking at a pattern and making our definition fit the pattern. This process may now be extended to negative integer exponents.

Negative Integer Exponents

Let's look at the chart of the powers of 2 in a different way. Starting with $2^4 = 16$, we notice that when we reduce the exponent by 1, the result is half of the previous result:

$$2^4 = 16$$

$$2^3 = 8 = \tfrac{1}{2} \text{ of } 16$$

$$2^2 = 4 = \tfrac{1}{2} \text{ of } 8$$

$$2^1 = 2 = \tfrac{1}{2} \text{ of } 4$$

Now we continue the pattern to include zero and negative integer exponents:

$$2^0 = 1 = \tfrac{1}{2} \text{ of } 2$$
$$2^{-1} = \tfrac{1}{2} = \tfrac{1}{2} \text{ of } 1$$
$$2^{-2} = \tfrac{1}{4} = \tfrac{1}{2} \text{ of } \tfrac{1}{2}$$
$$2^{-3} = \tfrac{1}{8} = \tfrac{1}{2} \text{ of } \tfrac{1}{4}, \text{ etc.}$$

We note that according to this pattern, $2^{-3} = \frac{1}{8}$. Because $8 = 2^3$, we have $2^{-3} = \frac{1}{2^3}$. This pattern generalizes and gives us our definition for negative integer exponents.

Definition Negative Integer Exponents

Let a be any number except 0 and let n be any positive integer. Then

$$a^{-n} = \frac{1}{a^n}$$

EXAMPLE 3 Find each value without using a calculator.

a. 3^{-2}
b. 12^{-1}
c. 4^{-3}
d. 2^{-6}
e. 891^0

Solution

a. $3^{-2} = \dfrac{1}{3^2} = \dfrac{1}{9}$

b. $12^{-1} = \dfrac{1}{12^1} = \dfrac{1}{12}$

c. $4^{-3} = \dfrac{1}{4^3} = \dfrac{1}{4 * 4 * 4} = \dfrac{1}{64}$

d. $2^{-6} = \dfrac{1}{2^6} = \dfrac{1}{2 * 2 * 2 * 2 * 2 * 2} = \dfrac{1}{64}$

e. $891^0 = 1$ by the definition of the zero exponent

Powers of Ten and Scientific Notation

Figure 17 Place value in the decimal numeration system.

Our decimal numeration system is based on powers of ten, as shown in Figure 17.

In this example,

$$\begin{aligned} 87{,}215.9346 &= 80{,}000 + 7{,}000 + 200 + 10 + 5 + .9 + .03 + .004 \\ &\quad + .0006 \\ &= 8 * 10^4 + 7 * 10^3 + 2 * 10^2 + 1 * 10^1 + 5 * 10^0 + 9 * 10^{-1} \\ &\quad + 3 * 10^{-2} + 4 * 10^{-3} + 6 * 10^{-4} \end{aligned}$$

Note that positive powers of 10 and 10^0 are located to the left of the decimal point. Negative powers of ten are located to the right of the decimal point. Adding the suffix "th" to the name for a power of ten is like taking the opposite of the exponent. For instance, one hundred is 10^2, whereas one-hundredth is 10^{-2}. Powers of 10 are used to express numbers in scientific notation.

Definition Scientific Notation

A number is written in scientific notation if it is of the form $a * 10^b$, where a is an integer or decimal such that $1 \leq a < 10$ and b is an integer.

The exponent b gives an idea of approximately how large the number is.

To change a number in decimal form to scientific notation, move the decimal point so that you have a number between 1 and 10. Then count the number of places the decimal point has moved. Remember that a whole number has a decimal point immediately to its right.

EXAMPLE 4 Write each number in scientific notation.

a. The speed of light is 186,000 miles per second.
b. In 1999 the world's population reached 6 billion.
c. The radius of a hydrogen atom is .000000000053 m.

Solution

a. $186{,}000. = 1.86 * 10^5$ We moved the decimal point 5 places. We know that the exponent is positive because 186,000 is greater than 1. We could have written $1.86000 * 10^5$, but we usually drop unnecessary zeros after the decimal point.

b. $6{,}000{,}000{,}000. = 6 * 10^9$ We moved the decimal point 9 places. Again, the exponent is positive.

c. $.000000000053 = 5.3 * 10^{-11}$ We moved the decimal point 11 places. Because the number is less than 1, the exponent must be negative.

Note: In this book we use an asterisk (∗) for multiplication. In many texts you'll see a × used in scientific notation. Example 4(a) then would be written 1.86×10^5.

Calculators display scientific notation with an uppercase E. When you encounter E, read it as "times 10 to the power."

EXAMPLE 5 Use a calculator to find the answer. Then write the answer displayed by the calculator in both scientific notation and decimal notation.

a. Suppose that every day each of the world's 6 billion people had a nutritionally adequate diet with about 2500 calories. How many calories per day would the human population consume?

b. If a glacier moves .3 cm in 1 year, how far will it move in 1 day?

Solution

a. Enter 6000000000 × 2500. The calculator shows 1.5E13, which is $1.5 * 10^{13}$. Move the decimal point 13 places to the right because the number is greater than 1: that gives 15000000000000 calories. Inserting commas makes the number easier to read: 15,000,000,000,000 calories, or 15 trillion calories.

b. Enter .3 ÷ 365 because there are 365 days in a year. The calculator displays 8.219178082E$^-$4. That is approximately $8.22 * 10^{^-4}$. Move the decimal point 4 places to the left because the number is less than 1 when the exponent is negative. The glacier moves .000822 cm in 1 day. You would need a microscope to observe this small distance.

Squares and Square Roots

The area of a figure is measured in square units. For instance, a square room that has a length of 20 feet and a width of 20 feet has an area of 400 square feet. Because feet are multiplied by feet we can also express square feet as feet^2. Similarly, when we find the volume of three-dimensional figures, we use cubic units. Cubic feet may be abbreviated feet^3.

Because these exponents refer to squares and cubes, we often use the terms "squared" and "cubed" when raising numbers to the second and third powers. Thus we may say "x squared" for x^2 and "y cubed" for y^3.

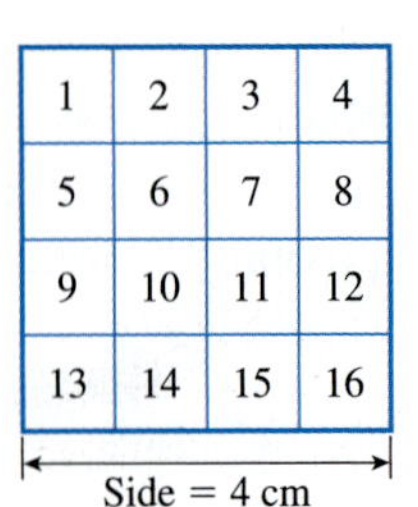

Figure 18

The square in Figure 18 has an area of 16 cm^2. The length of one side is 4 cm. We say "4 squared is 16"; using exponents, $4^2 = 16$. If we know the area of a square and want to know the length of the side, we can work backward. We reverse the process and say, "A **square root** of 16 is 4," or $\sqrt{16} = 4$.

A positive number such as 16 actually has two square roots, one positive (4) and one negative ($^-4$). In this case, since the side of a square cannot be negative, we speak only of the positive root. The symbol $\sqrt{\ }$ is called a **radical** and is used to indicate the positive square root of a positive number. You will learn more about radicals later in the course.

The number 16 is called a **perfect square** because its positive square root is an integer. Other perfect squares with which you should be familiar are 1, 4, 9, 25, 36, 49, 64, 81, and 100.

EXAMPLE 6 Use mental math to find each of these square roots.

a. $\sqrt{36}$
b. $\sqrt{81}$
c. $\sqrt{25}$
d. $\sqrt{100}$

Solution

a. $6^2 = 36$; therefore, $\sqrt{36} = 6$.
b. $9^2 = 81$; therefore, $\sqrt{81} = 9$.
c. $5^2 = 25$; therefore, $\sqrt{25} = 5$.
d. $10^2 = 100$; therefore, $\sqrt{100} = 10$.

EXAMPLE 7 The shaded square in Figure 19 is formed by joining the midpoints of the sides of the square in Figure 18.

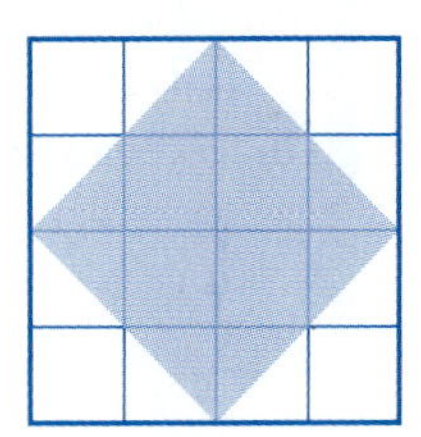

Figure 19

a. Find the area of this square.
b. Find the length of one side of this square.

Solution

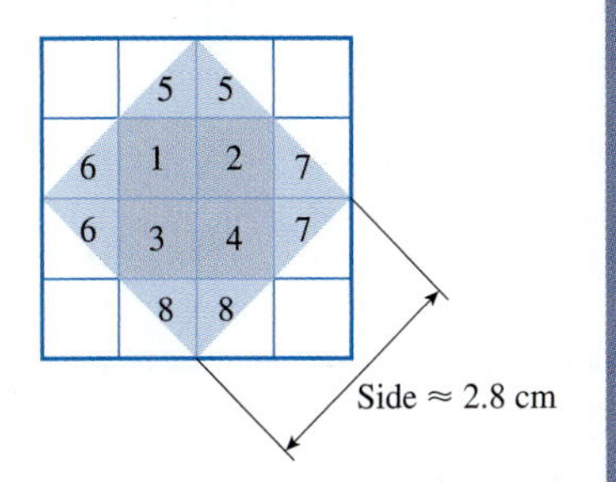

Figure 20

a. The area of the shaded square is exactly half of the original square. You can verify this by counting square centimeters from the original grid, as shown in Figure 20. There are four complete squares (1–4) and four pairs of triangles (5–8), each of which can be combined to form 1 cm^2. The area of the square is 8 cm^2.

b. You can estimate the length of a side of the square by drawing it on a centimeter square grid and using a ruler. The side is approximately 2.8 cm in length. The length of the side, which is $\sqrt{8}$, is not an integer because 8 is not a perfect square. Because 8 lies between two perfect squares, 4 and 9, it stands to reason that $\sqrt{8}$ is between $\sqrt{4}$ and $\sqrt{9}$, that is, greater than 2 but less than 3. A more precise approximation may be found with a calculator: $\sqrt{8} \approx 2.828427125$.

Irrational Numbers

The number $\sqrt{8}$ is called an **irrational** number. This means that it cannot be expressed as the ratio of two integers and there is no pattern of repeating digits in its decimal representation. The square root of any whole number that is not a perfect square is an irrational number. Thus, the following numbers are irrational: $\sqrt{2}$, $\sqrt{3}$, $\sqrt{5}$, $\sqrt{6}$, $\sqrt{7}$, and so on. Calculators can be used to get approximations of irrational numbers.

Another irrational number is $\pi \approx 3.141592654$. This number, called *pi*, is used to find the circumference and area of a circle, as you will see in the next section.

Although most square roots are irrational, if you know the perfect squares up through 100, you should be able to estimate the square root of any number between 0 and 100. You can check your estimate using a calculator.

EXAMPLE 8 Estimate these square roots.

a. $\sqrt{50}$

b. $\sqrt{20}$

c. $\sqrt{95}$

Solution

a. 50 is slightly greater than 49, so $\sqrt{50}$ is a bit more than 7. A good estimate is 7.1.

b. 20 is about halfway between 16 and 25. A good estimate is 4.5.

c. 95 is between 81 and 100 but closer to 100. A good estimate would be 9.7 or 9.8.

Exercises 1.3

1. Do not use a calculator.

a. Use repeated addition to find $3 * 4$.

b. Use repeated multiplication to find 3^4.

2. A student reasons that 5^3 is $5 * 3$. What mistake is the student making?

3. Find each value without using a calculator.

a. 2^4 **b.** 1^{15} **c.** $\left(\frac{1}{4}\right)^2$

d. $(^-3)^2$ **e.** $(^-3)^3$

4. Evaluate $(^-1)^2$, $(^-1)^3$, $(^-1)^4$, and $(^-1)^5$. What pattern do you notice?

5. On your calculator, experiment raising different numbers (except 0) to the zeroth power.

a. What is the result?

b. What do you conclude about raising any number except 0 to the zeroth power?

6. Without using a calculator, write each answer as a fraction and a decimal.

a. 10^0 **b.** 10^{-1} **c.** 10^{-2}

d. 10^{-3} **e.** 10^{-4}

7. Refer to Exercise 6.
   a. Plot each number on a number line.
   b. Continue evaluating negative powers of 10 (10^{-5}, 10^{-6}, 10^{-7}, etc.). To what number do you get closer?
   c. If you continue the process in (b), will you ever get a negative number? Explain.

8. Find each value without using a calculator.
   a. 5^{-2} b. 23^{-1} c. 3^{-4} d. 2^{-5}

9. Write each number in scientific notation.
   a. At freezing, the speed of sound is 1,089 feet/second.
   b. Jupiter's mean distance from Earth is 484 million miles.
   c. An aerosol particle has a diameter of .000000002 m.

10. a. Which number is greater, 1.2×10^{7} or 1.2×10^{-15}?
   b. Write both numbers from part (a) in decimal notation.
   c. Does your answer in part (b) confirm your choice of the greater number?

11. Use a calculator to find each answer; then write the answer displayed by the calculator in both scientific notation and decimal notation.
   a. Some scientists predict the temperature of the earth's atmosphere will warm 3.5°F in the next 100 years. How much will it warm in 1 day?
   b. A typical daily diet for each of the 270 million U.S. residents should include 51 g of protein. How much protein should the entire population of the United States consume in one day? (http://www.foodstandards.gov)

12. Find each value without using a calculator.
   a. $\sqrt{4}$ b. $\sqrt{49}$ c. $\sqrt{4900}$ d. $\sqrt{.49}$

13. Which of the following statements does not describe $\sqrt{144}$?
   a. A number multiplied by itself is 144.
   b. A number squared is 144.
   c. The square root of 144 is 144 divided by 2.
   d. 144 divided by a number is the same number.

14. The square has an area of 36 cm^2.

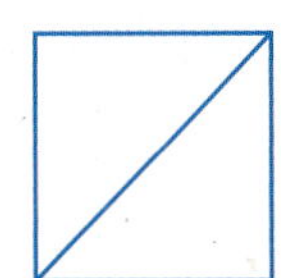

   a. Find the length of each side of the square.
   b. Find the area of each triangle inside the square.
   c. Describe how you found the area of the triangles.

15. Estimate each square root and check your answer with a calculator.
   a. $\sqrt{10}$ b. $\sqrt{42}$ c. $\sqrt{80}$

Skills and Review 1.3

16. Round 64.268 tons to the nearest hundredth.

17. Express each number of minutes as a fraction of an hour and as a decimal. If possible simplify your fractions.
   a. 15 minutes b. 0 minutes
   c. 30 minutes d. 60 minutes

18. Place each number on a number line. (*Hint:* It may be helpful to write each number in the same form in order to determine placement.)
   a. $\frac{3}{2}$ b. .25 c. 212.5%
   d. $1\frac{1}{3}$ e. .875

19. There are 2.54 cm per 1 inch.
   a. How many centimeters are there in 7.6 inches?
   b. Show how you performed the calculation.

20. Write each number in the specified forms.
   a. $\frac{5}{6}$ as a decimal and a percent
   b. .3 as a fraction and a percent
   c. 20% as a fraction and a decimal

21. Simon can detail 10 cars in 5 days. What is his work rate in cars per day?

22. The first inch of a tape measure has been enlarged. Label each of the ticks with an appropriate fraction.

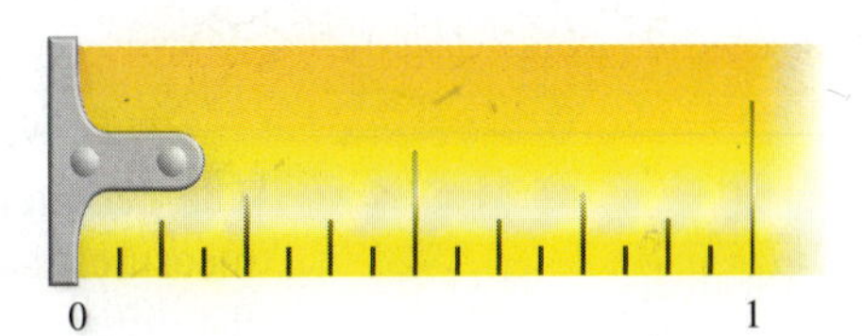

23. Without using a calculator, perform each calculation and include the appropriate units.
   a. $5 \text{ m/s} * 4.2 \text{ s}$
   b. $7 \text{ feet} \div \left(\frac{1}{3}\right) \text{ minute}$
   c. $10 \text{ cm/h} * 92.76 \text{ h}$
24. Write each verbal phrase as a numerical phrase and then evaluate.
   a. The sum of 2 and $^{-}5$
   b. The difference of 2 and $^{-}5$
   c. 1 decreased by 3
   d. 1 less than 3
   e. $\frac{1}{4}$ of 8
   f. 6 more than twice 3
25. The lowest elevation in the state of Louisiana is $^{-}8$ feet in New Orleans. The highest elevation is 535 feet on Driskill mountain. Find the difference in elevations.

1.4 Perimeter and Area

After Studying This Section You Will Be Able To

- Find the perimeter of a two-dimensional figure.
- Apply formulas to find areas of rectangles, triangles, parallelograms, and trapezoids.
- Find the circumference and area of a circle.

In the previous section you counted small squares to find the area of a larger square. The same technique applies to rectangles. In Figure 21 a rectangle has been formed with 1-inch-square tiles. By counting the tiles we can determine that the area is 18 square inches. You may also find the area of the rectangle by multiplying the length (6 inches) by the width (3 inches). In fact, the formula for the area of a rectangle is given by area = length $*$ width, or $a = l * w$.

Figure 21

There are two measurements related to a two-dimensional geometric figure. The **perimeter** measures the distance around the outside of the figure. The **area** measures the amount of space inside the figure. Our first two examples illustrate the difference between these two concepts.

EXAMPLE 1 You move into a new house with an L-shaped living room that has the dimensions shown in Figure 22. You need to put a strip of molding along each wall. Find the total length of the molding you need.

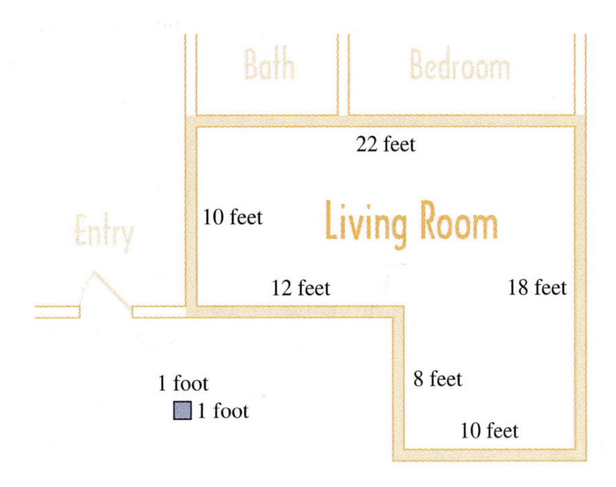

Figure 22

Solution Add the lengths of the walls to find the perimeter:

$$22 \text{ feet} + 18 \text{ feet} + 10 \text{ feet} + 8 \text{ feet} + 12 \text{ feet} + 10 \text{ feet} = 80 \text{ feet}$$

You will need 80 feet of molding.

EXAMPLE 2 You plan to cover the floor of the living room in Example 1 with square tiles that are 1 foot on each side. How many tiles will you need? (See Figure 23.)

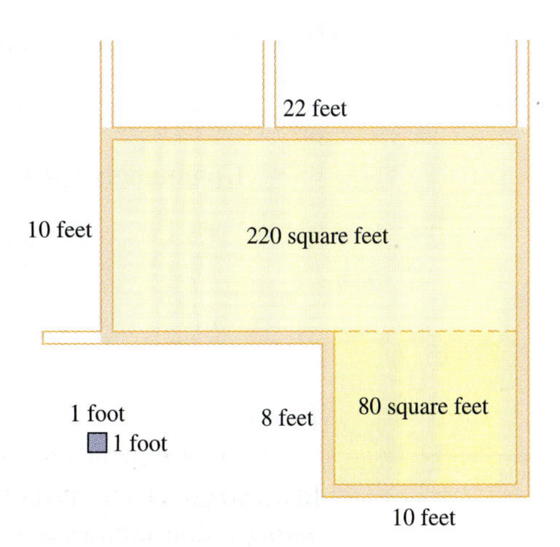

Figure 23

Solution You could cover the room with squares that are 1 foot by 1 foot and count to find out how many there are. That would take a long time. A more efficient method is to split the room into two or more rectangular regions and find the area of each by multiplying length times width. The area of the larger rectangle is 220 square feet. The area of the smaller rectangle is 80 square feet. The total area of the room is 300 square feet. You will need 300 tiles.

The Commutative Properties of Addition and Multiplication

In Example 1, the order in which we add the sides does not make a difference. For instance, instead of starting with 22 feet and going around the figure in a clockwise direction, we could have started with 8 feet and gone in a counterclockwise direction:

$$8 \text{ feet} + 10 \text{ feet} + 18 \text{ feet} + 22 \text{ feet} + 10 \text{ feet} + 12 \text{ feet} = 80 \text{ feet}$$

We get the same sum. In other words, the *order* in which we add numbers does not affect the result. This is called the **commutative property of addition**.

Multiplication is also commutative. For instance, in calculating the area of the larger rectangle in Figure 23 you could multiply 22 feet by 10 feet or 10 feet by 22 feet. In either case, the area is 220 square feet. This is the **commutative property of multiplication**.

The commutative properties may be stated as follows:

The Commutative Properties

- Commutative property of addition:

  For any two numbers a and b, $a + b = b + a$.

- Commutative property of multiplication:

  For any two numbers a and b, $a * b = b * a$.

Formulas

In Example 2 you found the area of a rectangle by multiplying length times width. Formulas for the areas of other familiar geometric figures—triangles, parallelograms, and trapezoids—are shown in the following box.

Area Formulas

Figure	Formula	Abbreviation
Rectangle	area = length $*$ width	$a = l * w$
Triangle	area = $\frac{1}{2}$ $*$ base $*$ height	$a = \frac{1}{2} * b * h$
Parallelogram	area = base $*$ height	$a = b * h$
Trapezoid	area = $\frac{1}{2} * (\text{base}_1 + \text{base}_2)$ $*$ height	$a = \frac{1}{2} * (b_1 + b_2) * h$

Note: In the last formula, the 1 and 2 are called **subscripts**. Base_1 is read "base sub 1" and refers to the first base. Base_2 is read "base sub 2" and refers to the second base.

The following examples illustrate the use of these formulas.

EXAMPLE 3 ■ Use Figure 24 to explain the formula for the area of a triangle. (*Hint:* The height is always **perpendicular** to the base. Two lines are perpendicular when they form **right angles**.)

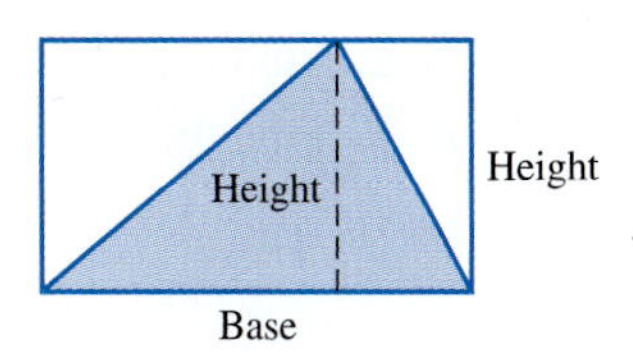

Figure 24

Solution The area of the triangle is half the area of the rectangle that surrounds it. Notice that the base of the triangle is the same as the length of the rectangle. Also, the height of the triangle is the same as the width of the rectangle. Since the area of the rectangle is length $*$ width, the area of the triangle must be $\frac{1}{2}$ $*$ base $*$ height. ■ ■

EXAMPLE 4 ■ Use Figure 25 to explain the formula for the area of a parallelogram.

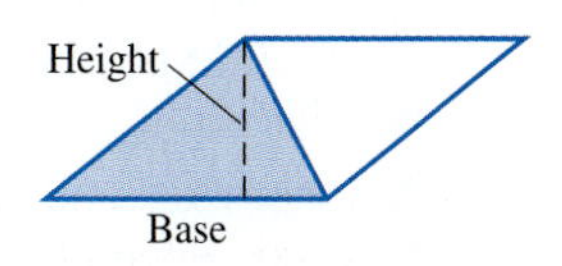

Figure 25

Solution The parallelogram is made up of two triangles that are the same size and same shape. Each triangle has an area equal to $\frac{1}{2}$ * base * height. Therefore the area of the parallelogram is $2 * \frac{1}{2}$ * base * height, or simply base * height.

EXAMPLE 5 Use Figure 26 to explain the formula for the area of a trapezoid.

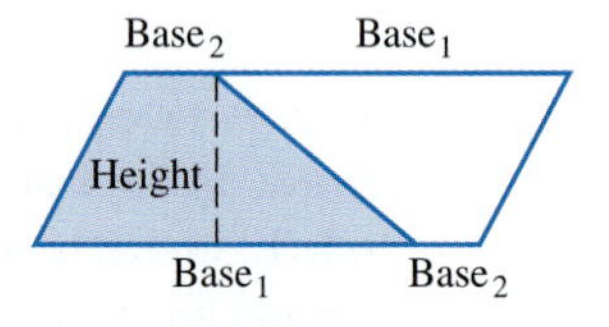

Figure 26

Solution Two trapezoids fit together to form a parallelogram. The area of the parallelogram is $(\text{base}_1 + \text{base}_2) * \text{height}$. Therefore, the area of the trapezoid is $\frac{1}{2} * (\text{base}_1 + \text{base}_2) * \text{height}$, half the area of the parallelogram.

EXAMPLE 6

a. Find the perimeter of the trapezoid in Figure 27.

b. Find the area of the trapezoid in Figure 27.

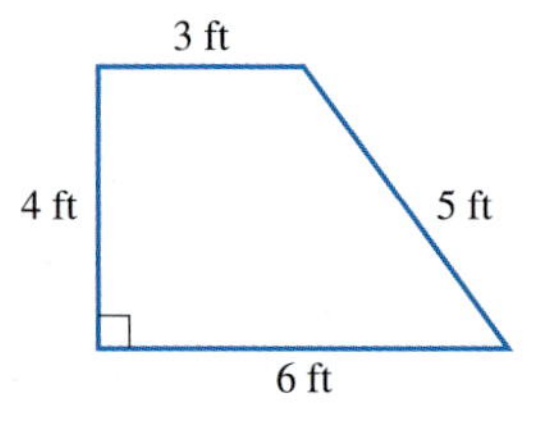

Figure 27

Solution

a. To find the perimeter, add the lengths of the sides: 3 feet + 5 feet + 6 feet + 4 feet = 18 feet.

b. To find the area of the trapezoid, use the formula $A = \frac{1}{2} * (\text{base}_1 + \text{base}_2) * \text{height}$. Note that the small square in the lower-left corner indicates that two sides are perpendicular, so 4 feet is the height. The bases are 3 feet and 6 feet. Thus, $\text{base}_1 = 6$ feet, $\text{base}_2 = 3$ feet, and height = 4 feet.

Substitute these values into the formula:

$$A = \frac{1}{2} * (\text{base}_1 + \text{base}_2) * \text{height}$$

$$A = \frac{1}{2} * (6 + 3) * 4 \qquad \textit{Substituting}$$

$A = \frac{1}{2} * 9 * 4$ *Performing the operation inside parentheses first*

$A = 18$

The area is 18 square feet.

Circles

A circle is the set of all points in a plane that are a given distance from a point called the *center*. A **radius** is a line segment joining the center to a point on the circle. A **diameter** is a line segment that passes through the center and joins two points on the circle (see Figure 28).

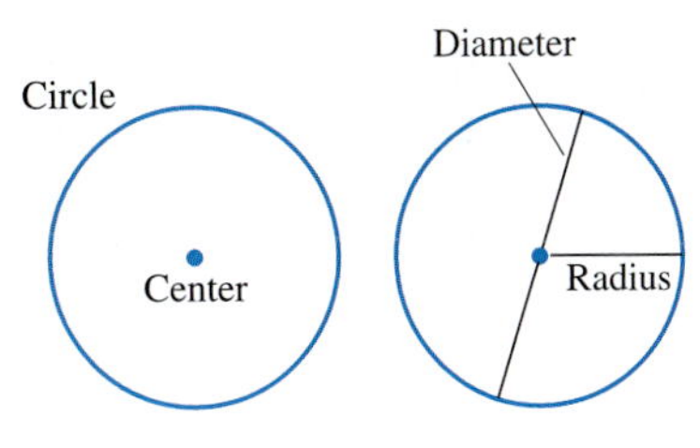

Figure 28

The perimeter of a circle is called the **circumference**. The ratio of the circumference to the diameter is the same for all circles. This ratio is a special number called pi, symbolized by the Greek letter π.

According to a calculator, $\pi \approx 3.141592654$. Recall that pi is an irrational number, so its digits continue forever with no apparent pattern. A commonly used approximation for π is 3.14.

Formulas for the circumference and area of a circle are given in the following box.

Circumference and Area of a Circle

Formula	**Abbreviation**
circumference $= \pi *$ diameter $= 2 * \pi *$ radius	$C = \pi * d$ or $C = 2 * \pi * r$
area $= \pi *$ radius $*$ radius $= \pi *$ radius^2	$A = \pi * r * r$ or $A = \pi * r^2$

EXAMPLE 7 Carl is going to work out by running around a circular track with a radius of 70 yards. His goal is to run 5 miles. How many laps should he run (see Figure 29)? (1 mile = 1760 yards)

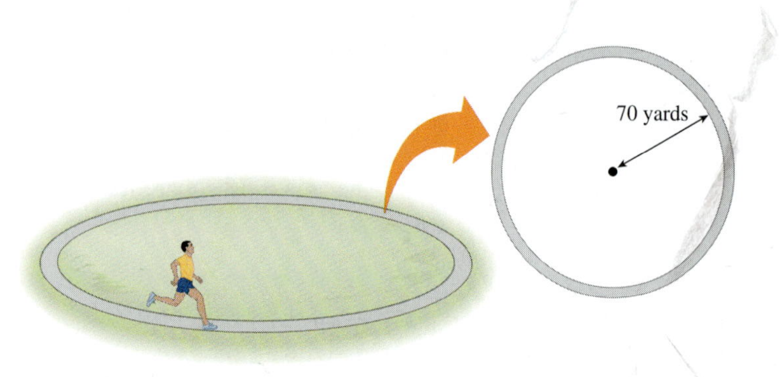

Figure 29

Solution In order to solve this problem, Carl needs to know the length of one lap—that is, the circumference of the circle.

We can use the formula circumference $= 2 * \pi *$ **radius**. Using the approximation $\pi \approx 3.14$, $2 * \pi \approx 6.28$. Multiply 6.28 by **70** yards to get circumference ≈ 439.6 yards, or 440 yards.

If Carl wants to run 5 miles, that's $5 * 1760 = 8800$ yards. Because $8800 \div 440 = 20$, he'll have to run 20 laps to make his goal.

Note: Here is another method. Because the radius is 70 yards, the diameter must be twice that distance, or 140 yards. Using circumference $= \pi *$ diameter and $\pi \approx 3.14$, we get circumference ≈ 439.6 yards, or 440 yards, to the nearest yard.

Exercises 1.4

1. In Example 1 you found the perimeter of the figure. In Example 2 you found the area. Explain the difference between the two concepts.

2. In the following figure, find each value.

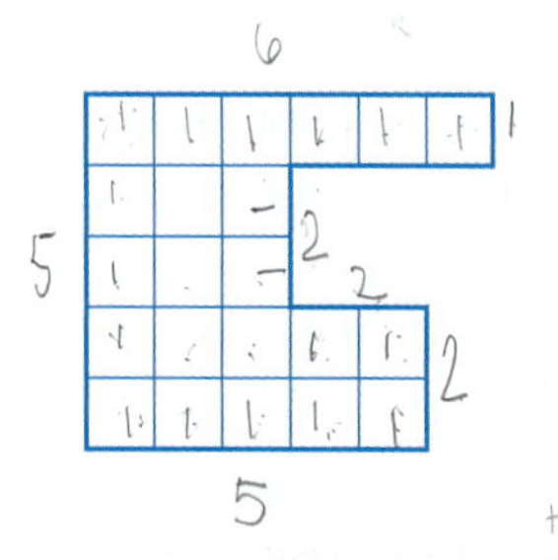

a. Area

b. Perimeter

3. Use grid paper to draw two rectangles, each with an area of 24 square units but with different perimeters.

4. A rectangular field has a perimeter of 600 feet. Find two possible fields with different areas and find the areas.

5. The length of a rectangle is 4.5 cm and its width is 2.5 cm.

a. Find its perimeter.

b. Draw the rectangle on grid paper and estimate its area.

c. Use the formula area = length $*$ width to find its area exactly.

d. Compare your answer in (c) with the estimate from part (b).

6. You are told that the length of a rectangle is twice its width and that the width is 50 yards.

a. Find the area of the rectangle.

b. Find the perimeter of the rectangle.

c. Suppose that the length of another rectangle is 10 yards more than twice its width. Assume that the width of this rectangle is also 50 yards. Answer parts (a) and (b) for this rectangle.

7. The Chang family owns a farm in the shape of a rectangle $\frac{1}{2}$ mile long and $\frac{1}{4}$ mile wide (see the figure).

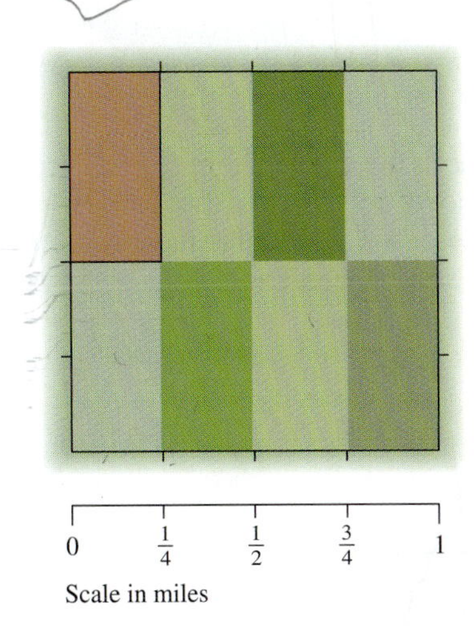

a. What portion of the figure is shaded pink?

b. Use the formula $A = \text{length} * \text{width}$ to find the area of the farm. Compare your answer with your answer in (a).

c. Find the perimeter of the farm.

d. Find the area of the farm in square rods and the perimeter in rods. (*Hint:* 1 mile = 320 rods.)

e. Find the area of the farm in acres. (*Hint:* 1 acre = 160 square rods.)

f. Check the answer in (e) by using the fact that there are 640 acres in 1 square mile.

8. Find the area and the perimeter of a rectangle that is $\frac{3}{4}$ mile long and $\frac{1}{3}$ mile wide.

9. A triangular patio has the dimensions shown in the figure.

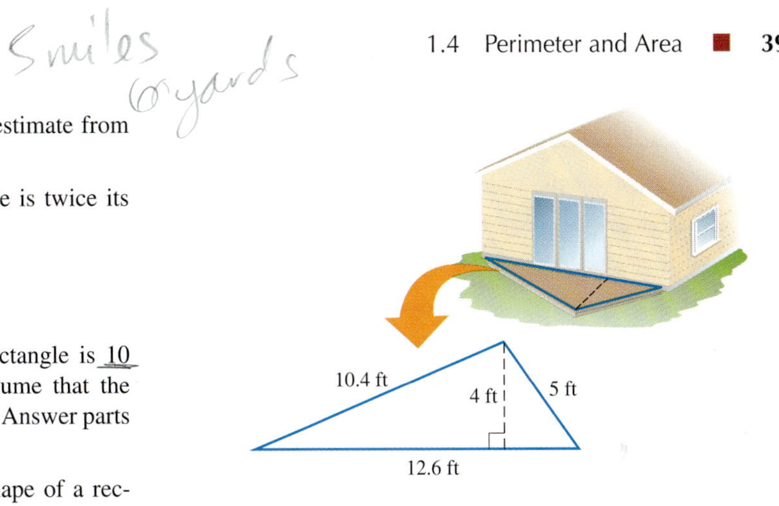

a. Find the area of the patio.

b. Find the perimeter of the patio.

10. In order to reduce the noise from above, the ceiling shown in the figure is covered with acoustic tiles.

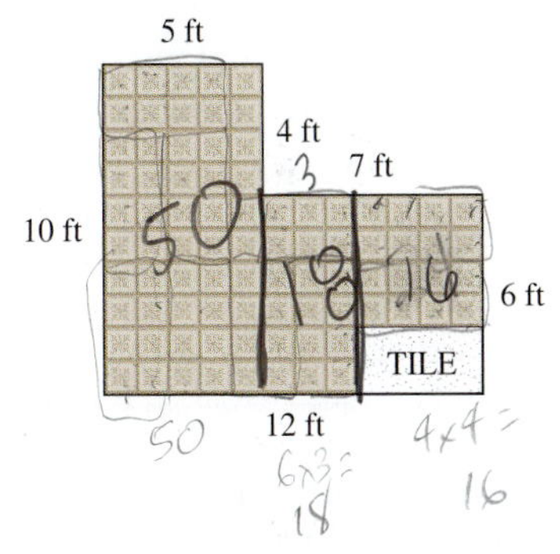

a. What is the area of the ceiling?

b. Each tile measures 2 feet by 4 feet. How many tiles are necessary to cover the ceiling?

11. Find the area of the trapezoid (see the figure).

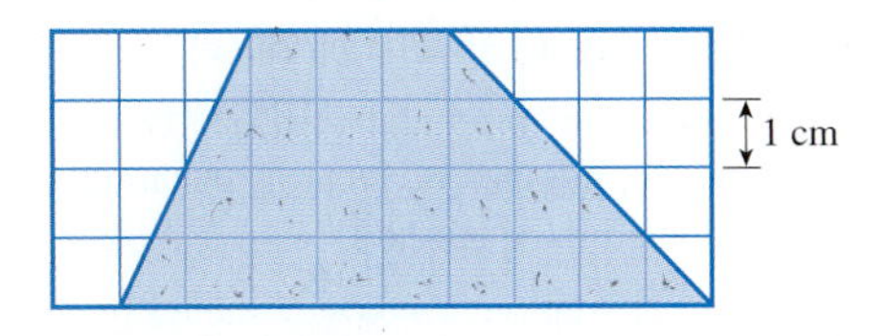

a. Count squares (and parts of squares).

b. Use a formula.

12. Which of the figures have the same area? Which have the same perimeter? Explain.

a.

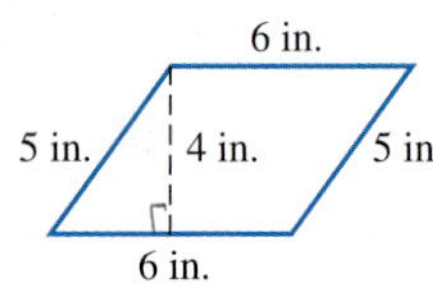

b.

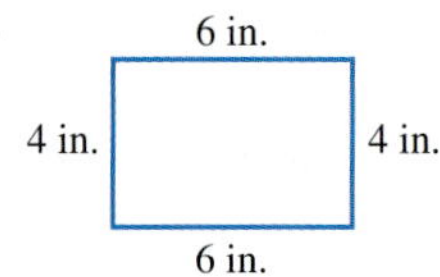

c.

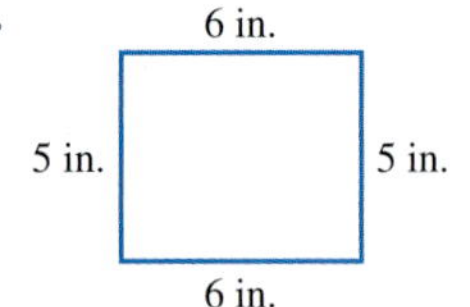

d.

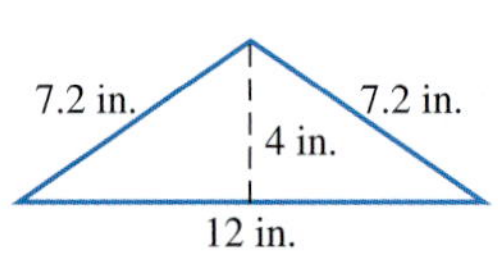

13. In Example 7 why did Carl need to know the circumference of the circle rather than the area?

14. A couple owns a circular swimming pool with a radius of 10 feet (see the figure).

a. They want to buy a cover for the pool. If pool covers are measured in square feet, what size cover should they purchase? Use $\pi \approx 3.14$.

b. The couple also wants to place a fence around the pool. The fence is 4 feet from the edge of the pool (see figure). How many feet of fencing will they need to surround the pool?

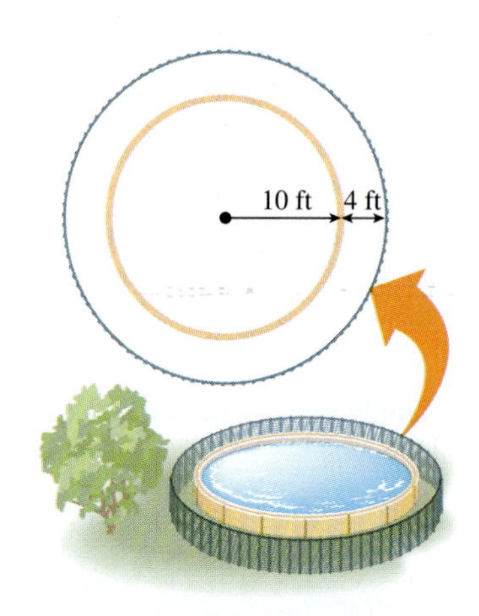

15. Two students use different formulas to calculate the circumference of a circle. Lynn uses $C = 2 * \pi * r$. George uses $C = \pi * d$. Show that these two formulas produce the same circumference for a circle with a radius of 8.5 m.

Skills and Review 1.4

16. Sammy performs the sum $5 + 3 + 5 + 6 + 7 + 4$ mentally by finding pairs of numbers that add to 10.

a. Find the sum by Sammy's method.

b. Explain why Sammy's method works.

17. Suppose a young student knows that $7 * 5$ is 35 but doesn't know the product $5 * 7$.

a. How can the student use the commutative property of multiplication to find the product?

b. Explain how the commutative property can be used to reduce the number of facts in the multiplication table that need to be learned.

18. Perform the calculation $7 * 6 * 5$ mentally using the commutative property of multiplication to reorder for easier multiplication.

19. Find each value without using a calculator.

a. $\sqrt{36}$ **b.** $\sqrt{100}$ **c.** $\sqrt{10{,}000}$

d. $\sqrt{.81}$ **e.** $\sqrt{\frac{9}{25}}$

20. Evaluate without using a calculator and change your answers to scientific notation. Be sure to include the appropriate units.

a. 700 feet $*$ 2000 feet $*$ 5000 feet

b. $\dfrac{48{,}000{,}000 \text{ cm}^2}{60{,}000 \text{ cm}}$

21. What is the value of the digit 7 in the number 1,247,650.12?

22. Find each value without using a calculator.

a. 4^3 **b.** $\left(\frac{5}{6}\right)^2$ **c.** 1^{12}

d. $({}^{-}3)^2$ **e.** 7^0

23. Without using a calculator, find 5^{-2}. Write your answer as a decimal.

24. Nicole reads 300 words per minute. What is her reading rate in words per second?

25. Which of the following numbers is not equal to 1.25?

a. $1\frac{1}{4}$ **b.** $\frac{1.25}{1}$ **c.** $\frac{5}{4}$

d. $1 \div \frac{4}{5}$ **e.** $\frac{7}{5}$ **f.** $\frac{10}{8}$

CHAPTER 1 ▪ KEY CONCEPTS

Adding Signed Numbers To add (or combine) two signed numbers, if the numbers have the *same* sign, *add* their absolute values and keep the sign. If the numbers have *opposite* signs, *subtract* their absolute values, and take the sign of the number with the *larger* absolute value. (Page 3)

Subtracting Signed Numbers To subtract two signed numbers, change the sign of the second number and change the operation to addition. Then follow the rules for the addition of signed numbers. (Page 7)

Number Line Models for Addition and Subtraction (Page 6)

$$\text{start} + \text{change} = \text{final}$$

$$\text{final} - \text{start} = \text{change}$$

Multiplying and Dividing Signed Numbers (Page 9)

positive ∗ positive = positive	positive ÷ positive = positive
positive ∗ negative = negative	positive ÷ negative = negative
negative ∗ positive = negative	negative ÷ positive = negative
negative ∗ negative = positive	negative ÷ negative = positive

Multiplying and Dividing with Zero 0 multiplied by any number is 0. 0 divided by any positive or negative number is 0. Division by 0 is undefined. (Page 10)

Distance, Rate, and Time $\text{rate} = \dfrac{\text{distance}}{\text{time}}$. (Page 13).

Work, Rate, and Time $\text{work rate} = \dfrac{\text{work}}{\text{time}}$. (Page 15)

Rational Numbers A rational number is one that can be expressed as the quotient or ratio of two integers. Rational numbers include integers and fractions and may be expressed as decimals or percents. (Page 17)

Common Fraction-Decimal-Percent Equivalents These equivalents are shown in the table. (Page 19)

Positive Integer Exponents Positive integer exponents may be used to represent repeated multiplication. For example, 5^3 means the same as $5 * 5 * 5$. (Page 22)

Zero Exponent Let a be any number except 0. Then $a^0 = 1$. (Page 24)

Negative Integer Exponents Let a be any number except 0 and let n be any positive integer. Then

$$a^{-n} = \frac{1}{a^n}$$

(Page 26)

Scientific Notation A number is written in scientific notation if it is of the form $a * 10^b$ where $1 \leq a < 10$ and b is an integer. (Page 27)

Square Root If we know the area of a square, the length of its side is called a square root of the area. For example, $\sqrt{16} = 4$ because $4^2 = 16$. (Page 28)

Irrational Numbers An irrational number cannot be expressed as the ratio of two integers and there is no pattern of repeating digits in its decimal representation. (Page 30)

Perimeter and Area The perimeter measures the distance around the outside of a two-dimensional figure. The area measures the amount of space inside the figure. (Page 32)

The Commutative Properties Addition: For any two numbers a and b, $a + b = b + a$. Multiplication: For any two numbers a and b, $a * b = b * a$. (Page 34)

Area Formulas (Page 35)

Figure	**Formula**	**Abbreviation**
Rectangle	area = length $*$ width	$a = l * w$
Triangle	area = $\frac{1}{2}$ $*$ base $*$ height	$a = \frac{1}{2} * b * h$
Parallelogram	area = base $*$ height	$a = b * h$
Trapezoid	area = $\frac{1}{2}$ $*$ ($\text{base}_1 + \text{base}_2$) $*$ height	$a = \frac{1}{2} * (b_1 + b_2) * h$

Circumference and Area of a Circle (Page 37).

Formula	**Abbreviation**
circumference = π $*$ diameter = $2 * \pi *$ radius	$C = \pi * d$ or $C = 2 * \pi * r$
area = π $*$ radius $*$ radius = π $*$ radius^2	$A = \pi * r * r$ or $A = \pi * r^2$

Exploration

The week of June 3, 2002, was not a good one for Wall Street. The changes in the Dow Jones Industrial Average (DJIA) for each day were as follows: Monday, $^-215.46$; Tuesday, $^-21.95$; Wednesday, 108.96; Thursday, $^-172.16$; Friday, $^-34.97$.

a. Find the net change in the DJIA for the week.

b. When the market opened on Monday, June 3, the DJIA stood at 9925.25. What was the DJIA at the close of trading on Friday, June 7?

c. Explain your answer to part (b) using a number line.

Chapter 1 ▪ Review Exercises

Section 1.1 Positive and Negative Numbers

1. Use a charged particle model to find each sum.

a. $^+3 + {}^+4$ **b.** $^+2 + {}^-5$

c. $^-1 + {}^+3$ **d.** $^-4 + {}^-2$

2. Use the addition rules for signed numbers to find each sum.

a. $^+71.2 + {}^-2.5$ **b.** $^-15.1 + {}^-56.76$

3. Use the rule for subtraction to find each difference.

a. $^{+}6 - {}^{+}5$ b. $^{+}18 - {}^{-}7$

c. $\frac{-3}{8} - \frac{+1}{4}$ d. $^{-}12.1 - {}^{-}4.3$

4. At 8 A.M. a balloonist was floating over Badwater Basin at an elevation of 106 feet below sea level. By 11 A.M. the balloonist had risen 151 feet. What was the balloonist's elevation at 11 A.M.?

5. At 2 P.M. the balloonist in Exercise 4 was at an elevation of 86 feet. By 5 P.M. the balloonist had fallen to an elevation of $^{-}34$ feet. By how much had the balloonist's elevation changed?

6. Find each product.

a. $^{+}3 * {}^{+}2$ b. $^{+}25 * {}^{-}8$

c. $\frac{-4}{5} * \frac{+5}{6}$ d. $^{-}1.6 * {}^{-}5$

7. Find each quotient.

a. $^{+}8 \div {}^{+}4$ b. $^{+}102 \div {}^{-}2$

c. $\frac{-1}{10} \div \frac{+3}{2}$ d. $^{-}3.9 \div {}^{-}1.3$

Section 1.2 Rates and Ratios

8. Find each rate in kilometers per hour.

a. Kim drives 140 km in 2 h.

b. Phil jogs 3 km in $\frac{1}{4}$ h.

c. Laura walks 6 km in $\frac{3}{4}$ h.

9. Find each work rate.

a. Ed can carpet 400 square feet in 50 minutes. What is his work rate in square feet per minute?

b. Ed can complete a carpet job in 3 hours. What is Ed's work rate in jobs per hour?

c. Debbie can complete the same carpet job in 4 hours. What is Debbie's work rate in jobs per hour?

d. What is Ed and Debbie's combined work rate?

10. Express each fraction as a decimal. Round any decimals with more than three decimal places to the nearest thousandth.

a. $\frac{9}{5}$ b. $\frac{17}{30}$ c. $\frac{2}{35}$

11. Express each of the decimals in Exercise 10 as a percent.

12. Place each number on the number line.

a. $\frac{3}{4}$ b. 20% c. $1\frac{3}{8}$ d. 1.2 e. $\frac{20}{5}$

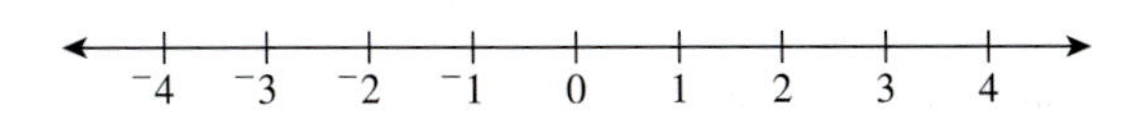

Section 1.3 Exponents and Radicals

13. Find each value without using a calculator.

a. 4^3 b. 3^4 c. 8^1

d. $\left(\frac{1}{2}\right)^5$ e. 1^{36}

14. Find each value without using a calculator.

a. $(^{-}3)^1$ b. $(^{-}3)^2$ c. $(^{-}3)^3$ d. $(^{-}3)^4$

15. Find each value without using a calculator.

a. 2^{-4} b. 15^{-1} c. 58^0

16. Write each number in scientific notation.

a. 348,000,000 b. .00074 c. 5,772

17. Use a calculator to find each answer. Write your answer in both scientific notation and decimal notation.

a. $561{,}000 * 24{,}800$ b. $2.5 \div 5000$

18. Find each value without using a calculator.

a. $\sqrt{9}$ b. $\sqrt{64}$ c. $\sqrt{36}$

19. Estimate each square root.

a. $\sqrt{8}$ b. $\sqrt{37}$ c. $\sqrt{72}$

20. The area of the square is 48 square inches. Estimate the length of one side of the square.

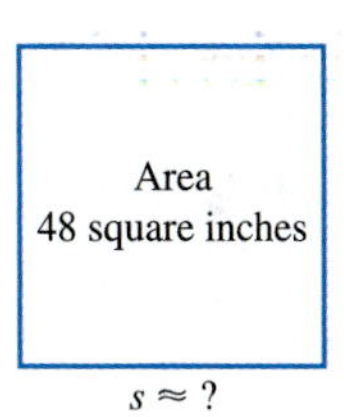

Section 1.4 Perimeter and Area

In Exercises 21–23, use each figure to find the perimeter and area. Assume each square within the figures measures 1 cm by 1 cm.

21.

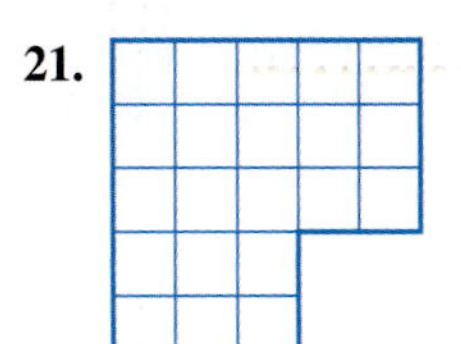

22.

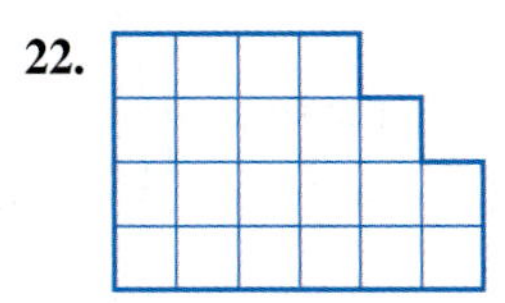

23.

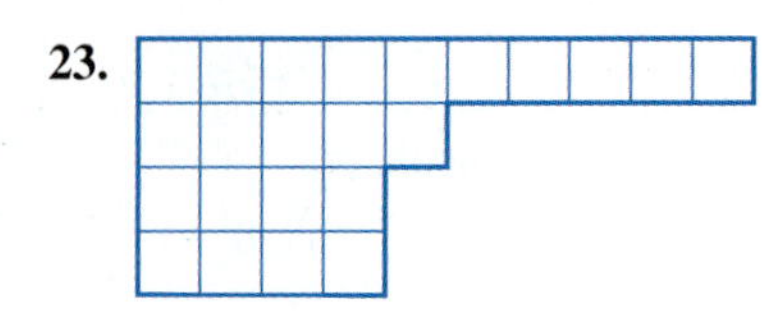

24. Use the commutative property of addition to reorder the following sum. Choose an order that will allow you to perform this sum mentally.

$$3 + 18 + 27 + 2$$

25. Use the commutative property of multiplication to reorder the multiplication. Choose an order that will allow you to perform this multiplication mentally.

$$2 * 16 * 5$$

26. Find the perimeter and area of the following parallelogram.

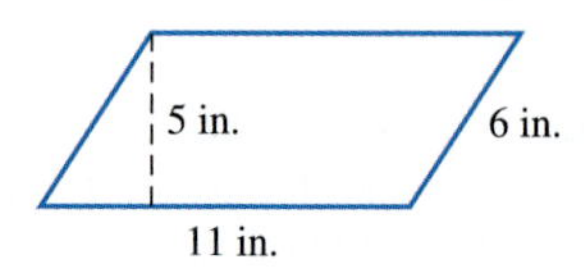

27. Find the circumference of a circle with a radius of 7 feet. Use $\pi \approx 3.14$.

CHAPTER 1 ■ TEST

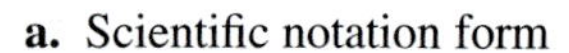

1. Determine the sum $^{-}8 + 5$.
2. Determine the difference $^{-}7 - 3$.
3. Suppose the temperature at the North Pole went from $^{-}23°$C at 3 A.M. to 2°C at noon. By how much had the temperature changed?
4. Determine each product or quotient.
   a. $12 \div {}^{-}3$ **b.** $^{-}3 * 2$ **c.** $^{-}3 \div {}^{-}12$
5. Larry drinks 64 ounces of water per day. Find his rate of drinking water in ounces per hour.
6. Express each fraction as a decimal and as a percent.
   a. $\frac{2}{5}$ **b.** $\frac{3}{4}$ **c.** $\frac{3}{10}$ **d.** $\frac{1}{3}$ **e.** $\frac{9}{4}$
7. Order the numbers from smallest to largest:
   $1\frac{5}{8}$; 1.512; $\frac{3}{2}$; 145%; 2^0
8. Find each value without using a calculator.
   a. 2^3 **b.** $(^{-}2)^3$ **c.** 2^{-3}
   d. 2^0 **e.** 1^{72}
9. Express each number in scientific notation.
   a. 115,000 **b.** .0000074
10. On your calculator, find .000000000032 mm $*$ 52,000 mm. Include the proper units and express the result as specified.
    a. Scientific notation form
    b. Decimal form
11. Estimate the square root $\sqrt{52}$ without using a calculator.
12. Determine the perimeter and area of the given figure. Assume each square within the figure measures 1 cm by 1 cm.

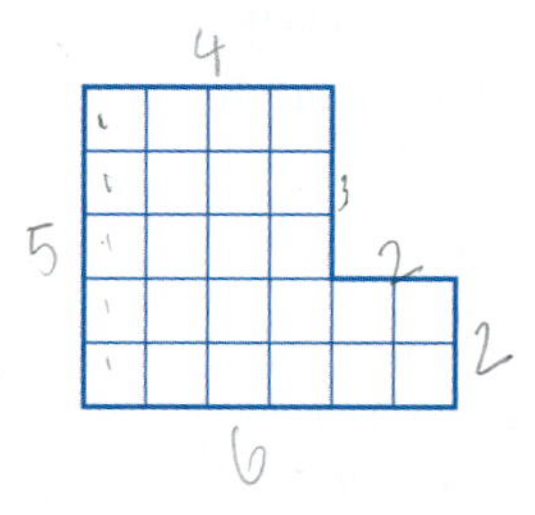

13. Use the commutative property of addition to reorder $14 + 33 + 2 + 16 + 7 + 8$ so that the sum can be found mentally.
14. The radius of a tire is 280 mm. Find the circumference of the tire. Use $\pi \approx 3.14$.
15. The area of a rectangular living room is 112 square feet and the room is 10 feet wide. What is the length of the living room?

CHAPTER

2

Algebraic Expressions

Where on Earth Are We?

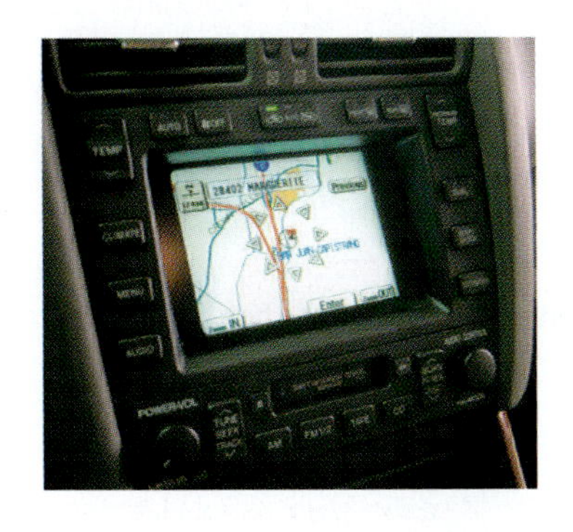

Global positioning system (GPS) technology has revolutionized the art of navigation. This system is based on 24 satellites, which orbit the earth at an altitude of 11,000 miles and send signals to receivers on or near the earth below. These signals are then used to determine the distance from a satellite to a receiver. Once the distance from a receiver to three or four of the satellites is known, a computer can find the location of the receiver within 100 m and with 95% accuracy.

GPS receivers are now standard equipment on commercial and military aircraft. They have helped explorers in remote regions of the world such as Antarctica and were used in the construction of the tunnel beneath the English Channel, which links England and France. The technology is now available for use in private cars in conjunction with a database of road maps. With GPS in your car you may never need to ask for directions!

GPS is based on a three-dimensional coordinate system similar to the two-dimensional system you study in this chapter. (See Exploration on p. 84.)

Need help? For on-line resources, visit this web site: **math.college.hmco.com/students.**

2.1 Grouping and the Order of Operations

AFTER STUDYING THIS SECTION YOU WILL BE ABLE TO

- Apply the commutative, associative, and distributive properties.
- Apply the order of operations.
- Apply special properties of 0, 1, and $^-1$.

In Chapter 1 you observed that letters may be used to represent numbers. In the formula for the circumference of a circle, $C = 2 * \pi * r$, for example, C represents the circumference and r represents the radius. Both C and r are **variables**, that is, they may assume many different values. On the other hand, the Greek letter π is a **constant**. It always represents the same number, approximately 3.14. Of course, 2 is also a constant, because its value is fixed.

The focus of this chapter is algebraic expressions. An **expression** is a meaningful arrangement of constants, variables, and special symbols.

The special symbols include operations such as addition, subtraction, multiplication, and division, as well as parentheses, ( ), used for grouping. In this section you learn how grouping is used in expressions.

The Associative Properties of Addition and Multiplication

In addition to being commutative, addition and multiplication are also *associative*.

EXAMPLE 1 You need to build a fence around a triangular piece of land with the dimensions shown in Figure 1. Find the length of the fence.

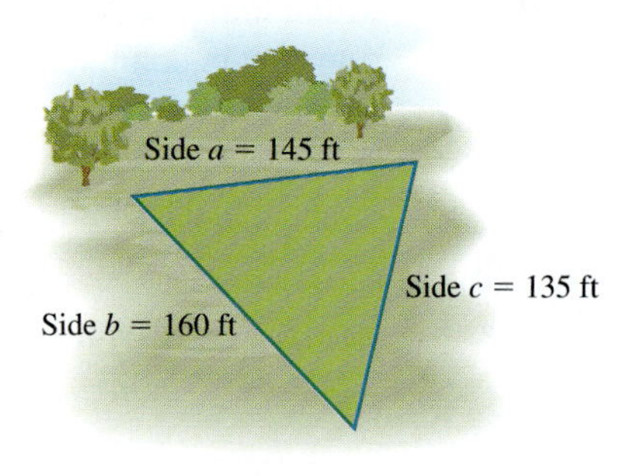

Figure 1

Solution The length of the fence is the perimeter of the triangle, which is the sum of the lengths of the three sides. There are several ways to find the perimeter. Here are two of them:

1. First add a and b then add the result to c.

$$\begin{aligned}(a + b) + c &= (145 \text{ feet} + 160 \text{ feet}) + 135 \text{ feet} \\ &= 305 \text{ feet} + 135 \text{ feet} \\ &= 440 \text{ feet}\end{aligned}$$

2. First add b and c and then add the result to a.

$$\begin{aligned}a + (b + c) &= 145 \text{ feet} + (160 \text{ feet} + 135 \text{ feet}) \\ &= 145 \text{ feet} + 295 \text{ feet} \\ &= 440 \text{ feet}\end{aligned}$$

Example 1 shows that three numbers added together may be grouped with parentheses in two different ways to get the same sum. In other words, $(a + b) + c = a + (b + c)$. This is called the **associative property of addition**, because the parentheses determine which numbers are to be associated. There is also an **associative property for multiplication**.

The Associative Properties

- Associative property of addition: For any numbers a, b, and c, $(a + b) + c = a + (b + c)$.
- Associative property of multiplication: For any numbers a, b, and c, $(a * b) * c = a * (b * c)$.

The Distributive Property

Another important property is the *distributive property*, as illustrated in Example 2.

EXAMPLE 2 A farmer has a rectangular field with dimensions 42 m and 20 m (Figure 2).

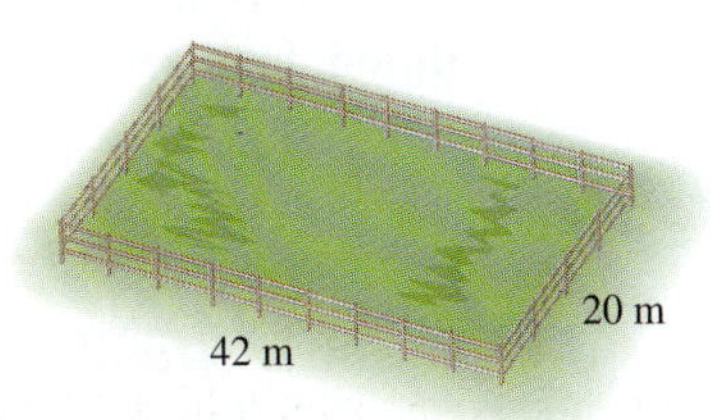

Figure 2

He decides to build a fence parallel to one of the sides, as shown in Figure 3. Find the area of the original field and of each of the two new fields.

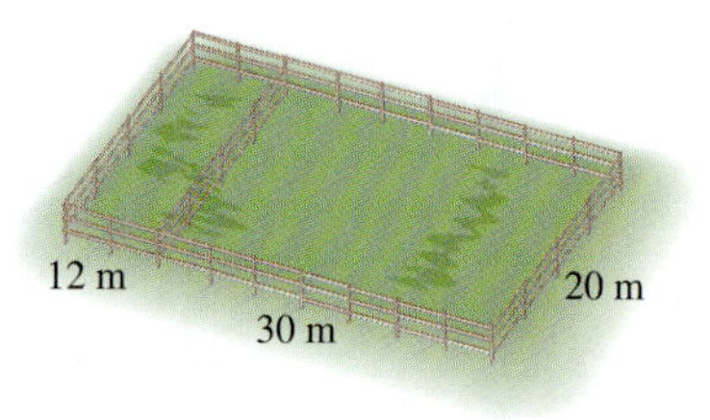

Figure 3

Solution Use the formula for area of a rectangle, area = **length** $*$ **width**.

In Figure 2,

$$\text{Area of original field} = \mathbf{20} * \mathbf{42} = 840 \text{ m}^2$$

In Figure 3,

$$\text{Area of field on left} = \mathbf{20} * \mathbf{12} = 240 \text{ m}^2$$

$$\text{Area of field on right} = \mathbf{20} * \mathbf{30} = 600 \text{ m}^2$$

Check: Because the total area has not changed, the sum of the areas of the two new fields should be the same as the area of the original field:

$$240 \text{ m}^2 + 600 \text{ m}^2 = 840 \text{ m}^2$$

Example 2 illustrates an important property of multiplication and addition. Notice that the length of the original field, 42 m, is the sum of the lengths of the two new fields, 12 m + 30 m. Thus we can write $20 * (12 + 30) = 840$ and $(20 * 12) + (20 * 30) = 840$. Because the expressions are both equal to 840, they must be equal to each other:

$$20 * (12 + 30) = (20 * 12) + (20 * 30)$$

This is an example of the **distributive property of multiplication over addition**. In general we can say that $a * (b + c) = (a * b) + (a * c)$, as illustrated in Figure 4.

The Distributive Property of Multiplication over Addition

For all numbers a, b, and c, $a * (b + c) = (a * b) + (a * c)$.

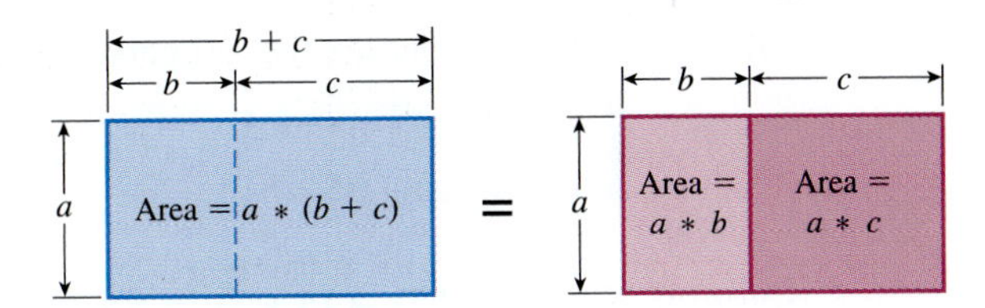

Figure 4

Here are some important facts about the distributive property.

1. The distributive property states that the *product*, $a * (b + c)$, is equal to the *sum*, $(a * b) + (a * c)$.
2. We use the words "expand" and "factor" in connection with the distributive property. If we start with the product, we **expand** $a * (b + c)$ to get $(a * b) + (a * c)$. If we start with the sum, we **factor** $(a * b) + (a * c)$ to get $a * (b + c)$.
3. We sometimes say that multiplication distributes over addition. Multiplication by a is distributed to both b and c.
4. Because subtracting a number is the same as adding its opposite (page 7), multiplication also distributes over subtraction; that is, $a * (b - c) = (a * b) - (a * c)$.
5. Because of the commutative property of multiplication, we can also write $(b + c) * a = (b * a) + (c * a)$. Thus, you can distribute from both the left and the right.

Examples 3 and 4 show how to use the distributive property.

EXAMPLE 3 Use the distributive property to expand each expression.

a. $4 * (20 + 5)$
b. $6 * (x + y)$
c. $3 * (a + b + c)$

Solution

a. $4 * (20 + 5) = (4 * 20) + (4 * 5)$. You can check this result by finding the value of each expression.

$$4 * (20 + 5) = 4 * 25 = 100$$

$$(4 * 20) + (4 * 5) = 80 + 20 = 100$$

b. $6 * (x + y) = (6 * x) + (6 * y)$

c. Multiplication distributes over all three terms, a, b, and c.

$$3 * (a + b + c) = (3 * a) + (3 * b) + (3 * c)$$

EXAMPLE 4 Use the distributive property to factor each expression.

a. $(5 * 97) + (5 * 3)$

b. $(8 * a) - (8 * b)$

Solution **a.** $(\mathbf{5} * 97) + (\mathbf{5} * 3) = \mathbf{5} * (97 + 3)$. You can check this result by finding the value of each expression.

$$(5 * 97) + (5 * 3) = 485 + 15 = 500$$
$$5 * (97 + 3) = 5 * 100 \quad = 500$$

b. $(\mathbf{8} * a) - (\mathbf{8} * b) = \mathbf{8} * (a - b)$

The Order of Operations

You may have noticed that we used a lot of parentheses in the preceding examples. Some of these parentheses are not necessary if we take advantage of a rule called the **order of operations**.

The Order of Operations

1. Do operations inside grouping symbols first.
2. Evaluate exponents and roots from left to right.
3. Perform multiplication and division from left to right.
4. Perform negations, then addition and subtraction from left to right.

This rule guarantees that whenever we evaluate an expression, everyone will arrive at the same answer. To be a successful algebra student, it is *extremely important* that you learn the order of operations, just as to be a successful driver you must learn the rules of the road. The order of operations is summarized in Figure 5.

First parentheses, then	( )
Exponents and roots, followed by	^ √
Multiplication and division, and finally	* /
Negation, addition, and subtraction	⁻ + −

Figure 5 Use this chart to remember the order of operations.

EXAMPLE 5 Apply the order of operations to find the value of each of these expressions.

a. $2 * 3 + \frac{12}{4}$

b. $(10 - 3 * 6)^2$
c. $^-2^4 + (^-3)^4$
d. $\frac{\sqrt{16}}{3+1}$

Solution

a. There are no grouping symbols, exponents, or roots. Step 3 says to perform multiplication and division from left to right. Then step 4 says to perform the addition.

$$\begin{aligned} 2 * 3 + \tfrac{12}{4} &= 6 + 3 \\ &= 9 \end{aligned}$$

b. Do the operations inside the parentheses first. Inside the parentheses multiplication (step 3) comes before subtraction (step 4). The result is then raised to the second power.

$$\begin{aligned} (10 - 3 * 6)^2 &= (10 - 18)^2 \\ &= (^-8)^2 \\ &= 64 \end{aligned}$$

c. There are two negation symbols, one inside parentheses and the other outside. Because exponents (step 2) come before negation (step 4), apply the exponents to positive 2 and negative 3. Then negate 16 and add the result to 81.

$$\begin{aligned} ^-2^4 + (^-3)^4 &= {}^-16 + 81 \\ &= 65 \end{aligned}$$

d. Remember that the horizontal fraction bar is a symbol of grouping. Apply operations in the numerator and denominator before performing the division.

$$\begin{aligned} \frac{\sqrt{16}}{3+1} &= \frac{4}{4} \\ &= 4 \div 4 \\ &= 1 \end{aligned}$$

Caution: Scientific and graphics calculators such as the TI-83 use the order of operations. Be careful, however—not all calculators do. If you have an unfamiliar calculator, test it first to see if it follows the rules. Enter $1 + 2 * 3$. If the answer is 7, the calculator is following the order of operations. If the answer is 9, it is just taking the operations in order from left to right.

Special Properties of 0, 1, and $^-1$

Chapter 1 introduced the commutative property of addition and multiplication. This section introduced the associative and distributive properties. These three properties will be extremely useful in your study of algebra. In addition, there are three special numbers, 0, 1, and $^-1$, with special properties that are important to know.

Special Properties of 0

a. The sum of a number and its opposite is zero. For any number a, $a + {}^-a = 0$. *Example:* $6 + {}^-6 = 0$.

b. Zero added to any number gives the same number. For any number a, $0 + a = a$ and $a + 0 = a$. *Example:* $3 + 0 = 3$.

c. Zero times any number gives zero. For any number a, $0 * a = 0$ and $a * 0 = 0$. *Example:* $4 * 0 = 0$.

d. You cannot divide by 0. For any number a, $\frac{a}{0}$ is undefined; that is, it has no meaning. *Example:* $\frac{7}{0}$ is undefined. When you enter $7 \div 0$ on a calculator, you should get an error message. Try it.

Special Properties of 1

a. The product of a number and its reciprocal is 1. For any number a (except 0), $a * (\frac{1}{a}) = 1$. *Example:* $3 * (\frac{1}{3}) = 1$.

b. One multiplied by any number gives the same number. For any number a, $1 * a = a$ and $a * 1 = a$. *Example:* $1 * 7 = 7$.

c. $1*$ may be inserted in front of a left parenthesis. Thus $(a + b) = 1 * (a + b)$. *Example:* $(6 + x) = 1 * (6 + x)$.

d. $*1$ may be inserted behind a right parenthesis. Thus $(a + b) = (a + b) * 1$. *Example:* $(6 + x) = (6 + x) * 1$.

Special Properties of $^-1$

a. Negative 1 multiplied by any number gives the opposite of the number. For any number a, $^-1 * a = {}^-a$ and $a * {}^-1 = {}^-a$. *Example:* $^-1 * 7 = {}^-7$.

(continued)

b. A negation symbol in front of a left parenthesis may be replaced by $^-1*$. Thus $^-(a + b) = {}^-1 * (a + b)$.
Example: $^-(6 + 1) = {}^-1 * (6 + 1)$.

c. A subtraction symbol in front of a left parenthesis may be changed to addition if $^-1*$ is inserted. Thus $a - (b + c) = a + {}^-1 * (b + c)$.
Example: $13 - (4 + 7) = 13 + {}^-1 * (4 + 7)$. Recall that subtraction is the same as adding the opposite.

Examples 6 and 7 illustrate these properties.

EXAMPLE 6 Identify the special property of 0, 1, or $^-1$ used in each instance.

a. $^-3 + 3 = 0$
b. $1 * 7 = 7$
c. $\left(\frac{1}{5}\right) * 5 = 1$
d. $^-(3 + x) = {}^-1 * (3 + x)$
e. $(6 + 3) = (6 + 3) * 1$

Solution

a. Special property of 0 (a on page 52)
b. Special property of 1 (b on page 52)
c. Special property of 1 (a on page 52)
d. Special property of $^-1$ (b on page 53)
e. Special property of 1 (d on page 52)

EXAMPLE 7 Use the special properties and the distributive property to remove parentheses.

a. $(x - y) + (z + 7)$
b. $4 - (a + b)$
c. $^-(x + 4) + y$

Solution

a. $(x - y) + (z + 7)$

$= \mathbf{1} * (x - y) + \mathbf{1} * (z + 7)$ *Using special property c of 1*

$= \mathbf{1} * x - \mathbf{1} * y + \mathbf{1} * z + \mathbf{1} * 7$ *Using the distributive property on page 48*

$= x - y + z + 7$ *Using special property b of 1*

b. $4 - (a + b)$

$= 4 + \mathbf{^-1} * (a + b)$ *Using special property c of $^-1$*

$= 4 + \mathbf{^-1} * a + \mathbf{^-1} * b$ *Using the distributive property*

$= 4 + {}^-a + {}^-b$ *Using special property a of $^-1$*

The last expression may also be written as $4 - a - b$ using the rule for subtraction on page 7.

c. $^{-}(x + 4) + y$

$= {}^{-}\mathbf{1} * (x + 4) + y$ *Using special property b of* $^{-}1$

$= {}^{-}\mathbf{1} * x + {}^{-}\mathbf{1} * 4 + y$ *Using the distributive property*

$= {}^{-}x + {}^{-}4 + y$ *Using special property a of* $^{-}1$

The last expression may also be written as $^{-}x - 4 + y$ using the rule for subtraction.

With a bit of practice you should be able to combine steps to write:

a. $(x - y) + (z + 7) = x - y + z + 7$

b. $4 - (a + b) = 4 - a - b$

c. $^{-}(x + 4) + y = {}^{-}x - 4 + y$

However, in order to avoid errors, particularly with minus signs, it is important that you be aware of the special properties when you apply the distributive property to remove parentheses.

Exercises 2.1

1. A triangle has sides that measure 7.64 mm, 5.2 mm, and 4.8 mm.

a. Make a sketch of the triangle and label the sides.

b. An expression for the perimeter of the triangle is (7.64 mm + 5.2 mm) + 4.8 mm. Use the associative property of addition to group with parentheses in a different way.

c. Evaluate the two expressions for the perimeter of the triangle to show that the sums are the same.

2. Recall from Section 1.4 that the area of a triangle is area = $(\frac{1}{2}) *$ base $*$ height.

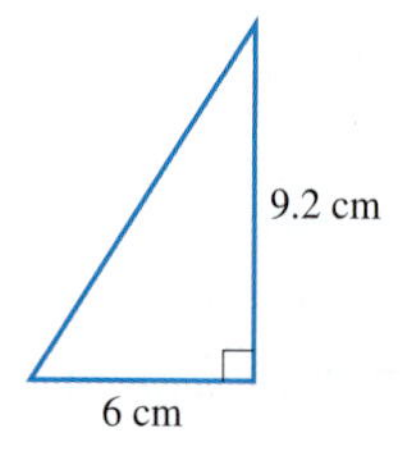

a. An expression for the area is $(\frac{1}{2} * 6) * 9.2$. Group with parentheses in a different way.

b. Do the two expressions give the same area? Explain.

c. What is the property called that allows grouping multiplication in different ways?

3. A land developer wants to subdivide a rectangular parcel of land 220 feet by 891 feet into three smaller rectangular lots. See the accompanying figure.

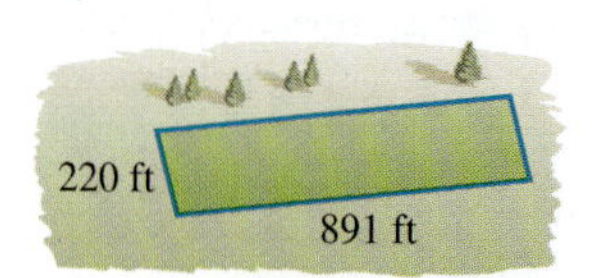

Use the distributive property to show that the area of the original parcel is the same as the sum of the areas of the three lots.

4. Use the distributive property to expand each expression.

a. $8 * (4 + 30)$

b. $6 * (200 - 5)$

c. Check your work by finding the value of each expression before and after rewriting.

5. Justin brags that he can perform many complicated multiplications in his head. For example, he says, "I can multiply $4 * 512$ by multiplying $4 * 500$ and adding $4 * 12$."

a. Use the distributive property to show what Justin did.

b. Use Justin's method to compute $11 * 403$ mentally.

6. Use the distributive property to expand each expression.

a. $2 * (L + W)$

b. $7 * (x - y)$

c. Substitute $x = 5$ and $y = 2$ into both expressions in part (b). Does your work check?

7. Factoring is the opposite of what process?

8. Use the distributive property to factor each expression.

a. $(4 * 61) + (4 * 9)$ **b.** $(9 * x) + (9 * y)$

c. $(23 * s) - (23 * t)$

9. Consider $3 - 4 \div 2$. Explain why it is necessary to follow the order of operations to evaluate this expression correctly.

10. Use the order of operations to evaluate each expression.

a. $3 + 5 * 2$ **b.** $6 - 2^3 * 4$

c. $\dfrac{4-6}{^-8+2}$ **d.** $\sqrt{(5-2)^2}$

e. $(^-3)^2$ **f.** $^-3^2$

g. $\dfrac{5 + 43 * 1}{6}$ **h.** $13 - (2 + 9)$

11. a. Use the order of operations to evaluate

$$\frac{3 + 24 - 6}{6 - 2}$$

b. How would you enter this expression into your calculator to check your answer?

12. Identify the special property of 0, 1, or $^-1$ used in each instance.

a. $29 \div 0$ is undefined.

b. $^-3 * \left(\frac{^-1}{3}\right) = 1$

c. $(5 + x) = 1 * (5 + x)$

d. $^-4.2 + 4.2 = 0$

13. Demonstrate the special properties of 0, 1, and $^-1$ with numbers of your choice.

14. Which of the expressions does not equal $^-(x + 7)$?

a. $^-1 * (x + 7)$ **b.** $(x + 7) * {}^-1$

c. $^-1 * x + 7$ **d.** $^-1 * x + {}^-1 * 7$

e. $x * {}^-1 + 7 * {}^-1$

15. Use the special properties and the distributive property to write each expression without parentheses.

a. $(a + b) - (8 + c)$ **b.** $^-(y + 6) + z$

c. $3 - (s + t)$

Skills and Review 2.1

16. A rectangle has a width of 6 feet and a length of 30 feet.

a. Sketch a figure and label the width and length.

b. Find the perimeter.

c. Find the area.

17. Recall that a formula for the circumference of a circle is $2 * \pi * r$. Find the circumference of a circle with a radius of 5 inches. Use $\pi \approx 3.14$. *Hint:* You may use the commutative property of multiplication to find the circumference without a calculator.

18. Find each value without using a calculator.

a. $2 * 5$ **b.** 2^5

19. Find each value without using a calculator.

a. 9^2 **b.** $\left(\frac{1}{2}\right)^3$

c. 1^5 **d.** 23^0

e. $3^{^-4}$

20. Find each value without using a calculator.

a. $^-7^2$ **b.** $(^-7)^2$

21. Write $2^{^-10}$ in each format,

a. As a fraction with a positive exponent

b. In scientific notation

c. As a decimal

22. Find the missing decimal, fraction, or percent in each row of the table.

Decimal	Fraction	Percent
?	$\frac{1}{3}$	?
?	?	20%
.25	?	?

23. One wall of Cynthia's house has an area of 450 square feet. It takes Cynthia 9 hours to paint the wall.

a. What is her work rate in square feet per hour?

b. What is her work rate in square feet per minute?

c. What is her work rate in walls per hour?

24. Which is greater, $\frac{2}{5}$ or $\frac{3}{8}$? How did you decide?

25. Evaluate $16 - {}^{-}3$. Use words to describe a change-in-temperature situation involving these numbers.

2.2 Expressions and Formulas

After Studying This Section You Will Be Able To

- Find equivalent algebraic expressions by combining like terms.
- Translate verbal expressions into algebraic expressions.
- Apply the Pythagorean theorem to find the hypotenuse of a right triangle.

The statement that two expressions are equal is called an **equation**. A **formula** is an equation that shows how one variable depends upon one or more other variables. Formulas generally take the form: dependent variable = *expression*.

EXAMPLE 1 Use toothpicks to make a row of squares that share adjacent sides. See Figure 6. The number of toothpicks (T) you need depends upon the number of squares (S) in the row. Find a formula for T in terms of S.

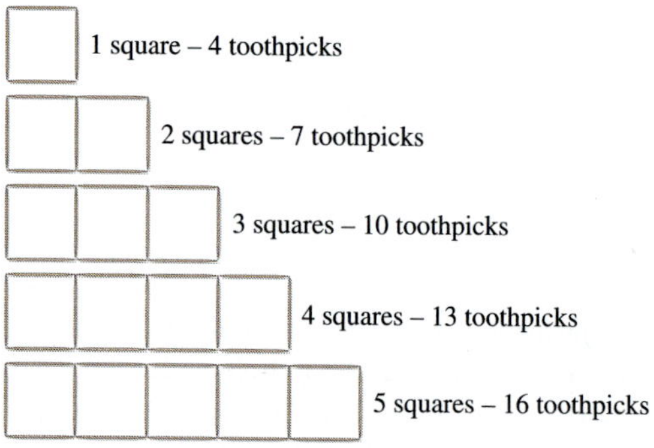

1 square – 4 toothpicks

2 squares – 7 toothpicks

3 squares – 10 toothpicks

4 squares – 13 toothpicks

5 squares – 16 toothpicks

Figure 6

Solution One formula is $T = 3 * S + 1$. Each time a new square is added, three more toothpicks are needed. If you have only $3 * S$ toothpicks, however, you will not be able to close up the left side, so one additional toothpick is needed. See Figure 7. You can check that this formula works by substituting values for S and

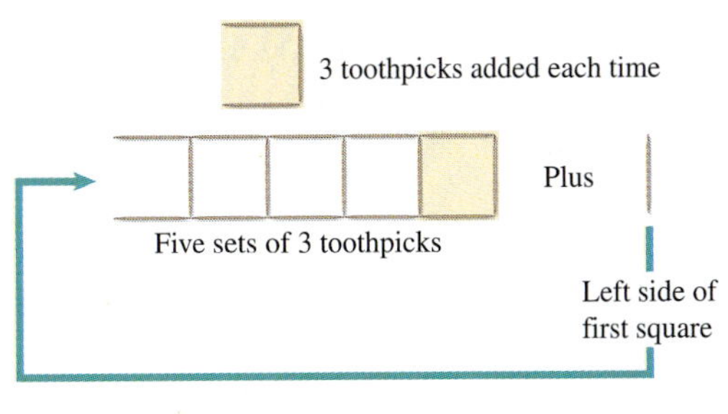

Figure 7

showing that you get the correct values for T:

when $S = 1$	when $S = 2$	when $S = 3$
$T = 3 * S + 1$	$T = 3 * S + 1$	$T = 3 * S + 1$
$T = 3 * 1 + 1$	$T = 3 * 2 + 1$	$T = 3 * 3 + 1$
$T = 3 + 1$	$T = 6 + 1$	$T = 9 + 1$
$T = 4$	$T = 7$	$T = 10$

Another formula is $T = (2 * S) + (S + 1)$. We obtain this formula by noticing that each square requires two horizontal toothpicks, one on top and one on the bottom. The number of vertical toothpicks is one more than the number of squares. So the total number of toothpicks is found by adding the number of horizontal toothpicks $(2 * S)$ to the number of vertical toothpicks $(S + 1)$. Again, this formula may be checked by substituting values for S:

when $S = 1$	when $S = 2$	when $S = 3$
$T = (2 * S) + (S + 1)$	$T = (2 * S) + (S + 1)$	$T = (2 * S) + (S + 1)$
$T = (2 * 1) + (1 + 1)$	$T = (2 * 2) + (2 + 1)$	$T = (2 * 3) + (3 + 1)$
$T = 2 + 2$	$T = 4 + 3$	$T = 6 + 4$
$T = 4$	$T = 7$	$T = 10$

Equivalent Expressions

In Example 1 two different formulas give the same result. That is because the expressions $3 * S + 1$ and $(2 * S) + (S + 1)$ are **equivalent**. Two expressions are equivalent if the results are the same no matter what values are substituted for the variable or variables.

The properties you have learned can be used to explain *why* the two expressions are equivalent. Start with

$(2 * S) + (S + 1)$	*Original expression*
$= ((2 * S) + S) + 1$	*Using the associative property of addition to change grouping*

$= ((2 * S) + (1 * S)) + 1$ — *Using a special property of 1 to rewrite S as $1 * S$*

$= ((2 + 1) * S) + 1$ — *S is a common factor; use the distributive property*

$= (3 * S) + 1$ — *Substituting 3 for $2 + 1$*

$= 3 * S + 1$ — *Removing the parentheses because they are not needed*

Putting all the steps together, we conclude that $(2 * S) + (S + 1) = 3 * S + 1$ for all values of S.

Combining Like Terms

To show that the two formulas in Example 1 are equivalent, we used the distributive property to show that

$$2 * S + S = 2 * S + 1 * S = (2 + 1) * S = 3 * S$$

$2 * S$ and S are called **like terms** because they contain the same variable. Like terms may be combined by adding (or subtracting) the **coefficients**, that is, the constants that are multiplied by the variables. In this case the coefficients are 2 and 1. The 1 did not appear in the original expression but can always be thought of as being there, using a special property of 1.

The procedure of combining like terms is used so frequently that you may soon be able to write $2 * S + S = 3 * S$ without going through the intermediate steps. Furthermore, the multiplication symbol ($*$) may be omitted between a coefficient and a variable or between two variables. Thus, the process may be further streamlined by writing $2S + S = 3S$. Be sure, however, that if you take this shortcut you know what you are doing! It is better to take an extra step and get the right result than to make a mistake.

EXAMPLE 2 Simplify these expressions by combining like terms:

a. $5x + 8x$
b. $y - 7y$
c. $4a + 5 - 3a + 7$
d. $2x^2 + 3x - x^2 - 5x$

Solution

a. $5x + 8x = 13x$ (*Think:* $5 + 8 = 13$.)
b. $y - 7y = {}^-6y$ (*Think:* $1 - 7 = {}^-6$.)
c. $4a + 5 - 3a + 7 = a + 12$ (*Think:* Combine the a terms: $4 - 3 = 1$. Combine constant terms: $5 + 7 = 12$.)
d. $2x^2 + 3x - x^2 - 5x = x^2 - 2x$ (*Think:* Combine the x^2 terms: $2 - 1 = 1$. Combine the x terms: $3 - 5 = {}^-2$.)

Caution: In the last expression x^2 and x are *not* like terms because for most values of x, x and x^2 are different numbers. Do not combine x and x^2 to get x^3.

Combining like terms is like combining fractions with the same denominator. For instance, you will recall that to add two fractions with the same denominator, you simply add the numerators:

$$\tfrac{2}{7} + \tfrac{3}{7} = \tfrac{2+3}{7} = \tfrac{5}{7}$$

When fractions do not have the same denominator, as in the case of $\frac{2}{5} + \frac{1}{4}$, however, you have to find a common denominator.

This principle extends to algebraic fractions, that is, those with variables in the numerator or denominator. The next example shows how the principle of combining like terms applies to fractions.

EXAMPLE 3 Simplify each of these expressions by combining like terms.

a. $\dfrac{5}{x} - \dfrac{2}{x}$

b. $\dfrac{a}{b} + \dfrac{3}{b}$

c. $\dfrac{x}{5} + \dfrac{2x}{5}$

d. $\dfrac{2}{5} + \dfrac{1}{4}$

e. $\dfrac{1}{x} + x + x^2 + x^{-1}$

Solution

a. The denominators are the same, so we combine like fractions.

$$\frac{5}{x} - \frac{2}{x} = \frac{5-2}{x} = \frac{3}{x}$$

b. The two fractions have the same denominator and can be combined. Because a and 3 are not like terms, they cannot be combined in the numerator.

$$\frac{a}{b} + \frac{3}{b} = \frac{a+3}{b}$$

c. The two fractions have the same denominator. The numerators are like terms and can be combined.

$$\frac{x}{5} + \frac{2x}{5} = \frac{x+2x}{5} = \frac{3x}{5}$$

d. The fractions do not have the same denominator. Once we find equivalent fractions with a common denominator, we can combine them.

$$\frac{2}{5} + \frac{1}{4} = \frac{8}{20} + \frac{5}{20} = \frac{13}{20}$$

e. Recall that $x^{-1} = \frac{1}{x}$. The two fractions with x in the denominator may be combined. As in Example 2(d), x^2 and x can't be combined because they are not like terms.

$$\frac{1}{x} + x + x^2 + x^{-1} = \frac{1}{x} + x + x^2 + \frac{1}{x} = \frac{1+1}{x} + x + x^2 = \frac{2}{x} + x + x^2$$

Algebraic and Verbal Expressions

In solving word problems you will often need to rewrite a verbal expression as an algebraic expression. When doing so, look for key phrases that indicate mathematical operations.

EXAMPLE 4 A real estate developer is building homes in a new subdivision. The lots are rectangles with the same width, w feet, but different lengths. See Figure 8.

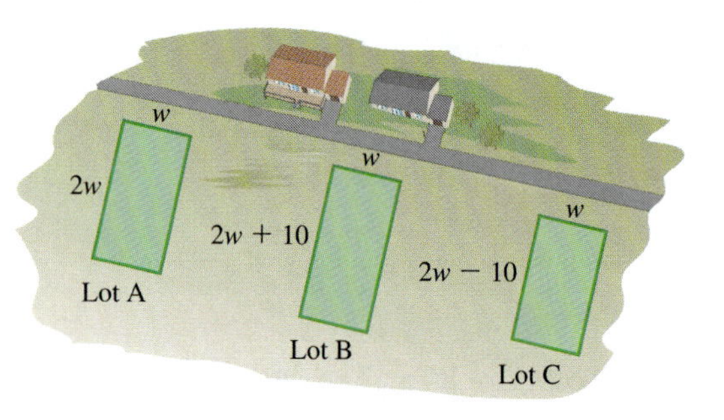

Figure 8

a. The length of lot A is twice the width. Find an expression for the perimeter of lot A.

b. The length of lot B is 10 feet more than twice the width. Find an expression for the perimeter of lot B.

c. The length of lot C is 10 feet less than twice the width. Find an expression for the perimeter of lot C.

d. Find the perimeter of each lot when $w = 120$ feet.

Solution

a. The key words are "twice the width." We are letting w represent the width, so we know that $2w$ represents twice the width. To find the perimeter add all the sides: $w + 2w + w + 2w$. Combine like terms to get $6w$ as a simplified expression for the perimeter of lot A.

b. The key words are "10 feet more than twice the width." Because $2w$ is twice the width, the length is $2w + 10$. To find the perimeter add all the sides:

$w + 2w + 10 + w + 2w + 10$. Combine like terms to get $6w + 20$ as a simplified expression for the perimeter of lot B.

c. The key words are "10 feet less than twice the width." Because $2w$ is twice the width, the length is $2w - 10$. To find the perimeter add all the sides: $w + 2w - 10 + w + 2w - 10$. Combine like terms to get $6x - 20$ as a simplified expression for the perimeter of lot C.

d. Substitute 120 feet for w in each expression for the perimeter.

$$\text{lot A: } 6w = 6 * 120 = 720 \text{ feet}$$

$$\text{lot B: } 6w + 20 = 6 * 120 + 20 = 720 + 20 = 740 \text{ feet}$$

$$\text{lot C: } 6w - 20 = 6 * 120 - 20 = 700 \text{ feet}$$

Caution: A common error made by many students in a problem such as Example 4(c) is to translate incorrectly the phrase "10 feet less than twice the width." Writing $10 - 2w$ instead of $2w - 10$ gives the opposite of the required expression. One way to avoid this error is to check the work by substituting a possible value of w into the expression.

For instance, suppose $w = 90$ feet. Then $2w - 10 = 2 * 90 - 10 = 180 - 10 = 170$ feet, a reasonable length. But $10 - 2w = 10 - 2 * 90 = 10 - 180 = {}^{-}170$. This isn't reasonable because the length of the lot can't be negative.

A Formula for Right Triangles

One of the most important formulas in geometry is named for the Greek mathematician Pythagoras, who lived in the sixth century B.C. However, the same formula was discovered by the Chinese and the Egyptians many centuries earlier.

In a **right triangle** the longest side is opposite the right angle and is called the **hypotenuse**. The Pythagorean theorem states that the area of a square built on the hypotenuse is equal to the sum of the areas of the squares built on the other two sides, as shown in Figure 9. We usually label the hypotenuse c and the other two sides a and b. Thus, $c^2 = a^2 + b^2$. This statement is called the Pythagorean theorem and is proved in Chapter 6. It leads us to a formula for the hypotenuse.

The Length of the Hypotenuse of a Right Triangle

In a right triangle with sides of length a, b, and c, if c is the length of the side opposite the right angle, then

$$c = \sqrt{a^2 + b^2}$$

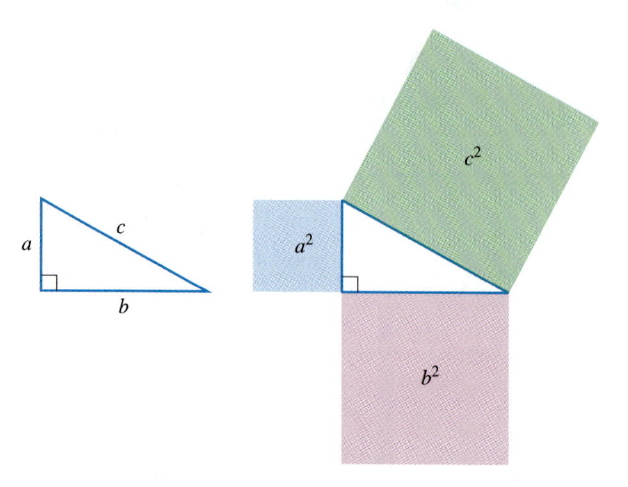

Figure 9

EXAMPLE 5 A rectangular lot is 120 feet long and 90 feet wide. Find the distance from one corner of the lot to the opposite corner.

Solution The diagram in Figure 10 shows that a right triangle is formed. We can use the formula for the hypotenuse to find c. Substitute $a = \mathbf{90}$ feet and $b = \mathbf{120}$ feet into the formula

$$c = \sqrt{a^2 + b^2} = \sqrt{90^2 + 120^2}$$

Here we must use the order of operations. The radical symbol groups the sum of a^2 and b^2 just like parentheses. So first we apply the powers and get $c = \sqrt{8100 + 14{,}400}$. Then we add $8100 + 14{,}400$, which gives $c = \sqrt{22{,}500}$. Finally, we find the square root of 22,500, which is $c = 150$. The distance between the two corners is 150 feet.

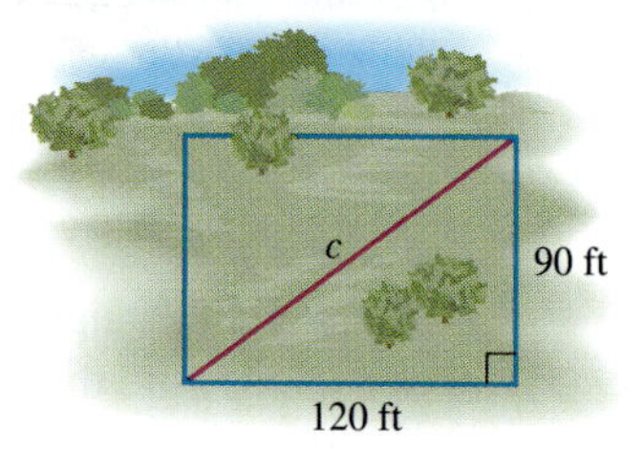

Figure 10

Exercises 2.2

1. Evaluate each expression for the given values of the variables.
   a. $3x - 5$ when $x = 2$
   b. $^-x + 1$ when $x = {}^-3$
   c. $4x^2$ when $x = {}^-1$
   d. $2\pi r$ when $r = \frac{3}{2}$ (use $\pi \approx 3.14$).
   e. $2L + 2W$ when $L = 4.5$ and $W = 3$
   f. $\sqrt{a^2 + b^2}$ when $a = 3$ and $b = 4$

2. Two formulas for the circumference of a circle are $C = 2 * \pi *$ radius and $C = \pi *$ diameter.
   a. Check that these formulas are equivalent by substituting three numerical values for the radius. (*Hint:* diameter $= 2 *$ radius.)
   b. Check that these formulas are equivalent by using the properties in Section 2.1 to derive one formula from the other.

3. Substitute the given values into each formula to find values of the indicated variables.
   a. $y = {}^-2x + 6$; find y when $x = 3$.
   b. $y = 3x^2 - x$; find y when $x = {}^-2$.
   c. The formula for the area of a rectangle is $a = l * w$. Find a when $l = \frac{5}{2}$ cm and $w = 8$ cm.
   d. The formula for the area of a trapezoid is $A = \frac{1}{2}(b_1 + b_2)h$. Find A when $b_1 = 5$ mm, $b_2 = 11$ mm, and $h = 6$ mm.
   e. The formula for the hypotenuse of a right triangle is $c = \sqrt{a^2 + b^2}$. Find c when $a = 6$ inches and $b = 8$ inches.
   f. The formula for the surface area of a cone is $S = \pi * r^2 + \pi * r * \sqrt{r^2 + h^2}$. Find S when $r = 3$ feet and $h = 4$ feet. Use $\pi \approx 3.14$.

4. The surface area of a rectangular box is the sum of the areas of its six rectangular faces, as shown in the figure. Let w = width of the box, l = length of the box, and h = height of the box.

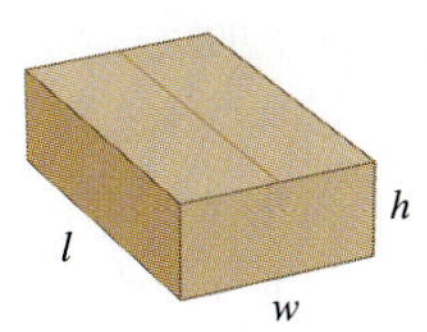

   a. Find a formula for the surface area (S) of the rectangular box in terms of w, l, and h.
   b. Find the surface area when $w = 8$ cm, $l = 10$ cm, and $h = 5$ cm.

5. The table shows the number of cups of flour (F) required to make S servings for a particular recipe.
   a. How many cups of flour are required to make 7 servings?
   b. How many cups of flour are required to make S servings?
   c. Write a formula for F in terms of S.

Servings (S)	Cups of Flour (F)
1	$\frac{1}{3}$
2	$\frac{2}{3}$
3	$\frac{3}{3} = 1$
4	$\frac{4}{3} = 1\frac{1}{3}$

6. Simplify each expression by combining like terms.
   a. $7a - 8a$
   b. $11b + b$
   c. $3x + 5 - 2x + 9$
   d. $6x^2 + 7x + x^2 - 6x$

7. A student concludes that $12x - 12x = 0x = x$. Demonstrate with $x = 1$ that $12x - 12x \neq x$.

8. Simplify each expression by combining like fractions.
   a. $\frac{3}{7} + \frac{2}{7}$
   b. $\frac{3}{x} - \frac{1}{x}$
   c. $\frac{a}{y} + \frac{8}{y}$
   d. $x^{-1} + \frac{2}{x}$

9. Assume the cost of operating a car is 6 cents per mile for fuel and an additional 25 cents per mile for wear and tear. Let M = number of miles driven.
   a. Write a formula for the cost of operating the car in terms of M.
   b. Write an equivalent expression to the one found in part (a).
   c. What is the cost for each mile driven?
   d. What is the cost for driving 10 miles?

10. A length of a rectangle is 4 feet more than three times the width.

a. Let w represent the width. Find an expression for the length of the rectangle in terms of the width.

b. Draw and label a diagram.

c. Find an expression for the perimeter of the rectangle.

d. Find the value of the expression when the width is 35 feet.

11. The height of a parallelogram is 5 inches less than the base.

a. Draw and label a diagram.

b. Assign a variable for the length of the base. Then, write an expression for the height in terms of the base.

c. Find an expression for the area of the parallelogram.

d. Find the value of the expression when the base is 20 inches.

12. A triangular banner has a height that is one-eighth of the base. Let b represent the base.

a. Find an expression for the height of the banner.

b. Find an expression for the area of the banner.

c. Find the area of the banner when the base is 80 cm.

13. The floor of a rectangular room measures 24 feet by 18 feet. What is the distance from one corner of the floor to the opposite corner?

14. A boat sails from Brisbane 40 km east and from that point, 30 km south.

a. Draw and label a diagram showing the path of the boat.

b. Find the boat's distance from Brisbane.

15. Suppose there are two alternative routes from Startville to Finishville. One route comprises two highways that form the legs of a right triangle and the other route is a secondary road, which is the hypotenuse of the triangle (see the figure).

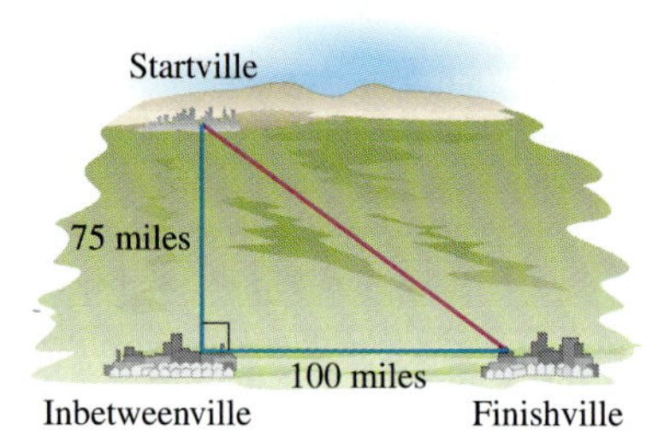

a. Find the length of the secondary road.

b. If you can average 50 miles per hour on the highways and 25 miles per hour on the secondary road, which route takes the least amount of time? Remember that time = distance/rate.

Skills and Review 2.2

16. a. Use the distributive property to expand $2(x - y)$.

b. Use distribution to factor $2x - 2y$.

c. How are expanding and factoring related?

17. Use the order of operations to evaluate $5 + 7 * (^-4)$.

18. Use the order of operations to evaluate

$$\frac{^-3 * 4^2}{2 - 6}.$$

19. Use the order of operations to evaluate

$$\frac{5 + \sqrt{(^-5)^2 - 4 * 1 * 6}}{2 * 1}$$

20. Use a special property of $^-1$ and the distributive property to write the expression without parentheses: $2 - (c + d)$.

21. The width of a rectangle is 5 inches. The length is 3 inches less than twice the width.

a. Write an expression for the length of the rectangle.

b. Evaluate your expression for the length.

c. Find the area.

d. Find the perimeter.

22. Recall that for any circle π = circumference/diameter.

a. Measure the circumference and diameter of a glass and use the formula above to approximate the value of π.

b. Is it possible to get an exact value for π by measurement? Why or why not?

23. Find each value without using a calculator.

a. 4^3 b. 3^0 c. 1^{39}

d. $5^{^-2}$ e. $(^-2)^4$

24. Find each absolute value.

a. $|9|$ **b.** $|^-9|$

c. $|0|$

25. Find each sum or difference.

a. $8 + {}^-5$ **b.** $^-7 - 2$

c. $^-6 + 4$ **d.** $3 - {}^-8$

2.3 The Coordinate Plane

AFTER STUDYING THIS SECTION YOU WILL BE ABLE TO

- Locate points in the coordinate plane.
- Apply the definitions of run and rise.
- Apply the distance formula.

In 1637 the French mathematician René Descartes invented a system for naming points in a plane. Today we refer to that system as the **coordinate plane**. We describe how the system works through a story about an imaginary town called Gridville.

In Gridville the roads form a rectangular grid. Roads that run from south to north are called "avenues." Those that run from west to east are called "streets." The two major roads, called Avenue 0 and Street 0, intersect at the center of town, which is also called the **origin**.

Avenues to the east of Avenue 0 have positive numbers. Avenues to the west of Avenue 0 have negative numbers. Streets to the north of Street 0 have positive numbers. Streets to the south of Street 0 have negative numbers (see Figure 11).

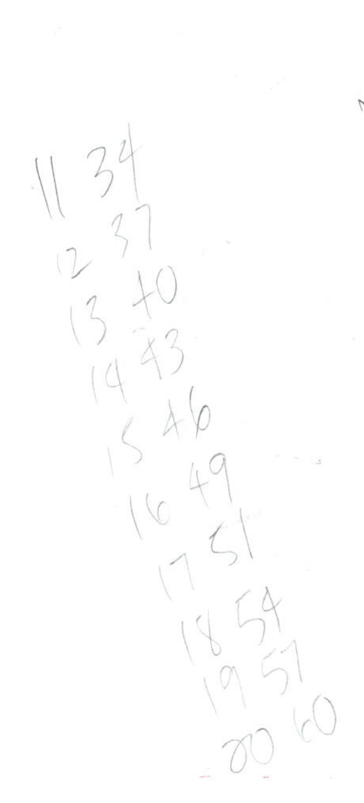

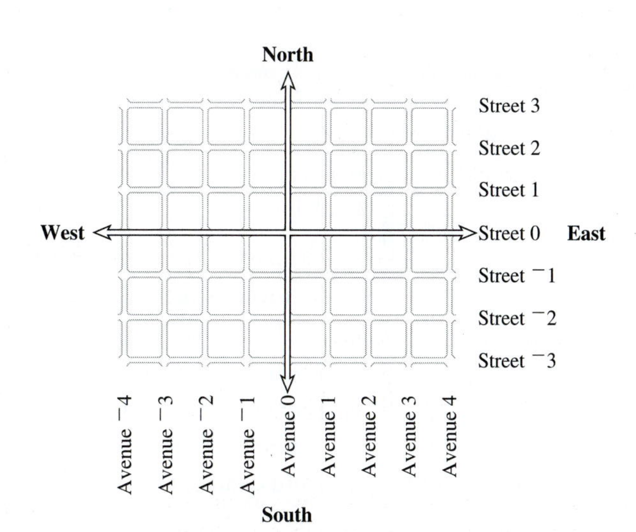

Figure 11

People who live in Gridville have an unusual method of describing locations. Each intersection of a street and an avenue is given a pair of numbers. The avenue number is always given first. Thus the intersection of Avenue $^{-}1$ and Street 2 is given by the pair $(^{-}1, 2)$. Instead of using house numbers, a resident of this town gives the pair for the nearest intersection. He or she figures that once you are at that intersection you should be able to find the house you are looking for quite easily.

The first number in the ordered pair is called the ***x*-coordinate**. The second number is the ***y*-coordinate**. Because the avenue (x) must be given before the street (y), order matters, and (x, y) is called an **ordered pair**. By knowing the ordered pair for a location, you also know what section of the town you are in. For instance, if both numbers are positive, your location is in the northeast quadrant, also known as the first quadrant. The names for the four quadrants are shown in Figure 12. The boundaries of the quadrants are Street 0, called the ***x*-axis**, and Avenue 0, called the ***y*-axis**.

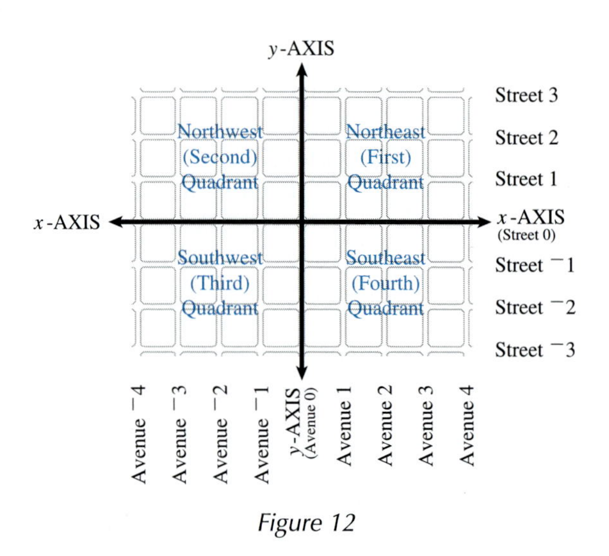

Figure 12

EXAMPLE 1 For each location, identify the quadrant or axis and locate the point on a map of Gridville.

a. The public library is located at $(^{-}4, ^{-}2)$.
b. A grocery store is located at $(^{-}1, 3)$.
c. The high school is located at $(3, 2)$.
d. The hospital is located at $(1, ^{-}2)$.
e. There is a video-rental store at $(0, ^{-}3)$.

Solution

a. Third quadrant
b. Second quadrant
c. First quadrant

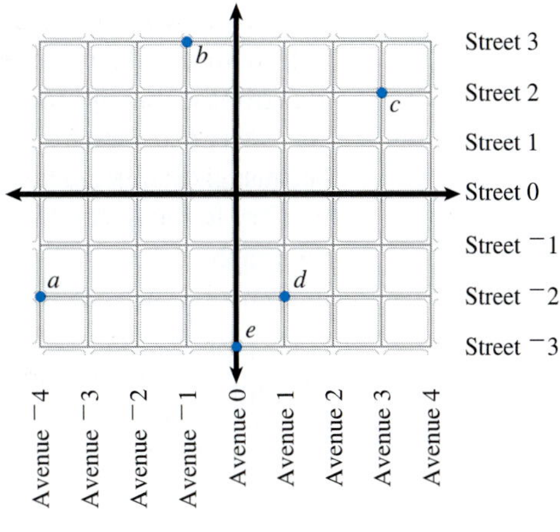

Figure 13

d. Fourth quadrant

e. y-axis

The locations are shown in Figure 13.

Coordinates on a Calculator

Throughout this book we use a graphing calculator to represent the coordinate plane. Let's begin by showing how to display the intersections in Gridville on the calculator screen.

First we must set the **window**. Press the WINDOW key and set the window variables for these values:

$$\text{Xmin} = {}^-4.7 \quad \text{Xmax} = 4.7 \quad \text{Xscl} = 1$$

$$\text{Ymin} = {}^-3.1 \quad \text{Ymax} = 3.1 \quad \text{Yscl} = 1$$

On a TI-83 a shortcut for this window is ZOOM 4 (ZDecimal).

Next, turn on the grid. On a TI-83 press 2ND FORMAT and select GRIDON. The screen should look like the screen on the left in Figure 14. Press GRAPH,

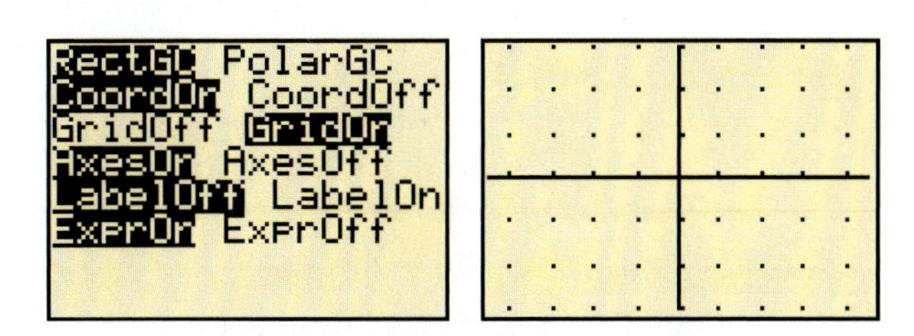

Figure 14

and you should get the picture on the right in Figure 14. Each dot on the grid represents the intersection of a street and an avenue. Use the four direction keys to move the cursor around the screen. Notice that every time you move the cursor, either the x-coordinate or the y-coordinate will change by .1. That happens because in the Decimal window, the distance between adjacent pixels is .1.

Try to locate the five points from Example 1 and read their coordinates from the graph. Take some time to become familiar with how coordinates are related to points on the screen. As you do so, notice the following:

- If you move the cursor up or down, you will find that the value of y changes while the value of x remains the same.
- If you move right or left, the value of x changes while the value of y remains the same.
- When you move the cursor along the x-axis, $y = 0$.
- When you move the cursor along the y-axis, $x = 0$.

Run and Rise

There is a taxi service in Gridville. The dispatcher receives calls from riders and relays information to the drivers. Instead of telling drivers where to go, the dispatcher tells how to get there. The dispatcher and the drivers use two code words to communicate. *Run* means to travel along a street, and *rise* means to go along an avenue.

For example, suppose John, who works at the grocery store (located at $(^-1, 3)$), needs a ride home and that the closest available cab is at $(2, ^-1)$. The dispatcher tells the driver "run $^-3$, rise 4." See Figure 15. Run $^-3$ means go three blocks in the negative direction (west) along a street. Rise 4 means go four blocks in the positive direction (north) along an avenue. A positive run would mean to go east. A negative rise would mean to go south.

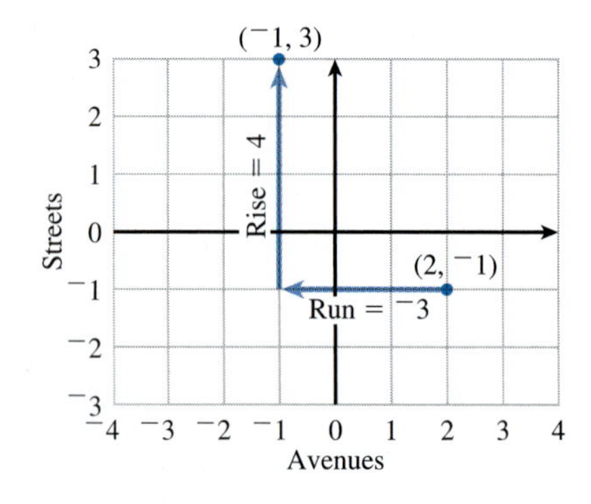

Figure 15

EXAMPLE 2 **a.** Suppose the driver is at $(^-2, 1)$ and the dispatcher tells her "run 5, rise $^-2$." To what location will the cab go?

b. One cab has just dropped off a teacher at the high school, (3, 2). Now another rider is waiting at ($^-1$, 1) to be picked up. What directions will the dispatcher give?

Solution

a. From Figure 16 you can see that the destination of the cab is (3, $^-1$). You can arrive at the same answer by using the number line model for addition (page 6). Recall that start + change = final. For the x-coordinate this means that $\mathbf{x_{start}} + \mathbf{run} = x_{\text{final}}$:

$$\mathbf{^-2} + \mathbf{5} = x_{\text{final}}$$

$$3 = x_{\text{final}}$$

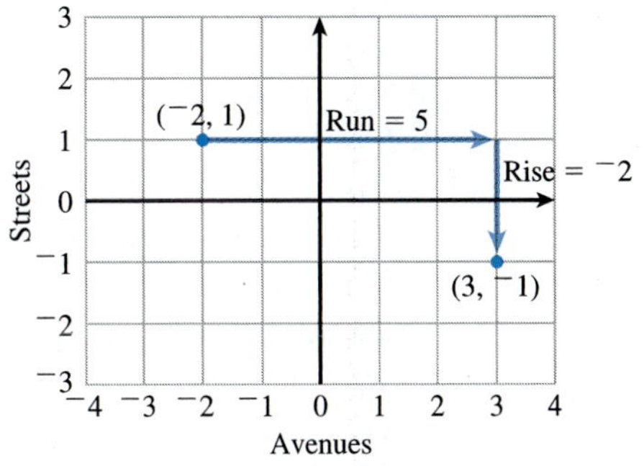

Figure 16

Similarly for y, $\mathbf{y_{start}} + \mathbf{rise} = y_{\text{final}}$:

$$\mathbf{1} + \mathbf{^-2} = y_{\text{final}}$$

$$^-1 = y_{\text{final}}$$

b. From Figure 17 you can see that run = $^-4$ and that rise = $^-1$. You can arrive at the same answer by using the number line model for subtraction (page 6).

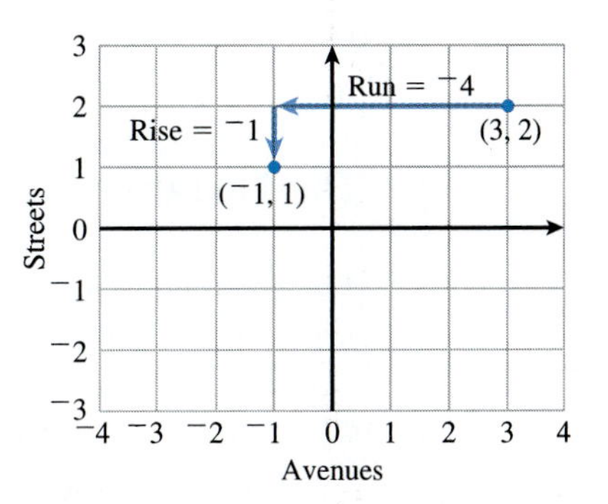

Figure 17

Recall that final − start = change. For the x-coordinates this means that

$$x_{\text{final}} - x_{\text{start}} = \text{run}$$

$$^{-}1 - 3 = \text{run} \qquad \textit{Substituting}$$

$$^{-}1 + {}^{-}3 = \text{run} \qquad \textit{Changing to addition}$$

$$^{-}4 = \text{run}$$

For the y-coordinates this means that

$$y_{\text{final}} - y_{\text{start}} = \text{rise}$$

$$1 - 2 = \text{rise} \qquad \textit{Substituting}$$

$$1 + {}^{-}2 = \text{rise} \qquad \textit{Changing to addition}$$

$$^{-}1 = \text{rise}$$

The solution to Example 2(b) suggests a general formula for run and rise. Call the starting point P, with coordinates (x_1, y_1), and the final point Q, with coordinates (x_2, y_2). Then the run is the difference in the x-coordinates, that is, $x_2 - x_1$, and the rise is the difference in the y-coordinates, that is, $y_2 - y_1$.

Definition **Formula for Run and Rise**

Suppose P and Q are two points in the coordinate plane with coordinates $P = (x_1, y_1)$ and $Q = (x_2, y_2)$. Then the **run** and **rise** from P to Q are

$$\text{run} = x_2 - x_1$$

$$\text{rise} = y_2 - y_1$$

(x_2, y_2)

(x_1, y_1) Rise

Run

The Distance Formula

The emergency medical service (EMS) team for Gridville owns a helicopter, which it uses for emergencies. Unlike the taxicabs, the helicopter does not have to travel along the streets and avenues of Gridville. Instead, the helicopter takes the shortest path, "straight as the crow flies."

The management of EMS prepares contingency plans for every foreseeable emergency. They know, for instance, that Ms. Robinson, who lives at the intersection of Avenue $^-4$ and Street 2, has a serious heart condition and may need to be rushed to the hospital. They want to know in advance the distance the helicopter would have to travel.

EXAMPLE 3 Find the distance (in blocks) from Ms. Robinson's home at $(^-4, 2)$ to the hospital at $(1, ^-2)$.

Solution Straight as the crow flies means that the helicopter will travel along the hypotenuse of a right triangle as shown in Figure 18. Using the formulas for run and rise we have

$$\text{run} = x_2 - x_1 = 1 - {}^-4 = 1 + 4 = 5$$

$$\text{rise} = y_2 - y_1 = {}^-2 - 2 = {}^-2 + {}^-2 = {}^-4$$

The run is 5 blocks and the rise is $^-4$ blocks.

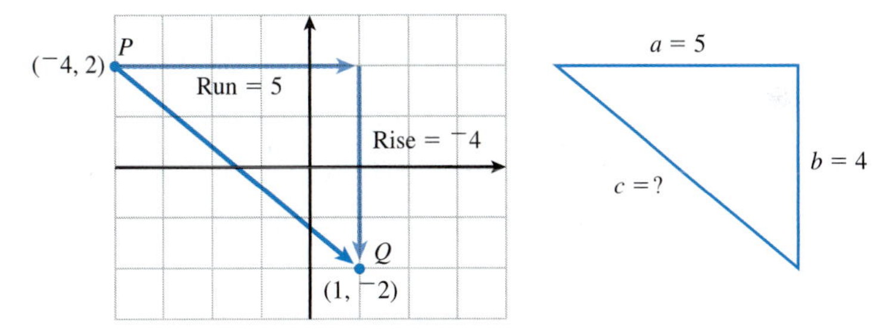

Figure 18

The length of a side of a triangle is always a positive number, so we can take the absolute values of the run and rise to find two sides of the right triangle. As shown in Figure 18 we have $a = 5$ and $b = 4$. Using the Pythagorean theorem (page 61), $c = \sqrt{a^2 + b^2}$. Substituting for a and b gives $c = \sqrt{5^2 + 4^2} = \sqrt{25 + 16} = \sqrt{41} \approx 6.403$. Thus, the helicopter will have to travel about 6.4 blocks to get Ms. Robinson to the hospital.

In Example 3 we were careful to distinguish between the rise of $^-4$ and its absolute value 4, which is the side b of the right triangle. Note that when we square 4 we get 16. If we square $^-4$ we get the same result because $(^-4)^2 = (^-4) * (^-4) = 16$. This suggests that we can find the distance between any two points in the coordinate plane directly from the run and the rise.

Definition The Distance Formula

Suppose P and Q are two points in the coordinate plane with coordinates $P = (x_1, y_1)$ and $Q = (x_2, y_2)$. Then the distance from P to Q is given by

(continued)

the formula

$$\text{distance} = \sqrt{\text{run}^2 + \text{rise}^2}$$

The distance may also be calculated directly from the coordinates:

$$\text{distance} = \sqrt{(x_2 - x_1)^2 + (y_2 - y_1)^2}$$

Notice that the distance formula is based on the Pythagorean theorem. If you have learned that $c = \sqrt{a^2 + b^2}$, then you should be able to remember that distance $= \sqrt{\text{run}^2 + \text{rise}^2}$.

EXAMPLE 4 Use the distance formula to find how far the helicopter will travel.

a. From (4, 2) to (1, $^-2$)

b. From ($^-4$, $^-3$) to (1, $^-2$)

c. From (3, 2) to ($^-3$, 2)

Solution

a.

$$\text{run} = x_2 - x_1 = 1 - 4 = {}^-3$$

$$\text{rise} = y_2 - y_1 = {}^-2 - 2 = {}^-4$$

$$\text{distance} = \sqrt{\text{run}^2 + \text{rise}^2} = \sqrt{({}^-3)^2 + ({}^-4)^2} = \sqrt{9 + 16} = \sqrt{25} = 5$$

Caution: Once again, remember that when you square a negative number, you must put it in parentheses: $({}^-3)^2 = 9$, but ${}^-3^2 = {}^-9$, which is incorrect.

b.

$$\text{run} = x_2 - x_1 = 1 - {}^-4 = 1 + 4 = 5$$

$$\text{rise} = y_2 - y_1 = {}^-2 - {}^-3 = {}^-2 + 3 = 1$$

$$\text{distance} = \sqrt{\text{run}^2 + \text{rise}^2} = \sqrt{5^2 + 1^2} = \sqrt{25 + 1} = \sqrt{26} \approx 5.099$$

c.

$$\text{run} = x_2 - x_1 = {}^-3 - 3 = {}^-6$$

$$\text{rise} = y_2 - y_1 = 2 - 2 = 0$$

$$\text{distance} = \sqrt{\text{run}^2 + \text{rise}^2} = \sqrt{({}^-6)^2 + 0^2} = \sqrt{36 + 0} = \sqrt{36} = 6$$

Note: In this case, although the distance formula works, rise = 0, so it may be easier just to take the absolute value of the run.

Exercises 2.3

1. For each ordered pair, identify the quadrant or axis and locate the point on a graph of the coordinate plane.

a. ($^-3$, 1) **b.** ($^-2$, $^-4$) **c.** (3, 0)

d. (2, 1) **e.** (1, $^-3$)

2. Give an ordered pair in each location.

a. Quadrant I

b. Quadrant II

c. Quadrant III

d. Quadrant IV

e. On the x-axis

3. Find the coordinates for the point on Street 4 that is halfway between Avenue $^{-}4$ and Avenue $^{-}6$.

4. Consider Avenue 2.

a. List four ordered pairs on Avenue 2.

b. What is the same about these ordered pairs?

c. What is different about them?

5. Consider Street $^{-}3$.

a. List four ordered pairs on Street $^{-}3$.

b. What is the same about these ordered pairs?

c. What is different about them?

6. On a graphing calculator, the cursor is located at $(^{-}2, 1)$. Suppose you move the cursor down.

a. Which variable, x or y, changes?

b. Does this variable increase or decrease?

7. On a graphing calculator, the cursor is located at $(^{-}3, ^{-}2)$. Suppose you move the cursor to the right.

a. Which variable, x or y, changes?

b. Does this variable increase or decrease?

8. Suppose you start at $(^{-}1, 2)$. Use the given run and rise to locate your final location.

a. run 4, rise 2 **b.** run 0, rise $^{-}3$

c. run 2, rise 0 **d.** run $^{-}3$, rise $^{-}4$

9. What are the run and rise needed to get from $(2, ^{-}5)$ to $(3, 4)$? What are the run and rise needed to get back to the starting position?

10. Plot $(0, 1)$. Use run 1 and rise 2 to plot the next point. Continue plotting new points using the same run and rise. What do you notice about the plotted points?

11. Given a starting point of $(1, ^{-}3)$ and a final point of $(^{-}2, ^{-}4)$,

a. Graph the points on a coordinate plane.

b. Find the run and rise and label your graph.

c. Find the distance between the two points.

12. Find the distance between $(^{-}2.3, 4)$ and $(1.2, ^{-}3.4)$.

13. A point on a circle is $(0, 2)$, and the center of the circle is $(^{-}1, 6)$. Use the distance formula to find the radius of the circle.

14. Suppose a hospital wants to estimate the time it will take a helicopter to bring an injured patient to the hospital. The patient is at $(60, ^{-}20)$ and the hospital is at $(20, 10)$.

a. Find the distance to the hospital.

b. If the helicopter's average speed is 120 blocks per hour, find the time (in minutes) it will take to reach the hospital. (*Recall:* time = distance ÷ rate.)

15. Find the area of the triangle shown.

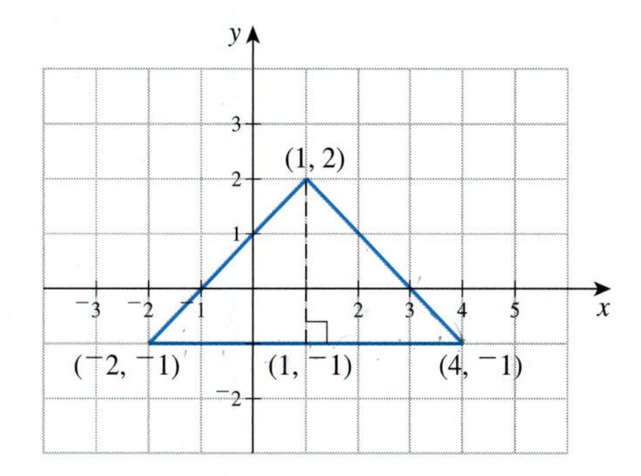

Skills and Review 2.3

16. For each statement, find an expression for the length in terms of the width. Let x represent the width.

a. The length is 7 more than the width.

b. The length is 3 times the width.

c. The length is 4 less than twice the width.

d. The length is 8 more than one-half the width.

17. Simplify each expression by combining like terms.

a. $3y + 9y$ **b.** $6x - x$

c. $^{-}2x - 4 + 5x + 7$ **d.** $5x^2 - x - 7x^2 + 2x$

18. Simplify the expression by first using the distributive property and then combining like terms: $^{-}(x + 2) - 6x$.

19. Evaluate $^{-}\frac{b}{2a}$ when $a = {}^{-}2, b = 6$.

20. Recall that the area of a trapezoid is $A = \frac{1}{2}(b_1 + b_2)h$. Find the area when $h = 7$ cm, $b_1 = 3$ cm, and $b_2 = 8$ cm.

21. Use the order of operations to evaluate $(1 - 3 * 7)^2$.

22. Suppose your calculator shows 2.73E^{-}5. Write this number in each form.

a. Scientific notation

b. Decimal notation

23. Find each value without using a calculator.

a. 2^5 **b.** $(^{-}3)^2$ **c.** $^{-}3^2$

d. $\left(\frac{1}{5}\right)^2$ **e.** $(^{-}1)^{23}$

24. Find each quotient or product.

a. ${}^{-}48 \div {}^{-}6$ **b.** $7 * {}^{-}8$

c. ${}^{-}105 \div 5$ **d.** $\frac{{}^{-}3}{4} * \frac{5}{3}$

25. Find each sum or difference.

a. $5 - 8$ **b.** ${}^{-}5 + 8$

c. $5 - {}^{-}8$ **d.** ${}^{-}5 - 8$

2.4 Representing Functions

After Studying This Section You Will Be Able To

- Represent functions in four ways: verbally, symbolically, numerically, and graphically.
- Use a calculator to display tables and graphs.

We now look more closely at how two variables may be related to each other. We begin with an example.

EXAMPLE 1 Jamie has joined a health club. In addition to a monthly fee of $20, she will be charged $3 every time she uses the facility to work out. She plans to work out about 12 times each month. How much will it cost her to use the club for 1 month?

Solution The cost of belonging to the health club for 1 month has two parts, an amount that varies according to how many times she uses the club during the month and a monthly membership fee that is fixed. These two components may be called *variable cost* and *fixed cost*, respectively. Jamie is interested in the total cost for the month, which may be computed using this relationship:

total cost = **variable cost** + **fixed cost**

The variable cost may be found by multiplying the number of times she uses the club, 12, by the cost for each visit, $3. Thus, the variable cost is **$36**. The fixed cost is given to be **$20**.

total cost = **$36** + **$20** = $56

Jamie will spend $56 for 1 month's membership in the club.

Functions

Of course, Jamie may not use the club exactly 12 times every month. Her actual cost will depend upon the number of times she uses the club. We can express this situation as a relationship between two variables.

EXAMPLE 2 Let y represent the total cost (in dollars) of 1 month's membership in the health club. Let x represent the number of times Jamie uses the health club. Write a formula that shows how y depends upon x.

Solution Again, use the relationship total cost = **variable cost** + **fixed cost**.

In Example 1, we found the variable cost by multiplying 12 visits times the rate $3 per visit. In this problem we don't know the number of visits, but we are representing it by the variable x, so x visits times $3 per visit gives a variable cost of $3x$. We know that the fixed cost is the same as before, $20, and that the variable y represents total cost. The formula $y = \mathbf{3x} + \mathbf{20}$ shows how y depends upon x.

When the value of one variable depends upon the value of another variable, we have a **function**. In this case y is a function of x. The value y is called the **output** (or **dependent**) **variable**, and the value x is the **input** (or **independent**) **variable**. The equation $y = 3x + 20$ is one way of representing this function. Because the equation takes the form dependent variable = expression, we may consider this equation to be a formula. (See Section 2.2.)

EXAMPLE 3 Complete this table using the formula $y = 3x + 20$ to show how y depends upon x.

x	y
0	
2	
4	
6	
8	
10	
12	

Solution In each case the value of y may be found by substituting for x in the formula.

x	y	
0	20	$= 3 * \mathbf{0} + 20$
2	26	$= 3 * \mathbf{2} + 20$
4	32	$= 3 * \mathbf{4} + 20$
6	38	$= 3 * \mathbf{6} + 20$
8	44	$= 3 * \mathbf{8} + 20$
10	50	$= 3 * \mathbf{10} + 20$
12	56	$= 3 * \mathbf{12} + 20$

Most graphing calculators will display a table for you. On a TI-82 or TI-83, use 2ND TBLSET to access the TABLE SETUP menu. Enter these values:

TblStart = 0	(TblMin on the TI-82)	*This starts the table at $x = 0$.*
Δ Tbl = 2		*This increments x by 2.*
Indpnt: Auto Depend: Auto		*These commands tell the calculator to make the table automatically.*

Then enter Y1 = 3 ∗ X + 20 in the Y= menu. 2ND TABLE will give you the desired table.

Note that the table illustrates an important property of functions. For each value of the independent variable, x, there is exactly one value of the dependent variable y. In other words, the value of x determines uniquely the corresponding value of y.

EXAMPLE 4 ■ Display the data in the table in Example 3 in a graph.

Solution Because neither x nor y can take on negative values, you need to show only the *first quadrant*. Choose a range and scale that will allow you to display values of x from 0 to 12 and values of y from 0 to 80.

On the left in Figure 19, paper with a centimeter grid is used to make the graph. On the x-axis 1 cm represents 2 visits. On the y-axis 1 cm represents 10 dollars.

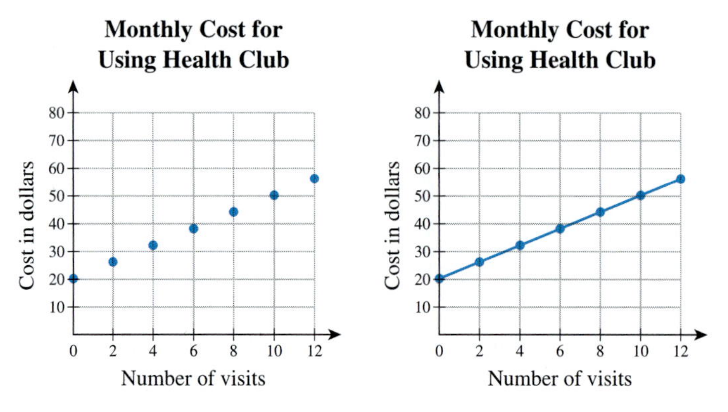

Figure 19

The x-axis is labeled with the name of its variable, Number of visits. Similarly, the y-axis is labeled Cost in dollars. The title, Monthly Cost for Using Health Club, appears above the graph.

From the graph in Figure 19, it appears that the data points lie in a line. In the graph on the right the line has been drawn. Chapter 4 provides more detail about equations of lines.

You may also use a calculator to display the graph drawn in Figure 19. Use the Y= menu. Enter 3X + 20 for Y1. Use WINDOW to set the range and scale.

One set of values that will work is:

$$\text{Xmin} = 0,\ \text{Xmax} = 47,\ \text{Xscl} = 5$$

$$\text{Ymin} = 0,\ \text{Ymax} = 80,\ \text{Yscl} = 10$$

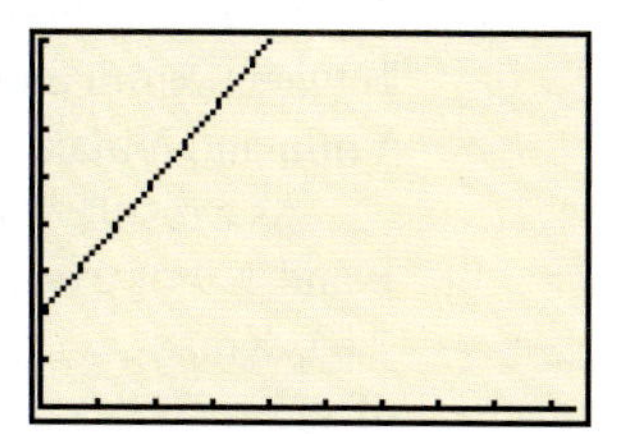

Figure 20

Figure 20 shows the graph that is displayed when GRAPH is pressed. Because 47 creates a *friendly window* on the TI-82 and TI-83, 47 was chosen for Xmax. Friendly windows minimize the number of decimal places displayed when the TRACE feature is used. Numbers that can be used to create friendly windows are shown in the table in Figure 21.

PIXELWIDTH	X_{min}	X_{max}	X_{scl}	Shortcut for Square Window
0.05	0	4.7	0.5	
0.1	$^-4.7$	4.7	1	ZOOM 4 ZDecimal
0.1	0	9.4	1	
0.2	$^-9.4$	9.4	2	ZOOM 4 ZDecimal ZOOM 3 Zoom Out ENTER
0.2	0	18.8	2	
0.5	$^-23.5$	23.5	5	ZOOM 8 ZInteger ZOOM 2 Zoom In ENTER
0.5	0	47	5	
1	$^-47$	47	10	ZOOM 8 ZInteger
1	0	94	10	
2	$^-94$	94	20	ZOOM 8 ZInteger ZOOM 3 Zoom Out ENTER
2	0	188	20	
5	$^-235$	235	50	
5	0	470	50	
10	$^-470$	470	100	
10	0	940	100	

Figure 21 "Friendly" windows for TI-82 and TI-83 calculators.

To see how the graph is related to the table, press the TRACE key. The calculator will display the point where X = 23.5 and Y = 90.5. Move the left arrow and watch the values of X and Y change. You should be able to find the ordered pairs (12, 56), (10, 50), etc., just as they appeared in the table in Example 3.

Pixels and Friendly Windows

The graphing screen on a calculator is made up of hundreds of pixels (*picture elements*), which may be turned on or off. The screen on a TI-83 is 95 pixels wide and 63 pixels high. Consequently, the width of one pixel is $\frac{1}{94}$ of the distance between Xmin and Xmax. The height of one pixel is $\frac{1}{62}$ of the distance between Ymin and Ymax.

As a result, some windows will give awkward-looking decimals for values of x and y. For example, when you set the Standard Window (ZOOM 6), Xmin = $^{-}10$, Xmax = 10, Ymin = $^{-}10$ and Ymax = 10. In this window the width of one pixel is approximately 0.2127659574 and the height is approximately 0.3225806452.

You may prefer to create a **friendly window** that will give pixel dimensions of at most one or two decimal places. The values of Xmin and Xmax suggested in Figure 21 will do just that.

The table in Figure 21 shows both four-quadrant windows in which the origin is at the center of the screen and first-quadrant windows in which the origin is at the lower-left corner of the screen. You can use a shortcut to get some of the windows with the ZOOM menu. Zoom In and Zoom Out will work if ZOOM FACTORS are both set at 2. (To set ZOOM FACTORS, use ZOOM MEMORY 4). ZInteger will always give a screen with PIXELWIDTH = 1 and PIXELHEIGHT = 1. The center of the screen depends upon the last location of the cursor. If necessary, activate ZOOM 6 Zstandard before pressing ZOOM 8 ZInteger.

Zoom 4 ZDecimal and Zoom 8 ZInteger produce square windows in which the width and height of the pixels are the same. An example of a square window is the one used in Figure 14 on page 67 to display street intersections in Gridville.

In the health club example shown in Figure 20, a square window is not appropriate. This is true in most problems involving applications. Once you have chosen friendly values for Xmin and Xmax, you can choose any appropriate values of Ymin and Ymax.

At first you may want to refer to Figure 21 when choosing a window. After a while you will probably remember that values of X like 4.7, 9.4, 47, and 94 will give you friendly windows.

EXAMPLE 5 Chose a friendly window that will include each set of points.

a. Data from the toothpick problem (Example 1, Section 2.2), where x is the number of squares and y is the number of toothpicks: (1, 4), (2, 7), (3, 10), (4, 13), (5, 16).

b. Data from the real estate problem (Example 4, Section 2.2), where x is the width of lot A in feet and y is the perimeter of the lot: (100, 600), (120, 720), (140, 840), (150, 900).

c. Points that satisfy the equation $y = 3x + 5$: $(^{-}4, ^{-}7)$, $(^{-}2, ^{-}1)$, (0, 5), (2, 11), (4, 17).

Solution

a. All values of x and y are positive, so a first-quadrant window is appropriate. The largest value of x is 5 and the largest value of y is 16. For a friendly window we should have Xmax = 4.7 or Xmax = 9.4. Because 4.7 is too small, let's choose 9.4. One possible friendly window is Xmin = 0, Xmax = 9.4, Ymin = 0, Ymax = 20.

b. Again, all values are positive so we may choose a first-quadrant window. Because both x- and y-values are in the hundreds, we must be sure that Xmax is at least 150 and Ymax is at least 900. One possible friendly window is Xmin = 0, Xmax = 188, Ymin = 0, Ymax = 1000.

c. Here we have both positive and negative values of x and y, so a four-quadrant window is in order. One possibility is Xmin = $^{-}4.7$, Xmax = 4.7, Ymin = $^{-}10$, Ymax = 20. Just make sure that all the values of x and y that you want to show lie within the window.

Four Representations for Functions

You have seen four ways of representing the function that shows how the cost for 1 month's membership at the health club depends upon the number of times Jamie uses the facility:

- Verbally: A description of the situation in words: She is charged a monthly fee of \$20 and \$3 for each visit.
- Symbolically: A formula, $y = 3x + 20$, as found in Example 2
- Numerically: A table of values, as found in Example 3
- Graphically: A graph, as found in Example 4

The **rule of four** states that any function may be represented in these four ways: verbally, symbolically, numerically, and graphically. In our study of functions throughout this course, we will be using these four different representations and exploring the connections among them.

EXAMPLE 6

To convert from Celsius to Fahrenheit temperatures, multiply the Celsius temperature by 1.8. Then add 32°. Show how the rule of four can be applied to this function.

Solution

The situation has already been described *verbally*. The rule works because 0° on the Celsius scale is the freezing point of water, which corresponds to 32° on the Fahrenheit scale. Furthermore, each Celsius degree is 1.8 times as great as a Fahrenheit degree.

For a formula, let y represent Fahrenheit temperature and x represent Celsius temperature. The rule described in words can be represented *symbolically* by the formula $y = 1.8 * x + 32$.

To represent the function *numerically*, you may create a table.

x (Celsius)	y (Fahrenheit)
$^{-}10$	14
0	32
10	50
20	68
30	86

A traveler from the United States to a country where Celsius is used might want a copy of the table for quick reference. Or he or she could learn this poem:

```
30 is hot
20 is pleasing
10 is cool, and
0 is freezing!
```

Finally, we may represent the function *graphically* as in Figure 22. Notice that the graph includes points outside the first quadrant because both Celsius and Fahrenheit temperatures can take on negative values. An appropriate friendly window for the calculator would be Xmin = $^{-}94$, Xmax = 94, XScl = 20, Ymin = $^{-}100$, Ymax = 100, and YScl = 20.

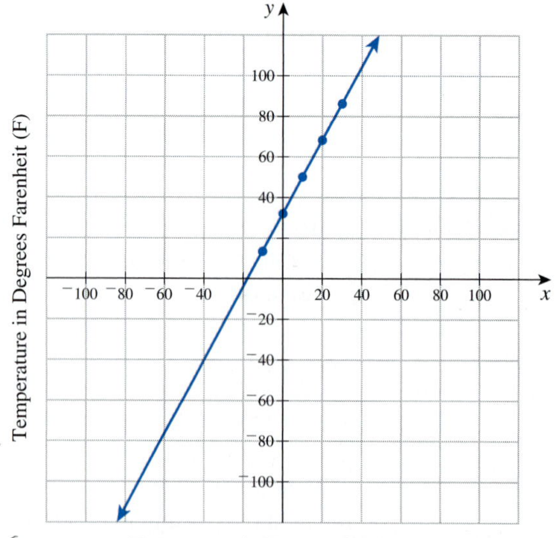

Figure 22

Exercises 2.4 

1. From your reading of Section 2.4, explain the meaning of the terms *fixed cost* and *variable cost.*
2. Gertrude needs to rent a car for 1 day. She will be charged a daily fee of \$40.00 in addition to 6 cents for every mile she drives. Assign the variables by letting x = the number of rental miles she drives and y = the total cost. This situation can modeled by the equation $y = .06x + 40$.
   - **a.** Which part of the formula represents the variable cost?
   - **b.** Which part of the formula represents the fixed cost?
   - **c.** What is the total cost if Gertrude drives 100 miles?
3. A plumber charges \$75.00 for house calls plus \$42.00 for every hour he works.
   - **a.** Let t represent the time (in hours) the plumber works and c represent the cost for his services.
   - **b.** Write a formula for c in terms of t.
   - **c.** Is this formula a function? Explain.
   - **d.** What is the output if the input is 2 hours?

Use the following situation for Exercises 4–6: An office supply company delivers paper that costs \$10.00 for each box. They charge \$25.00 for delivery.

4. **a.** Let x represent the number of boxes and y be the total cost. Write a formula for y in terms of x.
   - **b.** Which is the independent variable and which is the dependent variable?
   - **c.** Make a table by hand for values of 0, 2, 4, 6, and 8 boxes of paper.
   - **d.** Enter the formula from part (a) into the [Y =] menu on the calculator. Press [2ND] TBLSET. Start the table at 10 and increment the number of boxes by 2. Press [2ND] TABLE to display a table, and record your calculator's table values in the table you started in part (c).
5. **a.** Do x-values of 0 or less make sense in this situation? Explain.
   - **b.** Use the values in the table from Exercise 4(d) to make a graph by hand. Be sure to label the axes and title the graph.
   - **c.** Do the ordered pairs on your graph appear to fall on a line? If so, connect the points.
   - **d.** What does the ordered pair (14, 165) represent?
6. Now display the graph on your calculator.
   - **a.** Enter the formula from Exercise 4(a) into the [Y =] menu on the calculator.
   - **b.** Enter the following values in the [WINDOW] menu: Xmin = 0, Xmax = 47, Xscl = 5, Ymin = 0, Ymax = 500, Yscl = 100.
   - **c.** Press [GRAPH]. Use the [TRACE] feature to find the cost when 15 boxes of paper are shipped.
   - **d.** Replace your current-window Xmax setting with double its value and view the graph. Why does the graph appear different?
7. Consider the function $y = 2x + 5$. We wish to produce a calculator graph that will include the following table of values.

x	y
$^-15$	$^-25$
$^-10$	$^-15$
$^-5$	$^-5$
0	5
5	15
10	25
15	35

   - **a.** Pick a friendly window for viewing these values. (See the chart in Figure 21 on page 77.)
   - **b.** Display the graph and describe what it looks like.
   - **c.** Use [TRACE] to find the value of y when $x = {}^-4$.
8. Display the function $y = 2x + 5$ from Exercise 7 with the Standard Window, Xmin = $^-10$, Xmax = 10, Xscl = 1, Ymin = $^-10$, Ymax = 10, Xscl = 1. *Hint:* You may use Zoom 6 to obtain this window.
   - **a.** Describe in what ways the graph is similar to the graph from Exercise 7 and in what ways it is different.
   - **b.** Press [TRACE] and move the cursor to the right. What do you notice?
   - **c.** Is it possible to use [TRACE] to find the exact value of y when $x = {}^-4$? Explain.
   - **d.** Why is the Standard Window not considered to be friendly?
9. Example 6 gave a formula for converting Celsius temperatures to Fahrenheit temperatures. Here is a verbal description of how to convert from Fahrenheit to Celsius: To find a Celsius temperature, subtract 32° from

the Fahrenheit temperature and multiply the result by $\frac{5}{9}$. Now represent this situation three other ways:

a. Write a formula for Celsius temperature in terms of Fahrenheit temperature.

b. Make a table of values by hand that includes Fahrenheit temperatures of 10°, 30°, 50°, 70°, and 90°. Round Celsius temperatures to the nearest .1°.

c. Make a graph by hand.

10. Represent the same situation as Exercise 9 with your calculator.

a. Produce a table of values with Fahrenheit temperatures of 10°, 30°, 50°, 70°, and 90°.

b. Give a friendly window setting for graphing the function.

c. Draw the calculator's graph with scales and labels.

11. The simple interest on an investment is 6% of the principal invested. Recall that to find 6% of a number you may multiply by .06.

a. Assign variables to represent the principal and the interest.

b. Represent this function symbolically with a formula giving interest in terms of the principal.

c. Represent this function numerically with a table of values for investments of $500, $1000, $1500, $2000, and $2500.

d. Represent this function graphically.

12. The number of gallons of paint required to cover the walls of a room is found by dividing the square footage of the walls by 350. Represent this function by a formula, a table of values, and a graph for square footages of 1000, 2000, 3000, and 4000. Be sure to assign your variables before you write a formula.

13. Assume a new product costing $10,000 depreciates the same amount each year. Its remaining value can be found by subtracting the product of $1500 and the years of service from $10,000.

a. Represent this function by a formula, a table of values, and a graph.

b. What range of years makes sense in this situation?

c. What is the range of values for the product?

d. How do your answers from parts (b) and (c) help you choose window settings?

14. An art collector has a $300 painting, which she believes will appreciate 10% of its original value each year. The value of the painting, then, is $300 plus $30 for each year she owns the painting.

a. Write a formula for the value of the painting.

b. Make a table of values.

c. Draw a graph by hand or create a graph on your calculator.

15. a. Describe the difference between a four-quadrant window and a first-quadrant window.

b. For which of these functions is a four-quadrant window appropriate? For which is a first-quadrant window appropriate?

- The cost of office supplies (Exercises 4–6)
- $y = 2x + 5$ (Exercises 7–8)
- The formula for converting temperatures from Fahrenheit to Celsius (Exercises 9–10)
- The paint required to cover the walls of a room (Exercise 12)

Skills and Review 2.4

16. Graph, by hand, the ordered pairs $(^-1, 3)$ and $(2, ^-4)$ and connect the points.

a. Find the run and rise from $(^-1, 3)$ to $(2, ^-4)$. Label your graph.

b. Find the distance between the ordered pairs.

c. How do the run and rise relate to the distance between the ordered pairs?

17. Start at $(^-4, 6)$. Use a run of 2 and a rise of $^-3$ to find your next location.

18. Find the hypotenuse of a right triangle whose other sides measure 30 feet and 40 feet.

19. A handheld computer has a rectangular viewing screen. The length is 1.4 inches more than the width.

a. Let x represent the width of the computer screen, in inches. Find an expression for the perimeter of the screen.

b. Find the perimeter of the computer screen when the width is 3.8 inches.

20. Evaluate $(a + 3)(6 - b)$ when $a = 4$ and $b = 8$.

21. Consider the expressions $50 - 4x$ and $^-4x + 50$.

a. Show that the expressions are equivalent by the commutative property.

b. Show equivalence by substituting three values into each expression.

22. Use the order of operations to evaluate

$$\frac{2 + 5 * 4^2}{6 - 8}.$$

23. Estimate $\sqrt{41}$.

24. Write $^{-}6.15 * 10^{^{-}4}$ in decimal notation.

25. Find the rate in miles per hour for each situation.

a. Uta walks 6 miles in 2 hours.

b. Ben jogs 6 miles in 1 hour.

c. Tanya runs 6 miles in $\frac{2}{3}$ hour.

d. Roberto bicycles 6 miles in $\frac{2}{5}$ hour.

CHAPTER 2 ▪ KEY CONCEPTS

Expression An expression is a meaningful arrangement of constants, variables, and symbols for operations and grouping. (Page 46)

The Associative Properties Addition: for any numbers a, b, and c, $(a + b) + c = a + (b + c)$. Multiplication: for any numbers a, b, and c, $(a * b) * c = a * (b * c)$. (Page 47)

The Distributive Property For all numbers a, b, and c, $a * (b + c) = (a * b) + (a * c)$. (Page 48)

The Order of Operations (Page 50).

a. Do operations inside grouping symbols first.

b. Evaluate exponents and roots from left to right.

c. Perform multiplication and division from left to right.

d. Perform negation, then addition and subtraction, from left to right.

Special Properties of 0 (Page 52).

a. For any number a, $a + {}^{-}a = 0$.

b. For any number a, $0 + a = a$ and $a + 0 = a$.

c. For any number a, $0 * a = 0$ and $a * 0 = 0$.

d. For any number a, $\frac{a}{0}$ is undefined; that is, it has no meaning.

Special Properties of 1 (Page 52).

a. For any number a (except 0), $a * (\frac{1}{a}) = 1$.

b. For any number a, $1 * a = a$ and $a * 1 = a$.

c. $1*$ may be inserted in front of a left parenthesis. Thus $(a + b) = 1 * (a + b)$.

d. $*1$ may be inserted behind a right parenthesis. Thus $(a + b) = (a + b) * 1$.

Special Properties of $^{-}1$ (Page 52).

a. For any number a, $^{-}1 * a = {}^{-}a$ and $a * {}^{-}1 = {}^{-}a$.

b. A negation symbol in front of a left parenthesis may be replaced by $^{-}1*$. Thus $^{-}(a + b) = {}^{-}1 * (a + b)$.

c. A subtraction symbol in front of a left parenthesis may be changed to addition if $^{-}1*$ is inserted. Thus $a - (b + c) = a + {}^{-}1 * (b + c)$.

Equations and Formulas The statement that two expressions are equal is called an equation. A formula is an equation that shows how one variable depends upon one or more other variables. (Page 56)

Equivalent Expressions Two expressions are equivalent if the results are the same no matter what values are substituted for the variable or variables. (Page 57)

Like Terms Like terms contain the same variable raised to the same power. Like terms may be combined by adding (or subtracting) the **coefficients**, that is, the constants that are multiplied by the variables. For example, $3x + 4x = 7x$. *Caution:* x and x^2 are *not* like terms. (Page 58)

The Pythagorean Theorem The Pythagorean theorem states that in a right triangle with sides of length a, b, and c, if c is the length of the side opposite the right angle, then $c^2 = a^2 + b^2$ and $c = \sqrt{a^2 + b^2}$. (Page 61)

The Coordinate Plane Every point in the plane is represented by an ordered pair of numbers (x, y). The first number is the x-coordinate; the second number is the y-coordinate. (Page 65)

Run and Rise Suppose P and Q are two points in the coordinate plane with coordinates $P = (x_1, y_1)$ and $Q = (x_2, y_2)$. Then the run and rise from P to Q are

$$\text{run} = x_2 - x_1$$

$$\text{rise} = y_2 - y_1 \qquad \text{(Page 70)}$$

The Distance Formula If P and Q are two points in the coordinate plane with coordinates $P = (x_1, y_1)$ and $Q = (x_2, y_2)$, the distance from P to Q is given by the formula

$$\text{distance} = \sqrt{\text{run}^2 + \text{rise}^2} \qquad \text{(Page 71)}$$

Functions on a Graphing Calculator Use [Y =] to enter the function in symbolic form. Use [2ND] TBLSET and [2ND] TABLE to display the function numerically. Use [WINDOW] and [GRAPH] to display the function graphically and [TRACE] to show the values of x and y for points on the graph. (Page 76)

Friendly Windows Friendly windows for the TI-83 calculator are shown in Figure 21. (Page 77)

The Rule of Four Any function may be represented in these four ways: verbally, symbolically, numerically, and graphically. (Page 79)

Exploration

An automobile near Chambersburg, Pennsylvania, is equipped with a GPS receiver. It receives a signal from a satellite located above the Pacific Ocean midway between Los Angeles, California, and Honolulu, Hawaii. The center of the earth is at the origin of a coordinate system with distances measured in miles. The coordinates of the automobile are $x_1 = 661$, $y_1 = {}^{-}3033$, and $z_1 = 2522$. The coordinates of the satellite are $x_2 = {}^{-}9654$, $y_2 = {}^{-}8692$, and $z_2 = 7500$.

a. Find the differences in the x-, y-, and z-coordinates:

$$x_2 - x_1 = ?$$

$$y_2 - y_1 = ?$$

$$z_2 - z_1 = ?$$

b. Use the three-dimensional distance formula

$$\text{distance} = \sqrt{(x_2 - x_1)^2 + (y_2 - y_1)^2 + (z_2 - z_1)^2}$$

to find the distance from the automobile to the satellite.

c. Given the fact that the satellite is 11,000 miles directly above the surface of the earth, is your answer to part (b) reasonable? Explain.

Chapter 2 ▪ Review Exercises

Section 2.1 Grouping and the Order of Operations

1. Use the associative property of addition to regroup with parentheses and find the sum. (Do not use a calculator.)

a. $6 + (124 + 58)$ **b.** $(189 + 43) + 17$

2. Does the associative property hold for subtraction? Give an example to support your answer.

3. Use the associative property of multiplication to regroup with parentheses and find the product. (Do not use a calculator.)

a. $(7 * 8) * 5$ **b.** $\frac{1}{3} * (3 * 467)$

4. Does the associative property hold for division? Give an example to support your answer.

5. Use the distributive property to expand.

a. $3 * (40 + 7)$ **b.** $4 * (90 - 2)$

c. $6 * (8 + 10)$ **d.** $2 * (x + y)$

e. $9 * (a - b)$ **f.** $8 * (x - y)$

6. Use the distributive property to factor.

a. $(4 * 52) + (4 * 8)$ **b.** $(3 * 9) + (3 * 81)$

c. $(5 * x) + (5 * y)$ **d.** $(7 * a) - (7 * b)$

7. Apply the order of operations to evaluate each expression.

a. $8 + 4 \div 4$ **b.** $3 + 5^2 - (7 + 2)$

c. $\frac{\sqrt{5+20}-7}{2}$

8. Write the keystrokes necessary to correctly evaluate the given expression on your calculator:

$$\frac{5 + 11}{6 - 8}$$

9. Fill in the blanks.

a. $4 + ? = 0$ **b.** $\frac{23}{?}$ is undefined

c. $? * 17 = 17$

10. Name the special property of 1 or $^{-}1$ that makes each equation true.

a. $(x + y) = 1 * (x + y)$

b. $x - (y + z) = x + {}^{-}1 * (y + z)$

11. Demonstrate, with numbers, each of the four special properties of 0.

12. Demonstrate, with numbers, each of the four special properties of 1.

13. Demonstrate, with numbers, each of the three special properties of $^{-}1$.

Section 2.2 Expressions and Formulas

14. Evaluate each formula with the given substitution.

a. $C = 2 * \pi * r$ for $r = 2$ inches. You may use $\pi \approx 3.14$.

b. $r = \frac{d}{t}$ for $d = 3$ feet, $t = 6$ seconds.

c. $A = \pi * r^2$ for $r = 4$ feet.

d. run $= x_2 - x_1$ for $x_2 = 5, x_1 = {}^{-}1$

e. distance $= \sqrt{(x_2 - x_1)^2 + (y_2 - y_1)^2}$ for $x_2 = {}^{-}6$, $x_1 = 0$, $y_2 = 5$, $y_1 = {}^{-}3$

15. Determine if each pair of expressions is equivalent by substituting the values $x = 4, 3, {}^{-}2$.

a. $5 * x - 2$ and $6 * x - (x + 2)$

b. $\sqrt{x^2 + 2 * x + 1}$ and $x + 1$

c. $(x - 1) * (x + 1)$ and $x^2 - 1$

16. One formula for the area of a trapezoid is $A = \frac{1}{2} * (b_1 + b_2) * h$; another formula is $A = \frac{1}{2} * b_1 * h + \frac{1}{2} * b_2 * h$. Use the distributive property to show that the two expressions are equivalent.

17. Simplify each expression by combining like terms.

a. $6y + 9y$ **b.** $4x - x$

c. $a^2 - 3a + 2a^2 - 9a$

18. Simplify each expression by combining like terms.

a. $\frac{3}{a} + \frac{8}{a}$ **b.** $\frac{5}{a} - \frac{2}{3} + \frac{7}{a} + \frac{1}{9}$

c. $\frac{2}{y} + \frac{x}{y}$ **d.** $x - \frac{1}{x^2} + x^2 + x^{-2}$

19. Rewrite each verbal expression as an algebraic expression. Let x represent the unknown number.

a. A number increased by 5

b. 4 more than a number

c. 4 less than a number

d. The difference between a number and 2

e. Twice a number

f. One-third of a number

g. The product of 7 and a number

h. The quotient of a number and 6

i. A number divided by 3

j. 4 more than 3 times a number

k. 4 less than 3 times a number

20. A rectangle's length is 3 yards more than 4 times its width.

a. Let x represent the width of the rectangle, in yards. Find an expression for the perimeter of the rectangle.

b. Find the value of this expression when $x = 15$ yards.

21. Use the Pythagorean theorem to find the hypotenuse of a right triangle whose base is 6 cm and height is 17 cm.

Section 2.3 The Coordinate Plane

22. On the coordinate plane, locate the following points and identify the quadrant or axis where each lies.

a. $(^{-}2, 3)$ **b.** $(0, 5)$ **c.** $(1, 4)$

d. $(^{-}1, ^{-}2)$ **e.** $(2, ^{-}4)$

23. Explain why the x-coordinates of all the ordered pairs on a vertical line are the same. *Hint:* Draw a vertical line and label several ordered pairs.

24. Draw a horizontal line through the y-axis at $(0, 3)$; what can you say about the y-coordinates of each ordered pair on that line?

25. Compute the run and rise between the given ordered pairs.

a. $(1, ^{-}2)$ and $(^{-}3, 4)$ **b.** $(2, 6)$ and $(^{-}2, ^{-}3)$

26. Start at the point $(^{-}1, 0)$. If you move a run of 5 and a rise of $^{-}2$, at what point will you end?

27. Apply a run of $^{-}2$ and a rise of 1 to find the ending location if you start at $(3, ^{-}4)$.

28. Use the shortcut ZOOM 4 ZDecimal to find the value of the WINDOW variables Xmin = ?, Xmax = ?, Xscl = ?, Ymin = ?, Ymax = ?, and Yscl = ?

29. Describe what the WINDOW settings Xmin $= ^{-}10$, Xmax $= 10$, Ymin $= ^{-}10$, and Ymax $= 10$ will do to the window of your GRAPH screen.

30. Find the distance between the points $(2, 0)$ and $(^{-}1, ^{-}4)$.

31. Given the two ordered pairs $(2, ^{-}1)$ and $(^{-}6, 7)$:

a. Write the keystrokes necessary to enter your distance calculation on your calculator.

b. Approximate the distance shown on your calculator.

Section 2.4 Representing Functions

For Exercises 32–36, use the following situation: The Nomystery Airline Company charges $100 plus .25 per mile of travel.

32. How much will it cost to travel 800 miles on Nomystery Airline?

33. Let x represent the number of miles of travel and let y represent the cost (in dollars) of the travel.

a. Write a formula that shows how y depends upon x.

b. What is the independent variable?

c. What is the dependent variable?

34. By hand, create a table of values with inputs of $x = 0$, 100, 200, 300, and 400 miles.

35. By hand, create a graph that includes the values from Exercise 34.

36. Consider representing the same function on your calculator.

a. How is the formula from Exercise 33(a) entered into your Y = menu?

b. Choose table settings in 2ND TBLSET so that the table shown in 2ND TABLE includes your table of values from Exercise 34.

c. Choose a friendly window for viewing a graph that includes the table of values from Exercise 34.

CHAPTER 2 ▪ TEST

1. Use the associative property of addition or multiplication to regroup with parentheses and find the sum. (Do not use a calculator.)

a. $^{-}8 + (^{-}2 + 51)$ **b.** $(9 * 5) * 6$

c. $(9 + ^{-}73) + 73$

2. Use the distributive property to either expand or factor the following expressions.

a. $5 * (3 - 8)$ **b.** $(6 * a) + (6 * b)$

c. $7 * (a + b)$ **d.** $(2 * x) - (2 * y)$

3. Apply the order of operations to evaluate each of the expressions.

a. $\sqrt{16-7}+3$ **b.** $(12-9\div 3)^2$

c. $(^-2)^4-\sqrt{4}*6$

4. Use the special properties of 0, 1, or $^-1$ to fill in the blanks.

a. $?+{}^-7={}^-7$ **b.** $?*6=0$

c. $5*?=1$ **d.** $6*?={}^-6$

5. Evaluate each formula with the given substitution.

a. $y=5*x+6$ for $x=3$

b. $A=\frac{1}{2}*b*h$ for $b=4$ cm, $h=10$ cm

6. Determine if the expressions $x+4*x$ and $x*(1+4)$ are equivalent by substituting the values $x=5,\ 2,\ {}^-3$.

7. One formula for the amount of money in an account is $A=P*(1+r*t)$. Use the distributive property to find an equivalent formula.

8. Simplify each expression by combining like terms.

a. $11x-5x$ **b.** $\frac{1}{2}+\frac{4}{x}-\frac{3}{8}-\frac{1}{x}$

c. ${}^-6a+3+7a-8$ **d.** $\frac{3}{a}+\frac{8}{a}$

9. Soccer fields are rectangular in shape. The length of the field is 135 feet more than the width of the field.

a. Let w represent the width of the field, in feet. Find an expression for the perimeter of the field.

b. Find the perimeter of the soccer field if the width is 225 feet.

10. Suppose a cornerback has scored a touchdown by returning an interception from the corner of his goal line to the goal line at the opposite corner of the field. Use the dimensions in the figure in the next column to find the distance he ran.

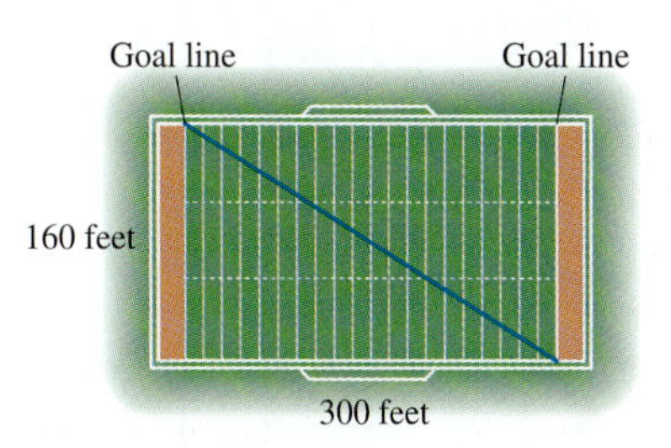

11. On the coordinate plane, locate the following points and identify their quadrants or axes.

a. $(^-4, 1)$ **b.** $(2, 0)$

c. $(1, {}^-3)$

12. Compute the run and rise between the given ordered pairs.

a. $(0, 3)$ and $(^-1, 5)$ **b.** $(^-3, 4)$ and $(1, 5)$

13. What is the distance between the ordered pairs $(^-1, {}^-5)$ and $(5, 3)$?

For Exercises 14–16, use the following situation: A phone company provides long-distance service for $4.95 per month plus $.07 per minute for each long distance call.

14. Write a formula for the monthly cost of long-distance service in terms of the number of minutes of calls. *Note:* Be sure to assign the variables you use.

15. On your calculator, create a table of values that includes inputs of $x=0$, 20, 40, 60, and 80 minutes.

16. On your calculator, create a graph that includes the table of values from Exercise 15.

CHAPTER

3 Linear Equations

A Matter of Balance

Photosynthesis is the process by which plants harness the sun's energy to produce sugar from water and carbon dioxide. It is one of the most important biological phenomena, for without photosynthesis, life as we know it would not exist. This process may be described as a chemical equation in words: water + carbon dioxide = oxygen + glucose.

Scientists prefer to write this equation using chemical formulas indicating the elements comprising each of these substances. Water is H_2O, because one molecule of water contains 2 hydrogen atoms and 1 oxygen atom. Carbon dioxide is CO_2, because its molecule contains 1 carbon atom and 2 oxygen atoms. Oxygen molecules each contain 2 atoms of the element oxygen, so oxygen is written O_2. Glucose, the most simple of sugars, has the most complex formula: $C_6H_{12}O_6$. The equation for photosynthesis may then be written $H_2O + CO_2 = O_2 + C_6H_{12}O_6$.

There is something wrong with this equation, however. There are only 2 hydrogen atoms on the left side but 12 hydrogen atoms on the right side. Chemical equations must be balanced, just as we maintain the balance of an algebraic equation when we perform the same operation on both sides. In this chapter you will learn more about how the principle of balance is used to solve algebraic equations. (See Exploration on p. 118.)

Need help? For on-line resources, visit this web site: **math.college.hmco.com/students.**

3.1 Solving Simple Equations

AFTER STUDYING THIS SECTION YOU WILL BE ABLE TO

- Solve simple linear equations with a variable on one side of the equation.
- Solve simple linear equations with a variable on both sides of the equation.
- Solve word problems leading to simple linear equations.

In Chapter 2 you learned to use variables to form algebraic expressions. In this chapter you learn how to solve equations. Recall from Section 2.2 that an *equation* is a statement that two *expressions* are equal. Solving an equation involves finding what value or values of the variables make that statement true. Example 1 shows a problem situation that leads to an equation.

EXAMPLE 1 John's mechanic is replacing the timing chain on his car. The part costs \$95 and she charges \$60 per hour for labor. If the total cost for the repair was \$200, how long did the mechanic work? Write an equation to solve the problem.

Solution In order to write an equation, we need an expression for the cost of repairs. To write an expression we must answer the question, "What should the variable be?" Look at what the problem is asking: "How long did the mechanic work?" This suggests a variable will be needed for time. We let t represent the number of hours she worked.

Recall from Section 2.4 that **total cost** = **variable cost** + **fixed cost**. The total cost is known to be **\$200**. The variable cost is the hourly rate times the number of hours she works, or **60*t***. The fixed cost is the cost of the part, **\$95**. Putting all this together gives an equation:

$$\$200 = \$60 \text{ per hour} * t + \$95$$

or, more simply,

$$200 = 60t + 95.$$

■ ■

Once we have an equation, the next step is to solve it. We need to use the following fact.

> **Basic Principle of Equation Solving**
>
> If we are given an equation, we can add the same term on both sides or multiply by the same number on both sides and still have a valid equation.

This principle works because if two things are equal, we think of them as *the same*, and if we do the same thing to things that start the same, they will end up the same. The only exception is division by zero, which is never allowed.

EXAMPLE 2 ■ Solve the equation in Example 1.

Solution Start with the equation

$$200 = 60t + 95$$

Our goal is to get a valid equation with t all by itself on one side. Think about the order of operations. The expression on the right side says "take t, multiply it by 60 and then add 95." To find t, reverse the steps, first eliminating 95 and then eliminating 60 from the right side of the equation.

To eliminate 95 from the right side of the equation, use the **basic principle of equation solving** and *add* $\mathbf{^-95}$ on both sides:

$$200 + \mathbf{^-95} = 60t + 95 + {}^-95$$

Combining constant terms on each side gives

$$105 = 60t$$

To eliminate 60 from the right side of the equation, we again use the basic principle of equation solving and *multiply* on both sides by the reciprocal of 60, that is, $\mathbf{\frac{1}{60}}$:

$$\mathbf{\frac{1}{60}} * 105 = \mathbf{\frac{1}{60}} * 60t$$

and because

$$\frac{1}{60} * 60t = \frac{60}{60}t = t$$

we get $t = \frac{105}{60} = 1.75$ hours. The answer is in hours, because that is how we measured time.

Note: We could have described the first step as *subtracting* 95 on both sides as well as adding $^-95$. Likewise we could have described the second step as *dividing* by 60 as well as multiplying by $\frac{1}{60}$. You may think of subtraction as adding an opposite and division as multiplying by a reciprocal.

■ ■

EXAMPLE 3 ■ Suppose we place equilateral triangles that are 2 feet on each side together, as in Figure 1, to make quadrilaterals. How many triangles will it take to get a quadrilateral with a perimeter of 54 feet?

Solution Notice that only two sides from each new triangle are added to the perimeter, and one side of a previous triangle is lost, so we add 2 to the perimeter for each new

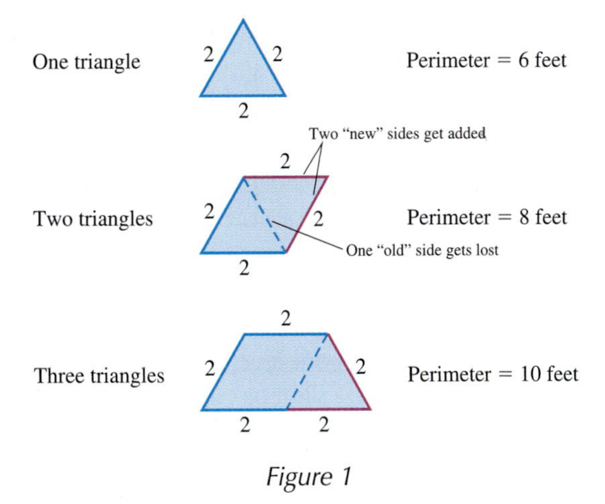

Figure 1

triangle. We are asked to solve for the number of triangles, so we should give that a variable name. Let N represent the number of triangles. Because we add 2 to the perimeter for each new triangle and the first perimeter is 6 feet, we can check the formula

$$\textbf{perimeter} = 4 + 2 * N$$

against the values in Figure 1 and see if it works.

We want to know how many triangles will give a **perimeter** of **54** feet. Substituting 54 for perimeter gives us the equation

$$\mathbf{54} = 4 + 2 * N$$

We can solve the same way we solved the equation in Example 2, by getting a constant term on one side and the variable on the other.

$$
\begin{aligned}
54 &= 4 + 2N \\
\underline{+{}^{-}4} &= \underline{+{}^{-}4} && \textit{Adding } {}^{-}4 \textit{ on both sides} \\
50 &= 2N && \textit{Combining constant terms} \\
\tfrac{1}{2} * 50 &= \tfrac{1}{2} * 2N && \textit{Multiplying by } \tfrac{1}{2} \textit{ on both sides} \\
25 &= N && \tfrac{1}{2} * 50 = 25 \textit{ and } \tfrac{1}{2} * 2 = 1
\end{aligned}
$$

We can check our work using a calculator. We enter the left and right sides of the equation separately in the [Y=] screen (Figure 2), where X stands in for N. Because we suspect that $N = 25$ and because Y1 = 54, we need to find a friendly window from Figure 21 on page 77 that will work; it looks as if Xmin = 0, Xmax = 47 will work, and we can let Ymin = 0 and Ymax = 100. Figure 3 shows the resulting graph.

Press the [TRACE] key and move the cursor to where the two lines intersect. When you reach that point you will see X = 25 and Y = 54 displayed at the

Figure 2

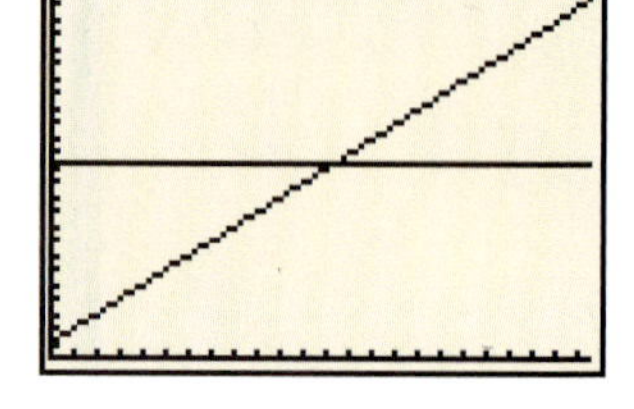

Figure 3

bottom of the screen. This shows that the point (25, 54) lies on both lines and that X = 25 is a solution to the equation 54 = 4 + 2X. See Figure 4.

It looks as if it will take 25 triangles to build a parallelogram with a perimeter of 54 feet. We can check by substituting 25 into our formula for perimeter:

$$\text{perimeter} = 4 + 2 * N = 4 + 2 * \mathbf{25}$$
$$= 4 + 50$$
$$= 54$$

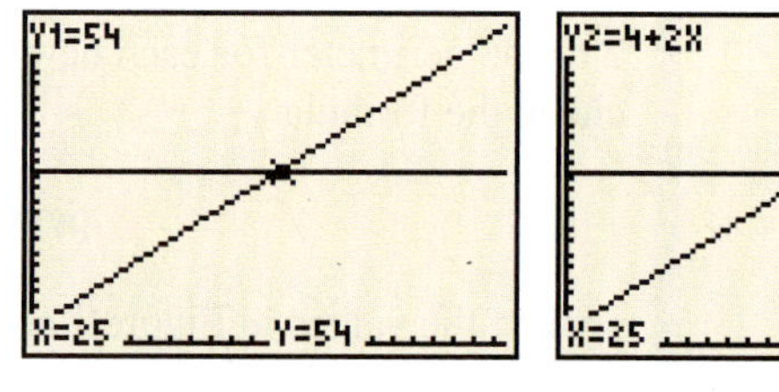

Figure 4

In Examples 1–3 our procedure was first to add (or subtract) the same number on both sides and then to multiply (or divide) the same number on both sides. When the variable appears on both sides of an equation, however, we have an additional first step, as Example 4 shows.

EXAMPLE 4 Recall Jamie's health club from Examples 1 and 2 of Section 2.4. An expression for the amount (in dollars) she will have to pay for one month is $3x + 20$, where x is the number of times she visits the club. There is another health club she is considering where she would pay a monthly fee of \$2 and a charge of \$6 for each visit. The cost of belonging to that club is $6x + 2$. Knowing how many times per month she would use a health club, Jamie figures that the cost of the two clubs will be the same. How many times per month does Jamie plan on using a health club?

Solution In effect, we need to find a value of x that makes the two expressions $3x + 20$ and $6x + 2$ equal each other. Thus, we have the equation

$$3x + 20 = 6x + 2$$

To solve this equation, we first add an x-term to both sides to eliminate $3x$. The x-term we add is $^{-}3x$, that is, the opposite of $3x$.

$3x + 20 = 6x + 2$	*Original equation*
$\underline{+^{-}3x} \quad \underline{+^{-}3x}$	*Adding $^{-}3x$ on both sides*
$20 = 3x + 2$	*Combining like terms*
$\underline{+^{-}2} \quad \underline{+^{-}2}$	*Adding $^{-}2$ on both sides*
$18 = 3x$	*Combining constant terms*
$\frac{1}{3} * 18 = \frac{1}{3} * 3x$	*Multiplying by $\frac{1}{3}$ on both sides*
$6 = x$	$\frac{1}{3} * 18 = 6$, *and* $\frac{1}{3} * 3 = 1$

Jamie plans to visit a health club about 6 times a month.

We can check our work by substituting $\mathbf{x = 6}$ into both sides of the equation. When checking an equation, we write a question mark (?) above the equal sign until we reach a statement where the left and right sides are equal.

$3\mathbf{x} + 20 = 6\mathbf{x} + 2$	*Original equation*
$3 * \mathbf{6} + 20 \stackrel{?}{=} 6 * \mathbf{6} + 2$	*Substituting 6 for x*
$18 + 20 \stackrel{?}{=} 36 + 2$	
$38 = 38$	*OK!*

We can also check our answer using a calculator, as in Example 3, by entering Y1 = 3X + 20 and Y2 = 6X + 2 and using [TRACE] to find the point of intersection.

Some equations involve fractions. The next example shows how these may be solved.

EXAMPLE 5 Solve the equation

$$\frac{2}{3}x + \frac{1}{2} = \frac{11}{2}$$

Solution Don't let the fractions scare you! Think of this as an equation similar to Example 3, $54 = 4 + 2N$. The first step is to add (or subtract) the same number on both sides. In this case the logical choice is to eliminate $\frac{1}{2}$ from the left side.

$\frac{2}{3}x + \frac{1}{2} = \frac{11}{2}$	*Original equation*
$\underline{+^{-}\frac{1}{2}} \quad \underline{+^{-}\frac{1}{2}}$	*Adding $^{-}\frac{1}{2}$ on both sides*
$\frac{2}{3}x = \frac{10}{2}$	*Combining constants*

To get x alone, we can multiply on both sides by the reciprocal of $\frac{2}{3}$, that is, $\frac{3}{2}$.

$$\frac{3}{2} * \frac{2}{3}x = \frac{3}{2} * \frac{10}{2}$$ *Multiplying by $\frac{3}{2}$ on both sides*

$$\frac{6}{6}x = \frac{30}{4}$$ *Multiplying numerator times numerator and denominator times denominator*

$$x = \frac{15}{2} = 7\frac{1}{2} = 7.5$$

The solution to this equation may be checked by substituting $\frac{15}{2}$ for x in the original equation:

$$\frac{2}{3}x + \frac{1}{2} = \frac{11}{2}$$ *Original equation*

$$\frac{2}{3} * \frac{15}{2} + \frac{1}{2} \stackrel{?}{=} \frac{11}{2}$$ *Substituting $\frac{15}{2}$ for x*

$$5 + \frac{1}{2} \stackrel{?}{=} \frac{11}{2}$$

$$5\frac{1}{2} = 5\frac{1}{2}$$ *OK!*

Exercises 3.1

1. In Example 2 we solved for t in the equation $200 = 60t + 95$. If the total cost of the repair is \$245, we get a slightly different equation. Solve for t in the equation $245 = 60t + 95$ to find the time the mechanic worked with the new total cost.

2. In Example 3 we solved for N and checked our answer to the equation $54 = 4 + 2N$. Solve for N and check your answer to the equation $108 = 4 + 2N$.

3. Solve $8 = 2x - 6$ for x.

4. Solve $\frac{1}{2}z - 5 = 6$ for z.

5. Solve $1.8b - 3.4 = {}^{-}10.2$ for b. Write your repeating decimal as a fraction by using the fraction feature on your calculator. (See Exercise 14 of Section 1.2).

6. A camper has collected 25 logs for an overnight fire. Suppose he burns 4 logs every hour. A formula for L, the number of logs remaining after t hours, is $L = 25 - 4t$.
   - **a.** Enter the equation Y1 = 25 − 4X in your calculator's [Y=] menu. Note that Y takes the place of L and X takes the place of t.
   - **b.** Use the window settings Xmin = 0, Xmax = 9.4, Xscl = 1, Ymin = 0, Ymax = 30, and Yscl = 5. Press [GRAPH] and observe that the graph is a line.
   - **c.** Press [TRACE] and move the cursor to estimate for what value of X (time) the number of logs (Y) is zero.
   - **d.** Solve the equation $0 = 25 - 4t$ to find exactly how much time it takes for the logs to run out.
   - **e.** The camper begins his fire at 11:00 P.M. At what time of day does he run out of wood?

7. Solve $11x = 20x - 18$ for x.

8. A carpenter and his helper decide to have a nail hammering contest. The carpenter is so confident that he can beat his helper that he lets him count the 30 nails he hammered in practice. They set a timer and both begin hammering nails. The carpenter hammers nails at

the rate of 50 per minute, whereas the helper hammers only 40 nails per minute.

a. Complete the table of values.

Time (minutes)	Nails Hammered by Carpenter	Nails Hammered by Helper
0	0	30
1	50	70
2	100	110
3		
4		
5		

b. Let t represent the number of minutes that have passed since they start the timer. Write an expression for the number of nails hammered by each person in terms of t.

c. Set the expressions in (b) equal to each other and solve to find the time it will take the carpenter to equal the helper's nail hammering.

d. If you were the helper, how long would you want the contest to last to guarantee your victory?

9. In Example 4, we solved the equation $3x + 20 = 6x + 2$. The basic principle of equation solving (p. 89) says that as long as we add the same amount on both sides of an equation or multiply by the same amount on both sides we still have a valid equation. Therefore, we can arrive at the same solution by performing the steps in a different order. Describe what was done next to each step in solving this equation.

$3x + 20 = 6x + 2$	*Original equation*
$3x + 20 + {}^{-}20 = 6x + 2 + {}^{-}20$	*?*
$3x = 6x + {}^{-}18$	*Combining like terms*
$3x + {}^{-}6x = 6x + {}^{-}18 + {}^{-}6x$	*?*
${}^{-}3x = {}^{-}18$	*Combining like terms*
$\frac{{}^{-}3x}{{}^{-}3} = \frac{{}^{-}18}{{}^{-}3}$	*?*
$x = 6$	*Solution*

What does the solution $x = 6$ stand for in this real-life problem?

10. Solve ${}^{-}7x + 3 = 4x - 5$ for x.

11. Solve $\frac{1}{3}x - 7 = 9 + x$ for x.

12. Show that $x = 3$ checks as the solution to the equation $60 + 5x = 20x + 15$.

13. Solve $\frac{1}{3}x - \frac{2}{3} = x - \frac{1}{2}$ for x.

14. An employee is offered a choice between two vacation plans. The number of days off is given by the expressions $.5m + 1$ for plan A and $.4m + 3$ for plan B, where m is the number of months employed.

a. Write an equation by setting the two expressions equal to each other.

b. Solve the equation in order to find the number of months the employee needs to work before the plans give the same number of days off.

c. Which plan would you prefer if you worked less than 12 months at the company? Explain.

15. The monthly charges for an Internet provider are \$12 plus \$2 for each hour of use. Another provider charges \$15 plus \$1.50 for each hour of use.

a. Write an expression for each provider's charges.

b. Write an equation by setting the two expressions equal to each other.

c. Solve the equation.

d. Interpret the solution in the context of the problem.

Skills and Review 3.1

16. Make a table of values and graph the function $y = 2x + 1$. Do not use a calculator.

17. The amount a person spends at the movies is \$7 for a ticket plus \$5 for each pound of candy. Let sweets represent the number of pounds of candy and cost represent the amount spent for a ticket and candy. A formula for the cost in terms of sweets is cost = 7 + 5 * sweets.

a. How is this function entered into the [Y=] menu?

b. Use the Table feature on your calculator to determine the cost when sweets = $\frac{1}{4}$, $\frac{1}{2}$, $\frac{3}{4}$, 1, and 2 pounds. Organize your results in a table.

c. Use the table values from part (b) to draw a graph by hand.

18. Data from the previous movie problem are: (.25, 8.25), (.5, 9.5), (.75, 10.75), (1, 12), and (2, 17).

a. In what quadrants do these points appear?

b. Choose a friendly window that includes these points.

19. Find the run and rise when you start at $(2, {}^{-}5)$ and end at $({}^{-}1, 3)$.

20. Find the distance between the points (0, 4) and (3, 9).

21. Evaluate the expression $3x - 4y$ when $x = \frac{-1}{3}$ and $y = 2$.

22. Evaluate without using a calculator:

$$\frac{12 + 4}{7 - {}^{-}5}$$

23. Evaluate $3 + 8 \div 2 * \frac{1}{4} - 2^3$ without a calculator.

24. Evaluate $5 + 3 * \sqrt{36}$ without a calculator.

25. Which expression is not equal to $5^{{}^{-}2}$?

a. $\frac{1}{25}$ **b.** ${}^{-}25$

c. .04 **d.** $\frac{1}{5^2}$

3.2 Simplifying Expressions to Solve Equations

AFTER STUDYING THIS SECTION YOU WILL BE ABLE TO

- Simplify expressions by removing parentheses and combining like terms.
- Solve equations that require the expression on one or both sides to be simplified.

In the previous section you learned to use the basic principle of equation solving to solve equations. In this section you solve more complex equations. These require that the expressions on one or both sides of the equation be simplified before adding the same number or multiplying by the same number on both sides of the equation. Example 1 illustrates this situation.

EXAMPLE 1 A train heading east passes a train heading west on a parallel track, and both engineers start blowing their whistles to say hello. The eastbound train is going 75 miles/hour and the westbound train is going 60 miles/hour. If the sound of the whistles carries for 20 miles, how long will it be before the engineers cannot hear each other's whistles?

Solution Because the whistles can be heard for 20 miles, the question is really asking how long until the trains are 20 miles apart. If we draw a picture (Figure 5), we can

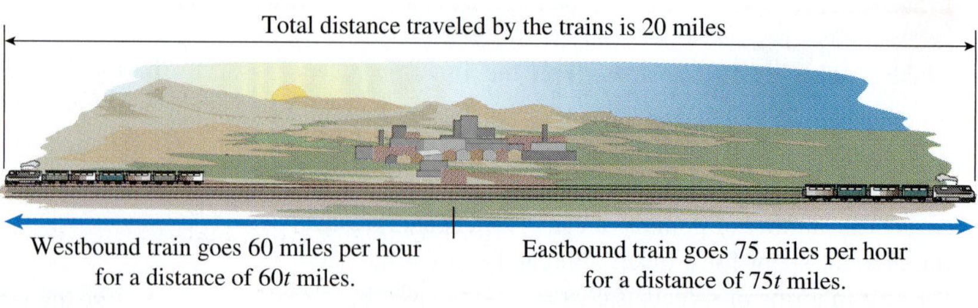

Figure 5 The eastbound train travels farther in the same amount of time.

see that this will happen when the *sum* of the distances traveled by the two trains is 20 miles. The unknown is the time until the distance equals 20 miles, so we should give that a variable name, say, t.

We know distance = **rate** $*$ **time**, so in t hours the eastbound train travels **75*t*** miles and the westbound train travels **60*t*** miles. We want the sum of these distances to equal 20 miles, which gives the equation

$$20 = 75t + 60t$$

The expression on the right side must be simplified. We combine the t terms and get

$$20 = 135t$$

Then we divide both sides by 135 and get

$$t = \frac{20}{135} = \frac{4}{27} \approx .15$$

So it will be about .15 hours, or 9 minutes, until the trains are 20 miles apart and the engineers cannot hear each other's whistles.

In Example 1 we simplified the expression on the right side by combining like terms. In Example 2 we need to use the distributive property to remove parentheses on one side of the equation.

EXAMPLE 2 One cellular phone plan charges a flat fee of \$10 per month plus 5 cents per minute, and another adds 100 minutes to the number of minutes you use and then charges 7 cents per minute. For how many minutes of use will the cost of the two plans be equal?

Solution The formula for the cost of the first plan is

$$10 + .05t$$

where t is the number of minutes used. The cost of the second plan is

$$.07(t + 100)$$

To find out for how many minutes of use the cost will be the same, we equate the two costs:

$$10 + .05t = .07(t + 100)$$

To solve this equation, we need to simplify the expression on the right side.

$10 + .05t = .07(t + 100)$	*Original equation*
$10 + .05t = .07t + 7$	*Simplifying the right side by removing parentheses*

$\underline{+^{-}.05t \quad +^{-}.05t}$	*Adding* $^{-}.05t$ *on both sides*
$10 = .02t + 7$	*Combining t terms*
$\underline{+^{-}7 \quad +^{-}7}$	*Adding* $^{-}7$ *on both sides*
$3 = .02t$	*Combining constant terms*
$\dfrac{3}{.02} = \dfrac{.02t}{.02}$	*Multiplying on both sides by* $\frac{1}{.02}$ *(or dividing by .02)*
$150 = t$	$3 \div .02 = 150$ *and* $.02 \div .02 = 1$

If the phone is used for 150 minutes, the cost of the two plans is the same.

In both Examples 1 and 2 we found it necessary to simplify an expression on one side of the equation before applying the basic principle of equation solving. In Example 1, we simplified the expression $75t + 60t$ by combining like terms. We replaced $75t + 60t$ with the *equivalent expression* $135t$. In Example 2, we simplified the expression $.07(t + 100)$ by applying the distributive property to remove parentheses. We replaced $.07(t + 100)$ with the equivalent expression $.07t + 7$.

Example 3 shows an expression that needs several steps to simplify.

EXAMPLE 3 ■ Is there a shorter way to write the expression $7(2x + 3) - 2x + 4 - 4x$?

Solution First, use the distributive property to remove the parentheses around $2x + 3$. Then simplify.

$7(2x + 3) - 2x + 4 - 4x$

$= 14x + 21 - 2x + 4 - 4x$	*Using the distributive property*
$= 14x - 2x - 4x + 21 + 4$	*Grouping like terms using the commutative property*
$= 8x + 25$	$14x - 2x - 4x = 8x$ *and* $21 + 4 = 25$

In Example 3, nothing more can be combined in the expression $8x + 25$. We say that $8x + 25$ is in **simplified form**. In manipulating an expression to put it in simplified form, the special properties of 1 and $^{-}1$ are useful, as Example 4 shows.

EXAMPLE 4 ■ Simplify the expression $3(x + 2) - (x - 1) + 5$.

Solution First, use the distributive property to eliminate any parentheses involved with a multiplication. Remember that subtracting an expression in parentheses is really like multiplying each term by $^{-}1$.

$3(x + 2) - (x - 1) + 5$

$= 3x + 6 + {}^{-}x + 1 + 5$ *Using the distributive property*

$= 3x - x + 6 + 1 + 5$ *Using the commutative property to collect like terms*

$= 2x + 6 + 1 + 5$ *Combining the variable terms*

$= 2x + 12$ *Combining the constant terms*

The expression $2x + 12$ is in simplified form because $2x$ and 12 cannot be combined.

We may want to write fewer steps.

$$3(x + 2) - (x - 1) + 5 = 3x + 6 + {}^{-}x + 1 + 5$$
$$= 2x + 12$$

If we do so, we still need to be conscious of the steps left out.

Our solution states that the expressions $3(x + 2) - (x - 1) + 5$ and $2x + 12$ are equivalent. This means that they have the same value no matter what values are substituted for X. You may demonstrate this with the TABLE feature of the calculator.

In the [Y=] menu, enter the expressions as Y1 and Y2:

$$Y1 = 3(X + 2) - (X - 1) + 5$$
$$Y2 = 2X + 12$$

Then press [2ND] TBLSET and let TblStart = 0 and ΔTbl = 1. Independent and dependent variables should both be set at AUTO. Finally press [2ND] TABLE and notice that for all the values of X displayed, the expressions are equal (Figure 6).

X	Y1	Y2
0	12	12
1	14	14
2	16	16
3	18	18
4	20	20
5	22	22
6	24	24

Figure 6

We may now summarize what we have learned in the previous examples as a procedure for simplifying expressions.

Procedure for Simplifying an Expression

Step 1 Use the special properties of 1 and $^{-}1$ if necessary.
Step 2 Remove all parentheses using the distributive property.
Step 3 Group like terms using the commutative property.
Step 4 Combine like terms.

Now that we know how to simplify an expression, we may apply this skill in solving a more complex linear equation. Example 5 requires that both sides be simplified.

EXAMPLE 5 ■ Solve for x in this equation: $2(9 + x) = 4(2x + 1) - (7 - x)$.

Solution

$2(9 + x) = 4(2x + 1) - (7 - x)$	*Original equation*
$18 + 2x = 4(2x + 1) - (7 - x)$	*Simplifying the left side using the distributive property*
$18 + 2x = 4(2x + 1) + {}^{-}1(7 + {}^{-}1x)$	*Simplifying the right side using the special properties of* $^{-}1$
$18 + 2x = 8x + 4 - 7 + x$	*Simplifying the right side by removing parentheses using the distributive property*
$18 + 2x = 8x + x + 4 - 7$	*Simplifying the right side by grouping like terms*
$18 + 2x = 9x - 3$	*Simplifying the right side by combining like terms*
$\underline{+{}^{-}2x} \quad \underline{+{}^{-}2x}$	*Adding* $^{-}2x$ *on both sides*
$18 = 7x - 3$	
$\underline{+3} \quad \underline{+3}$	*Adding 3 on both sides*
$21 = 7x$	
$\frac{1}{7} * 21 = \frac{1}{7} * 7x$	*Multiplying on both sides by* $\frac{1}{7}$
$3 = x$	

This may seem like a lot of steps to solve one equation. With practice, we may leave out some steps when we write our solution, but we must still be aware of each step and the justification for each. We can help ourselves do this with the following abbreviations:

LS Simplifying the expression on the left side of the equation.

RS Simplifying the expression on the right side of the equation.

A Adding the same quantity on both sides (or subtracting, because adding a negative is the same as subtracting).

M Multiplying both sides by the same quantity (or dividing, because multiplying by a reciprocal is the same as dividing).

Here is a streamlined version of the preceding solution using these abbreviations.

$2(9 + x) = 4(2x + 1) - (7 - x)$	
$18 + 2x = 8x + 4 - 7 + x$	*LS, RS*
$18 + 2x = 9x - 3$	*RS*

$$\underline{-2x} \quad \underline{-2x}$$

$$18 = 7x - 3 \qquad A$$

$$\underline{+3} \quad \underline{+3}$$

$$21 = 7x \qquad A$$

$$\frac{21}{7} = \frac{7x}{7} \qquad M$$

$$3 = x \qquad LS,\ RS$$

You should check that $x = 3$ is indeed a solution by substituting 3 for x in the original equation.

We summarize the steps needed to solve linear equations with the following procedure.

Procedure for Solving Linear Equations

Step 1 If necessary, simplify the expression on the left side.

Step 2 If necessary, simplify the expression on the right side.

Step 3 If necessary, add the same terms on both sides until you have a variable term on one side and a constant term on the other side.

Step 4 If necessary, multiply both sides by the reciprocal of the coefficient of the variable in order to get the variable alone.

Let's apply this procedure to a word problem that requires simplification on both sides of an equation.

EXAMPLE 6 A rectangle and an equilateral triangle have the same perimeter. The length of the rectangle is three times the width. Each side of the triangle is 10 feet more than the width of the rectangle. Find the dimensions of both figures.

Solution Let w represent the width of the rectangle (in feet). Then $3w$ is an expression for the length. The perimeter of the rectangle is $w + 3w + w + 3w$. Each side of the triangle is $w + 10$ feet. Because the three sides are equal, the perimeter of the triangle is $3(w + 10)$. See Figure 7.

Because the two perimeters are equal, we have the equation $w + 3w + w + 3w = 3(w + 10)$. To solve this equation, we need to simplify the expressions on both sides of the equation.

Figure 7

$$
\begin{aligned}
w + 3w + w + 3w &= 3(w + 10) & \\
8w &= 3(w + 10) & \textit{LS} \\
8w &= 3w + 30 & \textit{RS} \\
-3w \quad & \quad -3w & \\
5w &= 30 & \textit{A} \\
\frac{5w}{5} &= \frac{30}{5} & \textit{M} \\
w &= 6 & \textit{LS, RS}
\end{aligned}
$$

The width of the rectangle is 6 feet. The length of the rectangle is $3w$, or 18 feet. Therefore, the perimeter of the rectangle is $6 + 18 + 6 + 18 = 48$ feet. Each side of the triangle is $w + 10 = 16$ feet. So, the triangle's perimeter is $3 * 16 = 48$ feet. Our solution to the equation checks with the conditions given in the problem.

Exercises 3.2

1. What property is shown in this figure?

2. Consider the four rectangles in this figure.

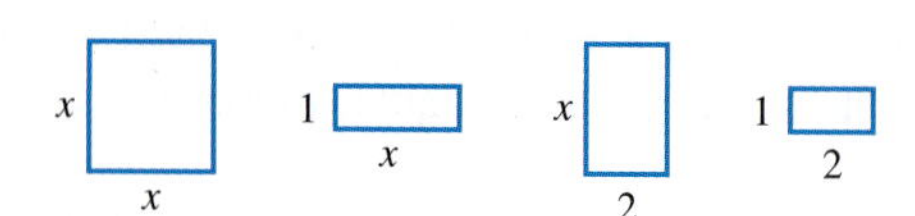

   a. Write an expression for the area of each rectangle.

   b. Write the areas of the rectangles as a sum.

   c. Simplify to find an equivalent sum. Be sure to combine only like terms.

3. Group the like terms and simplify the expression:

$$\frac{3}{5}x + 2 - \frac{1}{10}x - 6$$

4. Identify and correct the error in each of the simplifications.

   a. $5 + 3(x - 11) = 5 * 3x - 33$
   $= 15x - 33$

   b. $^{-}2(b + 4) = {}^{-}2b + 4$

5. Simplify $6(x + 5) + 2x - 7$.

6. Review the procedure for simplifying an expression and describe each step used to simplify $7a - (2a + 5) + 1$.

$7a - (2a + 5) + 1$	*Original expression*
$7a + {}^-1(2a + 5) + 1$	?
$7a + {}^-2a + {}^-5 + 1$	?
$5a + {}^-4$	?
$5a - 4$	*Using the rule for subtraction*

7. Simplify each expression.

a. $c^2 + 3c - 2c^2 + 8c - 5$

b. ${}^-(x + 2) - 3x + 7$

c. 2 inches^2 + 5 inches + 7 inches^2

d. $2(3x - 7) - (x - 4) + 5$

e. $\frac{5}{3}(x - 6) + x - 4$

8. Solve each equation for x. Use the abbreviations on page 100 to keep track of what you did at each step.

a. $8(4 - 2x) + 10 = 5x$

b. $(2x - 1) - (4x + 3) = 8$

9. Solve each equation for x.

a. $2 - 3x + 15 = 4 + x - 6$

b. $7 = 4x - 3 - 7x - 11$

c. $.6x + 2 - 1.1x = 2.6$

d. $\frac{1}{4}(2x + 5) - x = \frac{1}{2}(5x - 2)$

10. Show that $x = {}^-3$ checks as a solution to the equation $3x + 5 - 6x = 2(4 - x)$.

11. Suppose two cars leave the same spot traveling in opposite directions. One car travels 40 miles per hour and the other travels 55 miles per hour. Use (a)–(c) to find how long it will be before the cars are 57 miles apart.

a. Let t represent the number of hours of travel. Write an expression for the distance each car travels.

b. Write an equation showing that the sum of the distances is 57 miles.

c. Solve the equation for t to find the time that these cars travel.

12. Two insects are 100 feet apart at 11:00 A.M. The insects move toward each other, with the junior insect traveling 8 feet per minute and the older insect traveling 12 feet per minute. At what time will they meet?

In Exercises 13–15, use the fact that the sum of the measures of the three angles in any triangle is 180°.

13. In a triangle, the measure of angle B is 10° more than that of angle A and the measure of angle C is twice that of angle A.

a. Let x represent the measure of angle A. Write expressions for the measures of angle B and angle C. Label the diagram in the figure with your expressions for angles A, B, and C.

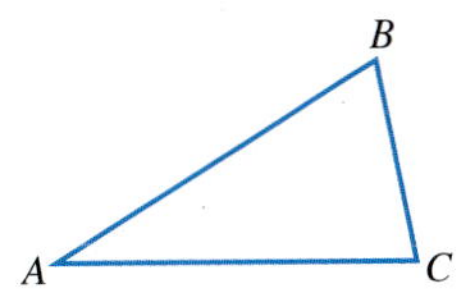

b. Write an equation showing that the sum of the measures of these three angles is 180°.

c. Solve the equation for x to find the measure of angle A.

14. In a triangle often used in trigonometry, angle B measures twice angle A and angle C measures three times angle A.

a. Let x represent the measure of angle A. Draw a diagram and label each angle with an expression for its measure.

b. Write an equation for the sum of the angles.

c. Solve the equation and find the measure of each angle.

15. In another triangle, angle B measures 5° more than twice angle A and angle C measures 15° less than angle A. Find the measure of each angle.

Skills and Review 3.2

16. Solve for x: $5x - 6 = 2x + 12$.

17. Verify that $x = 1$ is a solution to the equation $2x - 5 = {}^-10x + 7$.

18. The weekly savings (S) for an employee is given by the formula $S = .03P + 20$, where P is his weekly pay. Find his pay if he saves \$35.00.

19. Are the expressions $\frac{1}{3}x + \frac{8}{5}$ and $\frac{5x+24}{15}$ equivalent? Explain.

20. Let $y = {}^-\frac{2}{3}x + 20$.
    a. Use the table feature on your calculator to find five ordered pairs that satisfy this equation.
    b. Find a friendly window that will include the ordered pairs from part (a).
    c. Use the friendly window from part (b) to graph the function on your calculator. Copy the graph.

21. Recall from Exercise 9 on page 81 that the formula

$$C = \frac{5}{9}(F - 32)$$

converts Fahrenheit temperatures to Celsius. Find the Celsius temperature when the Fahrenheit temperature is 14°.

22. Evaluate $\frac{1}{3}x^2 + 5$ when $x = 6$.

23. Use the order of operations to evaluate

$${}^-8 + \frac{({}^-1)^{19} + 6}{9 - 10}$$

How you would enter this expression on your calculator to check your work?

24. Write the decimal number .000521 in scientific notation.

25. Below are steps for simplifying the expression

$$\frac{{}^-16 \text{ ft}}{\text{sec}^2}(2 \text{ sec})^2 + \frac{50 \text{ ft}}{\text{sec}}(2 \text{ sec}) + 150 \text{ ft}.$$

Explain what was done at each step.

$$\frac{{}^-16 \text{ ft}}{\text{sec}^2}(2 \text{ sec})^2 + \frac{50 \text{ ft}}{\text{sec}}(2 \text{ sec}) + 150 \text{ ft} \quad \textit{Original expression}$$

$$= \frac{{}^-16 \text{ ft}}{\text{sec}^2}(4 \text{ sec}^2) + \frac{50 \text{ ft}}{\text{sec}}(2 \text{ sec}) + 150 \text{ ft} \quad ?$$

$$= \frac{{}^-64 \text{ ft} * \text{sec}^2}{\text{sec}^2} + \frac{100 \text{ ft} * \text{sec}}{\text{sec}} + 150 \text{ ft} \quad ?$$

$$= {}^-64 \text{ ft} + 100 \text{ ft} + 150 \text{ ft} \quad ?$$

$$= 186 \text{ ft}$$

3.3 The Problem-Solving Process

AFTER STUDYING THIS SECTION YOU WILL BE ABLE TO

- Describe a seven-step process for solving problems.
- Apply the seven-step process to a variety of problems.

A Guide for Solving Problems

Throughout this book we emphasize that algebraic equations may be used to model real-world applications. Knowing how to solve equations, however, is just part of the problem solving process. An effective problem solver needs to be able to read a problem and set up an appropriate equation or equations. It is helpful, therefore, to have a *rough guide*, a series of steps that can help us organize our work toward a solution.

Seven Steps for Problem Solving

Step 1 *Understand* the problem. The best way to do this is to read the statement of the problem *carefully* before you begin to solve it.

Step 2 *Visualize* the problem. The right way to carry out this step depends on what type of problem you are trying to solve. For some problems, especially those involving geometry and distance-rate-time problems, you will probably want to draw a picture of some sort. It might not be possible to draw a picture that will help you, but you can still try to *visualize* the problem by thinking of it as happening right in front of you.

Step 3 *Assign* variable(s). If you are going to use algebra to help you solve a problem, you will need to have variables to work with. You should assign a variable to the quantity for which you are solving (which will usually be easy to spot as a question in the original problem) and possibly to other quantities that are unknown at the start of the problem. A picture might help you work out what the variables should be.

Step 4 *Write equation(s).* Use any information from the first two steps to write one or more equations involving the variables you assigned in step 3.

Step 5 *Solve* equation(s). Use the techniques we have discussed so far to solve any equations you wrote in step 4.

Step 6 *Answer* the question. Go back to the original problem and answer the question. It is important to remember that just writing $x = 3$, for example, will probably not answer the question asked in the original problem.

Step 7 *Check* your answer. Does your answer make sense? Go back to the original problem and make sure your answer works.

Applying the Seven Steps

Examples 1–4 illustrate how to use the seven steps for problem solving. Example 1 is an application of the distance formula.

EXAMPLE 1 Suppose a helicopter in Gridville (see Section 2.3) travels from the police station on Avenue $^{-}2$ Street 1 to pick up an officer who lives on the corner of Avenue 2 and Street 3. From there, it picks up a second officer at Avenue 1, Street $^{-}1$ and then returns to the police station. How far did the helicopter fly?

Solution ■ Use the seven steps to help solve this problem.

Step 1 To *understand* this problem, we go back to Section 2.3 where Gridville is described and read the description carefully. Recall that the streets and avenues of Gridville are laid out in a grid on the coordinate plane, with avenues running north-south and streets running east-west. Each corner is labeled with two numbers, the first for the avenue and the second for the street. A helicopter flies in a straight line, so it does not fly along avenues and streets but takes the shortest path. The distance it travels is given by the distance formula, which is

$$\text{distance} = \sqrt{\text{run}^2 + \text{rise}^2}$$

Step 2 Now that we understand the problem, we can try to *visualize* the problem. In this case we draw a figure, making sure to include all the information we have (see Figure 8). Drawing a picture lets us see that the helicopter's flight path is a triangle and that the total distance the helicopter flies is the *perimeter* of this triangle.

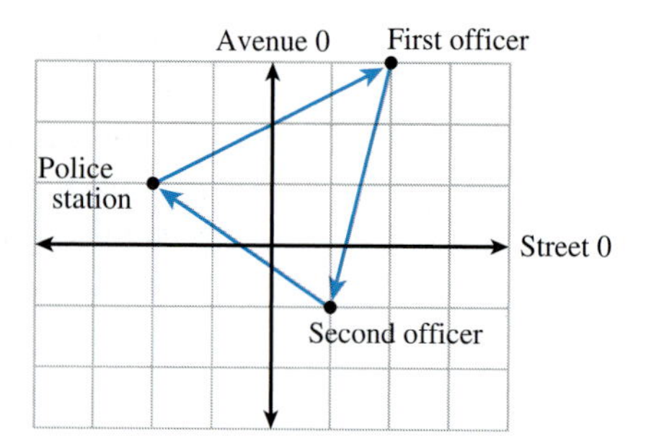

Figure 8 The arrows indicate the flight path of the helicopter.

Step 3 The next step is to *assign* variables. Because we are asked to find the distance, it might seem that the best assignment would be d equals the distance the helicopter travels, but if we look carefully at the way the distance formula works, we see that it can be used only to compute the distance traveled for each leg of the helicopter's flight and not for all of them together. So, let's call the three distances d_1, d_2, and d_3, with d_1 the distance from the police station to the first officer, d_2 the distance from the first officer to the second officer, and d_3 the distance from the second officer back to the police station.

Step 4 Now we are ready to use the distance formula to *write equations* for the three variables for which we want to solve. Looking back at Figure 8 to work out the run and rise for each trip, we have

$$d_1 = \sqrt{4^2 + 2^2}$$

$$d_2 = \sqrt{(^-1)^2 + (^-4)^2}$$

$$d_3 = \sqrt{(^-3)^2 + 2^2}$$

Step 5 The next step, *solving* the equations, is straightforward. We need only to simplify inside the square roots and then compute the square roots.

$$d_1 = \sqrt{4^2 + 2^2} = \sqrt{20} \approx \mathbf{4.5}$$

$$d_2 = \sqrt{(^-1)^2 + (^-4)^2} = \sqrt{17} \approx \mathbf{4.1}$$

$$d_3 = \sqrt{(^-3)^2 + 2^2} = \sqrt{13} \approx \mathbf{3.6}$$

Step 6 The second-to-last step is to *answer* the question. You might think you have already done this, because you have written down three distances, d_1, d_2, and d_3. But if you look back at the original question, you were asked to find how far the helicopter flew and not how far the helicopter flew for each leg of its trip. The answer to the question is given by the *total* distance flown, so we need to add the distances d_1, d_2, and d_3, which gives

$$\text{total distance} \approx \mathbf{4.5} + \mathbf{4.1} + \mathbf{3.6} = 12.2$$

The solution can be stated, The helicopter flew approximately 12.2 blocks.

Step 7 Finally, you should complete the last step by *checking* your answer. In this case, you can use Figure 8 to estimate the length of each side of the triangle and check that your answer looks right.

Our second example is a problem involving perimeter.

EXAMPLE 2 The organizers of the county fair want to fence in a rectangular area for vendors. Each vendor will set up a booth on the inside of the fence. They need 900 feet of fence to accommodate all the vendors and they want the rectangle to be twice as long as it is wide. What should the dimensions of the rectangular area be?

Solution

Step 1 To *understand* this problem, we need to decide exactly what we are being asked. Because the fence will form a rectangle, 900 feet will be the perimeter of the rectangle. The dimensions are the length and the width, so we need to be able to find a rectangle with length twice its width and perimeter 900 feet.

Step 2 To *visualize* the situation, we draw a picture of a rectangle and move immediately to step 3.

Step 3 We can *assign* variables by labeling the rectangle (see Figure 9).

We let x represent the width of the rectangle. Because the length of the rectangle must be twice the width, we may let $2x$ represent the length.

Step 4 We may now *write an equation*. The perimeter of 900 feet is the sum of all the sides, so $900 = x + 2x + x + 2x$.

Figure 9 The length of the rectangle should be twice the width.

Step 5 To *solve* the equation, we combine like terms and then divide on both sides by 6.

$$900 = 6x$$
$$\frac{900}{6} = \frac{6x}{6}$$
$$150 = x$$

Step 6 To *answer* the question, we need to give both the length and width of the rectangle. Because $\mathbf{x = 150}$, and x stands for width, we know the width is 150 feet. The length is $2x = 2 * \mathbf{150} = 300$, so the length is 300 feet. Both answers must include the unit (feet) to have any meaning.

Step 7 Let's *check*: does this answer make sense? If we add all the sides, the perimeter is $150 + 300 + 150 + 300 = 900$ feet, so we have found the correct answer.

Example 3 is an application of fixed and variable costs.

EXAMPLE 3 Rashid subscribes to a cell-phone service that charges a flat fee of $12.95 per month plus 2 cents for each minute he uses the phone. He knows that he can afford only $30 per month for his telephone bill. How much time will he be able to spend each month on the phone?

Solution

Step 1 Let's see if we *understand* the problem situation. Rashid has $30 to spend. The more he talks, the more he spends. We are asked to find how much time he will be able to spend on the phone. Because he is charged by the minute, the answer should be in minutes. Even if he never picks up the phone, he still has to pay the flat fee of $12.95.

Step 2 Can we *visualize* this problem? If a picture does not immediately come to mind, we may be able to relate this situation to a similar one, Jamie's health club (Section 2.4). There we used the formula

$$\text{total cost} = \text{variable cost} + \text{fixed cost}$$

represented by the picture in Figure 10.

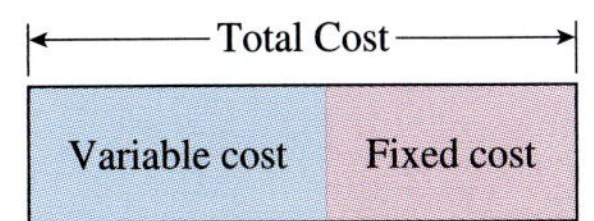

Figure 10

Step 3 We know we have to find time, so we may *assign* the variable t to represent the number of minutes Rashid uses the phone in one month.

Step 4 We may *write an equation* with the information in steps 1, 2, and 3. The total cost is \$30. The fixed cost is the flat fee, \$12.95. The variable cost depends upon t. Because he is charged 2 cents (\$.02) per minute, the cost is $.02t$ for t minutes. Putting it all together we have

$$\textbf{total cost} = \textbf{variable cost} + \textbf{fixed cost}$$

$$\mathbf{30 = .02}\boldsymbol{t}\mathbf{ + 12.95}$$

Step 5 We may now *solve* the equation.

$$30 = .02t + 12.95$$

$$\underline{^{-}12.95} + \underline{^{-}12.95} \qquad \textit{Adding } ^{-}12.95 \textit{ on both sides}$$

$$17.05 = .02t \qquad \textit{Combining constant terms}$$

$$\frac{17.05}{.02} = \frac{.02t}{.02} \qquad \textit{Dividing on both sides by .02}$$

$$852.5 = t \qquad 17.05 \div .02 = 852.5$$

Step 6 The *answer* to our question is that Rashid will be able to use the phone 852.5 minutes each month. For a month with 30 days, that works out to about 28.4 minutes a day, just under half an hour. Rashid will have to budget his time carefully!

Step 7 We may *check* our answer by going back to the original problem. If he talks for 852.5 minutes at 2 cents per minute, the variable cost will be $\$.02 * 852.5 = \$17.05$. Adding that to the fixed cost of \$12.95 gives the \$30 Rashid can afford.

Our last example is a distance-rate-time problem.

EXAMPLE 4 Ruth leaves her house at 9 A.M., driving west along Route 7, which runs right past her house. She drives at a steady 40 miles per hour. An hour after she leaves, her husband Jacob realizes that she forgot the slides she will need for the presentation, so he takes off after her, driving at a steady rate of 60 miles per hour. When will Jabob catch Ruth?

Solution

Step 1 Here is what we should *understand* from reading the problem: We are asked to find *when* Jacob catches Ruth, so we need to solve for time. We know that Jacob drives faster than Ruth, but he starts later so he has less time. They both start from the house, and when he catches up to her they will both have traveled the same distance.

Step 2 To *visualize* this problem you may want to act it out with a classmate. Have Ruth leave the house before Jacob does and have Jacob move at a faster speed until they meet. The diagram in Figure 11 represents the situation.

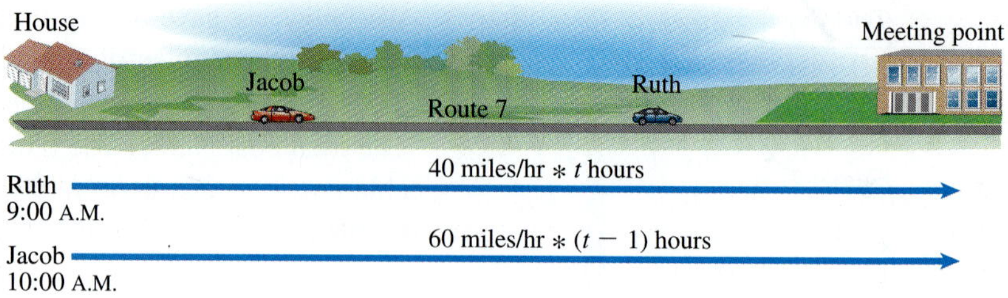

Figure 11

Step 3 Let's *assign* t as the variable to represent Ruth's time, that is, time elapsed since 9 A.M. Jacob leaves 1 hour later, so his time is 1 hour less, that is, $t - 1$.

Step 4 The diagram in Figure 11 may help us to *write an equation*. distance = ***rate*** * ***time***, so Ruth's distance is **$40t$** and Jacob's distance is **$60(t - 1)$**. Because these two distances are the same, we have the equation

$$40t = 60(t - 1).$$

Step 5 Simple algebraic manipulation lets us *solve* this equation for t.

$60(t - 1) = 40t$	*Original equation*
$60t - 60 = 40t$	*Distributive property*
$\underline{-60t} \quad \underline{-60t}$	*Subtracting 60t on both sides*
$^{-}60 = {}^{-}20t$	*Combining t terms*
$\dfrac{^{-}60t}{^{-}20} = \dfrac{^{-}20t}{^{-}20}$	*Dividing by $^{-}20$ on both sides*
$3 = t$	$\dfrac{^{-}60t}{^{-}20} = 3$ *and* $\dfrac{^{-}20t}{^{-}20} = t$

Step 6 The question asked what time it would be when Jacob catches Ruth, not how long it would take. So to *answer* the question, we need to convert the elapsed time, $t = 3$ hours on Ruth's stopwatch, to the actual time of day. Ruth left at 9 A.M., so 3 hours later it is noon.

Step 7 We can *check* this answer calculating the distances Ruth and Jacob travel. Ruth traveled for $t = 3$ hours at a rate of 40 miles per hour for a distance of $40 * 3 = 120$ miles. Jacob traveled $t - 1 = 2$ hours at a rate of 60 miles per hour for a distance of $60 * 2 = 120$ miles. As expected, they traveled the same distance.

Exercises 3.3

1. Summarize the seven steps for problem solving.
2. Review Example 1 and outline the work performed for each of the seven steps for problem solving.
3. Review Example 2 and outline the work performed for each of the seven steps for problem solving.
4. In Example 3, how many hours per month will Rashid use the phone?
5. In Example 4, does the solution $t = 3$ complete the problem-solving process? Explain.

Exercises 6 and 8 explore various approaches to visualizing a problem. Exercises 7 and 9 then ask you to apply those results to continue solving the problems.

6. The sum of two numbers is 11. Ten times the smaller number minus 5 times the larger number is 8. What are the two numbers?
   a. Step 1: Understand the problem by completely reading the problem. Then state what the problem is looking for.
   b. Step 2: Visualize the problem by using guess and check on pairs of numbers that sum to 11. Make a table. Do your guesses come close to meeting the second condition of the problem? Explain.
   c. Step 3: Let x represent the smaller number. Explain why $11 - x$ represents the larger number.
7. Continue the seven steps for problem solving to find the two unknown numbers from Exercise 6.
8. The length of a rectangle is four times the width. What are the dimensions of the rectangle if the perimeter is 100 m?
   a. Complete step 2 of the problem-solving process by drawing a picture of the information given in the problem.
   b. What additional information did your picture help you gather?
9. Complete the problem-solving steps to find the dimensions of the rectangle in Exercise 8.
10. The perimeter of a rectangle is 98 feet. The width of the rectangle is 5 feet more than one-third the length. Use the seven steps for problem solving to find the width and length of the rectangle.
11. Steve wants to record his music professionally. A studio charges an up-front fee of \$500 plus \$12 for each minute of recording. How many minutes can he record for \$1200?
12. Each month Rachel has spring water delivered to her apartment. She rents a water dispenser for \$6.25 per month. Each jug of water costs \$3.70. She budgets \$25 per month for water and the dispenser. How many jugs of water can she buy each month?
13. Two trains pass each other going in opposite directions. The first train travels 5 miles per hour faster than the second train. What is the average speed of each train if they are 370 miles apart after 2 hours?
   a. Complete step two of the seven steps for problem solving by completing the table below.

	Rate (miles/hour)	Time (hours)	Distance (miles)
Train A			
Train B			
Total			

   b. Complete the remaining problem-solving steps to find the average speed of each train.
14. Two cyclists begin from the same place at the same time riding toward the same final destination. One cyclist averages 20 miles per hour and the other averages 26 miles per hour. After how many minutes will the two cyclists be 10 miles apart?
15. Kyle and Missy buy a total of 52 lottery tickets. Missy buys 8 less than twice the number of tickets that Kyle buys. How many tickets did each buy?

Skills and Review 3.3

16. Simplify.
   a. $2x - (x + 6)$
   b. $5(3x - 7) - 4x - 2$
   c. $(x^2 - 5) - 4(2x + 7) + (x^2 - 6x)$
   d. $x + x^{-2} + 5 + \frac{3}{x^2} - 6x$

17. Simplify

$$6\left(x - \frac{9}{2}\right) - (4 - 1)$$

18. Solve for x if $2x + 3 = 11$.

19. Solve for x: $^{-}(3 + x) - 6 = 1 + (4x - 30)$.

20. Solve for x: $\frac{1}{3}x - 5 = \frac{1}{2}x + 4$.

21. Solve for y when $x = {}^{-}4$ if $5x - 7y = 22$.

22. Suppose a person travels 10 miles/hour in one direction and another person travels 12 miles/hour in the opposite direction. How long will it be before they are 110 miles apart?

23. Evaluate $x^2 + 2x - 3$ when $x = {}^{-}1$.

24. Evaluate using the order of operations.

a. $\dfrac{^{-}2 - (^{-}5)}{^{-}3 - 6}$

b. $\dfrac{\sqrt{(^{-}3)^2 - 4 * 1 * (^{-}4)}}{2 * 1}$

25. Find the run and rise going from $(2, {}^{-}6)$ to $(7, {}^{-}3)$.

3.4 Arithmetic Sequences (Optional)

After Studying This Section You Will Be Able To

- Determine if a sequence of numbers is an arithmetic sequence.
- Find the formula for the *n*th term in an arithmetic sequence.
- Use the formula to find terms in an arithmetic sequence.

A sequence is a list of numbers that show some pattern or rule. The numbers in a sequence are called terms. If you can figure out the rule, you can find more terms.

EXAMPLE 1 Find the pattern for each sequence and then find the next three terms.

a. 12, 15, 18, 21, . . .

b. 100, 94, 88, 82, . . .

c. 3, 6, 12, 24, . . .

d. 8, 10, 8, 10, . . .

Solution

a. The rule appears to be add 3 to each term to find the next term. The next three terms are 24, 27, and 30.

b. The rule appears to be subtract 6 from each term to find the next term. The next three terms are 76, 70, and 64.

c. The rule appears to be multiply each term by 2 to find the next term. The next three terms are 48, 96, and 192.

d. The rule appears to be add 2, then add $^{-}2$, then add 2, then add $^{-}2$, etc. The next three terms are 8, 10, and 8.

The sequences in Example 1(a) and (b) above are called **arithmetic sequences**. In these sequences a term is found by adding or subtracting the same number to the previous term. The number added or subtracted is called the **com-**

mon difference. In (a) the common difference is 3, and in (b) the common difference is $^-6$.

The sequences in (c) and (d) are not arithmetic sequences. In (c) each term is found by multiplying the previous term by 2. It is an example of a **geometric sequence**. In (d), the terms alternate between 8 and 10, and the differences flip back and forth between 2 and $^-2$. Because the differences are not the same, (d) does not qualify as an arithmetic sequence.

Formulas for Arithmetic Sequences

Variables may be used to describe an arithmetic sequence. The terms are usually designated by the letter a. Because there is more than one term, subscripts are used: a_1 represents the first term, a_2 represents the second term, a_3 represents the third term, and so on, until a_n represents the nth term. The letter d is used to represent the common difference.

EXAMPLE 2 ■ Find a formula for the nth term of an arithmetic sequence.

Solution Start with the first term and keep adding the common difference.

first term a_1

second term $a_2 = a_1 + d$

third term $a_3 = a_2 + d = a_1 + d + d = a_1 + 2d$

fourth term $a_4 = a_3 + d = a_1 + 2d + d = a_1 + 3d$

fifth term $a_5 = a_4 + d = a_1 + 3d + d = a_1 + 4d$

It appears that each term may be found by starting with a_1 and adding a specified number of d's. To determine how many d's to add, look more closely at the pattern:

a_1

$a_2 = a_1 + d$

$a_3 = a_1 + 2d$ *2 is one less than 3*

$a_4 = a_1 + 3d$ *3 is one less than 4*

$a_5 = a_1 + 4d$ *4 is one less than 5*

In the last three rows, the coefficient of d is one less than the number of the term. This pattern also holds for the first two rows if we write

$a_1 = a_1 + 0d$ *0 is one less than 1*

$a_2 = a_1 + 1d$ *1 is one less than 2*

We are now ready to write a general formula for the nth term, a_n.

$$n\text{th term } a_n = a_1 + (n-1)d$$

Substituting 1, 2, 3, 4, and 5 in the formula will give the preceding results.

Formula for an Arithmetic Sequence

Let a_1 be the first term and d be the common difference for an arithmetic sequence. Then the nth term is given by the formula

$$a_n = a_1 + (n-1)d$$

We may remember this formula by recalling that something must be added to a_1 to get a_n. A common mistake is to add $n * d$ rather than $(n-1)d$. Example 3 shows a way to remember why it should be $n-1$.

EXAMPLE 3 Your backyard is 120 feet wide. You want to build a fence and will need to place posts 10 feet apart. How many posts will you need?

Solution The easy answer is wrong! You can't simply divide 120 by 10 and get 12 as your answer. Figure 12 illustrates why 12 does not work. From Figure 12 we can see that 13 fence posts are needed in order to create 12 spaces of 10 feet each between the posts.

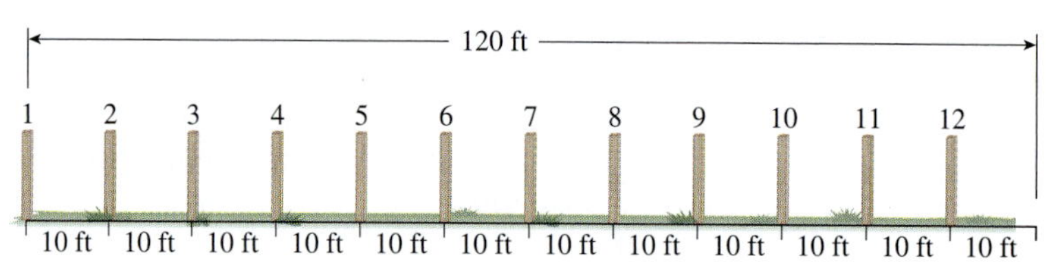

Figure 12 Twelve fence posts are not enough. Another post is needed at the end.

The Fence Post Principle

n fence posts create $n-1$ spaces between the posts.

In Example 3, $n = 13$ and $n - 1 = 12$. In the formula for an arithmetic sequence, think of the terms of the sequence as fence posts and the spaces between the terms as common differences. According to the fence post principle, you will need $n-1$ differences. Therefore, the formula is $a_n = a_1 + (n-1)d$.

Applying the Arithmetic Sequence Formula

Example 4 shows how to apply the arithmetic sequence formula.

EXAMPLE 4 Find the fourteenth term of the arithmetic sequence: 4, 7, 10, 13, ...

Solution Because $7 - 4 = 3$, $10 - 7 = 3$, and $13 - 10 = 3$, there is a common difference of 3. We now know that $a_1 = 4$ and $d = 3$. We want the fourteenth term, so $n = 14$:

$$a_n = a_1 + (n - 1)d \quad \text{Arithmetic sequence formula}$$

$$a_{14} = 4 + (14 - 1) * 3 \quad \text{Substituting for } a_1, d, \text{ and } n$$

$$a_{14} = 4 + (13) * 3 \quad \text{Applying the order of operations}$$

$$a_{14} = 4 + 39$$

$$a_{14} = 43$$

The fourteenth term of this sequence is 43.

EXAMPLE 5 The first term of an arithmetic sequence is 53. The ninth term is 93. Find the common difference.

Solution We know that $a_1 = 53$, $n = 9$, and $a_9 = 93$. We again use the formula. This time we will have an equation to solve.

$$a_n = a_1 + (n - 1)d \quad \text{Arithmetic sequence formula}$$

$$93 = 53 + (9 - 1)d \quad \text{Substituting 53 for } a_1, \text{ 9 for } n, \text{ and 93 for } a_n$$

$$93 = 53 + 8d \quad \text{RS (Simplifying parentheses)}$$

$$40 = 8d \quad \text{A (Adding } {}^{-}53 \text{ on both sides)}$$

$$\frac{40}{8} = \frac{8d}{8} \quad \text{M (Dividing on both sides by 8)}$$

$$5 = d$$

The common difference is 5.

EXAMPLE 6 Suppose we have the sequence 1000, 991, 982, 973, ...

a. Does the number 100 appear in this sequence? If so, which term is it?
b. Does the number 13 appear in this sequence? If so, which term is it?

Solution The common difference appears to be $991 - 1000 = {}^{-}9$. We check that the sequence is arithmetic by adding $^{-}9$ to 991 to get 982 and then again to 982 to get 973.

Thus we have $a_1 = 1000$ and $d = {}^{-}9$. For each example we know the value we want for a_n but we don't know if we can choose an n to get that value.

a. If 100 appears in the sequence, $a_n = 100$. We substitute into the formula:

$a_n = a_1 + (n-1)d$	*Arithmetic sequence formula*
$100 = 1000 + (n-1) * {}^{-}9$	*Substituting into the formula*
$100 = 1000 - 9n + 9$	*RS (Applying the distributive property)*
$100 = 1009 - 9n$	*RS (Combining the constant terms)*
$^{-}909 = {}^{-}9n$	*A (Adding $^{-}1009$ on both sides)*
$\frac{^{-}909}{^{-}9} = \frac{^{-}9n}{^{-}9}$	*M (Dividing both sides by $^{-}9$)*
$101 = n$	

When we solved for n, we got a whole number. We can conclude that 100 does appear in the sequence as the 101st term.

b. If 13 appears in the sequence, $a_n = 13$. We substitute into the formula:

$a_n = a_1 + (n-1)d$	*Arithmetic sequence formula*
$13 = 1000 + (n-1) * {}^{-}9$	*Substituting into the formula*
$13 = 1000 - 9n + 9$	*RS (Applying the distributive property)*
$13 = 1009 - 9n$	*RS (Combining the constant terms)*
$^{-}996 = {}^{-}9n$	*A (Adding $^{-}1009$ on both sides)*
$\frac{^{-}996}{^{-}9} = \frac{^{-}9n}{^{-}9}$	*M (Dividing by $^{-}9$ or multiplying by $^{-}\frac{1}{9}$ on both sides)*
$110.667 \approx n$	

When we tried to solve for n, we got a fraction. We conclude that 13 is not a term in the sequence.

Exercises 3.4

1. Find the pattern for each sequence and then find the next three terms.

   a. 1, 2, 4, 8, . . .

   b. 13, 23, 33, 43, . . .

   c. 41, 39, 37, 35, . . .

   d. 50, 51, 49, 52, 48, 53, . . .

2. Which sequences in Exercise 1 are arithmetic?

3. For each arithmetic sequence, determine a_1 and d.

a. 17, 20, 23, 26, . . .

b. 85, 80, 75, 70, . . .

c. 15, 17.5, 20, 22.5, . . .

d. 1, $^-2$, $^-5$, $^-8$, . . .

4. Find the nth term for each arithmetic sequence with the given information.

a. $a_1 = 14, d = 4, n = 10$

b. $a_1 = 30, d = {}^-3, n = 7$

c. $a_1 = {}^-500, d = 19, n = 35$

5. How many fence posts are needed to make a fence 210 feet long with the posts 15 feet apart?

6. The first term of an arithmetic sequence is 11. The 18th term is 130. Find the common difference.

7. Does the number 1000 appear in the sequence 25, 38, 51, 64, . . .? If so, which term is it?

8. The 100th term of an arithmetic sequence is 0 and the common difference is $^-8$. Find the first term.

CHAPTER 3 ▪ KEY CONCEPTS

Basic Principle of Equation Solving If we are given an equation, we can add the same term on both sides or multiply by the same number on both sides and still have a valid equation. (Page 89)

Equivalent Expressions Equal quantities written in different form: for example, $x + x$ and $2x$ are the same for any x, so they are equivalent, and we may write $x + x = 2x$. (Page 98)

Simplified Form An expression is in simplified form if it has no parentheses and no terms can be combined. (Page 98)

Procedure for Simplifying an Expression (Page 99).

1. Use the special properties of 1 and $^-1$ if necessary.
2. Remove all parentheses using the distributive property.
3. Group like terms using the commutative property.
4. Combine like terms.

Procedure for Solving Linear Equations (Page 101).

1. If necessary, simplify the expression on the left side.
2. If necessary, simplify the expression on the right side.
3. If necessary, add the same terms on both sides until you have a variable term on one side and a constant term on the other side.
4. If necessary, multiply on both sides by the reciprocal of the coefficient of the variable in order to get the variable alone.

Seven Steps for Problem Solving These steps provide a logical way to set up and work through word problems. (Page 105)

Step 1 *Understand* the problem. The best way to do this is to read the statement of the problem *carefully* before you begin to solve it.

Step 2 *Visualize* the problem. The right way to carry out this step depends on what type of problem you are trying to solve. For some problems, especially those involving geometry and distance-rate-time problems, you will probably want to draw a picture of some sort. It might not be possible to draw a picture that will help you, but you can still try to *visualize* the problem by thinking of it as happening right in front of you.

Step 3 *Assign* variable(s). If you are going to use algebra to help you solve a problem, you will need to have variables to work with. You should assign a variable to the quantity for which you are solving (which will usually be easy to spot as a question in the original problem) and possibly to other quantities that are unknown at the start of the problem. A picture might help you work out what the variables should be.

Step 4 *Write equation(s).* Use any information from the first two steps to write one or more equations involving the variables you assigned in step 3.

Step 5 *Solve* equation(s). Use the techniques we have discussed so far to solve any equations you wrote in step 4.

Step 6 *Answer* the question. Go back to the original problem and answer the question. It is important to remember that just writing $x = 3$, for example, will probably not answer the question asked in the original problem.

Step 7 *Check* your answer. Does your answer make sense? Go back to the original problem and make sure your answer works.

Exploration

Recall the equation for photosynthesis from the chapter opener, $H_2O + CO_2 = O_2 + C_6H_{12}O_6$. Balance the equation by following these steps:

a. Find out how many molecules of H_2O will be needed to provide the hydrogen (H) for one molecule of $C_6H_{12}O_6$.

b. Find out how many molecules of CO_2 will be needed to provide the carbon (C) for one molecule of $C_6H_{12}O_6$.

c. Find out how many extra oxygen atoms you will have. These extra atoms will make how many molecules of O_2?

d. Now fill in the blanks with the numbers of each molecule you will need:

$$\_\_H_2O + \_\_CO_2 = \_\_O_2 + 1C_6H_{12}O_6.$$

e. Check to see that the equation balances:

Element	Atoms on Left Side	Atoms on Right Side
Carbon (C)	__	__
Hydrogen (H)	__	__
Oxygen (O)	__	__

f. How is checking a balanced chemical equation similar to checking an algebraic equation? How are the procedures different?

CHAPTER 3 ▪ REVIEW EXERCISES

Section 3.1 Solving Linear Equations in One Variable

1. Solve each equation for x.

a. $3x + 8 = 11$

b. $7 + x = 3 - 4x$

c. $5x + 7 = {}^{-}23$

d. $1.2 - x = .7 - .5x$

e. $6 = 2x + 9$

f. $7x - 8 = 5x + 13$

g. $\frac{1}{2}x + 6 = \frac{7}{2}x - 21$

h. $\frac{2}{3}x - 1 = 6 + \frac{1}{3}x$

2. Show that $x = 9$ is a solution to $2x - 3 = x + 6$ by substituting into both sides of the equation.

3. Suppose you leave home traveling at a rate of 50 miles per hour.

   a. Find the distances in the following table.

Time (hours)	Distance (miles)
1	50
2	?
3	?
⋮	⋮
t	?

   b. In the last row, we don't know the amount of time, so we let t represent the number of hours of travel. Explain why the expression $50t$ represents the distance (in miles) that is traveled.

   c. Suppose you travel 120 miles. What does the solution to the equation $120 = 50t$ represent?

   d. Solve the equation from part (c).

4. A formula for temperature in Kelvins in terms of Celsius temperature is $K = C + 273°$. Water boils at a temperature of 373 Kelvins. What is the equivalent Celsius temperature?

   a. In this example $K = 373$. Substitute this value for K and write the equation.

   b. Solve the equation for C to find the Celsius temperature.

5. Recall, from Example 1 of Section 2.2, the formula $T = 3S + 1$ for finding the number of toothpicks T required to make S squares in a row. How many squares are in a row made of 28 toothpicks? Substitute this value for T and solve the equation to find the number of squares.

6. Suppose you are considering joining a music club. At one club, membership fees are \$4 plus \$1 for each CD. At the other club, membership fees are \$2.50 plus \$1.50 for each CD. Find the number of CDs for which these two clubs cost the same.

   a. Let d represent the number of CDs purchased. An expression for the cost of the first club is $4 + 1d$. Write an expression for the cost of the other club.

   b. Write an equation showing that the two expressions are equal because we want their costs to be equal.

   c. Solve the equation to find the number of CDs.

Section 3.2 Simplifying Expressions to Solve Linear Equations

7. Simplify each expression.

   a. $3x + 2 - x - 5$

   b. $2(x + 1) + 3x - 4$

   c. $x - (6 - 4x) + 2x$

   d. $3(2x - 4) + x^2 - 12$

   e. $3y - x^2 - y + 4x^2$

   f. $\frac{1}{3} + x + \frac{1}{6} + 2x$

   g. $\frac{1}{2}x - \frac{1}{3}x$

   h. $\frac{3}{4}(2x + 8) - \frac{1}{2}x$

8. Explain how the properties of $^-1$ can be used to simplify $3x - (3 - x) + 3$.

9. Solve each equation for x. Next to each step, write **A** if you added (or subtracted) the same quantity on both sides of the equation, **M** if you multiplied (or divided) both sides of the equation by the same quantity, **LS** if you simplified the expression on the left side of the equation, and **RS** if you simplified the expression on the right side of the equation.

   a. $2x + 8 + x = 32$

   b. $5(x - 2) + 9 = 7(x + 3)$

10. Solve each equation for x.

    a. $x + 8 + x + 8 = 26$

    b. $4x - 3 + 6x = 5 + 3x - 1$

    c. $x + (x + 30) + (2x + 10) = 180$

    d. $4(x - 5) + x = 10 + 6x - 70$

    e. $1 - (x + 5) = 3(x - 2) + 4x$

    f. $\frac{1}{4}(8x + 2) = 6(\frac{1}{4}x - \frac{2}{3})$

    g. $2 + 3(4x + 6) = x + 7 - (2x - 8)$

    h. $\frac{2}{3}(x - 9) + \frac{1}{3}x = 7x - 15$

11. Recall that a formula for the perimeter of a rectangle is $P = 2W + 2L$. Suppose a rectangle has a perimeter of 20 inches and a width of 4 inches. Substitute these values into the formula and solve the formula for L in order to find the length of this rectangle.

12. A formula for finding the area of the rectangle in the figure is $A = 4(x + 3)$. Find x when the area is 15 cm^2.

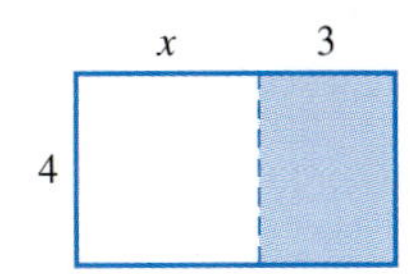

13. At 2:00 P.M., two trains leave the station in Hartford, one traveling north toward Springfield at a speed of 45 miles per hour and the other traveling south toward New Haven at 55 miles per hour. At what time will the two trains be 36 miles apart?

a. Let t represent the time in hours that the trains travel. Do you expect t to be greater than 1 or less than 1? Explain.

b. The southbound train travels a distance of $55t$ miles. Write an expression for the distance traveled by the northbound train.

c. Write an equation showing that the sum of the distances traveled by the two trains is 36 miles.

d. Solve the equation and find out (to the nearest minute) when the trains will be 36 miles apart.

In Exercises 14–16, use the fact that the sum of the measures of the three angles in any triangle is 180°.

14. In a triangle, angle B measures 20° more than angle A, and angle C measures 70° more than angle A.

a. Let x represent the measure of angle A. What is the measure of angle B? What is the measure of angle C?

b. Write an equation showing that the sum of the measures of these angles is 180°.

c. Solve the equation for x.

d. What are the measures of each angle?

15. In another triangle, one angle measures 90° and the other two angles have equal measures.

a. Let x represent the measure of one of the unknown angles. Write an equation using the fact that the sum of the three angles is 180°.

b. Solve the equation for x.

c. What is the measure of the two unknown angles?

16. The measures of all three angles of an equilateral triangle are equal.

a. Draw and label a diagram.

b. Write and solve an equation to find the measures of the angles.

Section 3.3 The Problem-Solving Process

17. A pilot in Gridville (see Section 2.3) flies from (2, 3) to ($^-5$, 1) and then to (1, $^-2$). Find the total distance this pilot has flown.

a. Understand the problem by writing what the problem is asking for.

b. Visualize the problem by drawing a picture.

c. Assign variables for the unknown(s).

d. Write equations using the variables.

e. Solve or simplify the equations.

f. Answer the question.

g. Check that your answer makes sense.

18. The in-bound lines on a tennis court form a rectangle. The length is 6 feet more than twice the width. The perimeter is 228 feet. Use the seven steps for problem solving to find the dimensions of the tennis court.

19. Sue has $40 to spend on her phone service each month. She pays a flat fee of $11.30 per month plus 6 cents for each minute she uses the phone. How much time can she spend on the phone each month?

20. Suppose Marion Jones gives her opponent a .3-second head start in a 200-m race. Jones' speed is 8.8 m/s and her opponent's is 8.6 m/s.

a. How long does it take Jones to catch her opponent?

b. At what distance does Jones catch her opponent?

c. Who wins the race?

Section 3.4 Arithmetic Sequences (Optional)

1. Find the next three terms in each sequence and determine which sequences are arithmetic.

a. 6, 10, 14, 18, . . .

b. 5, 15, 45, 135, . . .

c. 2, 0, 7, 0, 12, . . .

d. 19, 17, 15, 13, . . .

2. Find the missing piece of information for each arithmetic sequence.

a. $a_1 = 1$ and $d = 7$, find a_{10}

b. $a_1 = {}^-3$ and $a_9 = 5$, find d

c. $d = 11$ and $a_{14} = 151$, find a_1

CHAPTER 3 ▪ TEST

1. Solve for x if $7x + 5 = 2x - 5$.
2. Solve for x if
$$6 - \left(\tfrac{4}{5}x + 2\right) = 9 + \tfrac{2}{5}x - 3$$
3. Show that $x = \frac{3}{2}$ is a solution to the equation $6x - 1 = 2x + 5$.
4. Write and solve an equation to find the value of x that makes the expressions $2x + 9 - 5x$ and $4(x - 3)$ equal.
5. A formula for Fahrenheit temperature in terms of Celsius temperature is
$$F = \tfrac{9}{5}C + 32$$
Find the Celsius temperature when the Fahrenheit temperature is 90°.
6. An SUV holds 20 gallons of gasoline and gets 13 miles per gallon.
   a. Write an expression for the number of gallons in the SUV's gas tank if the SUV has traveled x miles.
   b. How much gas is left in the tank after 5 miles?
   c. How far can the SUV travel on a full tank of gas?
7. At 12:25 A.M. the Carpathia steamed toward the Titanic at a rate of 15.5 miles per hour. Suppose the Titanic was also drifting toward the Carpathia at a rate of 1.5 miles per hour. If the boats started 58 miles apart, at what time did they meet? (See the figure.)
   a. Identify the variable.
   b. Write an equation.
   c. Solve the equation and answer the question.
8. In triangle ABC, the measure of angle B is twice the measure of angle A. The measure of angle C is equal to the measure of angle B.
   a. Identify a variable and use it to write an equation.
   b. Solve the equation by any suitable method.
   c. Find the measure of each angle.
9. Doug is faster than Lance, so he decides to give Lance a 15-minute head start in a foot race. Doug runs 9 miles per hour and Lance runs 7.8 miles per hour.
   a. How long after he begins running does Doug catch Lance?
   b. Does your answer seem reasonable? Explain.
10. The perimeter of an isosceles triangle (two equal legs) is 20 cm. The remaining side is 4 cm more than three times the length of a leg. What is the length of each side of the triangle?

Figure for Exercise 7

CHAPTER

4

Systems of Linear Equations

For Safety's Sake

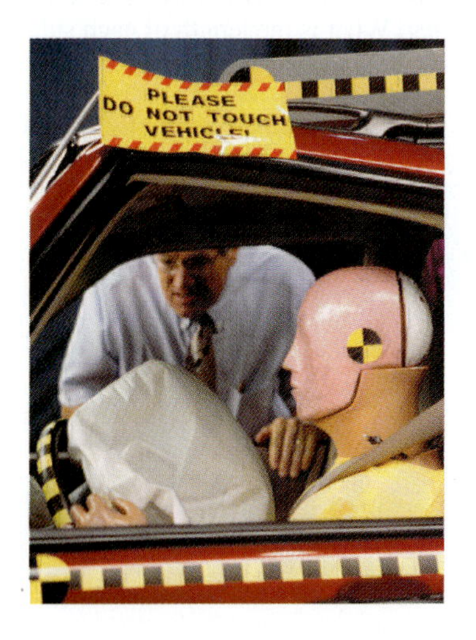

Beginning in 1998 airbags became standard equipment on all new cars sold in the United States. As the table shows, the number of lives saved by airbags steadily increased throughout the 1990s (*Source:* National Highway Traffic Safety Administration).

Year	Number of lives saved
1993	169
1994	276
1995	470
1996	686
1997	842
1998	1042
1999	1263
2000	1584

Transportation officials use data like these to predict future trends. In this chapter you learn how to use an equation to predict lives saved by airbags in the year 2006. (See Exploration on p. 169.)

Need help? For on-line resources, visit this web site: **math.college.hmco.com/students.**

4.1 Slope and Rate of Change

AFTER STUDYING THIS SECTION YOU WILL BE ABLE TO

- Find the slope of a line given two points.
- Find the slope of a line from its equation.
- Interpret slope as a rate of change and the y-intercept as an initial value.

Linear Equations in Two Variables

In the first example of Chapter 3 John had to take his car to a mechanic to replace the timing chain. He was charged $200 for the repair and wanted to find out how long the mechanic worked. Now let us look at the same situation from the mechanic's point of view.

EXAMPLE 1 John's mechanic is replacing the timing chain on his car. The part costs $95 and she charges $60 per hour for labor. Find an equation to show how the total cost of the repair depends upon the amount of time it takes to replace the chain. Use your equation to make a table and a graph that represent this situation.

Solution We use the formula

$$\text{total cost} = \textbf{variable cost} + \textbf{fixed cost}$$

first introduced in Section 2.4.

The fixed cost is the cost of the part or **$95**. The variable cost depends upon how many hours the job takes and is $\mathbf{60}t$, where t is the time measured in hours.

Substituting into the formula we have

$$\text{total cost} = \mathbf{60}t + \mathbf{95}$$

Substituting the values 1, 2, 3, and 4 for t gives us a table.

Time (hours)	Total Cost ($)
1	155
2	215
3	275
4	335

The table gives the mechanic a fairly good idea of how cost depends upon time. We can graph the four pairs of values, as shown in Figure 1. Notice that each tick mark on the vertical axis represents \$50. We chose this scale so that all four data points could be displayed.

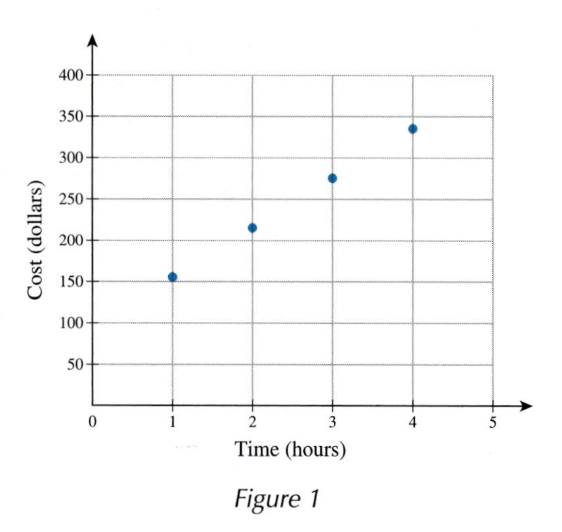

Figure 1

From Figure 1 we can see that the four points appear to lie on a straight line. We can confirm that it is a straight line by using a calculator. To do that, we need to name our variables x and y.

The equation

$$\textbf{total cost} = \mathbf{60}t + 95$$

becomes

$$y = \mathbf{60}x + 95$$

We enter the equation in the Y= menu. A friendly setting for the window is Xmin = 0, Xmax = 9.4, Xscl = 1, Ymin = 0, Ymax = 500, Yscl = 50.

The graph as it appears on the calculator screen is shown in Figure 2. If you use TRACE on your calculator, you will find all four points on this line, because we are using a friendly window. The final equation we found in Example 1 is called a *linear equation*.

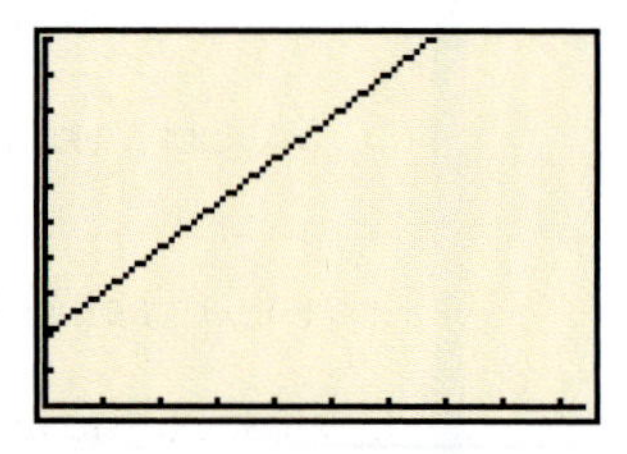

Figure 2 $Y1 = 60X + 95(0 \le X \le 9.4, 0 \le Y \le 500)$

Definition Linear Equation in Two Variables

A **linear equation in two variables** is any equation with simple terms involving two different variables and possibly a constant term. If we let x stand for one variable and y for the other, then a linear equation in two variables can almost always be put in the form $y = \text{coefficient} * x + \text{constant}$. An equation in this form is also called a **linear function**.

The following are important facts about linear functions.

Facts about Linear Functions $y = \text{coefficient} * x + \text{constant}$

Fact 1 The constant term on the right-hand side is an **initial value**. In Example 1 the initial value is the cost for the timing belt, $95.

Fact 2 The coefficient of the variable on the right is a **rate**. In the example the rate is what she charges for labor, $60 per hour.

Fact 3 The variable on the left is the **dependent**, or **output, variable**. The letter y is usually used for this variable. In the example y is the total cost. It depends upon how many hours the mechanic works.

Fact 4 We compute the value of the dependent variable using the value of the **independent**, or **input, variable** on the right. The letter x is usually used for this variable. In the example x is the number of hours the mechanic works. We put a value for the input variable, x, into the equation and out comes a value for the output variable, y.

There is another extremely important fact about linear equations in two variables. You have already observed this in Figures 1 and 2.

Graphs of Linear Functions

The graph associated with any *linear equation in two variables* (*linear function*) is a straight line.

We can learn some important information about an equation from its graph.

EXAMPLE 2 Trace the graph in Figure 2 all the way to the left so that $x = 0$.

a. What do you notice about y?

b. What does this value represent?

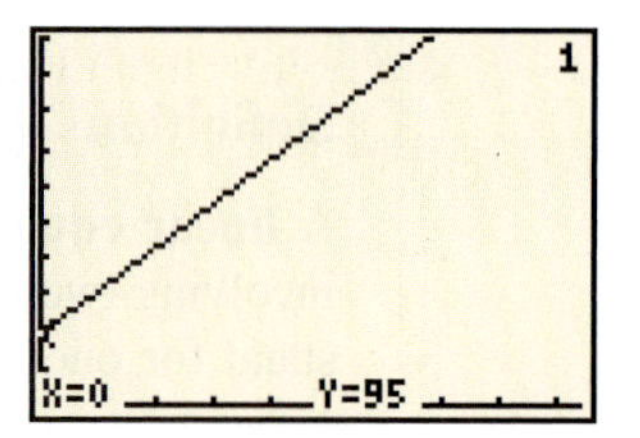

Figure 3 Trace to the point where X = 0.

Solution

a. When you trace to the point where $x = 0$, you will find that $y = 95$, as shown in Figure 3. You can also find this value by substituting $x = 0$ into the equation $y = 60x + 95$.

b. $95 is the cost of the timing belt. It is also the **initial value**.

From Example 2 we see that the initial value of a linear function shows up as the y-coordinate of the point where the line crosses the y-axis. This point is called the ***y*-intercept**.

Here is another example of a linear equation in two variables.

EXAMPLE 3 Suppose we start a stopwatch when a sprinter is running at a constant rate of 30 feet/second and is 140 feet from the finish line. Find an equation that shows how his distance from the finish line depends upon the time shown on the watch. Draw a graph, make a table, and use them to find the y-intercept.

Solution We use the formula distance = rate ∗ time, but we need to be careful. First, the initial value is 140, because we start 140 feet from the finish line. Second, the distance *decreases* as time increases. An appropriate equation is

$$\text{distance} = 140 - 30 * \text{time}$$

The graph is given in Figure 4.

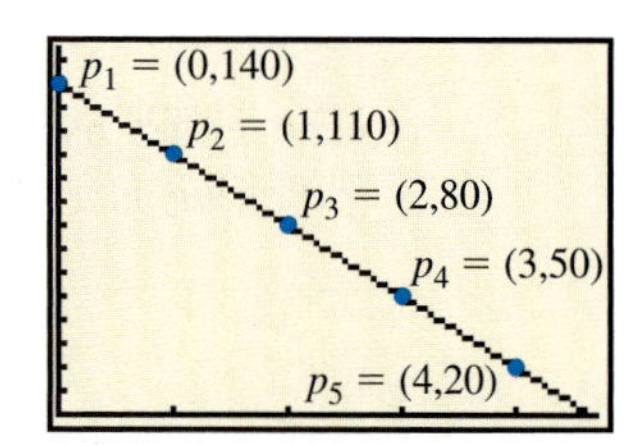

Figure 4 Graph of Y1 = 140 − 30X. We used a friendly window with Xmin = 0, Xmax = 4.7, Ymin = 0, and Ymax = 155.

We can enter Y1 = 140 − 30X in our calculator and make a table. Go to the TblSet menu and make sure that ΔTbl = 1. Let's set TblStart = 0, because negative input values certainly don't make any sense for this problem. Now when you press 2nd TABLE, you should see a table like in Figure 5.

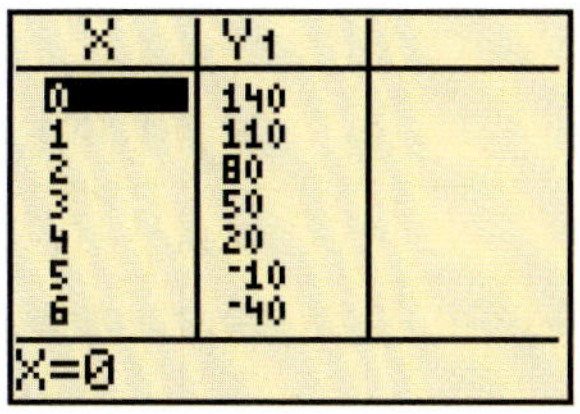

X	Y1	
0	140	
1	110	
2	80	
3	50	
4	20	
5	-10	
6	-40	

X=0

X	Y1
0	140
1	110
2	80
3	50
4	20
⋮	⋮

Figure 5

Now we can find the y-intercept by looking in the table for the value of y when $x = 0$. Clearly it is 140 feet, the distance the sprinter is from the finish line when we start the watch. This should not surprise us because the starting distance is the *initial value*. We may find the same result by tracing along the graph.

■ ■

The Slope of a Line

Compare the graphs in Figures 2 and 4. Reading from left to right, each graph starts at a point on the y-axis, called the y-intercept. As you move to the right the line in Example 1 rises, but the line in Example 3 falls. This suggests that we may be able to describe each line in terms of the *run* and *rise* introduced in Section 2.3. Run and rise are used to compute a value called the *slope of a line*.

Definition Slope of a Line

$$\text{slope} = \frac{\text{change in output}}{\text{change in input}} = \frac{\text{rise}}{\text{run}}$$

To compute the **slope of a line**, pick two points (x_1, y_1) and (x_2, y_2) on the line. Then **rise** $= \mathbf{y_2 - y_1}$ and **run** $= \mathbf{x_2 - x_1}$, so

$$\text{slope} = \frac{\textbf{rise}}{\textbf{run}} = \frac{y_2 - y_1}{x_2 - x_1}.$$

EXAMPLE 4 Show how to use the definition for slope to compute the slopes of the lines in each example.

a. Example 1

b. Example 3

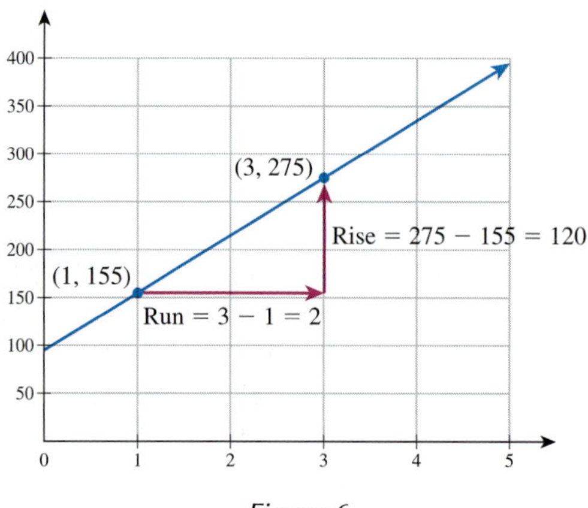

Figure 6

Solution

a. Use the table of values in Example 1 to pick two points on the line, as shown in Figure 6. From these points compute **run** = **2** and **rise** = **120**. Slope is then found to be the ratio

$$\frac{\textbf{rise}}{\textbf{run}} = \frac{\textbf{120}}{\textbf{2}} = 60$$

b. Use the table of values in Figure 5 to pick two points on the line shown in Figure 4. We'll use several different pairs to show we get the same result each time.

First, let's use $p_1 = (\textbf{0}, \textbf{140})$ and $p_2 = (\textbf{1}, \textbf{110})$. Substitute to get

$$\text{slope} = \frac{\textbf{rise}}{\textbf{run}} = \frac{y_2 - y_1}{x_2 - x_1} = \frac{\textbf{110} - \textbf{140}}{\textbf{1} - \textbf{0}} = {}^{-}30$$

Now let's use $p_1 = (\textbf{0}, \textbf{140})$ and $p_3 = (\textbf{2}, \textbf{80})$:

$$\text{slope} = \frac{\textbf{rise}}{\textbf{run}} = \frac{y_2 - y_1}{x_2 - x_1} = \frac{\textbf{80} - \textbf{140}}{\textbf{2} - \textbf{0}} = \frac{{}^{-}\textbf{60}}{\textbf{2}} = {}^{-}30$$

So far, so good. Now let's make the second point to the *left* of the first point. We'll use $p_5 = (\textbf{4}, \textbf{20})$ and $p_2 = (\textbf{1}, \textbf{110})$:

$$\text{slope} = \frac{\textbf{rise}}{\textbf{run}} = \frac{y_2 - y_1}{x_2 - x_1} = \frac{\textbf{110} - \textbf{20}}{\textbf{1} - \textbf{4}} = \frac{\textbf{90}}{{}^{-}\textbf{3}} = {}^{-}30$$

It still works! In fact, this formula for slope will work no matter what two points we choose on the line and in either order we subtract.

Example 4 illustrates an important concept.

Slopes of Lines

Straight lines have constant slope.

We may make another observation. The slope we found in Example 4(a) is 60. This is the same as the coefficient of x in the equation $y = 60x + 95$. It is the *rate* the mechanic charges for labor, $60 per hour.

In part (b) the slope is $^{-}30$. This is the same as the coefficient of x in the equation $y = 140 - 30x$. Note that we can write the equation as $y = {}^{-}30x + 140$. The slope is also a rate, $^{-}30$ feet/second. It is a negative rate, because the distance to the finish line is decreasing as time increases.

Meaning of Slope

The slope of a line represents the **rate of change** of the output variable with respect to the input variable. In an application the rate may be expressed as units of y *per* units of x.

When interpreting slope in an application problem, keep the rate-of-change concept in mind. Think of the units for slope as something per something else. In the preceding examples we had dollars *per* hour and feet *per* second. Finally, we noticed that in both cases, the slope is the *coefficient* of x.

Finding the Slope

For the line given by the equation y = coefficient $* \, x$ + constant, the coefficient of x is the slope of the line.

EXAMPLE 5 Graph these equations using your calculator and describe how changing the slope affects the steepness of a line. Notice that all these lines have the same y-intercept, (0, 1).

$$y = 2x + 1$$

$$y = 1x + 1$$

$$y = \tfrac{1}{2}x + 1$$

$$y = 0x + 1$$

$$y = {}^{-}\tfrac{1}{2}x + 1$$

$$y = {}^{-}1x + 1$$

$$y = {}^{-}2x + 1$$

Solution Figure 7 shows all seven equations graphed in the ZoomDecimal Window (${}^{-}4.7 \le x \le 4.7, {}^{-}3.1 \le y \le 3.1$). You may want to look at only one of them at a time.

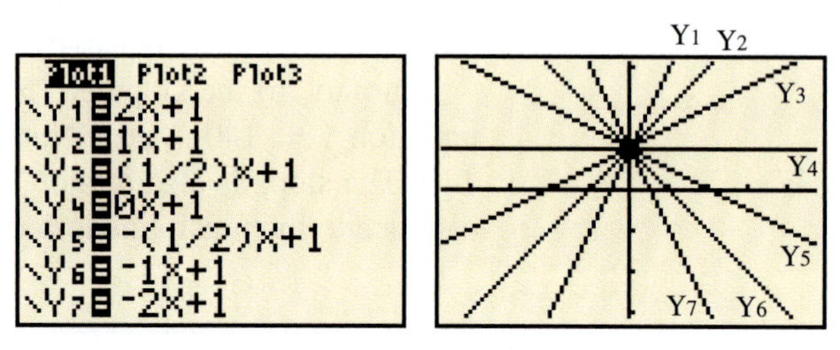

Figure 7 All seven equations are on the left, with the graphs on the right. Notice that we put parentheses around fractional coefficients of X.

We can make these observations:

1. The lines Y1, Y2, and Y3, which all have positive slope, rise as we move to the right.
2. The lines Y5, Y6, and Y7, which all have negative slope, fall as we move to the right.
3. The line Y4, which has zero slope, is horizontal.
4. The larger the *absolute value* of the slope (in other words the bigger the number, if you forget the minus sign), the steeper the rise or fall of the line.

We can summarize the previous example by noting the following facts.

How Do Graphs of Lines with Different Slopes Look?

Fact 1 Lines with positive slopes go to the right and up. The larger the positive value, the steeper the line.

Fact 2 A line with zero slope is horizontal.

Fact 3 Lines with negative slopes go to the right and down. The larger the negative slope in *absolute value*, the steeper the line.

EXAMPLE 6 Find the slope of the line passing through each pair of points.

a. $({}^{-}4, 5), ({}^{-}1, {}^{-}2)$

b. $(1, {}^{-}2), (4, 3)$

c. $(^-3, 4), (5, 4)$
d. $(3, ^-2), (3, 1)$

Figure 8 shows the pairs of points and the lines passing through them.

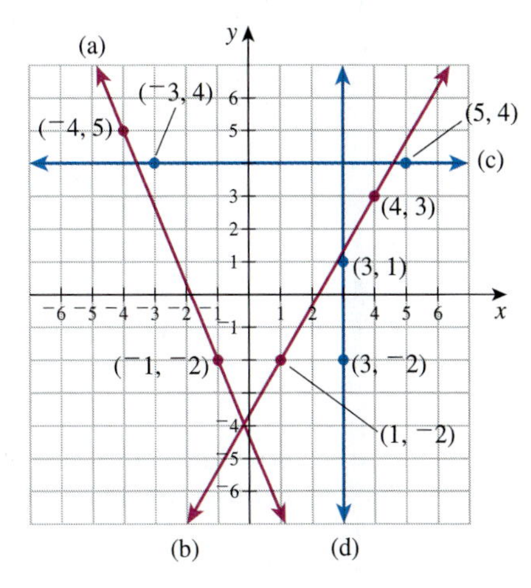

Figure 8

Solution

a. $$\text{slope} = \frac{\text{rise}}{\text{run}} = \frac{y_2 - y_1}{x_2 - x_1} = \frac{^-2 - 5}{^-1 - ^-4} = \frac{^-2 + ^-5}{^-1 + 4} = \frac{^-7}{3}$$

The line connecting $(^-4, 5)$ and $(^-1, ^-2)$ has a negative slope. It goes to the right and down.

b. $$\text{slope} = \frac{\text{rise}}{\text{run}} = \frac{y_2 - y_1}{x_2 - x_1} = \frac{3 - ^-2}{4 - 1} = \frac{3 + 2}{4 + ^-1} = \frac{5}{3}$$

The line connecting $(1, {}^{-}2)$ and $(4, 3)$ has a positive slope. It goes to the right and up.

c. $\text{slope} = \dfrac{\text{rise}}{\text{run}}$

$$= \frac{y_2 - y_1}{x_2 - x_1}$$

$$= \frac{4 - 4}{5 - {}^{-}3}$$

$$= \frac{0}{5 + 3}$$

$$= \frac{0}{8}$$

$$= 0$$

The line connecting $({}^{-}3, 4)$ and $(5, 4)$ has zero slope. It goes from left to right without rising or falling. It is horizontal.

d. $\text{slope} = \dfrac{\text{rise}}{\text{run}}$

$$= \frac{y_2 - y_1}{x_2 - x_1}$$

$$= \frac{1 - {}^{-}2}{3 - 3}$$

$$= \frac{1 + 2}{0}$$

$$= \frac{3}{0}$$

We have a problem here because division by 0 is not allowed. We say that the slope is *undefined*. The line goes straight up and down. It is vertical.

EXAMPLE 7 Michael and Sarah start a business selling their shareware computer program over the Internet. Their web host charges them a flat fee of \$150 per month and then charges them \$.10 for every megabyte their customers download. The encoded file containing their program is 7.5 megabytes. They are hoping that because they got a good review on a popular website, their downloads will increase next month by 2000 units. How much will their cost go up if this happens?

Solution If we think of the number of programs downloaded as an input and the cost per month as the output, then the question they need to answer can be rephrased as, If the input goes up by 2000, by how much will the output go up? To answer this question we can use the slope of the linear equation relating input to output.

Each program is 7.5 megabytes and each megabyte costs 10 cents, so they pay 75 cents, or $\frac{3}{4}$ of a dollar, per download. Because the slope of a line is the *rate of change* of the output with respect to the input (page 129), the slope of the line is .75.

We are given the change in input, or run, which is **2000**. We need to find the change in output, that is, the rise. Using the formula

$$\textbf{slope} = \frac{\text{rise}}{\textbf{run}}$$

we have

$$\textbf{.75} = \frac{\text{rise}}{\textbf{2000}}.$$

Solving for rise gives 1500. Because the output is expressed in dollars, their cost will go up by $1500.

Notice that to solve this particular problem we did not need to use the flat fee nor did we need to know how many downloads they are currently selling each month.

Exercises 4.1

1. For each situation, identify the variable cost, the fixed cost, and the rate of change.

   a. The charge for a telephone call from the United States to Japan is $1.25 plus 25 cents per minute.

   b. In 2003, first-class postage costs 14 cents plus 23 cents for every ounce a letter weighs.

   c. To rent a car from the Pinnacle Car Company, you must pay 25 cents for each mile driven plus an 8 dollar service fee.

2. Given the equations

   a. $y = 60x + 95$ **b.** $y = 2x + 4$

   c. $y = 3x$ **d.** $y = 800x + 1750$

   (1) What is the output for each equation when the input is zero?

   (2) Give another name for the constant term.

3. Suppose you are driving up a hill of steady incline. We describe the pitch, or steepness, of the hill algebraically as the slope. Let $y = .02x + 1000$ model this situation, where y is the elevation and x is the horizontal distance traveled, both measured in feet.

   a. Use this function to generate two ordered pairs called (x_1, y_1) and (x_2, y_2).

   b. Find the slope using the slope formula.

   c. Generate two more ordered pairs and calculate the slope again.

   d. Compare the slopes from parts (b) and (c). What does this imply about the rate of incline of the hill?

4. Find the slope of the line given by each equation.

   a. $y = 60x + 95$ **b.** $y = 2x + 4$

   c. $y = 3x$ **d.** $y = 800x + 1750$

   e. $y = {}^{-}5x + 1$ **f.** $y = \frac{4}{3}x - 7$

5. The figure shows a graph of four linear equations. Identify the slope of each line as a positive, zero, or negative slope.

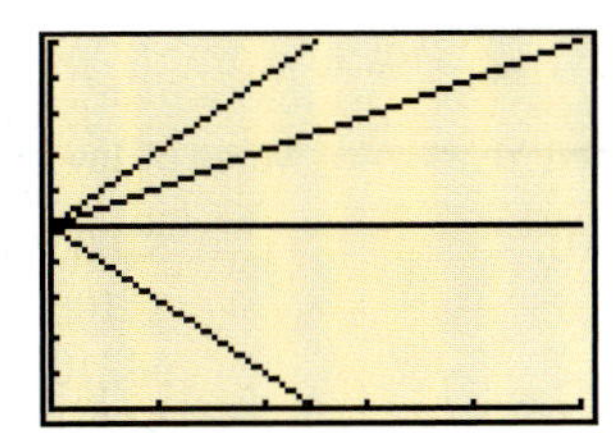

6. Given the three ordered pairs $A = (^-1, ^-3)$, $B = (0, ^-1)$, and $C = (1, 7)$:

a. Graph the ordered pairs and connect the points.

b. Does the graph appear to have a constant slope?

c. Compute the slope between A and B and then between B and C.

d. Do the slopes found in part (c) support your conclusion from part (b)?

7. In Section 2.3 you were introduced to Gridville, where the run and rise were used to find a new location.

a. If you begin at (0, 4) and know run = 5 and rise = 3 (so the slope = $\frac{3}{5}$), where will you end up?

b. Begin at $(1, ^-3)$ and apply a slope of $\frac{^-3}{2}$ to find a new location.

c. Any whole number such as 4 can be expressed as a fraction $\frac{4}{1}$. Starting at the point (1, 5) and following a line with a slope of 4, find the next location.

d. Start at $(^-1, 3)$ and use a slope of $\frac{1}{2}$ to find two new locations.

8. Show that a slope of $\frac{^-1}{2}$ is equivalent to the slope $\frac{1}{^-2}$ by using the following steps:

a. Plot a starting point (0, 3) on a graph.

b. Use the slope of $\frac{\text{rise}}{\text{run}} = \frac{^-1}{2}$ to find a new point.

c. Again, start at (0, 3) and use the slope of $\frac{\text{rise}}{\text{run}} = \frac{1}{^-2}$ to find a new point.

d. Are all three points on the same line? What does this imply about the slopes $\frac{^-1}{2}$ and $\frac{1}{^-2}$?

9. Find the slope of the line between each pair of points.

a. $(^-4, 1)$ and $(2, ^-5)$ **b.** $(3, ^-5)$ and $(^-2, 6)$

c. $(2, 4)$ and $(^-2, 1)$ **d.** $(0, ^-3)$ and $(5, ^-3)$

10. An investment of \$1000 increases at the rate of \$100 per year.

a. What is the initial value?

b. Use the rate of change to find the investment's worth after 1 year and 2 years.

c. Does this investment's growth represent a linear relationship? Explain.

11. A different \$1000 investment is appreciating according to the following table.

Elapsed Years	Value of Investment
0	\$1000
1	\$1100
2	\$1210
3	\$1331

a. What is the investment's initial value?

b. Is this investment's rate of change constant? Explain.

c. Does this investment's growth represent a linear relationship? Explain.

d. Which investment would you rather own, this one or the one in Exercise 10? Why?

12. The following table shows the dividend per share paid by the XYZ corporation.

Year	Years since 2000	Dividend per Share
2000	0	.25
2001	1	.25
2002	2	.25
2003	3	.25

Let x be the number of years since 2000 and y be the dividend per share.

a. Draw a graph.

b. Find the slope.

c. Interpret the slope.

13. The Super Light Soft-Drink Company claims its soda is less dense than water. An independent lab wishes to check the truth of the company's claim. They place a beaker weighing 50 g on a scale and weigh the beaker when it contains 10 mL, 20 mL, 30 mL, 40 mL and 50 mL. The results are presented graphically in the figure.

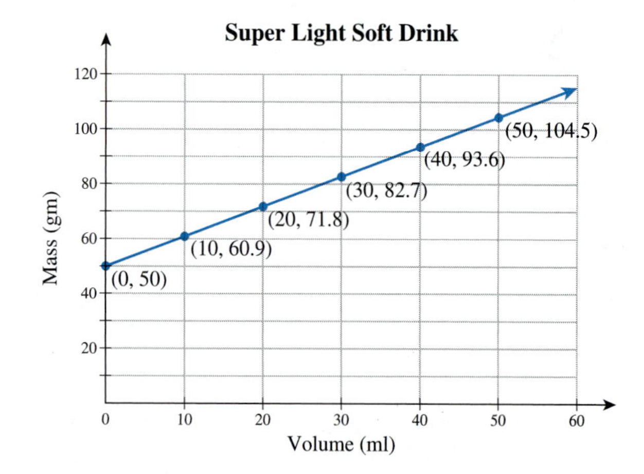

a. Calculate the density of Superlight by finding the slope of the line.

b. Compare Superlight's density with that of pure water (1.00). Did the company tell the truth?

14. Digitech computer software company can sell 400 disks a day at a price of \$60. The company can, however, increase its sales by 28 disks a day for each \$2 reduction in price.
    a. What is the rate of change?
    b. Interpret the rate of change.
    c. Use the rate of change to find additional entries for the table.

x (Price)	y (Number of Disks Sold)
60	400
58	428
56	456
⋮	⋮

15. The following table shows basal metabolic rates (BMR) for 22-year-old men and women of different weights and average height. BMR is the number of calories per day the body requires while at rest. [Harris-Benedict equation (http://www.weight-free-lifestyles.com)]

Weight	Female BMR	Male BMR
100	1283	1412
120	1371	1536
140	1459	1660
160	1547	1784
180	1635	1908
200	1723	2032

a. Draw a graph using the BMR column for your gender.

b. Find the slope.

c. Interpret the slope.

Skills and Review 4.1

16. Point A is located at $(^-2, 1)$ and point B is located at $(3, ^-5)$. Find the distance traveled from point A to point B. Outline your work using the seven steps for problem solving.
17. Suppose you are making a choice between two cell-phone plans. Plan A charges \$22.95 plus 7 cents for each minute of calling. Plan B charges \$28.45 plus 3 cents for each minute of calling. After how many minutes would the two plans cost the same? Outline your work using the seven steps for problem solving.
18. Solve for x if $11 - x = 8(x + 4) - 3$.
19. Verify that $x = 5$ is a solution to the equation $^-4x + 23 = 2x - 7$.
20. Solve $5x - (6 - x) = 4 + 2x$ for x.
21. A formula for the circumference of a circle is $C = 2\pi r$. Find the radius of the circle if the circumference is $(\frac{11}{6})\pi$ cm.
22. The perimeter of an equilateral triangle is three times the length of a side. Let P represent the perimeter of the triangle and let s represent the length of a side of the triangle.
    a. Write a formula that shows how P depends upon s.
    b. Complete a table of values using your formula from part (a).
    c. Display your data from part (b) in a graph.
23. Simplify the expression $4c^2 - 5(d - 2) + 7c^2 + 6 - d$.
24. Evaluate $x + \frac{2}{3} - 4 + (^-3)^4$ for $x = 11$.
25. Convert 20 feet/minute into inches/second.

4.2 Forms of Linear Equations

AFTER STUDYING THIS SECTION YOU WILL BE ABLE TO

- Find an equation of a line given a slope and y-intercept, one point and the slope, or two points.
- Write the equation of a line in slope-intercept form.
- Use the x- and y-intercepts to sketch the graph of a linear equation.

Slope-Intercept Form

Any equation involving simple terms with two variables (say x and y) and possibly a constant term may be put in the form

$$y = \text{coefficient} * x + \text{constant}$$

using simple algebraic manipulations. In Section 4.1 we learned that the graph of this equation is a straight line. Linear equations in this form are so common that there is a special formula for them.

> **Definition Slope-Intercept Form**
>
> A linear equation in two variables is in **slope-intercept form** if it is written
>
> $$y = mx + b$$
>
> where x and y can be replaced by any two variable names. The *slope* of the line is given by m, and the y-intercept is the point $(0, b)$.

The coefficient m and the constant b have important graphical meanings.

- b is the y-value at which the line crosses the y-axis, called the y-intercept. When we substitute $x = 0$ into the formula $y = mx + b$, we get

  $$y = m * 0 + b = b$$

  We saw in Section 4.1 that b may also be considered the *initial value* of the output variable.

- m is the *slope* of the line, which you can think of as the *rate of change*. It tells us how much y changes if x increases by 1.

EXAMPLE 1 Michelle is going to start a business renting DVDs. She wants to make a bulk purchase of disks to start. A company offers to sell her disks for \$22 each, with a fixed shipping charge of \$400. Write a linear equation and draw a graph that Michelle can use to study this offer.

Solution The input is the number of DVDs Michelle purchases, so we call that x. The output, y, is her cost. In this case, even if Michelle bought 0 DVDs (but had the empty carton shipped!), she would spend \$400, so the line should cross the

y-axis at the point (0, 400). Therefore, $\boldsymbol{b} = \mathbf{400}$. Each extra DVD costs her \$22, so the slope $\boldsymbol{m}$ is **22**.

We substitute for m and b in the formula $y = \boldsymbol{m}x + \boldsymbol{b}$. This gives Michelle the equation she wants:

$$y = \mathbf{22}x + \mathbf{400}$$

Figure 9 shows the graph of this equation. We check the slope by choosing two points on the line. The slope is

$$\frac{\textbf{rise}}{\textbf{run}} = \frac{\mathbf{4400}}{\mathbf{200}} = 22$$

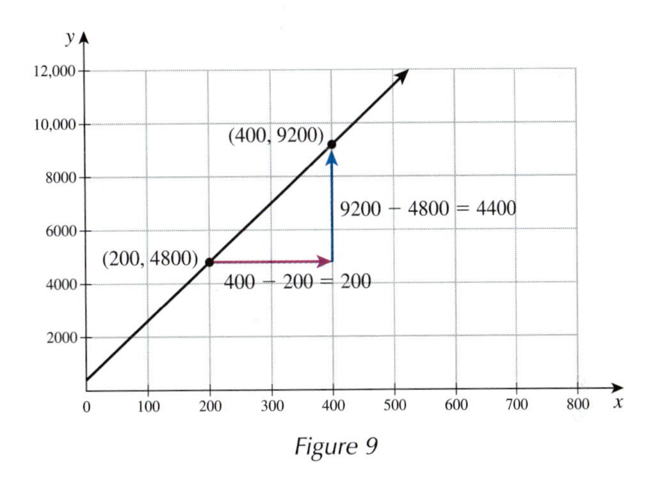

Figure 9

EXAMPLE 2 An airplane is flying at 30,000 feet and begins to descend to the airport at a constant rate of 250 feet/minute. What linear equation models the height of the airplane in terms of the number of minutes since it began descending?

Solution Once again we are given a physical problem that lends itself to the *slope-intercept formula* very well. We know the initial value is 30,000 feet, and the rate of change (slope) is $^-250$ feet/minute. Notice that we made the slope a negative number because the airplane is losing altitude.

We let y represent the altitude (in feet) and x the time (in minutes) after the plane starts dropping. Again, we use the formula $y = \boldsymbol{m}x + \boldsymbol{b}$. We substitute $\boldsymbol{m} = \mathbf{^-250}$ and $\boldsymbol{b} = \mathbf{30{,}000}$ to obtain the equation

$$y = \mathbf{^-250}x + \mathbf{30{,}000}$$

Let's check to see if this equation is a good model. If $x = 0$—in other words, if the plane is about to start descending—then

$$y = {}^{-}250 * 0 + 30{,}000 = 30{,}000$$

which is correct. To check the rate, let's see how high the plane is after **1** minute:

$$y = {}^{-}250 * \mathbf{1} + 30{,}000 = 29{,}750$$

The plane started at 30,000 feet and in 1 minute dropped to 29,750 feet, so it is descending 250 feet/minute, which is also correct. See Figure 10.

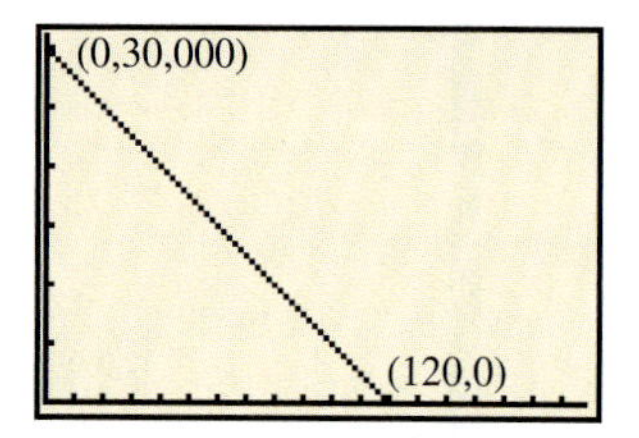

Figure 10 Graph of the altitude function Y1 = ⁻250X + 30,000. We have used the friendly window Xmin = 0, Xmax = 188, Ymin = 0, and Ymax = 31,000.

If we look carefully at Figure 10, we notice that the graph crosses not only the y-axis but the x-axis as well. The y-intercept is the initial altitude (the altitude when $x = 0$), but does the **x-intercept** have any special significance? If we look at the graph on our calculator, we can use TRACE to find the coordinates of the x-intercept. We get the values given in Figure 10, (120, 0).

Does the point (120, 0) have some special meaning? This point indicates that 120 minutes after the plane has started to descend, the altitude y is 0. That means the plane is on the ground. This model for the altitude of the plane is good only for 120 minutes, because the plane won't get below the ground (we hope!).

We can solve for the x-intercept algebraically using the fact that it is the place where y must be 0. Setting **y** to **0** in the equation, we get

$$\begin{aligned} \mathbf{0} &= {}^{-}250x + 30{,}000 \\ 250x &= 30{,}000 \\ \frac{250x}{250} &= \frac{30{,}000}{250} \\ x &= 120 \end{aligned}$$

which is just what the calculator graph gave us. In this case we were lucky that the graph gave us the *exact* value for the x-intercept. Sometimes tracing will give us only an approximation.

As in Example 2, we are often interested in both the x-intercept and the y-intercept. Here's how we find them.

Finding Intercepts

For any linear equation in two variables,

- We find the y-intercept by setting $x = 0$. The y-intercept is always (0, some number).
- We find the x-intercept by setting $y = 0$. The x-intercept is always (some number, 0).

Standard Form

Not all linear equations are written in slope-intercept form. In Example 3 we encounter a linear equation in **standard form**.

Definition Standard Form of a Linear Equation

A linear equation in two variables is in **standard form** if it is written

$$Ax + By = C$$

A, B, and C may be any numbers as long as A and B are not both zero, and x and y can be replaced by any variable names.

Example 3 shows how to interpret the standard form of an equation.

EXAMPLE 3 Suppose there are two dams on a lake. In order to keep the lake level, the dams must release 1000 gallons of water per second. The first dam releases 8 gallons per second for each click its gates are opened, and the second dam releases 10 gallons per second for each click its gates are opened. The dam keeper needs an equation that will let him work out how much to open the second dam if he knows how far open the first dam is.

Solution Let's let x be the number of clicks that the first dam is open and y be the number of clicks the second dam is open. We can compute how many gallons will be released in 1 second with the equation

$$\text{water flow} = 8x + 10y$$

We know that **water flow** has to be **1000**, so we can substitute:

$$\mathbf{1000} = 8x + 10y$$

Now we have an equation with two variables. We can get some idea what the graph of this equation looks like by finding both the x- and y-intercepts.

To find the x-intercept, we set $\mathbf{y} = \mathbf{0}$ and substitute:

$$1000 = 8x + 10 * \mathbf{0}$$
$$1000 = 8x$$
$$\frac{1000}{8} = \frac{8x}{8}$$
$$125 = x$$

The x-intercept is (125, 0).

To find the y-intercept, we set $\mathbf{x} = \mathbf{0}$ and substitute:

$$1000 = 8 * \mathbf{0} + 10y$$
$$1000 = 10y$$
$$\frac{1000}{10} = \frac{10y}{10}$$
$$100 = y$$

The y-intercept is (0, 100).

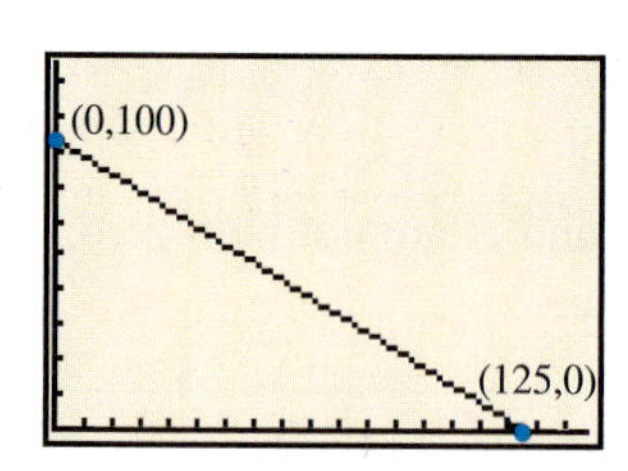

Figure 11

We can now sketch the graph in the first quadrant using the two intercepts. See Figure 11. Because the line appears to have a negative slope, the dam keeper can see that as x increases, y decreases. This makes sense to him, because if he opens the first gate a little more, he can close the second gate a bit and still get the same amount of water flow. If the dam keeper wants to be able to determine quickly how to adjust the second dam based on how open the first dam is, standard form is not very useful. Having an equation in slope-intercept form expresses the variable for the second dam, y, as a function of the variable for the first dam, x. So he needs to solve for y *in terms of* x.

Start with the equation $1000 = 8x + 10y$. Don't worry about the fact that we don't know the value of $8x$. Treat it as if it were a known quantity, and subtract it from both sides of the equation to isolate y. This gives us

$$1000 - 8x = 10y$$

We now know that we can solve for y if we divide by 10 on both sides.

$$\frac{1000 - 8x}{10} = \frac{10y}{10}$$

Remember that dividing by 10 is like multiplying by $\frac{1}{10}$, so the distributive property applies.

$$\frac{1}{10} * 1000 - \frac{1}{10} * 8x = \frac{1}{10} * 10y$$
$$100 - .8x = y$$

This equation may be rearranged so that it is in slope-intercept form:

$$y = {}^{-}.8x + 100$$

The dam keeper is satisfied. He may now take any value for x, substitute it into the equation, and find the corresponding value of y.

Although the dam keeper is happy with the equation $y = {}^{-}.8x + 100$, we might benefit by taking another look at it in connection with the graph in Figure 11. If we do that, we may observe the following.

- Using the formula $y = mx + b$, we know that $m = {}^{-}.8$. This gives us a negative slope, which is what we should expect from the line we see in Figure 11. Furthermore, if we go from the y-intercept to the x-intercept, we have **run** = **125** and **rise** = **⁻100**. So

$$\text{slope} = \frac{\textbf{rise}}{\textbf{run}} = \frac{\mathbf{{}^{-}100}}{\mathbf{125}} = {}^{-}.8$$

  That's the same as m. So, we know we are on the right track.
- Using our formula $y = mx + b$, we know that $b = 100$. This gives us a y-intercept of (0, 100), which is what is shown in Figure 11.
- Slope-intercept form is exactly what our calculator needs to draw a graph. We can enter the equation $y = {}^{-}.8x + 100$ in the Y = menu and find a suitable window (for example, $0 \le X \le 188$, $0 \le Y \le 120$). We should get a graph that looks a lot like the one in Figure 11.

EXAMPLE 4 Jessica's father gives her a piggy bank when she turns 12 and tells her, "I put some money in the bank to get you started, but now you have to deposit some of your allowance into the bank each week to make it grow." Jessica deposits \$5 into the bank each week. At the end of one year she breaks open the bank and discovers that she has \$360. How much money did Jessica's father deposit initially?

Solution The information we are given is the rate at which the money is increasing (\$5 per week) and the amount of money after 52 weeks (1 year). Because the rate never changes, we may use a linear equation to model the data. Let y be the amount in the savings account and x be the time in weeks since the account was opened. The rate is the slope, and the amount of money after 52 weeks gives us one point on the line.

Let's see what we can do with our favorite formula, $y = mx + b$.

We know the slope, m, is 5. And we know that when $\mathbf{x = 52}$, $\mathbf{y = 360}$. Let's substitute to get

$$\mathbf{y} = mx + b$$

$$\mathbf{360} = 5 * \mathbf{52} + b$$

$$360 = 260 + b$$

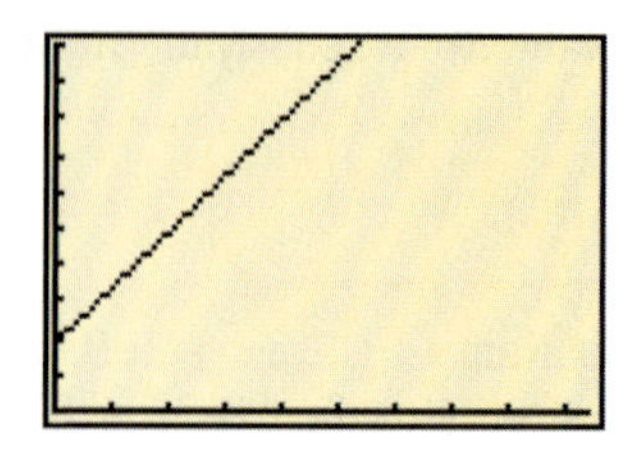

Figure 12 Jessica's father initially deposited \$100 into the piggy bank. We have used the friendly window $0 \leq X \leq 94$, $0 \leq Y \leq 500$.

Now we have an equation with only one variable, b. We may solve for b by subtracting 260 on both sides to obtain

$$100 = b$$

Because the y-intercept is also the initial amount deposited, we can answer the question. Jessica's father deposited \$100 initially.

We may also write an equation that shows how y depends upon x by substituting for both m and b in the formula $y = mx + b$.

$$y = mx + b$$

$$y = 5x + 100$$

A graph of this equation is shown in Figure 12.

EXAMPLE 5 PCBs were dumped into the Hudson river for many years. The United States Environmental Protection Agency extensively studied the concentration of these chemicals in various fish as part of the process of determining whether to force the General Electric company to dredge the river bottom. The approximate concentrations of PCBs in brown bullheads were as follows:

Year	Concentration (mg/kg)
1984	46
1985	42.5
1986	39

These points look as if they might lie on a straight line (Figure 13). Assuming they do, we should be able to find an equation that will allow us to determine the PCB concentration in past or future years and answer the following questions.

a. What was the concentration in 1980?

b. When will the concentration be 0?

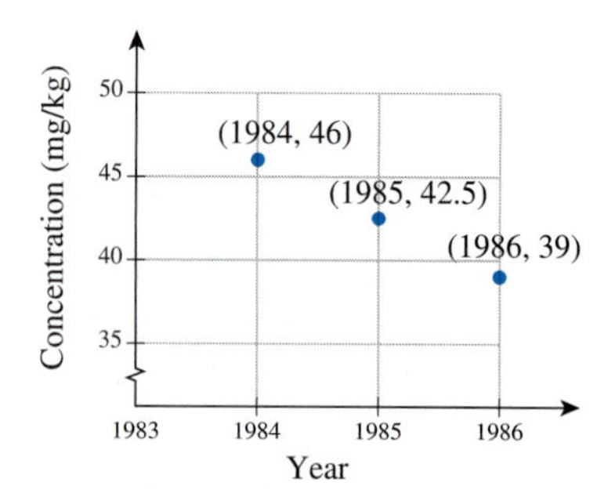

Figure 13 Data from the EPA (http://www.epa.gov/hudson).

Solution This example is very different from our previous ones. We are *not* given an initial value, and we aren't even given a rate of change. All we have are the coordinates of some points on our line. Because the first question we want to answer is what is the concentration in 1980, we let 1980 be year 0. Our table of values is now as follows.

Years Since 1980, x	Concentration (mg/kg), y
4	46
5	42.5
6	39

In order to answer parts (a) and (b), we need to write an equation relating x and y. From the table we observe that an increase of 1 in x results in a decrease of 3.5 in y (we have approximated the actual concentrations; in most real-world situations you would *not* expect even three points to lie exactly on a line). In other words, when run = 1, rise = $^{-}3.5$. Therefore, we will choose a line with

$$\text{slope} = \frac{\text{rise}}{\text{run}} = {}^{-}3.5$$

a. We can use the slope-intercept formula for the equation of a line and substitute what we do know. We can use any of the first three pairs of values for x and y, and we know that the slope is $m = {}^{-}3.5$. Let's use the pair (4, 46). We get

$$y = mx + b$$

$$46 = {}^{-}3.5 * 4 + b$$

$$46 = {}^{-}14 + b$$

$$60 = b \qquad \textit{Adding 14 on both sides of the equation}$$

In slope-intercept form, b is the y-intercept, the value of y when $x = 0$. In this case $b = 60$ means the concentration of PCBs was 60 mg/kg in 1980, if our linear model is correct.

b. Now we have an equation in slope-intercept form:

$$y = {}^{-}3.5x + 60$$

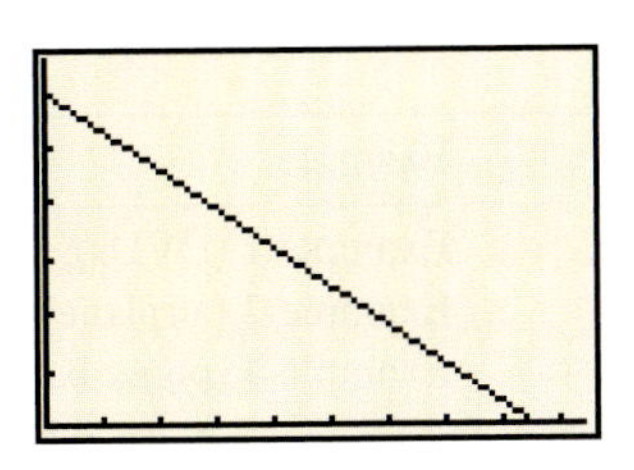

Figure 14

We find the x-intercept of the graph, which corresponds to the point where the concentration is 0. We can enter the equation $y = {}^{-}3.5x + 60$ into our calculator and estimate the x-intercept, as shown in Figure 14. We can compute the approximate intercepts by using TRACE in the friendly window Xmin = 0, Xmax = 18.8, Ymin = 0, and Ymax = 62; the x-intercept is about 17, so if our model is correct, the PCBs should have disappeared in 1980 + 17 = 1997.

We can also find the solution algebraically. We set $y = 0$ and solve for x.

$$0 = {}^{-}3.5x + 60$$

$${}^{-}60 = {}^{-}3.5x \qquad \textit{Subtracting 60 on both sides}$$

$$\frac{^{-}60}{^{-}3.5} = \frac{^{-}3.5x}{^{-}3.5}$$ *Dividing on both sides by* $^{-}3.5$

$$17.1 \approx x$$

It will take about 17 years for the PCBs to disappear completely.

If this model were accurate, the EPA would have a very tough time convincing anyone that dredging is necessary. In fact, any traces would be gone in only 17 years. In general, however, linear models are *not* well suited to situations like this. We can see this by looking at the actual PCB concentration in brown bullheads over a longer time period (Figure 15). The straight line works for a little while, but it quickly fails to model the actual data. The real concentration in fish drops slowly toward zero, whereas the linear equation implies that by the year 2000 the PCB concentration would be negative.

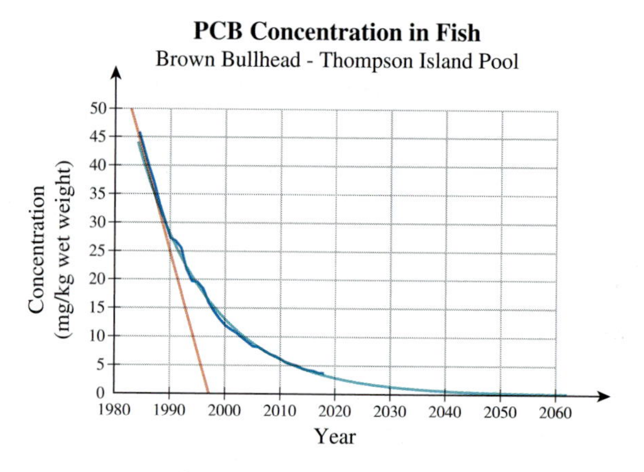

Figure 15 Taken from http://www.epa.gov/hudson.

The following chart summarizes what we have learned about linear equations in this section.

How to Find an Equation of a Line

Given	Here's What to Do	Example
The slope, m, and the y-intercept, b	Substitute into the equation $y = mx + b$.	Example 1 (DVDs), Example 2 (airplane)
One point (x, y) and the slope, m	First substitute for x, y, and m in $y = mx + b$. Solve this equation for b. Then substitute m and b in $y = mx + b$.	Example 4 (piggy bank)
Two points, (x_1, y_1) and (x_2, y_2)	First find the slope $= \frac{\text{rise}}{\text{run}} = \frac{y_2 - y_1}{x_2 - x_1}$. Next substitute for x, y, and m in $y = mx + b$. Solve this equation for b. Then substitute m and b in $y = mx + b$.	Example 5 (PCBs)

How to Graph an Equation of a Line (Using a Calculator)

The equation is in slope-intercept form.	You may enter it directly into the Y= menu.	Examples 1, 2, 4, and 5
The equation is not in slope-intercept form.	First solve for y. Then enter the equation into the Y= menu.	Example 3 (dam keeper)

Exercises 4.2

1. Write each equation in the alternative form.

Slope-intercept Form	Standard Form
$y = 8x - 10$	?
?	$x + y = 7$
?	$6x - 2y = 7$

2. Find the slope and y-intercept for each of the equations.
   a. $y = .36x - .1$ **b.** $y = 4x + 9$
   c. $y = \frac{1}{3}x - 5$ **d.** $x + 4y = 7$
3. Consider the equation $y = {}^{-}3x + 5$.
   a. Is this equation in slope-intercept form or standard form?
   b. What is the slope, or rate of change?
   c. What is the y-intercept, or initial value?
   d. What is the x-intercept?
4. Given the graph shown:

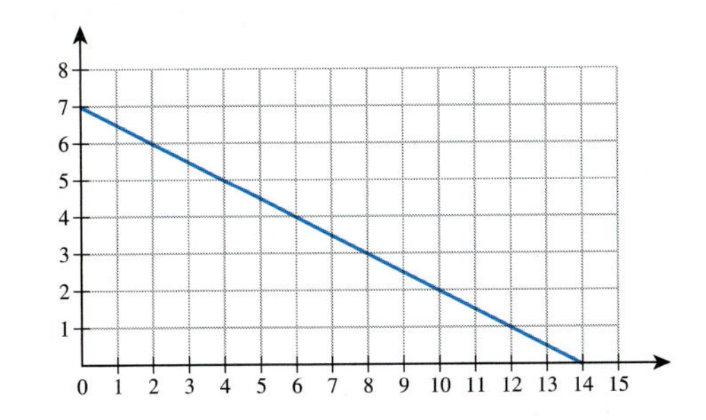

   a. Label the x- and y-intercepts on the graph.
   b. Use the x- and y-intercepts to find the slope.
   c. Use the slope and y-intercept to write an equation for the line.
5. Given the following table of values:

x	y
0	$^{-}4$
1	$^{-}2$
2	0

   a. What are the x- and y-intercepts?
   b. Use the intercepts to draw a graph by hand.
   c. Find the slope.
   d. Use the slope and y-intercept to write an equation.
6. Given the equation $4x + 5y = 20$:
   a. Find the x- and y-intercepts.
   b. Use the intercepts to draw a graph by hand.
7. Write an equation of the line with given slope and y-intercept.
   a. $m = \frac{3}{5}$, y-intercept $(0, 4)$
   b. $m = 2$, y-intercept $(0, {}^{-}1)$
   c. $m = 0$, y-intercept $(0, 7)$
   d. $m = \frac{^{-}1}{2}$, y-intercept $(0, 0)$
8. Write an equation of the line with given slope and containing the given point.
   a. $m = 2$, $(5, 1)$ **b.** $m = \frac{^{-}1}{3}$, $(6, 7)$
   c. $m = {}^{-}4$, $({}^{-}3, 5)$ **d.** $m = 0$, $(1, 2)$
9. Write an equation of the line containing the given points.
   a. $(1, 3)$ and $(2, 2)$ **b.** $({}^{-}2, 1)$ and $(3, 6)$
   c. $(4, {}^{-}5)$ and $(0, {}^{-}3)$ **d.** $(3, 4)$ and $({}^{-}1, 4)$
10. At a local bank, the monthly service fee on a checking account depends on the number of checks written. The fees are \$4.95 plus 10 cents for every check written.

a. What is the rate of change?

b. What is the initial value, or y-intercept?

c. Write an equation.

d. Draw a graph and label the information from parts (a) and (b).

11. A parachutist opens her chute at 3500 feet above the ground and falls at a rate of 22 feet/second after that. We wish to find a formula that models the parachutist's height in feet above the ground in terms of the time in seconds after she opens her chute. Let x be the time in seconds since she opened her chute and let y be her height in feet above the ground.

a. What is the value of y when $x = 0$?

b. If the value of x increases by 1, what will happen to the value of y?

c. Use the information in parts (a) and (b) to write a formula for the height, y, in terms of the time since the chute was opened, x.

12. Now we will model the situation from Exercise 11 with a table of values and a graph.

a. Make a table of values with inputs of 0, 40, 80, 120, and 160 seconds.

b. Draw a graph using a friendly window of Xmin = 0, Xmax = 188, Ymin = 0, Ymax = 3600.

c. Trace the graph to find an approximation for the x- and y-intercepts.

d. Find the exact x- and y-intercepts.

e. Interpret the x- and y-intercepts in relation to the real-world problem.

f. Find and interpret the slope.

13. Suppose you want to know the low temperature at a vacation spot. At 10 P.M. you hear the temperature is 75°F. Assume the temperature drops at a steady rate of 1.5°F each hour from the high at 2 P.M. until the low at 4 A.M. We can represent this problem with a table of values, an equation, and a graph.

a. Complete the table to find the low temperature.

Hour of Day	Elapsed Hours Since 2 P.M.	Temperature (°F)
10 P.M.	8	75
11 P.M.	9	73.5
12 A.M.	10	72
1 A.M.	?	?
2 A.M.	?	?
3 A.M.	?	?
4 A.M.	?	?

b. Use the rate of change and one of the ordered pairs from the table to write an equation for the temperature in terms of the hours since 2 P.M.

c. Find a friendly window to graph the equation. Trace the graph to approximate the low temperature. *Note:* The Y= menu requires that your equation be solved for y.

d. Because the low temperature occurs 14 hours after the high, let $x = 14$ hours and solve the equation from part (b) for y. This will give an exact low temperature.

14. Recall that the basal metabolic rate (BMR) determines the number of calories per day a person needs while at rest. The table shows BMR values for a 22-year-old male of average height. [Harris-Benedict equation (http://www.weight-free-lifestyles.com)]

Weight in Pounds	BMR
100	1412
120	1536
140	1660
160	1784
180	1908

a. Find the rate of change, or slope.

b. Write an equation for BMR in terms of weight.

c. Find the BMR for a 150-pound male.

15. Paul and Joan have a house-cleaning business. Paul can complete one job in 3 hours and Joan can complete one job in 2 hours. Let x represent the number of hours Paul works and y represent the number of hours Joan works.

a. Write an expression for the number of jobs Paul completes in terms of x.

b. Write an expression for the number of jobs Joan completes in terms of y.

c. Write an equation that shows how x and y are related if they complete a total of 6 jobs.

d. Write the equation in slope-intercept form.

e. Find a friendly window and make a graph.

f. Interpret the x- and y-intercepts.

g. Interpret the slope.

Skills and Review 4.2

16. Find m in the formula for the slope of a line,

$$m = \frac{y_2 - y_1}{x_2 - x_1},$$

when $x_1 = 5$, $x_2 = {}^-2$, $y_1 = {}^-1$, and $y_2 = 4$.

17. Shown here are the graph of a line and two points.

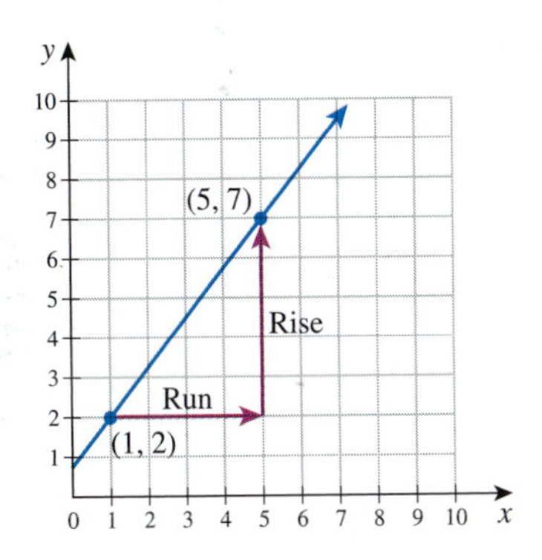

a. Find the run and rise.

b. Find the slope of the line.

c. The slope of a line is the steepness of the line connecting the two points, whereas the distance is the length of the line segment connecting the two points. Find the distance between the two points using the formula:

$$\text{distance} = \sqrt{\text{run}^2 + \text{rise}^2},$$

or

$$\sqrt{(x_2 - x_1)^2 + (y_2 - y_1)^2}.$$

18. Suppose two people start from the same place and travel in opposite directions. One person travels at 50 miles/hour. The other person travels at 60 miles/hour. How long will it be before they are 385 miles apart? Outline your work using the seven steps for problem solving.

19. Solve for x if $1 - \frac{1}{3}(x + 9) = 4(5 - \frac{1}{4}x)$.

20. A formula for the area of a trapezoid is $A = (\frac{1}{2})(b_1 + b_2)h$.

a. Find b_1 if $A = 13 \text{ cm}^2$, $b_2 = 7$ cm, and $h = 2$ cm.

b. Evaluate the expression for the values of b_1, b_2, and h to check the accuracy of your work in part (a).

21. Find the length of the diagonal of the rectangle in the figure.

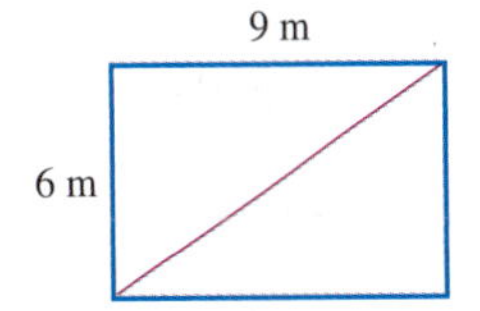

a. Give an approximate answer.

b. Give an exact answer.

22. Evaluate ${}^-\sqrt{b^2 - 4ac}$ when $a = 3$, $b = {}^-4$, and $c = {}^-1$.

23. Use the distributive property (Section 2.1) to factor $(4 * 83) - (4 * 3)$. Simplify the result mentally using the order of operations.

24. Mentally evaluate

$$\frac{4^2 - 2^4}{3^{-1}}$$

Check your work on a calculator.

25. Mentally evaluate ${}^-24 - {}^-8 - 6 + 22$.

4.3 Graphical Solution of Systems of Equations

AFTER STUDYING THIS SECTION YOU WILL BE ABLE TO

- Solve a system of linear equations in two variables by graphing.
- Identify an inconsistent system of linear equations in two variables.

Systems of Equations

You have now seen several examples of **linear equations in two variables**. We now examine what happens when we have two such equations. First, let's take a closer look at the relationship between an equation and the points that lie on its line.

Consider the graph in Figure 16. Each of the points labeled is on the line and must, therefore, make the equation that corresponds to the line $y = 3x + 5$ true. We can check two: p_1 is the point $(^-1, 2)$, so $x = {}^-1$ and $y = 2$. If we substitute these values for x and y, we get

$$2 \stackrel{?}{=} 3(^-1) + 5$$

$$2 = 2 \qquad \textit{OK!}$$

p_2 is the point $(1.5, 9.5)$, and we can check again:

$$9.5 \stackrel{?}{=} 3(1.5) + 5$$

$$9.5 = 4.5 + 5 \qquad \textit{OK!}$$

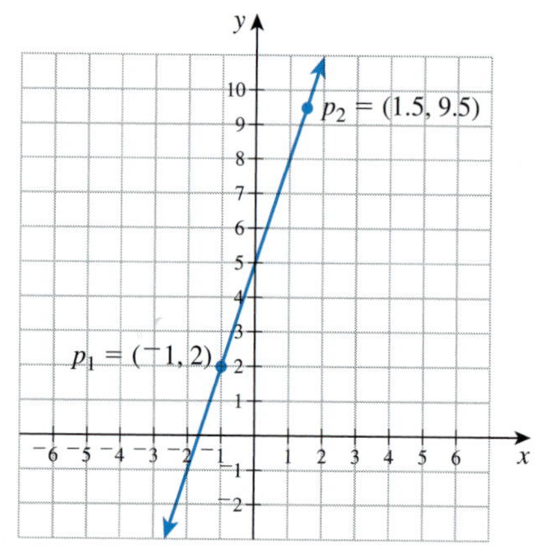

Figure 16 Any point on the line satisfies the equation $y = 3x + 5$.

The same thing will happen with *every* point on the line when we substitute its coordinates in the equation. Consequently, there are an *infinite* number of points that satisfy the given equation.

> **Definition Solution of a Linear Equation in Two Variables**
>
> A **solution** of a linear equation in two variables is an ordered pair of numbers (x, y) that satisfies the equation.

Many real-life problems have lots of solutions. In Example 3 in Section 4.2, there are many pairs of values for x and y that satisfy the equation $y = {}^{-}.8x + 100$. These values give the dam keeper what he wants: a combination of clicks on the gates for the dams that will produce a flow of 1000 gallons each second. The only limitations are that x and y must be integers and neither of them can be negative. For instance, he can't open a dam 4.5 clicks or ${}^{-}3$ clicks. Some examples of pairs that do satisfy the equation are (0, 100), (25, 80), (50, 60), and (125, 0). There are other cases in real life, however, where there is only one solution. Often these cases involve *two* linear equations in two variables. We look for a pair of values (x, y) that satisfy *both* equations.

EXAMPLE 1 Suppose we are given the equations

$$4 = 3x - y$$
$$y = {}^{-}3x + 2$$

Can we find values of x and y that will satisfy *both* equations?

Solution We will write the first equation in slope-intercept form so that we can graph it easily.

$$4 = 3x - y$$
$$4 + y = 3x$$
$$y = 3x - 4$$

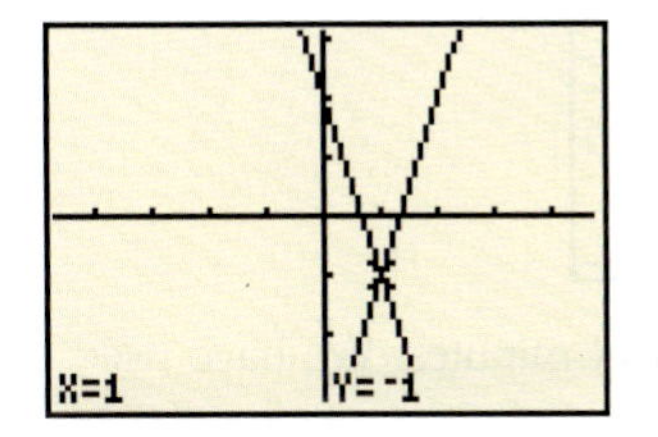

Figure 17 The point on both lines satisfies both equations. We use the zoom decimal window ${}^{-}4.7 \le X \le 4.7$, ${}^{-}3.1 \le Y \le 3.1$.

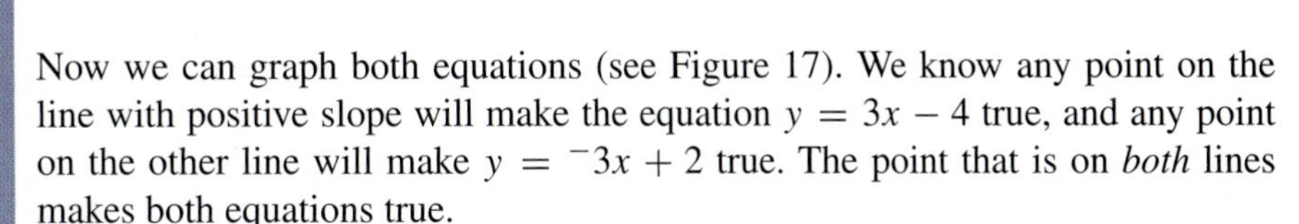

Now we can graph both equations (see Figure 17). We know any point on the line with positive slope will make the equation $y = 3x - 4$ true, and any point on the other line will make $y = {}^{-}3x + 2$ true. The point that is on *both* lines makes both equations true.

From our graph, the solution is the point $(1, {}^{-}1)$.

What makes Example 1 different from our previous examples is that we have *two* different linear equations. Instead of a variety of possible solutions, there is only one solution. It is found at the *intersection point* of the two lines.

Definition System of Linear Equations in Two Variables

Two linear equations in two variables are called **a system of linear equations in two variables**. Any point that is a solution to both equations is a solution to the system of equations.

Because a solution to a system of linear equations is a point that is a solution to *both* equations, it occurs at a point on the lines of both equations in the system.

Finding the Solution to a System of Linear Equations

We can find the solution to a system of linear equations by finding the **intersection point** of the two lines corresponding to the equations in the system (if such a point exists). This can be done either graphically or algebraically.

EXAMPLE 2 Suppose that one long-distance phone company charges 3 cents per minute plus a monthly fee of \$4, and another company charges 6 cents per minute with no monthly fee. How many minutes of phone calls could you make in a month to have the same charge from each company?

Solution We write equations for the cost of long distance calls per month for each company. If we let x be the number of minutes of calls, then the first company charges $y = 3x + 400$, because \$4 is 400 cents. The second company charges $y = 6x$. We can enter these two equations in the calculator as Y1 and Y2. A graph of the two equations, in a first quadrant window, looks like the one in Figure 18.

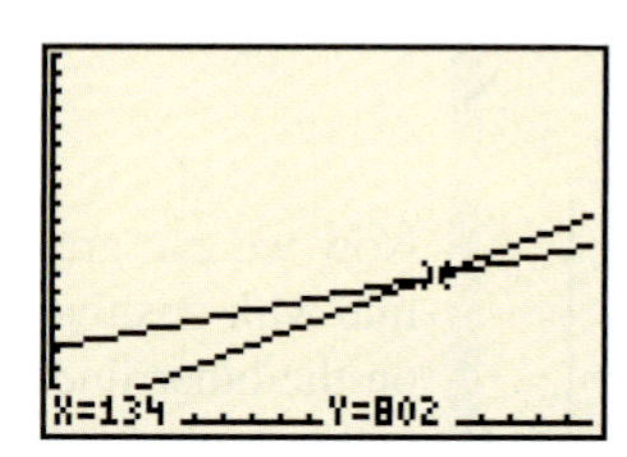

Figure 18 The two lines meet when X is about 134 minutes. We have used the friendly window $0 \leq X \leq 188, 0 \leq Y \leq 2000$.

The intersection point is the solution because at that point the minutes are the same and the cost is the same. We estimate the coordinates by tracing along one of the lines until we come as close as we can to the intersection point (Figure 18). The point seems to be about (134, 802).

We can get a more precise result using a calculator feature called *intersect*. This is how it works on a TI-83. First trace along one of the lines until the cursor is near the intersection point. Then press [2nd] CALC. Scroll down to 5:intersect. Press [ENTER] three times, and the coordinates of the intersection point will appear on the screen: $X \approx 133.33333$, $Y \approx 800$ (Figure 19). We conclude that after about 133 minutes of phone calls, the cost of the two plans is the same. ■

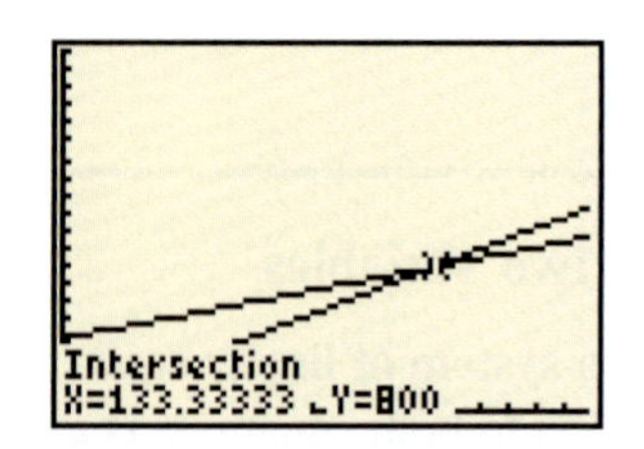

Figure 19 The two lines meet when X is about 133 minutes.

EXAMPLE 3 A hare is going to race a tortoise to prove that slow and steady *doesn't* always win. The hare will be a good sport and give the tortoise a head start of 10 miles. The hare can run at a speed of 18 miles/hour. The tortoise moves along at 2 miles/hour. How long will it take for the hare to catch the tortoise?

Solution This is a distance-rate-time problem. If we let t represent the time (in hours) that they race, the hare travels a distance of $18t$ miles and the tortoise travels only $2t$ miles.

Figure 20 shows the distance covered by each runner and the tortoise's 10-mile head start. Let y represent the distance of each runner from where the hare started. For the hare, this is just how far he runs, so $y = 18t$. For the tortoise, we need to include the 10-mile head start, so $y = 2t + 10$.

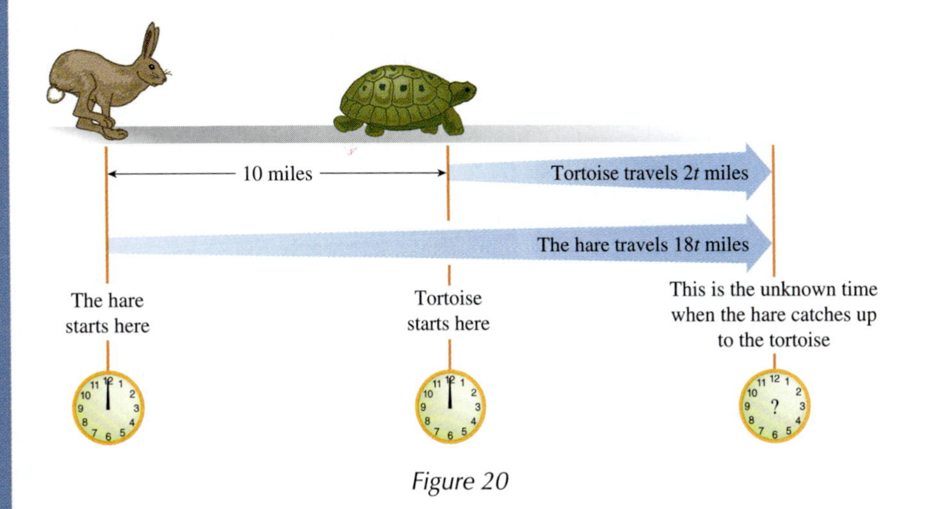

Figure 20

We may graph these equations to find the point of intersection. We'll have to use X in place of t for our calculator, so we have

$$Y1 = 18X$$

and

$$Y2 = 2X + 10$$

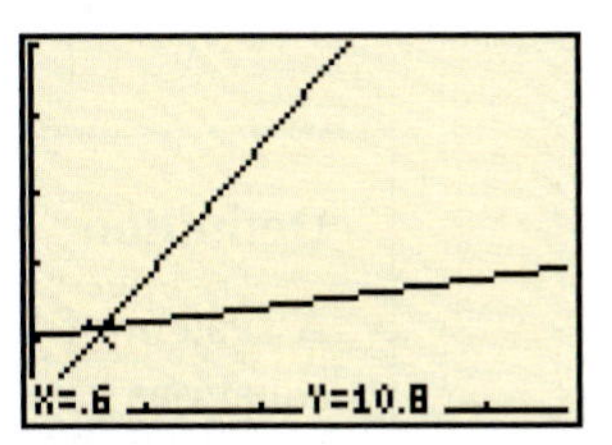

Figure 21 A graph with the hare's and the tortoise's distances versus time in the friendly window $0 \le X \le 4.7, 0 \le Y \le 50$.

From the graph in Figure 21 you can see that the line for the hare has a steeper slope and that he catches up to the tortoise in about .6 hour.

Inconsistent Systems

EXAMPLE 4 The hare is so proud of the fact that he caught up with the tortoise in less than one hour that he challenges a roadrunner to a race. He gives the roadrunner the same 10-mile head start as he gave the tortoise. Little does he realize that the roadrunner can also run at 18 miles/hour. Make a graph showing the result of this race.

Solution By the same reasoning we used in Example 3, we have the equations

$$Y1 = 18X$$

$$Y2 = 18X + 10$$

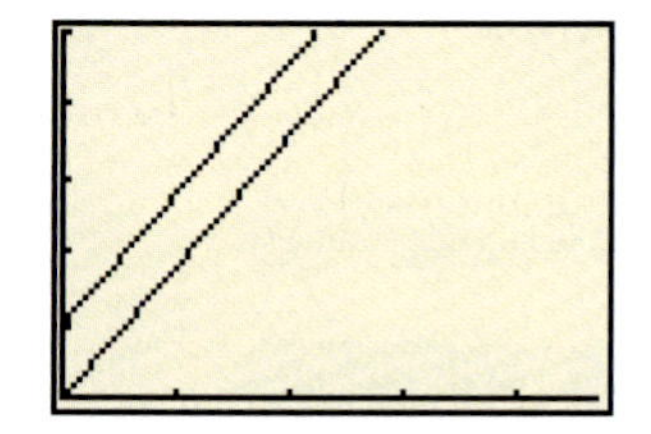

Figure 22 The hare and the roadrunner never meet.

When we graph this system of equations, we get two **parallel lines**, as shown in Figure 22.

It appears that these lines will never meet. The hare can't catch up with the roadrunner since they are both traveling at the same speed. Because there is no intersection point, the system of equations has no solution.

Notice in Example 4 that the two lines have the same slope. This should not be surprising because you know that the slope of a line is the *rate of change* of the variables, and in this case the rate is the same, 18 miles per hour. This observation leads us to an important fact about parallel lines.

Slopes of Parallel Lines

Two different lines with the same slope are parallel.

We were unable to find a solution to the system of equations in Example 4. There is a name for such a system of equations.

Definition Inconsistent System

A system of linear equations in two variables is **inconsistent** if the two equations have no common solution.

Exercises 4.3

1. The figure shows the graphs of two linear equations, with several ordered pairs on each line. What is the intersection of the two lines? Verify that the ordered pair at the intersection of the two lines is a solution to the system of equations: $y = {}^-x$ and $y = x + 2$.

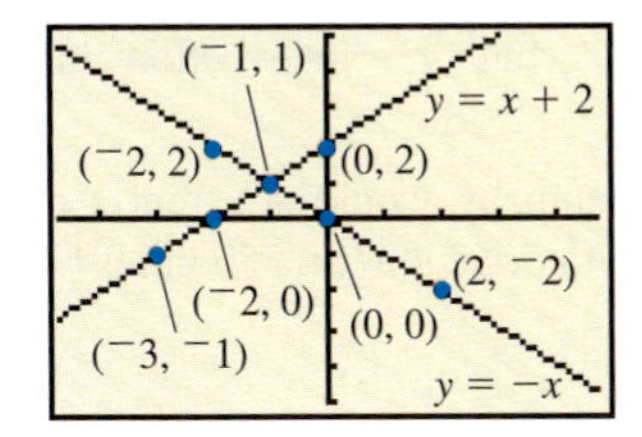

2. The sum of two numbers is 6. The difference of the same two numbers is 4.
   a. Let x = one of the numbers and y = the other number. Write a separate equation for each statement.
   b. From the set $\{1, 2, 3, 4, 5, 6\}$, find pairs of numbers that satisfy each statement.
   c. Pick the pair of numbers that satisfy both statements; this is the solution of the system of equations.

3. The sum of two numbers is 10. The difference of the same two numbers is 5.
   a. List three pairs of numbers that satisfy each statement.
   b. Plot the pairs of numbers that satisfy the first statement on a coordinate plane (Section 2.3) and connect the points.
   c. Plot the pairs of numbers that satisfy the second statement on the same coordinate plane and connect only the new points.
   d. Because one line represents all the pairs of numbers whose sum is 10 and the other line represents all the pairs of numbers whose difference is 5, what does the intersection of the two lines represent?
   e. Use the graph to find the coordinates of the intersection point.

4. If the graph of a system of equations forms parallel lines, how many solutions will the system have? What is the special name for this type of system of equations?

In Exercises 5–12, use a graphing method to find the approximate solution(s), if any, to the given systems of equations. Tell which systems of equations are inconsistent. *Note:* Begin with a ZOOM 8: ZInteger window and adjust if necessary.

5. $y = {}^-x + 1$
   $y = x - 3$

6. ${}^-x + y = {}^-6$
   $2x + 5 = y$
   a. In what form are the original equations?
   b. What equations did you enter into the Y= screen?
   c. In what form are the equations for the Y= screen?

7. $x = y - 1$
   $2 = x - y$
   What do the slopes of these equations tell you about the solution(s)?

8. $x - 2y = 6$
   $y = 5$

9. $3y = 4x - 1$
   $y = 2x - \frac{1}{2}$

10. $x + y = 2$
    ${}^-y = {}^-2x + \frac{1}{4}$

11. $.7x - .1y = 0$
    $.2x + .6y = 4.4$

12. $x + y = \frac{13}{15}$
    $2x - 3y = \frac{11}{15}$
    Is the solution shown on the calculator exact? Explain.

13. The Pinnacle car company brags about its low rental car costs by offering the following cost comparisons against a competitor.

Miles Driven	Pinnacle's Cost	Competitor's Cost
0	$8.00	$20.00
10	10.50	20.50
20	13.00	21.00
30	15.50	21.50
40	18.00	22.00

a. Continue the cost comparison values for 50, 60, and 70 miles driven by noting the change in cost for each additional 10 miles driven. Recall that this is the rate of change discussed in Section 4.1.

b. At what number of miles do the two rental cars have the same cost?

c. On a graph of the data, what is this point called?

14. In the previous problem we found that each car's rental cost changed at a constant rate. Thus, each car's cost may be expressed as a linear equation. For Pinnacle, the cost equation is $C = .25d + 8$, and for their competitor, the equation is $C = .05d + 20$.

a. Approximate the solution of this system of equations by using the TRACE or 2nd CALC 5:intersect feature on your graphing calculator.

b. How does this solution compare with your answer from Exercise 13(b)?

15. Suppose you are considering renting a car from one of two possible companies. Udrive's rental charges may be modeled by the equation $C = .10d + 35$, where d is the miles driven and C is the cost in dollars. Mile Rent-a-Car calculates its charges by the formula $C = .15d + 10$.

a. Graphically estimate the intersection of these two rental-cost equations.

b. Describe what the intersection in this problem means.

c. Which company would you rent a car from if you expected to drive at least 550 miles? Why?

Skills and Review 4.3

16. Write an equation of a line from the given information.

a. Slope of $^{-}\frac{3}{4}$, y-intercept of (0, 5)

b. Slope of $\frac{1}{2}$, containing the point $(4, {}^{-}1)$

c. Containing the points (5, 2) and $(0, {}^{-}1)$

17. Refer to the table of values for Pinnacle Rent-a-Car presented in Exercise 13.

a. Explain how you could find the y-intercept (initial value) for Pinnacle Rent-a-Car.

b. Show the calculation for finding the slope, or rate of change, for Pinnacle.

18. Write an equation of a line containing the points $(0, {}^{-}2)$ and $({}^{-}1, 4)$.

a. Write your equation in slope-intercept form.

b. Write your equation in standard form.

19. Find the slope and y-intercept for each line.

a. $y = {}^{-}3x + 2$

b. $3x - 2y = 6$

20. Consider the equation $2x + 3y = 18$.

a. Find the x-intercept.

b. Find the y-intercept.

c. By hand, draw a graph of the equation.

21. The traffic on a highway decreases constantly between the hours of 8 P.M. and midnight. Use the data in the table to find the rate of change in the number of vehicles per hour.

Time	Number of Vehicles
8	5000
9	4250
10	3500
11	2750
12	2000

22. Determine whether the line in each graph has a positive, negative, zero, or undefined slope.

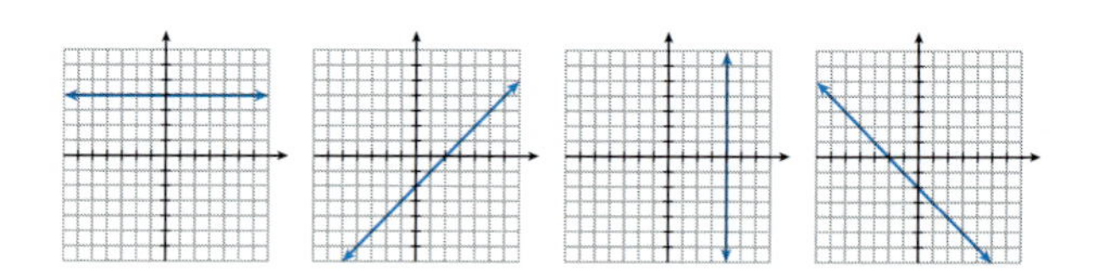

23. Find W in the formula for the perimeter of a rectangle, $P = 2L + 2W$, when $L = 4.5$ cm and $P = 15$ cm.

24. Recall from Section 2.4 that a function may be represented in four ways. Consider the problem situation where the number of words Ramona reads is a function of the time she is reading. Suppose she reads 300 words per minute. Express this function in each way.

a. As a table

b. As a graph

c. As a formula

25. Consider the expression $3x + 6 - (8x - 4)$.

a. Simplify the expression.

b. Evaluate the original expression for $x = {}^{-}3$.

c. Evaluate your simplified expression for $x = {}^{-}3$.

d. Do the evaluations agree? If not, go back and check your work.

4.4 Algebraic Solution of Systems of Equations

AFTER STUDYING THIS SECTION YOU WILL BE ABLE TO

- Solve a system of linear equations in two variables by the substitution method.
- Solve a system of linear equations in two variables by the elimination method.

Solving by Substitution

In the last section you used graphs to solve systems of linear equations. In this section you learn methods that do not require a graph. One method for solving a system of equations algebraically is called the **method of substitution**.

The Method of Substitution

Step 1 Put one of the equations in slope-intercept form, $y = mx + b$.

Step 2 Substitute the expression $mx + b$ for y in the other equation.

Step 3 Use algebraic manipulation to solve the new equation for x.

Step 4 Substitute this value of x back into the first equation to find the value of y.

Example 1 illustrates this method.

EXAMPLE 1 Show how the method of substitution may be used to solve the problem of the hare and the tortoise (Example 3, Section 4.3). A hare is going to race a tortoise to prove that slow and steady *doesn't* always win. He will be a good sport and give the tortoise a head start of 10 miles. The hare can run at a speed of 18 miles per hour. The tortoise moves along at 2 miles per hour. How long will it take for the hare to catch the tortoise?

Solution As we did before, we let y represent the distance from where the hare started and x be the time in hours until the hare catches up. We have a system of two equations:

$$y = 18x$$
$$y = 2x + 10$$

The first step is to put one of the equations in slope-intercept form, which is not necessary because both equations are already in that form.

The second step is to substitute the expression $mx + b$ from one equation into y for the other equation. If we substitute $2x + 10$ from the second equation for y in the first equation, we have

$$2x + 10 = 18x$$

The third step is to solve this equation for x:

$$\begin{aligned} 2x + 10 &= 18x \\ 10 &= 16x \\ \frac{10}{16} &= \frac{16x}{16} \\ .625 &= x \end{aligned}$$

Our solution is $x = .625$, which means that the hare overtakes the tortoise after .625 hours (which is $.625 * 60 = 37.5$ minutes). Note that this solution is more precise than the estimate we got from the graph in Example 3 in Section 4.3.

We could substitute $x = .625$ back into the equation $y = 2x + 10$ to determine how far the hare will have traveled when he catches the tortoise. We get $y = 2 * .625 + 10 = 11.25$, which means the hare will have gone 11.25 miles. The ordered pair (.625, 11.25), or $x = .625$, $y = 11.25$ is the solution to the system of linear equations, but the problem asked only for how long the hare took, not how far he ran.

We can always use the method of substitution to solve a system of linear equations algebraically. It is easiest if at least one of our two equations is already solved for a particular variable, for instance, when it is in slope-intercept form.

EXAMPLE 2 Solve the following system of equations using the method of substitution.

$$y = 5x - 3$$
$$2x + y = 4$$

Solution Because the equation $y = 5x - 3$ tells us the value of y in terms of x, we may substitute this into the equation $2x + y = 4$ and remove the y, leaving a linear equation in one variable.

$$2x + y = 4$$

$2x + (\mathbf{5x - 3}) = 4$ — *Putting $5x - 3$ in parentheses to indicate that the entire expression is substituted for y*

$7x - 3 = 4$ — *Removing parentheses (special property of 1) and combining like terms*

$$7x = 7$$

$$\mathbf{x = 1}$$

Once we know the value of one variable, we can *substitute* again into the equation $y = 5x - 3$ to get the value of y:

$$\begin{aligned} y &= 5\mathbf{x} - 3 \\ &= 5 * \mathbf{1} - 3 \\ &= 2 \end{aligned}$$

So, the solution is the ordered pair $(1, 2)$ because $x = 1$ and $y = 2$.

If neither equation is in slope-intercept form, then we must follow the first step in the list on page 155.

EXAMPLE 3 Solve the following system of equations using the method of substitution.

$$\begin{aligned} 3x + y &= 3 \\ 2x - 2y &= 10 \end{aligned}$$

Solution Neither equation is in slope-intercept form, so we must manipulate one equation until it is in that form and then do substitution. We can manipulate the equation $3x + y = 3$ like this:

$$\begin{aligned} 3x + y &= 3 \\ \mathbf{y} &= \mathbf{{}^{-}3x + 3} \end{aligned}$$

Now, we can substitute for y in the equation $2x - 2y = 10$:

$$2x - 2y = 10$$

$2x - 2(\mathbf{{}^{-}3x + 3}) = 10$ — *Don't forget the parentheses*

Once we have reduced our system of equations in two variables to a single equation in one variable, we can solve using the steps we learned in Chapter 3:

$2x - 2(\mathbf{{}^{-}3x + 3}) = 10$ — *Original equation*

$2x + 6x - 6 = 10$ — *Using the distributive property to remove parentheses on left side*

$$8x - 6 = 10 \quad \textit{Combining like terms on left side}$$

$$8x = 16 \quad \textit{Adding 6 on both sides}$$

$$x = \frac{16}{8} \quad \textit{Dividing on both sides by 8}$$

$$\mathbf{x = 2}$$

Next, we substitute $\mathbf{x} = \mathbf{2}$ into the equation $y = {}^{-}3x + 3$ to compute the value of y,

$$y = {}^{-}3\mathbf{x} + 3$$

$$y = {}^{-}3 * \mathbf{2} + 3$$

$$y = {}^{-}6 + 3$$

$$y = {}^{-}3$$

The solution to this system is $x = 2$ and $y = {}^{-}3$.

Solving by Elimination

Example 4 shows how to solve a system of equations when neither is in slope-intercept form.

EXAMPLE 4 Solve the following system of linear equations.

$$2x + 3y = 6$$

$$5x - 3y = 8$$

Solution We could solve it as we did the system in Example 3, but this would involve a series of steps: First, write one of the equations in slope-intercept form; next, substitute the resulting expression for one variable into the other equation; and then solve the new equation. There is an easier way.

We know that we may add the same quantity to both sides of an equation and still have a valid equation. The equation $5x - 3y = 8$ tells us that $5x - 3y$ and 8 are the *same quantity*. So we may add $5x - 3y$ to the left side of the equation $2x + 3y = 6$ and 8 to the right side of the equation $2x + 3y = 6$.

$$\begin{array}{r} 2x + 3y = 6 \\ +\ 5x - 3y = 8 \\ \hline 7x = 14 \end{array}$$

The sum of ${}^{+}3y$ and ${}^{-}3y$ is 0, so we get a simple linear equation in one variable, $7x = 14$. Then $\mathbf{x} = \frac{\mathbf{14}}{\mathbf{7}} = \mathbf{2}$. We can substitute this value of x into either equation to find the value of y. We use the equation $2x + 3y = 6$:

$$2x + 3y = 6$$
$$2 * 2 + 3y = 6$$
$$4 + 3y = 6$$
$$3y = 2$$
$$y = \frac{2}{3}$$

The solution is $x = 2$ and $y = \frac{2}{3}$, or $(2, \frac{2}{3})$.

We call this method of solving a system of linear equations the **elimination method**, because we *eliminate* one of the variables. We were able to use the elimination method to easily solve the system of equations in Example 4 because $3x$ and ^-3x canceled out, *eliminating* the variable x. We can use the elimination method to solve other systems.

First we need to arrange the equations so that x-terms, y-terms, and constant terms line up with each other. This is easy if both equations are in standard form (see page 139).

Then, we need to make sure that the coefficient of one of the variables in the first equation is the *opposite* of the coefficient of the same variable in the second equation; when we add the corresponding sides of the two equations, that variable will then be eliminated.

Method of Elimination

Step 1 Put both equations in standard form, $Ax + By = C$.

Step 2 If necessary, multiply one or both equations by the same number on both sides so either the x-terms or the y-terms are opposites.

Step 3 Add the two equations to get an equation with only one variable. Solve for that variable.

Step 4 Substitute this value of x or y back into one of the original equations to find the value of the other variable.

EXAMPLE 5 Solve the system of equations:

$$2x + 4y = 32$$
$$3x + 4y = 28$$

Solution We need to find a way to eliminate one of the variables from the system. Because the coefficient of y in both equations is 4, if we multiply by $^-1$ on both sides of one equation, then the coefficients of y will be opposites; we can *add the equations* (left side to left side, and right side to right side) like this:

$$\begin{array}{rl} 2x + 4y = 32 & \\ {}^{-}3x - 4y = {}^{-}28 & \textit{Multiplying on both sides of the second equation by } {}^{-}1 \\ \hline {}^{-}1x = 4 & \textit{Adding the equations} \\ x = {}^{-}4 & \textit{Multiplying by } {}^{-}1 \textit{ on both sides} \end{array}$$

Now, we can substitute $\mathbf{x = {}^{-}4}$ back into the equation $2x + 4y = 32$ to solve for y:

$$2x + 4y = 32$$
$$2 * \mathbf{{}^{-}4} + 4y = 32$$
$${}^{-}8 + 4y = 32$$
$$4y = 40$$
$$y = 10$$

The solution is $x = {}^{-}4$ and $y = 10$, or $({}^{-}4, 10)$.

Example 6 shows another example of using the elimination method.

EXAMPLE 6 You decide to enter the Iditarod sled dog race in Alaska. After extensive research into assembling a dog team, you decide to use two different types of dog. One type is a dog with very good endurance but who can only pull 55 pounds. The second type is a very strong dog who can pull 85 pounds, but has less endurance than the first type of dog. You will need 15 dogs for the team, and the team should be able to pull 975 pounds.

Solution In order to figure out how many of each type of dog you will use, we may let x be the number of type 1 dogs and y the number of type 2 dogs. Because we need 15 dogs total,

$$x + y = 15$$

The other information in the problem is the number of pounds each type of dog can pull and the total number of pounds you need the team to pull. Because each type 1 dog can pull 55 pounds, x type 1 dogs can pull $55x$ pounds. Likewise, y type 2 dogs can pull $85y$ pounds. You need the team to pull 975 pounds, so

$$55x + 85y = 975$$

We now have a system of two equations in x and y. Both equations are in standard form, but neither the x- nor the y-coefficients are equal. Not to worry. We could multiply both sides of the first equation by 55 so that its x-term has the same coefficient as the x-term in the second equation. Better yet, if we multiply by $^{-}55$, the x-terms will be *opposites*, and we will be all set to use the method of elimination. So our first equation becomes

$$^{-}55x - 55y = {}^{-}55 * 15$$
$$^{-}55x - 55y = {}^{-}825$$

We now have

$$\begin{array}{r} ^{-}55x - 55y = {}^{-}825 \\ 55x + 85y = 975 \\ \hline 30y = 150 \end{array}$$

The x-terms have been eliminated. We now solve for y:

$$\frac{30y}{30} = \frac{150}{30}$$
$$y = 5$$

Now that we have y, we can substitute into one of the equations and solve for x. In this case our best choice is $x + y = 15$, where we quickly find that $x = 10$.

You will need 10 type 1 dogs and 5 type 2 dogs for your Iditarod team.

In using the method of elimination, it is sometimes necessary to multiply *both* equations in order to eliminate one of the variables.

EXAMPLE 7 Solve the system of equations

$$4x + 3y = 29$$
$$3x - 2y = {}^{-}8$$

Solution In order to eliminate the y-terms we must make their coefficients opposites of each other. One way to do this is to find the least common multiple of 2 and 3, that is, 6. We can get $6y$ to appear in the first equation if we multiply by 2 on both sides. We can get $^{-}6y$ to appear in the second equation if we multiply by 3 on both sides.

We now have

$$\begin{array}{rr} 4x + 3y = 29 \rightarrow & 8x + 6y = 58 \\ 3x - 2y = {}^{-}8 \rightarrow & 9x - 6y = {}^{-}24 \\ \hline & 17x = 34 \\ & x = 2 \end{array}$$

Once we have found $\mathbf{x = 2}$, we substitute into one of the original equations to find y.

$$4 * 2 + 3y = 29$$
$$8 + 3y = 29$$
$$3y = 21$$
$$y = 7$$

Check by substituting $x = 2$ and $y = 7$ into the second equation:

$$3x - 2y = {}^{-}8$$
$$3 * 2 - 2 * 7 \stackrel{?}{=} {}^{-}8$$
$$6 - 14 \stackrel{?}{=} {}^{-}8$$
$${}^{-}8 = {}^{-}8 \qquad \textit{OK!}$$

Choosing a Method

You have learned three methods for solving systems of two linear equations in two variables. Often it will be up to you to choose which method to use. The chart in the following table gives you some guidance.

Method	Form of Equations	Advantages	Disadvantages
Graphing	Both equations must be in the form $y = mx + b$.	It quickly gives an estimate without much algebraic manipulation.	It may not give exact answers.
Substitution (an algebraic method)	At least one equation should be solved for one variable in terms of the other, as in $y = mx + b$.	It gives an equation in one variable very quickly.	It may become messy if the answers are fractions.
Elimination (an algebraic method)	Both equations should be in the form $Ax + By = C$.	If both equations are in standard form to begin with, the solution may be found in fewer steps than with substitution.	You need to think carefully about how to eliminate one variable.

Exercises 4.4

1. Which of the following are complete solutions to the given system of equations? *Note:* there may be more than one answer.

$$x - 5y = 16 \qquad 2x + y = {}^{-}1$$

a. $x = 1$

b. $(11, {}^{-}1)$

c. $x = 1$ and $y = {}^{-}3$

d. $(1, {}^{-}3)$

e. $y = {}^{-}3$

2. Verify by hand that $(^-2, 5)$ is a solution to the following system.

$$\frac{1}{2}x + \frac{1}{5}y = 0$$
$$y = {}^-3x - 1$$

3. Given is a system of equations and a partial solution to them:

$3x + 2y = 3$

$4x + y = {}^-1$

$y = {}^-4x - 1$ *Solving the second equation for y*

$3x + 2({}^-4x - 1) = 3$ *Substituting $^-4x - 1$ for y into the equation $3x + 2y = 3$*

$3x - 8x - 2 = 3$ *Distributing to clear parentheses*

${}^-5x = 5$ *Solving for x*

$x = {}^-1$

a. Explain why this system is not yet completely solved.

b. Complete the solution to the system and show it as an ordered pair.

In Exercises 4–6, use substitution to solve each system of equations.

4. $y = 2x - 7$
$y = {}^-x + 8$

5. $3y = x + \frac{6}{5}$
$4x + 15y = {}^-12$

6. $x + 2y = 6$
$y = \frac{3}{2}$

7. A mistake was made using elimination to solve the given system of equations.

$$4x - 3y = 22 \rightarrow 4x - 3y = 22 \rightarrow 4x - 3y = 22$$
$$3x - y = 14 \rightarrow 3(3x - y) = 3(14) \rightarrow 9x - 3y = 42$$
$$13x = 64$$
$$x = \frac{64}{13}$$
$$3\left(\frac{64}{13}\right) - y = 14$$
$$\frac{192}{13} - y = 14$$
$$y = \frac{10}{13}$$

Incorrect solution: $\left(\frac{64}{13}, \frac{10}{13}\right)$

a. Locate the error.

b. Correct the error and continue the process of elimination to find the correct answer.

In Exercises 8–10, use elimination to solve each system of equations.

8. $x + y = {}^-3$
$x - y = {}^-5$

9. $2x + 5y = 24$
$x - 3y = 1$

10. $7x - 3y = 13$
$5x - 2y = 27$

In Exercises 11–12, suppose you are equally comfortable with substitution or elimination to solve a system of two equations.

11. a. Explain which method requires fewer steps to solve the system:

$$7x - 3y = 4$$
$${}^-2x - 3y = 4$$

b. Use the method you chose to solve the system.

12. a. Explain which of the two algebraic methods requires fewer steps to solve the system:

$$y = 5x + 1$$
$$y = x - 5$$

b. Use the method you chose to solve the system.

In Exercises 13–15, solve each system of equations using either substitution or elimination.

13. $x + 2y = 4$
${}^-x = y + 3$

Check by substituting your solution in both equations.

14. $16x + 12y = 15$
$12x - 10y = 16$

15. $y = 2x - 1$
${}^-4x + 2y = {}^-5$

Your algebra probably gave you a strange result. Use your grapher to determine what kind of solution this system has.

Skills and Review 4.4

16. Use a graph to approximate the solution(s) to the system of equations.

$$x + 5y = {}^{-}13$$
$$y = 3x - 9$$

17. Use a graphing method to approximate the solution to the system of equations:

$$y - 5.3 = .21x$$
$$.72x + y = 6.4$$

18. Write an equation of a line from the given information:

a. Slope of $\frac{5}{2}$ and y-intercept of $(0, {}^{-}1)$

b. Slope of 3 and contains the point $({}^{-}4, 5)$

c. Contains the points $({}^{-}2, 4)$ and $(5, {}^{-}1)$

19. Determine the x- and y-intercepts of the linear equation $2x - 3y = 6$. Use the intercepts to sketch a graph by hand.

20. Describe the slope of each linear equation as positive, negative, zero, or undefined. Explain how you arrived at your conclusions.

a. $y = {}^{-}3$ **b.** $y = 2x$ **c.** $y = {}^{-}x + 5$

21. Which road is steeper, a road with constant slope $^{-}5$ or a road with constant slope 3? Explain your answer.

22. Company A charges \$40 plus \$.10 for each mile to rent an economy car for the day. Company B charges \$30 plus \$.15 for each mile to rent the same type of car. Use a graphing method to approximate the number of miles at which the cars cost the same. Use the seven steps for problem solving.

23. Recall from Section 2.3 that the distance formula is

$$\text{distance} = \sqrt{\text{run}^2 + \text{rise}^2}$$

a. Find the distance between the two points $({}^{-}4, 3)$ and $(2, {}^{-}1)$. Leave the radical so that your answer is exact.

b. Estimate the square root from part (a) (see Section 1.3).

24. Evaluate:

$$\left(\frac{x-5}{6-y}\right)^2 \quad \text{when} \quad x = 7, y = {}^{-}1.$$

25. If necessary, review exponents so that you can find the value of each expression without a calculator.

a. 2^1 **b.** 2^0 **c.** 2^{-1}

d. $({}^{-}4)^1$ **e.** $1.5 * 10^{-1}$

4.5 Relationships Between Lines (Optional)

In Example 4 in Section 4.3, we saw that two lines with the same slope are parallel and never meet. Any system of linear equations that consists of two *different* lines with the same slope is an *inconsistent system* (page 152) and has no solution. Because the hare gives the roadrunner a head start and they move at the same speed, the hare *can't* catch the roadrunner. There is one type of system with lines with the same slope that is different, however; we investigate it in the next example.

EXAMPLE 1 Suppose the hare realizes that he can't ever catch the roadrunner with the race setup in Example 4, page 152. Instead, the hare proposes giving the roadrunner an 18-mile head start but allows himself to start 1 hour ahead of the roadrunner. What happens in this race?

Solution We can use the same reasoning we did in Example 3 in Section 4.3 to work out formulas for the distance each racer is from the starting line. If we let x be the number of hours since the hare starts running, then the hare is $y = 18x$ miles from the starting line after x hours. Because the roadrunner doesn't start running for an hour, he will be running for only $x - 1$ hours, but he gets an 18-mile head start. Therefore, his distance from the starting line is $y = 18(x - 1) + 18$ miles. We have a system of equations

$$y = 18x$$
$$y = \mathbf{18}(x - 1) + 18 = \mathbf{18}x - \mathbf{18} + 18 = 18x$$

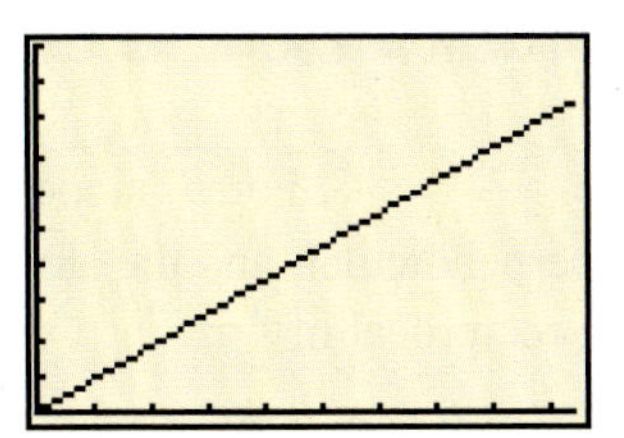

Figure 23

Both equations are the *same*. If we try to plot this, we get Figure 23. What happened? After the hare has been running for 1 hour, he reaches the roadrunner. Then, the roadrunner starts running at the same speed as the hare, and they stay right next to each other. In this system, instead of having two lines with the same slope that will never meet, we have two copies of the *same line*. In this case we say the lines are **coincident** and the system of equations is **dependent**. There are an infinite number of points that satisfy both equations.

Another special example of pairs of lines are lines that are **perpendicular**—in other words, lines that meet at a right (90°) angle. Parallel lines are easy to identify, because they are lines with the same slope (which are not the same line—the lines in Example 1 are really the same and not parallel). There is an easy way to tell that lines are perpendicular by using the slope.

EXAMPLE 2 Find a relationship between the slopes of the legs of the right triangle in Figure 24, which meet at a 90° angle.

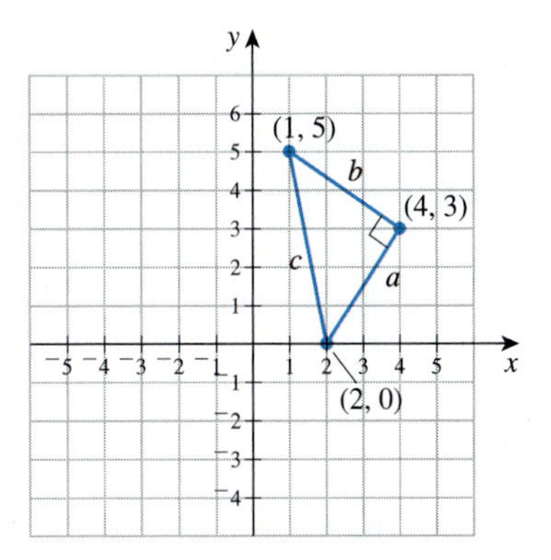

Figure 24

Solution First, let's check to make sure this really is a right triangle. We will use the distance formula (page 71) to find the lengths of the sides, a, b, and c.

$$a = \sqrt{(4-2)^2 + (3-0)^2} = \sqrt{2^2 + 3^2} = \sqrt{13}$$
$$b = \sqrt{(4-1)^2 + (3-5)^2} = \sqrt{3^2 + (^-2)^2} = \sqrt{13}$$
$$c = \sqrt{(2-1)^2 + (0-5)^2} = \sqrt{1^2 + (^-5)^2} = \sqrt{26}$$

Now, let's see if these lengths work in the **Pythagorean theorem** (page 61).

$$a^2 + b^2 \stackrel{?}{=} c^2$$
$$(\sqrt{13})^2 + (\sqrt{13})^2 = 13 + 13 = 26 = (\sqrt{26})^2 \quad \textit{OK!}$$

Recall that slope $= \frac{\text{rise}}{\text{run}}$, so for a we get slope $= \frac{3-0}{4-2} = \frac{3}{2}$. For b, slope $= \frac{3-5}{4-1} = \frac{^-2}{3}$. In this case, the two line segments are perpendicular, and the slopes are *negative reciprocals of each other*. The negative reciprocal of $\frac{3}{2}$ is

$$\frac{^-1}{\frac{3}{2}} = \frac{^-2}{3}$$

This relationship between the slopes of perpendicular lines is always true.

Slopes of Perpendicular Lines

Two lines with slopes m_1 and m_2 are *perpendicular* if their slopes are *negative reciprocals* of each other, in other words, if

$$m_1 = \frac{^-1}{m_2}$$

EXAMPLE 3 Which pairs of lines are perpendicular?

a. $y = 3x - 5$ and $y = {^-3}x + 2$
b. $y = 5x + 2$ and $y = {^-\frac{1}{5}}x + 17$
c. Street 2 and Avenue $^-3$ in Gridville (Section 2.3)

Solution **a.** The two slopes are $m_1 = 3$ and $m_2 = {^-3}$. Taking the negative reciprocal,

$$\frac{^-1}{m_1} = \frac{^-1}{3} \neq m_2$$

so the lines are *not* perpendicular.

b. The two slopes are $m_1 = 5$ and $m_2 = {}^{-}\frac{1}{5}$. Taking the negative reciprocal,

$$\frac{^{-}1}{m_1} = \frac{^{-}1}{5} = m_2$$

so the lines *are* perpendicular.

c. Street 2 is parallel to the x-axis, and the equation for it is $y = 2$, so it has slope $m_1 = 0$. Avenue ${}^{-}3$ has *undefined slope*, because every point on the avenue has the same x-coordinate. Hence the *run* is always 0, and we are not allowed to divide by 0. If we try to take the negative reciprocal of m_1, we get

$$\frac{^{-}1}{m_1} = \frac{^{-}1}{0} = \text{undefined}$$

which agrees with m_2, so the lines *are* perpendicular.

It is important to realize that checking whether lines are perpendicular is very difficult on a graphing calculator, because the *shape* of the window matters. Only *square* windows (where the pixel height of the window and the pixel width are the same) work well; other windows distort the picture (see Figure 25).

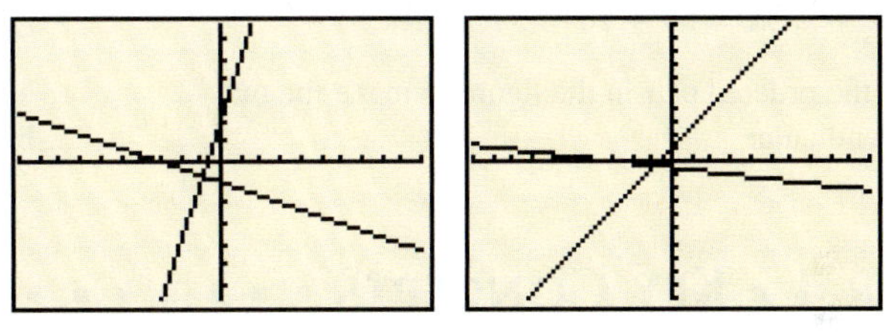

Figure 25 The lines $y = 3x + 2$ and $y = {}^{-}\frac{1}{3}x - 1$, which are perpendicular, in a square window (left) and a window that is not square (right). Notice that in the square window, the lines do look perpendicular, but in the other window they don't.

Exercises 4.5

1. Show that the two lines in the figure are parallel by comparing their slopes.

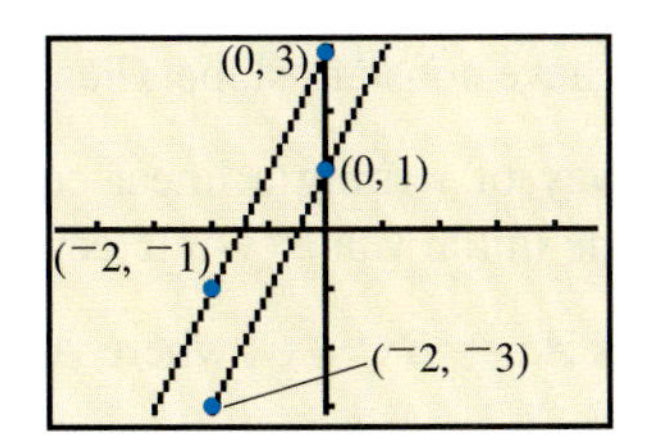

2. Show that the two lines in the figure are perpendicular by comparing their slopes.

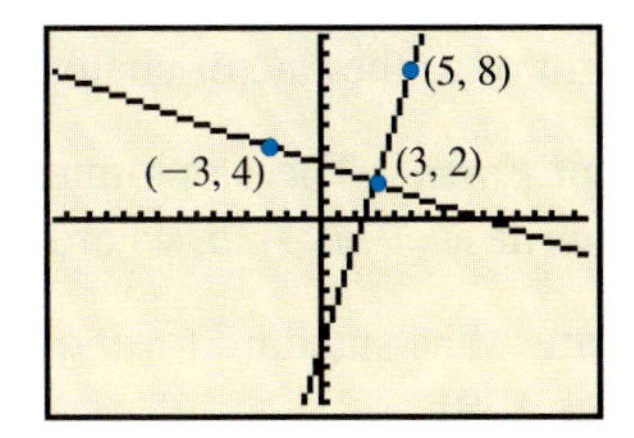

3. A line contains the points $(4, {}^{-}1)$ and $(6, 1)$. Another line contains the points $({}^{-}2, {}^{-}7)$ and $(2, {}^{-}3)$. Are the lines parallel, the same, perpendicular, or none of these? Explain your choice.

4. Classify each pair of linear equations as parallel, the same, perpendicular, or none of these.

   a. $y = 4x - 1$, $y = \frac{-1}{4}x + 5$

   b. $y = 2x$, $y = 6(\frac{1}{3}x)$

   c. $y = \frac{5}{3}x + 1$, $y = \frac{3}{5}x + 1$

   d. $y = x - 2$, $y = x - 4$

5. Classify each pair of linear equations as parallel, the same, perpendicular, or none of these. Recall that the slope is easiest to recognize when the linear equation is written in slope-intercept form, $y = mx + b$.

   a. $3x + 2y = {}^{-}1$, $3x - y = 4$

   b. $4y = x - 12$, $y - 3 = \frac{1}{4}x$

   c. $2x + y = 1$, $2x - 4y = {}^{-}5$

   d. $y = \frac{3}{5}x + 2$, $3x = 5y - 10$

6. Complete the ordered pair in the figure to make the two lines perpendicular.

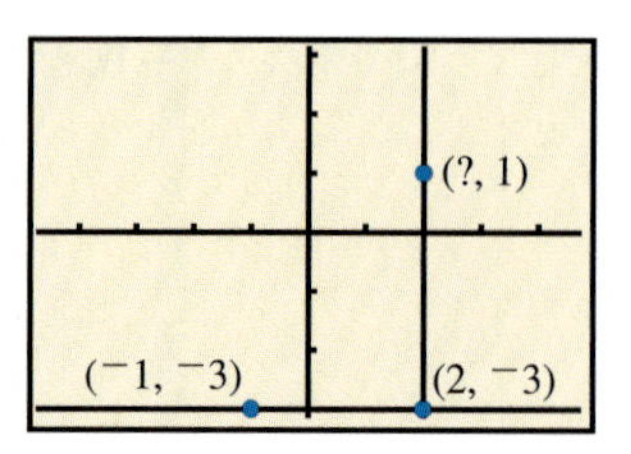

7. Will the two lines in the figure ever intersect? Explain your reasoning.

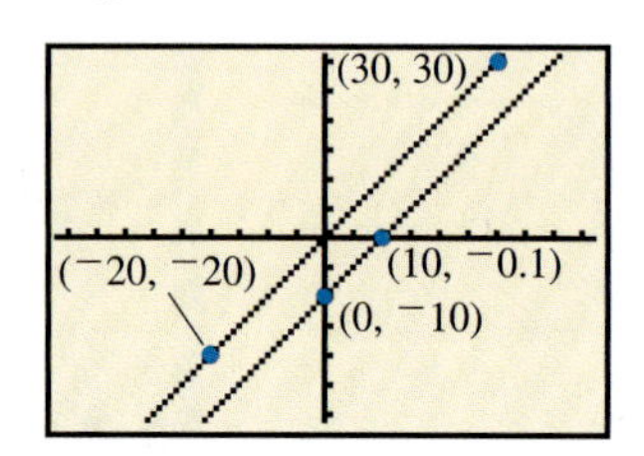

8. Suppose Miguel and Sofia run at the same speed of 10 feet per second. Miguel decides to give Sofia a 3-second head start in a race. The distance y in feet that Miguel runs is given by the equation $y = 10x$. The distance Sofia runs is given by the equation $y = 10(x + 3)$, where x represents Miguel's elapsed running time in seconds.

   a. Will Miguel ever catch Sofia? Explain.

   b. How do the linear equations support your answer?

CHAPTER 4 ▪ KEY CONCEPTS

Linear Equation in Two Variables A linear equation in two variables is any equation with simple terms involving two different variables and possibly a constant term. If we let x stand for one variable and y for the other, then a linear equation in two variables can almost always be put in the form $y = \text{coefficient} * x + \text{constant}$. An equation in this form is also called a **linear function**. (Page 125)

Initial Value, or y-intercept The y-intercept is the point where the graph of a linear equation in two variables crosses the y-axis. This point is the value of y when $x = 0$, and hence it is also called the **initial value**. (Page 125)

Slope The slope of a line is computed using the formula

$$\text{slope} = \frac{\text{rise}}{\text{run}} = \frac{\text{change in output}}{\text{change in input}}$$

It can be thought of as the rate of change of y in relation to x. Straight lines have constant slope. (Page 127)

Slope-intercept Form The slope-intercept form of an equation is one way of writing a linear equation in two variables. We write $y = mx + b$, where m is the slope of the line, and b is the initial value (y-intercept). (Page 136)

Standard Form The standard form of a linear equation in two variables is $Ax + By = C$, where A, B, and C are constants. (Page 139)

System of Linear Equations Two linear equations in two variables make a system of equations. The solution to the system, if it exists, is the intersection point of the two lines. This point is a point that satisfies *both* equations in the system. (Page 149)

Inconsistent System An inconsistent system of equations is a system of linear equations that has *no* solutions. (Page 152)

Methods for Solving Systems The three methods for solving a system of linear equations are graphing, substitution, and elimination. The chart on page 162 can be used to help choose a method. (Page 162)

Exploration

The table in the chapter opener shows the number of lives saved by airbags for the years 1993 to 2000. Let x represent the number of years since 1993. (For 1993, $x = 0$, for 1994, $x = 1$, etc.) Let y represent the number of lives saved by airbags. The linear function $y = 199x + 95$ is considered by statisticians to be a good model for this set of data.

a. Graph the function $y = 199x + 95$ in an appropriate window.

b. What is the slope of the line?

c. On average, approximately how many additional lives were saved by airbags each year?

d. Use the function to predict how many lives will be saved by airbags in the year 2006.

e. Use the function to predict how many lives were saved by airbags in the year 1990. What is wrong with this prediction?

Chapter 4 • Review Exercises

Section 4.1 Slope and Rate of Change

1. Find the

$$\text{slope} = \frac{\text{rise}}{\text{run}} = \frac{y_2 - y_1}{x_2 - x_1}$$

of the line that contains each pair of points.

a. (1, 2) and (2, 6)

b. ($^-5$, $^-6$) and ($^-8$, 2)

c. ($^-2$, 3) and (1, 0)

d. (4, $^-1$) and (7, $^-1$)

e. ($^-3$, 2) and ($^-7$, 4)

2. Suppose there is no snow on the ground and it begins to snow at the rate of 2 inches per hour.

a. How much snow is on the ground at the beginning of the storm? This is the initial value.

b. What is the rate of change in inches of snow per hour?

c. Use the rate of change to find how much snow has fallen in 1 hour, 2 hours, and 3 hours. Present these findings in a table with time (in hours) as the input and the amount of snow (in inches) as the output.

d. Plot the table values on a coordinate plane. The points form a line. What is the slope of the line and how does it compare to the rate of change?

3. Vitomax can sell 5000 supplements a day at a price of \$20. They can, however, increase their sales by 100 a day for each \$3 reduction in price.

a. What is the rate of change in supplements sold with respect to the price?

b. Explain how the rate of change is used to find the second ordered pair in the table below.

c. Use the rate of change to find a third ordered pair.

d. What is the slope of the line connecting these points?

Input (Price in Dollars)	Output (Number of Supplements Sold)
20	5000
17	5100
?	?

4. What is the slope of the line containing the ordered pairs in the following table?

Input, x	Output, y
0	5
2	8
4	11
6	14

5. Match each equation with its graph in the figure.

a. $y = 3x + 5$ **b.** $y = x$

c. $y = {}^{-}x + 2$ **d.** $y = {}^{-}4$

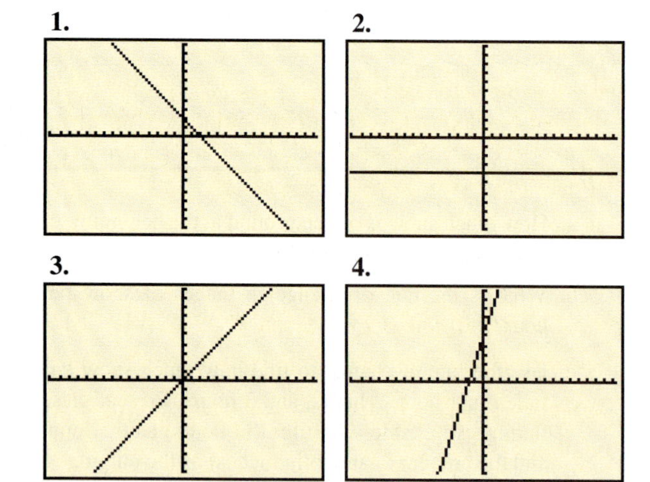

6. A line contains the point (0, 3) and has a slope of $\frac{3}{2}$. By hand, sketch a graph of this line.

Section 4.2 Forms of Linear Equations

7. Suppose there are 5 inches of snow on the ground when it begins snowing at the rate of 2 inches per hour.

a. What is the initial value?

b. What is the rate of change in inches of snow per hour?

c. We now have the slope and y-intercept. Use the slope-intercept formula to write an equation for the amount of accumulated snow in terms of the number of hours of snowfall.

8. Suppose each day you drink 300 mL from a container that initially holds 1800 mL of liquid. Let x represent the number of days you drink from the container and let y represent the amount of liquid (in mL) left in the container after x days.

a. Write an equation for y in terms of x.

b. Compare ordered pairs that satisfy your equation with those in the following table. They should be the same.

x (Days)	y (mL)
0	1800
1	1500
2	1200
3	900
4	600
5	300
6	0

9. Write an equation of the line with given slope and y-intercept.

a. $m = 6$, y-intercept (0, 2)

b. $m = \frac{1}{3}$, y-intercept (0, 4)

c. $m = 2$, y-intercept (0, 1)

d. $m = {}^{-}5$, y-intercept $(0, \frac{1}{2})$

e. $m = 1$, y-intercept $(0, {}^{-}3)$

f. $m = {}^{-}1$, y-intercept (0, 5)

10. Determine the slope and y-intercept for each equation. Some equations will need to be solved for y.

a. $y = 4x + 3$ **b.** $y + 2 = {}^{-}(x - 1)$

c. $y - 1 = 3(x - 2)$ **d.** $x - y = 3$

e. $2x + 3y = 12$ **f.** $y = {}^{-}x + 2$

11. Solve the equation $3x - 4y = 12$ for y in terms of x.

12. Write an equation of the line with given slope and containing the given point.

a. $m = 5, (^-2, 1)$ **b.** $m = 1, (^-4, 5)$

c. $m = ^-3, (1, 4)$ **d.** $m = \frac{1}{2}, (2, ^-3)$

13. Write an equation of the line containing each pair of points.

a. $(1, 4)$ and $(2, 6)$ **b.** $(4, ^-1)$ and $(5, 3)$

c. $(2, 6)$ and $(6, ^-8)$ **d.** $(0, 4)$ and $(^-9, 7)$

14. Two musicians are writing songs for an album. One musician writes 3 songs for each month she works and the other writes 4 songs for each month he works. Their album contains 12 songs.

a. The first musician's work rate in songs per month is $\frac{3}{1}$. What is the second musician's work rate?

b. Let x represent the number of months that the first musician writes songs for the album and let y represent the number of months that the second musician writes songs for the album. Then, an expression for the number of songs from the first musician is $3x$. Give an expression for the second musician's number of songs.

c. Write an equation showing that the total number of songs is 12.

d. Suppose the first musician writes songs for 1 month. For how many months must the other musician write songs in order to complete the album?

e. Mentally, find the intercepts using your equation from part (c) and interpret their meaning.

f. Use the intercepts to sketch a graph by hand.

15. Mentally, find the x- and y-intercepts for each equation. Sketch the graph for part (a) only.

a. $x + y = 5$

b. $x - 2y = 4$

c. $3x + 2y = 6$

d. $\frac{1}{2}x - 5y = 10$

e. $2x - \frac{1}{3}y = 4$

f. $4x + 5y = 9$

Section 4.3 Graphical Solution of Systems of Equations

16. The graph of the system of equations $y = 2x + 1$ and $y = ^-1x + 4$ is shown in the figure.

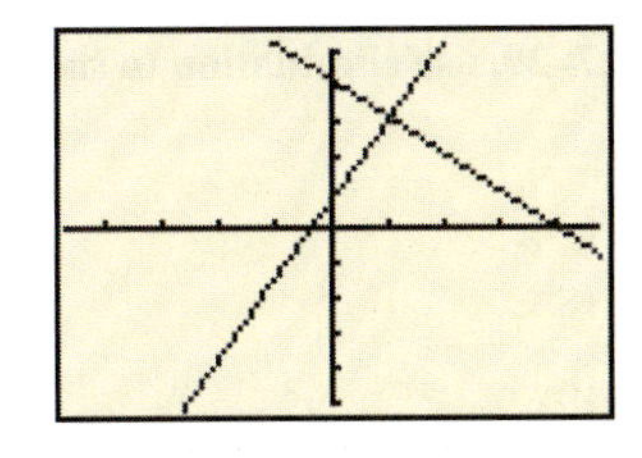

a. Mark the solution of the system on the graph.

b. What is this point on the graph called?

c. Estimate the solution of the system using the graph. Each tick represents one unit.

For Exercises 17–21, use a graphing method to find the solution of each system of equations. Note any systems that do not have a solution.

17. $y = 3x + 1$
$y = 2x + 2$

18. $y = x - 1$
$2x + y = ^-7$

19. $4x - 2y = 7$
$y = 2x + 3$

20. $3x + \frac{1}{7}y = 2$
$\frac{1}{2}x - y = 8$

21. $y = ^-x + 5$
$2x - y = 6$

22. The sum of two numbers is 78.2. The difference of the numbers is 29.3. Use a graph to find the two numbers.

23. Suppose you are comparing the cost of advertising in two local newspapers. The first charges \$7 and an additional \$.65 for each line of advertising. The second charges \$8.30 plus \$.52 for each line. For how many lines of advertisement will the cost for these two newspaper ads be the same? Use a graph.

Section 4.4 Algebraic Solution of Systems of Equations

24. Use substitution to find an exact solution to the system of equations in Exercise 21.

For Exercises 25–26, use substitution to solve each system of equations.

25. $y = 5x + 1$
$x + y = ^-11$

26. $y = 3x$
$y = 7x - 2$

For Exercises 27–29, use elimination to solve each system of equations.

27. $3x + 4y = 7$

$2x - 4y = 18$

28. $x - 5y = 10$

$2x + 6y = 4$

29. $3x + 7y = 11$

$4x - 5y = {}^{-}14$

For Exercises 30 and 31, use either elimination or substitution to solve the system of equations.

30. $y = 9x - 6$

$x + 2y = {}^{-}12$

31. $2x + 4y = {}^{-}5$

$5x - 8y = {}^{-}26$

In Exercise 32, use the fact that the sum of the measures of two supplementary angles is 180°.

32. Two angles are supplementary. The larger angle measures 30° less than twice the measure of the smaller angle. Complete (a) through (c) to find the measure of each angle.

a. Let x represent the measure of the smaller angle and y represent the measure of the larger angle. Write an equation involving the sum of the measures of these angles.

b. Write an equation for the measure of the larger angle in terms of the measure of the smaller angle.

c. The equations from parts (a) and (b) form a system of equations. Use algebra to solve this system.

33. A mouse runs past a cat. Then, $2\frac{1}{4}$ seconds later, the cat chases the mouse. The mouse runs 2 m/s and the cat runs 5 m/s. If both animals run in a straight line, how long will it take the cat to catch the mouse?

Section 4.5 Parallel and Perpendicular Lines

34. Classify each pair of linear equations as parallel, the same, perpendicular, or none of these.

a. $y = 5x - 2$ and $y = 5x + 1$

b. $y = \frac{1}{4}x$ and $y = {}^{-}4x + 7$

c. $4x + 2y = 6$ and $y = {}^{-}2x + 3$

d. ${}^{-}3x + y = 1$ and $x - 5y = 9$

35. A line contains the points $(0, {}^{-}3)$ and $(2, 9)$. Another line contains the points $(3, .5)$ and $(12, {}^{-}1)$. Are the lines parallel, the same, perpendicular, or none of these?

CHAPTER 4 ▪ TEST

1. Find the slope of the line that contains the points $({}^{-}6, 1)$ and $({}^{-}10, 2)$.

2. By hand, sketch the graph of a line with given slope passing through the origin.

a. $m = 0$ **b.** $m = \frac{4}{3}$

c. $m = 1$ **d.** $m = 2$

3. A type of tree grows 5 inches every 2 years.

a. What is the rate of change in inches with respect to the number of years?

b. An initial measurement of the tree is 250 inches. Use the rate of change to find two more ordered pairs in the following table.

Time (years)	Height (inches)
0	250
?	?
?	?

4. For Exercise 3, write a linear equation that models the height of the tree in terms of the number of years since its initial measurement.

5. Solve the equation $2x + 3y = 6$ for y in terms of x.

a. What is the slope of this equation?

b. What is the y-intercept?

6. Write an equation of a line with slope $\frac{-3}{4}$ and y-intercept $(0, 2)$.

7. Write an equation of a line with slope 5 containing the point $(1, {}^{-}2)$.

8. Mentally, find the x- and y-intercepts of the equation $4x - 5y = 20$. Use the intercepts to sketch a graph by hand.

9. Renting a truck from company A costs \$19.95 and an additional \$.08 for each mile driven. Renting from company B costs \$22.95 plus \$.05 for each mile. Use a graph to find the number of miles at which these two rental trucks cost the same.

10. Use a graph to find the approximate solution to the system of linear equations.

$$y = 3x - \tfrac{15}{7}$$
$$6x + 7y = 3$$

11. Use algebra to find an exact solution to Exercise 10.

12. Find an exact solution to the system of equations.

$$2x + 7y = {}^{-}5$$
$$3x + y = {}^{-}17$$

13. Two angles are supplementary. One-third of the larger angle is 10° less than the smaller angle. Find the measure of both angles.

14. Describe the graph of a system of equations that has no solutions.

CHAPTER

5

More Applications of Linear Equations

Coffee, Anyone?

The student center at a major university has just been renovated. There is now additional space to lease to commercial enterprises. A distributor of gourmet coffee is considering opening a franchise at the student center. However, the executives of this company want to know whether this is a good location for their business.

The coffee distributor hires a marketing firm to conduct a survey. The firm selects a random sample of students from the university and asks students in this sample whether or not they would patronize a gourmet coffee shop if one were opened on the campus. The results of this poll are used by the company to make its decision. In this chapter you learn how to apply proportions and percents to analyze information obtained from samples. (See Exploration on p. 206.)

Need help? For on-line resources, visit this web site: **math.college.hmco.com/students.**

5.1 Proportions

After Studying This Section You Will Be Able To

- Solve proportion equations.
- Set up proportions.
- Apply proportions to problems.

When two rates or ratios are equal to each other, the equation that expresses this relationship is called a **proportion**. In this section you learn how to solve proportions and how to set up problems that involve proportions.

One application of proportion is found in **similar figures**. Think of a photocopier with a setting that allows you to enlarge or reduce a picture. When such a setting is used, the photocopy is *similar* to the original picture. This means that the two figures have the same shape and that all the distances in the original figure have been changed by the same ratio. In Figure 1, the original house has been enlarged by a ratio of 3 to 2. We can work out the proportion between the houses by measuring any two corresponding parts of the house. For instance, if on the original picture the base of the house measures 5 inches, then on the photocopy it measures 7.5 inches.

Figure 1 The photocopy is $\frac{3}{2} = 1.5$ times as large as the original.

Example 1 deals with similar triangles. Two triangles are similar if their angles have the same measures and corresponding sides are proportional. Similar figures are just *scaled* versions of each other, like the houses in Figure 1.

EXAMPLE 1 At a certain time of the day, a 3-foot stick casts a 5-foot shadow. Find the height of a tree that casts a 30-foot shadow.

Solution Figure 2 shows how the stick, the tree, the shadows, and the rays of the sun form two triangles. Because the rays of the sun are parallel and both the stick and the tree form 90° angles with the ground, it can be shown, using theorems from geometry, that the two triangles are similar. Because corresponding sides of similar triangles have the same ratio, we can set up a proportion:

$$\frac{\textbf{height of stick}}{\textbf{length of stick shadow}} = \frac{\text{height of tree}}{\textbf{length of tree shadow}}$$

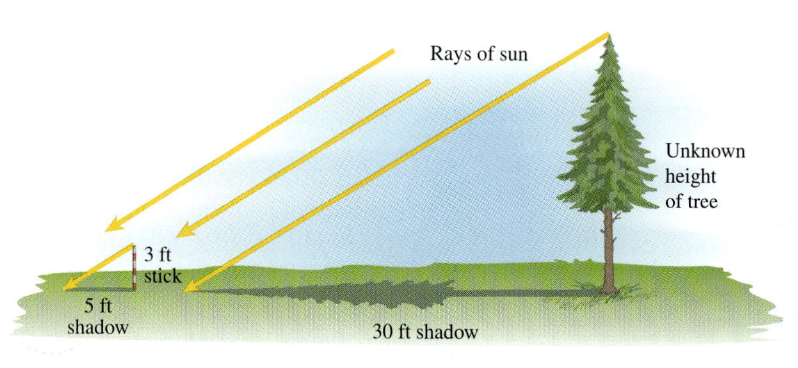

Figure 2

Let h represent the unknown height of the tree. Substituting the known quantities and h, we have

$$\frac{3}{5} = \frac{h}{30}$$

One way to solve this equation is to multiply both sides by 30. Then

$$30 * \frac{3}{5} = \frac{h}{30} * 30$$
$$18 = h$$

The tree is 18 feet tall.

■ ■

Direct Variation

In Example 1 we saw that the ratio of any vertical object to its shadow is the same at any particular time of day. The ratio itself depends upon how high the sun is in the sky. If we let x represent the length of the shadow and y represent the height of the vertical object, we can write an equation with the variables x and y. In Example 1, the ratio is $\frac{3}{5}$, or .6, so the equation becomes

$$\frac{y}{x} = .6$$

This ratio, .6, is called the *variation constant*.

Definition Direct Variation Equation

Two quantities, represented by the variables x and y, are said to be **directly proportional** if their ratio is constant:

$$\frac{y}{x} = k \qquad \text{or} \qquad y = kx$$

The **variation constant** k is the slope of the line passing through the origin that shows the relationship between x and y. When x and y are directly proportional, we can also say that y **varies directly as** x.

EXAMPLE 2 Roger bought 6 gallons of regular unleaded gasoline for \$9.00. How much will it cost him to fill his car's 20-gallon tank?

Solution The quantities are the amount of gasoline purchased and the total cost. They are directly proportional because their ratio is the price of regular unleaded gasoline in dollars per gallon. Let x represent the amount of gasoline purchased and y represent the cost. Then

$$\frac{y}{x} = k$$

We know that when $x = 6$ gallons, $y = \$9$. Substituting into the equation gives $k = \frac{9}{6} = 1.50$ dollars/gallon. In this case the variation constant is a rate involving two quantities, dollars and gallons.

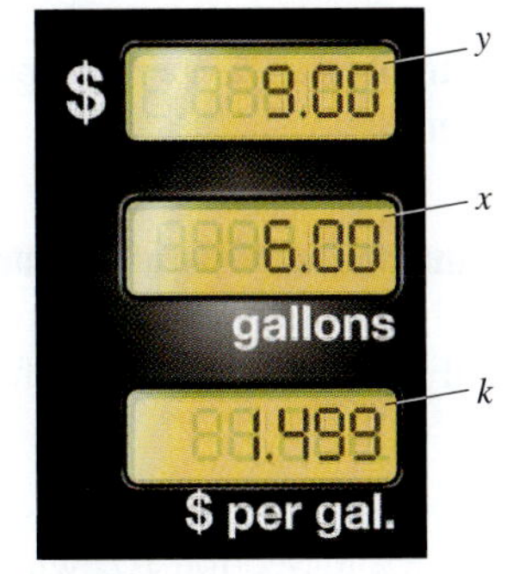

Figure 3

Roger observes this as he pumps the gas because the values of y, x, and k are displayed on the pump (Figure 3). Note that gasoline stations typically display the unit price ending in $\frac{9}{10}$ cent. Thus the price for 1 gallon appears as \$1.499, when to the nearest cent it is actually \$1.50.

Once Roger knows k, he can answer the question. The 20-gallon tank gives the value for x. Substituting into the equation we have

$$\frac{y}{20} = 1.50$$

Multiplying by 20 on both sides gives the solution $y = 30$. It will cost Roger \$30 to fill the tank.

The Cross-Multiplication Property

Example 2 illustrates one method to solve proportions—that is, to begin by finding the value of the variation constant k. Another common method for solving proportions relies on the following property.

Cross-Multiplication Property

$$\text{If } \frac{a}{b} = \frac{c}{d}, \quad \text{then } ad = bc$$

We find the cross-multiplication property of proportions by multiplying by the quantity bd on both sides of the original equation.

EXAMPLE 3 Show how to use the cross-multiplication property to solve the problem in Example 2.

Solution Let $c =$ the cost of filling the tank. Then, because the cost and amount of gasoline are directly proportional, we have

$$\frac{9}{6} = \frac{c}{20}$$

$$9 * 20 = 6c \qquad \textit{Using the cross-multiplication property}$$

$$180 = 6c$$

$$\frac{180}{6} = \frac{6c}{6} \qquad \textit{Dividing by 6 (or multiplying by } \tfrac{1}{6} \textit{ on both sides)}$$

$$30 = c \qquad \textit{Simplifying both sides}$$

It will cost Roger \$30 to fill the tank.

The next example shows how to use the cross-multiplication property as part of the seven-step process in solving a work-rate problem.

EXAMPLE 4 Rashid and Joshua decide to make some extra summer money painting cars. Rashid has done this before, so he is faster than Joshua. Rashid can paint an entire car in 3 hours, and Joshua takes 4. How long will it take Rashid and Joshua working together to paint a car?

Solution

Step 1 To *understand* this problem, review Example 2 in Section 1.2, which explains how work rates are computed. The most important thing to remember is that because Joshua and Rashid are working together, they should complete the job faster than *either* of them could do it alone.

Step 2 Drawing a picture can help you *visualize* the problem. Rashid paints a whole car in 3 hours, so if we represent a whole car by a box, then we need to split the box into thirds, each of which Rashid paints in an hour. Similarly, we split Joshua's box into fourths, each of which he paints in 1 hour (see Figure 4). So, adding the part of the car that

Figure 4 Each *large* rectangle represents one car to be painted.

Rashid paints in 1 hour to the part that Joshua paints in 1 hour, we can see that there is still almost half of a car left to paint after one hour.

Step 3 We *assign* the variable t to represent the time it will take them to paint together.

Step 4 To *write an equation*, use the work-rate formula from Section 1.2.

$$\text{rate} = \frac{\text{work}}{\text{time}}$$

$$\text{Rashid's rate} = \frac{1 \text{ car}}{3 \text{ hours}}$$

$$\text{Joshua's rate} = \frac{1 \text{ car}}{4 \text{ hours}}$$

You can add Rashid and Joshua's rates (like putting the boxes together in Figure 4) to get their combined rate:

$$\frac{1}{3} + \frac{1}{4} = \frac{7}{12} \text{ cars per hour}$$

which lets you write

$$\frac{\mathbf{7}}{\mathbf{12}} = \frac{\mathbf{1 \text{ car}}}{\boldsymbol{t}}$$

Step 5 Use the cross-multiplication property to *solve* for t.

$$\mathbf{7t = 12 * 1}$$

$$\frac{7t}{7} = \frac{12}{7}$$

$$t = \frac{12}{7}$$

Step 6 The *answer* to the question is that it will take them $\frac{12}{7} \approx 1.7$ hours to paint a car together.

Step 7 We can *check* that this makes sense because it is less than the time for either one alone but not too small, and it looks about right based on the last box in Figure 4.

Setting Up Proportions

Many students know how to use the cross-multiplication property effectively but are uncertain whether or not they have a correct proportion to begin with. Here is a technique that may prove to be helpful. If we have a situation involving a proportion, we should be able to label both ratios with an element they have in common. In addition, we should be able to do the same with the top row (the numerators) and the bottom row (the denominators). The next example illustrates this technique.

EXAMPLE 5 If a recipe that serves 8 people requires 3 cups of flour, then find how many cups of flour are needed to serve 20 people.

Solution Let f represent the amount of flour needed to serve 20 people. Then there are several proportions that can be used to solve this problem (see Figure 5). Upon cross multiplying, each of these proportions leads to the equation $8f = 60$, and dividing both sides by 8 gives $f = 7.5$. For 20 people, 7.5 cups of flour are needed.

	Recipe / Amount needed		Recipe / Amount needed		People / Cups of flour
People Cups of flour	$\frac{8}{3} = \frac{20}{f}$	Cups of flour People	$\frac{3}{8} = \frac{f}{20}$	Recipe Amount needed	$\frac{8}{20} = \frac{3}{f}$

Figure 5 Label the proportion to make sure the fractions relate the correct things; you can either have a fraction for the recipe amounts and one for the amount you will need or a fraction for the people and one for the cups of flour.

The labeling technique will guide us to one of several correct proportions. If we don't think carefully about labels, however, we might write an incorrect proportion such as

$$\frac{8}{20} = \frac{f}{3}$$

This is incorrect because recipe amount is in the numerator on the left and the denominator on the right.

Sampling

Proportions are found when a **sample** is used to draw a conclusion about an entire population. For instance, a polling organization might ask a random sample of about 1500 voters for whom they plan to vote in the next presidential election. From the information obtained in the sample, they predict the outcome of an election in which millions of people will vote.

Biologists use sampling to estimate the population of wildlife in a given environment. The technique is called "capture-recapture." Here's how the method works to estimate the number of fish of a certain species that live in one lake.

Step 1 Capture some fish and mark them with tags; then release them in the lake (Figure 6).

Figure 6 A sample of fish is captured and tagged.

Step 2 Allow the tagged fish enough time to swim around so that they are mixed up with the rest of the population (Figure 7).

Figure 7 The tagged fish are released into the lake.

Step 3 Return to the lake. Capture another sample of fish and count the number with tags (Figure 8).

Figure 8 A representative sample should have about the same proportion of tagged fish as the entire population.

Step 4 Use the number found in this sample to estimate the population of fish in the lake.

$$\frac{\textbf{tagged fish in sample}}{\textbf{total fish in sample}} = \frac{\textbf{tagged fish in lake}}{\textbf{total fish in lake}}$$

In these pictures we have the proportion

$$\frac{2}{6} = \frac{5}{15}$$

In reality, the numbers usually don't match perfectly. This technique just gives an estimate of the true population.

EXAMPLE 6 Biologists are studying small-mouth bass. They take **23** bass from Douglas Lake, tag them, and return them to the lake. Two weeks later they return and take a sample of **19** bass and find that **3** of them are tagged. Estimate the population of bass in Douglas Lake.

Solution Use the proportion

$$\frac{\textbf{tagged fish in sample}}{\textbf{total fish in sample}} = \frac{\textbf{tagged fish in lake}}{\text{total fish in lake}}$$

Let p represent the population of fish in the lake. Then

$$\frac{3}{19} = \frac{23}{p}$$

$3p = 437$ *Cross multiplying*

Because $3p = 437$, $p \approx 145.7$. Because we cannot have a fractional fish, we estimate the bass population of the lake to be 146.

Exercises 5.1

1. If a car goes 50 miles on 2 gallons of gas, it can go 150 miles on 6 gallons of gas. Set up a proportion and show that the cross-multiplication property holds.
2. Solve the proportion

$$\frac{x}{3} = \frac{5}{8}$$

3. Solve for x using the cross-multiplication property of proportions.

$$\frac{x+4}{8} = \frac{11}{3}$$

4. Suppose a 4-foot stick casts a 3-foot shadow. At the same time of day a flagpole casts a 36-foot shadow. How tall is the flagpole?
5. Note that triangle ABD is similar to triangle ECD in the figure. Set up and solve a proportion to find the length of AD.

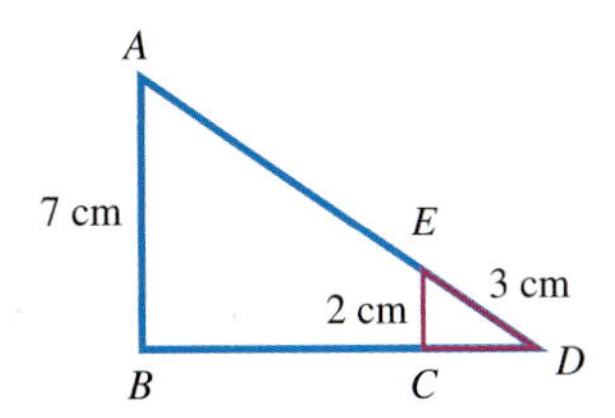

6. In Example 2 the equation $y = 1.5x$ gives the cost in dollars y of gasoline when x gallons are purchased.
   a. Draw a graph for this equation.
   b. What is the slope?
   c. What is the y-intercept?
   d. Is the slope or the y-intercept the same as the variation constant?

In Exercises 7–9 use the proportions A–D.

A. $\frac{6}{x} = \frac{7}{84}$

B. $\frac{6}{x} = \frac{84}{7}$

C. $\frac{x}{6} = \frac{7}{84}$

D. $\frac{x}{6} = \frac{84}{7}$

7. Which proportion has the same solution as proportion A?
8. Which proportion has the same solution as proportion B?
9. Write two other proportions with x, 6, 7, and 84 that have the same solution as proportion A.
10. Three bundles of asphalt roofing shingles cover 100 square feet. How many bundles do you need to cover 3000 square feet of roof?
11. Laura can pick up her toys in 5 minutes. Her mother can pick up the toys in 2 minutes. How long will it take Laura and her mother to pick up the toys together?
12. Running at a pace of one mile in 8 minutes, the average person burns 120 calories/mile. How many minutes would a person have to run to burn off a 600-calorie meal?
13. Latasha checks out a 145-page book from the library. She can read about 15 pages in 25 minutes. At this rate, how long will it take her to read the book?
14. An architect's drawing of a building has a scale of $\frac{1}{4}$ inch to 1 foot ($\frac{1}{4}$ inch on the drawing represents 1 foot of actual building length). If a wall measures 15 inches on the drawing, use a proportion to find the actual length of the wall.
15. Biologists capture 20 penguins and tag and release them. Later the biologists return and find that 4 are tagged in a sample of 17 penguins. Estimate the penguin population in the region.

Skills and Review 5.1

16. Solve for x and y by an algebraic method.

$$y = 2x - 8 \qquad 4x - y = 10$$

17. The sum of two supplementary angles is 180°. The larger angle is 10° more than 4 times the smaller. Find the measure of both angles.

18. A line has a slope of 1 and contains the point $(4, {}^{-}1)$.

a. Find an equation of the line.

b. Write your equation in standard form.

19. A line contains the points $(2, {}^{-}3)$ and $(5, 1)$.

a. Find the slope.

b. Find an equation of the line.

20. Two boats leave the same harbor heading in the same direction. The slower boat leaves at 10 A.M. and moves at a rate of 15 miles/hour. The faster boat leaves at noon and moves at a rate of 21 miles/hour. What time of day will the faster boat catch the slower boat?

21. Solve for x:

$$\frac{2x - 5}{3} = 5x$$

22. Solve for x: $x - 2(x + 3) + 7x + ({}^{-}4)^2 = 22$.

23. Recall that the area of a circle is πr^2. Find the exact area of a circle whose radius is 6 inches.

24. Evaluate $3 - 2 - 4 \div 2 + 5 * 2$.

25. Find the area of a parallelogram with base 11 mm and height 7 mm.

5.2 Percent and Percent Change

After Studying This Section You Will Be Able To

- Solve problems involving percents of a whole.
- Solve problems involving percent increase and decrease.

Using Percent to Compare a Part to a Whole

Percent may be used to compare a part to the whole. In Figure 9 part of the bar is shaded. The shaded part represents $x\%$ of the whole bar. The ratio of the part to the whole gives the **percent**. In other words,

$$\frac{\text{part}}{\text{whole}} = \text{percent}$$

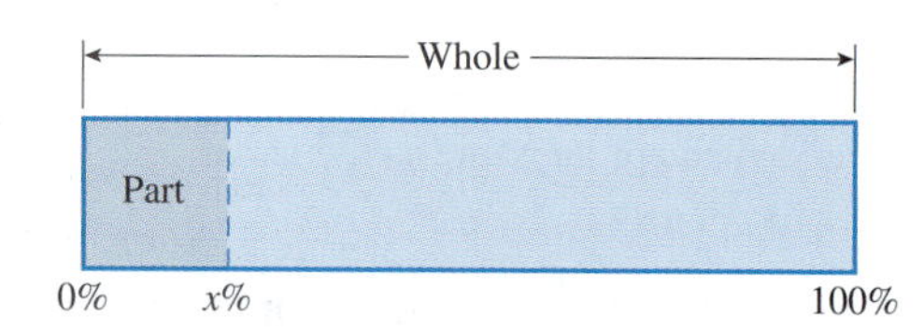

Figure 9

Multiplying both sides of this formula by the whole gives an equivalent formula:

$$\text{part} = \text{percent} * \text{whole}$$

The formula

$$\frac{\text{part}}{\text{whole}} = \text{percent}$$

may be used to find a percent when the whole and the part are known. The formula part = percent * whole may be used to find either the part or whole when the percent is known (the part can be found if you know the whole, and vice versa).

As the next three examples show, percents are used by nutritionists to help determine a balanced diet. For vitamins and minerals, the United States Department of Agriculture determines the recommended daily allowance (RDA), also called the recommended dietary intake or daily value (DV). References to RDA or DV are given on the labels of many food products and in nutritional guidebooks.

EXAMPLE 1 The RDA for vitamin C is 60 mg. One-half of a pink grapefruit contains 47 mg of vitamin C. What percent of the RDA is provided by this serving?

Solution Picture the situation with the diagram shown in Figure 10. The scale along the bottom of the bar shows percents. The scale on the top shows vitamin C in milligrams. More than half of the bar is shaded because **47** mg is more than half of **60** mg.

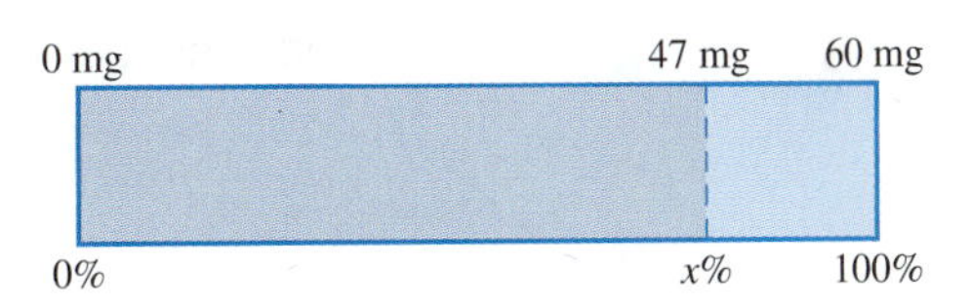

Figure 10

Both the part and the whole are known, so we may use the formula:

$$\text{percent} = \frac{\textbf{part}}{\textbf{whole}}$$

$$\text{percent} = \frac{\textbf{47 mg}}{\textbf{60 mg}} \approx .783 = 78.3\%$$

To change a decimal to percent, move the decimal point two places to the right and add a % sign, so we write .783 = 78.3%.

This problem may also be solved with a proportion. A percent is a ratio with 100 in the denominator, so we can write

$$\frac{47}{60} = \frac{x}{100}$$

Use the cross-multiplication property to get

$$100 * 47 = 60x$$

Solving for x gives $x \approx 78.3$. That means that $x\% = 78.3\%$. The serving of grapefruit provides about 78% of the RDA for vitamin C.

EXAMPLE 2 One serving of a breakfast cereal provides 9% of the RDA for sodium. The RDA for sodium is 2500 mg. How much sodium is provided in one serving of the cereal?

Solution Again, the scale along the bottom of Figure 11 shows percents and the scale along the top shows milligrams. The unknown is the milligrams of sodium in one serving of cereal, that is, the part. To estimate, think of 9% as slightly less than 10% and 10% of 2500 mg is 250 mg. So, we should have about 250 mg. To find a precise answer, use the formula part = percent * whole.

$$\begin{aligned} \text{part} &= \textbf{percent} * \textbf{whole} \\ &= \mathbf{.09} * \mathbf{2500} \\ &= 225 \end{aligned}$$

Figure 11

If you prefer, you may use a proportion:

$$\begin{aligned} \frac{x}{\mathbf{2500}} &= \frac{\mathbf{9}}{\mathbf{100}} \\ 100x &= 9 * 2500 \\ x &= 225 \end{aligned}$$

One serving contains 225 mg of sodium.

EXAMPLE 3 One serving of orange juice contains 450 mg of potassium. The nutrition label says that this is 13% of the RDA. What is the RDA for potassium?

Solution This time the unknown is the whole, as shown in Figure 12. To estimate the whole, think of 13% as a little more than 10%, so 10% of the whole is about 400 mg. The entire whole (100%) is ten times as much, or about 4000 mg.

Figure 12

To find a more precise answer, use the formula part = percent * whole.

$$\text{part} = \text{percent} * \text{whole}$$
$$450 = .13x$$
$$\frac{450}{.13} = x$$
$$3462 \approx x$$

Another method is to use a proportion:

$$\frac{450}{x} = \frac{13}{100}$$
$$450 * 100 = 13x$$
$$3462 \approx x$$

The RDA for potassium is about 3462 mg.

In this situation 3462 mg may be *too* precise. The RDA for potassium is probably **3500** mg. The food-distribution company could have used the formula $\frac{\text{part}}{\text{whole}}$ = percent to calculate the percent as

$$\frac{450 \text{ mg}}{3500 \text{ mg}} \approx .129 = 12.9\%$$

Rounding to the nearest percent, they would then report 13% on the nutrition label.

Solutions

A special type of whole-part situation occurs when one substance is dissolved in another to form a uniform mixture called a **solution**. For instance, table salt (sodium chloride) is dissolved in water to form a solution of salt water. The sodium chloride is the **solute**. The water is the **solvent** (Figure 13).

The strength of the solution is defined by the ratio of the mass of solute to the mass of solution, expressed as a percent. This ratio can be represented by a whole-part diagram like the one in Figure 14.

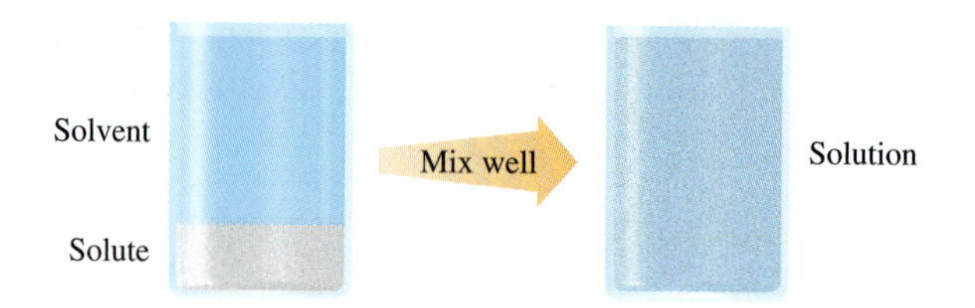

Figure 13

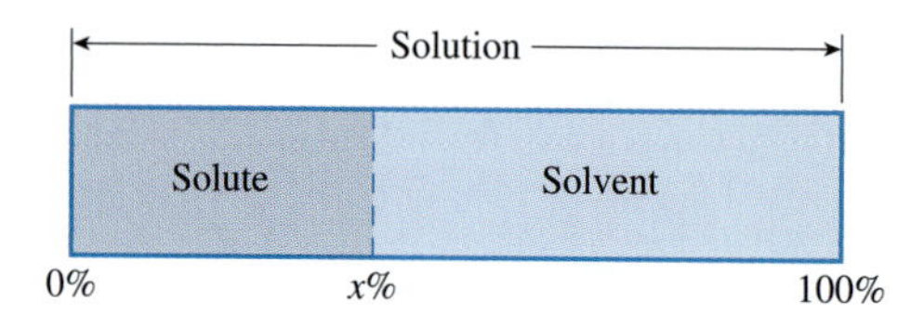

Figure 14

EXAMPLE 4 How much water and how much sodium chloride (NaCl) are needed to make 400 g of a 6% solution of sodium chloride?

Solution Because the whole (solution) is known to be 400 g and the percent is 6%, we need to find the part (solute). See Figure 15.

$$\text{part} = \textbf{percent} * \textbf{whole}$$

$$\text{solute} = \textbf{percent} * \textbf{solution}$$

$$x = \mathbf{.06} * \mathbf{400}$$

$$\boldsymbol{x} = \mathbf{24}$$

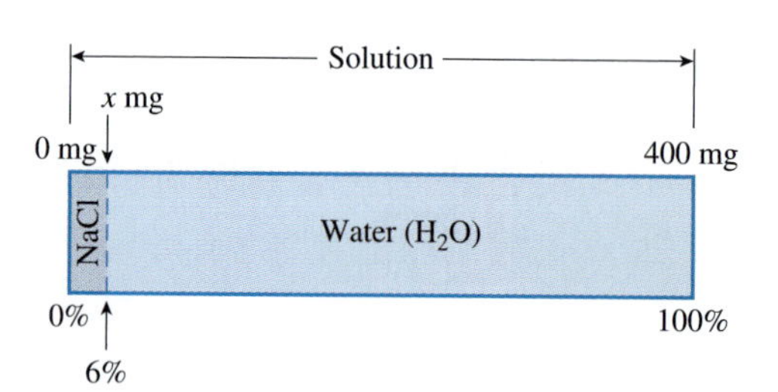

Figure 15

Thus we will need 24 g of solute, that is, sodium chloride (NaCl). To find how much water is needed, look at Figure 15:

$$\textbf{solution} = \textbf{solute} + \text{solvent}$$

$$\mathbf{400} = \mathbf{24} + \text{solvent}$$

$$376 = \text{solvent}$$

We will need to mix 376 g of water with 24 g of sodium chloride (NaCl) to produce 400 g of solution.

Definition

$$\text{Solute} = \text{percent} * \text{solution}$$

$$\text{Solution} = \text{solute} + \text{solvent}$$

Percent Change

Percent may also be applied to situations where something increases or decreases. For instance, the price of an item increases when a sales tax is added to it. Also, the price decreases when it is on sale. It is helpful in both situations to think of the original price as 100% and the change in price as a positive or negative quantity added to the original price. Recall the number line model for addition (Section 1.1):

$$\text{start} + \text{change} = \text{final}$$

See Figures 16 and 17.

Figure 16

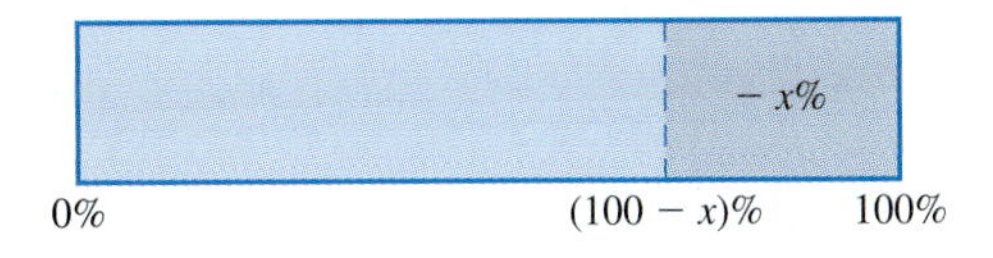

Figure 17

EXAMPLE 5 The sales tax rate in Connecticut is 6%. Mary bought an air conditioner priced at \$349.95. How much did she pay altogether?

Solution Picture this situation as a percent increase (Figure 18). The total price is 106% of the original price: **\$1.06** $*$ **\$349.95** = \$370.947. Rounding to the nearest cent, the total price is \$370.95.

Another way to solve this problem is to calculate the tax (6% of \$349.95 ≈ \$21.00) and add the tax to the original price (\$349.95 + \$21.00 = \$370.95).

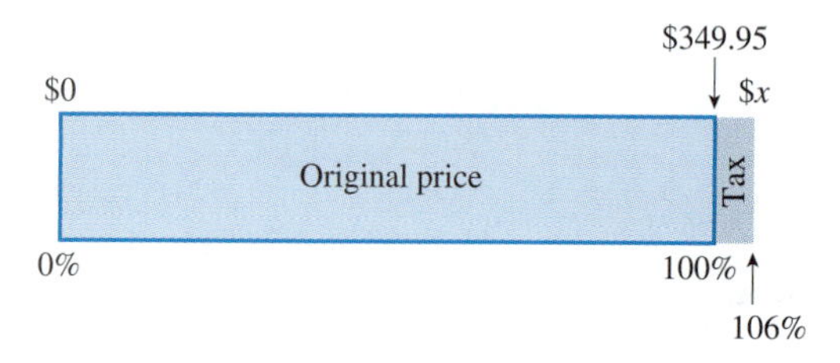

Figure 18

Example 6 shows a percent decrease.

EXAMPLE 6 ■ A coat that normally costs $150 is on sale for $120. Find the percent discount.

Solution Picture this situation as a percent decrease. See Figure 19.

$$x\% = \frac{\textbf{sale price}}{\textbf{original price}} = \frac{\$120}{\$150} = .8$$

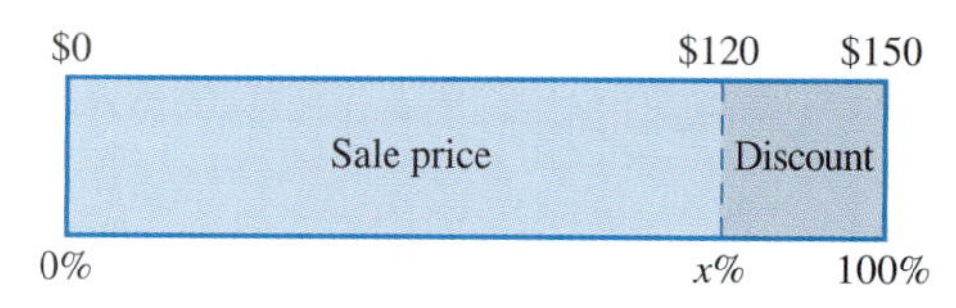

Figure 19

Thus the sale price is 80% of the original price. The discount is 20% of the original price.

Another way to solve this problem is to find the dollar amount of the discount ($150 − $120 = $30) and use it to calculate the percent discount directly:

$$\frac{\text{discount}}{\text{original price}} = \frac{30}{150} = .2$$

Again, the discount is 20% of the original price.

Often, statisticians use percent increase or decrease.

EXAMPLE 7 According to the 1990 census, the population of Nevada was 1,202,000, to the nearest 1000. In 2000, the U.S. Census Bureau found the state's population to be 1,998,000. Find the percent increase over this 10-year period.

Solution

$$\frac{\textbf{new population}}{\textbf{original population}} = \frac{\textbf{1998}}{\textbf{1202}} \approx 1.662$$

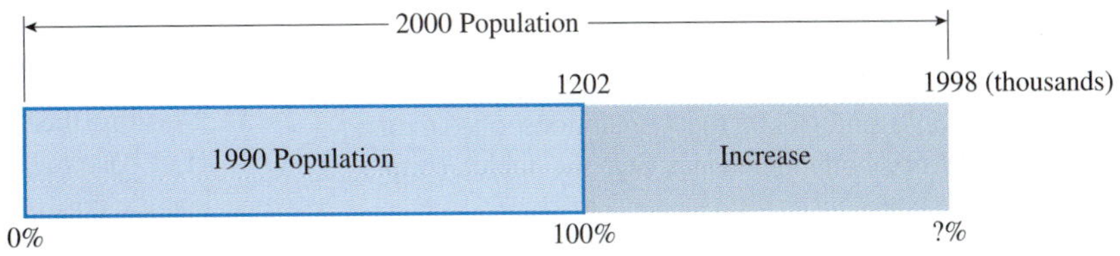

2000 Population

1202

1998 (thousands)

1990 Population

Increase

0%

100%

?%

Figure 20

The 2000 population is about 166% of the 1990 population. Nevada's population increased by about 66%. See Figure 20.

Caution: Notice that in order to compute percent change correctly, we must start with the *original* population. If we put the new (2000) population in the denominator, we have the ratio $\frac{1202}{1998}$, which is approximately 60%. This gives 40% for the change in population, which is very different from the correct figure, 66%. Regardless of whether a quantity is increasing or decreasing, consider the *original* quantity to represent 100%.

Exercises 5.2

1. 8 is 25% of what number?
   a. Represent the question in a bar diagram.
   b. Estimate the number from your diagram.
   c. Find the exact number with a formula for the whole.
2. 3% of 1400 is what number?
3. 130 is what percent of 270?

For Exercises 4–11, draw a diagram for each situation and then solve the problem.

4. The laboratory grade for a chemistry course is based on 150 points. Jason wants to earn a grade of at least 80%. How many laboratory points must he earn?
5. In the 2002 election, Congresswoman Nancy Johnson received 114,253 votes out of a total of 210,007 cast in Connecticut's fifth congressional District. What percent of the votes did Ms. Johnson receive? (http://www.jsonline.com)
6. An ice cream label states that one serving contains 120 mg of cholesterol. That is 40% of the recommended daily allowance (RDA). What is the RDA for cholesterol? (http://www.nutribase.com)
7. How much water and how much sodium chloride are needed to make 250 g of a 9% solution of sodium chloride?
8. What is the strength of the solution made by dissolving 45 g of sodium chloride in 1455 g of water?
9. A dealer offers to sell you a used automobile for \$6500. How much will you have to pay for the car when you include the 6% sales tax?
10. At a local gas station, the price of gasoline rose from \$1.69 per gallon to \$1.93 per gallon. Find the percent increase.
11. A clothing store has a sale in which all items are 40% off. How much would you have to pay for a pair of pants that originally sold for \$29.95?
12. Jamie has trouble remembering formulas. For whole-part situations he remembers only the formula

$$\frac{\text{part}}{\text{whole}} = \text{percent}$$

If necessary, he solves this formula for the item he needs.

a. Solve this formula for the part.

b. Solve this formula for the whole.

13. Between 1990 and 2000, the populations of Arizona and California both increased. See the table. (http://www.quickfacts.census.gov)

State	1990 Population	2000 Population	Change
Arizona	3,665,228	5,130,632	1,465,404
California	29,760,021	33,871,648	4,111,627

a. Find the percent increase for Arizona.

b. Find the percent increase for California.

c. Explain why California had a larger change in population but a smaller percent increase.

14. Assume the tip on a meal is 15% of the cost of the meal. Let base meal be the cost before the tip and final meal be the cost of food with the tip (excluding tax).

a. Write a formula that can be used to find the final meal cost.

b. Assume a base meal costs $22.15. How much does the final meal cost?

15. Some small engines require a mix (solution) of oil and gasoline. Assume 3% of a solution is oil.

a. How much oil is required for 4 gallons of gasoline?

b. A gallon is 128 ounces. How many ounces of oil from part (a) are needed?

Skills and Review 5.2

16. Solve for x if $\frac{1}{3}x + 2 = \frac{5}{6}$.

17. Consider the proportion

$$\frac{3}{4} = \frac{10}{x}$$

a. Write two other proportions with 3, 4, 10, and x that have the same solution.

b. Use the cross-multiplication property to find the solution.

18. Solve for x and y by an algebraic method.

$$2x - 3y = 11$$

$$x + 4y = {}^{-}11$$

19. Use a graph to approximate the solution(s) to the system of equations.

$$y = {}^{-}\tfrac{1}{2}x + \tfrac{3}{2}$$

$$y = \tfrac{4}{3}x - 2$$

20. Find an equation of the line containing the points $({}^{-}1, {}^{-}5)$ and $(3, 4)$.

21. Matt is asked to find the slope of the line $y = 3x - 7$. He finds two ordered pairs that satisfy the equation and then he calculates the slope. Describe another method for finding the slope of the line.

22. Solve this problem using the seven steps for problem-solving described in Section 3.3: The length of a rectangle is 4 feet less than twice the width. The perimeter is 64 feet. Find the dimensions of the rectangle.

23. A video rental store charges $3.00 to rent a movie plus $1.00 each day. Complete this table:

Days Rented	Charge
1	?
2	?
?	7

24. Use the table values from Exercise 23 to find an equation that shows how the charge for a video tape depends upon the number of days it is rented.

25. Evaluate

$$5 + 2(6 - 4)^3$$

5.3 Mixtures and Investments

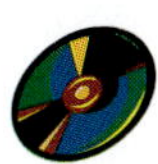

AFTER STUDYING THIS SECTION YOU WILL BE ABLE TO

- Solve problems involving mixtures.
- Solve problems involving investments.

As you will recall from Section 4.2, a linear equation in two variables, x and y, may be written in standard form as $Ax + By = C$. This form suggests that a certain amount of what x represents (Ax) is combined with a certain amount of what y represents (By) to produce a known quantity C. There are many applications that lead to equations of this form. In this section we learn how to solve a variety of problems that may be solved using a system of two equations in two variables, with one or both of the equations in standard form. Many of these equations involve applications of percents, which were introduced in Section 5.2.

EXAMPLE 1 At Kayla's Nut Shop, peanuts sell for \$2 per pound and cashews sell for \$5 per pound. A customer wants to buy a 10-pound mixture of peanuts and cashews and is willing to pay \$28. How many pounds of each type of nut should Kayla use to make the mixture?

Solution Before setting up this problem for an algebraic solution, let us observe Kayla as she goes through steps 1 and 2 of the problem-solving process, *understanding* and *visualization*.

Kayla thinks, "I need 10 pounds of nuts. If all 10 pounds are peanuts, the price for the mixture will be \$20, which is too low. On the other hand, if they are all cashews, the price will be \$50, too high. Suppose I make a half-and-half mixture—that would be 5 pounds of peanuts worth \$10 combined with 5 pounds of cashews worth \$25 for a total of \$35. That's too much, so let me try again."

Kayla is using the age-old problem-solving strategy called guess-and-check. She may eventually arrive at a solution, but it will take some time. One reason we learn algebra is to be able to solve problems like this more efficiently. We can gain some insight into the problem by examining the data Kayla has supplied.

Let us *assign* the variable x to represent the amount of peanuts, in pounds, and y to be the amount of cashews, in pounds, thus reaching step 3 of the problem-solving process. Kayla has given us three pairs of values for a table.

x	y	
10	**0**	(price of mixture = \$20)
0	**10**	(price of mixture = \$50)
5	5	(price of mixture = \$35)

If we graph these ordered pairs, they lie on a line with equation $x + y = 10$. We can call this equation the "weight equation." The weight equation line has (10, 0) as x-intercept and (0, 10) as y-intercept, as shown in Figure 21.

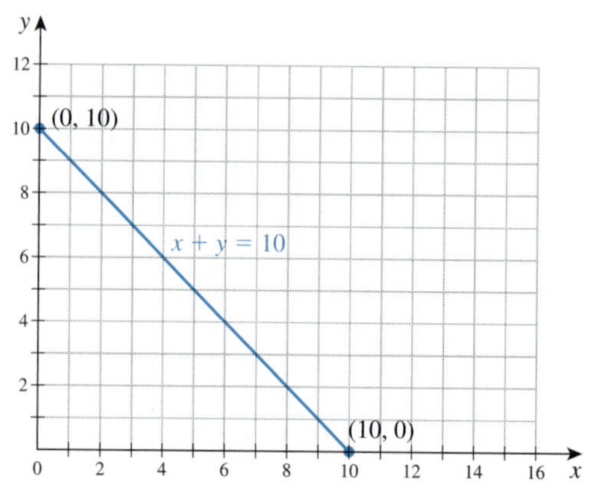

Figure 21

Kayla is not getting very far guessing and checking with this equation, so she tries another approach.

She reasons, "The customer wants to spend \$28. If she spends it all on peanuts at \$2 per pound she'll have 14 pounds of nuts—that's more than she wants. If she spends it all on cashews at \$5 per pound, she'll have—let me see, $28 \div 5$—that comes out to 5.6 pounds. That's less than the 10 pounds that she wants. What if she spent \$18 on peanuts and \$10 on cashews?"

Kayla isn't finding a solution this way either, but she has given us another table.

x	y	
14	**0**	(All money spent on peanuts)
0	**5.6**	(All money spent on cashews)
9	2	(\$18 on peanuts and \$10 on cashews)

The ordered pairs satisfy the equation money for peanuts + money for cashews = \$28, or, in terms of x and y, $2x + 5y = 28$. We have now completed step 4 of the problem-solving process by *writing* two *equations*.

The graph in Figure 22 shows the original weight equation, $x + y = 10$, graphed with the money equation, $2x + 5y = 28$. It appears that the two lines intersect. We can now *solve* this system of equations (step 5):

$$x + y = 10 \qquad \textit{Weight equation}$$

$$2x + 5y = 28 \qquad \textit{Money equation}$$

We may solve each equation for y in terms of x and graph

$$\text{Y1} = {}^{-}\text{X} + 10 \text{ (weight)} \qquad \text{Y2} = {}^{-}.4\text{X} + 5.6 \text{ (money)}$$

Using a graphing calculator, we find an approximate solution, $x \approx 7.33$ and $y \approx 2.67$.

We can also solve by an algebraic method such as substitution or elimination. Solving by elimination,

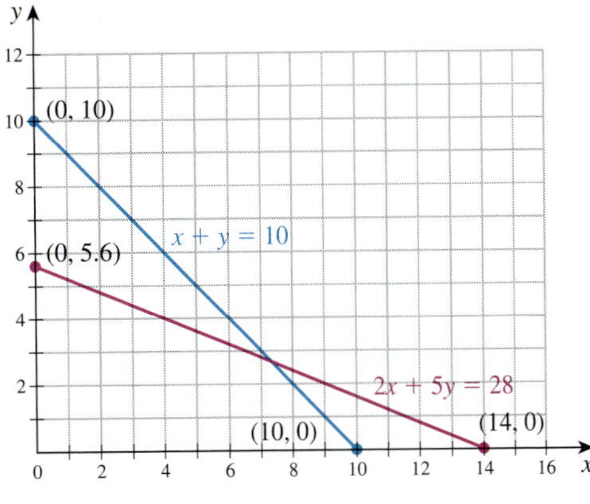

Figure 22

$$\begin{array}{rcl} x + y = 10 & \rightarrow & {}^{-}2x - 2y = {}^{-}20 \\ 2x + 5y = 28 & \rightarrow & 2x + 5y = 28 \\ \hline & & 3y = 8 \\ & & y = \frac{8}{3} = 2\frac{2}{3} \\ & & x + 2\frac{2}{3} = 10 \\ & & x = 7\frac{1}{3} \end{array}$$

The *answer* to our question (step 6) is that Kayla should mix $7\frac{1}{3}$ pounds of peanuts with $2\frac{2}{3}$ pounds of cashews to produce the desired mixture. We can *check* this result (step 7) by showing that the total weight is 10 pounds and the total cost is \$28.

Now let's apply the same technique to a situation involving chemical solutions.

EXAMPLE 2 A nurse needs 400 mL of a 35% solution of alcohol. She has one bottle which contains a 20% solution and another bottle which contains a 60% solution. How much of each solution should she mix together?

Solution Recall our discussion of solutions in Section 5.2. In Figure 23, the amount of alcohol (solute) in each container is represented by the shaded portion. We want to know how much of each solution to mix, so let x = the volume (in mL) of 20% solution to be used and y = the volume (in mL) of 60% solution. Our first equation is based on the relationship

volume of first solution + volume of second solution = volume of mixture,

$$x + y = 400$$

A second equation may be found by considering the amount of solute (in this case alcohol) in each container:

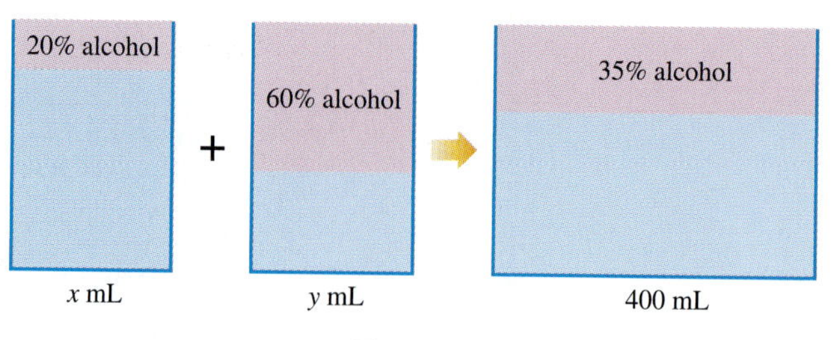

Figure 23

alcohol in first container + **alcohol in second container** = **alcohol in mixture**.

For each container we use the relationship part = percent $*$ whole to find the quantities of alcohol:

$$\mathbf{.20x} + \mathbf{.60y} = \mathbf{.35 * 400}$$

$$.20x + .60y = 140$$

We now have a system of two equations, which we can label the "solution equation" and the "alcohol equation":

$$x + y = 400 \qquad \textit{Solution equation}$$

$$.20x + .60y = 140 \qquad \textit{Alcohol equation}$$

This system can be solved by any method. Here we use substitution. Solve the first equation for y to get $\mathbf{y = 400 - x}$ and substitute into the second equation:

$$.20x + .60(\mathbf{400 - x}) = 140$$

$$.20x + 240 - .60x = 140$$

$$^{-}.40x + 240 = 140$$

$$^{-}.40x = {}^{-}100$$

$$\mathbf{x = 250}$$

$$y = 400 - \mathbf{250} = 150$$

She will need 250 mL of the 20% solution and 150 mL of the 60% solution.

EXAMPLE 3 Marcel has $2000 to invest in an IRA. His financial advisor suggests grade A bonds, which pay yearly interest at the rate of 5%, and grade B bonds, which pay yearly interest at the rate of 9.5%. Of course, he would prefer the higher interest rate, but grade B bonds are a riskier investment. His advisor suggests that he diversify by putting a portion of his $2000 in each of the two types of bonds. Marcel decides to do that but wants to make sure that he earns a 7.5% yield overall. How much should he invest in each grade of bond to meet this goal?

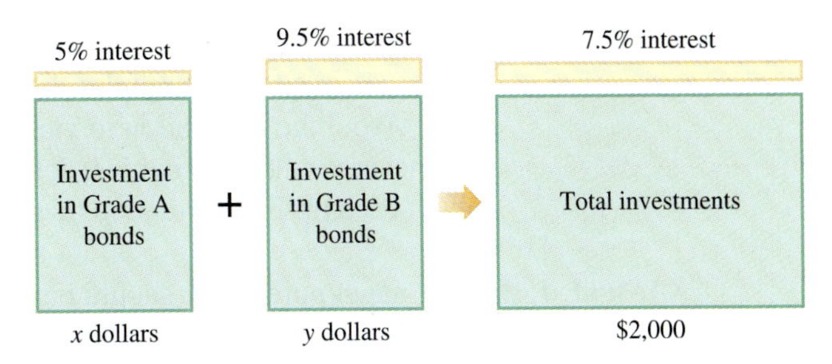

Figure 24

Solution Figure 24 shows a picture of this situation. Notice the similarity to the problem in Example 2, but also notice that in this case interest is not part of the original investment but rather is added onto it. Let x = amount of money (in dollars) invested in grade A bonds. Let y = amount of money (in dollars) invested in grade B bonds. Recall that interest = principal * rate * time.

The interest for 1 year earned on grade A bonds is $x * (.05) * 1 = .05x$. The interest for 1 year earned on grade B bonds is $y * (.095) * 1 = .095y$. The interest for 1 year earned on the total investment is $2000 * (.075) * 1 = 150$.

We have two equations, an "investment equation" and an "interest equation."

$$x + y = 2000 \qquad \textit{Investment equation}$$

$$.05x + .095y = 150 \qquad \textit{Interest equation}$$

Again, we may use any suitable method to solve this system. Solving by elimination we have

$$\begin{array}{rcl} x + y = 2000 & \rightarrow & {}^{-}.05x - .05y = {}^{-}100 \\ .05x + .095y = 150 & \rightarrow & .05x + .095y = 150 \\ \hline & & .045y = 50 \\ & & y \approx \mathbf{1111.11} \\ & & x + \mathbf{1111.11} \approx 2000 \\ & & x \approx 888.89 \end{array}$$

He should invest \$888.89 in grade A bonds and \$1111.11 in grade B bonds to meet his goal.

Note that in reality he probably won't be able to invest these amounts exactly, but this theoretical solution will help him come close to reaching his goal.

Although we used two variables to solve each of the previous problems, some students prefer to use only one variable. For instance, Example 3 could have been set up this way: Let x = amount of money (in dollars) invested in grade A bonds. Let $2000 - x$ = amount of money (in dollars) invested in grade B bonds. Then a single equation, $.05x + .095(2000 - x) = 150$ can be used to solve for x. You should convince yourself that this approach is very similar to solving the two equations in Example 3 with the substitution method.

Caution: All three examples included an equation of the form $x + y =$ something. The next two examples, however, demonstrate that this is not always the case. Avoid the temptation to view the previous examples as formulas that work whenever you have mixtures, solutions, or investments. Instead, think carefully about each problem situation to come up with appropriate equations.

EXAMPLE 4 How much pure water should be added to 200 mL of a 10% solution of sulfuric acid (H_2SO_4) to dilute its strength to 3%?

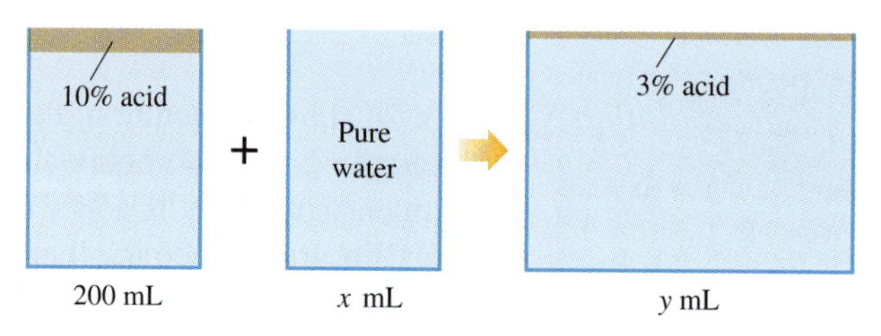

Figure 25

Solution As usual, a picture is helpful (see Figure 25). Let x = amount of pure water (in mL). Let y = amount of mixture (in mL). We have two equations based on

volume of first solution + volume of second solution = volume of mixture

acid in first container + acid in second container = acid in mixture

$$200 + x = y \quad \textit{Solution equation}$$

$$20 + 0 = .03y \quad \textit{Acid equation}$$

Solving the second equation for y, we have $y \approx 666.7$ mL. Substituting for y in the first equation gives $x \approx 466.7$ mL.

About 467 mL of pure water are needed. *Note:* As a safety precaution, the 10% acid solution should be poured into the container containing the pure water, not the other way around. Always add acid to water, never water to acid!

Our final example is a distance-rate-time problem that involves an equation in standard form, $Ax + By = C$. In this case the terms Ax and By represent the distances traveled by the two boys and C represents the total distance.

EXAMPLE 5 Jason sets out from home at 9:00 A.M. and walks toward Nathan's house at the rate of 4 miles/hour. At 9:30 A.M. Nathan leaves his house, riding his bike in the direction of Jason's house at the rate of 10 miles/hour. The two houses are 16 miles apart. When will Jason and Nathan meet?

Solution The situation is represented in Figure 26. There are several ways to approach this problem. One involves two variables.

Figure 26

Let x represent the time in hours Jason travels. Let y represent the time in hours Nathan travels. Use the relationship distance = rate $*$ time to find expressions for the distances both boys travel. Jason travels $4x$ miles and Nathan travels $10y$ miles. Because the total distance traveled is 16 miles, we have a *distance equation*:

$$4x + 10y = 16 \qquad \textit{Distance equation}$$

We also have a *time* equation, because Jason travels $\frac{1}{2}$ hour longer than Nathan does.

$$x = y + \tfrac{1}{2} \qquad \textit{Time equation}$$

Substituting x from the time equation into the distance equation gives us

$$4\left(y + \tfrac{1}{2}\right) + 10y = 16$$
$$4y + 2 + 10y = 16$$
$$14y + 2 = 16$$
$$14y = 14$$
$$y = 1$$
$$x = 1 + \tfrac{1}{2} = 1\tfrac{1}{2}$$

Nathan travels for 1 hour and Jason travels for $1\frac{1}{2}$ hours. They meet at 10:30 A.M.

Exercises 5.3

1. Often in linear combination problems, we can find a rough estimate for an answer before writing any equations. Consider Example 1, where a 10-pound mixture of nuts costs \$28.

   a. Find the cost per pound of the mixture.

   b. Was the cost per pound of the mixture closer to the cost per pound of the peanuts or the cashews?

c. Of which nut, then, should there be more in the mixture? Explain your reasoning.

2. In Example 2 a nurse made 400 mL of a 35% solution of alcohol from a combination of 60% and 20% solutions of alcohol. Follow the steps for a quick check of the work from Example 2.

a. Is the 35% for the mixture closer to 60% or 20%?

b. What does this imply about the relative quantities of each solution?

c. How can this method be used to determine if the algebraic results are reasonable?

3. A chemist needs to make a mixture containing 30 mL of a 25% solution of hydrochloric acid (HCl). She has a 15% solution and a 40% solution of HCl.

a. Model this situation with a diagram such as Figure 23.

b. From your diagram write a volume equation, where volume of first solution + volume of second solution = volume of mixture.

c. Write a solute equation where HCl in first solution + HCl in second solution = HCl in mixture.

d. Without solving this system of equations, describe the steps necessary to find the amount of each solution on a graphing calculator.

4. Maria has $3000 to invest between two accounts and wants an overall yield of 7.5%. One account earns 5% and the other earns 9%. Suppose a student writes the following two equations to represent this situation.

$$x + y = 3000 \qquad \textit{Investment equation}$$

$$.05x + .09y = .075 \qquad \textit{Interest equation}$$

Find the error in the equations and correct it.

5. In Example 5, you were given the two equations

$$4x + 10y = 16$$

$$x = y + \tfrac{1}{2}$$

Describe what the terms represent in each equation.

6. An amusement park sells children's tickets for $4 and adult tickets for $7. One day the park sold 1000 tickets for a total of $6100. Complete the steps to find the number of children's and adult tickets sold that day.

a. *Assign* the variables as x = number of children's tickets sold, and y = number of adult tickets sold. Complete the table.

	Children	Adults	Together
Rate (ticket)	$4	$7	—
Number of tickets	x		
Cost ($)	$4x$		$6100

b. From the table *write* two *equations*, each containing x and y.

c. *Solve* the system of equations to find the number of each type of ticket sold that day.

7. A gourmet coffee company sells a 3-pound mix of chocolate raspberry beans and French vanilla coffee beans for $13. If the chocolate raspberry is $4 per pound and the French vanilla is $6 per pound, how many pounds of each coffee are in the mix? Complete the steps to answer the question.

a. *Assign* the variable(s).

b. Complete a table similar to Exercise 6.

c. *Write* the *equation(s)*. *Note:* It is possible to have one equation in one unknown.

d. *Solve* the equation(s) algebraically.

e. How many pounds of each coffee are in the mix?

8. A chemist has two bottles of hydrochloric acid (HCl). Bottle A contains a 36% solution and bottle B contains a 12% solution. She needs solutions of different concentrations. Match each of these solutions with a description of how she would make it. Do not set up equations or make any calculation; just choose the most reasonable answer.

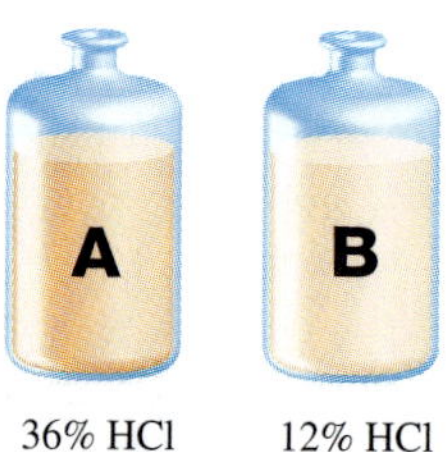

a. 18% solution 1. Mix equal amounts from bottle A and bottle B.

b. 24% solution 2. Mix a large amount from bottle A with a small amount from B.

c. 30% solution 3. Mix a small amount from bottle A with a large amount from B.

d. 48% solution 4. You can't make this solution with bottles A and B.

9. A chemist mixes two solutions of alcohol to make 600 mL of a 31% solution of alcohol. If the first solution is 10% alcohol and the second is 40% alcohol, how much of each solution is in the mixture? Complete the steps to answer the question.

a. Draw a picture.

b. *Assign* the variables.

c. *Write* a system of equation(s).

d. *Solve* the equations graphically.

e. *Answer* the question.

f. *Check* your answer from part (e) to see if it seems reasonable.

10. Alana has $5000 to invest. She wants to maximize her income but portion her investments to reduce risk. She decides to invest in a CD paying 6% interest per year and a lower-grade bond paying 11% per year. How much should she put into each investment in order to earn 8% overall? Complete the steps to answer the question.

a. Draw a picture.

b. *Assign* the variables.

c. *Write* the equation(s).

d. *Solve* the equation(s) algebraically.

e. *Answer* the question.

f. *Check* your answer for reasonableness.

11. Thelma and Louise traveled by car from Wichita, Kansas, to Saint Louis, Missouri, a distance of 446 miles. Thelma drove for the first part of the journey, averaging 61 miles/hour. Louise got impatient and asked to take over. She drove the remainder of the trip, averaging 72 miles/hour. The total driving time was 6 hours 30 minutes. How long was each of the drivers behind the wheel?

a. Draw a diagram.

b. *Assign* one or two variables and use them to *write* an *equation* or a system of equations.

c. *Solve* the equation(s) by any suitable method.

d. *Answer* the question.

12. A 5-pound mixture of peanuts and almonds sells for $15. If the peanuts sell for $1.50 per pound and the almonds sell for $4 per pound, how many pounds of each type of nut are in the mix?

13. A theater company wants to keep its ticket prices low so they sell most of the 600 seats at $4 per ticket and sell some special seats at $12 per ticket. If the company makes $3040, how many of each kind of seat does it sell? Assume all the seats are sold, and solve this problem graphically.

14. A restaurant has 2 cups of an oil and vinegar salad dressing that is 88% oil. How much vinegar should be added to dilute the dressing to 55% oil?

15. A couple needs to borrow $5600 for a home improvement project. They borrow part of the money at 7.5% simple interest and the rest at 11.25% simple interest. How much do they borrow from each source, if the interest for one year is $476.25?

Skills and Review 5.3

16. Answer each question about percents.

a. 140 is what percent of 250?

b. 15% of 12 is what number?

c. 35 is 60% of what number?

17. How much water and how much hydrochloric acid are needed to make 300 g of a 6% solution of hydrochloric acid?

18. In 1980 the U.S. budget deficit was $73.8 billion. In 1992 the budget deficit was $290.4 billion. (Congressional Budget Office)

a. How you can tell without doing any calculations that the percent change (from 1980 to 1992) was greater than 100%?

b. Find the percent change in the budget deficit from 1980 to 1992.

19. Solve the proportion.

$$\frac{x}{3.2} = \frac{^-24}{7}$$

20. Julie bought 75 gallons of heating oil for $78.75. How much will 120 gallons of heating oil cost her?

a. Set up a proportion.

b. Solve the proportion using the cross-multiplication property.

21. Find the x- and y-intercepts of the equation,

$$9x - 4y = 36$$

22. Solve this problem using the seven steps for problem-solving. Two cars start at 2:00 P.M., one from San Diego and the other from Las Vegas, and they travel toward each other. The car from San Diego travels at 70 miles/hour and the car leaving Las Vegas travels at 75 miles/hour. The distance between San Diego and Las Vegas is 338 miles. At what time will the two cars meet?

23. Solve the equation for x.

$$4 - (x + 3) = 2(5x - 1)$$

24. Evaluate the expression

$$\frac{2(4x - 5) + 7}{6}$$

for $x = {}^{-}3$.

25. Find the missing side on the right triangle shown.

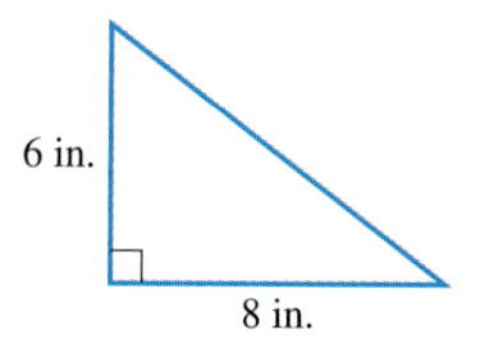

5.4 Systems of Three Equations (Optional)

In Sections 4.3 and 4.4 we solved systems of equations involving two variables: two variables and two equations. Many situations involve more than two unknowns, which can lead to systems of equations in *more than* two variables.

EXAMPLE 1 José and Valerie play a guessing game with coins. Valerie tells José that she has 15 coins in her pocket, some quarters, some dimes, and some nickels. The total value of the coins is \$2.40, and she has 2 more quarters than dimes. José has to guess how many of each type of coin she has.

Solution We can use the seven problem-solving steps to help us with this problem. First, we read carefully and see that what José needs to find out is how many quarters, dimes, and nickels Valerie has in her pocket. To *visualize* the problem, it might help to get some coins and try out various combinations. Next, we need to assign variables. Because we are asked to find the numbers of quarters, dimes, and nickels, we need a variable for each. Let x be the number of quarters, y be the number of dimes, and z be the number of nickels.

We have not encountered a problem like this before; we have *three* variables to work with. We can carry out step 4, writing equations, by looking back at the problem.

$x + y + z = 15$ *Valerie has 15 coins.*

$.25x + .10y + .05z = 2.40$ *The total value is \$2.40.*

$x = y + 2$ *She has 2 more quarters than dimes.*

We could try to solve this system of equations algebraically in the same way we solved systems of equations in Section 4.4, using substitution. Use the equation $x = y + 2$ to substitute for x in the equations $x + y + z = 15$ and $.25x + .10y +$

$.05z = 2.40$, which would give us two equations in two variables (y and z). Next, we could solve one of these new equations for y and substitute into the other equation, which would give an equation involving just z. We could solve this for z, and then substitute the value of z into the equation with y and z to find a value for y. Finally, we could substitute the values of y and z into any of the original equations to find the value of x.

Instead, we take advantage of our calculators, which can do the algebra for us in this case. First, we need to put the equations in *standard form*, with the variables on the left-hand side and any constant on the right-hand side.

$$x + y + z = 15$$

$$.25x + .10y + .05z = 2.40$$

$$x - y = 2$$

Now, we enter this system of equations into our calculator; first, choose 2nd MATRX, and then, EDIT. Choose one of the available names (say [A]) and press ENTER. The display will look like this:

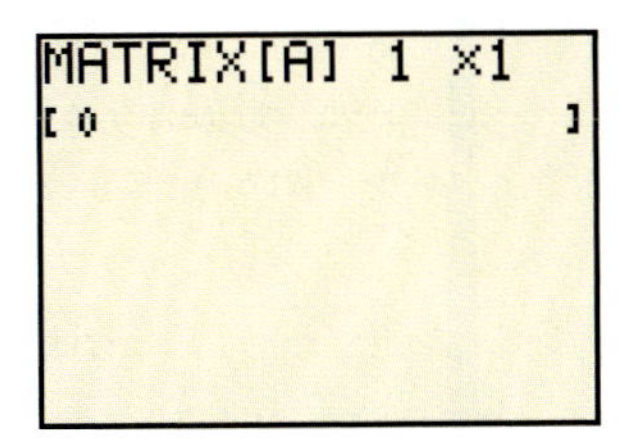

Push 3 ENTER 4 ENTER and the display becomes

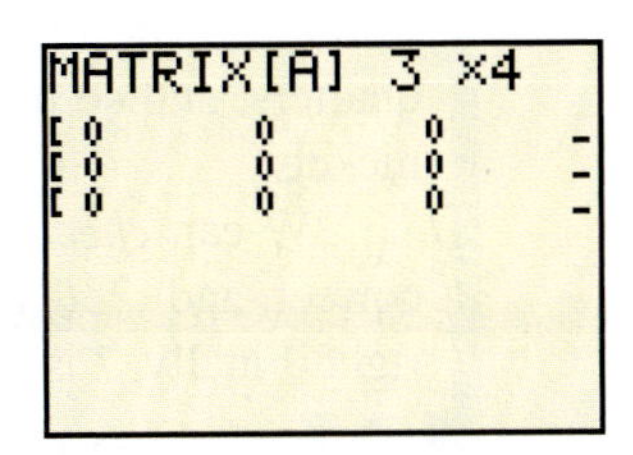

Next, we enter each coefficient to replace the zeros (. . . means there is another set of zeros off the screen). In this case, start with 1 ENTER, then 1 ENTER, then 1 ENTER, then 15 ENTER. This puts the first equation in. Then do .25 ENTER, .10 ENTER, and keep going until you have run out of equations. The display is now

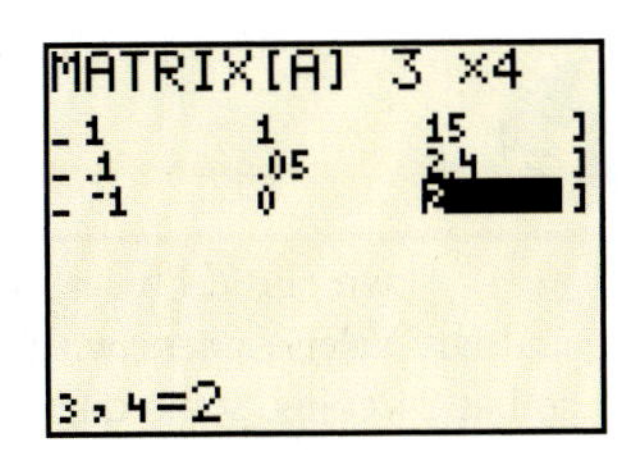

There is a zero on the third line because the coefficient of z in the last equation is zero.

To check your entry from the home screen, choose 2nd MATRX, and then NAMES. Select [A] and press ENTER. You should see the following:

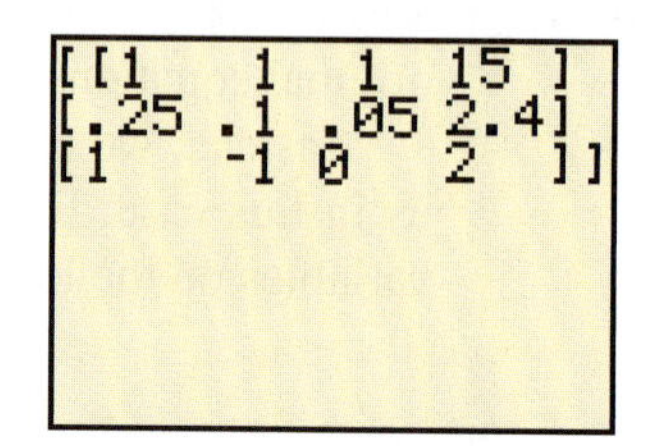

This is the **matrix** associated with our system of equations. The first column is all the coefficients of x, the second, of y, and the third, of z. The last column contains the numbers on the right-hand side of each equation.

Finally, choose 2nd MATRX again, then MATH, scroll down to rref, and press ENTER. Enter the name [A] from the MATRX menu, close parentheses, and press ENTER. Now the display is [A]

```
rref([A])
      [[1 0 0 7]
       [0 1 0 5]
       [0 0 1 3]]
```

We care only about the numbers in the right column, **7**, **5**, and **3**. These are the solutions. The first number corresponds to the first variable, $\boldsymbol{x}$, the number of quarters, and so on. The calculator says Valerie has 7 quarters, 5 dimes, and 3 nickels.

We can *check* this answer. $\mathbf{7} + \mathbf{5} + \mathbf{3} = 15$, so she has 15 coins, which is correct, and $.25(\mathbf{7}) + .1(\mathbf{5}) + .05(\mathbf{3}) = 1.75 + .5 + .15 = 2.40$, which is also right. Finally, **7** is 2 more than **5**, so Valerie has 2 more quarters than dimes.

This might seem like just as much work as doing the algebra, but with a little practice it goes quickly. You can also save time after the first problem by reusing A. When you choose EDIT the second time and pick A, all you need to do is change the numbers.

Exercises 5.4

1. José checks in his pocket and has Valerie guess what coins he has. He tells her he has 24 coins, some quarters, some dimes, and some pennies. The coins have a value of $3.84. He has twice as many dimes as pennies. How many of each coin does he have?

2. Solve the following system of three equations:

$$x + 2y + 3z = 4$$
$$5x - 5y + 7z = 8$$
$$10y + 11z = 12$$

a. Give the answers rounded to two decimal places.

b. If you use the full decimal as given by the calculator, is the answer exact? Explain.

3. A developer is building houses on a cul-de-sac. He builds three types of houses, small colonials, big colonials, and split-levels. Small colonials cost $135,000 to build and sell for $200,000. Big colonials cost $175,000 to build and sell for $250,000. Split-levels cost $150,000 to build and sell for $210,000. He will build 11 houses and plans on making a profit of $730,000. If he builds the same number of big colonials and split-levels, how many houses of each kind does he build?

CHAPTER 5 ▪ KEY CONCEPTS

Direct Variation Equation This equation involves two quantities, represented by the variables x and y, whose ratio is constant:

$$\frac{y}{x} = k \qquad \text{or} \qquad y = kx$$

The **variation constant** k is the slope of the line passing through the origin that shows the relationship between x and y. (Page 177)

Cross-Multiplication Property If $\frac{a}{b} = \frac{c}{d}$, then $ad = bc$. (Page 178)

Setting Up Proportions When setting up a **proportion** to solve a problem, check to make sure the fractions relate *corresponding* quantities. (Page 180)

Sampling In sampling, a sample is taken from a large quantity and proportions relate the sample to the whole quantity. For example, suppose 10,000 people vote in a local election and a reporter asks 100 people for whom they voted. If 65 people say candidate A, then the reporter is probably safe in assuming that candidate A won the election. (Page 181)

Comparing a Part to a Whole (Page 184)

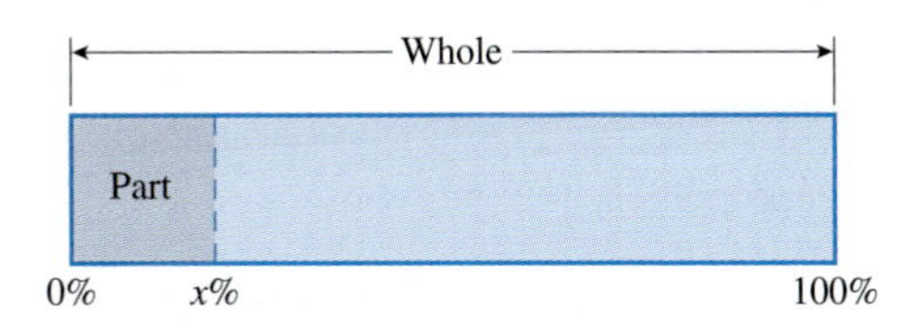

$$\frac{\text{part}}{\text{whole}} = \text{percent}$$
$$\text{part} = \text{percent} * \text{whole}$$

Solutions (Page 187)

$$\text{solute} = \text{percent} * \text{solution}$$
$$\text{solution} = \text{solute} + \text{solvent}$$

Percent Change Start + change = final. Consider the original quantity (start) to represent 100%. (Page 189)

Mixtures and Investments These problems often involve equations in standard form $Ax + By = C$. (Section 5.3)

Exploration

A sample of 500 students at a university are polled to see if they would patronize a proposed gourmet coffee shop. Of the students polled, 119 say yes. The total student population of the university is 18,500.

a. Set up and solve a proportion to estimate how many students would patronize the gourmet coffee shop.

b. What percent of the student population does the poll indicate would patronize the coffee shop?

c. All polls have a *margin of error*. The actual percentage in the population is likely to fall within a range determined by the sample percentage plus or minus the margin of error. For this poll the margin of error is 3.7%. Find the range of percents for this poll.

d. Given the margin of error, find the maximum and the minimum number of students (from the population) who are likely to patronize the coffee shop.

Chapter 5 • Review Exercises

Section 5.1 Proportions

1. Use the cross-multiplication property to solve each proportion.

a. $\frac{10}{4} = \frac{x}{18}$ **b.** $\frac{9}{x} = \frac{2}{3}$

c. $\frac{x}{5} = 3$ **d.** $\frac{5}{2} = \frac{8}{x+1}$

2. In Example 3 of Section 5.1, we found the proportion

$$\frac{9 \text{ dollars}}{6 \text{ gallons}} = \frac{c \text{ dollars}}{20 \text{ gallons}}$$

Write two other proportions that are also correct. *Note:* See Setting Up Proportions on page 180 for guidelines on how to write a proportion.

3. A formula for Celsius temperature in terms of Fahrenheit temperature is

$$C = \frac{F - 32}{1.8}$$

Notice that this equation is a proportion. Find the Fahrenheit temperature when the Celsius temperature is 20°.

4. The cost of heating oil is directly proportional to the number of gallons purchased. The Richards spent $210 on 175 gallons of heating oil. How much does 1 gallon cost? Use this unit price to find the cost of 150 gallons of heating oil.

Use proportions to solve Exercises 5–8.

5. At a certain time of day a 5-foot person casts a 4-foot shadow. How tall is a building that casts a 17-foot shadow?

6. At a local gas station, 15 gallons of regular gasoline cost $25. How many gallons of regular gasoline may be bought for $18?

7. Sid can perform a job in 2 hours. Joanne isn't nearly as skillful, so the same job takes her 6 hours. Suppose they work together on this job.

a. Without performing any calculations, how do you know that Sid and Joanne can finish this job in less than 2 hours?

b. Exactly how many hours does this job take if Sid and Joanne work together?

8. In a certain forest 42 deer are caught and tagged. A month later naturalists return to the same area and capture 22 deer, of which 9 have tags. Estimate the deer population in this forest.

Section 5.2 Percent and Percent Change

9. The formula for percent is $\frac{\text{part}}{\text{whole}} = \text{percent}$.

a. Give a formula for the *part*.

b. Give a formula for the *whole*.

10. Use a bar diagram such as the one shown in Figure 9 (page 184) to represent the following question: 14 is what percent of 30?

11. Use your bar diagram from Exercise 10 to estimate the percent.

12. Use the formula from Exercise 9 to find the exact percent in Exercise 10.

13. 21 is what percent of 80?

a. Represent the question in a bar diagram.

b. Estimate the percent from your diagram.

c. Find the exact percent using the formula.

14. 15% of 72 is what number?

15. 12 is 6% of what number?

16. The RDA for calcium is 1000 mg. A cup of lactose-reduced milk contains 300 mg of calcium. What percent of the RDA for calcium is provided in a cup of this milk?

17. The RDA for saturated fat is 20 g. A serving of peanut butter is 15% of the RDA for saturated fat. How many grams of saturated fat are in one serving of peanut butter?

18. One serving of sun-dried raisins contains 2000 mg of fiber. That amount of fiber is 8% of the RDA for fiber. What is the RDA for fiber? (http://www.nutribase.com)

19. Give a definition for each term.

a. Solute

b. Solvent

c. Solution

20. What is the strength of a 200-g solution of sodium chloride that contains 18 g of sodium chloride?

21. Suppose the taxes and surcharges on a recent phone bill were 12.5% of the charges for phone calls. If the phone calls alone were $29.65, how much was paid altogether? Round your answer to the nearest penny.

22. An airline ticket on a major carrier was recently reduced from $420 to $305. To the nearest hundredth, what was the percent discount?

23. The population of a town increased 17% to 12,500 people. What was the original population?

Section 5.3 Mixture and Investments

24. At the Luscious Candy Shop, peanut butter fudge sells for $5 per pound and vanilla fudge sells for $7 per pound. A customer wants to buy a 2-pound mixture of the fudge and is willing to pay $13. How many pounds of each type of fudge can this customer purchase?

a. Assign any variables.

b. Write an equation or system of equations.

c. Solve the equation(s) for either variable so that they may be entered into the Y = menu on your calculator.

d. Use a graph to approximate the number of pounds of each type of fudge.

25. A chemist requires 200 mL of a 22% solution of hydrochloric acid. The chemist has a 10% solution and a 30% solution. How much of each solution should the chemist mix together?

26. A checking account pays an annual interest rate of 2.3% and a savings account pays an annual interest rate of 5.4%. A total investment of $700 paid $34.30 in interest for the year. How much was deposited in each account?

5.4 Systems of Three Equations

27. Solve the following system of three equations and round your answers to two decimal places.

$$3y + z = 5$$
$$^{-}x + 2y + 4z = {}^{-}6$$
$$2x + 5y - 3z = 8$$

28. Giovanni has 31 coins in a jar containing pennies, nickels, and quarters. There are four times as many nickels as pennies. The coins have a value of $1.51. How many of each coin is in the jar?

CHAPTER 5 ▪ TEST

1. Solve the proportion

$$\frac{10}{x} = \frac{3}{8}$$

2. When traveling abroad people often exchange currencies. Suppose 50 U.S. dollars buy 42.9 euros. How many U.S. dollars can be purchased with 200 euros?

3. Ladeen and Larry have a snow-plowing business. Using her truck alone, Ladeen can finish the plowing in 6 hours. Larry needs 8 hours with his truck. Using both trucks, how long will it take Ladeen and Larry to finish plowing?

4. In an artic region 15 polar bears are caught and tagged. A couple of months later, in the same region, 10 polar bears are caught, and 4 of them have tags. Estimate the polar bear population for this region.

For Exercises 5–7, use a bar diagram to support your answers.

5. 18 is what percent of 40?

6. 85% of 200 is what number?

7. 60 is 105% of what number?

8. A basketball player averages 20 points per game. One night she scores 16 points. What percent of her usual output was this?

9. Due to a shortage of memory chips a computer price was increased to \$1800 from \$1600. What was the percent increase on the price of this computer?

10. Which solution contains more acid, 150 mL of a 12% solution of sulfuric acid (H_2SO_4) or 130 mL of a 15% solution of H_2SO_4? Explain your answer.

11. A nut shop sells almonds for \$7.75 per pound and peanuts for \$3.25 per pound. In order to sell an overstock of peanuts, the shop owner decides to mix the peanuts with almonds. How many pounds of each kind of nut are in a 4-pound mixture that costs \$19?

12. Fred has \$5000 to invest between AAA bonds and a certificate of deposit (CD). The bonds pay a yearly rate of 7.5% interest and the CD pays a yearly rate of 5% interest. How much should Fred place in each investment in order to earn a 6.5% yield overall?

13. A juice company decides to charge more for its product by adding water to increase the volume. The current product contains 2 L and is 15% juice. How much water is added if the concentration is diluted to 8% juice?

CHAPTER

6

Exponents and Factoring

Multiplying with Squares

In the late twentieth century the handheld calculator revolutionized many aspects of computation. Similarly, in fifteenth-century Europe the switch from Roman to Hindu-Arabic numerals made multiplication and division much easier.

Previously various algorithms were used to multiply with Roman numerals. One algorithm dating back to ancient Babylonia, involved looking up numbers in a table of squares. This method, outlined here, works if both numbers are odd or both are even. The introduction of Hindu-Arabic numerals made it possible for people to use the paper-and-pencil algorithms that your parents learned in school. One author has suggested that this change in number systems was a significant influence on the growth of commerce in the late middle ages, and thus changed the course of history.

Multiplying Using Squares

Step 1 Add the two numbers and divide by 2. Call the result x.

Step 2 Subtract the smaller number from the larger number and divide by 2. Call the result y.

Step 3 Find the squares of x and y in a table. Subtract y^2 from x^2. The result is the product of x and y. (See Exploration on p. 252.)

Need help? For on-line resources, visit this web site: **math.college.hmco.com/students.**

6.1 The Laws of Exponents

AFTER STUDYING THIS SECTION YOU WILL BE ABLE TO

- Explain the laws of exponents.
- Apply the laws of exponents to simplify expressions.

Discovering the Laws of Exponents

In this section you will learn some shortcuts that allow you to simplify expressions with exponents more easily. Let us begin by applying what we already know about exponents. Recall from Section 1.3 that repeated multiplication may be represented by positive integer exponents. We can use this definition to simplify expressions.

EXAMPLE 1 Simplify these expressions:

a. x^3x^4

b. $(y^3)^5$

c. $(xy)^6$

Solution

a. $x^3x^4 = (x * x * x)(x * x * x * x)$

$= x * x * x * x * x * x * x$ *Using the associative property of multiplication*

$= x^7$

b. $(y^3)^5 = (y * y * y)^5$

$= (y * y * y)(y * y * y)(y * y * y)(y * y * y)(y * y * y)$

$= y * y * y * y * y * y * y * y * y * y * y * y * y * y * y$

$= y^{15}$ *Using the associative property of multiplication*

c. $(xy)^6 = (xy)(xy)(xy)(xy)(xy)(xy)$

$= x * y * x * y * x * y * x * y * x * y * x * y$

Using the associative property of multiplication

$= (x * x * x * x * x * x) * (y * y * y * y * y * y)$

Using the commutative property of multiplication

$= x^6y^6$ ■■

Each of these results from Example 1 may be checked by assigning specific values to x and y. (If you do this, pick values other than 1 or 0, because 1 raised to any power is 1 and 0 raised to any positive power is 0.) For example, let $x = 3$ and let $y = 4$. Check each of the results.

EXAMPLE 1(a)

$$x^3x^4 = x^7 \quad \textit{Original equation}$$
$$3^33^4 \stackrel{?}{=} 3^7 \quad \textit{Substituting 3 for x}$$
$$27 * 81 \stackrel{?}{=} 2187$$
$$2187 = 2187 \quad \textit{OK!}$$

EXAMPLE 1(b)

$$(y^3)^5 = y^{15} \quad \textit{Original equation}$$
$$(4^3)^5 \stackrel{?}{=} 4^{15} \quad \textit{Substituting 4 for y}$$
$$64^5 \stackrel{?}{=} 1{,}073{,}741{,}824$$
$$1{,}073{,}741{,}824 = 1{,}073{,}741{,}824 \quad \textit{OK!}$$

EXAMPLE 1(c)

$$(xy)^6 = x^6y^6 \quad \textit{Original equation}$$
$$(3 * 4)^6 \stackrel{?}{=} 3^6 * 4^6 \quad \textit{Substituting 4 for y}$$
$$12^6 \stackrel{?}{=} 729 * 4096$$
$$2{,}985{,}984 = 2{,}985{,}984 \quad \textit{OK!}$$

Example 1 illustrates that you can apply the definition of positive integer exponents to simplify expressions. But when you did that, you had to write a lot of x's and y's. Not only is that time consuming, but it is easy to make a mistake in counting. Therefore, it would be nice if we had shortcuts to aid in simplifying expressions with exponents.

Begin by making these observations. In Example 1(a), we found the product of x^3 and x^4. Notice that the exponent in the product is the sum of the two exponents ($7 = 3 + 4$). This leads us to the product law for exponents.

Product Law for Exponents

Let a be any number and let m and n be integers. Then

$$a^m a^n = a^{m+n}$$

In Example 1(b), we found the result when we raised one power (y^3) to another power. Notice that the exponent in the result is the product of the two exponents ($15 = 3 * 5$). This leads us to the power-of-a-power law for exponents.

Power-of-a-Power Law for Exponents

Let a be any number and let m and n be integers. Then

$$(a^m)^n = a^{mn}$$

In Example 1(c), we found the result when we raised the product of two numbers (x and y) to the same power. Notice that each of the two numbers is raised to that power and the results are then multiplied together. This leads us to the power-of-a-product law for exponents.

Power-of-a-Product Law for Exponents

Let a and b be any numbers and let m be an integer. Then

$$(ab)^m = a^m b^m$$

Applying the Laws of Exponents

Now that you have discovered these three laws of exponents, you can apply them.

EXAMPLE 2 Use the laws of exponents to simplify these expressions. State which law you are using and identify the values or expressions assigned to a, b, m, and n.

a. $(z^4)^7$
b. $2x^5 * 3x^4$
c. $(3y)^2$
d. $(xz^3)^4$

Solution

a. $(z^4)^7 = z^{4*7}$ *Using the power-of-a-power law with $a = z$, $m = 4$, and $n = 7$*

$= z^{28}$

b. $2x^5 * 3x^4 = 2 * 3 * x^5 * x^4$ *Using the commutative property of multiplication*

$= 6x^{5+4}$ *Using the product law with $a = x$, $m = 5$, and $n = 4$*

$= 6x^9$

Notice that we multiply the coefficients, 2 and 3, because they are factors. We *add* the 5 and the 4 because they are exponents for factors with the same base, x.

c. $(3y)^2 = 3^2y^2$ *Using the power-of-a-product law with $a = 3$, $b = y$, and $m = 2$*

$= 9y^2$

d. $(xz^3)^4 = x^4(z^3)^4$ *Using the power-of-a-product law with $a = x$, $b = z^3$, and $m = 4$*

$= x^4z^{12}$ *Using the power-of-a-power law with $a = z$, $m = 3$, and $n = 4$*

EXAMPLE 3 The diameter of the sun is approximately $1.39 * 10^6$ km. Find the volume of the sun in cubic kilometers (km^3). Use this formula for the volume of a sphere: volume $= \frac{\pi}{6} * \text{diameter}^3$.

Solution

$$\text{volume} = \frac{\pi}{6} * \text{diameter}^3$$

$$= \frac{\pi}{6} * (1.39 * 10^6 \text{ km})^3 \quad \textit{Substituting into the formula}$$

$$= \frac{\pi}{6} * 1.39^3 * (10^6)^3 \text{ km}^3 \quad \textit{Using the power-of-a-product law}$$

$$= \frac{\pi}{6} * 1.39^3 * 10^{6*3} \text{ km}^3 \quad \textit{Using the power-of-a-power law}$$

$$= \frac{\pi}{6} * 1.39^3 * 10^{18} \text{ km}^3$$

$$\approx 1.41 * 10^{18} \text{ km}^3 \quad \textit{Using a calculator to find } \frac{\pi}{6} * 1.39^3$$

Note that the units (km) may be treated like a variable in applying the power laws, and that the answer is in cubic kilometers (km^3), an appropriate unit for volume.

We can check this answer by entering the entire expression in the calculator as $\frac{\pi}{6} * (1.39 * 10 \wedge 6) \wedge 3$. The display reads 1.40618682E18. Remember that E18 indicates *times* 10 *to the power* 18.

Negative Integer Exponents

The laws developed here apply to negative integer exponents as well as positive integer exponents. (In Chapter 9 you will learn that they apply to fractional exponents as well.) Recall the definition of negative exponents from Section 1.3.

Definition Negative Integer Exponents

Let a be any number except 0 and let n be any positive integer. Then

$$a^{-n} = \frac{1}{a^n}.$$

This definition is used in the following example.

EXAMPLE 4 Use the laws of exponents to simplify these expressions. Express the final answer using only positive exponents.

a. x^5x^{-3}
b. $x^{-5}x^3$
c. $(y^4)^{-2}$
d. $(x^2y^{-1})^{-3}$

Solution

a. $x^5x^{-3} = x^{5+(-3)}$ *Using the product law*

$= x^2$

b. $x^{-5}x^3 = x^{(-5)+3}$ *Using the product law*

$= x^{-2}$

$= \frac{1}{x^2}$ *Using the definition of negative integer exponent*

c. $(y^4)^{-2} = y^{4*(-2)}$ *Using the power-of-a-power law*

$= y^{-8}$

$= \frac{1}{y^8}$ *Using the definition of negative integer exponent*

d. $(x^2y^{-1})^{-3} = (x^2)^{-3}(y^{-1})^{-3}$ *Using the power-of-a-product law*

$= x^{2*(-3)} * (y^{-1*(-3)})$ *Using the power-of-a-power law*

$= x^{-6}y^3$

$= \left(\frac{1}{x^6}\right)\left(\frac{y^3}{1}\right)$ *Using the definition of negative integer exponent*

$= \frac{y^3}{x^6}$ *Multiplying fractions*

Here is an opportunity to apply negative exponents and the laws you have learned to a problem involving distance, rate, and time.

EXAMPLE 5 Light travels at a speed of $3 * 10^8$ m/s. One nanosecond (ns) is one billionth of a second, or 10^{-9} s. How far does light travel in 1 ns?

Solution We know the rate ($3 * 10^8$ m/s) and the time (10^{-9} s), so we can find the distance.

$$\begin{aligned} \text{distance} &= \text{rate} * \text{time} \\ &= 3 * 10^8 \text{ m/s} * 10^{-9} \text{ s} \\ &= 3 * 10^{8+(-9)} \text{ m} \\ &= 3 * 10^{-1} \text{ m} \\ &= 0.3 \text{ m} \end{aligned}$$

Light travels .3 m, or about 1 foot, in 1 ns. That's a short distance in an extremely small amount of time.

Quotient Laws for Exponents

Example 1 showed the first three laws of exponents. These laws may be extended to include quotients as well as products. Let's see if we can discover the quotient laws by working Example 6.

EXAMPLE 6 Use the definition of positive integer exponents to simplify these expressions:

a. $\frac{x^8}{x^5}$

b. $\frac{x^5}{x^8}$

c. $\left(\frac{x}{y}\right)^7$

Solution

a.
$$\frac{x^8}{x^5} = \frac{x * x * x * x * x * x * x * x}{x * x * x * x * x}$$
$$= \left(\frac{x}{x}\right)\left(\frac{x}{x}\right)\left(\frac{x}{x}\right)\left(\frac{x}{x}\right)\left(\frac{x}{x}\right)\left(\frac{x}{1}\right)\left(\frac{x}{1}\right)\left(\frac{x}{1}\right)$$
$$= \frac{x^3}{1}$$
$$= x^3$$

b.
$$\frac{x^5}{x^8} = \frac{x * x * x * x * x}{x * x * x * x * x * x * x * x}$$
$$= \left(\frac{x}{x}\right)\left(\frac{x}{x}\right)\left(\frac{x}{x}\right)\left(\frac{x}{x}\right)\left(\frac{x}{x}\right)\left(\frac{1}{x}\right)\left(\frac{1}{x}\right)\left(\frac{1}{x}\right)$$
$$= \frac{1}{x^3}$$
$$= x^{-3}$$

c.
$$\left(\frac{x}{y}\right)^7 = \left(\frac{x}{y}\right)\left(\frac{x}{y}\right)\left(\frac{x}{y}\right)\left(\frac{x}{y}\right)\left(\frac{x}{y}\right)\left(\frac{x}{y}\right)\left(\frac{x}{y}\right)$$
$$= \frac{x^7}{y^7}$$

The results of Example 6 may be generalized.

Quotient Law for Exponents

Let a be any number except 0 and let m and n be integers. Then

$$\frac{a^m}{a^n} = a^{m-n}$$

Power-of-a-Quotient Law for Exponents

Let a be any number, let b be any number except 0, and let m be an integer. Then

$$\left(\frac{a}{b}\right)^m = \frac{a^m}{b^m}$$

Caution: There are three power laws for exponents: power-of-a-power, power-of-a-product, and power-of-a-quotient laws. There is, however, no "power-of-a-sum" law. Don't make the mistake of thinking that $(x + y)^m = x^m + y^m$.

EXAMPLE 7 Use all five laws of exponents to simplify these expressions. Express the final answer using only positive exponents.

a. $\dfrac{x^5y^2}{xy^7}$

b. $\left(\dfrac{3x}{y^2}\right)^4$

c. $\left(\dfrac{x^2y^3}{xy^4}\right)^{-1}$

Solution

a. $\dfrac{x^5y^2}{xy^7} = \dfrac{x^5y^2}{x^1y^7}$ *Recognize that the x in the denominator may be written x^1*

$= \left(\dfrac{x^5}{x^1}\right)\left(\dfrac{y^2}{y^7}\right)$

$= x^{5-1}y^{2-7}$ *Applying the quotient law to both x- and y-factors*

$= x^4y^{-5}$

$= \left(\dfrac{x^4}{1}\right)\left(\dfrac{1}{y^5}\right)$ *Using the definition of negative exponent to write y^{-5} as a fraction*

$$= \frac{x^4}{y^5}$$

b. $\left(\frac{3x}{y^2}\right)^4 = \frac{(3x)^4}{(y^2)^4}$ *Using the power-of-a-quotient law*

$= \frac{3^4x^4}{y^{2*4}}$ *Using the power-of-a-product law in the numerator, power-of-a-power law in the denominator*

$= \frac{81x^4}{y^8}$

c. $\left(\frac{x^2y^3}{xy^4}\right)^{-1} = \left(\frac{x^2y^3}{x^1y^4}\right)^{-1}$ *Think of x as x^1*

$= (x^{2-1}y^{3-4})^{-1}$ *Applying the quotient law to both x- and y-factors*

$= (x^1y^{-1})^{-1}$

$= x^{1(-1)}y^{(-1)(-1)}$ *Applying the power-of-a-product law and the power-of-a-power law*

$= x^{-1}y^1$

$= \left(\frac{1}{x}\right)\left(\frac{y}{1}\right)$ *Rewriting x^{-1} as a fraction*

$= \frac{y}{x}$

Negative Exponents in Denominators

With some practice you should be able to apply these laws to any situation involving exponents. The most challenging problems may be those in which a negative exponent appears in the denominator. One more property of exponents comes in handy here.

Negative Exponent in the Denominator

Let a be any number except 0 and let n be a positive integer. Then

$$\frac{1}{a^{-n}} = a^n$$

This property can be demonstrated by considering the fact that $a^0 = 1$ and using the quotient law.

$$\frac{1}{a^{-n}} = \frac{a^0}{a^{-n}} = a^{0-(-n)} = a^{0+n} = a^n$$

In effect, a factor with a negative exponent may be moved from the denominator to the numerator or from the numerator to the denominator if the sign of the exponent is changed.

EXAMPLE 8 Simplify

$$\left(\frac{x^2}{x^{-1}y^3}\right)^{-2}$$

Solution

$$\left(\frac{x^2}{x^{-1}y^3}\right)^{-2} = \left(\frac{x^{2-(-1)}}{y^3}\right)^{-2}$$ *Using the quotient law to subtract the powers of x*

$$= \left(\frac{x^3}{y^3}\right)^{-2}$$

$$= \frac{(x^3)^{-2}}{(y^3)^{-2}}$$ *Using the power-of-a-quotient law*

$$= \frac{x^{-6}}{y^{-6}}$$ *Using the power-of-a-power law*

$$= \frac{y^6}{x^6}$$ *Moving factors with negative exponents from numerator to denominator and vice versa, changing the signs of the exponents*

Exercises 6.1

1. Simplify the expressions using repeated multiplication. See Example 1.

   a. x^2x^5 **b.** $(y^4)^3$ **c.** $(xy^2)^6$

2. Check your results from Exercise 1 by letting $x = 2$ and $y = 3$ and simplifying.

3. Use the product law for exponents, the power-of-a-power law for exponents, or the power-of-a-product law for exponents to simplify the expressions. State which fact you used and identify the values assigned to a, b, m, and n.

   a. $3a^4 * 4a^5$ **b.** $(b^3)^6$

   c. $(a^2b^4)^7$ **d.** $(4b)^3$

4. The earth is a sphere with an approximate diameter of $1.27 * 10^7$ m. Find the surface area of the earth using the formula surface area $= \pi *$ diameter^2. Perform your work in scientific notation.

5. Use the laws of exponents to simplify each expression and the definition of negative exponents (page 213) to write the final answer with only positive exponents.

a. $k^{-2}k^{6}$ **b.** $k^{2}k^{-6}$

c. $(j^{5})^{-3}$ **d.** $(j^{-5}k^{2})^{-4}$

6. Check your answers to Exercise 5 by picking values for j and k and evaluating the expressions on your calculator.

7. Simplify the expressions and write with only positive exponents.

a. $\dfrac{y^7}{y^3}$ **b.** $\dfrac{y^2}{y^8}$ **c.** $\left(\dfrac{y}{z}\right)^9$

8. A garden snail can move at a rate of $3.1 * 10^{-2}$ miles/hour. At this rate, how long would it take the snail to travel a distance of $1.55 * 10^3$ miles? Perform your work without using a calculator and write your answer in scientific notation. (World Almanac)

9. Correct the error in each line of the simplification of

$$\frac{(2x^3y^4)^3}{x^{-2}y^{12}}$$

$$\frac{(2x^3y^4)^3}{x^{-2}y^{12}} = \frac{6x^9y^{12}}{x^{-2}y^{12}}$$

$$= 6x^7y^0$$

$$= 6x^7(0)$$

$$= 0$$

10. Simplify the expressions and write the final answer using only positive exponents.

a. $\dfrac{xy^6}{x^3y^2}$ **b.** $\left(\dfrac{4y}{z^3}\right)^2$

c. $\left(\dfrac{a^{-3}b^2}{a^2b^5}\right)^3$

For Exercises 11–13, simplify the expressions and write the final answer using only positive exponents.

11. $\left(\dfrac{x^4y^{-3}}{x^5}\right)^{-1}$

12. $\dfrac{-6c^5d^{-2}}{(2c^{-3}d)^2}$

13. $\dfrac{4(x^7y^{-3})^0}{x^{-2}y}$

14. A formula for the volume of a sphere is $V = (\frac{4}{3})\pi r^3$, where r is the radius of the sphere. Find the exact volume of a sphere with a radius of $\frac{1}{2}$ m.

15. Find an integer value for n that will make the expression $\frac{x^3}{x^{-n}}$ equal to x.

Skills and Review 6.1

16. At a local candy shop, plain chocolate sells for \$3 per pound and deluxe chocolate sells for \$5 per pound. Suppose a 6-pound mixture of the two types of chocolate costs \$27.

a. What is the cost per pound of the mixture?

b. Does your answer from part (a) suggest that there is more plain chocolate or deluxe chocolate in the mixture? Explain your reasoning.

17. We wish to find the exact amount of each type of chocolate in the previous exercise.

a. Let x represent the number of pounds of plain chocolate and y the pounds of deluxe chocolate. Then a weight equation is $x + y = 6$ and a money equation is $3x + 5y = 27$. Explain what each term in the equations represents.

b. Solve the system of equations in order to find the exact amount of each chocolate in the mixture.

18. A pair of sneakers is marked up from \$60 to \$80. Find the percent increase.

19. Solve the proportion

$$\frac{3}{10} = \frac{x}{12}.$$

20. Given the equation

$$y = \frac{x}{4} - 5.$$

a. Find the slope.

b. Find the y-intercept.

c. Find the x-intercept.

21. Solve for x if $x - (7 - 3x) = 6x + 5$.

22. Verify the solution in Exercise 21.

23. Recall that a formula for the area of a triangle is

$$A = \frac{1}{2}bh.$$

Find the height of a triangle whose area is 20 cm^2 and base is 10 cm.

24. Write an algebraic expression to represent each statement.

a. 5 more than a number

b. 2 less than a number

c. 12 minus a number

d. The quotient of a number and 6

25. Evaluate $\sqrt{64} + 5 * 3^2$.

6.2 Products and Factors

AFTER STUDYING THIS SECTION YOU WILL BE ABLE TO

- Expand the product of a monomial and a binomial.
- Factor the product of a monomial and a binomial.
- Apply the zero-product property to solve equations of the form $ax^2 + bx = 0$.

Polynomials

Recall the distributive property, which states that for all numbers a, b, and c, $a * (b + c) = a * b + a * c$. This property is illustrated with an area model in Figure 1. In Section 2.1 we showed that the distributive property can be used in two ways:

1. We can expand $a(b + c)$ to get $ab + ac$.
2. We can factor $ab + ac$ to get $a(b + c)$.

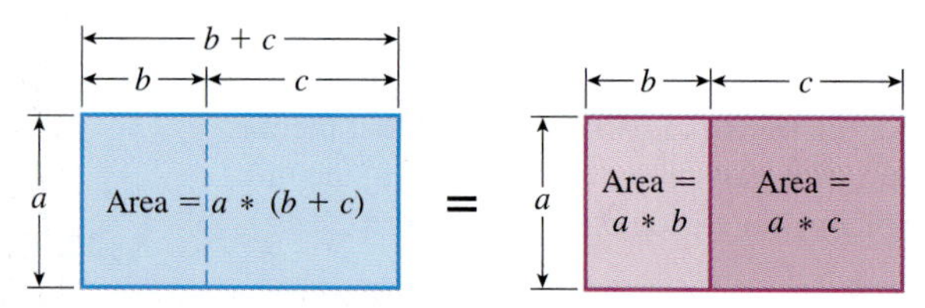

Figure 1 We can think of the rectangle as having one piece with area $a(b + c)$ or two pieces, one with area ab and one with area ac.

In this section, we extend our use of the distributive property by letting a, b, and c represent any algebraic expression involving the products of numbers and variables raised to positive integer powers. These expressions are called

monomials, indicating that they have one (mono-) term (-nomial). Here are some examples of monomials:

$3x$	a monomial in one variable, x
x^4y^2	a monomial in two variables, x and y
^-2xyz	a monomial in three variables, x, y, and z
5	a constant

Monomials may be joined by addition and subtraction to form expressions with several terms called **polynomials**, indicating that they have many (poly) terms. Polynomials with two terms are called **binomials**, and polynomials with three terms are called **trinomials**.

For example, $3x + 2$ is a binomial with two terms, $3x$ and 2; $4x^2 - 7xy + y^2$ is a trinomial with three terms, $4x^2$, ^-7xy, and y^2. The middle term is considered to be negative because subtracting $7xy$ is the same as adding ^-7xy. We use these vocabulary words extensively in the next three sections of this book. Our work with polynomials relies on the laws of exponents, which we learned in the previous section.

Multiplying a Monomial by a Polynomial

Our first examples involve expanding. That is, we start with two factors, one monomial and one polynomial, and we multiply them together to find their product.

EXAMPLE 1 ■ Expand the product of $3x^2$ and $6x^3 + 5xy^2$.

Solution An area model, as shown in Figure 2, can be used.

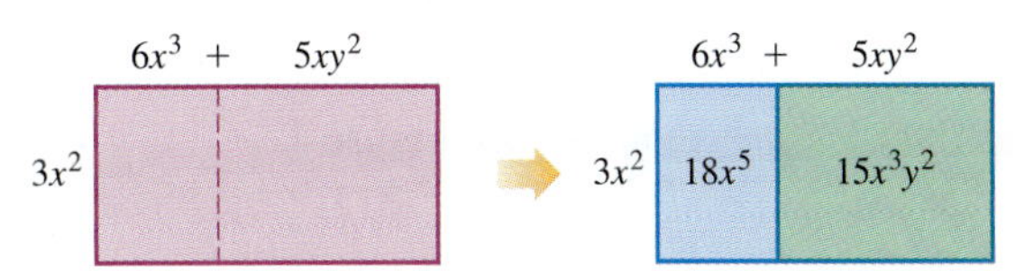

Figure 2 The area of the rectangle has two parts, $18x^5$ and $15x^3y^2$.

a. Start with a diagram showing the factors with $3x^2$ on one side of the rectangle and $6x^3 + 5xy^2$ on the other. Draw a line to show that the large rectangle is divided into two smaller rectangles.

b. Find the area of each of the smaller rectangles, using the product law for exponents:

$$3x^2 * 6x^3 = 18x^5$$

$$3x^2 * 5xy^2 = 15x^3y^2$$

c. Join the two terms to form the sum: $18x^5 + 15x^3y^2$.

The whole problem may be written as follows:

$$\mathbf{3x^2}(6x^3 + 5xy^2) = \mathbf{3x^2} * 6x^3 + \mathbf{3x^2} * 5xy^2$$
$$= 18x^5 + 15x^3y^2$$

Example 2 shows several other examples.

EXAMPLE 2 Expand these products:

a. $4xy(3x + 5y + 2z)$

b. $10x(7x - 6)$

c. $^-(x^2 - 3x - 4)$

Solution We show each example with an area model and with the steps written out symbolically.

a. This example shows the product of a monomial and a trinomial. See Figure 3.

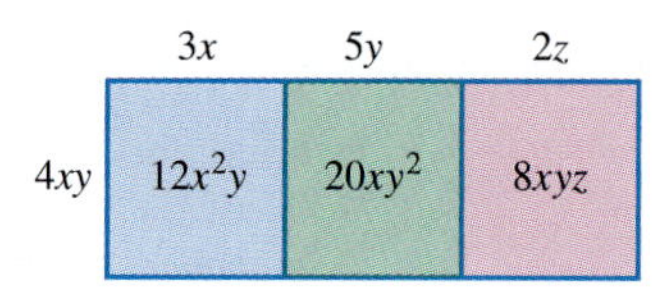

Figure 3

$$\mathbf{4xy}(3x + 5y + 2z) = (\mathbf{4xy} * 3x) + (\mathbf{4xy} * 5y) + (\mathbf{4xy} * 2z)$$
$$= 12x^2y + 20xy^2 + 8xyz$$

b. See Figure 4.

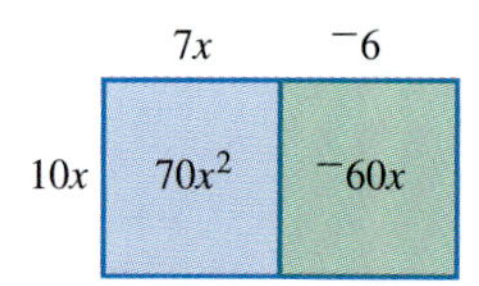

Figure 4

Note: Strictly speaking, there is no such thing as a negative length or a negative area. The area model, however, still works with negative quantities.

$$\mathbf{10x} * (7x - 6) = (\mathbf{10x} * 7x) + (\mathbf{10x} * {}^-6)$$
$$= 70x^2 + {}^-60x$$
$$= 70x^2 - 60x$$

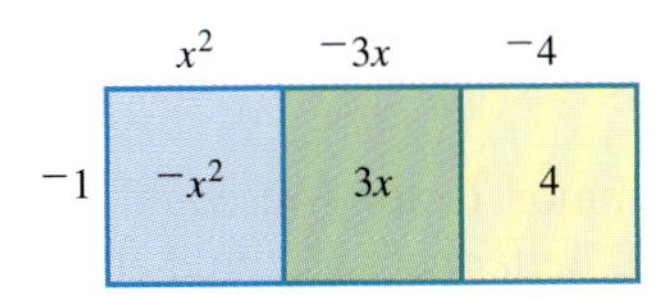

Figure 5

c. This example shows that the opposite of a polynomial may be thought of as a product of $^-1$ times the polynomial. (Recall the special properties of $^-1$, page 52.) See Figure 5.

$$
\begin{aligned}
^-(x^2 - 3x - 4) &= \mathbf{^-1}(x^2 - 3x - 4) \\
&= (\mathbf{^-1} * x^2) + (\mathbf{^-1} * {^-3x}) + (\mathbf{^-1} * {^-4}) \\
&= {^-x^2} + 3x + 4
\end{aligned}
$$

Reversing the Process: Finding Factors

Recall from Section 2.1 that factoring is the reverse process of expanding. We are given a polynomial. We try to find two factors, one a monomial and the other a polynomial, which multiplied together give us the original polynomial. Again, an area model can help us visualize the process.

EXAMPLE 3 ■ Use an area model to find factors that give the product $6x^2 - 15x$.

Solution Write $6x^2 - 15x$ as a sum, $6x^2 + {^-15x}$. Let the two terms, $6x^2$ and ^-15x, both represent areas of rectangles. Draw a diagram showing the two areas, as in Figure 6.

Figure 6 What can the height be?

We need to find an expression for the height of the rectangle, one that will allow us to find values that will fit along the top. In other words, we must find a *common factor* for $6x^2$ and ^-15x.

We could try x. Then we have the situation shown in Figure 7.

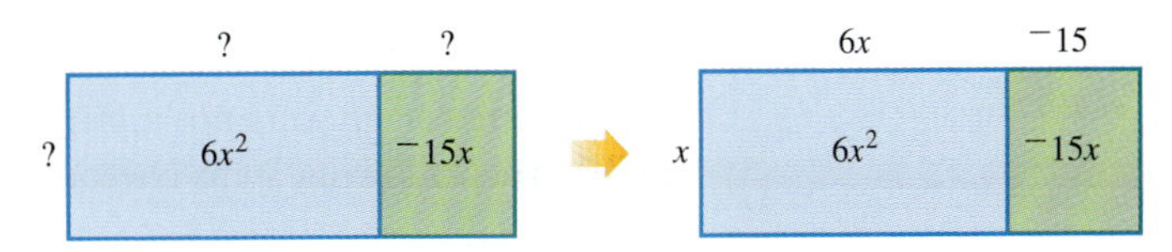

Figure 7 x is a common factor of $6x^2$ and ^-15x.

We think: x times what gives us $6x^2$? The answer is $6x$. We then think: x times what gives us ^-15x? The answer is $^-15$. We write $6x$ and $^-15$, as shown on the right in Figure 7. In symbols, we write

$$
\begin{aligned}
6x^2 - 15x &= 6x^2 + {^-15x} \\
&= x(? + ?) \\
&= x(6x + {^-15}) \\
&= x(6x - 15)
\end{aligned}
$$

More simply, we can write $6x^2 - 15x = x(6x - 15)$.

Greatest Common Factor

In the solution to Example 3 we chose x as a common factor. We could have chosen 3 because it is a factor of both $6x^2$ and $^-15$. Better yet, we could choose $3x$ as a common factor. We then have the situation shown in Figure 8. In symbols, we can write $6x^2 - 15x = 3x(2x - 5)$. When factoring a polynomial as the product of a monomial and a binomial, it is customary to find the **greatest common factor (GCF)** of the terms in the polynomial. We know we have found the greatest common factor if, after factoring out, there is no common factor left in the terms of the polynomial. Prime numbers are used in a procedure to find the GCF of the terms of a polynomial.

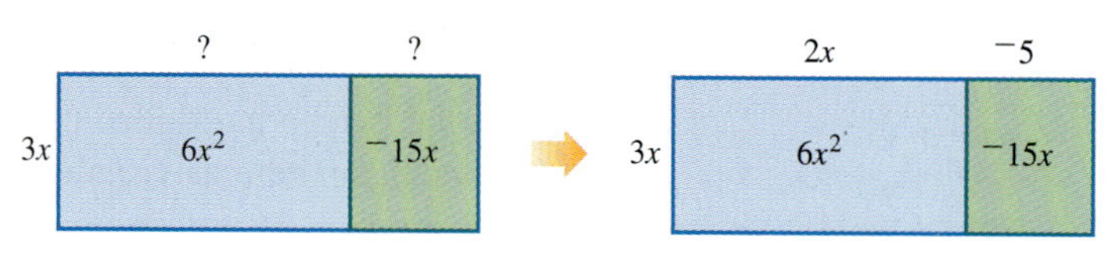

Figure 8

> **Definition Prime Number**
>
> A prime number is a positive integer with exactly two factors, the number itself and 1.

For example, 7 is a prime number because it has exactly two factors, 7 and 1. On the other hand, 6 is not a prime number because it has four factors, 6, 3, 2, and 1. The following steps let you find the GCF of any two monomials.

> **Finding the Greatest Common Factor (GCF)**
>
> Step 1 List all the variables and all the prime factors of the numerical coefficients.
>
> Step 2 When a variable or prime number does not appear, include that factor to the zeroth power. Remember that anything (except zero) raised to the zeroth power is 1.
>
> Step 3 Take the smallest power of each variable factor and each prime factor to form the GCF.

Example 4 illustrates finding the GCF of two monomials.

EXAMPLE 4 Find the GCF of these monomials.

a. x^7y^2 and xy^6
b. $14x^3y$, $2x^6y^4$, and $12y^5$
c. $84x$ and $140y$

Solution

a. The powers of x are x^7 and x^1. The smaller power is x^1. The powers of y are y^2 and y^6. The smaller power is y^2. Taking the smaller powers, the GCF is x^1y^2, or xy^2.

b. The powers of x are x^3, x^6, and x^0. The smallest power is x^0. The powers of y are y^1, y^4, and y^5. The smallest power is y^1. We now know that the GCF contains y^1. We can leave out x because its smallest power is $x^0 = 1$.

You may be able to see by inspection that the GCF of 14, 2, and 12 is 2. So, the GCF of the three monomials is $2y$. The procedure in step 2 may also be used to get the same result:

Factor the coefficients into prime factors: $14 = 2 * 7$; $2 = 2$; and $12 = 2^2 * 3$.

- The powers of 2 are 2^1, 2^1, and 2^2. The smallest power is 2^1.
- The powers of 3 are 3^0, 3^0, and 3^1. The smallest power is 3^0.
- The powers of 7 are 7^1, 7^0, and 7^0. The smallest power is 7^0.

Taking all the smallest powers, the GCF is $2^1 * 3^0 * 7^0 * x^0 * y^1$, or $2y$.

c. The smallest powers of x and y are x^0 and y^0, so the GCF does not contain any variable factors. Because the numerical coefficients are relatively large, you may want to use **factor trees** (as shown in Figure 9) to find the prime factors.

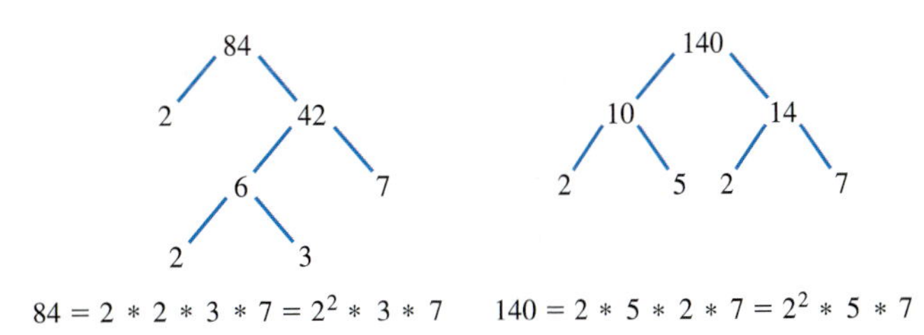

Figure 9

- The powers of 2 are 2^2 and 2^2. The smallest power is 2^2.
- The powers of 3 are 3^1 and 3^0. The smallest power is 3^0.
- The powers of 5 are 5^0 and 5^1. The smallest power is 5^0.
- The powers of 7 are 7^1 and 7^1. The smallest power is 7^1.

Taking all the smallest powers, the GCF is $2^2 * 7^1 = 28$.

In Example 5, we find the GCF of several terms.

EXAMPLE 5 ■ Factor completely: $8x^4y^2z - 6x^5yz^2 - 4x^3y^3$.

Solution The first step is to find the GCF of the monomials. By inspection the greatest common factor for the numerical coefficients is 2. To find the GCF of the variable factors, proceed as in Example 4:

- The powers of x are x^4, x^5, and x^3. The smallest power is x^3.
- The powers of y are y^2, y^1, and y^3. The smallest power is y^1.
- The powers of z are z^1, z^2, and z^0. The smallest power is z^0.

The GCF is $2x^3y$.

You are now ready to factor the polynomial:

$$8x^4y^2z - 6x^5yz^2 - 4x^3y^3 = 2x^3y(? - ? - ?)$$

- To find the coefficient in the first term, think: 2 times what gives 8? Answer: 4.
- To find the power of x in the first term, think: x^3 times what gives x^4? Answer: x^1.
- To find the power of y in the first term, think: y times what gives y^2? Answer: y.

Don't forget to include z in the first term. So the first term inside parentheses is $4xyz$:

$$8x^4y^2z - 6x^5yz^2 - 4x^3y^3 = 2x^3y(4xyz - ? - ?)$$

Do the same for each of the other terms:

$$8x^4y^2z - 6x^5yz^2 - 4x^3y^3 = 2x^3y(4xyz - 3x^2z^2 - 2y^2)$$

That's it!

Check the answer by expanding: $2x^3y(4xyz - 3x^2z^2 - 2y^2) = 8x^4y^2z - 6x^5yz^2 - 4x^3y^3$. Get in the habit of always checking your answer to the factoring problem. If you do, you will usually catch any mistakes. ■ ■

Factoring Out a Negative Number

Sometimes it is desirable to have a positive coefficient for the first term in the polynomial factor. If the first coefficient is negative, you may use a negative number for the GCF.

EXAMPLE 6 Factor these expressions so that the first term of the polynomial factor has a positive coefficient.

a. $^-12x^2 + 33x$

b. $^-16y - 7z$

Solution

a. The GCF of $^-12x^2$ and $33x$ is $3x$. Take ^-3x as the common factor:

$$^-12x^2 + 33x = {}^-3x(? + ?)$$

- Think: ^-3x times what gives $^-12x^2$? Answer: $4x$ (negative times positive = negative).
- Think: ^-3x times what gives $33x$? Answer: $^-11$ (negative times negative = positive).

$$^-12x^2 + 33x = {}^-3x(4x + {}^-11) = {}^-3x(4x - 11)$$

Again, you should check the answer by expanding $^-3x(4x - 11)$.

b. $16y$ and $7z$ appear to have no common factor, so their GCF is 1. We can take $^-1$ as the common factor.

$$^-16y - 7z = {}^-1(? + ?)$$

- Think: $^-1$ times what gives ^-16y? Answer: $16y$ (negative times positive = negative).
- Think: $^-1$ times what gives ^-7z? Answer: $7z$ (negative times positive = negative).

$$^-16y - 7z = {}^-1(16y + 7z)$$

Common Binomial Factors

In all the cases we have seen so far, the common factor has been a *monomial*. In the next two sections we encounter situations in which there is a common *binomial* factor. The next example illustrates this.

EXAMPLE 7

a. Factor $7xz - 4z$.

b. Factor $7x(y + 5) - 4(y + 5)$.

c. Show how (a) and (b) are related to each other.

Solution

a. The GCF of the two terms is z. So $7xz - 4z = z(? + ?) = z(7x - 4)$.

b. A common factor is the expression in parentheses, which is $y + 5$. The greatest common factor is also $y + 5$ because x does not appear in the second term and 4 and 7 have no common factor other than 1.

$$7x(y + 5) - 4(y + 5) = (y + 5)(? + ?) = (y + 5)(7x - 4)$$

To check this result, you need to expand the product of two binomials. You will learn how to do that in the next section.

c. (a) and (b) are essentially the same problem if we let z replace the binomial $y + 5$ in (b). When factoring out a common binomial factor, you may want to think of representing the binomial $(y + 5)$ as another variable, such as z.

The Zero-Product Property

If the product of two numbers is zero, then at least one of them must be zero. This is called the zero-product property, and it can be used to solve certain types of equations.

> **The Zero-Product Property**
>
> Suppose $a * b = 0$. Then either $a = 0$ or $b = 0$.

You can convince yourself that the zero-product property is true if you think of factor pairs that give a product of zero: for example, $4 * 0 = 0$, $0 * 5 = 0$, $^{-}3 * 0 = 0$, and $0 * 0 = 0$. In all cases, either the first or the second factor is zero. In the last case both factors are zero. On the other hand, if neither factor is zero, you get either a positive product (if both factors are positive or both are negative) or a negative product (if one factor is positive and the other factor is negative).

The zero-product property may be used to solve some equations.

EXAMPLE 8 ■ Solve the equation $x^2 - 6x = 0$.

Solution The expression on the left side of the equation has a common factor, x, and so it may be rewritten $x(x - 6) = 0$. Now we can apply the zero-product property. The factors are $a = x$ and $b = x - 6$. So, either $x = 0$ or $x - 6 = 0$. This means that the equation has two solutions, $x = 0$ and $x = 6$. Up until this point most equations you have seen have had just one solution. In the next chapter you will learn that this is an example of a quadratic equation. Some quadratic equations have two solutions.

You may check this result with a calculator. Enter Y1 = X ∧ 2 − 6X and make a table of values with TblStart = 0 and ΔTbl = 1. You will find that both $x = 0$ and $x = 6$ give values of 0 for Y1.

Exercises 6.2

1. Use an area model, such as the one shown in Figure 4, to show that the expansion of $3(x + 5)$ equals $3x + 15$.
2. Find the product of $2a$ and $4a^2 + 7a^3b$.

Expand the products in Exercises 3–5.

3. $8z(5z - 3)$
4. $6a^2b(2a + 3b - 7c)$
5. $^-(^-x^2 + 9x - 5)$
6. Recall that a prime number is divisible only by one and itself.
   a. Find six prime numbers.
   b. Factor 90 into a product of prime factors.
   c. Rewrite your prime factors of 90, showing all duplicate factors as a single factor with an exponent.
   d. Write 75 as a product of its prime factors.
   e. Find the greatest common factor (GCF) of 75 and 90.
7. **a.** Use the figure to show the factorization of $10x - 6$.

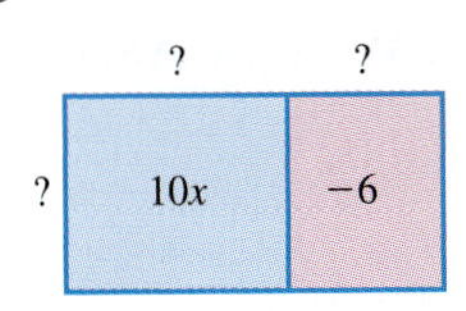

   b. Find the GCF of $54y$ and 72, then factor completely $54y + 72$.
   c. Find the GCF and factor completely: $14x^3 + 7xy - 42x^2$.
   d. Verify your factorization from part (c) by expanding using the distributive property.
8. Consider the monomials, x^2 and x^3.
   a. Use the steps on page 224 to find the greatest common factor (GCF) of the two monomials.
   b. Use the laws of exponents to show that x^3 is divisible by x^2.
9. Factor completely:
   a. $18x^5 + 6x^3$ **b.** $21x^2y^3 - 15xy^4$
10. Find the GCF and factor completely, $3xy^4z^2 - 12x^3y^2z + 9x^5z^3$. Expand your result to check your answer.
11. The surface area of a right circular cylinder is $2\pi rh + 2\pi r^2$ (see the figure). Factor out the GCF to find an equivalent expression for the surface area.

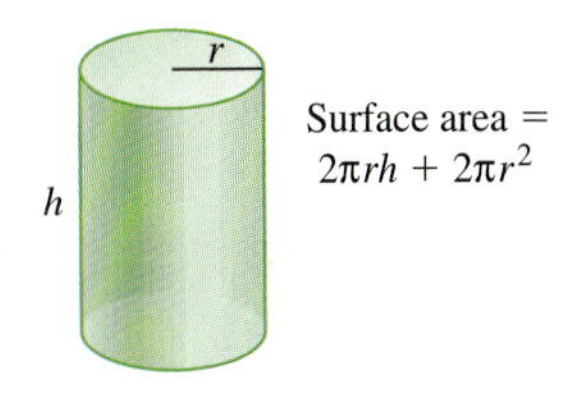

12. Factor out $^-1$ from each binomial.
   a. $^-x - 5$ **b.** $^-3x + 8$
13. Consider the expression $(x + 3)4y + (x + 3)9$. This expression has two terms.
   a. What is the common binomial factor in the two terms?
   b. Factor out the common binomial (GCF).
14. **a.** Factor $xz - 5z$.
   b. For the expression $x(2x - 3) - 5(2x - 3)$, what could you replace $(2x - 3)$ with in order to make it identical to the binomial in part (a)?
   c. Factor $x(2x - 3) - 5(2x - 3)$.
15. Factor the expression on the left side of each equation and then use the zero-product property to solve each equation.
   a. $x^2 + 4x = 0$ **b.** $x^2 - 11x = 0$

Skills and Review 6.2

16. Expand the product and write your final answer with only positive exponents.

$$2x^3y(5x^5 - x^4y^{-1} + 3z^{-2})$$

17. Use the rules of exponents to simplify. Write your answer with only positive exponents.

$$\left(\frac{6xy^4}{2x^5y^4}\right)^{-2}$$

18. Given the expression $\frac{5+x^4}{x^3}$.

a. Is $5+x$ or $\frac{5}{x^3}+x$ equivalent to the given expression?

b. Evaluate the expressions when $x = {}^-2$.

c. Does your evaluation support your answer from part (a)? Explain.

19. How much alcohol is in 2 L of a 25% solution of alcohol?

20. A chemist needs 2 L of a 25% solution of alcohol. She has a bottle that contains a 20% solution and another bottle that contains a 50% solution. She wants to know how much of each solution she should mix together.

a. Write a solution equation.

b. Write an alcohol equation.

c. Solve the system of equations.

d. Answer the question.

21. A 3-inch by 5-inch picture is enlarged so that the length of the enlargement is 6 inches more than its width. Find the width and length of the enlargement using either a proportion or some other method of your choice.

22. Solve the system of equations by an algebraic method.

$$5x - 2y = {}^-22$$

$$4x - 3y = {}^-19$$

23. Given an equation in standard form, $5x + 6y = 70$.

a. Find the slope.

b. Find the x- and y-intercepts.

24. Find the equation of the line passing through the points $({}^-1, 5)$ and $(2, {}^-4)$. See Section 4.2.

25. Find the distance between the points $({}^-6, 3)$ and $(6, 8)$.

6.3 Factoring by Grouping

AFTER STUDYING THIS SECTION YOU WILL BE ABLE TO

- Expand the product of two polynomials.
- Square a binomial.
- Use grouping to factor a polynomial with four terms.

Multiplying Polynomials

In the last section you saw how an area model represents the product of a monomial and a polynomial. The same method may be used to show multiplication by polynomials, each having two or more terms.

EXAMPLE 1 ■ Expand $(x + 3)(2x^2 + 4x + 5)$.

Solution Start with a diagram showing the factors $(x + 3)$ on one side of a rectangle and $(2x^2 + 4x + 5)$ on an adjacent side (Figure 10). Then find the area of each of the smaller rectangles (Figure 11).

The individual terms shown in Figure 11 can now be written as a polynomial with six terms:

$$(x + 3)(2x^2 + 4x + 5) = 2x^3 + 4x^2 + 5x + 6x^2 + 12x + 15$$

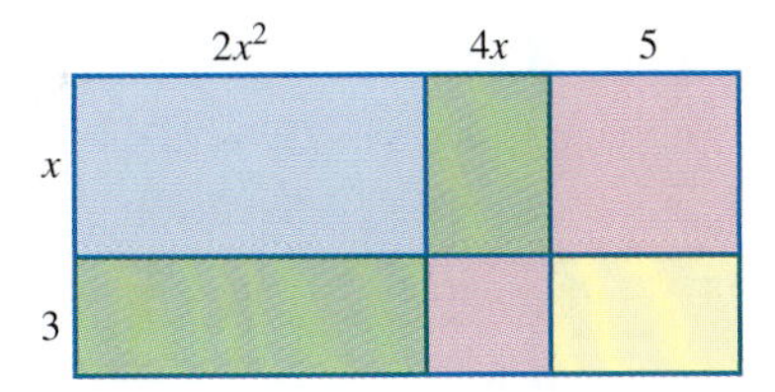

Figure 10 Each monomial in a factor goes with a row or column.

	$2x^2$	$4x$	5
x	$2x^3$	$4x^2$	$5x$
3	$6x^2$	$12x$	15

Figure 11

Combine like terms to get a polynomial with four terms:

$$(x + 3)(2x^2 + 4x + 5) = 2x^3 + 10x^2 + 17x + 15$$

You may always use an area model to expand the product of polynomials. With some practice, however, you may prefer not to draw a diagram but just to write out the results. When you do that, remember that you are really using the distributive property twice:

$$\begin{aligned}(x + 3)(2x^2 + 4x + 5) &= x(2x^2 + 4x + 5) + 3(2x^2 + 4x + 5) \\ &= 2x^3 + 4x^2 + 5x + 6x^2 + 12x + 15 \\ &= 2x^3 + 10x^2 + 17x + 15\end{aligned}$$

Note: In Example 1, both factors, $x + 3$ and $2x^2 + 4x + 5$, and the expanded polynomial $2x^3 + 10x^2 + 17x + 15$ are written in **descending order**: that is, the highest power of x comes first, followed by the second highest power, and so on, with the constant term last. In the remainder of this book we will usually write polynomials in descending order, because this makes it easier to keep track of like terms.

EXAMPLE 2 ■ Expand $(x - 7)(3x + 4)$.

Solution ■ Think of $x - 7$ as $x + {}^-7$. Then draw an area model (Figure 12).

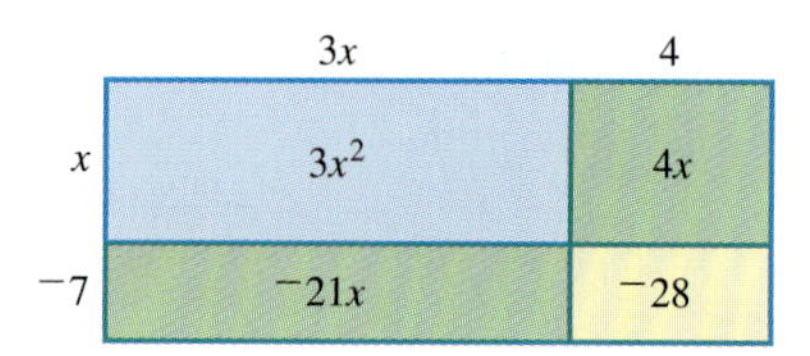

Figure 12

Combine terms to obtain

$$(x - 7)(3x + 4) = 3x^2 + 4x - 21x - 28$$
$$= 3x^2 - 17x - 28$$

Squaring a Binomial

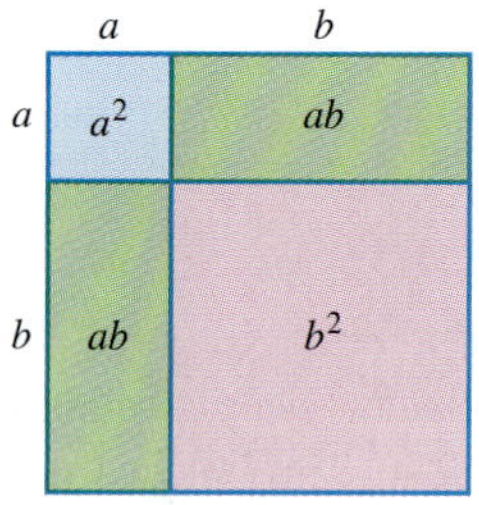

A special application of polynomial multiplication occurs when a binomial is squared. Figure 13 shows an area model for $(a + b)^2$.

Figure 13

Square of a Binomial

$$(a + b)^2 = a^2 + 2ab + b^2$$

Think: The square of $a + b$ is the square of the first term (a^2) plus twice the product of the two terms ($2ab$) plus the square of the second term (b^2).

Caution: Don't forget the middle term: $(a + b)^2 \neq a^2 + b^2$! This formula for $(a + b)^2$ may be used in solving quadratic equations, as you will see in Chapter 7. It may also be used to prove the Pythagorean theorem (see Exercise 15).

EXAMPLE 3 Expand each binomial square.

a. $(x + n)^2$

b. $(2x + 3y)^2$

c. $(x - 5)^2$

Solution

a. $(x + n)^2 = x^2 + 2xn + n^2$ *Squaring the binomial with $a = x$ and $b = n$*

b. $(2x + 3y)^2 = (2x)^2 + 2 * 2x * 3y + (3y)^2$ *Squaring the binomial with $a = 2x$ and $b = 3y$*

$= 4x^2 + 12xy + 9y^2$ *Using the power-of-a-product law*

c. $(x - 5)^2 = (x + {}^-5)^2$ *Subtracting is the same as adding the opposite*

$$= x^2 + 2 * x * {}^-5 + ({}^-5)^2$$
$$= x^2 - 10x + 25$$

Factoring the Product of Polynomials

Now we come to a greater challenge. We can always expand the product of two polynomials by creating an area model. But can we always factor a polynomial into two or more polynomial factors with integer coefficients? The answer is no, not all polynomials may be factored. Such polynomials are called **prime**, just as a prime number is an integer whose only factors are itself and 1.

There are techniques, however, for factoring some polynomials. We discuss some of those techniques in this course. A key strategy for these factoring problems is called **grouping**, as Example 4 shows.

EXAMPLE 4

a. Use an area model to factor $5y - 6x - 15 + 2xy$.

b. Use symbolic manipulation to factor $5y - 6x - 15 + 2xy$.

Solution

a. The only strategy we have learned so far is to look for a common factor. Unfortunately $5y$, ${}^-6x$, ${}^-15$, and $2xy$ have no factor other than 1 in common. Observe, however, that $5y$ and ${}^-15$ do have a common factor, 5. This suggests that $5y$ and ${}^-15$ may be grouped together. An area model may be used to show the grouping (Figure 14). The GCF for these two terms is 5, so we write 5 on the common side. We think: 5 times what is $5y$ and 5 times what is ${}^-15$? We write y and ${}^-3$, as shown in Figure 15 at the right.

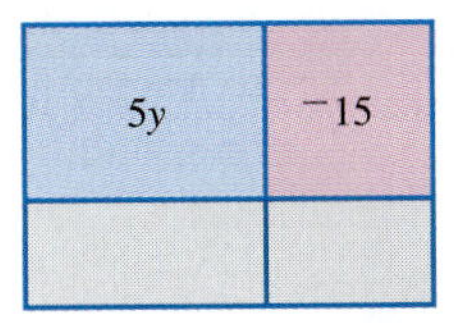

Figure 14 Because $5y$ and ${}^-15$ have a common factor, we put them in the same row.

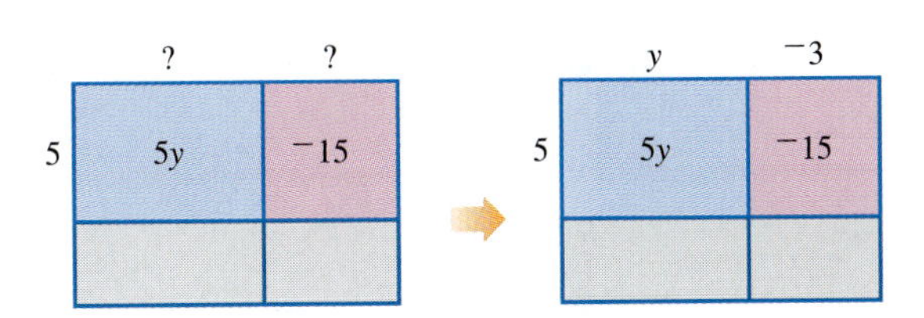

Figure 15

We now need to place $^{-}6x$ and $2xy$ in the area model. Two possibilities are shown in Figure 16. We ask whether or not y and $^{-}3$ are factors of the products with which they are aligned. We choose the placement shown in Figure 16 on the right.

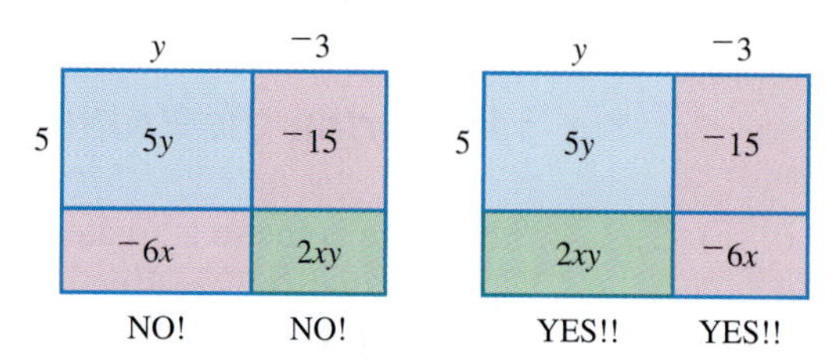

Figure 16

The final step is to find the remaining factor: $2x$ is the GCF of $2xy$ and $^{-}6x$, as shown in Figure 17.

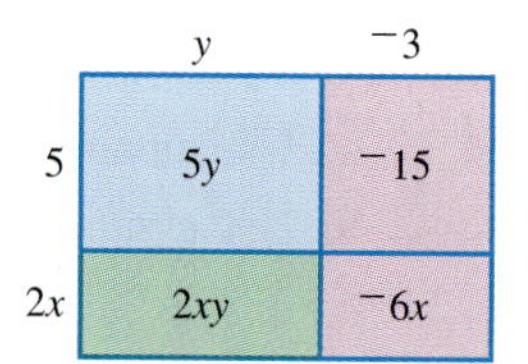

Figure 17

We can now write

$$5y - 6x - 15 + 2xy = (5 + 2x)(y - 3)$$

We have found two binomial factors for our original polynomial. The diagram in Figure 17 can be used to check our work by verifying that each of the four terms in the rectangles is equal to the product of the two terms on its sides.

b. You can factor by grouping without an area model if you prefer. You may have to rearrange the terms to accomplish this. Here is how you can use symbolic manipulation to factor $5y - 6x - 15 + 2xy$.

$5y - 6x - 15 + 2xy = 5y - 15 + 2xy - 6x$ *Using the commutative property to rearrange terms*

$= (5y - 15) + (2xy - 6x)$ *Forming groups of two terms*

$= 5(\mathbf{y - 3}) + 2x(\mathbf{y - 3})$ *Finding the GCF for each group*

$= (\mathbf{y - 3})(5 + 2x)$ *Factoring out the common binomial factor, $y - 3$*

A Test for Factorability

Factoring by grouping will work for some polynomials but not all. The next example leads to a way of testing a polynomial with four terms to see whether or not it can be factored by the grouping method.

EXAMPLE 5

a. Use an area model to factor $ac + ad + bc + bd$.

b. Find the product of each pair of diagonally opposite terms in the area model.

Solution

a. The first two terms, ac and ad, have a common factor, a. Once they are grouped together, it is natural to group bc and bd, as shown in Figure 18. Therefore, $ac + ad + bc + bd = a(c + d) + b(c + d) = (a + b)(c + d)$.

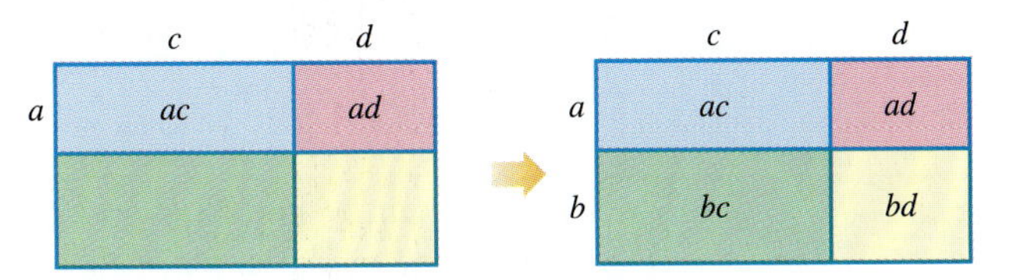

Figure 18 $ac + ad + bc + bd = (a + b)(c + d)$

b. Along each diagonal of the large rectangle in Figure 18, there are two pairs of terms. Going from northwest to southeast, we have ac and bd, with a product of $acbd$. Going from northeast to southwest, we have ad and bc, with a product of $adbc$. Because of the commutative property of multiplication, $acbd = adbc$, and we may conclude that the two diagonal products are equal. See Figure 19.

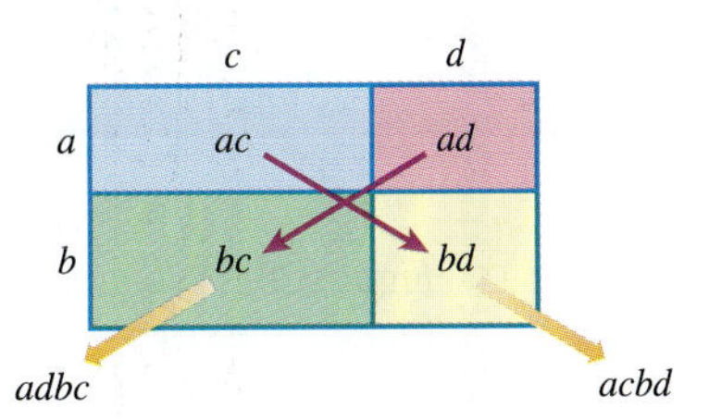

Figure 19 The products along the diagonals are equal.

The following test summarizes what we discovered in Example 5.

The Diagonal Test for Grouping

A polynomial with four terms may be factored by grouping if the product of one pair of terms is equal to the product of the other two terms. The

(continued)

polynomial may then be written in the form

$$ac + ad + bc + bd = (a + b)(c + d),$$

as in Figure 18.

EXAMPLE 6 ■ Factor $x^3 - 12x + 48 - 4x^2$.

Solution Use the diagonal test for grouping. Start with the first term, x^3, and see what other term it can be paired with:

$$x^3 * {}^-12x = {}^-12x^4 \neq 48 * {}^-4x^2 \qquad \textit{No}$$

$$x^3 * {}^-4x^2 = {}^-4x^5 \neq {}^-12x * 48 \qquad \textit{No}$$

$$x^3 * 48 = 48x^3 = {}^-12x * {}^-4x^2 \qquad \textit{Yes}$$

The last pair, x^3 and 48, works because its product is equal to ${}^-12x * {}^-4x^2$. Now use an area model to place each pair along a diagonal.

Figure 20 shows that there is more than one way to arrange the factors. In both cases x^3 and 48 appear along one diagonal and ${}^-4x^2$ and ${}^-12x$ along the other diagonal. The factorization is the same in both cases: $x^3 - 4x^2 - 12x + 48 = (x^2 - 12)(x - 4)$. If you prefer to use symbolic manipulation, rearrange the terms so that one pair, say, x^3 and 48, are the first and last terms and the other two terms are in the middle.

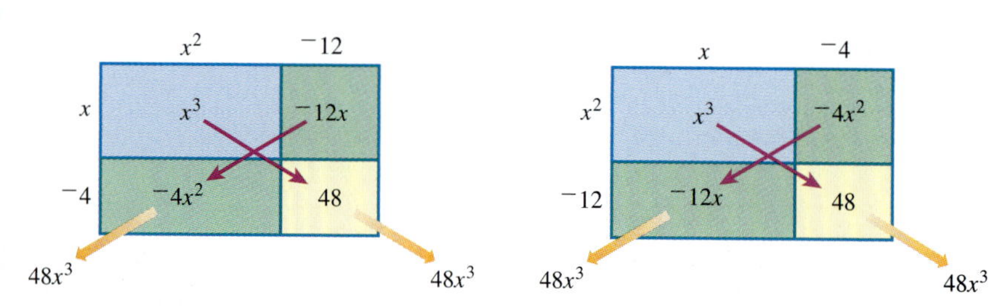

Figure 20 We get two pictures for the same factorization depending on where we place the terms.

$$x^3 - 4x^2 - 12x + 48 = x^2(\mathbf{x - 4}) + {}^-12(\mathbf{x - 4})$$ *Factoring out common factors for the first pair and second pair of terms: notice that the second common factor is negative*

$$= (\mathbf{x - 4})(x^2 - 12)$$ *Factoring out the common binomial factor*

EXAMPLE 7 ■ Use the diagonal test to factor $x^4 - x^2 + x + 3$.

Solution ■ Try to pair the first term with another term:

$$x^4 * 3 = 3x^4 \neq {}^-x^2 * x \qquad \textit{No}$$

$$x^4 * x = x^5 \neq {}^-x^2 * 3 \qquad \textit{No}$$

$$x^4 * {}^-x^2 = {}^-x^6 \neq x * 3 \qquad \textit{No}$$

Because none of these pairings works, the diagonal test fails and $x^4 - x^2 + x + 3$ cannot be factored. It is a prime polynomial.

The following procedure can be used to factor some polynomials with four terms.

Procedure for Factoring a Polynomial with Four Terms

Step 1 Apply the diagonal test for grouping. If the test works, go on to steps 2 and 3. If the test fails, then the polynomial cannot be factored.

Step 2 Arrange the terms so that the product of the first and last terms is equal to the product of the middle two terms.

Step 3 Use an area model or factor symbolically.

Exercises 6.3

1. **a.** Complete the area model in the figure to expand $(x + 6)(3x^2 + 1)$.

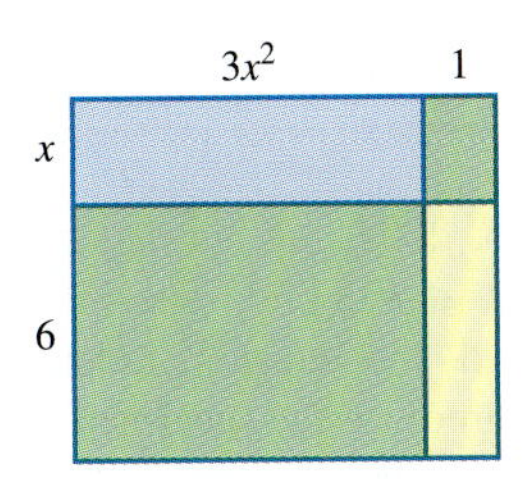

b. Without an area model, expand $(x + 6)(3x^2 + 1)$.

c. Check your answer by substituting a value for x into the original expression and into the expanded expression.

2. Use an area model to expand $(2x - 3)(x^2 - 5x + 4)$. On a calculator, enter the original expression as Y1 and the expanded expression as Y2. Check your answer by comparing your calculator's table values for the original expression to that for your expansion.

3. Expand $(x + 7)(4x - 9)$. On a calculator, enter the original expression as Y1 and the expanded expression as Y2. Use a window of ${}^-9.4 \leq X \leq 9.4$, ${}^-100 \leq Y \leq 100$, and check your expansion by comparing the graphs of Y1 and Y2.

4. Expand $(y + 3)(y^2 - 3y + 9)$.

5. The side of a square is $x + 4y$. Find its area.

6. In this section, you were shown a shortcut for squaring a binomial. Use this shortcut to perform the following operations.

a. $(a - b)^2$ **b.** $(2x + y)^2$ **c.** $(3x - 7y)^2$

7. Expansion is the reverse of what process?

8. Use an area model to show that the polynomial $6 + 12y - x - 2xy$ is equivalent to its factored form, $(6 - x)(1 + 2y)$.

9. Use an area model to factor the polynomial $2y^2 + 7x + 2xy + 7y$.

10. Use symbolic manipulation to factor $xy - 5x - 4y + 20$. See Example 4(b) for an illustration.

Exercises 11 and 12 concern the four groupings of the polynomial $3x - xy + 6 - 2y$ in the figure.

a.

$3x$	$^{-}2y$
6	^{-}xy

b.

$3x$	^{-}xy
6	$^{-}2y$

c.

$3x$	6
^{-}xy	$^{-}2y$

d.

6	$^{-}2y$
$3x$	^{-}xy

11. Use the diagonal test for grouping to determine which one grouping arrangement does not lead to a factorization.

12. In Exercise 11, you determined that one grouping arrangement does not lead to a factorization. Factor the remaining groupings to show that each will produce the same factorization of $3x - xy + 6 - 2y$.

For Exercises 13 and 14, use the diagonal test for grouping to determine if the polynomial can be factored, and, if so, factor it.

13. $6x^3 - 2x^2 - 15x + 5$

14. $2a + 2b + 5x + 5y$

15. Prove the Pythagorean theorem by following the steps indicated below.

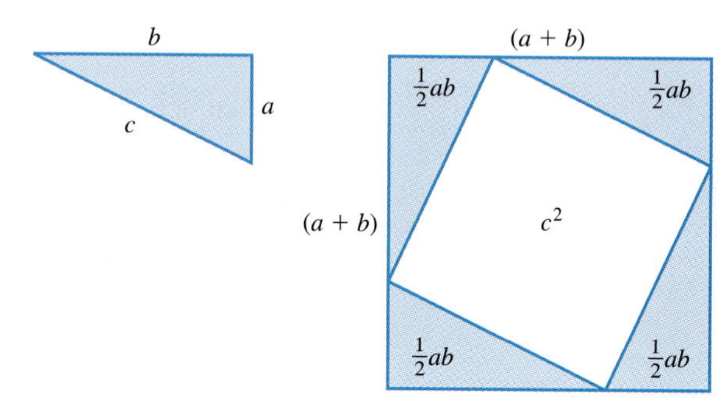

A square with side equal to $a + b$ is shown at the right of the figure. Four copies of a right triangle are arranged along the sides. The region in the middle of the large square is a smaller square whose side is of length c.

a. Find the area of one of the right triangles.

b. Find the total area of all four right triangles.

c. Find the sum of the area of the four right triangles and the square in the middle whose side is of length c.

d. Find the area of the large square using the expansion of $(a + b)^2$.

e. Write an equation with the results from (c) and (d).

f. Subtract the same quantity from both sides of the equation to arrive at $a^2 + b^2 = c^2$.

Skills and Review 6.3

16. Solve the equation $^{-}2x^2 + 14x = 0$.

17. Expand the product $6x(x^2 - 2x + 11)$.

18. Simplify and write without negative exponents: $a^4(3ab^2)^{-3}$.

19. Randy has \$5000 to invest. He invests some of the money in an annuity earning 8.2% annual interest and the remainder in a CD that pays 5% annual interest. How much should Randy invest in each account if he wants to earn a 7% yield overall?

20. On August 15, 2002, the Nasdaq composite index for stocks opened at 1334.30. The index gained 0.803% that day. Find the closing index. (*Note:* The index is measured to the nearest hundredth). (http://dynamic.nasdaq.com)

21. A beverage company found that in a sample of 1500 people, 276 would use their new power drink. Predict the number of people who will use their power drink in a market of 1,400,000 people.

22. Simplify $4x^3 + x^2(x - 2) + 3x^2$.

23. Solve for x if

$$\frac{2}{3}x - \frac{1}{4} = \frac{5}{6}$$

24. Verify that $x = 12$ is a solution to the equation $^-2x + 6x + 7 = 5(x - 1)$.

25. Evaluate $\frac{^-b}{2a}$ for $a = \frac{1}{5}$ and $b = {}^-\frac{3}{2}$.

6.4 Factoring Quadratic Trinomials

After Studying This Section You Will Be Able To

- Factor quadratic trinomials.
- Factor the difference of two squares.
- Completely factor polynomials requiring several steps.
- Apply the zero-product property to solve equations of the form $ax^2 + bx + c = 0$.

In the previous section you learned how to factor some polynomials with four terms by the method of grouping. We can now extend this method to **quadratic trinomials**. These polynomials have three terms (hence the name *trinomial*), and the term with the highest power of the variable contains a square (hence the adjective *quadratic*). Factoring quadratic trinomials is a skill that is used in solving some equations, as we see at the end of this section.

Splitting the Middle Term

The fundamental strategy we use to factor quadratic trinomials is to create four terms out of three and use the grouping method you already know. The next example shows how we can do this by *splitting the middle term*.

EXAMPLE 1 ■ Factor $x^2 + 8x + 12$.

Solution We can split the middle term, $\mathbf{8x}$, into two terms, $\mathbf{6x}$ and $\mathbf{2x}$. Then we rewrite the trinomial as $x^2 + \mathbf{6x} + \mathbf{2x} + 12$. Now we have a polynomial with four terms. We apply the diagonal test for grouping and find that the product of the first and last terms ($x^2 * 12 = 12x^2$) is the same as the product of the two middle terms ($6x * 2x = 12x^2$). We factor by grouping, as shown in Figure 21. We can conclude that

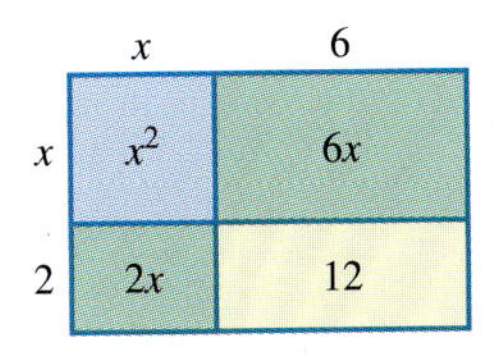

Figure 21

$$x^2 + 8x + 12 = (x + 2)(x + 6)$$

At this point you might reasonably ask how we knew to split $8x$ the way we did. Why didn't we say $8x = 3x + 5x$ or $8x = 4x + 4x$? That's a good question, and it must be answered if we are to succeed in factoring trinomials.

Let's look at the diagonal product $12x^2$ and ask ourselves how we can find two middle terms that will give $12x^2$ as a product. Obviously we are looking for two x terms, so our terms will have the form $\_\_x$ and $\_\_x$. We can fill in the blanks with pairs of numbers that have a product of 12. There are three such pairs:

$$1 * 12 = 12$$
$$2 * 6 = 12$$
$$3 * 4 = 12$$

We can make these factors coefficients of x-terms and add them together to see which give us the desired middle term $8x$.

$1x + 12x = 13x$	*No*
$\mathbf{2x + 6x = 8x}$	*Yes*
$3x + 4x = 7x$	*No*

Procedure for Factoring a Quadratic Trinomial

Step 1 Write the terms of the quadratic trinomial in descending order.
Step 2 Find the diagonal product of the first and last terms.
Step 3 Find all factor pairs for the diagonal product.
Step 4 Search systematically for a pair whose sum is the middle term. Use this pair to split the middle term.
Step 5 Factor with an area model or symbolically by grouping.

EXAMPLE 2 Factor these trinomials:

a. $2x^2 + 9x + 10$
b. $x^2 - 13x + 36$
c. $3x^2 + 20x - 7$
d. $10x^2 - 3 - x$

Solution **a.** $2x^2 + 9x + 10$ is already in descending order. The diagonal product is $2x^2 * 10 = 20x^2$. The middle term must be split so that $9x = \_\_x + \_\_x$. We need to find factor pairs for 20.

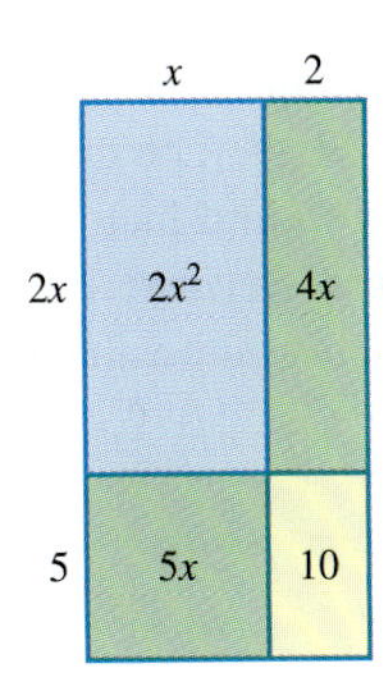

Figure 22

$1 * 20 = 20$	$1x + 20x = 21x$	*No*
$2 * 10 = 20$	$2x + 10x = 12x$	*No*
$\mathbf{4 * 5 = 20}$	$\mathbf{4x + 5x = 9x}$	***Yes***

We split the middle term $9x$ into two terms, $4x$ and $5x$. See Figure 22. Symbolically, we have

$$\begin{aligned} 2x^2 + 9x + 10 &= 2x^2 + 4x + 5x + 10 \\ &= 2x(\mathbf{x+2}) + 5(\mathbf{x+2}) \quad \textit{Factoring by grouping} \\ &= (\mathbf{x+2})(2x+5) \end{aligned}$$

b. $x^2 - 13x + 36$ is already in descending order. The diagonal product is $x^2 * 36 = 36x^2$. The middle term must be split so that $^-13x = \_\_x + \_\_x$. We need to find factor pairs for 36. Because the middle term is negative and 36 is positive, we look for two negative factors.

$^-1 * {}^-36 = 36$	$^-1x + {}^-36x = {}^-37x$	*No*
$^-2 * {}^-18 = 36$	$^-2x + {}^-18x = {}^-20x$	*No*
$^-3 * {}^-12 = 36$	$^-3x + {}^-12x = {}^-15x$	*No*
$\mathbf{^-4 * {}^-9 = 36}$	$\mathbf{^-4x + {}^-9x = {}^-13x}$	***Yes***
$^-6 * {}^-6 = 36$	$^-6x + {}^-6x = {}^-12x$	*No*

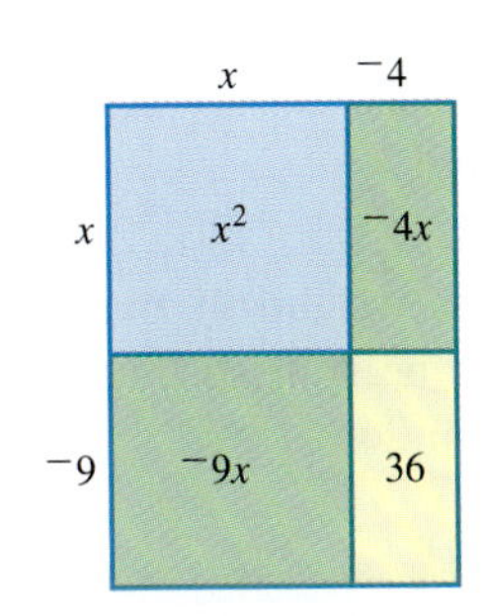

Figure 23

We split the middle term, ^-13x, into two terms, ^-4x and ^-9x. See Figure 23. Symbolically, we have

$$\begin{aligned} x^2 - 13x + 36 &= x^2 - 4x - 9x + 36 \\ &= x(\mathbf{x-4}) - 9(\mathbf{x-4}) \quad \textit{Factoring by grouping: } {}^-9 \textit{ is a negative common factor.} \\ &= (\mathbf{x-4})(x-9) \end{aligned}$$

c. $3x^2 + 20x - 7$ is already in descending order. The diagonal product is $3x^2 * {}^-7 = {}^-21x^2$. The middle term must be split so that $20x = \_\_x + \_\_x$. We need to find factor pairs for $^-21$, so we will need one positive and one negative factor. Since the middle term is positive, the positive factor must have larger absolute value than the negative factor.

$\mathbf{^-1 * 21 = {}^-21}$	$\mathbf{^-1x + 21x = 20x}$	***Yes***
$^-3 * 7 = {}^-21$	$^-3x + 7x = 4x$	*No*

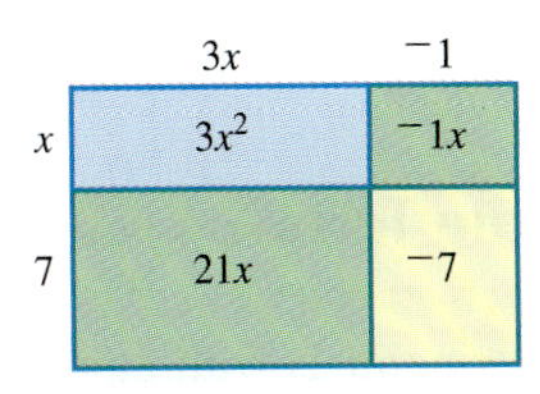

Figure 24

We split the middle term, $20x$, into two terms, ^-1x and $21x$. See Figure 24. Symbolically, we have

$$\begin{aligned} 3x^2 + 20x - 7 &= 3x^2 - 1x + 21x - 7 \\ &= x(\mathbf{3x-1}) + 7(\mathbf{3x-1}) \quad \textit{Factoring by grouping} \\ &= (\mathbf{3x-1})(x+7) \end{aligned}$$

d. First rewrite $10x^2 - 3 - x$ so that the powers of x are in descending order: $10x^2 - x - 3$. The diagonal product is $10x^2 * {}^-3 = {}^-30x^2$. The middle term must be split so that ${}^-1x = \_\_\_x + \_\_\_x$. We need to find factor pairs for ${}^-30$, so we need one positive and one negative factor. Because the middle term is negative, the negative factor must have larger absolute value than the positive factor.

$1 * {}^-30 = {}^-30$	$1x + {}^-30x = {}^-29x$	*No*
$2 * {}^-15 = {}^-30$	$2x + {}^-15x = {}^-13x$	*No*
$3 * {}^-10 = {}^-30$	$3x + {}^-10x = {}^-7x$	*No*
$\mathbf{5 * {}^-6 = {}^-30}$	$\mathbf{5x + {}^-6x = {}^-1x}$	***Yes***

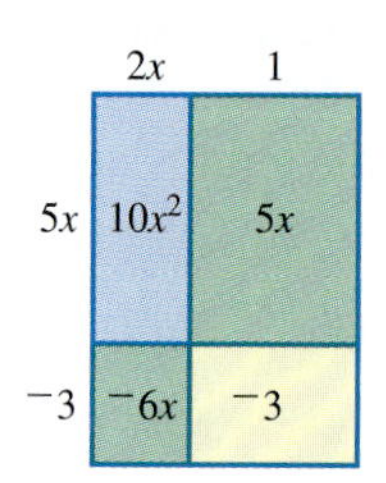

Figure 25

We split the middle term, ${}^-x$, into two terms, $5x$ and ${}^-6x$. See Figure 25. Symbolically,

$$10x^2 - x - 3 = 10x^2 + 5x - 6x - 3$$

$$= 5x(\mathbf{2x + 1}) - 3(\mathbf{2x + 1})$$ *Factoring by grouping: ${}^-3$ is a negative common factor.*

$$= (\mathbf{2x + 1})(5x - 3)$$

Example 3 shows a trinominal that cannot be factored.

EXAMPLE 3 ■ Show that the trinomial $x^2 - 7x + 24$ cannot be factored.

Solution Suppose $x^2 - 7x + 24$ could be factored. Proceed as we did in Example 2. The diagonal product is $24x^2$. The middle term must be split so that ${}^-7x = \_\_\_x + \_\_\_x$. We need to find two negative factors for 24.

${}^-1 * {}^-24 = 24$	${}^-1x + {}^-24x = {}^-25x$	*No*
${}^-2 * {}^-12 = 24$	${}^-2x + {}^-12x = {}^-14x$	*No*
${}^-3 * {}^-8 = 24$	${}^-3x + {}^-8x = {}^-11x$	*No*
${}^-4 * {}^-6 = 24$	${}^-4x + {}^-6x = {}^-10x$	*No*

We tried every possible pair and found that none of them work, so $x^2 - 7x + 24$ cannot be factored with integers. It is considered a prime polynomial.

Other Applications of Splitting the Middle Term

Splitting the middle term can be used to factor some trinomials with more than one variable or with higher powers of x.

EXAMPLE 4 Factor these trinomials.

a. $x^2 - 11xy - 42y^2$

b. $2x^4 + 9x^2 + 7$

Solution **a.** $x^2 - 11xy - 42y^2$ is written in descending order for powers of x. The diagonal product is $^-42x^2y^2$. The middle term must be split so that $^-11xy =$ ___$xy +$ ___xy. We need to find factor pairs of $^-42$, one positive and one negative. Because the middle term is negative, the negative factor must have larger absolute value than the positive factor.

$1 * {}^-42 = {}^-42$	$1xy + {}^-42xy = {}^-41xy$	*No*
$2 * {}^-21 = {}^-42$	$2xy + {}^-21xy = {}^-19xy$	*No*
$\mathbf{3 * {}^-14 = {}^-42}$	$\mathbf{3xy + {}^-14xy = {}^-11xy}$	***Yes***
$6 * {}^-7 = {}^-42$	$6xy + {}^-7xy = {}^-1xy$	*No*

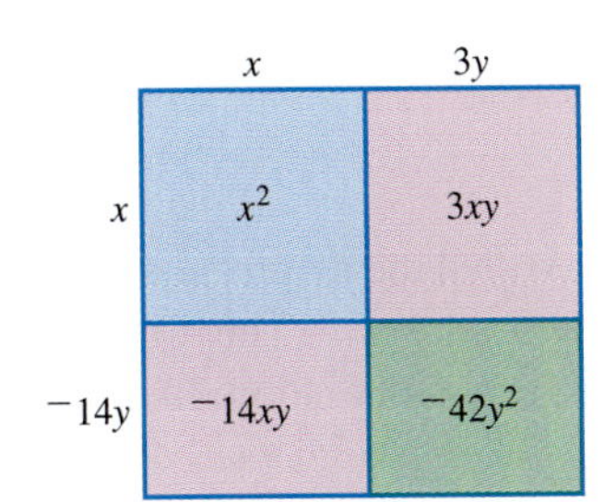

Figure 26

We split the middle term, ^-11xy, into two terms, $3xy$ and ^-14xy. See Figure 26. Symbolically,

$$x^2 - 11xy - 42y^2 = x^2 + 3xy - 14xy - 42y^2$$

$$= x(\mathbf{x + 3y}) - 14y(\mathbf{x + 3y}) \quad \textit{Factoring by grouping: } {}^-14y \textit{ is a negative common factor}$$

$$= (\mathbf{x + 3y})(x - 14y)$$

b. $2x^4 + 9x^2 + 7$ is written in descending order. The diagonal product is $14x^4$. The middle term must be split so that $9x^2 =$ ___$x^2 +$ ___x^2. We need to find two positive factors of 14.

$1 * 14 = 14$	$1x^2 + 14x^2 = 15x^2$	*No*
$\mathbf{2 * 7 = 14}$	$\mathbf{2x^2 + 7x^2 = 9x^2}$	***Yes***

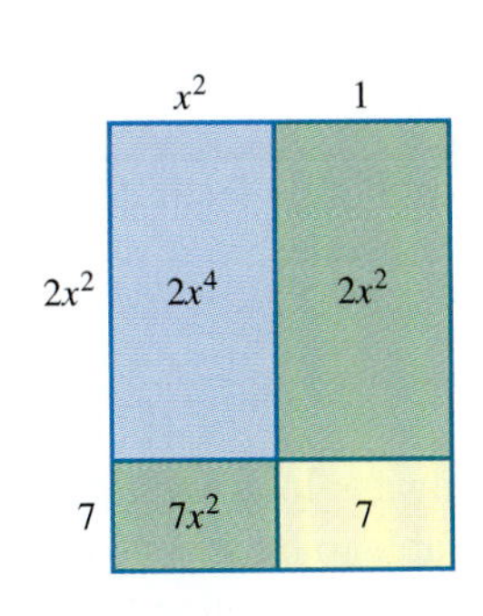

Figure 27

We split the middle term, $9x^2$, into two terms, $2x^2$ and $7x^2$. See Figure 27. Symbolically,

$$2x^4 + 9x^2 + 7 = 2x^4 + 2x^2 + 7x^2 + 7$$

$$= 2x^2(\mathbf{x^2 + 1}) + 7(\mathbf{x^2 + 1}) \quad \textit{Factoring by grouping}$$

$$= (\mathbf{x^2 + 1})(2x^2 + 7)$$

A Special Binomial

Example 5 considers a special binomial, the difference of two squares.

EXAMPLE 5 Factor $x^2 - 49$.

Solution $x^2 - 49$ is a binomial rather than a trinomial, but we can still split the middle term if we think of it as $0x$. The diagonal product is $^-49x^2$. The middle term must be split so that $0x =$ ___$x +$ ___x. We need to find factor pairs for $^-49$ and need one positive and one negative number because they must add to zero.

$1 * {}^-49 = {}^-49$	$1x + {}^-49x = {}^-48x$	*No*
$\mathbf{7 * {}^-7 = {}^-49}$	$\mathbf{7x + {}^-7x = 0x}$	***Yes***

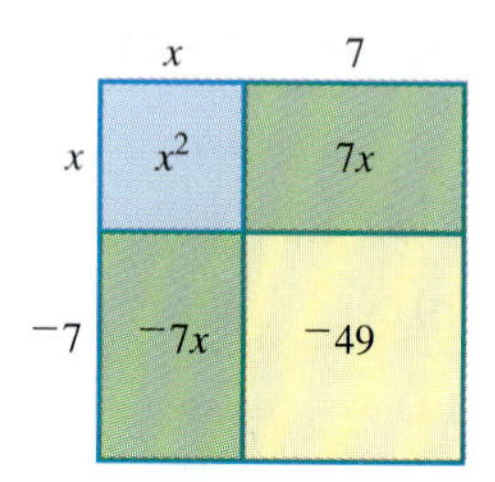

Figure 28

We split the middle term, $0x$, into two terms, ^-7x and $7x$. See Figure 28. Symbolically,

$$\begin{aligned} x^2 - 49 &= x^2 + 0x - 49 \\ &= x^2 - 7x + 7x - 49 \\ &= x(\mathbf{x - 7}) + 7(\mathbf{x - 7}) \\ &= (\mathbf{x - 7})(x + 7) \end{aligned}$$

The binomial $x^2 - 49$ is an example of a pattern called the **difference of two squares**. Note that x^2 is the square of x and 49 is the square of 7. The difference of two squares may always be factored by splitting a zero middle term. In general we have $a^2 - b^2$ with a middle term of $0ab$.

The Difference of Two Squares

$a^2 - b^2$ can be factored as $(a - b)(a + b)$.

Caution: Note that the sum of two squares, $a^2 + b^2$, cannot be factored.

Factoring Completely

Some polynomials may be factored in several steps. Always look for a common monomial factor first. If there is one, then you may be able to factor one or more of the resulting factors by grouping, splitting a middle term, or using the difference of two squares. Keep going until the polynomial is completely factored.

EXAMPLE 6 Factor each polynomial.

a. $12x^2 - 18x + 6$

b. $2xy^2 - 50x$

c. $x^4y^2 - 4x^3y^2 - 2x^4y + 8x^3y$

Solution

a. $12x^2 - 18x + 6 = 6(2x^2 - 3x + 1)$ *The common factor is 6*

$= 6(2x^2 - 2x - 1x + 1)$ *Splitting the middle term*

$= 6(2x(\mathbf{x-1}) + {}^-1(\mathbf{x-1}))$ *Factoring by grouping*

$= 6(\mathbf{x-1})(2x - 1)$

b. $2xy^2 - 50x = 2x(y^2 - 25)$ *The common factor is 2x*

$= 2x(y-5)(y+5)$ *Factoring as the difference of two squares*

c. $x^4y^2 - 4x^3y^2 - 2x^4y + 8x^3y = x^3y(xy - 4y - 2x + 8)$ *The common factor is x^3y*

$= x^3y((xy - 4y) + ({}^-2x + 8))$ *Grouping*

$= x^3y(y(\mathbf{x-4}) + {}^-2(\mathbf{x-4}))$

$= x^3y(\mathbf{x-4})(y + {}^-2)$

$= x^3y(x-4)(y-2)$

Solving Equations by Factoring

Section 6.2 introduced the zero-product property. We can use this property to solve quadratic equations in which a quadratic trinomial is equal to zero. In the next chapter we learn more about quadratic equations and different methods for solving them. In the meantime, here are examples of quadratic equations that can be solved by factoring.

EXAMPLE 7 Solve these equations.

a. $x^2 - 7x - 18 = 0$

b. $2x^2 + 5 = 11x$

Solution

a. $x^2 - 7x - 18 = 0$

$x^2 - 9x + 2x - 18 = 0$ *Splitting the middle term on left side*

$\mathbf{x}(\mathbf{x-9}) + \mathbf{2}(\mathbf{x-9}) = 0$ *Factoring by grouping*

$(\mathbf{x-9})(\mathbf{x+2}) = 0$

Now apply the zero-product property: Either

$\mathbf{x-9} = 0$ or $\mathbf{x+2} = 0$

$x = 9$ $\quad$ $x = {}^-2$

Typically for equations like this there is more than one solution. Both of them should check in the original equation.

Check for $\mathbf{x = 9}$

$\mathbf{x}^2 - 7\mathbf{x} - 18 = 0$

$\mathbf{9}^2 - 7 * \mathbf{9} - 18 \stackrel{?}{=} 0$

$81 - 63 - 18 = 0$ *OK!*

Check for $\mathbf{x = {}^-2}$

$\mathbf{x}^2 - 7\mathbf{x} - 18 = 0$

$(\mathbf{{}^-2})^2 - 7 * (\mathbf{{}^-2}) - 18 \stackrel{?}{=} 0$

$4 + 14 - 18 = 0$ *OK!*

b.

$$2x^2 + 5 = 11x$$

$$2x^2 - 11x + 5 = 0$$ *Making the right side zero by subtracting 11x on both sides*

$$2x^2 - 1x - 10x + 5 = 0$$ *Splitting the middle term*

$$x(2x - 1) - 5(2x - 1) = 0$$ *Factoring by grouping*

$$(2x - 1)(x - 5) = 0$$

Now apply the zero-product property: Either

$2x - 1 = 0$	or $x - 5 = 0$
$2x = 1$	
$x = \frac{1}{2}$	$x = 5$
Check for $x = \frac{1}{2}$	Check for $x = 5$
$2x^2 - 11x + 5 = 0$	$2x^2 - 11x + 5 = 0$
$2 * \left(\frac{1}{2}\right)^2 - 11 * \left(\frac{1}{2}\right) + 5 \stackrel{?}{=} 0$	$2 * 5^2 - 11 * 5 + 5 \stackrel{?}{=} 0$
$2\left(\frac{1}{4}\right) - \frac{11}{2} + 5 \stackrel{?}{=} 0$	$2 * 25 - 55 + 5 \stackrel{?}{=} 0$
$\frac{1}{2} - \frac{11}{2} + 5 \stackrel{?}{=} 0$	$50 - 55 + 5 = 0$ *OK!*
$-\frac{10}{2} + 5 \stackrel{?}{=} 0$	
$^{-}5 + 5 = 0$ *OK!*	

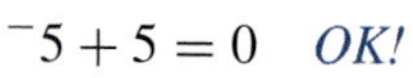

Exercises 6.4

1. Place the terms of each polynomial in descending order and determine which are quadratic trinomials.
   a. $x + x^2 - 6$ **b.** $x^3 - x + 6$
   c. $x^2 + 3$ **d.** $5 - x + 4x^2$
2. Follow these steps to factor the trinomial $x^2 + 18x + 32$.
   a. Find the diagonal product.
   b. List all the factor pairs for the diagonal product.
   c. Choose the factor pair that adds to equal the middle term.
   d. Rewrite the trinomial with a split middle term.
   e. Create an area model and factor your four-term polynomial from (d).

For Exercises 3–4, factor each trinomial.

3. $2x^2 + 7x + 3$
4. $4x^2 + 5x - 6$
5. George decided that $x^2 - 5x + 6$ wouldn't factor. His incorrect conclusion was based on this algebraic work:

$x^2 - 5x + 6 = x^2 - 3x - 2x + 6$ *Spliting the middle term*

$= (x^2 - 3x) - (2x + 6)$ *Forming groups of two terms*

$= x(x - 3) - 2(x + 3)$ *Finding the GCF of each group*

George stopped here because he couldn't find a GCF.

a. Use the diagonal test for grouping to show that $x^2 - 3x - 2x + 6$ does factor.

b. Use an area model to factor the polynomial.

c. Find the error in George's symbolic work and correct it.

For Exercises 6–7, factor each trinomial.

6. $x^2 - 12x + 35$

7. $^{-}11x + 6x^2 - 10$

For Exercises 8 and 9, show that each trinomial can't be factored. Do this by trying to split the middle term.

8. $x^2 + 4x + 6$

9. $5x^2 - 3x + 8$

For Exercises 10 and 11, use Example 4 as a model to factor the polynomial.

10. $x^2 + 2xy - 8y^2$

11. $2x^6 - x^3 - 28$

12. Factor each difference of two squares.

a. $x^2 - 16$ **b.** $4x^2 - y^2$ **c.** $36 - z^2$

13. Solve each equation by factoring.

a. $x^2 - 12x + 32 = 0$

b. $x^2 - 3x = 10$

c. $2x^2 + 7x + 3 = 0$

14. Factor the following polynomials in steps. First look for a common monomial factor. Then factor by grouping, by splitting a middle term if necessary, or by using the difference of two squares.

a. $2x^3 - 6x^2 - 36x$

b. $5xy - 20x - 10y + 40$

c. $6 - 24x^2$

15. The trinomial $x^2 + x - 2$ factors into $(x + 2)(x - 1)$. The two expressions look entirely different but are, indeed, equivalent.

a. Demonstrate their equivalence by evaluating both expressions for $x = {}^{-}2$ and $x = 1$.

b. Graph either expression in the [Y=] menu with a [ZOOM] 4: ZDecimal setting. [TRACE] to the points on the graph that correspond to $x = {}^{-}2$ and $x = 1$. What is the name for the line both points lie on?

c. Notice the choices of $^{-}2$ and 1 as values of x that produce an output of zero in the factored polynomial $(x + 2)(x - 1)$. What values of x do you suppose produce an output of zero in the factored polynomial $(x - 6)(x + 3)$?

d. Explain why it is helpful to factor a polynomial in order to find where the graph of a function crosses the x-axis. In the next chapter we examine this question in greater detail.

Skills and Review 6.4

16. Expand each product.

a. $(x - 6)(3x + 7)$

b. $(2x - 1)(x^2 + 4x + 3)$

17. Use an area model to show why $(x + 3)^2$ is not equal to $x^2 + 9$.

18. Expand each of the following.

a. $(2x + 5)^2$ **b.** $(3x - 4)^2$

19. Solve the equation $7x^2 - 21x = 0$.

20. Simplify the expression and write your answer with only positive exponents.

$$\left(\frac{x^{-2}y^4}{x}\right)^2$$

21. Linda and Lucy are $\frac{3}{4}$ mile apart. At 5 P.M. Linda starts running toward Lucy at the rate of $\frac{1}{8}$ mile/minute. At 5:03 P.M. Lucy begins riding her moped toward Linda at the rate of $\frac{1}{4}$ mile/minute. When will Linda and Lucy meet?

22. The price of a collectible went from \$42 to \$197. Find the percent increase.

23. Recall that a system of linear equations in two variables has no solutions if the lines are parallel. Show by demonstrating equal slopes or by an algebraic method (substitution or elimination) that the following system of two equations has no solutions.

$$x + \tfrac{1}{2}y = 6$$
$$y = {}^{-}2x + 10$$

24. Write the equation $2x - 4y = 8$ in slope-intercept form.

25. Solve the equation $11 - (x + 2) = 5(x - 7)$.

a. Write your solution as a simplified fraction.

b. Write your solution as a decimal rounded to the nearest thousandth.

6.5 Factoring the Sum and Difference of Two Cubes (Optional)

AFTER STUDYING THIS SECTION YOU WILL BE ABLE TO

- Factor the sum or difference of two cubes.
- Use the sum or difference of two cubes to factor more complex polynomials completely.

In Section 6.4 we observed that the difference of two squares may be factored using the pattern

$$a^2 - b^2 = (a - b)(a + b)$$

We also noticed that the *sum* of two squares $a^2 + b^2$ cannot be factored.

You may be surprised to learn that *both* the sum and the difference of two cubes may be factored. The technique of splitting zero middle terms that you used for the difference of two squares may be extended to expressions involving cubes. Figure 29 suggests how formulas for the sum and difference of two cubes may be found by splitting the terms a^2b and ab^2.

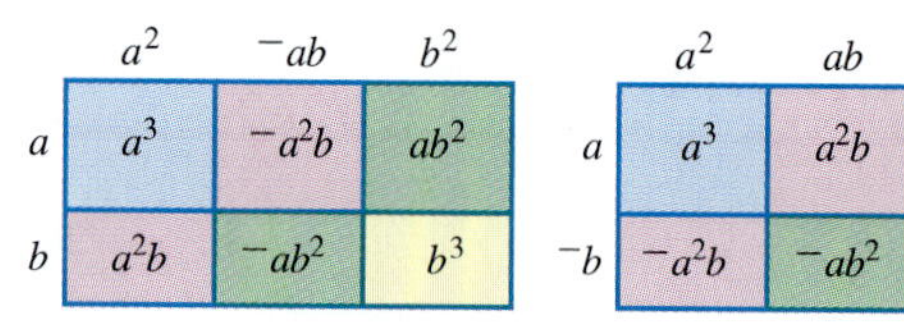

Figure 29

The Sum or Difference of Two Cubes

$a^3 + b^3$ can be factored as $(a + b)(a^2 - ab + b^2)$.
$a^3 - b^3$ can be factored as $(a - b)(a^2 + ab + b^2)$.

Observe that for the sum of two cubes, the binomial $a + b$ is also a sum. For the difference of two cubes, the binomial $a - b$ is a difference. The middle term of the second factor is $+ab$ or $-ab$. Its sign is the *opposite* of the sign of b in the first factor.

In order to use these patterns, you must recognize expressions that contain sums and differences of two cubes.

EXAMPLE 1 Factor these expressions.

a. $x^3 - 8$
b. $64x^3 + 125y^3$
c. $z^6 + 27$

Solution **a.** x^3 is the cube of x and 8 is the cube of 2. We have the difference of two cubes.

$$x^3 - 8 = \mathbf{x}^3 - \mathbf{2}^3$$

$$= (\mathbf{x} - \mathbf{2})(\mathbf{x}^2 + \mathbf{2x} + \mathbf{2}^2)$$ *Using the difference of two cubes with $a = x$ and $b = 2$*

$$= (x - 2)(x^2 + 2x + 4)$$

The work may be checked by expanding $(x - 2)(x^2 + 2x + 4)$ with an area model, as shown in Figure 30.

	x^2	$2x$	4
x	x^3	$2x^2$	$4x$
$^-2$	$^-2x^2$	^-4x	$^-8$

Figure 30

b. $64x^3$ is the cube of $4x$ and $125y^3$ is the cube of $5y$. We have the sum of two cubes.

$$64x^3 + 125y^3 = (\mathbf{4x})^3 + (\mathbf{5y})^3$$

$$= (\mathbf{4x} + \mathbf{5y})((\mathbf{4x})^2 - \mathbf{4x} * \mathbf{5y} + (\mathbf{5y})^2)$$ *Using the sum of two cubes with $a = 4x$ and $b = 5y$*

$$= (4x + 5y)(16x^2 - 20xy + 25y^2)$$ *Using the power-of-a-product law*

c. z^6 is the cube of z^2 and 27 is the cube of 3. We have the sum of two cubes.

$$z^6 + 27 = (z^2)^3 + 3^3$$

$$= (z^2 + 3)((z^2)^2 - 3z^2 + 3^2)$$ *Using the sum of two cubes with $a = z^2$ and $b = 3$*

$$= (z^2 + 3)(z^4 - 3z^2 + 9)$$ *Using the power-of-a-power law*

The sum or difference of two cubes may be used in connection with other factoring techniques. As a general rule, look for common factors or the difference of two squares before you apply the patterns for the sum and difference of two cubes.

EXAMPLE 2 Factor completely:

a. $24x^3 + 3$

b. $x^6 - 64$

Solution **a.** The two terms have a common factor, 3. Therefore, $24x^3 + 3 = 3(8x^3 + 1)$.

We now observe that the binomial factor, $8x^3 + 1$, is the sum of two cubes because $8x^3$ is the cube of $2x$ and 1 is the cube of 1. So,

$$24x^3 + 3 = 3(8x^3 + 1)$$

$$= 3((2x)^3 + 1^3)$$

$$= 3(2x + 1)((2x)^2 - 2x + 1^2)$$ *Using the sum of two cubes with $a = 2x$ and $b = 1$*

$$= 3(2x + 1)(4x^2 - 2x + 1)$$ *Using the power-of-a-product law*

b. The two terms have no common factor, but we may consider this expression as the difference of two squares. x^6 is the square of x^3 and 64 is the square of 8. Once we have factored the difference of two squares, we find that we have a sum of two cubes and a difference of two cubes.

$$x^6 - 64 = (x^3)^2 - 8^2$$

$$= (x^3 - 8)(x^3 + 8)$$ *Using the difference of two squares with $a = x^3$ and $b = 8$*

$$= (x^3 - 2^3)(x^3 + 2^3)$$

$$= (x - 2)(x^2 + 2x + 4)(x + 2)(x^2 - 2x + 4)$$ *Using the sum and difference of two cubes with $a = x$ and $b = 2$*

You may be surprised that this polynomial with two terms can be expressed as the product of four polynomials.

Exercises 6.5

1. Use the sum or difference of two cubes to factor.
   a. $y^3 + 125$
   b. $27 - 8x^3$
   c. $x^3 - 216$
   d. $1000z^3 + 1$
   e. $x^6 - y^3$
2. Factor completely.
   a. $2z^3 + 54$
   b. $40x^3 - 5y^3$
   c. $z^6 - 1$
   d. $x^6 - y^6$
3. Some of these expressions are prime—that is, they cannot be factored with integer coefficients. Identify those that are prime and factor those that can be factored.
   a. $x^2 + 25$
   b. $x^3 + 729$
   c. $x^4 + 16$
   d. $x^6 - 729$
   e. $x^6 + 512$

CHAPTER 6 ▪ KEY CONCEPTS

Checking with Specific Values This technique lets us check that our algebraic manipulations have been done correctly. If we substitute specific values for each variable in the original expression and in our simplified form, we should get the same result. If not, we have made a mistake somewhere. It is a good idea to choose small numbers to check with, but avoid 1 and 0. (Page 210)

Product Law for Exponents This rule lets us simplify expressions involving a product and exponents.

$$a^m a^n = a^{m+n} \qquad \text{(Page 211)}$$

Power-of-a-Power Law for Exponents This rule lets us simplify expressions involving a power of a power.

$$(a^m)^n = a^{mn} \qquad \text{(Page 212)}$$

Power-of-a-Product Law for Exponents This rule lets us simplify expressions involving a power of a product.

$$(ab)^m = a^m b^m \qquad \text{(Page 212)}$$

Negative Integer Exponents A negative integer exponent is another way to represent a quantity in the denominator of a fraction: $a^{-n} = \frac{1}{a^n}$. If a negative exponent appears in a factor in a denominator, then that factor can be moved to the numerator and the exponent becomes positive. (Page 213)

Quotient Law for Exponents This rule lets us simplify expressions involving quotients and exponents.

$$\frac{a^m}{a^n} = a^{m-n} \qquad \text{(Page 216)}$$

Power-of-a-Quotient Law for Exponents The rule which lets us simplify expressions involving the power of a quotient.

$$\left(\frac{a}{b}\right)^m = \frac{a^m}{b^m} \qquad \text{(Page 216)}$$

Monomials A monomial is any single algebraic expression involving a product of numbers and variables raised to positive integer powers. (Page 221)

Polynomials A polynomial is any combination of monomials by addition and subtraction. Polynomials with two terms are called **binomials** (page 221) and with three terms are called **trinomials**. (Page 221)

Area Models An area model is a useful tool to help expand or factor polynomials. The monomials in one factor are written as the rows on one side of a rectangle, and the monomials in the other factor are written at the tops of columns on the other side. The product of the two factors is the area of the rectangle. (Pages 221–223)

Greatest Common Factor (GCF) The GCF is the largest factor common to all of the monomials in a polynomial. This can be either the largest monomial common factor or a common binomial factor. (Page 224)

The Zero-Product Property If the product of two numbers is zero, then at least one of the numbers must be zero. This property can be used to solve **quadratic equations**. (Page 228)

The Square of a Binomial The square of a binomial has a distinctive form.

$$(a+b)^2 = a^2 + 2ab + b^2 \qquad \text{(Page 232)}$$

Factoring by Grouping A technique for factoring polynomials by grouping the monomials according to which pairs have common factors. (Page 233)

Prime Polynomial A polynomial is prime if it cannot be factored into two or more polynomial (or monomial) factors with only integer coefficients. (Page 233)

The Diagonal Test for Grouping This test tells us if a polynomial with four terms may be factored into two binomials with integer coefficients. If the product of two terms equals the product of the other two terms, then the polynomial may be factored; otherwise it cannot. (Page 235)

Splitting the Middle Term This technique allows us to factor trinomials by splitting the middle term into two terms in such a way that the diagonal test for grouping works and the polynomial can be factored.

$$\begin{aligned} x^2 + 5x + 6 &= x^2 + 2x + 3x + 6 && 2x * 3x = x^2 * 6, \textit{ so it factors} \\ &= x(x+2) + 3(x+2) \\ &= (x+2)(x+3) && \text{(Page 239)} \end{aligned}$$

The Difference of Two Squares This special binomial is the difference of two perfect squares, $a^2 - b^2$. It factors easily as $a^2 - b^2 = (a+b)(a-b)$. (Page 244)

Exploration

The method of multiplying using squares shown in the chapter opener can be used to find the product of two even numbers or two odd numbers. For example, you can multiply 23 * 55 using this method. (A table of squares is provided here.)

Step 1 $x = \dfrac{23+55}{2} = \dfrac{78}{2} = 39$

Step 2 $y = \dfrac{55-23}{2} = \dfrac{32}{2} = 16$

Step 3 $x^2 - y^2 = 1521 - 256 = 1265$

Number	Square	Number	Square	Number	Square
1	1	21	441	41	1681
2	4	22	484	42	1764
3	9	23	529	43	1849
4	16	24	576	44	1936
5	25	25	625	45	2025
6	36	26	676	46	2116
7	49	27	729	47	2209
8	64	28	784	48	2304
9	81	29	841	49	2401
10	100	30	900	50	2500
11	121	31	961	51	2601
12	144	32	1024	52	2704
13	169	33	1089	53	2809
14	196	34	1156	54	2916
15	225	35	1225	55	3025
16	256	36	1296	56	3136
17	289	37	1369	57	3249
18	324	38	1444	58	3364
19	361	39	1521	59	3481
20	400	40	1600	60	3600

a. Use the method of squares to multiply $82 * 46$.

b. Use algebra to explain why this method works. (*Hint:* Let the larger number be m and the smaller number be n. Show that $x + y = m$ and $x - y = n$.)

c. Why does this method work only if both numbers are odd or both are even?

d. The method can be modified to multiply an odd number by an even number. To multiply $77 * 44$, first multiply $76 * 44$ and then add 44 to the result. Explain why this works.

Chapter 6 ▪ Review Exercises

Section 6.1 The Laws of Exponents

1. Use repeated multiplication to simplify each expression.

a. $x^4 * x^5$ **b.** $(y^2)^4$

c. $(x * y)^5$

2. Use the laws of exponents to simplify each expression.

a. $(2xy^3)^6$ **b.** $4y * 7y^5$

c. $(z^2)^4$

3. The side of a cube measures $2.1 * 10^4$ inches. Find the volume of the cube in cubic inches. A formula for the volume of a cube is volume = side^3.

4. Use the laws of exponents to simplify each expression. Write your answer using only positive exponents.

a. x^4x^{-7} **b.** $x^{-4}x^7$

c. $(x^3)^{-3}$ **d.** $(x^{-2}y)^{-6}$

5. At room temperature sound travels at a speed of $1.13 * 10^3$ feet/second. One microsecond is one millionth of a second, or 10^{-6} seconds. How far does sound travel in 1 microsecond?

6. Simplify each expression and write with only positive exponents.

a. $\dfrac{x^7}{x^3}$

b. $\dfrac{x^3}{x^7}$

c. $\left(\dfrac{x}{y}\right)^6$

7. Match each expression (a)–(e) with its equivalent simplified expression (1)–(5). Also, name the law or laws of exponents used in simplifying each expression.

a. $\dfrac{x^3}{x^2}$ b. $(x^3)^2$ c. $\dfrac{x^2}{x^3}$ d. x^3x^2 e. $\dfrac{1}{x^{-3}}$

1. x^5 2. $\dfrac{1}{x}$ 3. x 4. x^3 5. x^6

8. Use the laws of exponents to simplify each expression. Write your answer using only positive exponents.

a. $\dfrac{x^5y}{x^2y^4}$

b. $\left(\dfrac{4x^2}{y}\right)^3$

c. $\left(\dfrac{x^3y^2}{x^7y}\right)^4$

d. $\left(\dfrac{2x^7}{3x^3y^2}\right)^{-3}$

Section 6.2 Products and Factors

9. Use an area model to find the product of $4x^3$ and $5x + 2x^4y$.

10. Find each product.

a. $5x(6x - 1)$

b. $2xz(4x + 7y - 3)$

c. $^{-}(x^2 - 8x + 2)$

11. Find the greatest common factor (GCF) for each group of expressions.

a. xy^4 and x^5y

b. $18x^2y$, $3x^3z^5$, and $9x^4$

c. $60x$ and $105y$

12. Use an area model to factor $8y^2 + 12y$.

13. Factor completely the expression $5x^2yz^3 - 15x^3y^4 + 10x^5y^2$.

14. Factor each expression so that the first term of the binomial factor has a positive coefficient.

a. $^{-}8x^2 + 10x$ **b.** $^{-}9x - 12y$

15. Factor each expression.

a. $4xz - 3z$

b. $4x(y + 7) - 3(y + 7)$

c. Show how (a) and (b) are related to each other.

16. Solve each equation by factoring and using the zero-product property.

a. $x^2 + 5x = 0$ **b.** $x^2 - 8x = 0$

Section 6.3 Factoring by Grouping

17. Use an area model to expand each product.

a. $(x + 4)(x + 2)$

b. $(x - 3)(4x^2 + 5x - 6)$

18. Factor each polynomial by grouping. Factor with an area model and then factor symbolically.

a. $2x - 9y + xy - 18$

b. $^{-}4y + 3z - 12yz + 1$

19. Use the diagonal test for grouping to determine if each polynomial can be factored, and if it can be, factor it.

a. $4x^3 + 10x^2 - 2x - 5$ **b.** $x^3 - 6x^2 - 5x + 2$

c. $3x + 8y + 6xy + 4$

20. In this section you learned a formula for squaring a binomial: $(a + b)^2 = a^2 + 2ab + b^2$. Use this formula to find each of the squares.

a. $(x + y)^2$ **b.** $(x + 2y)^2$

c. $(5x - 3y)^2$

Section 6.4 Factoring Quadratic Trinomials

21. Follow these steps below to factor the trinomial $x^2 + 11x + 24$.

a. Find the diagonal product.

b. List all the factor pairs of the diagonal product.

c. Choose the factor pair that has a sum equal to the middle term.

d. Rewrite the trinomial with a split middle term.

e. Create an area model and factor your four-term polynomial from (d).

22. Factor each trinomial by splitting the middle term.

a. $x^2 - 5x - 14$ **b.** $5x^2 + 9x - 2$

c. $3x^2 - 11x - 4$ **d.** $7x^2 - 8 + 26x$

23. Show that each trinomial cannot be factored by trying to split the middle term.

a. $x^2 - 2x + 5$ **b.** $2x^2 + x + 6$

24. Factor each trinomial. See Example 4.

a. $x^2 + 10xy + 21y^2$ **b.** $2x^4 - x^2 - 15$

25. Factor each difference of two squares.

a. $x^2 - 25$ **b.** $9 - y^2$

c. $4x^2 - 81y^2$

26. Factor the following polynomials in steps. First look for a common monomial factor. Then attempt more factoring by grouping, splitting a middle term, or using the difference of two squares.

a. $18x^2 - 36x + 16$

b. $xy^2 - xy + 2y^2 - 2y$

c. $3x^2y - 300y$

27. Solve each equation by factoring.

a. $x^2 + 13x + 40 = 0$ **b.** $x^2 + 2x = 3$

c. $3x^2 - 19x + 6 = 0$

Section 6.5 Factoring the Sum and Difference of Two Cubes (Optional)

28. Use the sum or difference of two cubes to factor:

a. $x^3 + 64$ **b.** $1 - y^3$

c. $8x^3 + 1000y^3$ **d.** $y^3 + 27z^6$

29. Factor completely:

a. $6 - 48z^3$ **b.** $250x^3 + 2y^3$ **c.** $64 - y^6$

CHAPTER 6 ▪ TEST

For Exercises 1–3, use the laws of exponents to simplify each expression. Write your answer using only positive exponents.

1. $(xy^3z^{-4})^2$

2. $\dfrac{6xy^{-3}}{20x^{-2}}$

3. $\left(\dfrac{3x^4y^2}{xy^2}\right)^{-3}$

4. Suppose the human brain sends $2.0 * 10^7$ signals through its neurons each second. How many signals are sent in one day? *Note:* There are $8.64 * 10^4$ seconds in one day. Perform your work using the laws of exponents. (http://vadim.www.media.mit.edu)

5. Find each product.

a. $3a * (a + 4)$

b. $^-6xy^2 * (^-x^2 - 2xy + 5y^2)$

6. Completely factor $^-6ab^2c^3 - 3bc^2 - 12a^3b^4c^3$.

7. Solve $2x^2 - 10x = 0$.

8. Expand each expression.

a. $(2x + 5)(x - 7)$

b. $(x - 4)(3x^2 - x + 6)$

9. Factor each polynomial.

a. $2x + 4xy + 3 + 6y$ **b.** $6x^3 - 9x - 2x^2 + 3$

10. Square the binomial $(2x - 7y)^2$.

11. Factor each trinomial.

a. $4x^2 - 23x - 6$ **b.** $x^2 + 2xy - 35y^2$

12. Factor the binomial $64x^2 - y^2$.

13. Completely factor each polynomial.

a. $2x^2y + 6xy^2 - 80y^3$ **b.** $12 - 3x^2$

14. Solve $6x^2 - 17x + 5 = 0$.

CHAPTER

7

Quadratic Equations and Graphs

The Right Price

A local charity is giving a benefit concert. They are trying to decide what price to charge for admission. Their goal is to maximize the revenue from ticket sales. Revenue is found by multiplying the number of tickets sold by the price of each ticket.

$$\text{revenue} = \text{number} * \text{price}$$

They know that the number of tickets sold will depend on the price, so the lower the price, the greater the number sold. In this chapter you learn a method that will help them decide exactly how much to charge for tickets in order to produce the maximum revenue for the charity. (See Exploration on p. 295.)

Need help? For on-line resources, visit this web site: **math.college.hmco.com/students.**

7.1 Solving Quadratic Equations

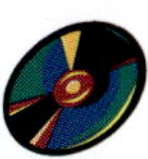

AFTER STUDYING THIS SECTION YOU WILL BE ABLE TO

- Apply the rule of four to a quadratic model.
- Solve a quadratic equation using the square root principle.
- Solve a quadratic equation using the zero-product property.

In the first six chapters of this book, we solved a wide range of problems. Most of these problems involved linear equations. Although it might be nice if all of our models could be developed using only linear formulas, unfortunately that is not the case. Many real-world situations result in formulas and equations involving more complicated algebraic expressions, such as the quadratic equations we saw in Chapter 6. In this chapter we continue our study of such equations.

A Quadratic Model

EXAMPLE 1 Rachel and Mike are planning to open a pizza parlor. Their recipe book tells them that they will need 1.5 g of cheese for every square inch of pizza.

a. They are considering making four sizes of pizza, with diameters of 8 inches, 12 inches, 16 inches, and 20 inches. Make a table showing how much cheese is needed for each of these pizzas.

b. Find a formula that shows how the amount of cheese depends upon the pizza's diameter.

c. Make a graph showing the relationship between the diameter of the pizza and the amount of cheese.

Solution **a.** Use the formula for the area of a circle, $A = \pi r^2$, with the approximation $\pi \approx 3.14$. Don't forget that the radius is half the diameter. When the diameter is 8 inches, $r = 4$ and we have

$$\begin{aligned} A &= \pi \mathbf{r}^2 \\ &= \pi \mathbf{4}^2 \\ &= 16\pi \\ &\approx 50.24 \text{ inches}^2 \end{aligned}$$

When the diameter is 12 inches, $r = 6$ and we have

$$\begin{aligned} A &= \pi \mathbf{r}^2 \\ &= \pi \mathbf{6}^2 \\ &= 36\pi \\ &\approx 113.04 \text{ inches}^2 \end{aligned}$$

When the diameter is 16 inches, $r = 8$ and we have

$$\begin{aligned} A &= \pi r^2 \\ &= \pi 8^2 \\ &= 64\pi \\ &\approx 200.96 \text{ inches}^2 \end{aligned}$$

When the diameter is 20 inches, $r = 10$ and we have

$$\begin{aligned} A &= \pi r^2 \\ &= \pi 10^2 \\ &= 100\pi \\ &\approx 314 \text{ inches}^2 \end{aligned}$$

Once we know the area in square inches, we multiply by 1.5 g/square inch to obtain the amount of cheese shown in the table. The numbers in the cheese column have been rounded to the nearest gram.

Diameter (inches)	Cheese (g)
8	75
12	170
16	301
20	471

b. To find a formula for cheese, think about the steps we followed in part (a). First, we divided the diameter by 2 to obtain the radius, that is, $r = \frac{d}{2}$. Then we substituted for r in the formula for the area of a circle.

$$\begin{aligned} A &= \pi r^2 \\ &= \pi \left(\frac{d}{2}\right)^2 \\ &= \pi \frac{d^2}{4} && \textit{Using the power-of-a-quotient law} \\ &\approx 3.14 \frac{d^2}{4} && \textit{Using } \pi \approx 3.14 \\ &= .785d^2 \end{aligned}$$

We then multiply by 1.5 g/square inch. Because $1.5 * .785 = 1.1775 \approx 1.2$, a good formula for the amount of cheese needed is $cheese = 1.2d^2$.

c. The independent variable is the diameter, d. The dependent variable is the amount of cheese. Using our traditional variables, x and y, we have $y = 1.2x^2$.

The table of values in (a) suggests a first-quadrant graph; after all, we cannot have a negative diameter. One suitable friendly window for the calculator is Xmin = 0, Xmax = 23.5, Ymin = 0, Ymax = 500. A graph is shown in Figure 1.

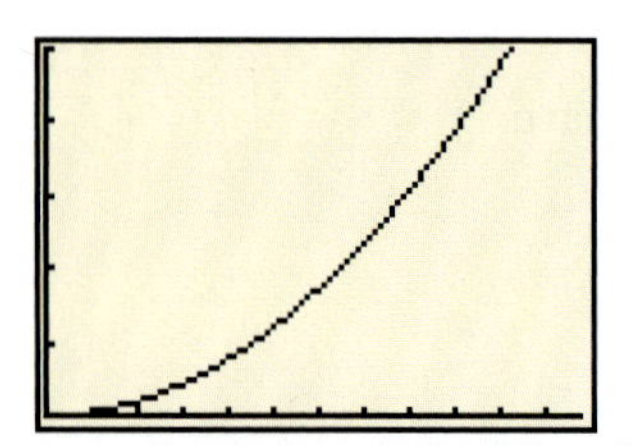

Figure 1

Rachel and Mike have found four representations of their pizza function (see the rule of four on page 79). The cookbook gave them a *verbal* description. They used the verbal description to represent the function *numerically* in a table of values. Then, they found a *symbolic* formula for the function; finally, they *graphed* the function. Each representation can be useful in different ways; for example, because they want to make only four sizes of pizza, the most useful representation for them is the *numerical* table of values.

EXAMPLE 2 The symbolic formula for the pizza function, cheese $= 1.2d^2$, is a quadratic monomial, different than the linear functions we studied in Chapters 3, 4, and 5. Linear equations have graphs with a constant slope, which represents the rate of change of one variable with respect to the other. How does the slope of a quadratic function behave?

Solution To compute slope, we need two points, which we can connect by a line. We choose several points on the graph of the pizza function and find the slopes between different pairs. The slopes of the lines connecting each pair of points from the table are different (see Figure 2).

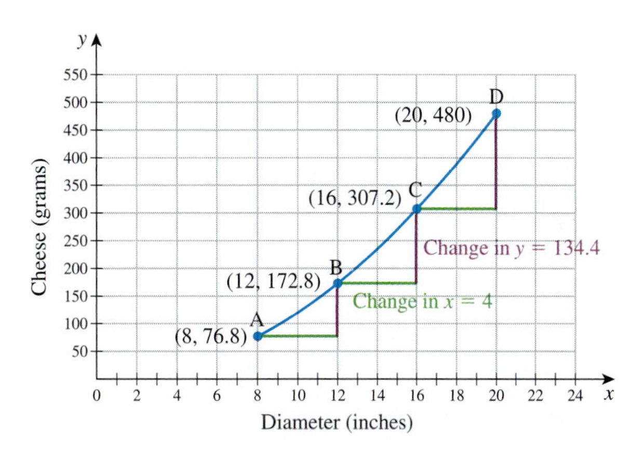

Figure 2

Using the formula for slope,

$$\text{slope} = \frac{\text{rise}}{\text{run}}$$

we get

$$\begin{aligned} \text{slope} &= \frac{172.8 - 76.8}{12 - 8} \\ &= \frac{96}{4} \\ &= 24 \end{aligned}$$

Slope from point A to point B

$$\text{slope} = \frac{307.2 - 172.8}{16 - 12}$$
$$= \frac{134.4}{4}$$
$$= 33.6 \qquad \textit{Slope from point B to point C}$$
$$\text{slope} = \frac{480 - 307.2}{20 - 16}$$
$$= \frac{172.8}{4}$$
$$= 43.2 \qquad \textit{Slope from point C to point D}$$

The slopes are increasing each time, and the points don't fit together to form a straight line. We can interpret this using rate of change: as the pizzas get bigger, the rate at which you need to add cheese *increases*.

EXAMPLE 3 Rachel and Mike think that if they use 250 g of cheese on a pizza, they could sell the pizza for $10 and make a profit.

a. Use a graph to estimate how big a pizza they can make with 250 g of cheese.
b. Solve an equation to find how big a pizza they can make with 250 g of cheese.

Solution

a. We can use our formula from Example 1(b). Substitute 250 into cheese = $1.2d^2$.

$$250 = 1.2d^2$$

You may enter the formula cheese = $1.2d^2$ into your calculator as Y1 = 1.2X ∧ 2, and the graph (Figure 3) should help you solve for x. You may use the window from Example 1(c): Xmin = 0, Xmax = 23.5, Ymin = 0, Ymax = 500.

Using the TRACE key to move along the graph until Y is about 250, you see that X should be between 14 and 15.

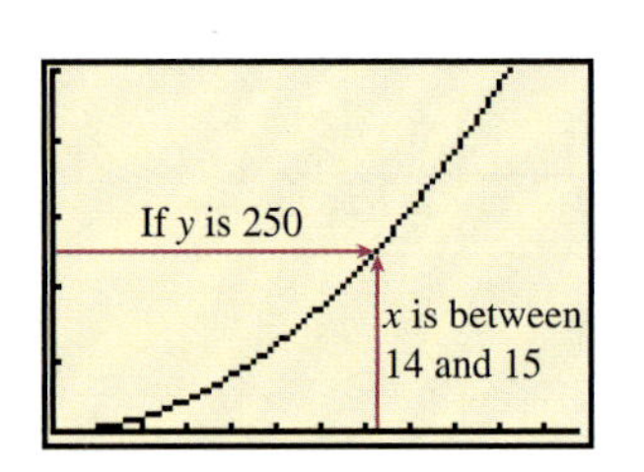

Figure 3 You can trace along the graph to find the approximate solution point.

b. We can get a more precise answer algebraically.

$$250 = 1.2d^2$$
$$\frac{250}{1.2} = d^2 \qquad \textit{Dividing by 1.2 on both sides}$$
$$208.3 \approx d^2 \qquad \textit{Rounding to the nearest .1}$$
$$\sqrt{208.3} \approx \sqrt{d^2} \qquad \textit{Taking a square root on both sides}$$
$$14.4 \approx d$$

Rachel and Mike decide to make pizzas about 14.4 inches in diameter.

In solving the equation $250 = 1.2d^2$, we applied the fundamental principle of equation solving when we divided by 1.2 on both sides. We did something similar when we took a square root on both sides. We need to be careful, however, because every positive number has *two* square roots, one positive and one negative. For example, the two square roots of 16 are 4 and $^-4$, because $4^2 = 16$ and $(^-4)^2 = 16$.

Square Root Principle for Solving Equations

In solving an equation, you may take square roots on both sides. When you do, you get two equations, one with a positive root and one with a negative root.

If $x^2 = a$, then

$$x = \sqrt{a} \quad \text{or} \quad x = {}^-\sqrt{a}$$

We may also write these two equations as a single equation using the symbol $\pm$, which stands for plus or minus.

$$x = \pm\sqrt{a}$$

The equation $250 = 1.2d^2$ in Example 3 actually has two solutions, $\pm 14.4 \approx d$. Because the diameter of a pizza cannot be negative, Rachel and Mike needed only the positive solution.

Quadratic Equations

We saw equations like $250 = 1.2d^2$ in Chapter 6.

Definition Quadratic Equation

A **quadratic equation** is an equation that can be put in the form

$$ax^2 + bx + c = 0$$

where x is the variable name used in the problem and a, b, and c are real numbers, called the *coefficients*, with $a \neq 0$.

It is not immediately clear that the equation $250 = 1.2d^2$ looks like the one in the definition, but we can do some simple algebraic manipulation to make it match.

$$250 = 1.2d^2$$

$$^{-}1.2d^2 + 250 = 0 \qquad \textit{Adding } ^{-}1.2d^2 \textit{ to both sides of the equation}$$

$$^{-}1.2d^2 + 0d + 250 = 0 \qquad \textit{Including 0d as the middle term}$$

The coefficients are $a = {}^{-}1.2$, $b = 0$, and $c = 250$.

The graphs associated with a quadratic equation are not straight lines, because the slope is always changing, and in order to solve these equations algebraically, we used a technique (taking a square root) that was never used to solve linear equations. Also, it seems that, theoretically at least, quadratic equations can have two solutions (the positive- and negative-diameter pizzas), even if only one of them actually works for the real-world problem.

Let's look at another example involving a quadratic equation.

EXAMPLE 4 An artist wants to create a tile consisting of four triangles arranged to form a square 19 inches on a side, as shown. To form the triangles she needs to make two smaller squares the same size and cut each in half along a diagonal (see Figure 4). How big should she make the smaller squares?

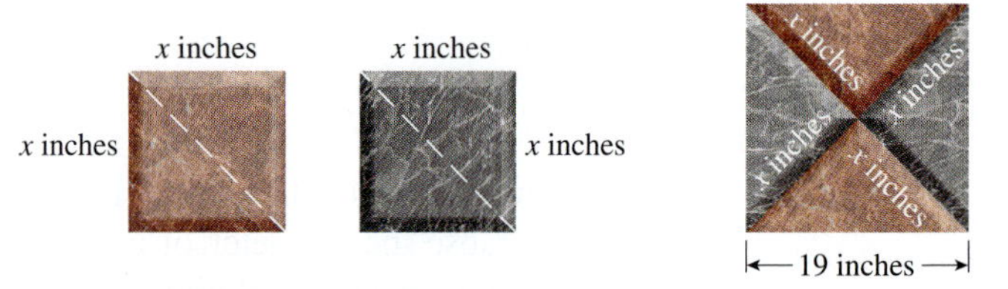

Figure 4 Cutting a small red tile and a small black tile to make a 19-inch square tile that is both red and black.

Solution We let x represent the length of the sides of the smaller squares. When the artist puts the tiles together to make the big square, the diagonals of the smaller tiles become the sides of the bigger square, so they must be 19 inches. Each diagonal is the hypotenuse of a right triangle, so we can use the Pythagorean theorem to get

$$x^2 + x^2 = 19^2$$

$$2x^2 = 361 \qquad \textit{Combining like terms and squaring}$$

$$x^2 = 180.5 \qquad \textit{Dividing by 2 on both sides}$$

$$x = \pm\sqrt{180.5} \qquad \textit{Taking square roots on both sides}$$

$$x \approx \pm 13.4$$

Because a tile cannot have negative dimensions, she needs to use small tiles that are about 13.4 inches on each side.

Solving Any Quadratic Equation

Both the pizza problem and the tile problem lead to quadratic equations that can be solved by taking square roots, because $b = 0$. What would happen if we end up with a different equation, such as

$$2x^2 + 5x - 361 = 0$$

Because the x-term is not zero, we cannot simply take square roots on both sides. There are several methods for solving such equations. One of them, using the zero-product property, was introduced in Chapter 6 and is illustrated in the next example. Other methods are introduced later in this chapter.

EXAMPLE 5 Solve the following quadratic equations.

a. $4x^2 - 22x = 0$

b. $x^2 + 2x - 15 = 0$

c. $6x^2 - 7x + 2 = 0$

d. $x^2 - 100 = 0$

Solution **a.** $4x^2 - 22x = 0$

$2x(2x - 11) = 0$ *Factoring out 2x, a common factor*

Apply the zero-product property: Either

$$2x = 0 \quad \text{or} \quad 2x - 11 = 0$$

$$x = 0 \qquad\qquad 2x = 11$$

$$x = \tfrac{11}{2} = 5\tfrac{1}{2}$$

b. To apply the zero-product property we need to factor $x^2 + 2x - 15$. Follow the steps on page 240. The terms of our quadratic polynomial are already in descending order, the x^2-term first, then the x-term, and then the constant. The diagonal product of the first and last terms is $^-15x^2$, so we need to find the factor pair of $^-15$ that sums to 2 and splits the middle term. Because 2 is positive and $^-15$ is negative, we need one negative term and one positive term, and the positive term should be larger.

$^-1 * 15 = {}^-15$	$^-1x + 15x = 14x$	*No, not 2x*
$^-3 * 5 = {}^-15$	$^-3x + 5x = 2x$	*Yes*

Now we can solve by grouping or with an area model, as shown in Figure 5.

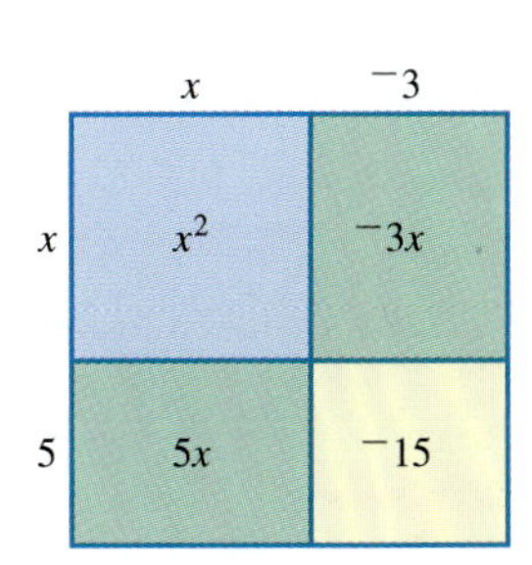

Figure 5

$$x^2 + 2x - 15 = 0$$

$$x^2 - 3x + 5x - 15 = 0 \quad \textit{Splitting the middle term}$$

$$x(x - 3) + 5(x - 3) = 0 \quad \textit{Factoring by grouping}$$

$$(x - 3)(x + 5) = 0$$

Apply the zero-product property: Either

$$x - 3 = 0 \qquad \text{or} \qquad x + 5 = 0$$
$$x = 3 \qquad\qquad x = {}^{-}5$$

c. The diagonal product of the first and last terms is $12x^2$, so we need to find the factor pair of 12 that sums to $^{-}7$ and splits the middle term. Because $^{-}7$ is negative and 12 is positive, we need two negative terms.

$$^{-}1 * {}^{-}12 = 12 \qquad {}^{-}1x + {}^{-}12x = {}^{-}13x \qquad \textit{No, not } {}^{-}7x$$
$$^{-}2 * {}^{-}6 = 12 \qquad {}^{-}2x + {}^{-}6x = {}^{-}8x \qquad \textit{No, not } {}^{-}7x$$
$$^{-}3 * {}^{-}4 = 12 \qquad {}^{-}3x + {}^{-}4x = {}^{-}7x \qquad \textit{Yes}$$

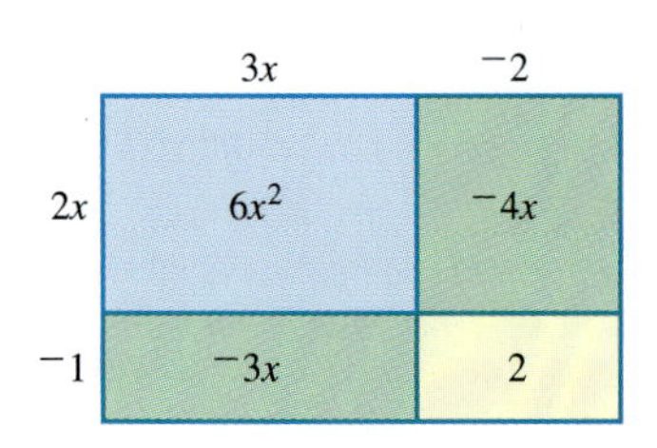

Figure 6

Solve by grouping or with an area model, as shown in Figure 6.

$$6x^2 - 7x + 2 = 0$$
$$6x^2 - 4x - 3x + 2 = 0 \qquad \textit{Splitting the middle term}$$
$$2x(3x - 2) - 1(3x - 2) = 0 \qquad \textit{Factoring by grouping (Note: } {}^{-}1 \textit{ is a common factor for the last two terms)}$$
$$(3x - 2)(2x - 1) = 0$$

Apply the zero-product property: Either

$$3x - 2 = 0 \qquad \text{or} \qquad 2x - 1 = 0$$
$$3x = 2 \qquad\qquad 2x = 1$$
$$x = \tfrac{2}{3} \qquad\qquad x = \tfrac{1}{2}$$

d. Because $b = 0$, we can use the square root principle.

$$x^2 - 100 = 0$$
$$x^2 = 100 \qquad \textit{Adding 100 on both sides}$$
$$x = \pm 10 \qquad x \textit{ is } {}^{+}10 \textit{ or } {}^{-}10\textit{; there are two solutions}$$

Another method is to factor the left side of the original equation as the difference of two squares.

$$x^2 - 100 = 0$$
$$x^2 - 10^2 = 0$$
$$(x - 10)(x + 10) = 0 \qquad \textit{Factoring as the difference of two squares}$$

Apply the zero-product property: Either

$$x - 10 = 0 \qquad \text{or} \qquad x + 10 = 0$$
$$x = 10 \qquad\qquad x = {}^{-}10$$

In this case, either method—taking square roots or applying the zero-product property—gives the same solution.

Exercises 7.1

1. Review the data in Example 1 on page 258.
   a. Plot the data on a graph with the diameter (in inches) on the x-axis and the grams of cheese on the y-axis.
   b. Compare your graph to those of Figures 1 and 2 on pages 258 and 259. Do the points line on a straight line?
   c. Does the slope of your graph remain constant? What does this imply about the data fitting a linear model?
2. Recall that a formula for the area of a circle is $A = \pi r^2$.
   a. Find the area of a circle with an 8-inch radius.
   b. Find the area of a circle with a 16-inch diameter.
   c. Compare the results in (a) and (b).
   d. Explain how $A = \pi r^2$ is helpful in finding a formula for the amount of cheese required on a pizza of given diameter.
3. Find the side of a square whose area is 64 square inches.
4. Estimate the side of a square whose area is 90 m². See the estimation of square roots in Section 1.3.
5. Solve each quadratic equation using the square root principle. Are your solutions approximate or exact? Explain.
   a. $9 = x^2$ b. $25 = s^2$ c. $26 = s^2$
6. Give the exact solutions to the quadratic equation $80 = 5x^2$.
7. Find the solutions to the equation $400 = 1.2d^2$. Are your solutions approximate or exact? Explain.
8. Graph the equation $y = .25x^2 + 1$. Does the shape of the graph indicate a linear or a quadratic equation? Why?
9. Use a graph to approximate the solutions to the quadratic equation $5 = .3x^2$, following these steps:
   a. Enter Y1 = .3X ∧ 2 and Y2 = 5.
   b. Set WINDOW variables to Xmin = $^-4.7$, Xmax = 4.7, Ymin = $^-10$, Ymax = 10.
   c. Press GRAPH and estimate the points of intersection.
10. Solve the equation $5 = .3x^2$ using the square root principle and compare your answers with the solutions found in the previous exercise.
11. Given the following linear equation and quadratic equation:

$$2x - 3 = 0 \qquad 2x^2 - 3 = 0$$

    Compare the steps needed to solve each equation algebraically. Which steps are similar? Which steps are different?
12. Use factoring to solve each quadratic equation. (See Section 6.2.)
    a. $x^2 + 5x = 0$ b. $x^2 - 7x = 0$
    c. $^-x^2 - 12x = 0$
13. Use factoring to solve each quadratic equation. (See Section 6.4.)
    a. $x^2 + 4x + 3 = 0$ b. $x^2 - 3x + 2 = 0$
    c. $y^2 + 5y - 14 = 0$ d. $x^2 - 11x = {}^-18$
    e. $2x^2 + 10x + 12 = 0$
14. Use the square root principle to solve for x when $(2x - 1)^2 = 100$.
15. Solve the equation $4x^2 - 49 = 0$ by two methods.
    a. Use the square root principle.
    b. Solve by factoring using the zero-product property.
    c. Compare your results in (a) and (b).

Skills and Review 7.1

16. Solve $x^2 - 8x + 15 = 0$ for x.
17. Solve $x^2 - 64 = 0$ for x.
18. Solve $3x^2 + 18x + 24 = 0$ for x.
19. Expand $(2x - 3)^2$.
20. A formula for the volume of a sphere is volume = $\frac{\pi}{6}$ diameter³. Use the laws of exponents to find the vol-

ume of the planet Venus if the diameter is $7.519 * 10^3$ miles. Use $\pi \approx 3.14$.

21. The sum of two numbers is 212 and their difference is 83.

a. Let x represent the first number and y the second number. Write an equation for the sum.

b. Write an equation for the difference.

c. Use a graph to solve the system of equations. (*Hint:* Your calculator window might need adjusting in order to show the point of intersection.)

22. Simplify

$$\frac{3(x-5)+3}{^-3+1}$$

23. Solve for z if

$$z - \frac{3}{4} = \frac{1}{3}z + \frac{1}{2}$$

Substitute your solution back into the original equation to check your answer.

24. Which of the following are equivalent to the expression $\frac{x+8}{2}$?

a. $\frac{x}{2} + \frac{8}{2}$ **b.** $4 + \frac{x}{2}$

c. $x + 4$

25. Evaluate

$$\frac{^-b}{2a} + \frac{\sqrt{b^2 - 4ac}}{2a}$$

for $a = 1$, $b = {}^-7$, and $c = 12$.

7.2 Parabolas

After Studying This Section You Will Be Able To

- Predict how changes in the values of a, b, and c affect the position of the parabola $y = ax^2 + bx + c$.
- Use a graph of $y = ax^2 + bx + c$ to determine the vertex, y-intercept, and x-intercepts (if any).

In the last section we noticed that the graph associated with a quadratic equation is not a straight line but rather a curve. This curve is given a special name, *parabola*.

Definition Parabola

The graph of $y = ax^2 + bx + c$ with $a \neq 0$ is called a **parabola**.

We can use the parabola associated with a quadratic equation to give us information about the solutions of the equation $ax^2 + bx + c = 0$. Let's explore different graphs generated by $y = ax^2 + bx + c$ as we vary the values of a, b, and c.

- Start with the simplest case, when a is 1 and b and c are 0. Let Xmin $= {}^-9.4$ and Xmax $= 9.4$ so we have a friendly window. Because $9^2 = 81$ we have to

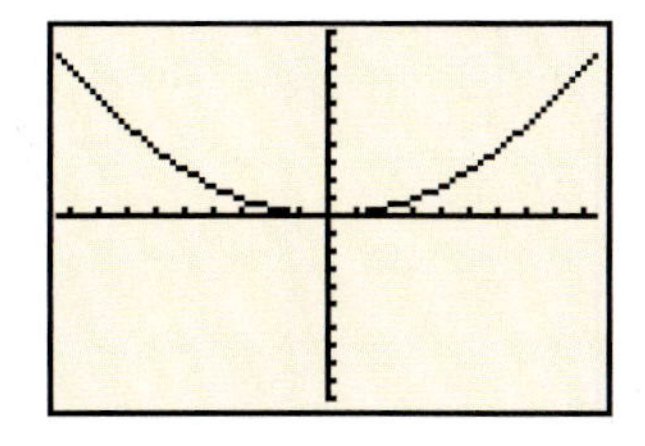

Figure 7 Graph of Y1 = X ∧ 2 with Xmin = ⁻9.4, Xmax = 9.4, Ymin = ⁻100, Ymax = 100.

let our y-values get close to 100, so we use Ymin = ⁻100 and Ymax = 100. That gives the four-quadrant window shown in Figure 7.

First, we see that when $x = 0$, $y = 0$, and in fact this is the lowest y-value the graph ever reaches. Unlike linear graphs, parabolas all *turn* somewhere, and thus there is a bottom (or maybe a top) for each parabola. In fact, the y-values are clearly dropping for x-values less than 0, something like a line with a negative slope, and rising for x-values bigger than 0, like lines with positive slope. So no single value represents the slope.

- Now let's switch a to ⁻1. We now have the graph of $y = {}^{-}x^2$ (Figure 8). The picture is *flipped*. This makes sense. Because x^2 must be a nonnegative number, the graph of $y = x^2$ doesn't ever get below the x-axis (see Figure 7). When we put a negative sign in front of the x^2, all the positive numbers become *negative*, so we get Figure 8.

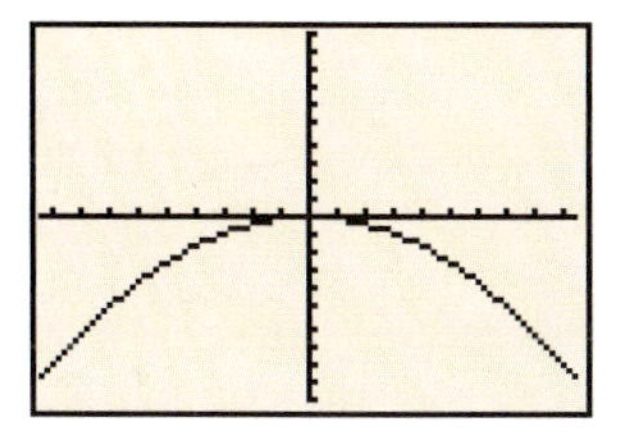

Figure 8 Graph of $y = {}^{-}x^2$

- What happens if we look at the graph of a more complicated quadratic? We can keep $a = 1$ and $b = 0$ and let c vary. See Figure 9.

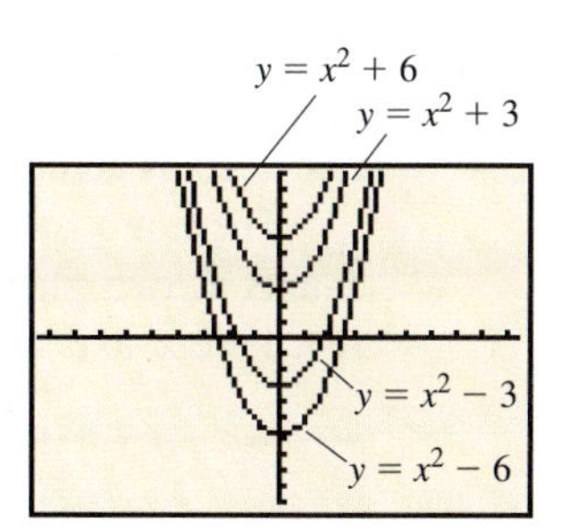

Figure 9 All these graphs are parabolas. As c changes, the graph moves up and down. The value of c is the y-coordinate of the y-intercept.

- In Figure 10, $a = 1$, $c = 0$, and b varies.
- In Figure 11, $a = {}^{-}1$ and b and c both vary.

You should start to notice some patterns. All the graphs (Figures 7, 8, 9, 10, 11) have the same basic shape; the biggest difference seems to be that some look like "cups," and others look like "hats." If we check the equations, it seems that if $a > 0$ we get a "cup," whereas if $a < 0$, we get a "hat."

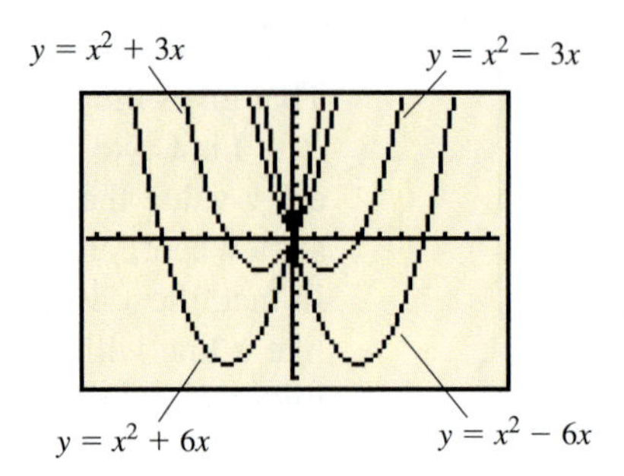

Figure 10 All these graphs are parabolas. As b changes, the graph moves left or right and up or down.

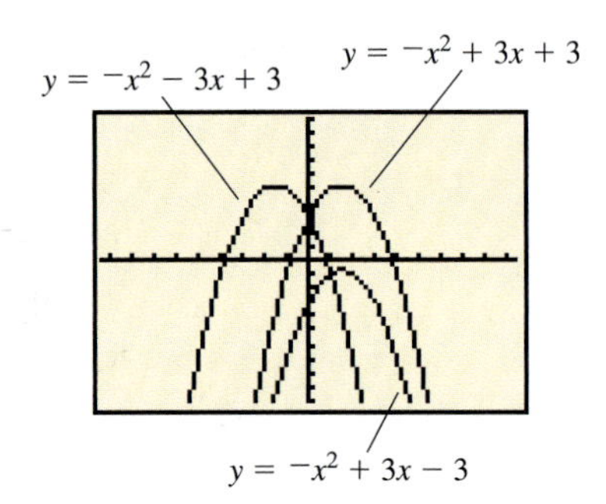

Figure 11 All these graphs are parabolas. As b and c change, the graph moves left or right and up or down.

Each graph has a point where it turns over (a *turning point*, also called the *vertex*), which gives either a maximum or minimum y-value. It also looks as though each graph is the same on either side of this turning point; if we fold the paper along a vertical line through this point, one side is a *mirror image* of the other. Each graph intersects the y-axis at one point; the y-value where the graph intersects is actually the value c. Can you see why from the formula?

There are some differences in the graphs. Some of the graphs intersect the x-axis at two places, some intersect the x-axis at one point, and some do not reach the x-axis at all.

Features of the Graph of $y = ax^2 + bx + c$

- If $a > 0$, then the parabola opens upward. Starting with a large (in absolute value) negative x-value and sliding right along the x-axis so the value of x gradually increases, the value of y drops, eventually reaching a minimum, and then begins to rise.
- If $a < 0$, then the parabola opens downward. Starting with a large negative x-value and sliding right along the x-axis so the value of x gradually increases, the value of y increases, eventually reaching a maximum, and then begins to drop.

(*continued*)

- The graph behaves the same to the right and left of the **turning point** (vertex). The **vertex** is the point (x, y) where y reaches its maximum (for a hat) or minimum (for a cup) value. If we fold the graph along a vertical line through this point, the right and left halves of the graph exactly match up.
- The graph can have zero, one, or two x-intercepts. The x-intercepts are the **roots** of the equation $ax^2 + bx + c = 0$.
- The graph crosses the y-axis at the point $(0, c)$, the y-intercept.

There are several points of special interest on a parabola.

Definition Interesting Points

In this book we often refer to the collection of points just described, the y-intercept, any x-intercepts, and the vertex, as the **interesting points** on the graph.

EXAMPLE 1 Two people are playing catch, and one tosses the ball to the other. The height of the ball y (in feet) x seconds after it is tossed is given by the formula

$$y = {}^{-}16x^2 + 44x + 5$$

a. Draw a graph of the equation $y = {}^{-}16x^2 + 44x + 5$.

b. Assume that the receiver misses the ball. When does it hit the ground?

Solution **a.** Enter Y1 = ⁻16X ∧ 2 + 44X + 5 in your calculator and graph it in the standard friendly window, Xmin = ⁻9.4, Xmax = 9.4, Ymin = ⁻10, Ymax = 10. Looking at this picture (Figure 12) we cannot see the turning point. We need to adjust the window.

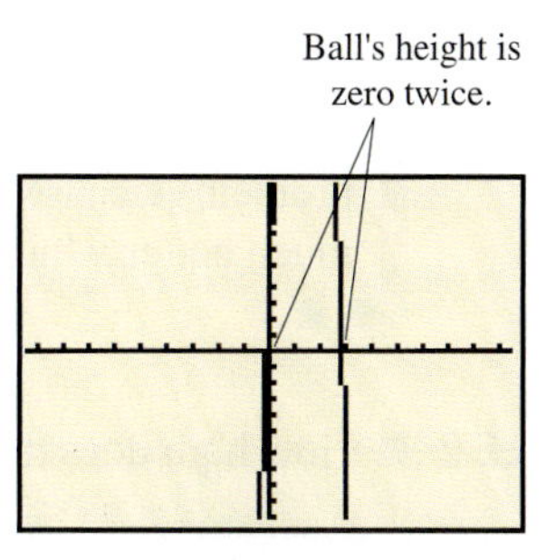

Figure 12 Y1 = ⁻16X ∧ 2 + 44X + 5 in the window Xmin = ⁻9.4, Xmax = 9.4, Ymin = ⁻10, Ymax = 10. The interesting points are not apparent.

We use the TABLE feature to get an idea what values of x and y we will need.

Time (seconds)	Height (feet)
0	5
1	33
2	29
3	$^-7$
4	$^-75$
⋮	⋮

This table suggests that we do not really need to go much beyond 3 on the x-axis but that our values on the y-axis should at least get into the 30s. So, let's change our window to Xmin $= {}^-4.7$, Xmax $= 4.7$, Ymin $= {}^-40$, Ymax $= 40$. Now all four interesting points are visible.

b. From the graph in Figure 13 we see that the height will be 0 at two different times, one positive and one negative. The positive time corresponds to the time that the ball will hit the ground, and if we use TRACE we can see that this x-intercept is between 2.8 and 2.9. Thus a good estimate for the time it hits is 2.85 seconds.

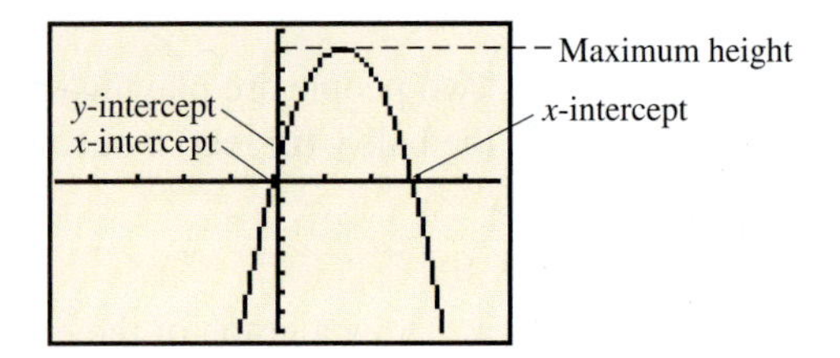

Figure 13 Y1 $= {}^-16X \wedge 2 + 44X + 5$ in the window Xmin $= {}^-4.7$, Xmax $= 4.7$, Ymin $= {}^-40$, Ymax $= 40$. This graph shows the vertex, both x-intercepts, and the y-intercept.

The x-intercept that occurs just to the left of the y-axis obviously does not correspond to the ball actually hitting the ground, because the ball could not hit the ground *before* it was thrown. We may safely ignore this point, because it does not relate to the physical problem.

In fact, it doesn't really make sense to use a four-quadrant window for this problem because negative values for *time* and *height* don't make any sense. A much better window is shown in Figure 14.

EXAMPLE 2 ■ How high does the ball get in Example 1?

Solution We can use the graph in Figure 14. If we trace along the graph, we find that the turning point lies somewhere between $x = 1.3$ and $x = 1.5$, with a y-value a

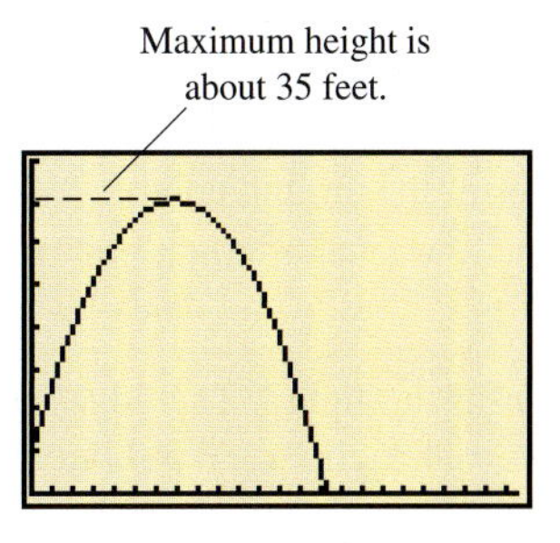

Figure 14 A first-quadrant window for Y1 = $^{-}16X \wedge 2 + 44X + 5$ with Xmin = 0, Xmax = 4.7, Ymin = 0, and Ymax = 40.

little over 35. So we estimate the maximum height of the ball to be just over 35 feet.

We may obtain a more precise answer by using the CALC menu. Begin by pressing 2ND CALC, then scroll down to maximum and press ENTER. Move the cursor to the left of the turning point and press ENTER to establish a left boundary. Then move the cursor to the right of the turning point and press ENTER again to establish a right boundary. Move the cursor close to the turning point and press ENTER to obtain a more precise estimate for the turning point. The calculator will indicate that $x \approx 1.375$ and $y \approx 35.25$.

We have seen how a calculator may be used to display a parabola and to estimate the interesting points. In the next section we learn algebraic techniques to find *exact* values for the x-intercepts and the coordinates of the vertex.

Exercises 7.2

1. What are the *interesting points* of a parabola?
2. On your grapher, experiment with different values of a in the quadratic equation $y = ax^2 + 4x - 3$. Try $a = {}^{-}2, {}^{-}1, {}^{-}\frac{1}{2}, \frac{1}{2}$, 1, and 2.
   a. What values of a make the parabola open downward (a hat)?
   b. Express your answer from part (a) in terms of an inequality involving a.
3. Experiment graphing the quadratic equation $y = 2x^2 + 5x + c$ with different values for c.
   a. Where does the value of c always appear on the graph?
   b. Compare the constant c from part (a) to the constant b in the linear equation $y = 3x + b$. Where does b always appear on the graph of a linear equation?
   c. Make a conjecture about what the value of d represents in the equation $y = x^3 + 4x^2 + 2x + d$.
   d. Refer to the last item of the features of the graph of $y = ax^2 + bx + c$ (page 268–269). What input is necessary to get an output equal to the constant in each of the three functions in this exercise?
4. Enter the equation Y1 = X ∧ 2 + 3X − 6 in your calculator.
   a. Graph the equation in the window Xmin = $^{-}9.4$, Xmax = 9.4, Ymin = $^{-}10$, Ymax = 10.
   b. What is the y-intercept?
   c. Use the graph to estimate the x-intercepts.
   d. Estimate the coordinates of the turning point.
5. Given $y = x^2 - 5x + 3$.

a. Find approximate solutions for $y = 0$.

b. What is the y-intercept?

c. Estimate the coordinates of the turning point.

d. At the turning point, does y reach a minimum or a maximum?

6. Consider the graph shown in the figure. The edges of the viewscreen are top: $y = 4$, right: $x = 5$, bottom: $y = {}^-2$, and left: $x = {}^-2$.

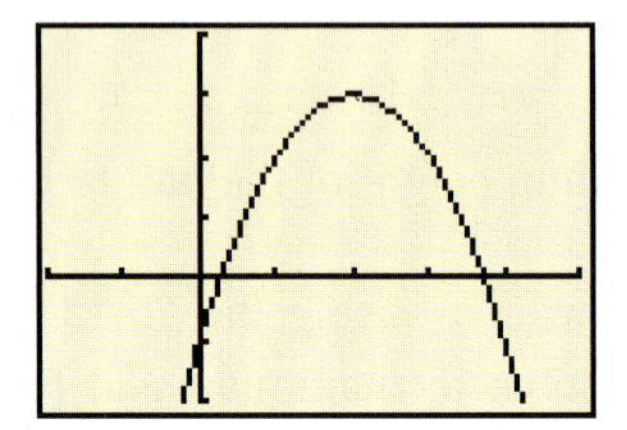

a. Estimate the interesting points of this parabola.

b. Label the interesting points on the graph of the parabola.

7. Graph each equation and visually determine whether the equation has zero, one, or two x-intercepts.

a. $y = x^2 + 4$ **b.** $y = x^2$

c. $y = x^2 - 5$

8. Use a graph to estimate the interesting points on the graph of the parabola $y = 2.5x^2 - 7x + 3$.

9. Let $y = {}^-16x^2 + 32x + 48$.

a. Factor the expression on the right-hand side completely. Use two steps:

$$y = {}^-16(x^2 + \_\_\_x + \_\_\_)$$

$$y = {}^-16(\_\_\_)(\_\_\_)$$

b. Use the factors to find values of x for which $y = 0$.

c. Use the results of (b) to determine Xmin and Xmax for your window.

d. Experiment with Ymin and Ymax to find a window that shows the turning point.

10. Let $h = {}^-4.9t^2 + 15t + 11.2$ model the situation where a ball is thrown in the air. After t seconds, h is the height of the ball in meters above the ground.

a. Although there are two t-intercepts for this parabola, why are we interested in only one of them?

b. What is the h-intercept?

c. On a calculator, h will be represented by Y and t by X. Find a window that includes the interesting points, and graph the equation.

d. Does the ball ever attain a height of 23 m? Explain.

11. Given the following table of values for a ball thrown in the air:

Time (seconds)	Height (feet)
0	5.1
1	69.1
2	101.1
3	101.1
4	69.1
5	5.1
6	$^-90.9$

a. Between what two whole seconds does the ball hit the ground?

b. Between what two whole seconds does the ball reach its maximum height?

12. Let $y = 3(x - 2.1)(x - 2.9)$.

a. How can you tell this is a quadratic equation without using your grapher?

b. At what values of x does $y = 0$?

c. In part (b), what are these values of x called?

d. We wish to get a table of values that will include the roots and the minimum. What should we use for TblSet settings to accomplish this?

13. Assume the function $h = {}^-16t^2 + 80t + 4$ describes the path of water shot from a firefighter's hose, where h is the height of the water in feet above the ground after t seconds. Assume the firefighter is directly below the fire.

a. Will the hose water reach a fire 120 feet above the ground?

b. Approximate how high on the building the water will reach.

14. Imagine Sammy Sosa has just hit a towering fly ball to center field. The outfield fence is 400 feet from home plate and the fence is 10 feet high. Assume

$$h = \frac{{}^-s^2}{500} + s + 4$$

traces the height, in feet, of the ball at s feet from home plate.

a. Find h when $s = 400$ feet.

b. Did Sosa's hit clear the fence for a home run? Why?

15. For a punter in the NFL it is advantageous to get the ball to remain in the air as long as possible (known as hang time). This gives his defense time to run down field before the ball reaches the opposing team. If $h = {}^{-}4.9t^2 + 22t + 2$ models the height of the football in meters after t seconds, how long does this punter's football remain airborne?

Skills and Review 7.2

16. A formula for the area, A, of an equilateral triangle is $A = (\frac{\sqrt{3}}{4})s^2$, where s is the length of a side of the triangle. Use the seven steps to problem solving (Section 3.3) to find the length of a side of the triangle when the area is 8 cm^2.

17. Let $y = 2x^2 + 7x - 15$.

a. Find the value of y if $x = {}^{-}1$.

b. Solve the equation for x if $y = 0$.

18. Find the exact solutions to the equation

$$7x^2 - x = 0$$

19. Find the exact solutions to the equation

$$x^2 + 3x - 28 = 0$$

20. Expand $(2x + 3)(x - 4)$.

21. Expand $(2y - 3)(2y + 3)$.

22. Use the laws of exponents (Section 6.1) to simplify and write $(2c^{-1}d^4)^{-3}$ without negative exponents.

23. Use the rules of exponents to simplify and write without negative exponents:

$$\frac{5xy^4z}{20x^3y^2z}$$

24. In the November 2002 elections, women won 60 of the 435 U.S. House seats. What percentage of the house seats is this? (*Source: Center for American Women and Politics, Rutgers University.*)

25. Letter-size copy paper measures $8\frac{1}{2}$ inches by 11 inches.

a. What is the area of the paper? Be sure to include units.

b. What is the perimeter of the paper? Again, include units.

7.3 Algebraic Techniques for Solving Quadratic Equations

After Studying This Section You Will Be Able To

- Complete the square for a quadratic expression.
- Use the quadratic formula to solve equations.
- Use algebraic techniques to find exact values for the vertex and x-intercepts of the graph of a quadratic function.

How can we find reasonable windows for more difficult quadratic functions? Suppose we need to know the *exact* value of an x-intercept of a parabola. Although we know how to use calculators to get *approximate* answers, we do not always get exact solutions this way.

In Section 7.1 we learned how to find exact solutions for some quadratic equations. If there is no x-term in the associated quadratic trinomial ($b = 0$), we can rearrange and get x^2 on one side of an equation and a number on the other side. Then, we can take square roots on both sides. If $b \neq 0$ and the coefficients are small, we can try to factor the associated quadratic trinomial and then use the zero-product property to find exact solutions. If neither of these techniques works, we need to use a different method. In this section we learn two new methods, completing the square and the quadratic formula, that enable us to solve *any* quadratic equation.

Completing the Square

In Example 3(a) of Section 6.3 (page 232) we saw that

$$(x+n)^2 = x^2 + 2nx + n^2$$

For example, when $n = 3$ we have

$$(x+3)^2 = x^2 + 6x + 9$$

We can take advantage of this fact to solve an equation with an x^2-term and a $2nx$-term by a technique called **completing the square**.

EXAMPLE 1 ■ Solve the equation $x^2 + 6x = 14$.

Solution The left side of this equation is almost the same as the right side of the equation in red above. All that's missing is the constant term, 9. So let's add 9 on both sides and see what happens.

$$x^2 + 6x = 14$$

$$x^2 + 6x + 9 = 14 + 9 \qquad \textit{Adding 9 on both sides}$$

$$(x+3)^2 = 23 \qquad \textit{Substituting } (x+3)^2 \textit{ for } x^2 + 6x + 9$$

$$x + 3 = \pm\sqrt{23} \qquad \textit{Using the square root principle}$$

$$x = {}^{-}3 \pm \sqrt{23}$$

This means we have two solutions, $x = {}^{-}3 + \sqrt{23}$ and $x = {}^{-}3 - \sqrt{23}$; 23 is not a perfect square but a calculator shows $\sqrt{23} \approx 4.796$. So our two solutions are approximately 1.796 and $^{-}7.796$.

■ ■

This technique can be used to solve a more general equation of the form

$$x^2 + 2nx - k = 0$$

The left side of this equation looks a lot like the right side of the earlier blue equation. Our problem is that we have ^{-}k instead of n^2. But we can rearrange like this:

$$x^2 + 2nx - k = 0$$

$$x^2 + 2nx = k \qquad \textit{Adding k on both sides}$$

$x^2 + 2nx + n^2 = n^2 + k$	*Adding n^2 on both sides so we get an expression like the right side of the blue equation*
$(x + n)^2 = n^2 + k$	*Substituting on the left side from the blue equation*
$x + n = \pm\sqrt{n^2 + k}$	*Using the square root principle*
$x = {}^{-}n \pm \sqrt{n^2 + k}$	*Adding ${}^{-}n$ on both sides*

Adding n^2 on both sides of the equation is called completing the square. We can see why if we look at a geometric square (Figure 15).

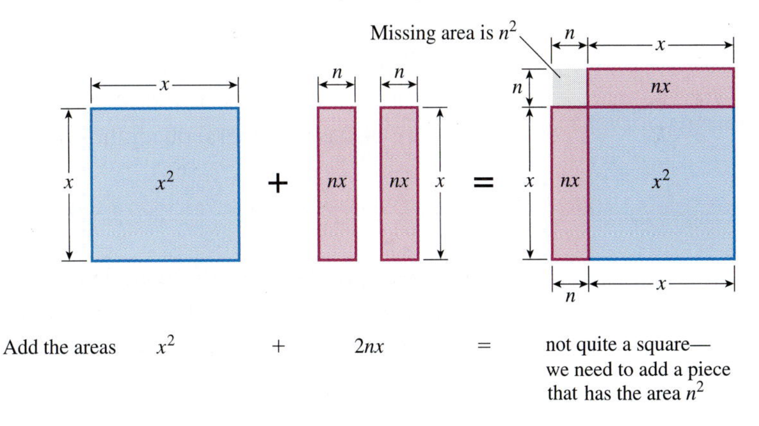

Figure 15 Completing the square.

In principle, any quadratic equation can be solved by completing the square, as we did in Example 1. In practice, it is easier to use a formula such as $x = {}^{-}n \pm \sqrt{n^2 + k}$.

EXAMPLE 2 Solve

$$x^2 + 3x - 10 = 0$$

using the formula $x = {}^{-}n \pm \sqrt{n^2 + k}$.

Solution We add 10 on both sides of the equation, which puts k on the right:

$$x^2 + 3x = 10$$

Therefore, $\mathbf{k = 10}$ and $2n = 3$, so $\mathbf{n = \frac{3}{2}}$. If we substitute this into the equation $x = {}^{-}\mathbf{n} \pm \sqrt{\mathbf{n}^2 + \mathbf{k}}$, we get

$$x = {}^{-}\frac{3}{2} \pm \sqrt{\left(\frac{3}{2}\right)^2 + 10}$$

$$x = {}^{-}\frac{3}{2} \pm \sqrt{\frac{9}{4} + 10} = {}^{-}\frac{3}{2} \pm \sqrt{\frac{49}{4}}$$

$$x = {}^{-}\frac{3}{2} \pm \frac{7}{2}$$ *Taking square roots of the numerator and denominator*

$$x = 2 \quad \text{or} \quad x = {}^{-}5$$

Notice that we were able to use the formula $x = {}^{-}n \pm \sqrt{n^2 + k}$ even though the coefficient of x wasn't an even number. When we solved for n, we got a fraction, but the formula still worked.

The Quadratic Formula

If we have a general quadratic equation

$$ax^2 + bx + c = 0$$

we can still solve by completing the square. Our problem is the fact that the coefficient of x^2 could be something other than 1. Because there is a zero on the right-hand side of the equation, we divide both sides by a:

$$ax^2 + bx + c = 0$$

$$x^2 + \frac{b}{a}x + \frac{c}{a} = 0$$

Now we can solve for n and k in the same way we did in Example 2.

$$2n = \frac{b}{a} \qquad {}^{-}k = \frac{c}{a}$$

$$\mathbf{n} = \frac{\mathbf{b}}{\mathbf{2a}} \qquad \mathbf{k} = {}^{-}\frac{\mathbf{c}}{\mathbf{a}}$$

Substituting into the formula, we have

$$x = {}^{-}\frac{\mathbf{b}}{\mathbf{2a}} \pm \sqrt{\left(\frac{\mathbf{b}}{\mathbf{2a}}\right)^2 - \frac{\mathbf{c}}{\mathbf{a}}}$$

In Chapter 11 you learn techniques to simplify the expression on the right side so that

$$x = {}^{-}\frac{b}{2a} \pm \frac{\sqrt{b^2 - 4ac}}{2a}.$$

This is one version of the *quadratic formula*.

Definition The Quadratic Formula

Any solutions to a quadratic equation $ax^2 + bx + c = 0$ are given by the **quadratic formula**,

$$x = \frac{^-b}{2a} \pm \frac{\sqrt{b^2 - 4ac}}{2a}$$

One important thing to notice is that by taking the square root in the quadratic formula, we have several different possibilities for how many solutions we get.

Definition Discriminant

The term $b^2 - 4ac$ found inside the radical in the quadratic formula is called the **discriminant**.

1. If the discriminant is positive, we get two real number solutions, or **roots**.
2. If the discriminant is zero, we get one real number solution, $x = \frac{^-b}{2a}$.
3. If the discriminant is negative, we get *no* real number solution, because we are not able to take the square root of a negative number.

Note: There is no *real number* solution if the discriminant is negative; however, if we add $i = \sqrt{^-1}$ to our number system, we will have imaginary, or complex, number solutions. Imaginary numbers are discussed in Section 7.5.

EXAMPLE 3 Use the quadratic formula to solve

$$2x^2 + 3x - 4 = 0$$

Solution $\boldsymbol{a} = \mathbf{2}$, $\boldsymbol{b} = \mathbf{3}$, and $\boldsymbol{c} = \mathbf{^-4}$, so the solutions are given by

$$x = \frac{^-b}{2a} \pm \frac{\sqrt{b^2 - 4ac}}{2a}$$

$$x = \frac{^-\mathbf{3}}{2 * \mathbf{2}} \pm \frac{\sqrt{\mathbf{3}^2 - 4 * \mathbf{2} * (\mathbf{^-4})}}{2 * \mathbf{2}}$$

$$x = \frac{^-3}{4} \pm \frac{\sqrt{9 + 32}}{4}$$

$$x = {}^{-}\frac{3}{4} \pm \frac{\sqrt{41}}{4}$$

$$x = {}^{-}\frac{3}{4} + \frac{\sqrt{41}}{4} \approx .85 \quad \text{or} \quad x = {}^{-}\frac{3}{4} - \frac{\sqrt{41}}{4} \approx {}^{-}2.35$$

Finding the Vertex

Notice that the right-hand side of the quadratic formula has two parts:

$$^{-}\frac{b}{2a}$$

and

$$\pm\frac{\sqrt{b^2 - 4ac}}{2a}$$

The second term tells us how many solutions (roots) the equation has. The first term is the same whether we add or subtract the second. We then use the fact that the graph of a parabola looks exactly the same on either side of the maximum or minimum point (the vertex) to realize the following: $x = {}^{-}\frac{b}{2a}$ is the x-coordinate of the point on the parabola with the maximum or minimum y-value, which is the turning point. Because the parabola must be symmetric around this point, the x-intercepts (if there are any) must be evenly spaced around it (see Figure 16). Think, "if the paper is folded on a vertical line through the vertex, then the x-axis is folded onto itself, and the x-intercepts must match."

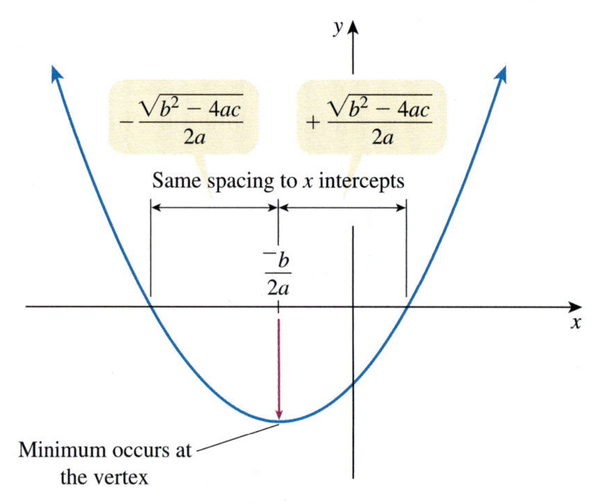

Figure 16 We can see the values of the parts

$$^{-}\frac{b}{2a} \quad \text{and} \quad \pm\frac{\sqrt{b^2 - 4ac}}{2a}$$

in the graph.

Definition Vertex of a Parabola

The **turning point** of a parabola $y = ax^2 + bx + c$ is called the **vertex**. The x-coordinate of this point is given by $x = {}^{-}\frac{b}{2a}$, and the y-coordinate can be obtained by substituting this x-value into the original equation. The graph of the parabola is symmetric around the vertical line $x = {}^{-}\frac{b}{2a}$. The vertex will give a maximum y-value if $a < 0$ and a minimum y-value if $a > 0$.

EXAMPLE 4 Let's look back at Example 1 on page 269. Find the exact value for the maximum height the ball will reach.

Solution The formula for the height of the ball is $y = {}^{-}16x^2 + 44x + 5$. The coefficients are

$$\mathbf{a = {}^{-}16}, \qquad \mathbf{b = 44}, \qquad \text{and} \qquad \mathbf{c = 5}$$

so the formula for the vertex tells us that the maximum height will occur when

$$\begin{aligned} x &= \frac{{}^{-}b}{2a} \\ &= \frac{{}^{-}44}{2({}^{-}16)} \\ &= \frac{{}^{-}44}{{}^{-}32} \\ &= \frac{11}{8} = 1\frac{3}{8} \text{ seconds} \end{aligned}$$

If we substitute this into our formula for height, we get

$$y = {}^{-}16\left(\frac{11}{8}\right)^2 + 44\left(\frac{11}{8}\right) + 5 = 35\frac{1}{4} \text{ feet}$$

Both the time and the height agree well with our initial approximations in Example 1.

EXAMPLE 5 Find the exact time the ball will hit the ground in Example 4.

Solution Remember, the equation is $y = {}^{-}16x^2 + 44x + 5$, where y is the height in feet, so the ball will hit the ground when $y = 0$. We have $\mathbf{a} = \mathbf{{}^{-}16}$, $\mathbf{b} = \mathbf{44}$, and $\mathbf{c = 5}$. Substitute these values into the quadratic formula:

$$x = \frac{{}^-b}{2a} \pm \frac{\sqrt{b^2 - 4ac}}{2a}$$

$$= \frac{{}^-44}{2 * ({}^-16)} \pm \frac{\sqrt{44^2 - 4 * ({}^-16) * 5}}{2 * ({}^-16)}$$

$$= \frac{{}^-44}{{}^-32} \pm \frac{\sqrt{1936 + 320}}{{}^-32}$$

$$= \frac{11}{8} \pm \frac{\sqrt{2256}}{{}^-32}$$

So we get two solutions to the equation $0 = {}^-16x^2 + 44x + 5$,

$$x = \frac{11}{8} - \frac{\sqrt{2256}}{32} \quad \text{and} \quad x = \frac{11}{8} + \frac{\sqrt{2256}}{32}$$

These are the exact solutions; once we use a calculator to compute the square roots, we write them as decimal approximations, $x \approx {}^-.109293$ and $x \approx 2.85929$, which agree with the answers we got in Section 7.2 when we used a calculator to approximate the answers with a graph. The advantage to the algebraic solution is twofold: we have the exact version, with square roots left in and available, and we can use that version to get *as many* decimal places of accuracy as we want, assuming we have a good enough calculator or computer program.

As in the previous section, however, only one of our solutions has any meaning in the context of this problem. The ball will hit the ground in approximately 2.86 seconds.

Note: In Chapter 9 you will learn that a simpler form for the solutions in Example 5 is $\frac{11}{8} \pm \frac{\sqrt{141}}{8}$. For now don't worry about simplifying the square root, because both forms of the solution are correct.

EXAMPLE 6 Suppose the profit (in millions of dollars) a company makes by selling widgets is given by $y = {}^-.3x^2 + 25x - 120$, where x is the price (in dollars) they charge for a box of widgets. How much should they charge for a box of widgets in order to make a profit of \$500 million?

Solution Setting up the equation gives

$${}^-.3x^2 + 25x - 120 = 500$$

$${}^-.3x^2 + 25x - 620 = 0$$

So $\boldsymbol{a} = {}^-\mathbf{.3}$, $\boldsymbol{b} = \mathbf{25}$, and $\boldsymbol{c} = {}^-\mathbf{620}$. Before we use the whole quadratic formula, we can check the discriminant to see if there are any solutions.

$$\begin{aligned} b^2 - 4ac &= \mathbf{25}^2 - 4(\mathbf{^-.3})(\mathbf{^-620}) \\ &= 625 - 744 \\ &= {}^-119 \end{aligned}$$

Because the discriminant is negative, there are *no* real solutions; in other words, the company cannot make a $500 million profit. You may check this result by graphing Y1 $= {}^-.3x^2 + 25x - 120$ in a suitable window and noticing that the maximum value of Y is less than 500.

EXAMPLE 7 Find the solutions to the equation

$$x^2 - 4x + 4 = 0$$

Solution Here, $\boldsymbol{a} = \mathbf{1}$, $\boldsymbol{b} = \mathbf{^-4}$, and $\boldsymbol{c} = \mathbf{4}$. If we check the discriminant,

$$\begin{aligned} b^2 - 4ac &= (\mathbf{^-4})^2 - 4(\mathbf{1})(\mathbf{4}) \\ &= 16 - 16 \\ &= 0 \end{aligned}$$

This means there is only one solution, which we can find using the quadratic formula (we can ignore the second half of the formula, because the square root is zero).

$$\begin{aligned} x &= \frac{{}^-b}{2a} \\ &= \frac{{}^-({}^-4)}{2} \\ &= 2 \end{aligned}$$

We can summarize what we know about quadratic equations:

- The quadratic formula

$$x = \frac{{}^-b}{2a} \pm \frac{\sqrt{b^2 - 4ac}}{2a}$$

gives the solutions (roots) to the equation $0 = ax^2 + bx + c$, *which are also* the x-intercepts for the graph of $y = ax^2 + bx + c$.

- The discriminant, $b^2 - 4ac$, tells you the number of solutions and x-intercepts.

a. If $b^2 - 4ac > 0$, there are two real number solutions (two x-intercepts).

b. If $b^2 - 4ac = 0$, there is one real number solution (x-intercept).

c. If $b^2 - 4ac < 0$, there is no real number solution to the equation and hence there is no x-intercept for the graph.

- $x = \frac{^-b}{2a}$ gives the x-coordinate of the vertex of the parabola associated with the quadratic equation.

Choosing a Method

If we are given a quadratic equation $ax^2 + bx + c = 0$, how can we tell whether it is better to try the factoring method or just to use the quadratic formula? There is no simple rule that will work all the time, but we can usually get a good idea just by looking at the coefficients. If either a or c is large, the diagonal product may have many factors and we'll be able to find a solution more quickly using the formula. If a and c are both small, we may be able to factor the quadratic trinomial; if not, we will have to use the formula anyway. If we are unsure, we can quickly check the sign of the discriminant, because if it is negative we can stop right there.

With a little practice it is possible to recognize quadratics that can be factored easily right away. We can summarize our methods for solving quadratic equations with the following table.

Solution Type	Quadratic	Example	Method
Approximate	Any	$0 = {}^-2.45x^2 + 6.33x + 4.2$	Graphing: TRACE
Exact	No x term, $b = 0$	$0 = 5x^2 - 4$	Square root principle
Exact	No constant term, $c = 0$	$0 = {}^-3x^2 - 7x$	Find the GCF and use the zero-product property
Exact	Many, integer coefficients	$0 = x^2 + 3x + 2$	Factoring, if it works (zero product property)
Exact	Any	$0 = 4x^2 - 5x - 8$	Quadratic formula

Exercises 7.3

1. As an aid, use the preceding table to determine the quickest method for finding the solutions of the following quadratic equations. (Do not solve.) Give the reasons for your choice.

a. Exact solution(s) for $42x^2 + 7x = 0$

b. Exact solution(s) for $x^2 + 7x + 12 = 0$

c. Approximate solution(s) for $3x^2 - 6x - 8 = 0$

d. Exact solution(s) for $5x^2 - 8 = 0$

e. Exact solution(s) for ${}^-2x^2 + 5x + 6 = 0$

2. Without finding the solution(s), pick a method for determining the solution(s) from the preceding table. Remember to make one side of the equation equal to zero before deciding.

a. Exact solution(s) for $1.5x^2 = {}^-7.2x + 5$

b. Approximate solution(s) for $34x^2 - 2x = 9$

c. Exact solution(s) for $2 = 5x^2$

d. Exact solution(s) for ${}^-x = 6 - x^2$

e. Exact solution(s) for $52x^2 = 4x$

3. Consider the quadratic equation $2x^2 + 12x - 5 = 0$.

a. Remember that the quadratic formula is used to solve the equations $ax^2 + bx + c = 0$. In this case, what are the values of a, b, and c?

b. Use the values of a, b, and c and the quadratic formula to solve for x.

4. What are the x-intercepts for the graph of $y = 5x^2 + 10x - 6.25$?

5. Find the x-coordinate of the turning point for $y = 5x^2 + 10x - 6.25$.

6. Use the quadratic formula to find the exact solutions of $2x^2 + 6x - 10 = 0$.

7. Consider the parabola with equation $y = 2x^2 + 6x - 10$.

a. Which part of the quadratic formula gives the x-coordinate of the vertex?

b. Find the x-coordinate of the vertex for $y = 2x^2 + 6x - 10$.

c. Find the y-coordinate of the vertex for $y = 2x^2 + 6x - 10$.

d. What points on the parabola are related to the solutions found in Exercise 6?

8. Consider the equation $x^2 + 10x = 15$.

a. What number must be added to $x^2 + 10x$ to complete the square?

b. Solve the equation $x^2 + 10x = 15$ by adding that number on both sides and using the method shown in Example 1.

c. Solve the equation $x^2 + 10x = 15$ using the quadratic formula.

d. Compare your answers in (b) and (c).

e. Express your answers as exact quantities and approximations to the nearest .01.

9. Consider the functions $y = x^2 + 2x + 1$, $y = x^2 + 2x + 3$, and $y = x^2 + 2x - 2$.

a. Graph each function.

b. Set $y = 0$ and calculate the discriminant part of the quadratic formula.

c. Match each statement in blue with a statement in red.

The equation has no real root.
The equation has one real root.
The equation has two real roots.

The discriminant is positive.
The discriminant is negative.
The discriminant is zero.

10. What is the maximum y-value on the graph of $y = {}^{-}x^2 - 7x + 9$?

11. A parabola has a minimum at $(3, {}^{-}1.5)$ and an x-intercept at $({}^{-}1, 0)$.

a. Sketch a graph.

b. What is the other x-intercept?

12. Find the solutions, if any, for $x^2 + 4.2x + 4.41 = 0$.

13. Find the solutions, if any, for ${}^{-}4x^2 - 2x - 7 = 0$.

14. Solve $x^2 + 8x = {}^{-}7$ in three ways.

a. Add the same number to both sides to solve by completing the square.

b. Add the same number to both sides to get 0 on the right side and solve by factoring.

c. Add the same number to both sides to get 0 on the right side and use the quadratic formula.

15. Solve for $5x^2 + 4x - 1 = 0$ in three ways.

a. Factor the left side and use the zero-product property.

b. Use the quadratic formula.

c. Graph the equation in the decimal window Xmin = ${}^{-}4.7$, Xmax = 4.7, Ymin = ${}^{-}3.1$, Ymax = 3.1, and approximate the solutions.

Skills and Review 7.3

16. The height of a ball y (in feet) x seconds after it is thrown is given by the formula $y = {}^{-}16x^2 + 30x + 5.5$. Use a graph to approximate the maximum height of the ball.

17. Use a graph to approximate the solutions of $.27x^2 - .96x + .2 = 0$.

18. Find the exact solutions to the equation $3x^2 - 2x - 1 = 0$.

19. At Elio's pizza house the amount of cheese on a circular pizza is 1.3 g/square inch multiplied by the diameter squared.

a. Represent this situation symbolically with a formula

for the amount of cheese in terms of the diameter d of the pizza.

b. Represent this situation numerically with a table of values for pizzas with 6-inch, 10-inch, and 14-inch diameters.

c. Represent this situation graphically.

20. Let $y = {}^{-}x^2 + 4x - 5$. Find y when $x = {}^{-}2$.

21. Simplify the expression and write without negative exponents:

$$\left(\frac{2x^{-2}y}{x^3}\right)^2$$

22. Solve the equation ${}^{-}5(x + 8) = {}^{-}7$.

23. The formula for the area of a trapezoid is

$$A = \tfrac{1}{2}(b_1 + b_2)h$$

Find b_1 if $A = \frac{15}{2}$ cm^2, $b_2 = 4$ cm, and $h = 2$ cm.

24. At what average rate does a plane travel if it flies 2317 miles in $4\frac{1}{2}$ hours?

25. Given the number .73.

a. Write this number as a common fraction.

b. Write this number as a percent.

7.4 Applications That Lead to Quadratic Equations

After Studying This Section You Will Be Able to

- Solve word problems with quadratic equations.
- Solve problems requiring that you find the maximum or minimum value of a function.

In this section you will apply what you have learned about parabolas and algebraic techniques for solving quadratic equations to a variety of problems.

EXAMPLE 1 The force due to gravity on the moon is weaker than on Earth, so the formula for the height of an object on the moon is a little different, but it is still a quadratic. Suppose two astronauts are playing catch with a baseball. The height of the ball is given by

$$h = {}^{-}2.7t^2 + 22t + 5$$

where h is in feet and t is the number of seconds after one astronaut throws the ball to the other.

a. How high does the ball go?

b. The second astronaut catches the ball when it is 4.5 feet off the ground. How long after it is thrown does the second astronaut catch the ball?

Solution **a.** Let's rewrite the equation with $y = h$ and $x = t$ so we can use a graphing calculator.

$$y = {}^{-}2.7x^2 + 22x + 5$$

To get an idea of what window to use, you may want to use the formula $x = \frac{{}^{-}b}{2a}$ to find the x-coordinate of the vertex.

$$
\begin{aligned}
x &= \frac{^{-}b}{2a} \\
&= \frac{^{-}22}{2(^{-}2.7)} \\
&= \frac{^{-}22}{^{-}5.4} \\
&\approx 4.07
\end{aligned}
$$

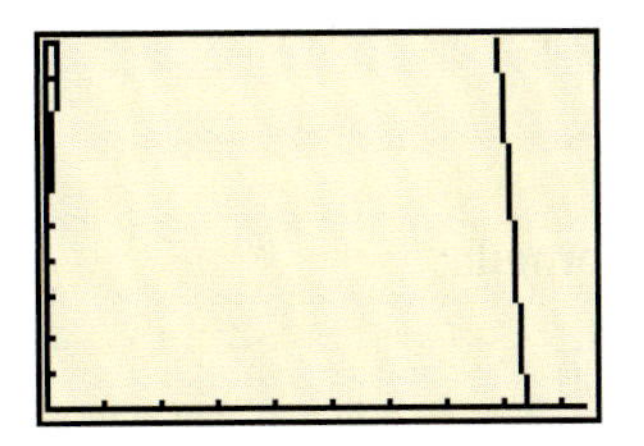

Figure 17

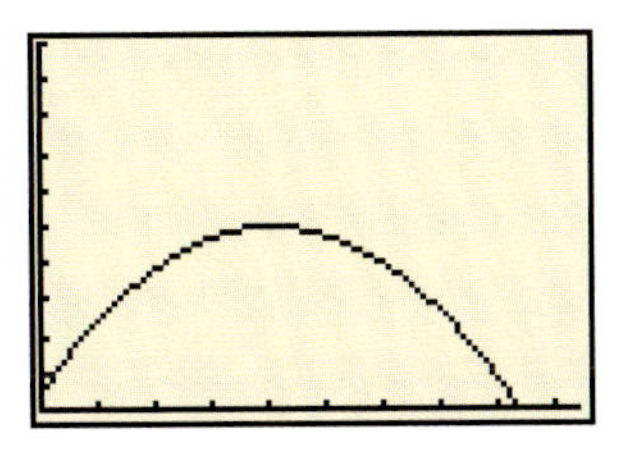

Figure 18

This suggests that a good friendly window to try is one with Xmin = 0 and Xmax = 9.4. If you set Ymin = 0 and Ymax = 10, you will get the graph in Figure 17. Because the vertex is nowhere in sight, we must increase Ymax. When Ymax = 100 you get the graph shown in Figure 18. Tracing along this graph, we find that the y-coordinate of the vertex is approximately 50. This means that the maximum height the ball reaches is about 50 feet.

We can substitute $x = 4.07$ in the equation $y = {}^{-}2.7x^2 + 22x + 5$ to find a more precise value for the maximum height.

b. Tracing along the graph we find that one x-intercept occurs when $x \approx 8.4$. However, the astronaut catches the ball at a height of 4.5 feet. Tracing back to where $y \approx 4.5$, we get $x \approx 8.2$. The astronaut catches the ball about 8.2 seconds after it is thrown.

We can find a more precise answer to this question using the quadratic formula to solve the equation ${}^{-}2.7x^2 + 22x + 5 = 4.5$.

EXAMPLE 2 Television manufacturers advertise their televisions in terms of the diagonal measurement, because that is longer than the length along either side of the screen. If a television is advertised as being a 25-inch screen and it is 5 inches wider than it is high, how high and wide are the screen?

Solution If we sketch a picture of the TV set and let x represent the height, then the width is $x + 5$ (Figure 19). The diagonal, which is 25 inches long, forms a right triangle with the two sides. Therefore, we can use the Pythagorean theorem (Section 2.2) and write a quadratic equation in x.

Figure 19 The width of the TV is 5 inches more than the height, and the diagonal is 25 inches.

$$x^2 + (x+5)^2 = 25^2$$
$$x^2 + x^2 + 10x + 25 = 625$$
$$2x^2 + 10x - 600 = 0$$

We can solve this equation by using the quadratic formula:

$$x = \frac{^-10}{4} \pm \frac{\sqrt{100 + 4800}}{4}$$
$$x = \frac{^-10}{4} \pm \frac{\sqrt{4900}}{4}$$
$$x = {}^-20 \quad \text{or} \quad x = 15$$

We can also solve by factoring

$$2x^2 + 10x - 600 = 0$$
$$2(x^2 + 5x - 300) = 0$$
$$2(x+20)(x-15) = 0$$
$$x = {}^-20 \quad \text{or} \quad x = 15$$

Clearly we are interested only in the positive solution. Because x represents the height, the television is 15 inches high and 20 inches wide.

EXAMPLE 3 You are going to supplement your income by opening a doggie day-care center at home. To make the dogs happy, you are going to make a rectangular dog run in your back yard. If you use the back of your house as part of the border, you will save money on the fencing.

You can afford 150 feet of fencing, and you also get a gate that is 5 feet wide. It is easiest to put the fence up in a rectangle.

a. What should the dimensions of the dog run be to maximize the area?

b. How large is the maximal dog run?

Solution **a.** We use the seven-step process to help us solve this problem (see page 105). We can *understand* the problem best if we know exactly what we are looking for, which is the maximum possible area, assuming the pen is a rectangle, built on the back of the house, with a 5-foot gate. A diagram will help *visualize* the problem (Figure 20). The diagram helps us to *assign variables*, in this case for the dimensions of the pen, length and width.

Next, we need to *write equations*. Because we want to find the largest possible area, we need to write a formula for the area; for a rectangle this is easy,

$$\text{area} = \text{length} * \text{width}, \quad \text{or} \quad A = lw$$

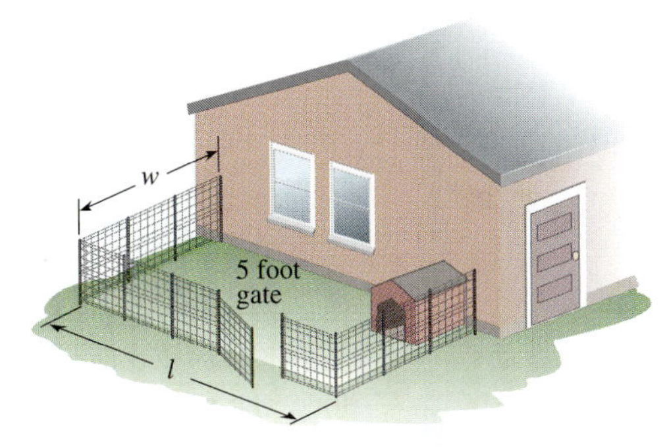

Figure 20 Notice that the gate is included in l, and we do not put any fence on the house side.

Obviously l and w can't be just anything since we have only 150 feet of fence (plus 5 feet of gate); if the length gets very large, then the width has to be small, and vice versa. If we look at the picture (Figure 20), we can see that the boundary of the run includes the house (which comes for free), the gate, and the fence. We need only the lengths of three sides to account for the gate and the fence. We have a second equation,

$$l + 2w = 155$$

Now, we need to *solve the equations*. First, we go back to the *understanding* step to work out what exactly we need to solve; because we are looking for the *maximum* area, we will need to use the equation $A = lw$. We cannot work with this equation yet, however, because it involves *three* variables, A, l, and w. The second equation, $l + 2w = 155$, uses only two variables, so we can solve for l to get

$$l = 155 - 2w$$

Now we substitute back into the first equation and get

$$\begin{aligned} A &= lw \\ &= (155 - 2w)w \\ &= 155w - 2w^2 \end{aligned}$$

Letting $\mathbf{x}$ be the width, $\mathbf{w}$, and $\mathbf{y}$ be the area, $\mathbf{A}$, we have a quadratic equation

$$y = {}^{-}2x^2 + 155x$$

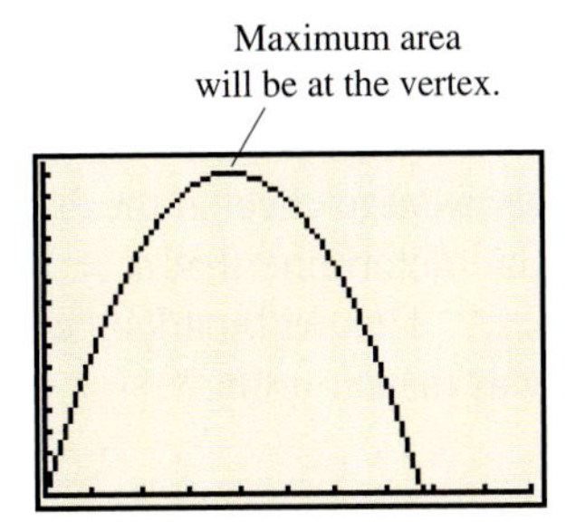

Figure 21 The x-coordinate of the vertex seems to be at a width a little less than 40 feet, and it seems to give an area of about 3000 square feet.

We want the maximum area, which occurs at the vertex. We can find the vertex by looking at the graph and approximating (Figure 21) or by using the algebraic formula for the vertex (see page 279). First, we find the x-coordinate of the vertex:

$$\begin{aligned} x &= \frac{{}^{-}b}{2a} \\ &= \frac{{}^{-}155}{{}^{-}4} \\ &= \mathbf{38.75} \end{aligned}$$

This does not *answer the question* in part (a); we want to know the dimensions (length and width) of the pen. Because x represents the width, we are half done; the width is **38.75** feet. We can find the length using the equation

$$\begin{aligned} l &= 155 - 2w \\ &= 155 - 2 * \mathbf{38.75} \\ &= 77.5 \text{ feet} \end{aligned}$$

We have already *checked* our answer using the graph in Figure 21. It certainly seems reasonable to think the pen should have these dimensions.

b. To determine the maximum area, we can substitute into the equation $y = {}^{-}2x^2 + 155x$ to find the y-coordinate of the vertex.

$$\begin{aligned} y &= {}^{-}2\mathbf{x}^2 + 155\mathbf{x} \\ &= {}^{-}2(\mathbf{38.75})^2 + 155(\mathbf{38.75}) \\ &= 3003.125 \end{aligned}$$

The largest possible area is 3003.125 square feet.

EXAMPLE 4 Insurance investigators who examine the scenes of car accidents have devised a formula that gives the distance it takes to stop a car in terms of the car's speed x in miles per hour. The formula is of the form

$$y = ax^2 + bx$$

where y is the stopping distance in feet, a depends on the car's weight and the condition of the tires and the road, and b depends on the reaction speed of the driver.

State police are investigating an accident, and they want to check if the driver was speeding. They examine the car and the road and determine that $\mathbf{a} = \mathbf{.04}$. After testing the driver's reaction, they estimate $\mathbf{b} = \mathbf{.2}$. The skid marks the car made while stopping were 200 feet long. How fast was the car going?

Solution If we substitute the police estimates for a and b, we have

$$y = \mathbf{.04}x^2 + \mathbf{.2}x$$

Because the skid marks were 200 feet long, we have $y = 200$, and our formula is

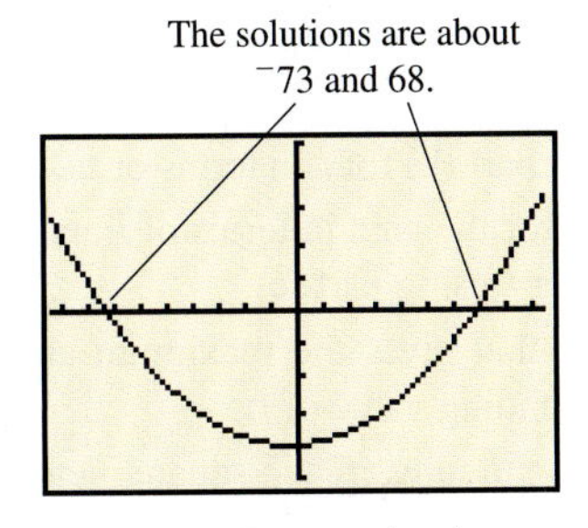

Figure 22 The graph of $y = .04x^2 + .2x - 200$ using the friendly window Xmin = ⁻94, Xmax = 94, Ymin = ⁻250, Ymax = 250. The solutions are about ⁻73 and 68.

$$200 = .04x^2 + .2x$$

$$0 = .04x^2 + .2x - 200$$

This equation won't factor easily, but we can look at a graph to estimate the solutions (Figure 22). The positive solution indicates that the car was traveling about 68 miles/hour. Because the speed limit was 55, the police gave the driver a ticket.

The police need a more accurate solution when they go to court, so they use the quadratic formula.

$$x = \frac{^-.2}{.08} \pm \frac{\sqrt{.04 + 32}}{.08}$$

$$x = \frac{^-.2}{.08} \pm \frac{\sqrt{32.04}}{.08}$$

$$x \approx {^-73.25} \quad \text{or} \quad x \approx 68.25$$

Exercises 7.4

1. Review Example 1 and explain why knowing the vertex is useful for picking a friendly window for viewing the graph.
2. Suppose the second astronaut in Example 1 throws the ball back to the first. Because her height and arm strength are different, the equation is different. The height of the ball is now given by the equation $h = {^-2.7}t^2 + 35t + 7$.
   a. Find the vertex.
   b. Use the vertex to pick a friendly window for viewing the graph.
3. In Example 2 the solutions to the equation $2x^2 + 10x - 600 = 0$ were $x = {^-20}$ and $x = 15$. Where do these solutions appear on the graph of $y = 2x^2 + 10x - 600$?
4. Outline the seven steps (*understand, visualize, assign* variable(s), *write* equation(s), *solve* equation(s), *answer* the question, *check* the answer) used to solve the problem in Example 3(a).
5. Why did we find the vertex in Example 3 and not the x-intercepts?
6. In Example 3 suppose we have 160 feet of fence and the gate is 4 feet wide. Use the seven-step process to find the dimensions of the dog run that maximizes the area.
7. Use the seven steps for problem solving to outline the work performed in Example 4.
8. Why did we find the x-intercepts in Example 4 and not the vertex?
9. Suppose the skid marks in Example 4 were 150 feet. How fast was the car going?
10. Assume a person can throw a baseball 25 m high on Earth. The same person's throw on the moon can be modeled by the equation $h = {^-.82}t^2 + 22t + 1.7$, where h is the height in meters above the moon after t seconds.
    a. How many times higher is the maximum height on the moon than on Earth?
    b. What does this indicate about the strength of gravity on the moon in relation to Earth?
11. Suppose an advanced civilization has a spaceship in orbit around our sun and they drop a probe into the sun to do research (their probe should withstand the heat of the sun). The equation for the height of the probe is $h = {^-448}t^2 + 1{,}056{,}000$, where h is the distance in feet from the sun after t seconds from release.
    a. How far is the spaceship from the sun at release?
    b. How many seconds would it take the probe to hit the sun?
    c. Suppose the probe is not as well designed as thought and would melt at 100 miles from the sun. How many seconds after release would the probe melt?

12. A television designer believes the dimensions (length and width) of a TV are as important to sales as the size (diagonal) of a TV. If the diagonal is 30 inches and the length is 6 inches more than the width, what are the dimensions of the TV?

a. Draw a picture and label the length and width.

b. Define the variable(s).

c. Write an appropriate equation.

d. Solve the equation. (This equation may be solved by factoring if you wish.)

e. Answer the question. Be sure to include the units.

13. A campground director has 30 m of rope with floats to make a rectangular swimming area. Notice in the figure that the beach and a 7-m dock partly enclose the swimming area. He wants to find the length of rope on each side necessary to maximize the swimming area.

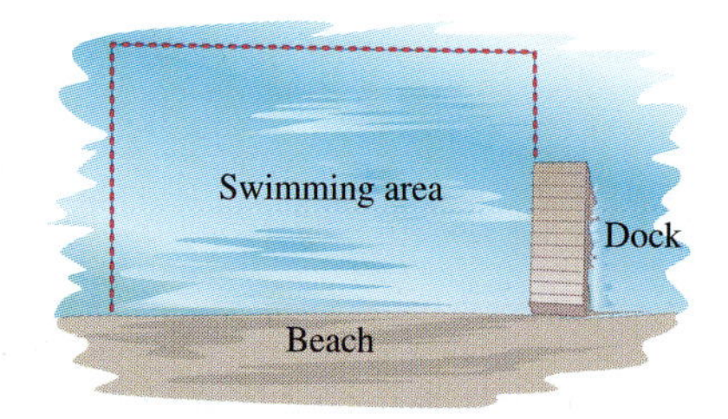

a. Draw a picture and label the length l and width w.

b. Write an equation in l and w that shows that the total amount of rope is 30 m.

c. Solve the equation in part (b) for l in terms of w.

d. Write an expression for the area in terms of l and w and then substitute for l using part (c).

e. Find the value of w that gives the maximum area; then find the corresponding value of l.

f. Find the maximum swimming area in square meters.

14. A detective is reviewing an accident report in order to calculate the speed of a car. He uses the general formula $y = ax^2 + bx$, where y is the stopping distance in feet and x is the speed of the car in miles per hour. For this particular accident the detective determines that $a = \frac{1}{20}$ and $b = .1$. If the car left a skid mark 180 feet long, how fast was the car going?

15. In Section 7.3 we introduced a business application of profit modeled by a parabola. Assume a company's revenue is modeled by the equation revenue $= {}^-2.9\text{ price}^2 + 80\text{ price}$, where price and revenue are in dollars for each item sold.

a. At what price(s) does the company generate no revenue?

b. At what should the company price its product to maximize its revenue?

c. At what price(s) does the company generate \$500.00 of revenue?

Skills and Review 7.4

16. Find the exact vertex for the graph of $y = 2x^2 - 12x + 1$.

17. Use the quadratic formula or the discriminant to show that the graph of the equation $y = 4x^2 + x + 5$ has no x-intercepts.

18. Let $y = x^2 + 8x + 6$.

a. Find the y-intercept.

b. Approximate the x-intercepts.

c. Find the exact vertex.

d. Use the information from the preceding parts to sketch a graph by hand.

19. Use a graph to approximate the solutions to the equation $x^2 + (x + 6)^2 = 196$.

20. Solve exactly the equation $4x^2 - 25 = 0$.

21. Simplify the expression and write without negative exponents

$$\left(\frac{x^5}{x^2 y}\right)^{-3}$$

22. A snowboard that normally sells for \$250 is on sale for \$210. Find the percent discount.

23. Solve the equation

$$\frac{x+5}{6} = \frac{1}{3}$$

24. Solve the system of equations by either elimination or substitution

$$x + 2y = 52$$
$$3x - y = 23$$

25. Locate $\frac{1}{3}$, 2.7, ${}^-4\frac{2}{5}$, and $\frac{11}{4}$ on a number line.

7.5 Complex Numbers (Optional)

Solving Cubic Equations

Knowledge of the quadratic formula (in some form) goes back several thousand years. It wasn't until the first half of the sixteenth century, however, that a similar formula was devised that allowed mathematicians to solve **cubic equations,**

$$ax^3 + bx^2 + cx + d = 0$$

It was first devised by Tartaglia (1500–1557) and published by Cardano (1501–1576) in his algebra book *Ars Magna*. The formula is too long for us to worry about here, but the length of the formula wasn't the only problem with it. We illustrate with an example.

EXAMPLE 1 Find the solutions to

$$x^3 - 6x^2 - 9x + 54 = 0$$

Solution Because we live in the 21st century, let's use our calculators to plot the function $y = x^3 - 6x^2 - 9x + 54$ and find the approximate solutions (see Figure 23). Looking at the graph, it is clear the equation has three solutions, which appear to be $x = {}^-3$, $x = 3$, and $x = 6$. We can confirm this by observing that when each of these values is substituted into $\mathbf{x}^3 - 6\mathbf{x}^2 - 9\mathbf{x} + 54 = 0$, the result is zero. For instance, when $\mathbf{x} = \mathbf{3}$, $(\mathbf{3})^3 - 6(\mathbf{3})^2 - 9(\mathbf{3}) + 54 = 27 - 54 - 27 + 54 = 0$. ■■

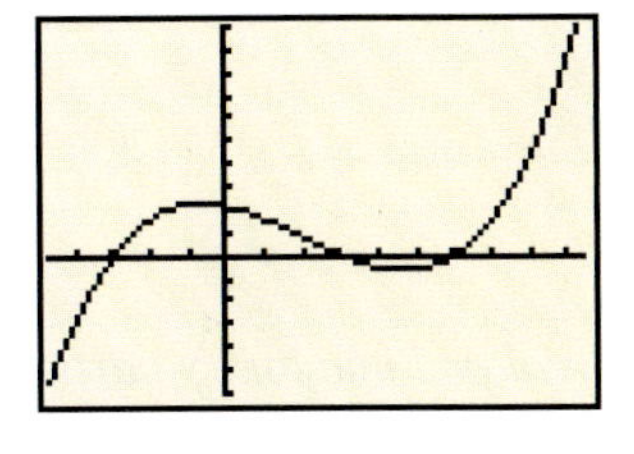

Figure 23

To solve this equation prior to the invention of the graphing calculator, one would have used the formula of Tartaglia and Cardano. After some calculation, this formula would tell you that one of the solutions is

$$x = 2 + \sqrt[3]{{}^-10 + \sqrt{{}^-243}} + \sqrt[3]{{}^-10 - \sqrt{{}^-243}}$$

That expression looks pretty scary, doesn't it? The radicals with the number 3 appearing above, $\sqrt[3]{}$, indicate **cube roots**, a topic we study in Chapter 9, so don't worry about them now. What is very interesting is that this expression contains the square root of a negative number, $\sqrt{{}^-243}$, which is not a real number.

We know from our calculators that *all* the solutions of the equation are real, so what is happening? Bombelli (1626–1672) realized that if you *allowed* yourself to manipulate the square roots of negative numbers as if they were OK, you could show that

$$2 + \sqrt[3]{{}^-10 + \sqrt{{}^-243}} + \sqrt[3]{{}^-10 - \sqrt{{}^-243}} = 6$$

which is one of the solutions we found from the graph. The fact that Tartaglia and Cardano's formula manipulated *square roots of negative* numbers in order to find *real solutions* to cubic equations motivated mathematicians to expand their notion of what a number was.

Definition **Complex Number**

A **complex number** is a number of the form $z = a + b\sqrt{^-1}$, where a and b are real numbers. It is customary to define $i = \sqrt{^-1}$ and write complex numbers as $z = a + bi$, where a is called the **real part** of z and bi is the **imaginary part** of z.

Complex numbers can be manipulated in much the same way as real numbers if you remember two simple rules:

1. When adding or subtracting complex numbers, add or subtract the real and imaginary parts separately.
2. When multiplying complex numbers, use an area model (Section 6.3) and remember that because $i = \sqrt{^-1}$, $i * i = i^2 = {^-1}$.

We have ignored *division* of complex numbers, which is more difficult than addition, subtraction, or multiplication. Division does not show up in any of the examples we cover here.

EXAMPLE 2 If $z_1 = 2 + i$ and $z_2 = 1 - 2i$, find each value.

a. $z_1 + z_2$

b. $z_1 - z_2$

c. $z_1 * z_2$

Solution

a. $z_1 + z_2 = (2 + i) + (1 - 2i)$

$= (2 + 1) + (1 - 2)i$ *Grouping the real and imaginary parts*

$= 3 - i$

b. $z_1 - z_2 = (2 + i) - (1 - 2i)$

$= (2 - 1) + (1 - {^-2})i$ *Grouping the real and imaginary parts*

$= 1 + 3i$

c. $z_1 * z_2 = (2 + i)(1 - 2i)$

$= 2 + 2 * {^-2i} + i * 1 + i * {^-2i}$ *Multiplying term by term*

$= 2 - 4i + i - 2 * i^2$

$= 2 - 3i - 2 * {^-1}$ *Using the fact that $i^2 = {^-1}$*

$= 4 - 3i$

An area model is shown in Figure 24.

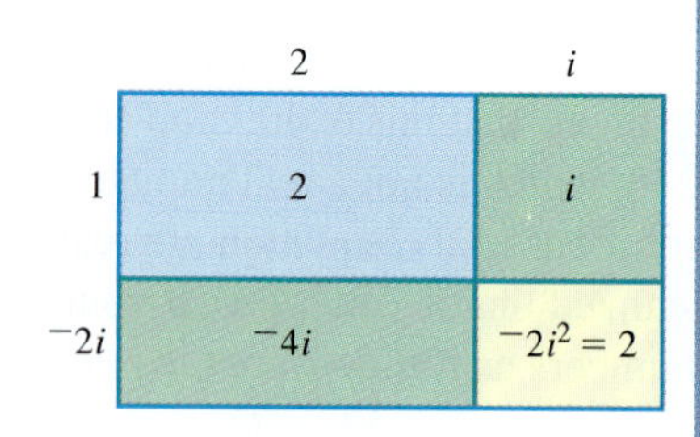

Figure 24

EXAMPLE 3 Find *any* solutions, real or complex, for each equation.

a. $x^2 + 4x + 8 = 0$

b. $x^2 - 3x + 4 = 0$

Solution Recall that in the quadratic formula, the discriminant $b^2 - 4ac$ is the term inside the square root, so we can always check the value of the discriminant; if it is *negative*, then the solutions are complex numbers, otherwise, the quadratic equation has either one or two real solutions. We can also check the graph to see if it crosses the x-axis (Figure 25).

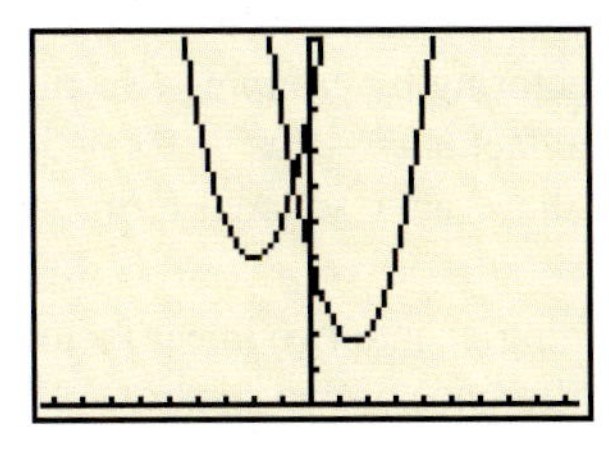

Figure 25 Xmin = $^-9.4$, Xmax = 9.4, Ymin = 0, Ymax = 10; neither graph, $y = x^2 + 4x + 8$ or $y = x^2 - 3x + 4$, intersects the x-axis.

a. Here, $a = 1$, $b = 4$, and $c = 8$, so the discriminant is $4^2 - 4 * 1 * 8 = 16 - 32 = {}^-16$, which is negative. The equation has two complex solutions given by the quadratic formula:

$$\begin{aligned} x &= \frac{^-4}{2} \pm \frac{\sqrt{^-16}}{2} \\ &= {}^-2 \pm \frac{\sqrt{^-1}\sqrt{16}}{2} \\ &= {}^-2 \pm \frac{4i}{2} \\ &= {}^-2 \pm 2i \end{aligned}$$

Notice that we simplified the square root by factoring out the imaginary part

b. Here, $a = 1$, $b = {}^-3$, and $c = 4$, so the discriminant is $(^-3)^2 - 4 * 1 * 4 = 9 - 16 = {}^-7$, which is negative. The equation has two complex solutions given by the quadratic formula:

$$\begin{aligned} x &= -\frac{(^-3)}{2} \pm \frac{\sqrt{^-7}}{2} \\ &= \frac{3}{2} \pm \frac{\sqrt{^-1}\sqrt{7}}{2} \\ &= \frac{3}{2} \pm \frac{\sqrt{7}i}{2} \\ &= \frac{3}{2} \pm \frac{\sqrt{7}}{2}i \end{aligned}$$

Exercises 7.5

1. In Example 1 we solved the equation $x^3 - 6x^2 - 9x + 54 = 0$ by graphing. Now follow these steps to solve the equation by factoring.
   a. Use the technique of factoring by grouping (Section 6.3) or an area model to factor $x^3 - 6x^2 - 9x + 54$. (*Hint:* One factor will have an x-squared term and the other, an x-term.)
   b. Break the factor containing x^2 into two linear factors using the rule for the difference of two squares.
   c. The left side of the equation should now be the product of three linear factors. Apply the zero-product property to find three solutions to the equation.
2. Suppose $z_1 = 4 + 3i$ and $z_2 = {}^-7 - 2i$. Find each value.
   a. $z_1 + z_2$
   b. $z_1 - z_2$
   c. $z_2 - z_1$
   d. $z_1 * z_2$
3. Use a calculator to show that the messy expression

$$2 + \sqrt[3]{({}^-10 + \sqrt{({}^-243)})} + \sqrt[3]{({}^-10 - \sqrt{({}^-243)})}$$

is actually equal to 6.

   a. Change MODE from Real to $a + bi$. Use the MATH menu to enter cube roots (MATH 4 on the TI-83), $\sqrt[3]{\ }$.
   b. Now enter the expression into the calculator, being careful to include appropriate parentheses:

$$2 + \sqrt[3]{\ }({}^-10 + \sqrt{\ }({}^-243)) + \sqrt[3]{\ }({}^-10 - \sqrt{\ }({}^-243))$$

Press ENTER and observe what happens.

Chapter 7 ■ Key Concepts

Slope Originally introduced in Section 4.1 for the graphs of linear equations, the slope of a **linear equation** is **constant** and can be defined as

$$\frac{\text{rise}}{\text{run}}$$

It tells you the rate of change of one variable in relation to the other. In this chapter we saw equations that are not linear, such as quadratic equations, which involve the square of one of the variables, have graphs with *changing* slopes. (Page 259)

Quadratic Equation A quadratic equation is any equation that can be put in the form

$$ax^2 + bx + c = 0$$

where x is a variable and a, b, and c are real numbers, called the coefficients. Quadratic equations can have up to two real solutions. (Page 261)

Zero-Product Property This rule says that if the product of two numbers is zero, then at least one of the numbers must be zero. (Page 263)

Factoring Factoring is another algebraic technique for getting exact solutions to quadratic equations $ax^2 + bx + c = 0$. It works by splitting the quadratic on the left side of the equation, $ax^2 + bx + c$, into two factors, such as

$$x^2 + 4x + 3 = (x + 3)(x + 1)$$

You can then use the zero-product property to say that if the product of the factors is zero, at least one of the factors must be zero. (Page 263)

Parabola The graph of a quadratic $y = ax^2 + bx + c$ is a parabola, which has changing slope. Every parabola has a **turning point**, called the **vertex**, where y achieves its maximum or minimum value. Parabolas can have 0, 1, or 2 x-intercepts (which are solutions to the equation $ax^2 + bx + c = 0$, called **roots**). (Page 266)

Vertex The vertex is also called the **turning point** in the graph of a parabola. The x-coordinate of the vertex is given by the first term in the quadratic formula, $\frac{^-b}{2a}$. The y-coordinate, which will be either the maximum or minimum y-value on the graph, can be obtained by substituting $x = \frac{^-b}{2a}$ into the original quadratic function. (Page 268)

Interesting Points The interesting points on the graph of a parabola are the x-intercepts (sometimes called **roots**), the y-intercept, and the turning point (vertex of the parabola). (Page 269)

Completing the Square Completing the square can be used to find the exact solutions of a quadratic equation $ax^2 + bx + c = 0$ by making the left-hand side a perfect square (the right-hand side will no longer be zero but will be some real number). Although completing the square works for any quadratic equation, it is usually much simpler to use the quadratic formula. (Page 274)

Quadratic Formula The quadratic formula gives the exact solutions to any quadratic equation $ax^2 + bx + c = 0$. The solution(s) are given by

$$x = \frac{^-b}{2a} \pm \frac{\sqrt{b^2 - 4ac}}{2a} \qquad \text{(Page 277)}$$

Discriminant The quantity $b^2 - 4ac$ under the square root in the quadratic formula is called the discriminant because it *discriminates* between equations with two real roots (if the discriminant is positive), one real root (if it is zero), and no real roots (if it is negative). (Page 277)

Exploration

Members of the planning committee for the benefit concert estimate that the following relationship exists between the price of each ticket and the number of tickets sold.

Ticket Price (dollars)	Number of Tickets Sold
20	100
15	300
10	500
5	700

Because the auditorium seats 1000 people, they realize they are not in danger of selling out. Use the table to help the charity decide how much to charge for tickets.

a. Let x represent the price of each ticket. Show that the expression $^-40x + 900$ gives the number of tickets sold for each value of x in the table.

b. Let y represent revenue. Use the formula revenue = number * price to express y as a function of x.

c. Find the value of x that gives the maximum value of y. What price should they charge in order to maximize revenue?

d. The cost of putting on the concert is about $1500. How much profit will the charity make after deducting their cost from the revenue?

CHAPTER 7 ▪ REVIEW EXERCISES

Section 7.1 Solving Quadratic Equations

1. Describe how the slope of a linear equation differs from the slope of a quadratic equation.

2. Identify each of the following equations as linear, quadratic, or other.

a. $3x - 2 = 5$ **b.** $x^2 = 5$

c. $x^3 + 2x^2 = 3$

3. Find the solutions to the quadratic equation $3x^2 - 1.1x - 3 = 0$ graphically.

For Exercises 4–8, use an algebraic method to find the solutions for each equation.

4. $.5x^2 = 4$

5. $7x^2 - 1 = 2$

6. $3x^2 + 4x = 0$

7. $2x^2 - 7x = 15$

8. $x^3 + 4x^2 + 4x = 0$

9. Assume the formula cheese = 1.05 * diameter^2 represents the grams of cheese required for a pizza of given diameter measured in inches.

a. Use a graphing method to find the approximate diameter of a pizza that uses 180 g of cheese.

b. Use an algebraic method to find the exact diameter.

Section 7.2 Parabolas

10. Describe each situation.

a. How you determine when a parabola opens upward and when it opens downward

b. How you determine where the turning point is on the graph of a parabola

c. How you determine whether the graph of a parabola has zero, one, or two x-intercepts

d. How you find the y-intercept of a parabola

For Exercises 11–13, use a graphing method to approximate the interesting points on the graph of each quadratic equation.

11. $y = 3x^2 + x - 4$

12. $y = {}^-x^2$

13. $y = 2x^2 + 10x + 15$

14. Suppose the height of a ball thrown in the air is modeled by the equation $y = {}^-16x^2 + 30x + 6.2$, where y is the height in feet after x seconds since release of the ball.

a. Use your calculator to create a table of values that has inputs beginning at 0 seconds and continuing in 1-second increments up to and slightly after the time it takes for the ball to hit the ground.

b. Use your table of values to find a friendly window for viewing the interesting points of this parabola.

c. Graphically, approximate the time it takes for the ball to hit the ground.

d. Approximate the turning point.

e. Find the y-intercept and interpret its meaning.

15. Consider a real-life problem where a ball is thrown or shot in the air. You wish to graph the equation that models the height of the ball after an elapsed time. Explain why knowing the x-intercepts and the vertex is enough to set a window for viewing the interesting points of the parabola.

Section 7.3 Algebraic Techniques for Solving Quadratic Equations

16. In Exercise 11 you found approximate x-intercepts on the graph of $y = 3x^2 + x - 4$.

a. Why is it necessary to let $y = 0$ and solve $0 = 3x^2 + x - 4$ in order to find the exact x-intercepts on the graph?

b. Find the exact x-intercepts on the graph of $y = 3x^2 + x - 4$.

17. Complete the square to solve $x^2 + 12x = 2$.

For Exercises 18 and 19, use the quadratic formula to find the exact roots of the given quadratic equations.

18. $0 = 3x^2 + x - 8$

19. $0 = 2x^2 + \frac{2}{3}x - \frac{1}{6}$

20. Given the quadratic equation $y = {}^{-}x^2 + 6x + 5$.

a. Find the exact vertex.

b. Use the quadratic formula to find the exact x-intercepts.

Section 7.4 Applications That Lead to Quadratic Equations

21. Find the length of a side of a square whose diagonal is 150 cm.

a. Draw a picture.

b. Identify any variables and write an equation.

c. Solve the equation by any suitable method.

d. Answer the question.

22. A rectangle has a length that is 1 m more than three times the width. The area of the rectangle is 10 m^2.

a. Use a graphing method to find the approximate dimensions of the rectangle.

b. Use an algebraic method to find the exact dimensions of the rectangle.

23. An astronaut throws a ball on the moon. Assume $h = {}^{-}.82t^2 + 10t + 4$ models the height of the ball in meters after t seconds from release.

a. Approximate the vertex by a graphing method.

b. Interpret the meaning of the vertex.

c. Approximate the positive t-intercept by a graphing method.

d. Interpret the meaning of the t-intercept.

24. Using the equation from Exercise 23, find the exact time(s) when the ball is at a height of 18 m.

25. A rectangular table top has a diagonal of 6 feet. The length is 16 inches more than the width. What are the dimensions of the table?

a. Draw a picture.

b. Identify a variable, label your picture, and write an equation.

c. Solve the equation by any suitable method.

d. Answer the question.

26. Suppose a family wants to enclose a rectangular space against their house (see the figure). They have 200 feet of fencing and a 4-foot gate. What should the dimensions be in order to maximize the area of the new space?

27. The formula $y = .045x^2 + .22x$ models the stopping distance, y, of a car in terms of the speed of the car, x, where y is in feet and x is in miles per hour. Suppose police measure a skid of 175 feet. What was the speed of the vehicle?

Section 7.5 Complex Numbers

28. Suppose $z_1 = 5 - 3i$, $z_2 = {}^{-}4 + 7i$, and $z_3 = 5 + 3i$, find

a. $z_1 + z_2$ **b.** $z_3 - z_2$

c. $z_2 * z_3$ **d.** $z_1 * z_3$

29. Find any solutions, real or complex, to

a. $x^2 + 4 = 0$ **b.** $x^2 - 6x + 25 = 0$

c. $x^2 + 5x + 9 = 0$ **d.** $x^2 + 6x + 27 = 0$

CHAPTER 7 ▪ TEST

For Exercises 1–5, choose from the square root principle, factor and zero-product property, completing the square, and the quadratic formula to find exact solution(s) to each quadratic equation.

1. $3x^2 = 12$

2. $2x^2 + 6x = 0$

3. $x^2 - 8x = -12$

4. $^{-}8x^2 - 18x + 5 = 0$

5. $x^2 + 4x = 5$

6. Use a graphing method to approximate the solutions of $0 = x^2 + 7.2x - 8$.

7. Consider the quadratic equation $y = {}^{-}3x^2 - x + 9$.

a. Does the parabola have a minimum or maximum? Explain.

b. Find the exact vertex.

c. Use the discriminant to determine whether there are zero, one, or two x-intercepts.

d. Find the approximate x-intercepts.

e. What is the y-intercept?

8. Refer to the quadratic's table of values.

x	y
0	400
10	0
20	$^{-}200$
30	$^{-}200$
40	0

a. Estimate the interesting points of this quadratic.

b. Pick suitable values for Xmin, Xmax, Ymin, and Ymax so that the interesting points of this quadratic may be viewed on one screen.

9. Let $h = {}^{-}4.9t^2 + 15t + 1.5$ model the height of water shot from a water pistol, where h is the height of the water in meters above the ground after t seconds.

a. What is the water's highest point?

b. What is the initial height of the water?

c. How long before the water hits the ground?

10. A big screen TV has a diagonal of 5 feet. The width is 1 foot more than the height. What are the dimensions of the TV?

11. Suppose the formula $R = {}^{-}5.2P^2 + 70P$ models the amount of revenue in dollars generated from a product priced at P dollars.

a. At what price (other than zero) does the product generate no revenue?

b. At what price does the product generate maximum revenue?

12. The formula $y = .05x^2 + .24x$ models the stopping distance y of a car in terms of the speed of the car (x), where y is in feet and x is in miles per hour. Skid marks of 190 feet were found at an accident site. Does the length of the skid suggest that the driver exceeded the 65 mile per hour speed limit? Explain.

CHAPTER

8 Functions

At the Click of a Mouse

The home page of Eastern Connecticut State University's website has links to 18 other pages. The home page is set up on a coordinate system with the lower-left corner corresponding to the origin and the upper-right corner corresponding to the point (180, 120). The links to other pages are activated when the mouse location has an x-coordinate between 110 and 170 and a y-coordinate between 30 and 90. When the mouse is clicked, the coordinates of its position are used to determine which web page, if any, to send the browser. The rule that assigns web pages to mouse locations is an example of a *function*. In this chapter you will learn more about functions. (See Exploration on p. 337.)

Need help? For on-line resources, visit this web site: **math.college.hmco.com/students.**

8.1 What Is a Function?

AFTER STUDYING THIS SECTION YOU WILL BE ABLE TO

- Apply the definition of function.
- Find the domain and range of a function.

This chapter is devoted to an extremely important idea, the concept of function. We encountered this concept in Section 2.4, where we described a function this way: "When the value of one variable depends upon the value of another variable, we have a function." We also learned that most functions may be represented in several ways: a verbal description, a formula, a table, and a graph (see page 79).

What Is a Function?

In Chapter 4 we studied linear functions, those that may be written in the form $y = mx + b$. In Chapter 7 we studied quadratic functions, those that may be written in the form $y = ax^2 + bx + c$. We are now ready to learn about functions in general and to become familiar with a wider variety of functions. Let us begin with an example that does not involve numbers.

EXAMPLE 1 Consider this situation: We are writing a computer program that allows a user to play solitaire. In order to draw a card on the monitor, we need to know what color to use based on the suit of the card. What we want is a function called *PickColor*, which takes as input the suit and returns as output the color. How can we represent this function?

Solution Because we are using a computer program, we can just set up a table listing each possible input and the output associated with it.

Input	Output
Heart	Red
Spade	Black
Diamond	Red
Club	Black

Once this information is programmed into the computer, the program will select the correct color, black or red, for each of the four possible suits. The output variable (color) *depends* upon the "value" of the input variable (suit). Thus, we may consider the relationship between these two variables to be a function.

We may use a picture to show that a *function* is just a rule that gives one output for each input. Think of a function as a collection of *links* connecting each input value to one output value (see Figure 1).

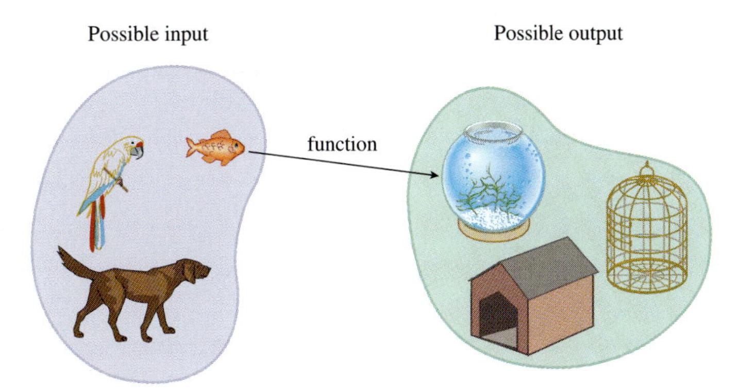

Figure 1 A function is a collection of arrows connecting input items in the left blob with output items in the right blob.

The blobs indicate the sets containing the inputs and outputs. In the previous example the input blob contains four elements (diamond, heart, spade, and club) and the output blob contains two (red and black) (see Figure 2). The two blobs in Figure 2 are given the special names *domain* and *range*.

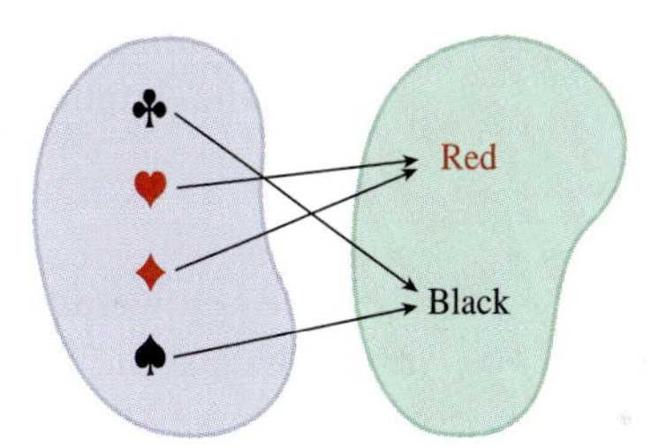

Figure 2 This is the connection picture for the function defined in Example 1. Notice that although only one arrow begins at each input, two arrows end at each output.

Definition Domain of a Function

The set of all allowable inputs to a function is called the **domain** of the function. The allowable inputs are those that correspond to an output.

In Example 1 the domain is the set {heart, spade, diamond, club}.

Definition Range of a Function

The set of possible outputs of a function is the **range** of the function.

In Example 1 the range is the set {black, red}.

In order to give a more formal definition of function, we need to know how the *domain* and *range* are linked. If we reverse the role of suit and color, we have a different table.

Input?	Output?
Red	Heart
Black	Spade
Red	Diamond
Black	Club

If we try to use a computer program with this table, we have a problem; suppose we input red. The computer has *two* choices for its output, heart and diamond. Computers aren't good at making choices, so we limit our definition of function.

Definition Function

A **function** is a rule that associates to an allowable input value one and only one output value.

The key words in this definition are *one and only one*. Each suit in the deck of cards has one and only one color, but each color does *not* identify one and only one suit.

Now let's turn to a more familiar setting to see how these concepts apply to functions involving numbers. Recall that in Section 7.1 we used a formula that gives the amount of cheese (in grams) needed for a pizza with a given diameter (in inches):

$$\text{cheese} = 1.2 * \text{diameter}^2$$

If the diameter is known, its value may be substituted into the formula to find the amount of cheese needed.

Definition Input Variable

The variable associated with the domain of a function is the **input variable**. It is also called the **independent variable**.

Diameter is the input variable for the pizza function. Suppose the diameter is 12 inches. If we substitute 12 for diameter in the formula, we get cheese $= 1.2 * 12^2 = 172.8$ g. 172.8 is the *output* of the formula.

Definition Output Variable

The variable associated with the range of a function is the **output variable**. It is also called the **dependent variable**.

Cheese is called the output (or dependent) variable for the pizza function.

We first analyzed this function by looking at a table similar to the one that we made for the card-color problem.

Diameter (inches)	Cheese (grams)
8	76.8
12	172.8
16	307.2
20	480

Unlike the previous example, however, the domain of this function contains more than four input values, so this table does not show us the entire domain.

EXAMPLE 2 Find the domain and range for the pizza function. Take into account that the largest pizza the oven can hold has a diameter of 42 inches.

Solution The diameter must be greater than 0 because otherwise we would not have a pizza. And we just learned that the maximum diameter is 42 inches, so it must be *less than or equal to* 42. We describe the domain with the sentence: $0 < \text{diameter} \leq 42$.

Similarly, the amount of cheese used must also be greater than 0. We know that as the diameter increases, so does the amount of cheese. So the maximum amount of cheese may be found by substituting the maximum diameter into the equation $\text{cheese} = 1.2 * \text{diameter}^2$. When we do that, we get

$$\text{cheese} = 1.2 * \text{diameter}^2$$

$$\text{cheese} = 1.2 * 42^2$$

$$\text{cheese} \approx 2117$$

We describe the range with the sentence $0 < \text{cheese} \leq 2117$.

Finding Domains and Ranges

When we use a graphing calculator for the function in the cheese equation, the input variable becomes x and the output variable becomes y. Thus we have

$$y = 1.2x^2$$

To see the entire graph of this function, we pick windows that include the domain and range. We then get a graph similar to the one shown in Figure 3.

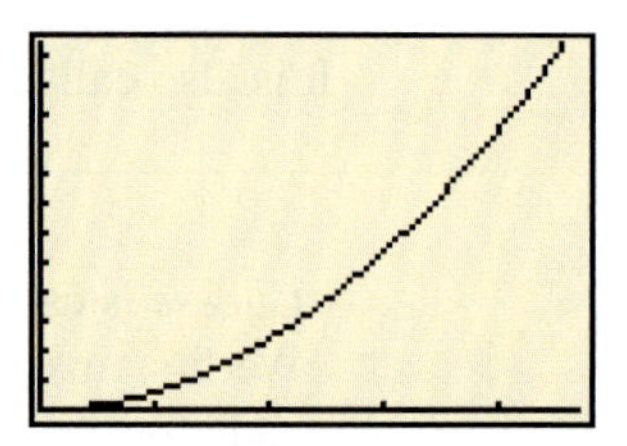

Figure 3 Xmin = 0, Xmax = 47, Ymin = 0, Ymax = 2500.

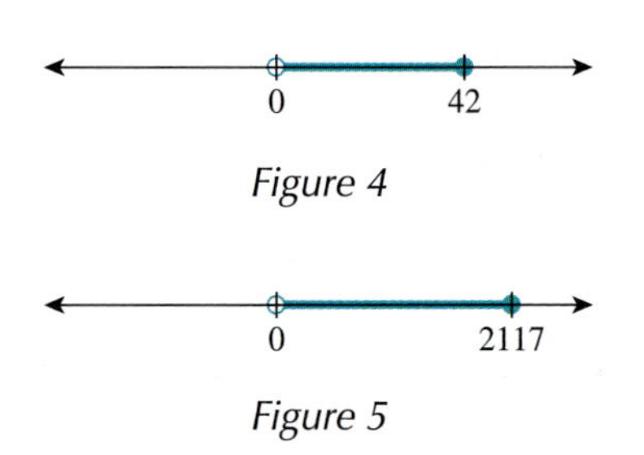

Figure 4

Figure 5

We can show the domain and range on number lines. In Figure 4, all numbers between 0 and 42 are shaded because they belong to the domain. A heavy dot at 42 indicates that 42 belongs to the domain. An open circle at 0 shows that 0 is not part of the domain. Similarly, the range is shown in Figure 5. We could also show the range on a vertical number line, but we show it horizontally here. The endpoints are 0 and 2117. Again, the right endpoint, 2117, is included and 0 is shown as an open circle.

Domains and ranges are usually sets of numbers, and shading number lines is a good way to represent them. However, it is sometimes helpful to express them as sets. We do this using set-builder notation.

> **Definition Set-Builder Notation**
>
> We write a set in **set-builder notation** by listing the conditions the elements of a set must meet:
>
> set = {variable name|conditions the variable must meet to be in set}

EXAMPLE 3 ■ Use set-builder notation to show the domain and range for the pizza function.

Solution

$$\text{domain} = \{x \mid 0 < x \leq 42\}$$

$$\text{range} = \{y \mid 0 < y \leq 2117\}$$

We read set-builder notation like this: domain equals the set of x *such that* 0 is less than x is less than or equal to 42. Read the range in the same way.

Each particular diameter determines one and only one amount of cheese required for the pizza. It is the uniqueness of each element in the range for a particular value in the domain that gives functions their special property—that the output variable *depends* upon the value of the input variable.

Writing Functions

We use special notation to write functions.

> **Definition Function Notation**
>
> When we have a function we write
>
> $$y = f(x)$$
>
> where x is the input (independent) variable, y is the output (dependent) variable, and f is the name of the function, or rule, that assigns to each x-value one and only one y-value. Read $f(x)$ as f of x.

Caution: $f(x)$ does *not* mean $f * x$. f is the name of a function, not a variable to be multiplied by x.

In our pizza example, for instance, we could name the rule "pizza." We would then write the rule as cheese = pizza (diameter) = 1.2 * diameter^2. Or, if we prefer to stick with letters for variables, we write $y = f(x) = 1.2x^2$.

In mathematics books f, g, and h are most often used for function names. In addition, when we use a calculator, we may refer to functions as Y1, Y2, etc. Functional notation also allows us to specify a particular value of the output when a particular value of the input is given. For example, if the diameter of the pizza is 12 inches, we can say that the amount of cheese is given by cheese = pizza (12). Pizza (12) tells us to evaluate the function *pizza* for an input of 12. As we have already seen,

$$\text{pizza}\ (\mathbf{12}) = 1.2 * \mathbf{12}^2 = 172.8 \text{ g}$$

The calculator gives the same result. Enter the pizza function as Y1 = 1.2 X ∧ 2. Then on the home screen, type Y1(12) and press enter. (We can get Y1 by going first to the Y-Vars menu and then to the Function submenu.) The function Y1 will be evaluated for X = 12, and the number produced is 172.8.

EXAMPLE 4 ■ Use the calculator to evaluate the function $f(x) = 1.2x^2$ for $\boldsymbol{x} = \mathbf{^{-}12}$.

Solution In the Y= screen, enter Y1 = 1.2X ∧ 2. On the home screen type Y1(**⁻12**) and press enter. Again, the number produced is 172.8.

■ ■

But now stop for a minute and think: how can a pizza have a diameter of ⁻12 inches? It can't; ⁻12 does not belong to the domain of the pizza function. But evidently ⁻12 *does* belong to the domain of the function

$$y = f(x) = 1.2x^2$$

Abstract and Concrete Functions

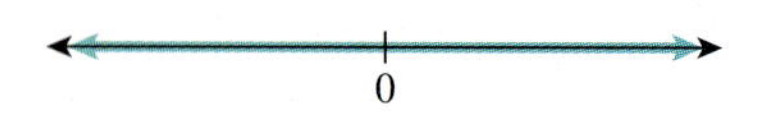

Figure 6 The domain of the *abstract* function $f(x) = 1.2x^2$ is all real numbers, which we can show by shading the number line.

This function exists apart from the context of the pizza problem. We may consider it an **abstract** function, as opposed to the **concrete** function of the pizza problem. The domain of the abstract function is not restricted. It may include values of x that give diameters too large to fit in the oven. It may even include negative numbers. When we think about it, we realize that for this abstract function, *any* real number may be substituted for x. The domain is the set of *all* real numbers. See Figure 6.

The arrows at each end of the line in Figure 6 indicate that the domain continues in both directions without any limitation. We sometimes say that the domain goes from negative infinity to positive infinity. Using the symbol ∞ for infinity, we express the domain in set-builder notation as $\{x \mid {}^{-}\infty < x < \infty\}$.

Finding the range is a bit trickier. Here a graph will help. Let's graph the function in the decimal window (${}^{-}4.7 \le \text{X} \le 4.7, {}^{-}3.1 \le \text{Y} \le 3.1$). The result is the familiar parabola shown in Figure 7.

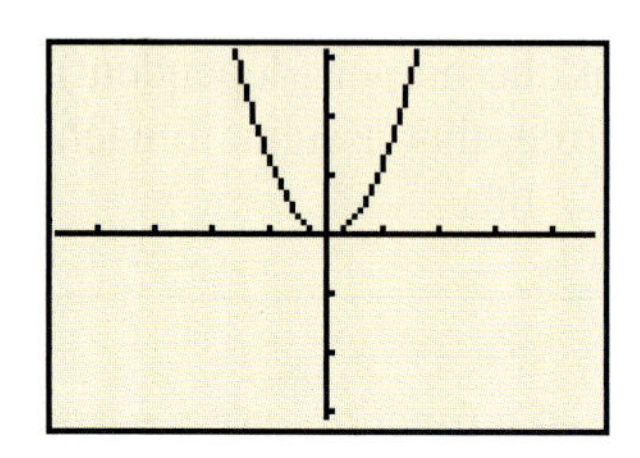

Figure 7

From the graph we can see that values of y are never negative. The smallest value of y is zero, and although the screen can't show the entire parabola, we know that it rises indefinitely on both sides of the y-axis. Therefore, the range of this function is shown on a number line in Figure 8. We give the range in set-builder notation as $\{y \mid 0 \le y < \infty\}$.

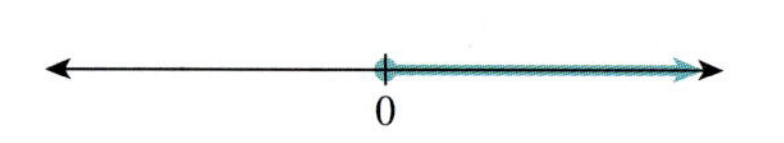

Figure 8 y cannot be negative.

We can also check to see if points on the output axis are in the range by drawing a horizontal line through them and checking to see if the graph meets the line. If it does, then clearly that point is in the range, because there is at least one value we can input to the function that will get us the desired output. Of course, we cannot do that for all output values, but by using basic properties of the graph and adjusting the window, we can usually work out the entire range. One way of thinking of this is to imagine that you are using a squeegee to clean off the graph.

Moving from left to right with the blade of the squeegee vertical gives you the domain on the x-axis, and moving from bottom to top with the blade horizontal gives you the range on the y-axis. The DRAW menu on the calculator allows you to simulate squeegees with vertical and horizontal lines.

EXAMPLE 5 Use the DRAW menu to show how a squeegee sweeps out the domain and range for the function $y = f(x) = 1.2x^2$.

Solution Start with the given function in Y1 and the Decimal window (${}^{-}4.7 \le \text{X} \le 4.7, {}^{-}3.1 \le \text{Y} \le 3.1$). Press GRAPH to draw the graph. Then press 2ND DRAW. Select Vertical. Use right and left arrows to move the squeegee back and forth. Notice that between $x = {}^{-}1.6$ and $x = 1.6$ the squeegee always intersects the graph. For $x < {}^{-}1.6$ and $1.6 < x$ we know that there the squeegee *would* intersect the graph if our screen extended upward far enough. So our experience with the squeegee should help convince us that the domain for this function is all real numbers.

Press [2ND] DRAW again and now select Horizontal. Use up and down arrows to move the squeegee. Notice that the squeegee starts out on the x-axis, where it intersects the vertex of the parabola, and that for positive values of y the squeegee intersects the parabola twice. When the squeegee is moved below the x-axis, however, it does not hit any point on the parabola. This experience should reinforce the conclusion that the range for this function is the set of all numbers greater than or equal to zero.

Exercises 8.1

1. Suppose a coin is used to settle a dispute. The input is either a head or tail. Depending on your choice, the output is either win or lose. The table summarizes the possibilities.

Input	Output
Heads	Win
Tails	Lose
Heads	Lose
Tails	Win

   a. Create a connection picture such as the one in Figure 2 and show there is not one and only one output for each input.
   b. Does this situation represent a function? Explain.

2. Determine if each table of inputs and outputs represents a function. For those that are functions, give the sets that describe their domains and ranges.

a.

Input	Output
a	1
c	3
g	7
z	26

b.

Input	Output
0	0
1	1
2	$\frac{1}{4}$
3	$\frac{1}{9}$

c.

Input	Output
5	0
5	1
5	2
5	3

3. Determine which of the connection pictures are functions and give the sets that describe their domains and ranges.

a.

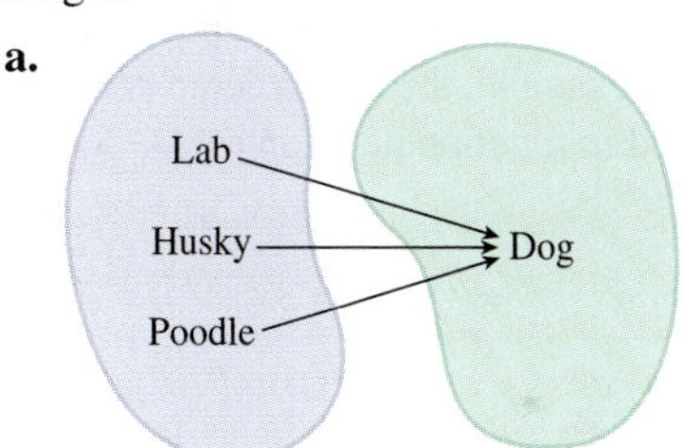

b.

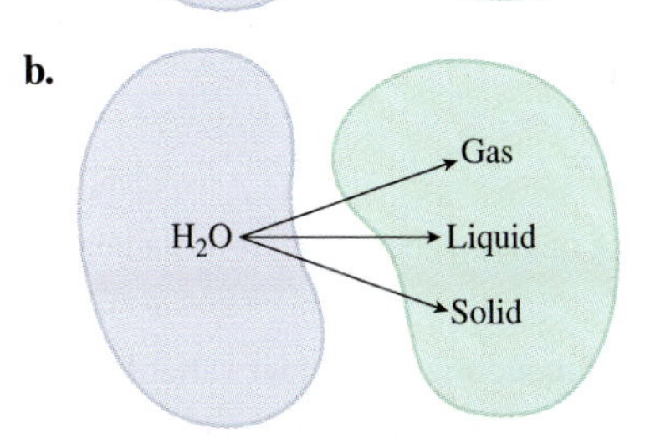

4. Consider the pizza function given by the formula cheese $= 1.6 * \text{diameter}^2$. Assume the largest pizza has a diameter of 38 inches.
   a. Evaluate the function for a diameter of 0 inches.
   b. Evaluate the function for a diameter of 38 inches.
   c. Use the information from parts (a) and (b) to determine the domain and range of this function. Express the domain and range on a number line.

5. Sandra makes a wish by dropping a rock down a 100-foot-deep well. The rock takes 2.5 seconds to hit the water. The rock's height above the water is a function of the time after it was dropped.

a. Between what two times does this function make sense? This is the domain.

b. Express the domain in set-builder notation and on a number line.

c. Between what two heights does this function make sense? This is the range.

d. Express the range in set-builder notation and on a number line.

e. Suppose we knew the formula for this function and wanted a graph. How would the domain and range help you determine an appropriate viewing window?

f. A graph of this function would appear in what quadrant? Explain your answer.

6. Write each equation in function notation and determine the independent and dependent variables.

a. $y = 6x - 5$; name the function Y_1.

b. $h = {}^{-}16t^2 + 50t + 200$; name the function h.

7. The table gives several ordered pairs that satisfy the function $f(x) = 3x^2$. Express each output as a function of the input. For example, the first ordered pair, $({}^{-}1, 3)$, can be expressed in function notation as $f({}^{-}1) = 3$.

x	$y = f(x) = 3x^2$
$^{-}1$	3
0	0
2	12
3	27

8. Recall the pizza function cheese $= 1.6 *$ diameter^2 from Exercise 4.

a. Rewrite the function with abstract variables x for diameter and y for cheese and name the function f.

b. Evaluate the function for a diameter of 8 inches. This is $f(8)$.

c. Evaluate $f(12)$.

9. Evaluate each function.

a. Evaluate $f(2)$ for $f(x) = x - 5$.

b. Evaluate $g({}^{-}4)$ for $g(x) = x^2 - 3x$.

c. Evaluate $h(3)$ for $h(t) = {}^{-}4.9t^2 + 10t + 5$.

10. Given the function $Y_1(x) = \frac{1}{5}x^2 - 8x$, use your calculator to evaluate $Y_1(.27)$.

11. Given the function $f(x) = 4x^3 + 5x^2$, use your calculator to evaluate $f({}^{-}2.15)$.

12. The domain and range of a function are shown in the figure with several of the inputs linked to their outputs.

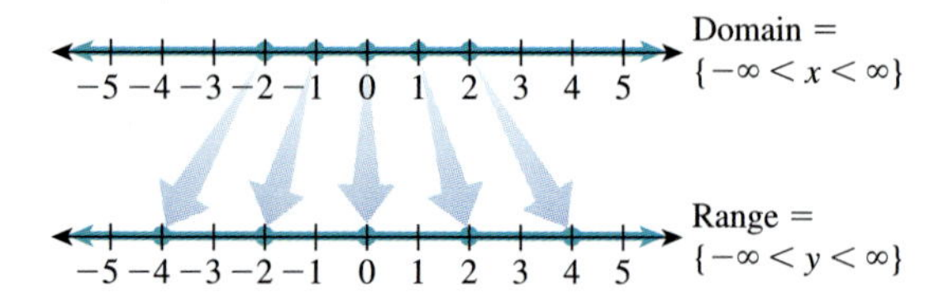

a. Express the linking of these inputs to outputs as ordered pairs.

b. Plot the ordered pairs on a graph.

c. You may have noticed that the rule for determining the outputs is twice the input. Use this rule to create additional links.

d. Because the domain includes all real numbers, there are an infinite number of links. How will these links appear on your graph?

13. Determine the domain and range of each function from its graph.

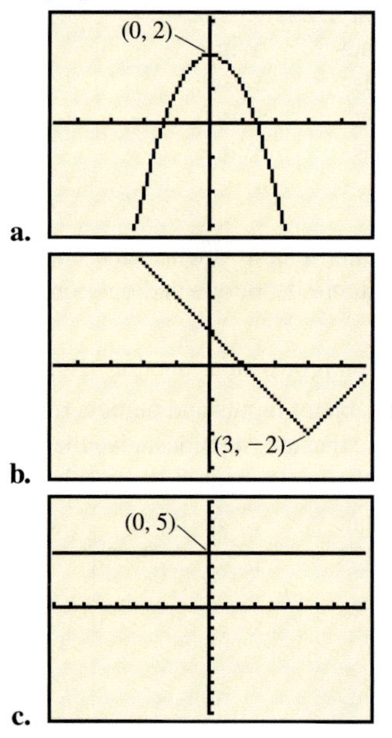

14. Graph each function with a ZOOM 4: ZDecimal setting and squeegee the graph (see Example 5) to determine the domain and range.

a. $f(x) = 2x + 1$

b. $g(x) = .3x^2$

c. $h(x) = {}^{-}1$

15. Answer the following.

a. The square of any real number input results in an output greater than or equal to zero. What does this imply about the range and graph of $f(x) = x^2$?

b. Explain why the range of the absolute value function, $h(x) = |x|$, is the same as the quadratic in part (a).

Skills and Review 8.1

Use the following situation for Exercises 16 and 17: The area of a rectangle is given by the formula $A = w(180 - 2w)$, where w is the width (in feet) of the rectangle.

16. Find the maximum area of the rectangle.

17. Find the dimensions of the rectangle that maximize the area.

18. Let $y = x^2 - 5x + 6$.

a. Find the exact values of the x-intercepts.

b. Find the the y-intercept.

19. Find the solutions to the quadratic equation $0 = x^2 + 8x + 16$ using each method.

a. Tracing the graph

b. Factoring

c. Using the quadratic formula

20. Let $y = {}^-4x^2 + 3x - 7$. Do not draw a graph.

a. Determine whether this parabola opens upward or downward.

b. Determine whether this parabola has zero, one, or two x-intercepts.

21. Solve the equation $\frac{1}{5}x^2 = 7$ for x.

22. Perform the following operations and simplify.

a. $3x + 3x$

b. $(3x)(3x)$

c. $(4x + 1)^2$

d. Is $2x^3$ equal to $(2x)^3$? Explain.

23. Write each expression in an equivalent form with a positive exponent.

a. $\dfrac{1}{x^{-2}}$ **b.** x^{-2}

24. Use the laws of exponents (Section 6.1) to simplify $(x^3y^4z)(x^{-5}y^2z^{-1})$.

25. In Section 5.1 we introduced the direct variation equation $y = kx$, where k is the variation constant. Find k if $x = {}^-4$ and $y = {}^-6$.

8.2 Graphs of Functions

After Studying This Section You Will Be Able To

- Determine the domain and range of a function from its graph.
- Determine the domain of a function from its equation.

We have seen that one way to represent a function is by its graph. Now consider this question: how can we tell if a graph is the graph of a function?

When Does a Graph Represent a Function?

Suppose we just scribble something on a set of axes, as in Figure 9. How can we tell which of these graphs are functions? As we have pointed out, what makes a function a function is that given an input, there is one and only one output. When listing pairs of inputs and outputs in a table of values, this translates to the fact that no two pairs can have the same first element. In a diagram like the one in Figure 2 (Section 8.1), we had to make sure that no two arrows start at the same place.

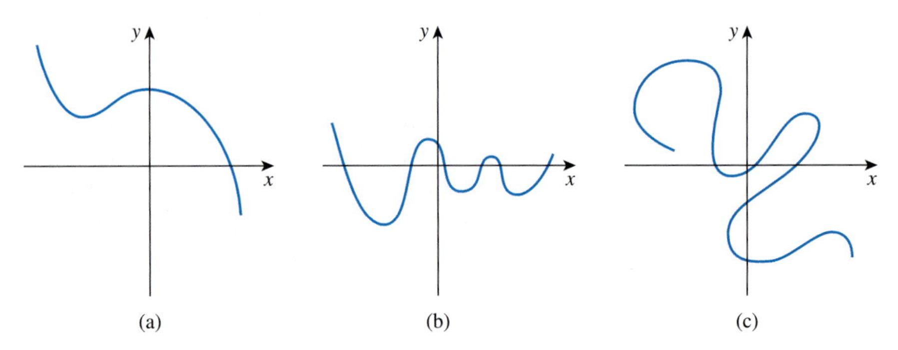

Figure 9 Do all these graphs represent functions?

On a graph, the first element in a pair may be thought of as a point on the x-axis. If a vertical line through that point intersects the graph in more than one place, then the graph cannot be a function, because some inputs will have more than one corresponding output. For example, in part (c) of Figure 10, the input x goes with outputs y_1 and y_2, thus violating the definition of a function.

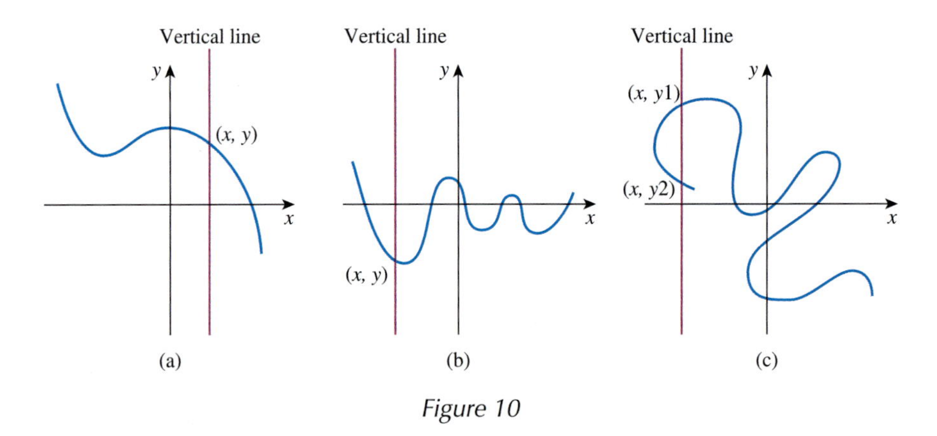

Figure 10

In the case of (a) and (b), however, each point on the input axis corresponds to at most one point on the graph, so these are graphs of functions.

In general, we imagine drawing vertical lines at every point on the input axis. If one of these lines intersects a graph more than once, it is not a function. We use a vertical line because all points on a vertical line have the same first (input) coordinate (see Figure 10).

Definition Vertical-Line Test

To test a graph to see if it is the graph of a function, draw a vertical line through every point on the input axis. If any of these lines intersect the graph at more than one point, the graph is not the graph of a function. If none do, the graph is the graph of a function.

Obviously, we can't draw all those lines. In practice, we can do the same thing we did to help us compute the domain of a function from its graph. Imagine holding a squeegee with the blade vertical on the left edge of the graph and wiping across left to right. If the blade hits the graph more than once at any time, we stop and say "not a function."

EXAMPLE 1

a. Show that any line of the form $y = mx + b$ is a function.

b. Explain why a vertical line is not a function.

Solution

a. Pick any values for m and b and draw the graphs.

There is no way that any vertical line will intersect such a line more than once; in fact every vertical line will intersect it *exactly* once. In Figure 11 several lines are drawn with different slopes (different values of m) and the same y-intercept, $b = 0$. The vertical line shown intersects each of these lines only once. And no matter where we move that vertical line, it will still intersect each line in only one place.

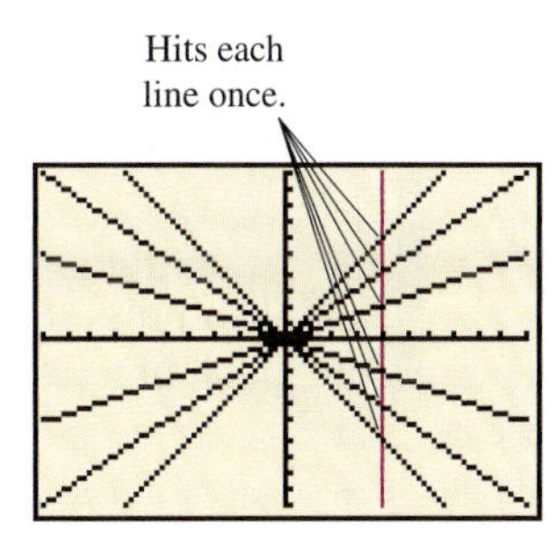

Figure 11 Each of these linear graphs represents a function.

To convince yourself of this fact, use the DRAW menu to make your own vertical line squeegee (as explained earlier). Move the vertical line back and forth to show that there is always just one point where it intersects the graph of the function $y = mx + b$.

b. Now suppose our first line is already vertical. Then almost every vertical line completely misses our original line, but when our squeegee is exactly lined up with our original line, it hits it an infinite number of times. Vertical lines have equations of the form $x =$ some number. On these lines y may take on any value, so y does not depend on, and is not a function of, x.

EXAMPLE 2 ■ Show that a parabola with equation $y = ax^2 + bx + c$ is the graph of a function.

Solution See Figure 12. Every vertical line will intersect a parabola exactly once. For a parabola, there are two inputs for almost every output (the exception occurs at the vertex), and so although each vertical line intersects the graph once, many horizontal lines intersect the graph twice (Figure 12). A graph may be intersected by a horizontal line several times and still be the graph of a function.

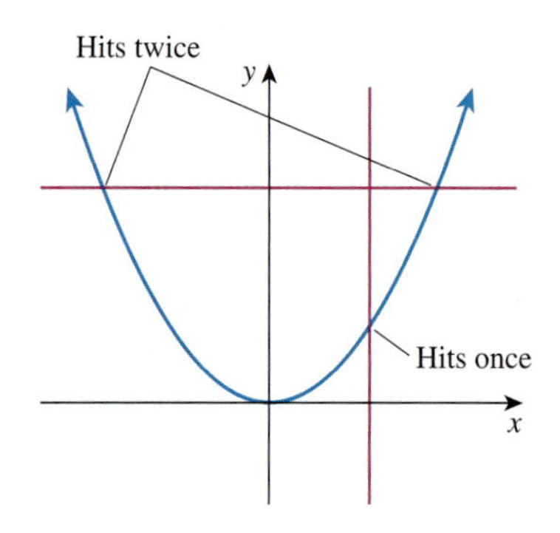

Figure 12 Notice that each output (except for the minimum) corresponds to two inputs.

EXAMPLE 3 ■ Use the ZoomDecimal window ($^-4.7 \leq X \leq 4.7, {}^-3.1 \leq Y \leq 3.1$). From the home screen press [2ND] DRAW and scroll down to select Circle. Enter two zeros and the number 3 separated by commas,

Circle (0, 0, 3)

and then press [ENTER]. You should get a circle, as shown in Figure 13. Is this the graph of a function?

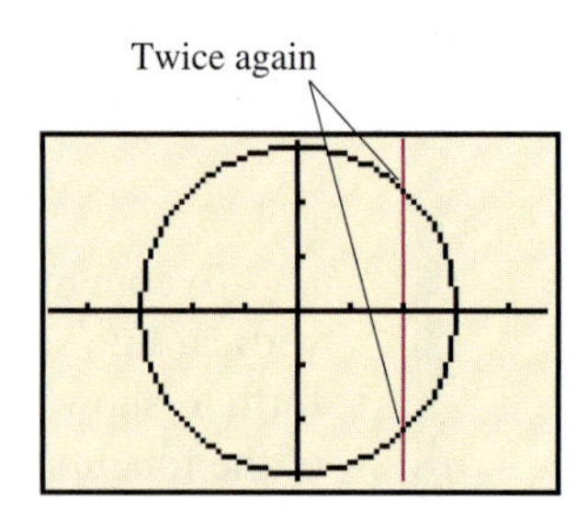

Figure 13 Any graph that looks like a circle does not represent a function. This should be a square window such as Xmin = $^-4.7$, Xmax = 4.7, Ymin = $^-3.1$, Ymax = 3.1.

Solution Using a vertical squeegee, you should see that many vertical lines intersect the circle in two places. A circle is not a graph of a function.

When you want to erase the circle, return to the DRAW menu and select ClrDraw.

EXAMPLE 4 Many graphs we see in everyday life do not look like most of the graphs in this book. Statistical information is often presented in a graphical format, because it is easy to scan a graph quickly while you are reading the newspaper but much more difficult to read through a table of numbers. For example, *bar graphs* are often used to display data, as in Figure 14. Does the bar graph in Figure 14 represent a function?

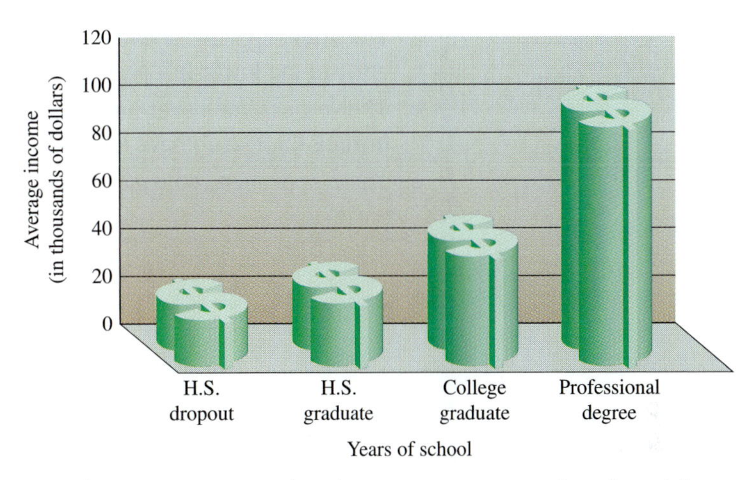

Figure 14 There are many graphical ways to represent a function. A bar graph is very useful when the input set is finite.

Solution The graph in Figure 14 shows how average annual income is related to educational level (data from the U.S. Census Bureau). To see that this is a function, let the domain be the set

$$\{\text{H.S. dropout},\ \text{H.S. graduate},\ \text{college graduate},\ \text{professional degree}\}$$

The elements in this set indicate categories based on the level of education completed.

The range of the function is a finite set of numbers,

$$\{\$18{,}900,\ \$25{,}900,\ \$45{,}400,\ \$99{,}300\}$$

indicating income in dollars. We could certainly have put this information in a table, but it is immediately obvious from the graph that income rises with schooling, and the pattern in that rise is also obvious.

From the graph we can confirm that it is indeed a function because any vertical line intersects the top of only one of the bars.

More on Finding Domains and Ranges

EXAMPLE 5 Find the domain and range for the function

$$y = f(x) = \frac{1}{x+1}$$

Solution Let's begin by drawing the graph in the ZoomDecimal window ($^-4.7 \le X \le 4.7$, $^-3.1 \le Y \le 3.1$). Use the DRAW menu to obtain a vertical squeegee and move this line to the right and left. You should notice at least three things:

a. The graph appears as two unconnected curves, one in the upper-right and the other in the lower-left portion of the window.

b. The vertical line never intersects the graph in more than one point (thus confirming that we do have a function).

c. There appears to be a gap between the two curves around $x = {}^-1$ (see Figure 15).

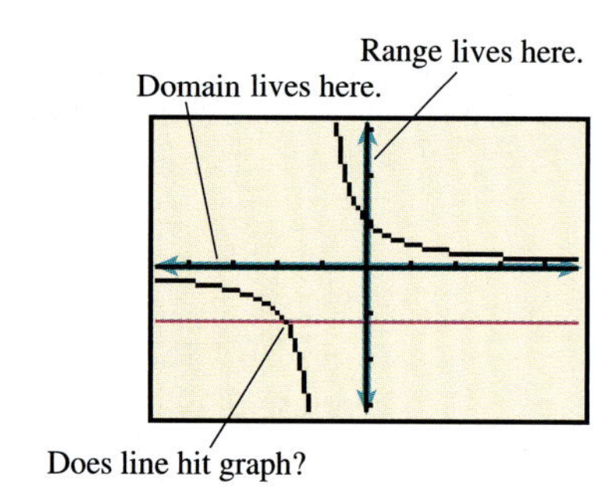

Figure 15 The function Y1 = 1/(X + 1) in the ZoomDecimal window.

The domain of this function is determined by the fact that the rule involves division, and we can never divide by zero. The only value of x that makes the denominator 0 is $x = {}^-1$, so the domain is all real numbers except $^-1$ (Figure 16).

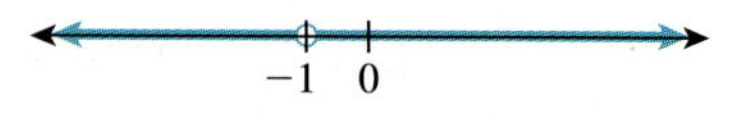

Figure 16 The number $x = {}^-1$ is missing from the domain of $f(x) = \frac{1}{x+1}$. We indicate this on the number line with a *hole* at $x = {}^-1$.

To find the range, we use a horizontal squeegee. Most output values seem to work. The graph of f goes both up toward ∞ and down toward $^-\infty$ for x-values near $^-1$. (If $x \approx {}^-1$, the denominator is very close to zero, and dividing 1 by a very small number gives a very large number.) So, the range extends from $^{-}\infty$ to ∞. The question is whether it misses any points in between. Looking carefully at the graph, we can see that the graph seems to miss the x-axis, which is the horizontal line corresponding to $y = 0$.

Does it make sense that $y = 0$ is missing from the range of the function $f(x) = \frac{1}{x+1}$? Because the numerator of the fraction defining f is 1, there is no value of x that will make the value of f zero; so $y = 0$ really is missing from the range, which means the range of $f(x) = \frac{1}{x+1}$ is missing $y = 0$ (Figure 17).

This represents the *range* of the function.

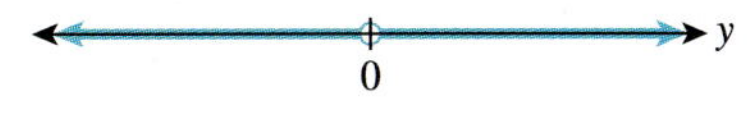

Figure 17

Using set-builder notation we write

$$\text{domain} = \left\{x \mid x \neq {}^{-}1\right\} = \left\{x \mid {}^{-}\infty < x < {}^{-}1 \quad \text{or} \quad {}^{-}1 < x < \infty\right\}$$

$$\text{range} = \{y \mid y \neq 0\} = \left\{y \mid {}^{-}\infty < y < 0 \quad \text{or} \quad 0 < y < \infty\right\}$$

Notice that we may describe these sets in terms of what does *not* belong using ($\neq$) or by describing the two separate pieces that *do* belong.

EXAMPLE 6 ■ Find the domain and range for the function $g(x) = \sqrt{x+1}$.

Solution The graph of this function is shown in Figure 18.

From this graph it appears that

$$\text{domain} = \{x \mid {}^{-}1 \leq x < \infty\}$$

$$\text{range} = \{y \mid 0 \leq y < \infty\}$$

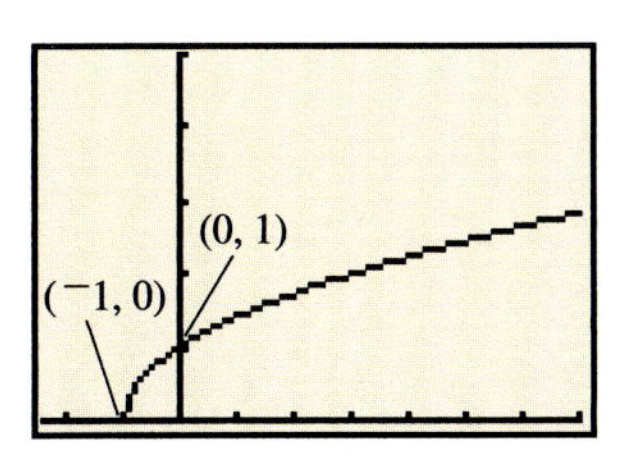

Figure 18 $y = g(x) = \sqrt{x+1}$.

As in the previous example, let's see if we can understand why. When we studied quadratic equations, we learned that every nonnegative number has two square roots, one positive and one negative. When we use the symbol $\sqrt{\ }$, we indicate only the positive root. Thus the range of the function cannot include negative numbers. Furthermore, we cannot take the square root of a negative number, so the smallest value $x + 1$ may assume is 0. This occurs when $x = {}^{-}1$.

Just to be sure, we should check the endpoints of our intervals: $g({}^{-}1) = \sqrt{{}^{-}1+1} = \sqrt{0} = 0$, which confirms that $^{-}1$ should be included in the domain and 0 in the range of $g(x)$.

At this point we might wonder if there is a systematic way of finding the domain of a function without drawing a graph. Indeed there is, if we think of the points as "innocent until proven guilty." Start by assuming that all possible input values belong to the domain. Then exclude input values that don't belong. In the examples we have studied, we have found three possible reasons for excluding values from the domain of a function.

1. The values don't make sense in the context of a problem situation. (Example: The pizza function does not allow for negative diameters or for diameters that exceed the size of the oven.)
2. The value of the input makes the denominator equal to zero. (Example: $f(x) = \frac{1}{x+1}$ for $x = {}^{-}1$.)
3. The value of the input would make us take the square root of a negative number. (Example: $g(x) = \sqrt{x+1}$ for values of x less than ${}^{-}1$.)

Just remember "innocent until proven guilty." Once you have excluded the guilty values of x, what remains is the domain.

Finding the range of a function may be more complicated. We always use a graph to get a fairly good idea of the range. Some functions, however, are so complicated that finding exact boundaries for the range requires methods that are taught only in more advanced courses.

Exercises 8.2

In Exercises 1–6, use the vertical-line test to determine which graphs represent functions.

1.

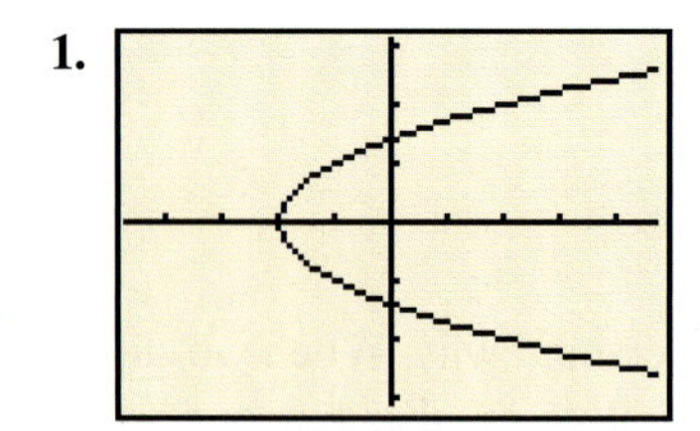

2.

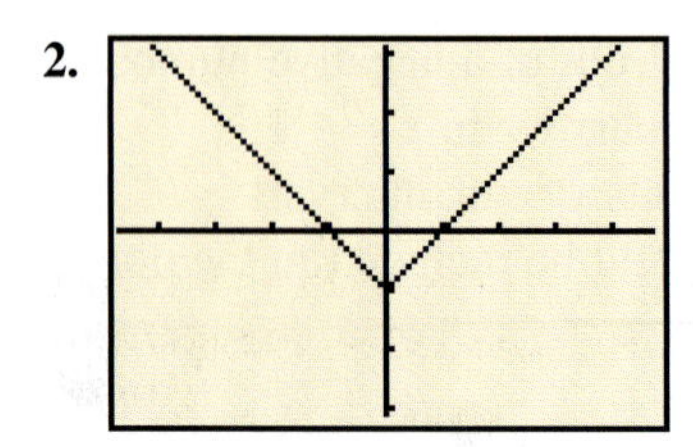

3.

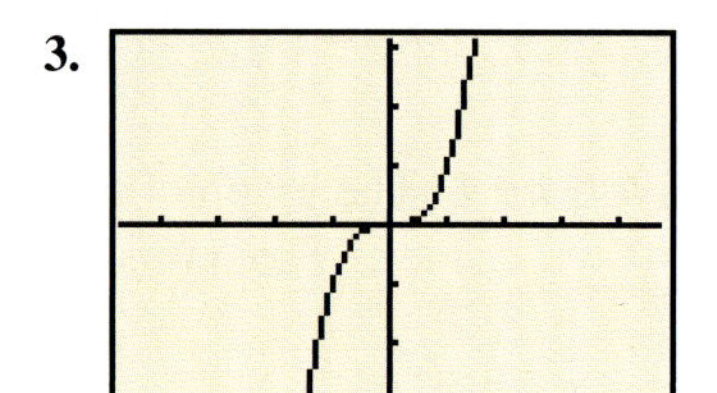

4.

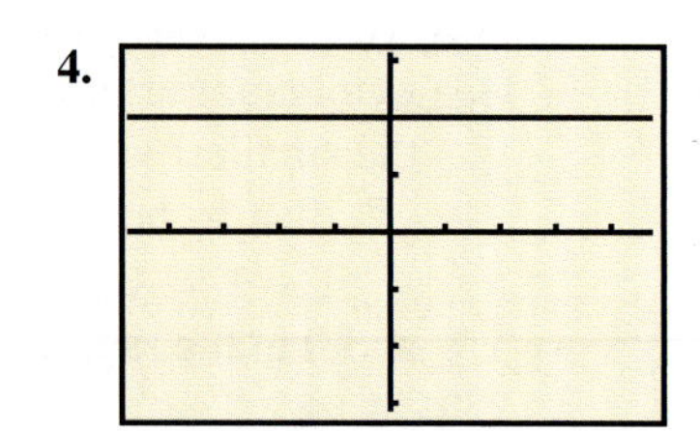

5. 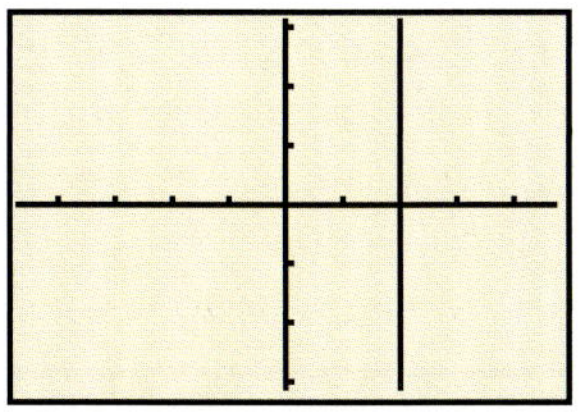

6. 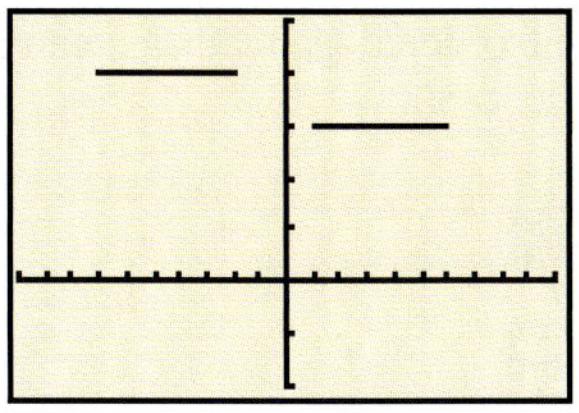

7. Let $^{-}y^2 = x$.

   a. Complete a table of values for $y = {}^{-}2, {}^{-}1, 0, 1, 2$.

   b. Draw a graph by hand.

   c. Is this a function? Explain.

8. Let $|y| = x$.

   a. Complete a table of values for $y = {}^{-}2, {}^{-}1, 0, 1, 2$.

   b. Draw a graph by hand.

   c. Is this a function? Explain.

9. Here is a bar graph of per capita carbonated soft-drink consumption for the years 1996 through 2001. (http://www.bevnet.com)

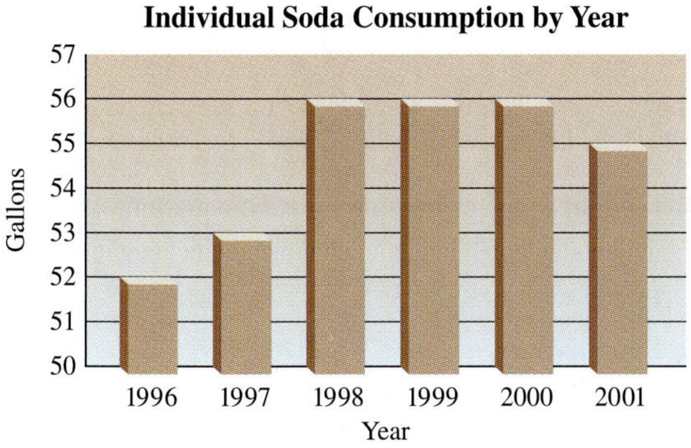

   a. Is this a graph of a function?

   b. Estimate the soda consumption for each year and write ordered pairs in the form of (year, soda consumption).

   c. Express the domain and range using set notation.

10. The dot plot shows the ages of children from three families labeled A, B, and C. Does the graph represent a function? Explain.

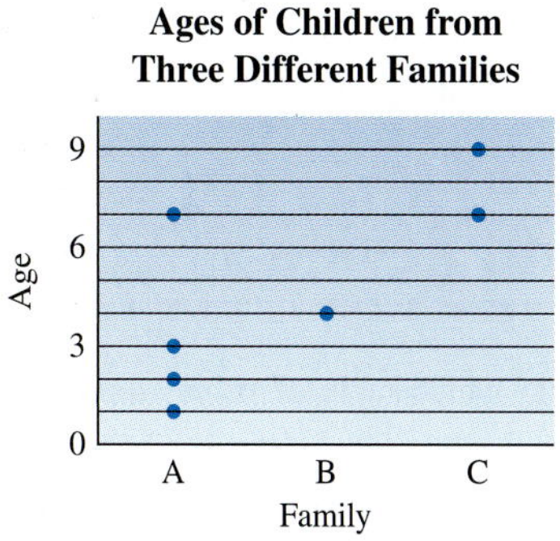

11. Match the domain and range with the appropriate graph.

   a. Domain $= \{x \mid {}^{-}\infty < x < \infty\}$;
      Range $= \{y \mid y \geq {}^{-}3\}$

   b. Domain = 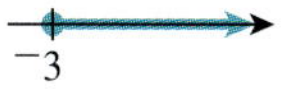

      Range = 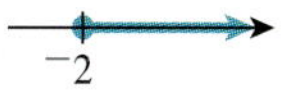

   c. Domain $= \{x \mid -\infty < x < \infty\}$;
      Range $= \{y \mid y = {}^{-}3\}$

1. 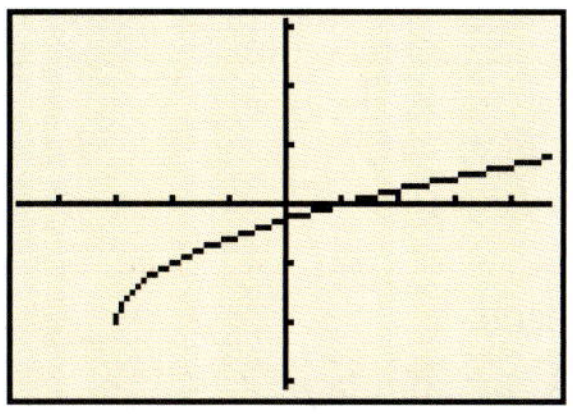

2. 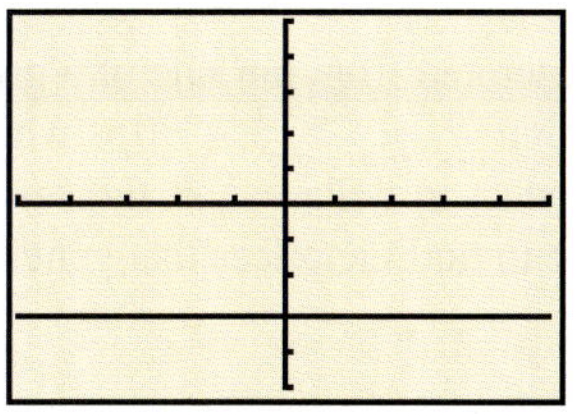

3. 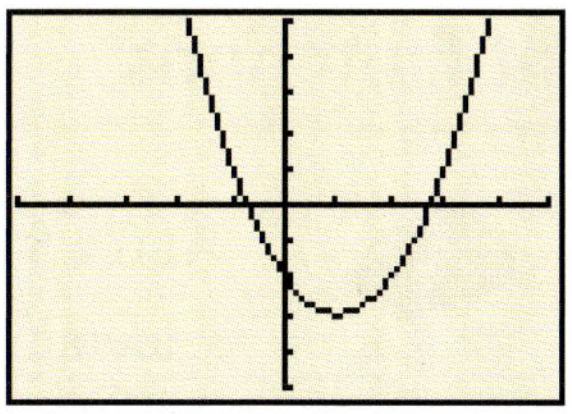

In Exercises 12–15, express the domain and range of each function on a number line and in set-builder notation.

12. $f(x) = \sqrt{x}$

13. $g(x) = \sqrt{x - 4}$

14. $h(x) = \frac{1}{x}$

15. $j(x) = \frac{1}{x+6}$

Skills and Review 8.2

16. Determine which table of values represents a function and explain your choice. (http://www.sportserver.com)

a.

Input?	Output?
31	Packers
32	Broncos
33	Broncos
34	Rams
35	Ravens

b.

Input?	Output?
Washington	1992
Dallas	1993
Dallas	1994
San Francisco	1995
Dallas	1996

17. State the domain and range of the function from Exercise 16.

18. Let

$$f(x) = \frac{\frac{1}{3}x + 8}{^{-}4}$$

a. Find $f(^{-}9)$ by hand.

b. Find $f(.16)$ on your calculator.

c. Find x if $f(x) = \frac{1}{4}$.

19. The revenue, R, generated from the sale of a product is given by the equation $R = {^{-}5}p^2 + 90p$, where p is the price of the product in dollars. Use the seven steps for problem solving to find the prices that generate zero revenue.

20. The table gives several ordered pairs that satisfy the function $y = {^{-}x^2} + 40x + 500$.

X	Y1	
-10	0	
0	500	
10	800	
20	900	
30	800	
40	500	
50	0	

Y1 = -X²+40X+500

a. From the table find the y-intercept, x-intercepts, and vertex.

b. Use the information from part (a) to pick a suitable window for graphing this function.

21. Solve the equations algebraically.

a. $7x - 6 = 8$

b. $2x^2 + 2x = 24$

22. Solve $9x = 6x^2 + 4$ for x.

23. Use the rules of exponents to simplify and write the expression with only positive exponents.

$$\left(\frac{5x^3}{x^{-1}y}\right)^2$$

24. Find the slope of the line for each equation.

a. $y = \frac{1}{2}x - 5$

b. $4x - 9y = 6$

25. Simplify $2a^2 - (5a - 3)^2 + 18a - 7$.

8.3 Important Functions

After Studying This Section You Will Be Able To

- Identify the graphs of basic linear, quadratic, cubic, square root, absolute value, and reciprocal functions.
- Find the composite of two functions.

Many of the functions for which we have an algebraic formula are closely related to each other: for example, all linear functions have common features; each has a constant slope, and each has one y-intercept. Any quadratic expression gives a function whose graph is a parabola, and all of these also share a common set of features.

Here is a list of some of the most important functions we encounter in the rest of this book.

Linear functions We have already studied linear functions of the form $f(x) = mx + b$. A special case (the basic linear function) occurs when $m = 1$ and $b = 0$.

Basic linear function The function

$$f(x) = x$$

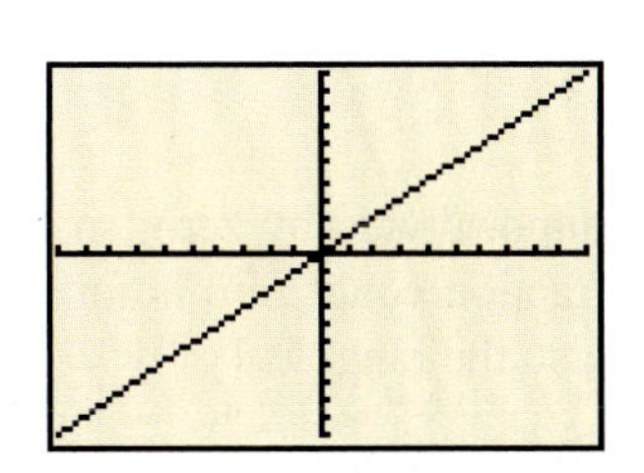

Figure 19 Graph of $f(x) = x$, the basic linear function.

is the basic linear function. The graph is a straight line through the origin with constant slope 1 (Figure 19). The domain is $\{x \mid {}^{-}\infty < x < \infty\}$ and the range is $\{y \mid {}^{-}\infty < y < \infty\}$.

Quadratic functions You have also studied quadratic functions of the form $f(x) = ax^2 + bx + c$. A special case (the basic quadratic function) occurs when $a = 1$, $b = 0$, and $c = 0$.

Basic quadratic function The function

$$f(x) = x^2$$

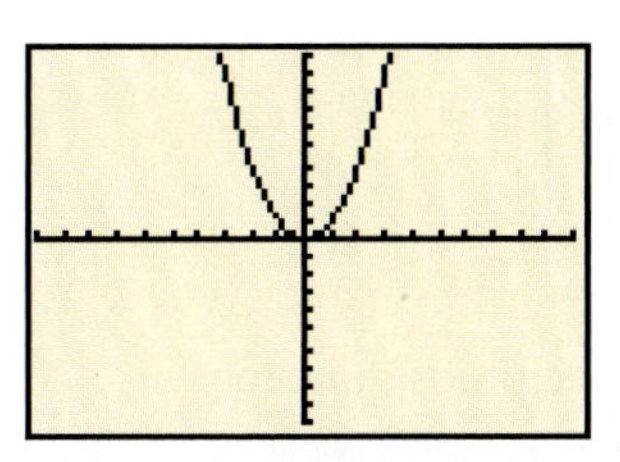

Figure 20 Graph of $f(x) = x^2$, the basic quadratic function.

is the basic quadratic function. Because the output is just the square of the input, the output can never be negative. The graph is a parabola that opens upward (a "cup"), with vertex ("bottom") at the origin (0, 0), which is also the only root (Figure 20). The parabola is thus symmetric about the y-axis. The domain is $\{x \mid {}^{-}\infty < x < \infty\}$; the range is $\{y \mid 0 \le y\}$.

Cubic functions These functions are similar to quadratic functions, except that the highest power of x is 3 rather than 2. Cubic functions may be written in the form $f(x) = ax^3 + bx^2 + cx + d$. We do not study cubic functions in as much detail as quadratics. However, we do need to know about the special case where $a = 1$, $b = 0$, $c = 0$, and $d = 0$. It is called the basic cubic function.

Basic cubic function The function

$$f(x) = x^3$$

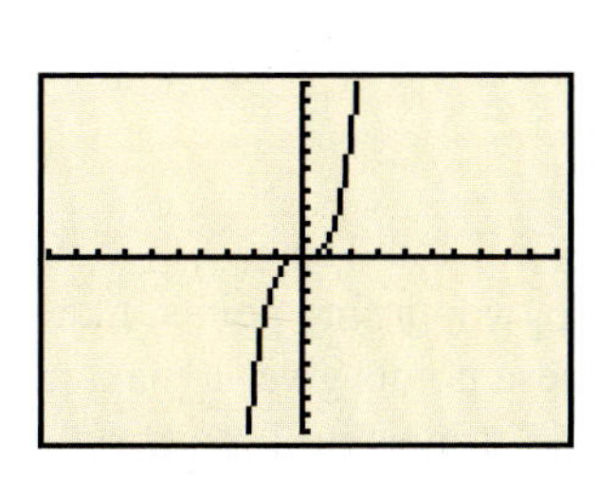

Figure 21 Graph of $f(x) = x^3$, the basic cubic function.

is the basic cubic function. The output is negative when the input is negative and positive when the input is positive. The graph of the function crosses the x-axis only once, at the origin, and it is always rising (Figure 21). The domain is $\{x \mid {}^{-}\infty < x < \infty\}$ and the range is $\{y \mid {}^{-}\infty < y < \infty\}$.

Basic square root function The function

$$f(x) = \sqrt{x}$$

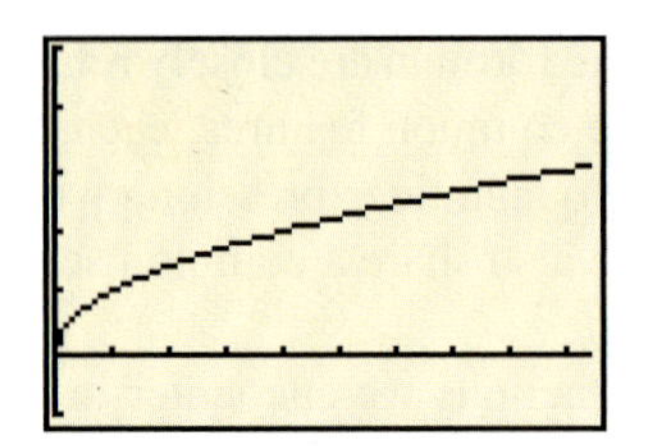

Figure 22 Graph of $f(x) = \sqrt{x}$, the basic square root function.

is the basic square root function. The graph begins at the origin and is always rising (Figure 22). The domain is $\{x \mid 0 \leq x\}$ and the range is $\{y \mid 0 \leq y\}$. We study this function in more detail in Chapter 9.

Basic absolute value function The function

$$f(x) = |x|$$

is the basic absolute value function. Because the absolute value of both a negative number and a positive number is positive, the graph of the absolute value function never goes below the x-axis. The graph touches the x-axis at the origin (Figure 23). The domain is $\{x \mid {}^{-}\infty < x < \infty\}$ and the range is $\{y \mid 0 \leq y\}$.

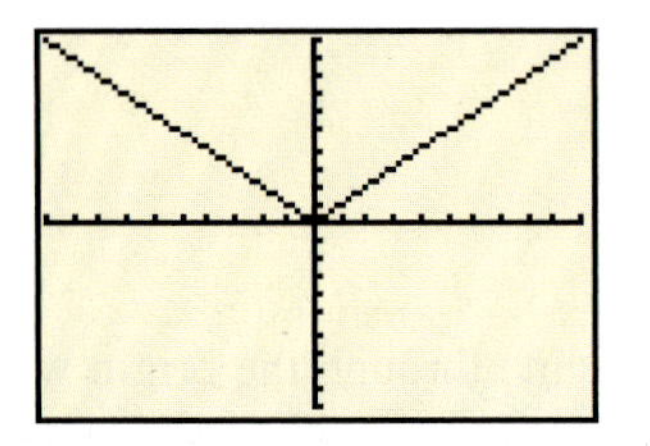

Figure 23 Graph of $f(x) = |x|$, the basic absolute value function. *Note*: Use abs when entering the function in the calculator.

Basic reciprocal function The function

$$f(x) = \frac{1}{x}$$

is the basic reciprocal function (Figure 24). We cannot divide by zero, so the domain is $\{x \mid x \neq 0\}$. The only way to make a fraction equal zero is for the numerator to equal zero, which cannot happen here, so the range is $\{y \mid y \neq 0\}$. We learn more about this function in Chapter 11.

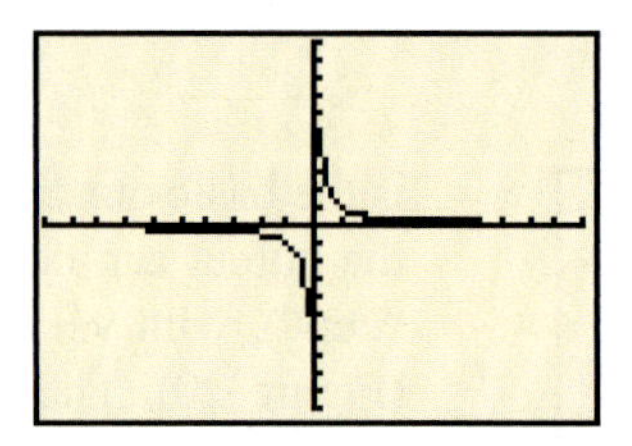

Figure 24 Graph of $f(x) = \frac{1}{x}$, the basic reciprocal function.

Moving Beyond the Basics

We can create many variations on this small set of basic functions. For example, not all cubics look like the basic cubic (see Figure 25).

EXAMPLE 1 ■ Describe the graph of $y = x^3 - 3x^2 - 3x + 1$.

Solution A good window to display the graph is ${}^{-}4.7 \leq \text{X} \leq 4.7$ and ${}^{-}15 \leq \text{Y} \leq 5$. One interesting new feature in this graph is that there are two turning points, like the vertex of a parabola. Unlike parabolas, neither of these points gives a maximum or minimum y-value; because the graph turns *twice*, it goes up forever on one side and down forever on the other (the range is $\{y \mid {}^{-}\infty < y < \infty\}$, as with the basic cubic).

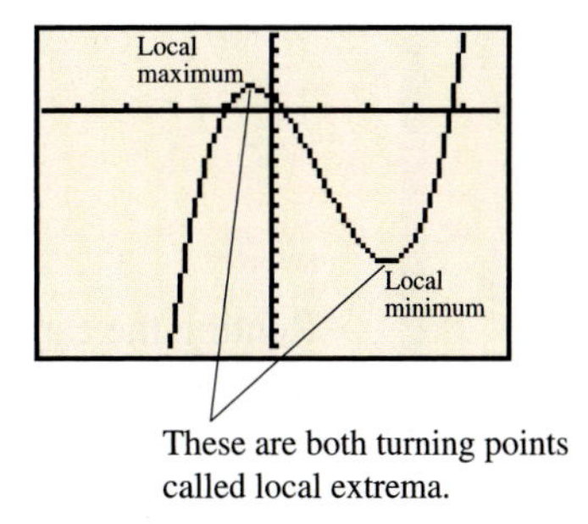

Figure 25 There are two points on the graph of $y = x^3 - 3x^2 - 3x + 1$ that look somewhat like the vertex of a parabola.

These points are places where the graph *turns*, either switching from increasing to decreasing or vice versa. You can think of the top of a mountain or the bottom of a valley. We call these points **local extrema**, because they are *extreme* values for the function, but only in the region of the graph immediately around themselves.

Another way to expand our library of functions is to combine two functions to create a more complicated function. The next example introduces a method called **function composition** for combining two functions.

EXAMPLE 2 The cost of renting a car for a weekend consists of the fixed fee paid to the rental agency ($60) and the cost of the gasoline purchased at the price of $1.50 per gallon. Thus, cost is a function of the number of gallons (g) the car consumes. (Don't forget that we have to return the car full or be charged extra).

$$\text{cost} = f(g) = 60 + 1.50g$$

The number of gallons you purchase depends upon how far you drive. The model you choose to rent gets 24 miles a gallon. If x is the number of miles driven,

$$\text{gasoline} = g(x) = \frac{x}{24}$$

a. Suppose we rent a car and drive it 900 miles. How much will it cost?

b. Suppose we rent a car and drive it x miles. How much will it cost?

Solution **a.** First, we need to find how many gallons of gas we use. Substitute 900 for x in the formula for $g(x)$.

$$g(900) = \frac{900}{24} = 37.5$$

The car will need 37.5 gallons. Now use the formula for $f(g)$ to find the cost of renting the car.

$$\begin{aligned} f(37.5) &= 60 + 1.50 * 37.5 \\ &= 60 + 56.25 \\ &= 116.25 \end{aligned}$$

Renting the car will cost \$116.25.

b. Substitute the formula for $g(x)$ for g in the formula for $f(g)$.

$$\begin{aligned} f(g) &= 60 + 1.50g \\ f(g(x)) &= 60 + 1.50g(x) \\ f(g(x)) &= 60 + 1.50 * \frac{x}{24} \end{aligned}$$

This last expression may be simplified to give

$$f(g(x)) = 60 + .0625x$$

Read the expression on the left side of this equation as f of g of x.

In effect we created a new function in part (b) of Example 2. We could call this function h and write $h(x) = f(g(x))$. The function $h(x)$ is called the *composite* of the functions f and g. The relationship among f, g, and h may be represented by a blob diagram, as shown in Figure 26.

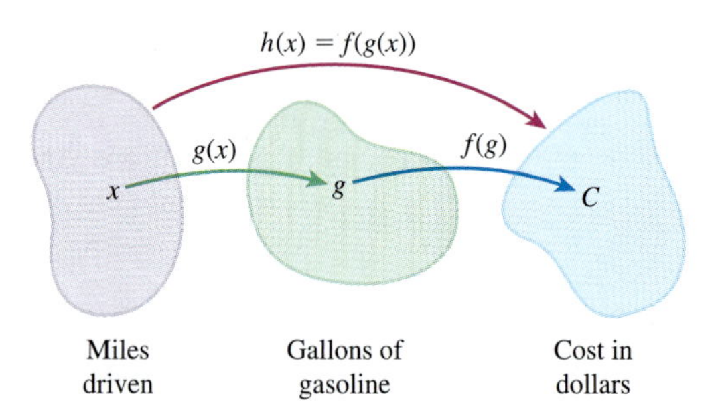

Figure 26 The first blob represents miles driven. The second blob represents gallons of gas. The third blob is cost in dollars. The arrow from blob 1 to blob 2 represents the function g. The arrow from blob 2 to blob 3 is f. The arrow from blob 1 to blob 3 is $h(x) = f(g(x))$.

Notice that in the example we gave the first function as $f(g)$ rather than $f(x)$. We did this to emphasize the fact that the output of function g is used as the input for function f. We could just as easily have described all three functions as functions of x.

Definition Composition of Functions

Given two functions $f(x)$ and $g(x)$, we can get a new function by taking f *composed with* g, written $f(g(x))$. This means that we substitute x into $g(x)$ and then substitute the output of $g(x)$ into $f(x)$ as input. We read $f(g(x))$ as f of g of x.

It is important to realize that it usually makes a difference which function we do *first*; $f(g(x))$ is almost never the same as $g(f(x))$. The next example illustrates this.

EXAMPLE 3 Suppose $f(x) = x^3$, the basic cubic function, and $g(x) = x + 1$.

a. Find $f(g(^-1))$ and $f(g(5))$.
b. Find $g(f(^-1))$ and $g(f(5))$.
c. Find a formula for $f(g(x))$.
d. Find a formula for $g(f(x))$.

Solution

a.
$$g(^-1) = {^-1} + 1 = 0$$
$$f(g(^-1)) = f(0) = 0^3 = 0$$

$$g(5) = 5 + 1 = 6$$
$$f(g(5)) = f(6) = 6^3 = 216$$

b.
$$f(^-1) = (^-1)^3 = {^-1}$$
$$g(f(^-1)) = g(^-1) = {^-1} + 1 = 0$$

Notice that for $x = {^-1}$, $f(g(x) = g(f(x))$.

$$f(5) = 5^3 = 125$$
$$g(f(5)) = g(125) = 125 + 1 = 126$$

Notice that for $x = 5$, $f(g(x)) \neq g(f(x))$.

c. Substitute the output of $g(x)$ as the input for f.

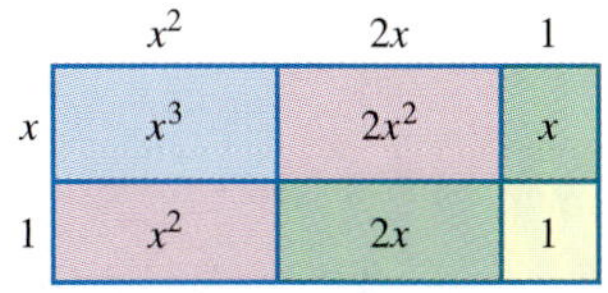

$$\begin{aligned} f(g(x)) &= f(x+1) \\ &= (x+1)^3 \\ &= (x+1)(x+1)(x+1) \\ &= (x+1)(x^2+2x+1) && \textit{Multiplying } (x+1)(x+1) \\ &= x^3 + 3x^2 + 3x + 1 && \textit{See Figure 27} \end{aligned}$$

	x^2	$2x$	1
x	x^3	$2x^2$	x
1	x^2	$2x$	1

Figure 27 We can use an area model to multiply $(x+1)(x^2+2x+1)$.

d. Substitute the output of $f(x)$ as the input for g.

$$g(f(x)) = g(x^3) = (x^3) + 1 = x^3 + 1$$

Notice that this expression is different from the one found in (c). *In general*, the functions $f(g(x))$ and $g(f(x))$ are not the same.

Exercises 8.3

1. In this section we discussed six basic functions: linear, quadratic, cubic, square root, absolute value, and reciprocal. It is important to remember these functions, because in the next section we modify them. Complete the chart below for the remaining five basic functions.

Name of the Basic Function	Function	Graph
Linear	$f(x) = x$	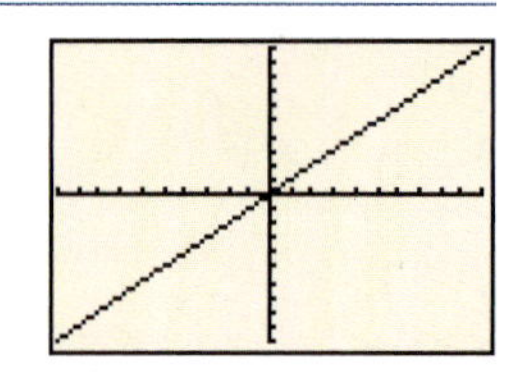
Quadratic		
Cubic		
Square root		
Absolute value		
Reciprocal		

2. Consider the basic linear function.

a. Find the slope.

b. Find the y-intercept.

3. Consider the basic quadratic function.

a. Evaluate the function for both positive and negative inputs. Why are the outputs always nonnegative?

b. Find the vertex (minimum) of the parabola.

c. Describe the symmetry of the parabola.

4. Notice that in the first quadrant, the graph of the cubic function looks similar to the quadratic. Why does the cubic turn down in the third quadrant instead of turning up like the quadratic? (*Hint:* Compare outputs of both functions for negative inputs.)

5. Consider the basic square root function.

a. Attempt to evaluate the function for a negative input. What happens?

b. How does this affect the domain?

6. Does the basic square root function have a minimum value? A turning point?

7. Compare the symmetry of the basic absolute value function to the symmetry of the basic quadratic function.

8. Consider the basic reciprocal function.

a. Are there any x- or y-intercepts? Explain.

b. Is there a minimum or maximum? Explain.

9. A student comments, "The basic reciprocal function has two identical halves." Interpret the meaning of this comment.

10. List the basic functions outlined in this section that have the given feature.

a. Constant slope

b. A minimum or maximum

c. No x-intercepts

d. No y-intercepts

e. Nonnegative outputs

11. The graph of a function with its local extrema is shown.

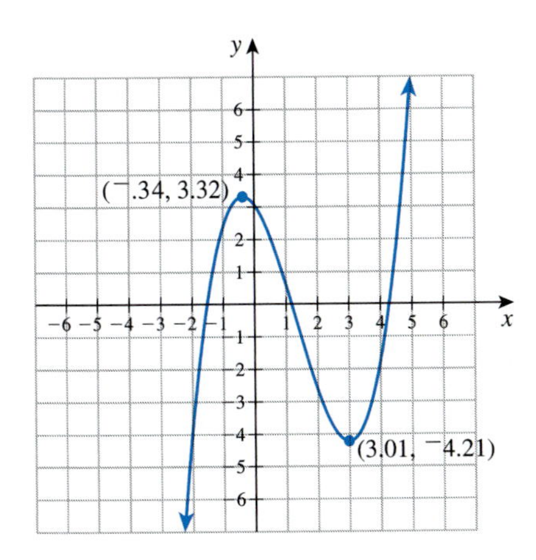

a. Give the local minimum.

b. Give the local maximum.

12. Use a graph to approximate the local extrema of the function $y = {}^-x^3 + 2x^2 - 2$.

13. The cost, $c(g)$, of renting a car is given by the function $c(g) = 50 + 1.65g$, where g is the number of gallons of gas used by the car. The number of gallons purchased depends upon the number of miles (x) driven and is given by the function $g(x) = \frac{x}{20}$.

a. How much will it cost to drive the car 500 miles?

b. How much will it cost to drive x miles?

14. Let $f(x) = 2x - 3$ and $g(x) = x + 1$. Find each composition.

a. $f(g(4))$ **b.** $f(g(x))$

c. $g(f({}^-2))$ **d.** $g(f(x))$

15. Let $f(x) = x^2 - 2x$ and $g(x) = x - 1$. Find each composition.

a. $f(g(3))$ **b.** $f(g(x))$

c. $g(f({}^-1))$ **d.** $g(f(x))$

Skills and Review 8.3

16. Use the vertical line test to determine which graphs represent functions.

a.

b.

c.

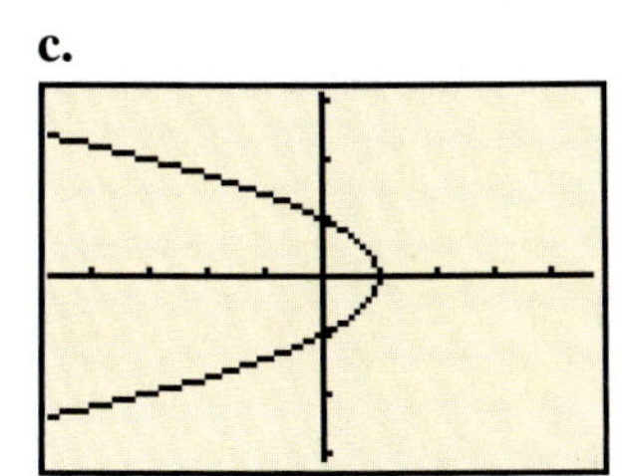

17. Find the domain and range for the function $f(x) = x^2 - 3$.

18. Let the function $f(x) = 1.4x^2$ represent the amount of frosting (in grams) on a cake of diameter x inches.

a. Use your calculator to find $f(24)$. (*Hint:* Enter the function in the Y= menu and find Y1(24) on the home screen using Y-VARS and the Function submenu.)

b. Suppose the maximum size cake is 24 inches. From part (a) we know the maximum amount of frosting. Show the domain and range on a number line.

19. Solve the equation $3x^2 - 18x = 0$ for x.

20. Solve the equation $7x^2 + 12x = 2$ for x.

21. Let $y = x^2 - 3x - 28$

a. Use the sign of the x^2-term to determine whether this parabola opens upward or downward.

b. Find the y-intercept.

c. Find the x-intercepts.

d. Find the vertex.

e. Use the information gathered to sketch a graph by hand.

22. Factor completely.

a. $x^2 - 4$ **b.** $10x^2 + 5x - 30$

c. $x^2 - 10x + 25$

23. Expand and simplify as necessary.

a. $x(3x^2 + 7x - 5)$ **b.** $(2x - 3)(x + 8)$

24. Simplify and write with only positive exponents.

$$\left(\frac{2x^4}{x^3z^5}\right)^{-1}$$

25. Solve the proportion,

$$\frac{1}{x+2} = \frac{5}{x}$$

8.4 Modifying Basic Functions

AFTER STUDYING THIS SECTION YOU WILL BE ABLE TO

- Apply these modifications to basic functions: stretching and compressing, reflecting across the x-axis, and vertical and horizontal shifts.
- Analyze a function in terms of the modifications introduced in this section.

In Section 8.3 you were introduced to several basic functions. In this section we explore ways these functions can be modified to produce new functions. Our strategy is to graph a basic function along with a modified one and observe the relationship between the two graphs.

Stretching and Compressing Graphs

EXAMPLE 1 Graph $f(x) = |x|$ and $g(x) = 3 * f(x)$ on the same graph. Observe what happens.

Solution In the Y= screen enter Y1 = abs(X). Note: On the TI-83 you will find abs under MATH. On the TI-82 it is the second function above the x^{-1} key.

Now enter Y2 = 3 ∗ Y1. You could have entered Y2 = 3 ∗ abs(X) to get the same result. However, in the next example we will keep Y2 the same but change the function in Y1. (See Figure 28.)

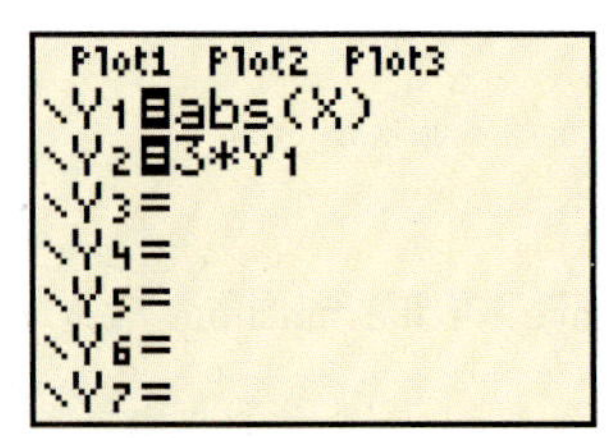

Figure 28

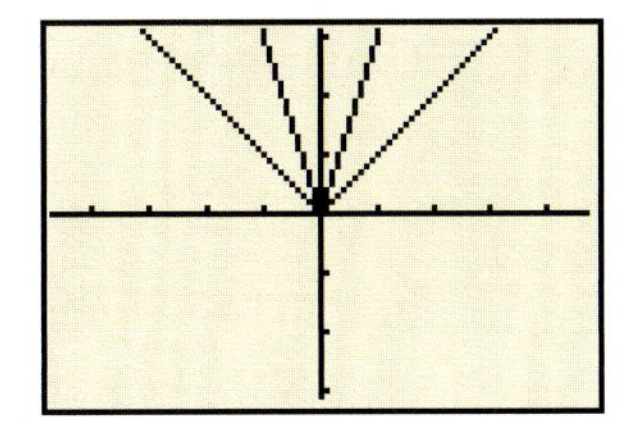

Figure 29 Y1 = *abs*(X), Y2 = 3 ∗ Y1.

Press GRAPH to see the result in Figure 29. Observe that the graph of Y2 is above the graph of Y1. We can imagine the graph of Y1 being stretched upward in such a way that the bottom of the *V* remains fixed at the origin.

EXAMPLE 2 Graph these pairs of functions on the same graph. Observe what happens in each case.

a. $f(x) = x^3$ and $g(x) = 3 * f(x)$

b. $f(x) = \sqrt{x}$ and $g(x) = 3 * f(x)$

Solution **a.** Enter Y1 = X ∧ 3 and Y2 = 3 ∗ Y1. The result appears in Figure 30. The original graph of $y = x^3$ appears to be stretched away from the x-axis in both the first and third quadrants.

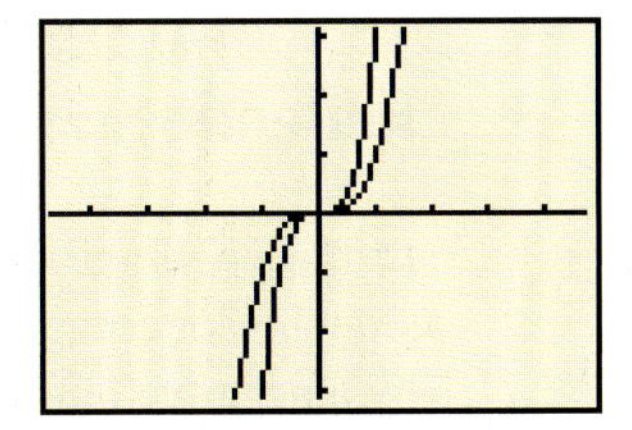

Figure 30

b. Enter Y1 = √(X). Keep Y2 the same. The result appears in Figure 31. Again the graph of the original function is stretched away from the x-axis.

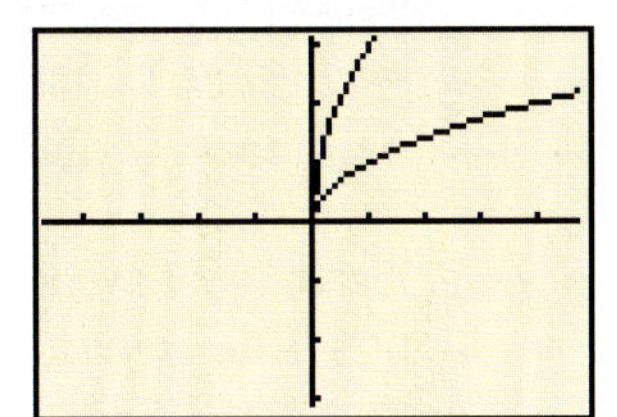

Figure 31

From these examples we conclude that when we multiply a function by a constant value greater than 1 (such as 3), our original graph is stretched away from the x-axis. The next example shows what happens if we multiply a function by a small positive number, such as $\frac{1}{2}$.

EXAMPLE 3 Graph these pairs of functions on the same graph. Observe what happens in each case.

a. $f(x) = |x|$ and $g(x) = \frac{1}{2} * f(x)$
b. $f(x) = x^3$ and $g(x) = \frac{1}{2} * f(x)$
c. $f(x) = \sqrt{x}$ and $g(x) = \frac{1}{2} * f(x)$

Solution We are starting with the same three functions we used in the two previous examples. What is different this time is that Y2 = (1/2) ∗ Y1. See Figure 32. The graphs of the three functions are shown in Figure 33. Notice that each of the original graphs is pulled *toward* the x-axis rather than away from it. We may think of this as shrinking, or *compressing*, rather than stretching.

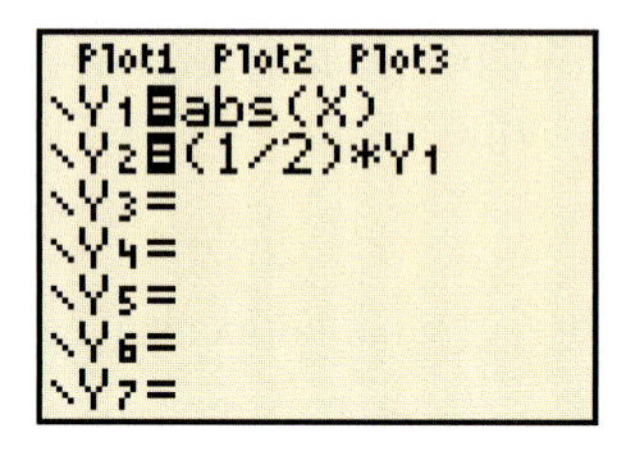

Figure 32

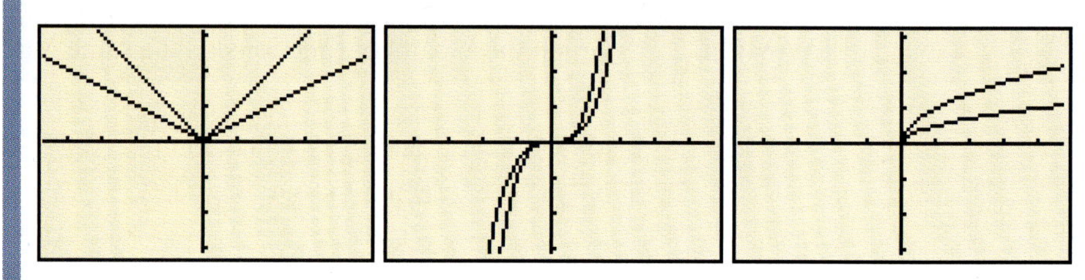

Figure 33

Reflecting a Graph Across the x-Axis

In the previous examples we multiplied a function by a positive number and stretched or compressed the graph. Now let's see what happens when we multiply by a negative number, such as $^-1$.

EXAMPLE 4 Graph these pairs of functions on the same graph. Observe what happens in each case.

a. $f(x) = |x|$ and $g(x) = {}^-f(x)$
b. $f(x) = x^3$ and $g(x) = {}^-f(x)$
c. $f(x) = \sqrt{x}$ and $g(x) = {}^-f(x)$

Solution Again, we have the same three functions. Now Y2 = $^-$Y1, so the second function is the *opposite* of the first. The graphs are shown in Figure 34.

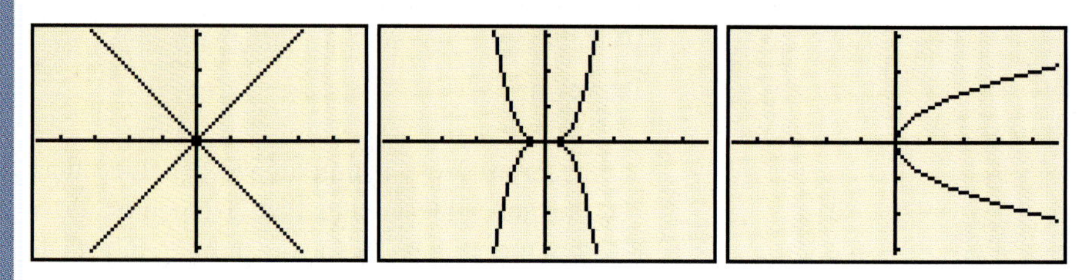

Figure 34

Observe that each of the original functions appears to be flipped, or *reflected* across the x-axis.

Moving a Graph Up or Down

Now let's see how we can move a function up or down.

EXAMPLE 5 Begin with the three basic functions used in the previous examples (absolute value, cube, and square root). Discover what happens in each case.

a. When 2 is added to the output

b. When 2 is subtracted from the output

Solution **a.** Enter the same three functions in Y1. Enter Y2 = Y1 + 2. The results are shown in Figure 35. The graphs of the original functions are moved up two units.

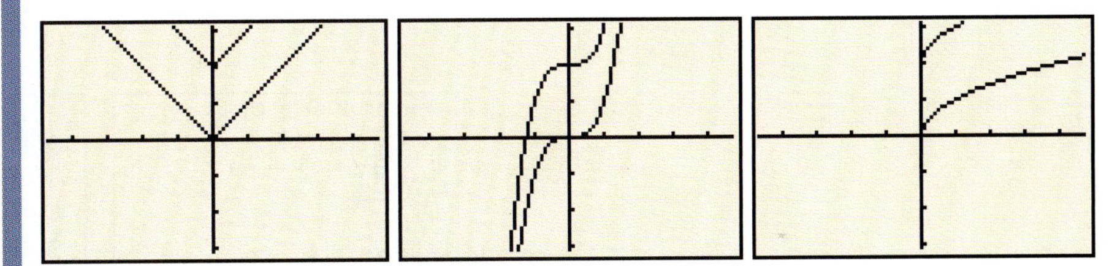

Figure 35

b. Change Y2 to Y1 − 2. The graphs of the original functions are now moved down two units. See Figure 36.

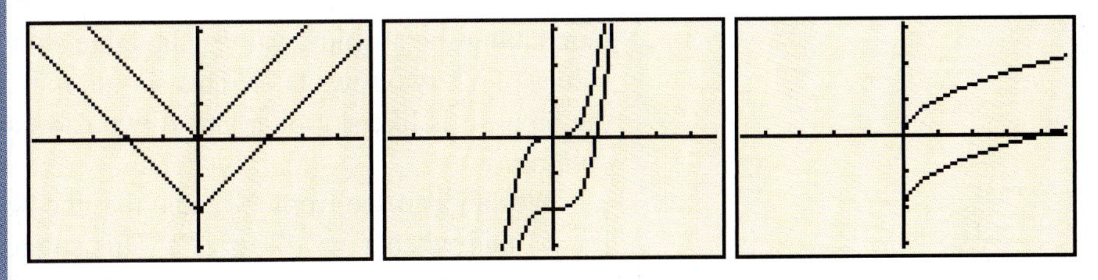

Figure 36

Example 5 illustrates what happens when we add or subtract a constant number (such as 2) from the output of a function. Another modification is to add 2 to the input before applying the particular function. The table shows the difference between these two methods for modifying a basic function.

Original Function	Adding 2 to output Symbolic representation	Verbal description	Adding 2 to input Symbolic representation	Verbal description
$f(x) = \|x\|$	$f(x) + 2 = \|x\| + 2$	Take the absolute value of x and then add 2.	$f(x+2) = \|x+2\|$	Add 2 to x and take the absolute value of the result.
$f(x) = x^3$	$f(x) + 2 = x^3 + 2$	Cube x and then add 2.	$f(x+2) = (x+2)^3$	Add 2 to x and cube the result.
$f(x) = \sqrt{x}$	$f(x) + 2 = \sqrt{x} + 2$	Take the square root of x and then add 2.	$f(x+2) = \sqrt{x+2}$	Add 2 to x and take the square root of the result.

Moving Graphs Left or Right

Finally, we look at how to modify a function in order to move its graph left or right.

EXAMPLE 6 Explain what the table shown in Figure 37 tells you about how the graph of Y2 is related to the graph of Y1.

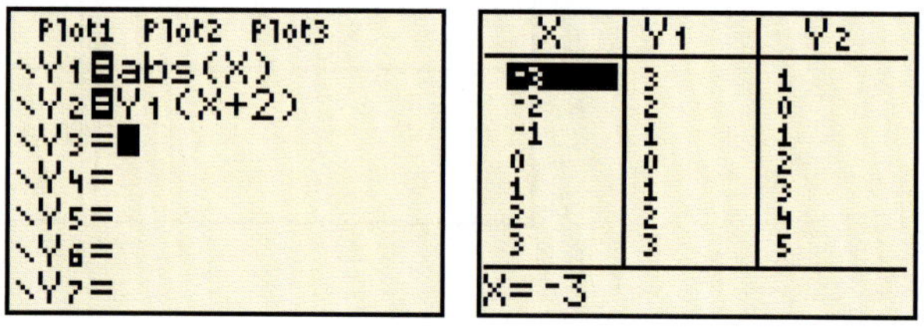

Plot1 Plot2 Plot3
\Y1=abs(X)
\Y2=Y1(X+2)
\Y3=
\Y4=
\Y5=
\Y6=
\Y7=

X	Y1	Y2
-3	3	1
-2	2	0
-1	1	1
0	0	2
1	1	3
2	2	4
3	3	5

X= -3

Figure 37

Solution Y1 is the absolute value function. Y2 is formed by adding 2 to the input, x, and then taking the absolute value. The table shows that values of Y2 correspond to values of Y1 two rows later. This should make sense because when you add 2 to the input, it is like skipping ahead two rows and finding a value for y in the Y1 column.

We can also see from the table that the bottom of the V occurs when $x = 0$ for Y1 and when $x = {}^-2$ for Y2. This suggests that the graph of Y1 is moved two units to the left to give the graph of Y2. The next example confirms this idea.

EXAMPLE 7 Start with the three basic functions used in the previous examples (absolute value, cube, and square root) and discover what happens in each case.

a. When 2 is added to the input

b. When 2 is subtracted from the input

Solution

a. Enter the same three functions in Y1. Enter Y2 = Y1(X + 2). The results are shown in Figure 38. The graphs of the original functions are moved to the left two units.

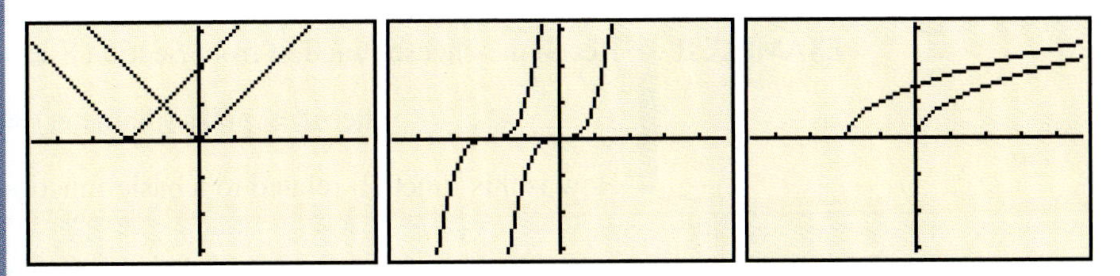

Figure 38

b. Change Y2 to Y1(X − 2). The graphs of the original functions are now moved to the right two units. See Figure 39.

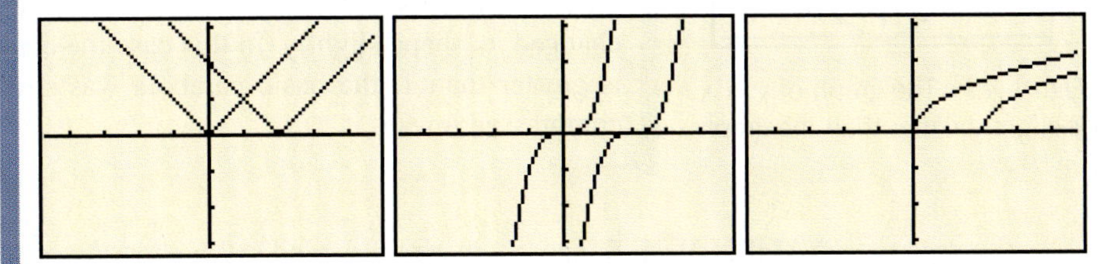

Figure 39

Caution: You may have been surprised in the previous example to discover that adding a positive number to the input shifts the graph to the left (which we consider the *negative* direction), whereas subtracting a positive number from the input shifts the graph to the right. This can be confusing, so try to keep this pattern straight in your mind when doing a problem involving modifications to inputs. You may always check your answer by drawing a graph or making a table.

The results of the previous examples are summarized in this table.

Function	Description	Effect on Graph
$f(x)$	Basic Function	
$c * f(x)$	Multiply the output by a positive number c.	If $c > 1$, it stretches graph away from the x-axis. If $0 < c < 1$, it compresses graph toward the x-axis.
$^-f(x)$	Take the opposite of the output.	It flips (reflects) the graph across the x-axis.
$f(x) + c$	Add a positive number c to the output.	It moves the graph up c units.
$f(x) - c$	Subtract a positive number c from the output	It moves the graph down c units
$f(x + c)$	Add a positive number c to the input.	It moves the graph left c units.
$f(x - c)$	Subtract a positive number c from the input.	It moves the graph right c units.

Relating Functions to the Basic Functions

Now let us see how these modifications may be applied to functions we have previously studied.

EXAMPLE 8 Recall the cheese function from Section 8.1:

$$\text{cheese} = \text{pizza (diameter)} = 1.2 * \text{diameter}^2$$

How is this function related to a basic function?

Solution This looks a lot like the basic quadratic function if we think of pizza as f and *diameter* as x:

$$f(x) = x^2$$

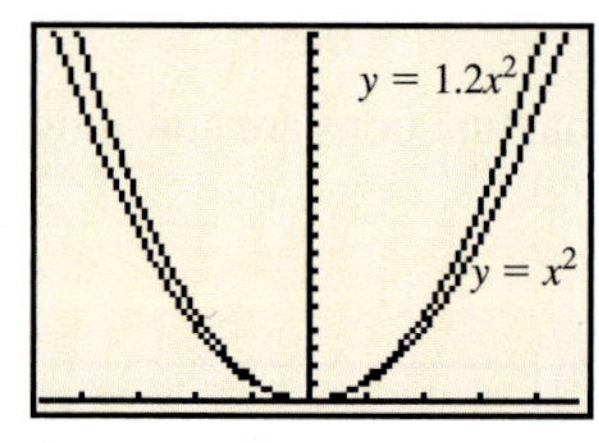

Figure 40 The graph of $y = x^2$ is slightly flatter than the graph of $1.2x^2$.

Let's examine how the graphs are related (see Figure 40).

Both are parabolas opening upward through the origin (a common vertex), but one is a little narrower than the other; in other words, it is steeper. Multiplying the x^2-term by a constant left the parabola with the same vertex and root but changed its shape slightly. In this case the constant is 1.2. Because the constant is greater than 1, the basic parabola was stretched away from the x-axis, not compressed toward it.

EXAMPLE 9 Suppose you set up a stand selling hot dogs outside the local softball game every Sunday. Supplies cost \$30 every week, and you sell the dogs for \$1 each.

a. What function tells how much profit you make each week?

b. How is this function related to a basic function?

Solution

a. The profit you make each week is given by a simple linear function,

$$P(n) = n - 30$$

where n is the number of hot dogs you sell that week.

b. The function in (a) looks just like the basic linear function, only we have subtracted 30. Figure 41 shows how the graphs are related.

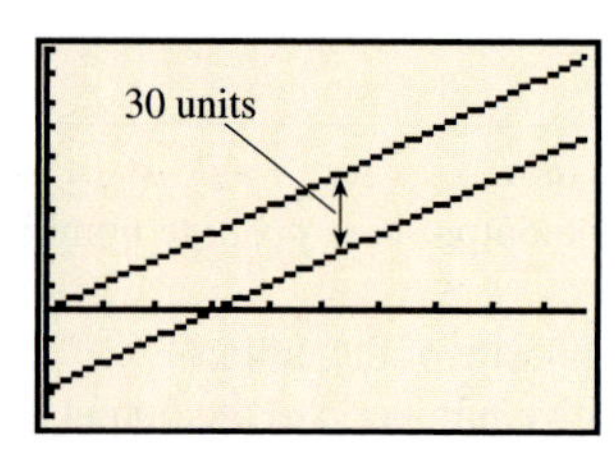

Figure 41 Subtracting 30 from the output moves each output point down. Because all output moves down by the same amount, the entire graph shifts down by 30 units.

Subtracting 30 shifted the line down everywhere. This makes sense, because we are just subtracting 30 from the value P would have from the basic linear function. That is, the function moves down 30 units on the P-axis.

EXAMPLE 10 A ball is dropped into a deep canyon. Let $d(t)$ be the distance (in feet) the ball has traveled t seconds after the moment it was dropped. The gravitational pull of the earth on the ball gives the equation $d(t) = 16t^2$.

Let the level of the canyon rim be considered zero and the position of the ball positive if it is above the rim or negative if it is below the rim.

a. Find a function $p(t)$ that describes the position of the ball t seconds after it is dropped.

b. How is this function related to a basic function?

Solution

a. As soon as the ball is dropped, it falls below its initial position of zero. Thus, its position is the *opposite* of the distance it has traveled.

$$p(t) = {}^{-}d(t) = {}^{-}16t^2$$

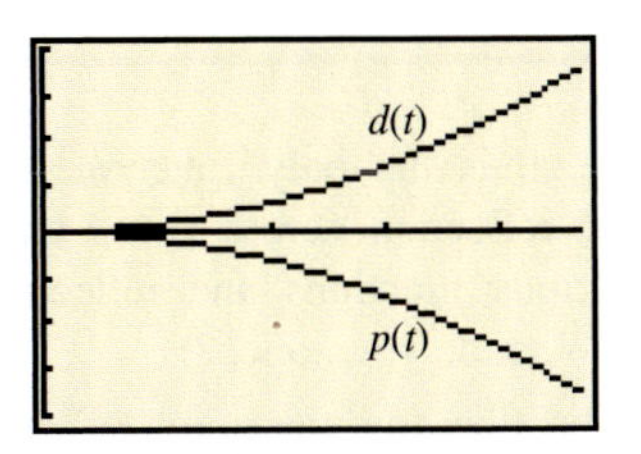

Figure 42 $p(t)$ is the reflection of $d(t)$ across the t-axis.

b. The graph of $p(t)$ is found by reflecting the graph of $d(t)$ across the x-axis, as shown in Figure 42. In this case we could think of the function $p(t)$ as the result of *two* modifications of the basic function $f(t) = t^2$. First, $f(t)$ is stretched by the factor 16 to give $d(t)$. Then $d(t)$ is reflected across the x-axis to give the function $p(t)$.

EXAMPLE 11 Consider the ball in Example 10. Now suppose we drop a second ball 2 seconds after we drop the first ball. Again, let t represent the time since the *first ball* was dropped.

a. Find a function for the position of the second ball in terms of t.

b. How are these functions related?

Solution

a. Because there is a 2-second delay, when $t = 2$ we will just have dropped the second ball, when $t = 3$ the second ball will have been moving for 1 second, when $t = 4$ the second ball will have been moving for 2 seconds, and so on. We are always subtracting 2 from the time, so we get a new function:

$$p_2(t) = p(t - 2) = {}^{-}16(t - 2)^2$$

The subscript 2 indicates this is the second ball's position

This gives us the distance the second ball has traveled in terms of the number of seconds t since the first ball was dropped. It is important to note that this function works only for t-values larger than 2, because the ball doesn't move at all before that; the part of the graph to the left of X = 2 on a calculator does not apply to the real-world problem.

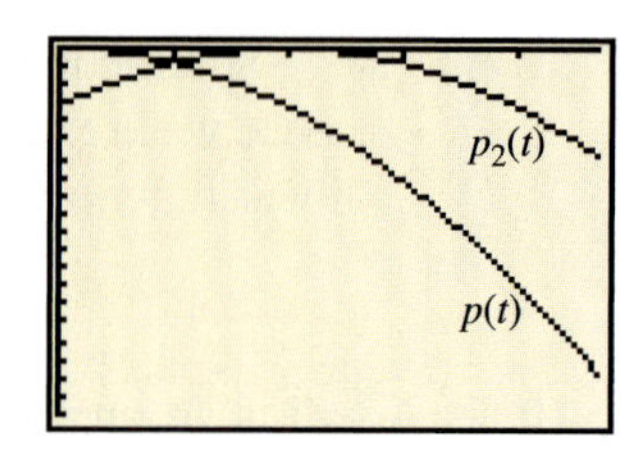

Figure 43 Subtracting from the input creates a change related to the input and not the output axis—the graph moves right.

b. The relationship between the two graphs is shown in Figure 43. Notice that the graph of $p(t)$ is shifted to the right 2 units to give the graph of $p_2(t)$.

Exercises 8.4

1. How is *stretching* a function different from *compressing* a function?

2. Determine whether each graph of $g(x)$ is a stretch or a compression of the graph of $f(x)$. Explain your reasoning.

a. $f(x) = x$ and $g(x) = 5 * f(x)$

b. $f(x) = x^2$ and $g(x) = \frac{1}{4} * f(x)$

c. $f(x) = \frac{1}{x}$ and $g(x) = .32 * f(x)$

3. Let $g(x) = cx^2$, where c is a constant greater than 0.

a. Determine the values of c that stretch the graph of the basic quadratic function.

b. Determine the values of c that compress the graph of the basic quadratic function.

c. Determine the value of c that produces no change in the graph of the basic quadratic function.

4. Use your knowledge of the basic functions from Section 8.3 along with stretching and compressing to draw by hand a graph of each function. (*Hint:* Draw $g(x)$ on the same graph as the basic function).

a. $g(x) = .2|x|$ **b.** $g(x) = 8.5x^3$

5. Describe what happens to the graph of a function when the output of the function is multiplied by $^-1$.

6. Let $f(x) = \frac{1}{x}$ and $g(x) = {}^-f(x)$. How is the graph of $g(x)$ related to the graph of $f(x)$?

7. Describe why $f(x) = 0$ is the only function that remains the same after reflection across the x-axis. (*Hint:* Try drawing various functions and reflecting them across the x-axis. What do you notice?)

8. For parts (a) and (b), describe how the graph of each function relates to the graph of the basic quadratic function.

a. $g(x) = x^2 + 5$ **b.** $g(x) = x^2 - 3$

c. Summarize what happens to the graph of a function when a positive constant is added to or subtracted from the output.

9. Look at the table of values shown in the figure and explain why the graph of the basic cubic function moves 1 unit right when we subtract 1 from the input.

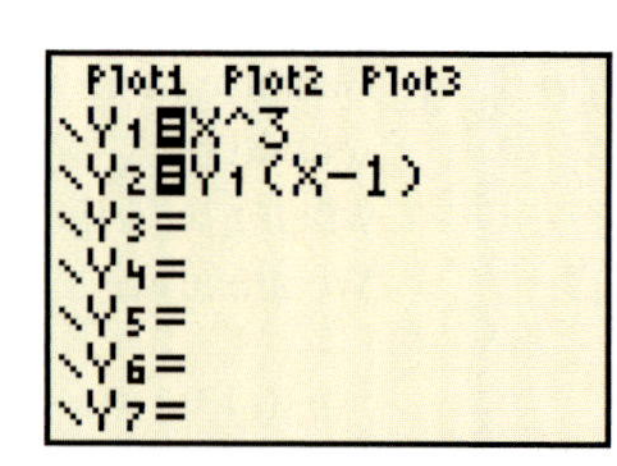

X	Y1	Y2
-3	-27	-64
-2	-8	-27
-1	-1	-8
0	0	-1
1	1	0
2	8	1
3	27	8

X= -3

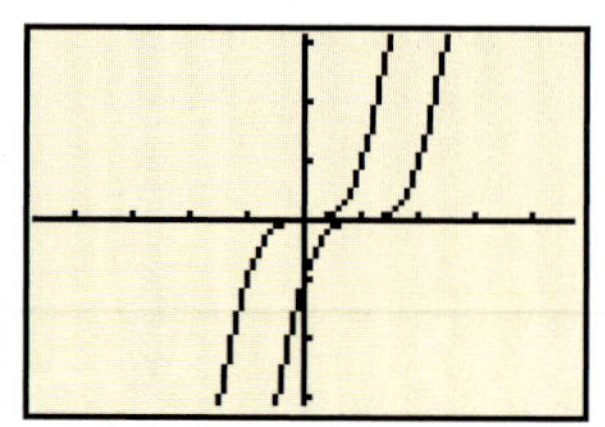

10. For parts (a) and (b), describe how the graph of each function relates to the basic square root function.

a. $g(x) = \sqrt{x - 4}$

b. $g(x) = \sqrt{x + 1}$

c. Summarize what happens to the graph of a function when a positive constant is added to or subtracted from the input.

11. Let $f(x)$ represent any one of the basic functions. Match each modification to the change in the graph of $f(x)$.

a. Add 3 to the input: $y = f(x + 3)$.

b. Multiply the output by $^{-}1$: $y = {}^{-}f(x)$.

c. Multiply the output by a constant $c > 1$: $y = c * f(x)$.

d. Subtract 2 from the output: $y = f(x) - 2$.

---

1. The graph moves down 2.
2. The graph moves left 3.
3. The graph flips across the x-axis.
4. Graph stretches away from the x-axis.

12. Match each modified quadratic function with the appropriate graph.

a. $g(x) = (x - 2)^2$

b. $g(x) = x^2 - 2$

c. $g(x) = 3x^2$

d. $g(x) = {}^{-}x^2$

1.

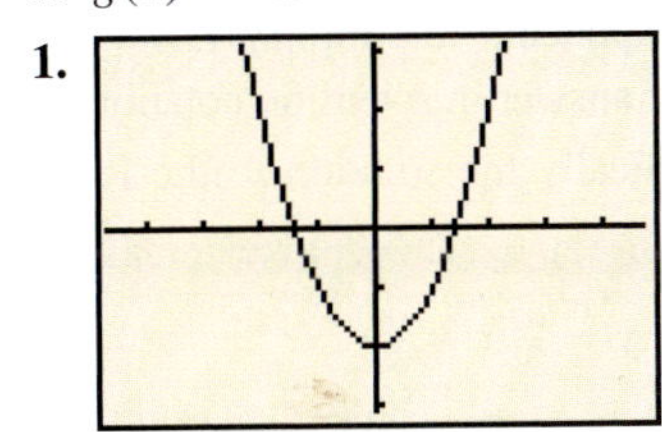

2.

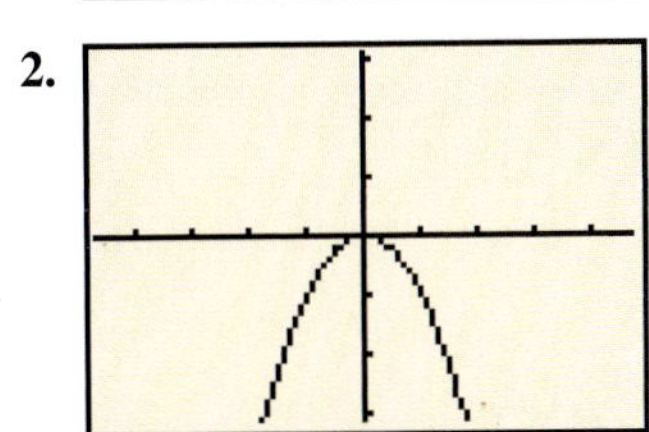

3.

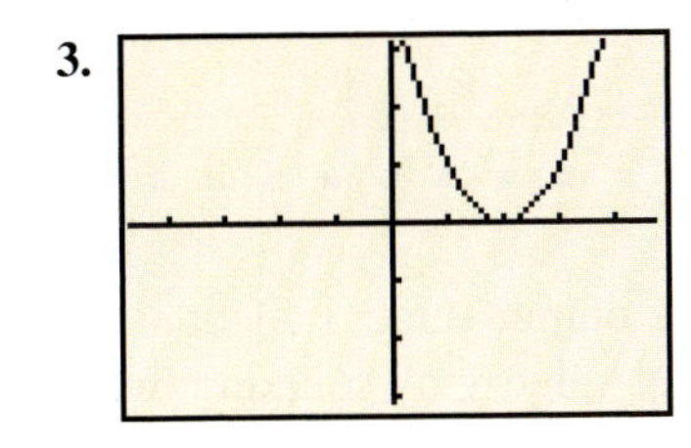

4.

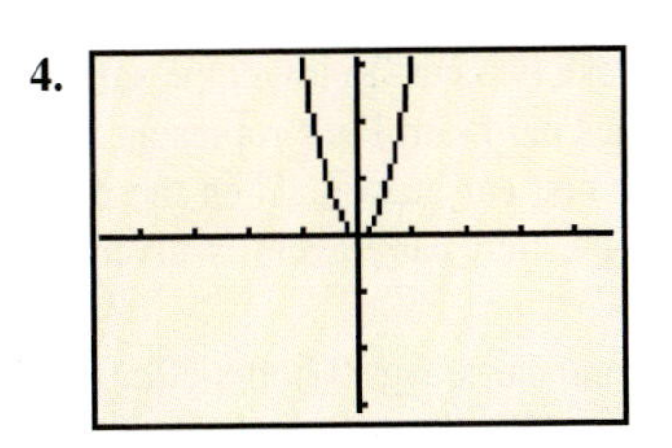

13. Each graph represents a modification of one of the following basic functions: linear, quadratic, cubic, square root, absolute value, or reciprocal. The modifications don't include stretching or compression and each tick on the axes represents one unit. Match each graph with the correct function from the choices given.

a.

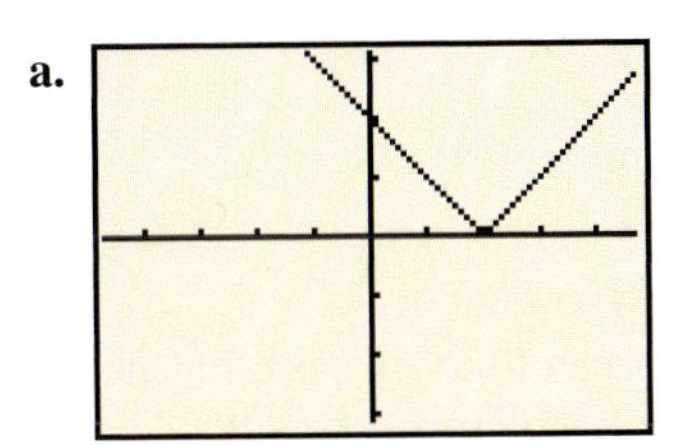

$g(x) = |x - 2|$, $g(x) = |x| + 2$, $g(x) = 2|x|$

b.

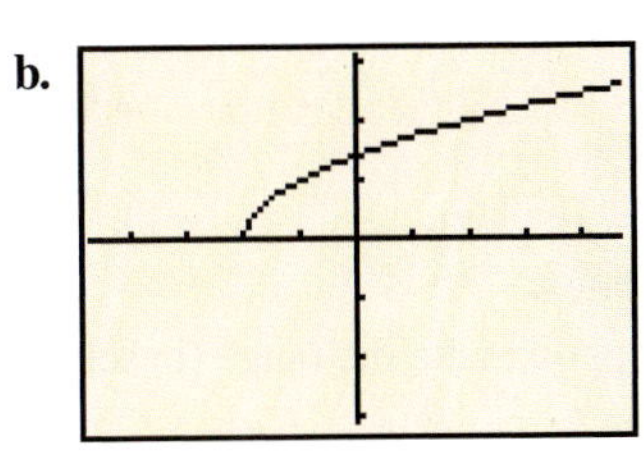

$g(x) = 2\sqrt{x}$, $g(x) = \sqrt{x + 2}$, $g(x) = \sqrt{x} - 2$

c.

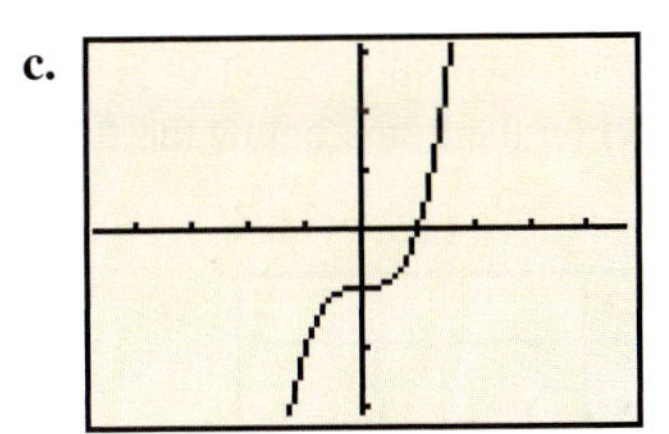

$g(x) = x^3 + 1$, $g(x) = (x - 1)^3$, $g(x) = x^3 - 1$

d.

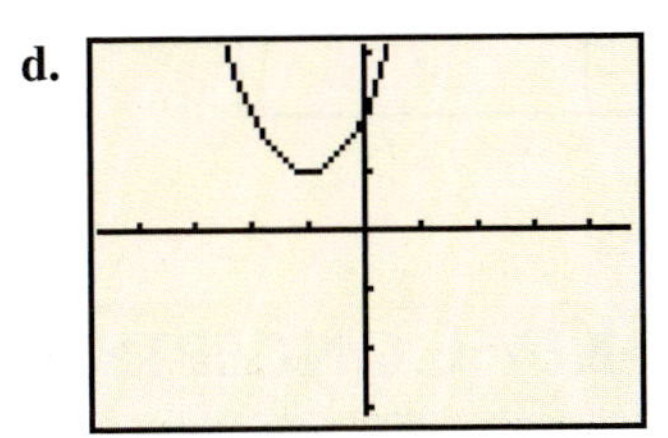

$g(x) = (x + 1)^2 - 1$, $g(x) = (x + 1)^2 + 1$, $g(x) = (x - 1)^2 + 1$

14. A group of friends take two cars to travel the same route at a speed of 50 miles per hour. Let t represent the time (in hours) since the first car started. Then the distance, $d(t)$, in miles that the first car travels is given by the function $d(t) = 50t$.

a. Suppose the second car leaves 3 hours after the first. Write a function for the distance the second car travels in terms of the time t.

b. How are the graphs of these two functions related?

15. The basic quadratic function, $f(x) = x^2$, has a vertex at $(0, 0)$. Use your knowledge of modifications to determine the exact vertex of the quadratic function $g(x) = (x - 9.5)^2 + 52$.

Skills and Review 8.4

16. Suppose $f(x) = x + 3$ and $g(x) = x^2 - 5$.

a. Find $f(g(2))$.

b. Find a formula for $f(g(x))$.

c. Find $g(f(^-4))$.

d. Find a formula for $g(f(x))$.

17. Sketch a graph of each of the six basic functions from Section 8.3. Next to each graph, show the domain and range in set-builder notation.

18. Why doesn't the range of $f(x) = |x|$ contain any negative values?

19. Let $y = x^2 - 5x + 9$.

a. Is this a function? Explain.

b. Find the exact vertex.

c. Use a graph to approximate the range (in set-builder notation).

20. Let $f(x) = 5 + \sqrt{^-x + 4}$. Find $f(1)$.

a. Give an exact answer.

b. Give an approximate answer.

21. Place a dot and label on the graph below the following points, if they exist.

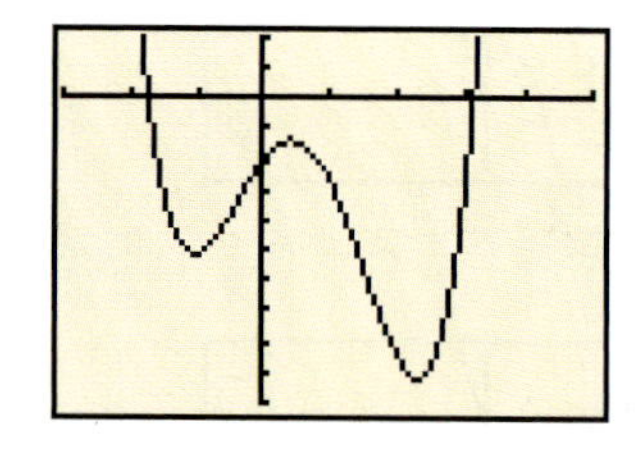

a. x-intercepts

b. y-intercept

c. Local minimum

d. Local maximum

22. The height of a falling acorn is given by the function $h = {^-16}t^2 + 80$, where h is in feet and t is the number of seconds after the acorn drops from the tree.

a. How long, to the nearest .1 second, does it take this acorn to hit the ground?

b. We know the initial height is the maximum. Verify this by finding the vertex.

c. Find the domain and range.

23. Factor the greatest common factor (GCF) out of the expression $72a^2b - 48b^3$.

24. Use the rules of exponents to simplify $(\frac{\pi}{2}) * (2.31 * 10^4 m)^3$. Write your answer in scientific notation.

25. Approximate graphically the solution to the following system of equations:

$$y = \frac{42}{37}x - \frac{65}{37}$$
$$y = {^-}\frac{75}{14}x - \frac{16}{7}$$

CHAPTER 8 ▪ KEY CONCEPTS

Function A function is a rule that associates to an input value one and only one output value. Functions can be specified using algebraic rules: $y = x^2$ takes an input value x and gives the output value y by squaring the input.

They can also be specified by drawing a graph, with the x-axis for input values and the y-axis for outputs, or by giving a table listing input and output values. (Page 302)

Dependent and Independent Variables, Function Notation The two variables we use when we write a function using function notation, $y = f(x)$. The input variable is the independent variable (x), the output variable is the dependent variable (y); f is the name of the rule that defines the function. (Page 302)

Domain The domain of a function is the set of allowable inputs. (Page 301)

Range The range of a function is the set of possible outputs. (Page 301)

Vertical Line Test This test indicates whether a given graph is the graph of a function. If any vertical line drawn hits the graph more than once, then the graph cannot be the graph of a function, since some input (where the vertical line hits the x-axis) would correspond to more than one output (where the line hits the graph, see Figure 10 in Section 8.2). (Page 311)

Specifying Domains and Ranges We can specify the domains and ranges for functions using different notation. Graphically, we can just draw them on a number line. Using more algebraic notation, we can use set-builder notation (page 304) and give conditions that a number must satisfy to be in the domain and range,

$$\{x \mid x < 3\} \qquad \textit{The set of } x\textit{'s less than } 3$$

Basic Functions Basic functions are elementary functions that can be used to examine the properties of more complicated functions. For example, the function $y = x^2 - 5x + 17$ shares many properties with the basic quadratic function $y = x^2$. (Section 8.3)

Composition of Functions A way of generating new functions from old ones by taking the output of one function as the input of another, $h(x) = f(g(x))$. Here, the function h is given by first using x as input to the function g, and then using the output of g as the input to f. (Page 323)

Modifying Functions Functions can be modified by making simple changes to either their input or output. (Section 8.4)

- Multiplying the input or output by a positive number changes the scale of the axes, *stretching* or *compressing* the graph.
- Changing the sign of the input reflects the graph across the x-axis.
- Adding or subtracting from the output moves the graph up or down.
- Adding or subtracting from the input of a function moves the graph left or right.

Exploration

The links from the home page to other pages on Eastern Connecticut State University's website are set up on a rectangular grid, as shown in the figure. As described in the chapter opener, the pages can be accessed when the mouse has a location with an x-coordinate between 110 and 170 and a y-coordinate between 30 and 90. For example, when the mouse is clicked at the location (135, 65), the browser links to the page for clubs and activities. When the mouse is clicked at the location (100, 40), the browser remains on the home page, because no link is associated with this mouse location.

	x = 110 to 130	x = 130 to 150	x = 150 to 170
y = 90 to 80	Academics	Administrative Offices	Admissions
y = 80 to 70	Alumni	Alphabetical Index	Athletics
y = 70 to 60	Calendars	Clubs and Activities	Continuing Education
y = 60 to 50	Directions and Maps	Faculty and Staff	Graduate Division
y = 50 to 40	Guestbook	Information Services	Library
y = 40 to 30	Online CSU	Online Services	Search Web and ECSU

a. What happens if the mouse is clicked when pointed to (154, 73)?

b. What happens if the mouse is clicked when pointed to (136, 97)?

c. What is the domain of this function?

d. What is the range of this function?

CHAPTER 8 ▪ REVIEW EXERCISES

Section 8.1 What Is a Function?

1. Define a function.

2. Define the domain and range of a function.

3. Determine if the table of inputs and outputs represents a function. Explain your answer.

Input	Output
α	1
β	2
γ	3
ϵ	1

a. Express the domain in a set.

b. Express the range in a set.

4. Does this connection picture represent a function? Explain.

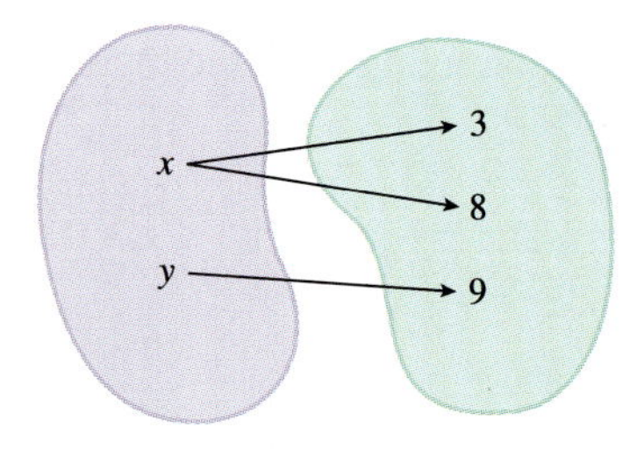

5. The formula sauce = $.2 * \text{diameter}^2$ gives the amount of tomato sauce (in ounces) needed for a pizza of given diameter (in inches). Assume the maximum diameter pizza is 35 inches.

a. Express the domain for this function on a number line.

b. Express the range on a number line.

6. Let f be the name of the function $y = x^2 - 3$.

a. Write f using function notation.

b. By hand, evaluate $f(^-2)$.

c. Evaluate $f(1.46)$ on your calculator.

7. Evaluate each function.

a. $f(x) = x + 1$; evaluate $f(3)$.

b. $g(x) = x^2 - x$; evaluate $g(2)$.

c. $h(t) = {}^-16t^2 + 25t + 6$; evaluate $h(^-1)$.

d. $p(x) = \sqrt{x+5}$; evaluate $p(4)$.

e. $q(x) = 4x^3 - 2x$; evaluate $q(^-2)$.

f. $r(x) = \frac{9}{x-4}$; evaluate $r(1)$.

Section 8.2 Graphs of Functions

For Exercises 8–10, use the vertical-line test to determine which graphs represent functions.

8.

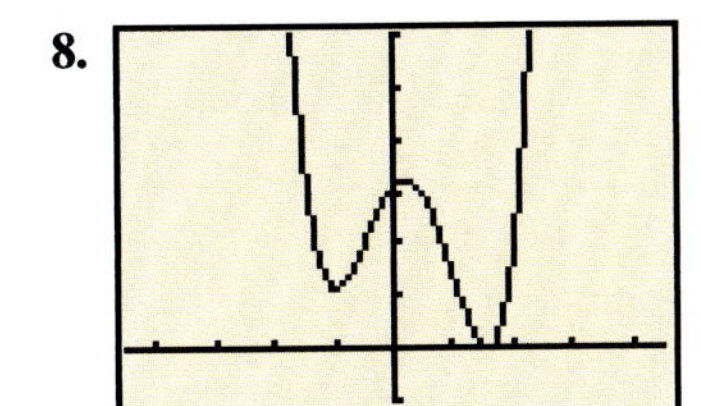

9.

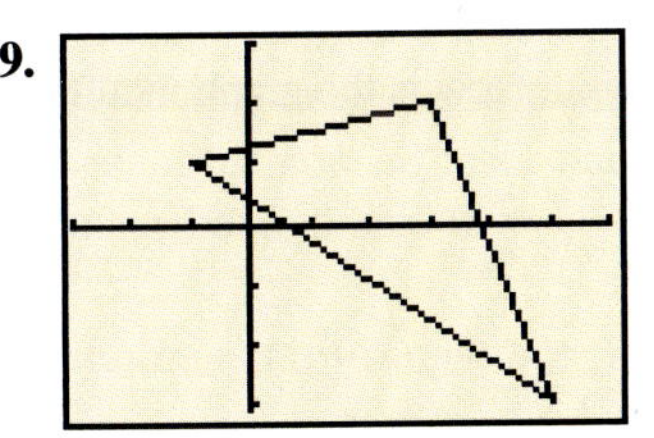

10.

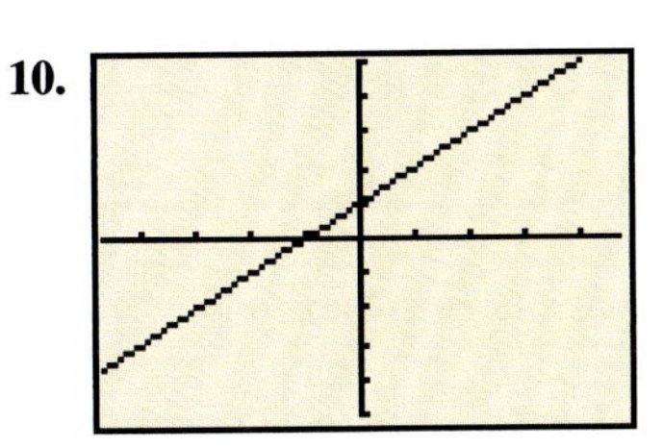

11. Use set notation to write the domain and range for the bar graph.

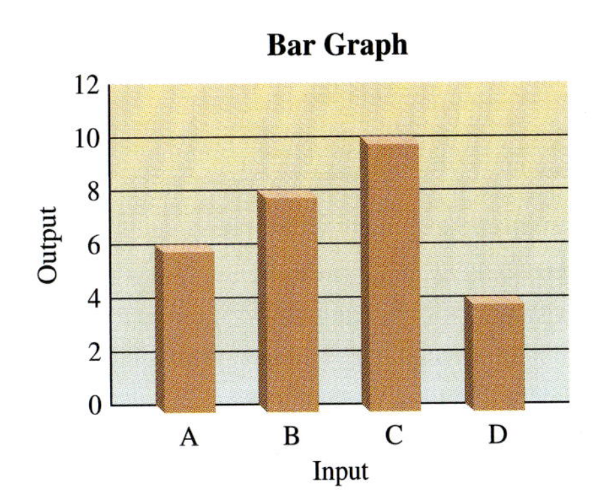

12. Use set notation to write the domain and range for the dot plot. Each tick is one unit.

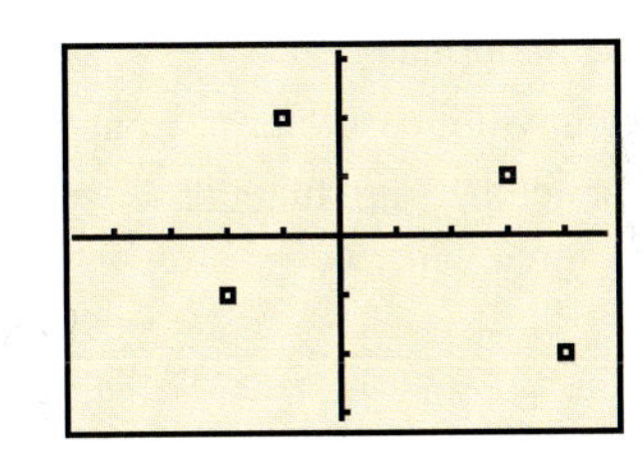

13. Let $f(x) = x^2 - 1$. Show the domain and range.

a. Use a number line.

b. Use set-builder notation.

14. Express the domain and range of each function in set-builder notation.

a. $f(x) = \sqrt{x-2}$

b. $g(x) = \sqrt{x+2}$

c. $h(x) = \dfrac{1}{x+2}$

d. $j(x) = \dfrac{1}{x-1}$

Section 8.3 Important Functions

15. Review the six basic functions introduced in this section.

a. Name the quadrants through which each function passes.

b. Which of the functions do not pass through the origin?

16. Use the graph to estimate the local extrema.

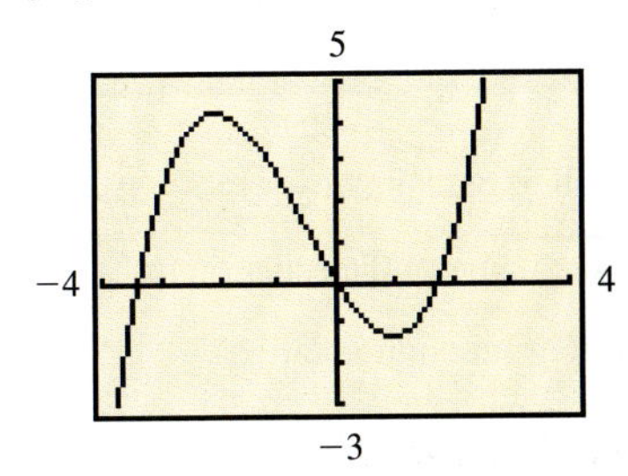

17. Graph

$$f(x) = \frac{^-x^3}{3} + x^2.$$

Approximate any local extrema.

18. It costs \$30 for a printer cartridge for a computer. Thus the cost (c) of operating the printer is a function of the number of cartridges (p) used and is given by $c(p) = 30p$. The number of cartridges used depends upon the number of pages (s) printed. Each cartridge lasts 1500 pages, so the number of printing cartridges is given by

$$p(s) = \frac{s}{1500}.$$

a. Suppose you print 4500 pages. How much will it cost to operate the printer?

b. How much will it cost to print s pages?

c. Write the composite function $c(p(s))$. Does this function agree with your answer from part (b)? Explain.

19. Let $f(x) = x + 2$ and $g(x) = x^2 - 4$. Find each composition.

a. $f(g(2))$ **b.** $f(g(3))$

c. $f(g(x))$ **d.** $g(f(0))$

e. $g(f(^-2))$ **f.** $g(f(x))$

20. Let $j(x) = \frac{1}{x}$ and $k(x) = x - 3$. Find each composition.

a. $k(j(^-1))$ **b.** $k(j(x))$

c. $j(k(2))$ **d.** $j(k(x))$

Section 8.4 Modifying Basic Functions

21. Match each modified absolute value function with the appropriate graph.

a. $y = |x| + 4$ **b.** $y = |x - 3|$

c. $y = ^-|x|$ **d.** $y = 4|x|$

e. $y = |x + 2| - 1$

1.

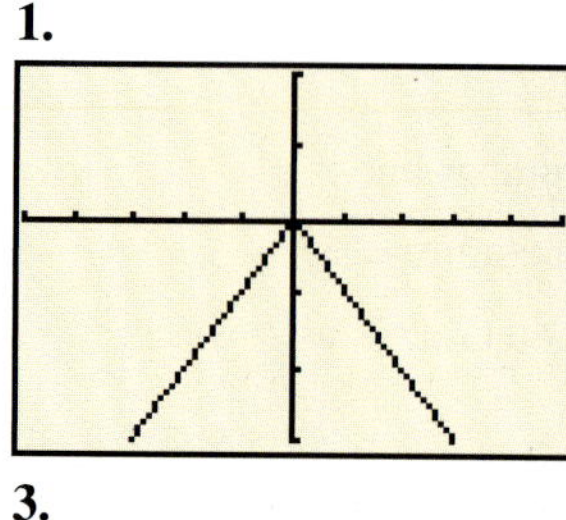

2.

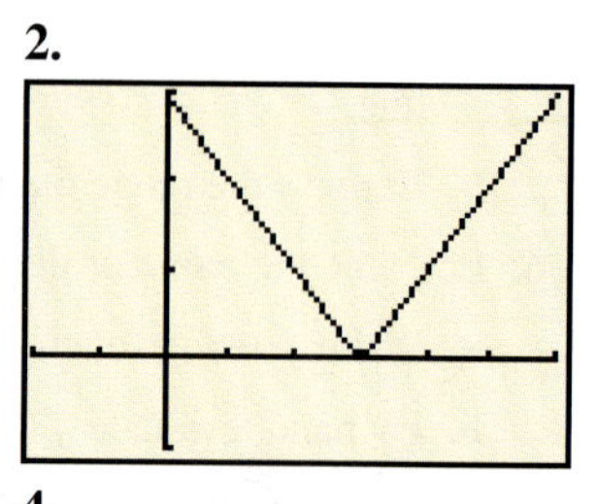

3.

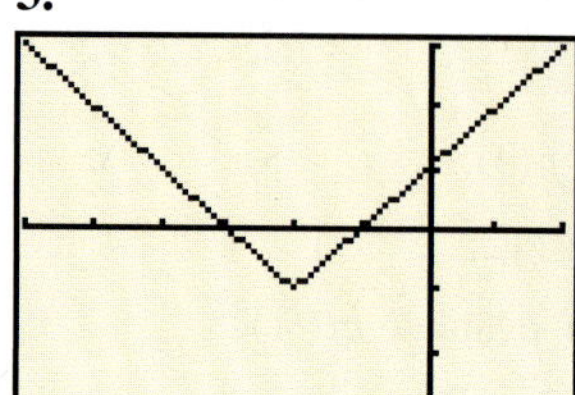

4.

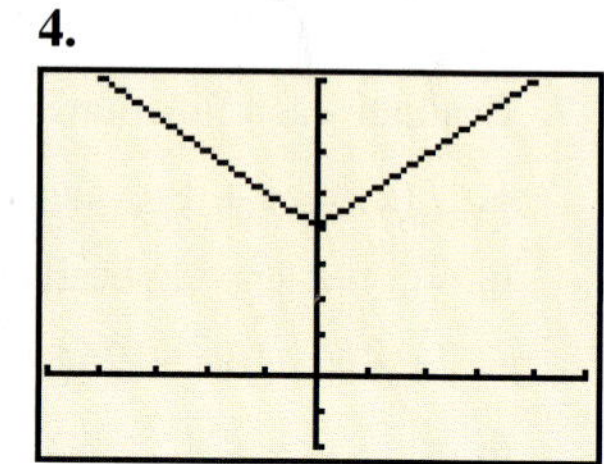

5.

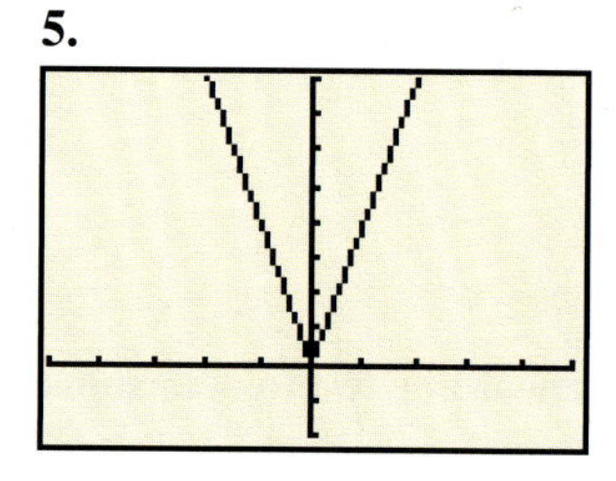

22. Describe how the graph of $f(x) = \frac{1}{x}$ is changed by each modification.

a. Add 3 to the input: $y = f(x + 3)$.

b. Subtract 3 from the output: $y = f(x) - 3$.

c. Multiply the output by $^-1$: $y = ^-f(x)$.

d. Multiply the output by $\frac{1}{10}$: $y = \frac{1}{10}f(x)$

23. Add 4 to the input and subtract 2 from the output of the basic square root function. How is the graph modified?

24. Review Examples 10 and 11 of Section 8.4 to describe how the graph of $f(x) = x^2$ is modified to become the graph of $g(x) = ^-16(x - 2)^2$.

25. How does the graph of $g(x) = (x + 1)^3 - 5$ relate to the graph of $f(x) = x^3$?

CHAPTER 8 ▪ TEST

1. Use a graph to explain why $y = x^3 - 3x^2$ is a function.

2. Let g be the name of the function in Exercise 1.

a. Write g in function notation.

b. Evaluate $g(2)$ by hand.

c. Evaluate $g(^-3.27)$ on your calculator.

3. Express the domain and range of the function $f(x) = \frac{1}{x+1}$ on a number line.

4. Use the table of values.

Input	Output
1	3
2	0
3	3
4	12

a. Draw a connection picture. Is this a function? Explain.

b. Plot the inputs and outputs as ordered pairs. Does the graph of these points support your conclusion from part (a)? Explain.

5. Is this graph a function? Explain.

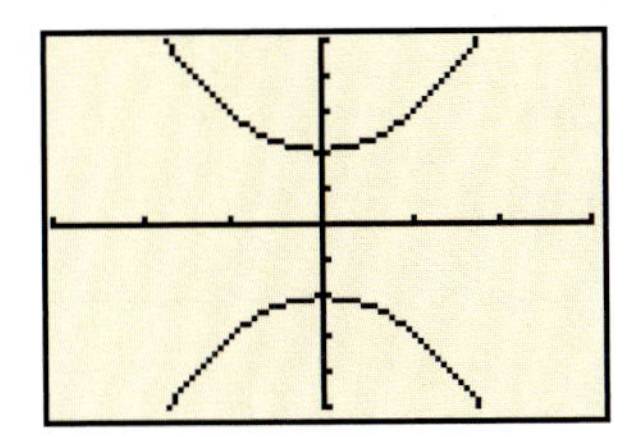

6. Here is a bar graph showing various types of Internet Access in the year 2000. (http://www.theinternet-analyst.com)

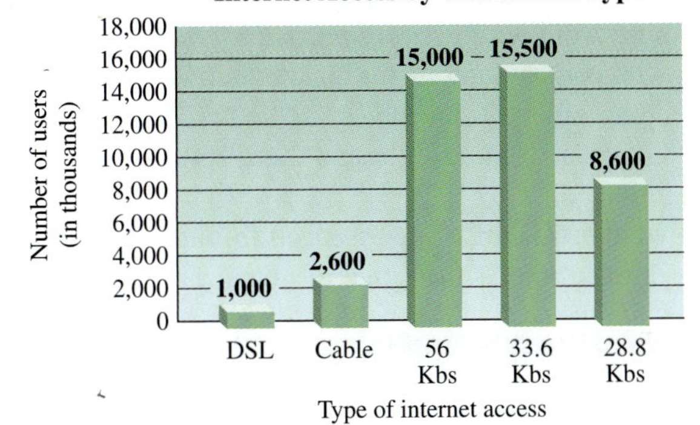

a. Does the bar graph represent a function? Explain.

b. Express the domain and range in a set.

7. Consider the basic cubic function, $f(x) = x^3$.

a. What are the x- and y-intercepts?

b. Are there any minimum or maximum points?

8. Sketch a graph of the modified absolute value function that has a maximum at $(3, {}^{-}4)$.

9. Suppose each cup of coffee requires 2 tablespoons of ground coffee. The number of cups (c) of coffee depends upon the number of tablespoons (t) of ground coffee and is given by $c(t) = \frac{t}{2}$. There are approximately 90 tablespoons of ground coffee per pound of beans. Thus, the number of tablespoons of coffee depends upon the pounds of coffee beans (b) and is given by $t(b) = 90b$.

a. Find the composite function $c(t(b))$ that determines the number of cups of coffee that can be made from b pounds of coffee beans.

b. How many cups of coffee can be made from $1\frac{3}{4}$ pounds of coffee beans?

10. Let $f(x) = \frac{1}{x}$ and $g(x) = x - 1$. Find each composition.

a. $f(g(4))$

b. $f(g(x))$

c. $g(f({}^{-}3))$

d. $g(f(x))$

11. How does the graph of $g(x) = 5\sqrt{x - 3}$ relate to the graph of $f(x) = \sqrt{x}$?

12. Describe how the graph of $f(x) = x$ is changed by each of the modifications.

a. Subtract 2 from the input: $y = f(x - 2)$.

b. Add 2 to the output: $y = f(x) + 2$.

c. Multiply the output by ${}^{-}1$: $y = {}^{-}f(x)$.

d. Multiply the output by 5: $y = 5f(x)$.

CHAPTER

9

Powers and Roots

Play That Tune

Mathematics has played an important role in the history of music. The first musical scales based on seven notes (the white keys on a piano) were studied by the followers of Pythagoras, who is most famous for his theorem about right triangles. The system was based on ratios of frequencies, so the two notes forming an octave (middle C to high C) are in a 1-to-2 ratio and those forming a perfect fifth (C to G) are in a 2-to-3 ratio.

A modification of the Pythagorean scale was used up through the Middle Ages, but the system had its drawbacks. As the range of notes was expanded to include the black keys on a piano, the scale could begin with any note. However, it became impossible to keep all the ratios consistent for all the scales and maintain intervals pleasing to the ear. This problem made the construction of keyboard instruments difficult. Organs were sometimes built with split black keys—for example, one for E flat and another for D sharp, two notes that are the same on modern keyboard instruments. But that arrangement still did not fully address the need for a consistent way to tune the instruments.

The eventual solution was to create a system of equal ratios between all 12 pairs of consecutive notes on the keyboard. In this chapter we study how mathematics was used to arrive at this solution, which musicians call *equal temperament*.

Need help? For on-line resources, visit this web site: **math.college.hmco.com/students.**

9.1 Rational Exponents and Radicals

AFTER STUDYING THIS SECTION YOU WILL BE ABLE TO

- Interpret the meaning of rational exponents.
- Use familiar powers to find values of rational exponent expressions without a calculator.
- Write an expression with square roots in simple radical form.

Extending the Definition of Exponent

In Chapter 2 we extended the definition of exponent to include zero and negative numbers. In Chapter 6 we developed the laws of exponents. These results are summarized as follows:

- *Zero exponent* Let a be any number except 0. Then $a^0 = 1$.
- *Negative integer exponents* Let a be any number except 0 and let n be any positive integer. Then

$$a^{-n} = \frac{1}{a^n}$$

- *Product law for exponents* Let a be any number and let m and n be integers. Then $a^m * a^n = a^{m+n}$.
- *Power-of-a-power law for exponents* Let a be any number, and let m and n be integers. Then $(a^m)^n = a^{m*n}$.
- *Power-of-a-product law for exponents* Let a and b be any numbers, and let m be an integer. Then $(a * b)^m = a^m * b^m$.
- *Quotient law for exponents* Let a be any number, and let m and n be integers. Then

$$\frac{a^m}{a^n} = a^{m-n}.$$

- *Power-of-a-quotient law for exponents* Let a be any number, let b be any number except 0, and let m be an integer. Then

$$\left(\frac{a}{b}\right)^m = \frac{a^m}{b^m}.$$

- *Negative exponents in the denominator* Let a be any number except 0, and let n be a positive integer. Then

$$\frac{1}{a^{-n}} = a^n$$

These rules tell us how to interpret many sorts of *integer* exponents but say nothing about exponents that are *rational numbers* (fractions). In this section we extend the definitions further to include all rational numbers. We begin by considering what it means to raise a number to the $\frac{1}{2}$ power.

EXAMPLE 1 ■ Find the value of $9^{1/2}$ and justify your answer.

Solution Let's start by looking at known powers of 9:

$$9^{-2} = \frac{1}{9^2} = \frac{1}{81} \approx .012$$
$$9^{-1} = \frac{1}{9^1} \approx .111$$
$$9^0 = 1$$
$$9^1 = 9$$
$$9^2 = 81$$
$$9^3 = 729$$

It is evident that as the exponent increases, 9 to the power of the exponent also increases. Because $\frac{1}{2}$ lies between 0 and 1, it is reasonable to conclude that $9^{1/2}$ must lie between 9^0 and 9^1; in other words, $1 < 9^{1/2} < 9$.

That narrows our range but still does not give us a solution. We are tempted to guess possible values for $9^{1/2}$ within our range. Here are some intelligent guesses students have made that turned out to be wrong.

1. Take half of 9. This gives $9^{1/2} = 4\frac{1}{2}$, which lies within our range. This is wrong, however. By the same reasoning we would have $9^{5/2} = \frac{5}{2} * 9 = 22\frac{1}{2}$. But from the chart above we can see that $9^{5/2}$ should lie between 81 and 729.
2. Take the average of 9^0 and 9^1. The average of 1 and 9 is 5. Again that lies within our range. This is also wrong. By that reasoning 9^1 should be the average of 9^0 and 9^2. The average of 1 and 81 is 41, not 9.

Neither of the methods proposed by the two incorrect guesses gives a result that is consistent when it is applied in other cases. To find a definition of $9^{1/2}$ that produces consistent results, we use the rules of exponents developed in Chapter 6 and listed at the beginning of this section.

Because we know that $9^1 = 9$, we can use the *power-of-a-power rule*, $(a^m)^n = a^{m*n}$ and think of the exponent as $1 = \frac{1}{2} * 2$. If we let $m = \frac{1}{2}$ and $n = 2$, then

$$(9^{1/2})^2 = 9^{(1/2)*2} = 9^1 = 9.$$

In other words, $9^{1/2}$ squared is 9, so

$$9^{1/2} = \sqrt{9} = 3$$

We might try $9^{1/2} = {}^-3$ because $({}^-3)^2 = 9$, but we choose the *principal square root* because ${}^-3$ is not within the range $1 < 9^{1/2} < 9$.

Our solution to Example 1 gives $9^{1/2} = 3$, the *principal* (nonnegative) square root of 9. We can generalize this for numbers other than 9, because for any nonnegative number a, $(\sqrt{a})^2 = a = (a^{1/2})^2$ (using the power-of-a-power rule.) Thus, it makes sense to adopt the following definition.

Definition Half-Power

For all nonnegative values of a ($a \geq 0$), $a^{1/2} = \sqrt{a}$.

Example 2 illustrates this definition.

EXAMPLE 2 Use the table feature of a graphing calculator to show that $\sqrt{x}$ and $x^{1/2}$ are the same function.

Solution In the Y= menu, set Y1 = $\sqrt{}$X and Y2 = X ∧ (1/2). In 2ND TBLSET, set TblStart = 0 and ΔTbl = 1. Then 2ND TABLE will give the following:

X	Y1	Y2
0	0	0
1	1	1
2	1.4142	1.4142
3	1.7321	1.7321
4	2	2
5	2.2361	2.2361
6	2.4495	2.4495

X=0

Scrolling down the table will give the same result: Y1 = Y2 for all values of x.

nth roots and $a^{1/n}$

The method we used to discover the meaning of $a^{1/2}$ can be extended to find $a^{1/n}$, where n is any positive integer. In this section we assume that a is a nonnegative number. In the next section we examine under what circumstances $a^{1/n}$ has meaning for negative values of a.

Again we use the power-of-a-power rule: $(a^{1/n})^n = a^{(1/n)*n} = a^1 = a$. This leads to the following definition:

Definition Unit-Fraction Power

For all nonnegative values of a ($a \geq 0$), $a^{1/n}$ is defined to be the nonnegative value of x that satisfies the equation $x^n = a$. When $n = 2$ we call this number the **principal square root** of a, and in general this number is called the **principal nth root** of a and is written $\sqrt[n]{a}$. The symbol $\sqrt{}$ is called a **radical**; n is called the **index** and a is called the **radicand**. When the index is 2, the radical is a square root, and the index may be omitted.

EXAMPLE 3 Evaluate the following expressions without using a calculator.

a. $32^{1/5}$
b. $125^{1/3}$
c. $\sqrt[4]{81}$
d. $\sqrt[6]{64}$

Solution You should not need a calculator because each of these problems involves a power that you should recognize. A table of familiar powers is given next.

Definition Familiar Powers

$2^2 = 4$	$2^3 = 8$	$2^4 = 16$	$2^5 = 32$	$2^6 = 64$
$3^2 = 9$	$3^3 = 27$	$3^4 = 81$	$3^5 = 243$	
$4^2 = 16$	$4^3 = 64$	$4^4 = 256$		
$5^2 = 25$	$5^3 = 125$	$5^4 = 625$		
$6^2 = 36$	$6^3 = 216$			
$7^2 = 49$				
$8^2 = 64$				
$9^2 = 81$				
$10^2 = 100$	$10^3 = 1000$	$10^4 = 10{,}000$	$10^5 = 100{,}000$	$10^6 = 1{,}000{,}000$

Using this table, we can easily solve each problem.

a. $2^5 = 32$, so $32^{1/5} = 2$.
b. $5^3 = 125$, so $125^{1/3} = 5$.
c. $3^4 = 81$, so $\sqrt[4]{81} = 81^{1/4} = 3$.
d. $2^6 = 64$, so $\sqrt[6]{64} = 64^{1/6} = 2$.

EXAMPLE 4 The twelfth root of 2 plays an important role in music. It represents the ratio of the frequencies of two adjacent keys on a piano. Evaluate $\sqrt[12]{2}$ to the nearest 0.0001.

Solution The simplest method is to use the definition of $a^{1/n}$ and the exponent key on the calculator.

Caution: Because of the rule for order of operations, the fractional exponent must be enclosed in parentheses. The keystrokes are 2 ∧ (1/12). The calculator gives $2^{1/12} \approx 1.0595$. An alternative is to use the xth-root feature under the MATH menu. The keystrokes are $12\ {}^x\!\sqrt{2}$. The calculator gives $\sqrt[12]{2} \approx 1.0595$.

Rational Exponents

We have found a meaning for fractional exponents when the exponent is a fraction with numerator 1, such as $\frac{1}{2}$ or $\frac{1}{5}$. We can generalize our definition to include any rational number as an exponent. Again, we use the power-of-a-power rule. Because $\frac{m}{n} = m(\frac{1}{n}) = (\frac{1}{n})m$, we can make the following definition.

Definition **Rational Power**

For all nonnegative values of a and for all positive integers m and n, $a^{m/n} = (a^m)^{1/n} = (a^{1/n})^m$.

EXAMPLE 5 Evaluate each of these expressions. For familiar powers, find an exact answer without a calculator. Otherwise, use a calculator to evaluate to the nearest .001.

a. $9^{5/2}$

b. $8^{2/3}$

c. $10{,}000^{3/4}$

d. $\left(\frac{27}{64}\right)^{2/3}$

e. $3^{5/6}$

f. $10^{3/2}$

Solution **a.** $9^{5/2} = (9^{1/2})^5 = 3^5 = 243$

Notice that $\frac{5}{2}$ is between 2 and 3 and that 243 lies between $9^2 = 81$ and $9^3 = 729$. Thus, using the same reasoning as in Example 1, we may conclude that 243 is a reasonable result.

b. $8^{2/3} = (8^{1/3})^2 = 2^2 = 4$

Because $\frac{2}{3} < 1$, it stands to reason that $8^{2/3} < 8^1$.

c. $10{,}000^{3/4} = (10{,}000^{1/4})^3 = 10^3 = 1000$

You may prefer to work an example like this by rewriting 10,000 as a power of 10 and applying the rules of exponents:

$$10{,}000^{3/4} = (10^4)^{3/4} = 10^{4*3/4} = 10^3 = 1000$$

d. $\left(\frac{27}{64}\right)^{2/3} = \left(\left(\frac{27}{64}\right)^{1/3}\right)^{2} = \left(\frac{3}{4}\right)^{2} = \frac{9}{16}$

e. Because 3 is not the sixth power of a rational number, we can't use a familiar power. We can find a decimal approximation with a calculator:

$$3 \wedge (5/6) \approx 2.498$$

To check for reasonableness, notice that $\frac{5}{6}$ is slightly less than 1 and 2.498 is slightly less than 3^1.

f. This is similar to part (e). The calculator gives

$$10 \wedge (3/2) \approx 31.623$$

Notice that $\frac{3}{2}$ lies between 1 and 2 and that 31.623 is between 10^1 and 10^2.

Simple Radical Form

Look at Example 5 (f) again: 31.623 is an approximation of the expression $10^{3/2}$ but not an *exact* value, because $10 \wedge (3/2)$ is an irrational number. To express the exact value of an irrational number such as $10^{3/2}$, it is customary to use radical notation.

EXAMPLE 6 ■ Express $10^{3/2}$ in radical notation.

Solution There are several ways to express $10^{3/2}$. Think about which properties of exponents are used in each case:

$$10^{3/2} = (10^{1/2})^3 = (\sqrt{10})^3$$
$$10^{3/2} = (10^3)^{1/2} = (1000)^{1/2} = \sqrt{1000}$$
$$10^{3/2} = (10^{1/2})^3 = (10^{1/2})^{2+1} = (10^{1/2})^2 * (10^{1/2})^1 = 10 * \sqrt{10}$$

Of the three expressions, the third one is preferred because it is in *simple radical form*. Writing an irrational number in simple radical form may be compared to writing a rational number as a simplified fraction. Recall that many different fractions may be used to express the same rational number. For instance, $\frac{12}{18}$ and $\frac{4}{6}$ are both equivalent to $\frac{2}{3}$; $\frac{2}{3}$ is the simplified fraction because the numerator and denominator have no common factor.

Similarly, we have a rule for simplifying irrational numbers that may be expressed with radicals.

Definition Simple Radical Form

An expression consisting of a single term is in **simple radical form** if the following are true.

- It contains only one radical.
- The radicand is an integer.
- The radical does not appear in the denominator of a fraction.
- The radical is not raised to a power.
- Both the radicand and the index take on the smallest possible absolute values.

Although this rule applies to all radical expressions, in this book we simplify only expressions involving square roots. It will take a little practice to manipulate an expression so that it appears in simple radical form. Here are a few examples to get you started. Notice that it helps to break an integer into prime factors. You may want to review how to make a factor tree on page 225.

EXAMPLE 7 Express each in simple radical form.

a. $\sqrt{75}$
b. $\sqrt{72}$
c. $\sqrt{21}$
d. $\frac{1}{\sqrt{3}}$
e. $\sqrt{\frac{5}{2}}$

Solution

a. $\sqrt{75} = 75^{1/2}$ *Rewriting with a fractional exponent*

$= (25 * 3)^{1/2}$ *Factoring $75 = 5 * 5 * 3 = 25 * 3$*

$= 25^{1/2} * 3^{1/2}$ *Using the power-of-a-product law*

$= 5\sqrt{3}$

This example shows how the power-of-a-product law may be applied to square roots. We can save a few steps by using radicals throughout.

$$\begin{aligned}\sqrt{75} &= \sqrt{25 * 3}\\ &= \sqrt{25} * \sqrt{3}\\ &= 5\sqrt{3}\end{aligned}$$

Notice that the factor 5 appeared twice to form a perfect square, 25. When there is a perfect square inside the radical, we may pull it out as we did here to get $5\sqrt{3}$.

b. $\sqrt{72} = \sqrt{2 * 2 * 2 * 3 * 3}$ *Writing the prime factors of 72*

$= \sqrt{4 * 9 * 2}$ *Arranging factor pairs to form perfect squares*

$= 2 * 3 * \sqrt{2}$ *Applying the power-of-a-product law for the power $\frac{1}{2}$*

$= 6\sqrt{2}$

c. $\sqrt{21} = \sqrt{3 * 7}$ *Writing the prime factors of 21*

We can stop here. No prime factor appears twice, so we cannot form a perfect square. $\sqrt{21}$ is already in simple radical form.

d. $\sqrt{3}$ is already in simple radical form, but it appears in the denominator. We can **rationalize** the denominator by multiplying both numerator and denominator by $\sqrt{3}$.

$$\frac{1}{\sqrt{3}} = \frac{1 * \sqrt{3}}{\sqrt{3} * \sqrt{3}} = \frac{\sqrt{3}}{3}$$

e. $\sqrt{\frac{5}{2}} = \left(\frac{5}{2}\right)^{1/2}$ *Rewriting with a fractional exponent*

$= \frac{5^{1/2}}{2^{1/2}}$ *Using the power-of-a-quotient law*

$= \frac{\sqrt{5}}{\sqrt{2}}$ *Rewriting with radicals*

$= \frac{\sqrt{5} * \sqrt{2}}{\sqrt{2} * \sqrt{2}}$ *Multiplying both numerator and denominator in order to rationalize the denominator*

$= \frac{\sqrt{10}}{2}$ *Simplifying numerator and denominator*

Each of these results may be checked using a calculator. For instance, for (e) the calculator gives $(\frac{5}{2})^{1/2} \approx 1.581$ and $\frac{\sqrt{10}}{2} \approx 1.581$.

In Chapter 7 we saw how to use the quadratic formula to find solutions to the equation $ax^2 + bx + c = 0$. When the discriminant, $b^2 - 4ac$, is not a perfect square, we can express the answer in simple radical form.

EXAMPLE 8 Find solutions to the equation $x^2 - 10x + 5 = 0$. Estimate the answers and express them in simple radical form.

Solution Substitute $a = 1$, $b = {}^-10$ and $c = 5$ in the quadratic formula.

$$\begin{aligned} x &= \frac{^-b}{2a} \pm \frac{\sqrt{b^2 - 4ac}}{2a} \\ &= \frac{^-(^-10)}{2*1} \pm \frac{\sqrt{(^-10)^2 - 4*1*5}}{2*1} \\ &= 5 \pm \frac{\sqrt{80}}{2} \end{aligned}$$

The two solutions to the equation are $5 + \frac{\sqrt{80}}{2}$ and $5 - \frac{\sqrt{80}}{2}$. Because $\sqrt{80} \approx \sqrt{81} = 9$, we may estimate the roots as approximately equal to $5 + \frac{9}{2}$ and $5 - \frac{9}{2}$, that is, 9.5 and .5.

To obtain simple radical form, we need to simplify $\sqrt{80}$. The prime factorization of 80 is $2^4 * 5^1$. Thus, we have

$$\sqrt{80} = \sqrt{16*5} = \sqrt{16} * \sqrt{5} = 4\sqrt{5}$$

Substituting into our answers, we get

$$5 + \frac{4\sqrt{5}}{2} = 5 + 2\sqrt{5}$$

and

$$5 - \frac{4\sqrt{5}}{2} = 5 - 2\sqrt{5}$$

Exercises 9.1

1. Find each power.

$$16^{-1} = ?$$
$$16^0 = ?$$
$$16^1 = ?$$

   a. Suppose you did not know how to find the value of $16^{1/2}$. Use your values from the list above to give a whole number interval in which $16^{1/2}$ must fall.
   b. Write $16^{1/2}$ in an equivalent form with a radical.
   c. Use your answer from part (a) to explain why $16^{1/2}$ equals 4 but not $^-4$.
   d. Are the solution(s) of $x^2 = 16$ the same as the solution(s) of $x = \sqrt{16}$? Explain.

2. Use the table feature on your calculator to show that $\sqrt[3]{x}$ and $x^{1/3}$ are the same. Describe the steps you used.

3. Write the exponential expressions as radical expressions or the radical expressions as exponential expressions.
   a. $7^{5/2}$ **b.** $13^{1/3}$
   c. $(\sqrt{5})^3$ **d.** $\sqrt[5]{11}$

4. **a.** Use the definition of unit-fraction power to identify the index, radical, and radicand of $\sqrt[3]{27}$.
   b. Write $\sqrt[3]{27}$ in an equivalent form with a rational exponent.
   c. Evaluate $\sqrt[3]{27}$ without a calculator.

5. The expressions $\sqrt{(^-5)^2}$ and $(\sqrt{^-5})^2$ appear to be similar. Use the order of operations (Section 2.1), to explain why $\sqrt{(^-5)^2}$ equals 5 but $(\sqrt{^-5})^2$ is undefined.

Why, however, are both $\sqrt[3]{(^-5)^3}$ and $(\sqrt[3]{^-5})^3$ defined and equal to $^-5$?

6. The prime factorization of 128 is 2^7. Use this fact to find $128^{1/7}$.

7. Write each base or radicand in factored form and evaluate each expression without a calculator.
   a. $16^{1/4}$ b. $216^{1/3}$
   c. $\sqrt[3]{64}$ d. $\sqrt[5]{32}$

8. Use your calculator to approximate the value of the sixth root of 10 to the nearest .001. Explain why your decimal answer is only an approximation.

9. A cube is a square box with equal length, width, and height. Suppose a cube has a volume of 20 cm^3; find the dimensions of the cube to the nearest .1 cm.

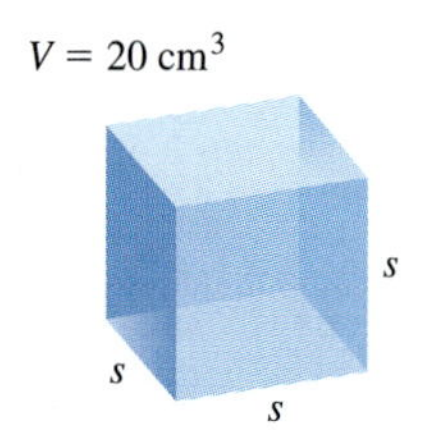

10. Use the definition of rational power to rewrite $4^{3/2}$ as a power-of-a-power. Evaluate the expression without a calculator.

11. Evaluate each expression. For familiar powers, find an exact answer without a calculator. Otherwise, use a calculator to evaluate to the nearest .01.
    a. $27^{2/3}$ b. $16^{3/4}$ c. $.04^{5/2}$
    d. $(\frac{9}{25})^{3/2}$ e. $7^{5/2}$

12. Use a factor tree to write each radicand as a product of primes. Then rewrite each expression in simple radical form.
    a. $\sqrt{18}$ b. $\sqrt{99}$ c. $\sqrt{112}$
    d. $\sqrt{500}$ e. $\sqrt{40}$ f. $\sqrt{35}$

13. Write these expressions in simple radical form without any radical in the denominator.
    a. $\frac{1}{\sqrt{5}}$ b. $\frac{3}{\sqrt{2}}$ c. $\frac{\sqrt{12}}{\sqrt{3}}$
    d. $\frac{\sqrt{14}}{\sqrt{5}}$ e. $\sqrt{\frac{3}{7}}$ f. $\sqrt{\frac{5}{6}}$

14. Solve these quadratic equations. Express the solutions as exact numbers in simple radical form.
    a. $2x^2 = 96$ b. $x^2 - 6x + 3 = 0$
    c. $2x^2 - 2x - 5 = 0$ d. $x^2 + 6x = 14$

15. In Exercise 11(e) we found an approximation for the expression $7^{5/2}$. Now we wish to find an exact expression.
    a. Refer to Example 6 and write $7^{5/2}$ in simple radical form.
    b. Find an approximation for your answer in part (a) to the nearest .01, and show that it is the same as your answer in Exercise 11(e).

Skills and Review 9.1

For Exercises 16–19, use the function $g(x) = \sqrt{x - 4}$.

16. How does the graph of $g(x)$ relate to the graph of $f(x) = \sqrt{x}$?

17. Evaluate each of the following.
    a. $g(4)$
    b. $g(3)$ (Why do you get an error?)

18. Show the domain of $g(x)$ in set-builder notation.

19. Show the range of $g(x)$ on a number line.

20. Let $f(x) = 4x^2$ and $g(x) = \sqrt{x}$.
    a. Find $f(g(9))$.
    b. Find $g(f(^-5))$.

21. Find the exact solutions of $c^2 + 8c = {}^-15$.

22. Solve for x: $(3x - 5)(x + 2) = 0$.

23. Factor each expression completely.
    a. $x^2 - 10x + 24$
    b. $^-4x^2 + 36$
    c. $10x^2 - 35x$

24. Expand.
    a. $(4 + x^3)^2$ b. $(4x^3)^2$

25. Solve for x: $6x - 7(1 + x) = 5 - 3x$.

9.2 Power Functions and Their Graphs

AFTER STUDYING THIS SECTION YOU WILL BE ABLE TO

- Use the odd- and even-root properties to solve equations.
- Use the odd- and even-power properties to solve equations.
- Identify extraneous roots of an equation.
- Demonstrate that two functions are inverses of each other.

Power Functions of the Form $y = f(x) = x^n$

Figure 1 shows graphs of $y = f(x) = x^2$, $y = f(x) = x^3$, $y = f(x) = x^4$, $y = f(x) = x^5$. These are called **power functions** because in each case x is raised to a known power. From these graphs we can draw the following conclusions about power functions, functions of the form $y = f(x) = x^n$, where n is a positive integer.

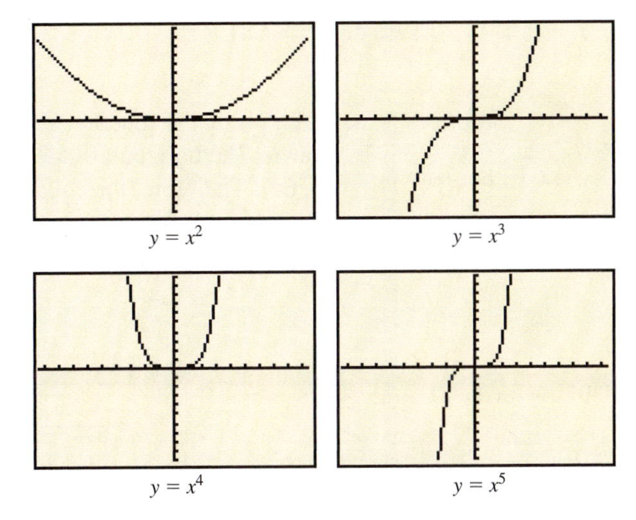

Figure 1 Windows: Xmin = $^-9.4$, Xmax = 9.4, Ymin = $^-100$, Ymax = 100

Definition Power Functions of the Form $f(x) = x^n$, Where n Is a Positive Integer

1. The domain of all these functions is the set of all real numbers.
2. When n is odd, the range is the set of all real numbers. When n is even, the range is the set of y such that $y \geq 0$. This makes sense, because

(continued)

when we raise a negative number to an even power, we get a positive number, but when we raise a negative number to an odd power, we get a negative number.

3. When n is odd, the graph of the function lies in the first and third quadrants. When n is even, the graph lies in the first and second quadrants.

Solving Equations of the Form $x^n = a$

Example 1 illustrates how to solve equations of the form $x^n = a$.

EXAMPLE 1 Solve each equation by graphing.

a. $x^3 = 64$
b. $x^3 = {}^-64$
c. $x^4 = 81$
d. $x^4 = {}^-81$

Solution **a.** In Figure 2, the functions Y1 = X ∧ 3, Y2 = 64, and Y3 = ⁻64 have been drawn. The horizontal line Y2 = 64 intersects the graph of Y1 = X ∧ 3 at (4, 64). Therefore, the equation $x^3 = 64$ has one solution, $x = 4$.

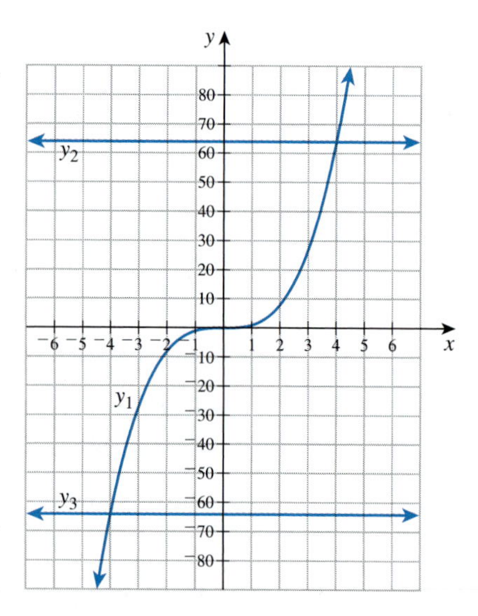

Figure 2

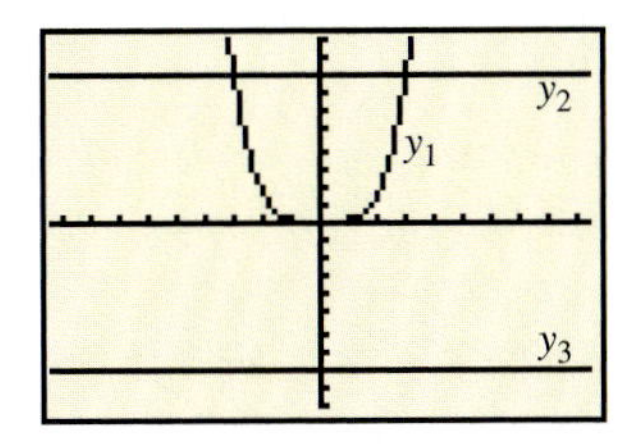

Figure 3

b. Likewise the horizontal line $Y3 = {}^-64$ intersects the graph of $Y1 = X \wedge 3$ at $({}^-4, {}^-64)$. Therefore, the equation $x^3 = {}^-64$ has one solution, $x = {}^-4$.

c. Figure 3 shows the functions $Y1 = X \wedge 4$, $Y2 = 81$, and $Y3 = {}^-81$. $Y2 = 81$ intersects the graph of $Y1 = X \wedge 4$ in two points, $(3, 81)$ and $({}^-3, 81)$. Therefore, the equation $x^4 = 81$ has two solutions, $x = 3$ and $x = {}^-3$.

d. $Y3 = {}^-81$ does not intersect the graph of $Y1 = X \wedge 4$. Therefore equation (d) has no solution.

According to the definition of unit-fraction power, the solution $x = 4$ to the equation $x^3 = 64$ may also be written as the principal third root of 64, or $\sqrt[3]{64}$. It may also be expressed with a fractional exponent: $x = 4 = 64^{1/3}$.

Because the equation $x^3 = {}^-64$ has the solution $x = {}^-4$, it *makes sense* to define $\sqrt[3]{{}^-64} = ({}^-64)^{1/3} = {}^-4$. Because the graph of $y = x^3$ extends into the third quadrant, we are able to find a real solution to equations of the form $x^3 =$ some negative number and thus define cube roots for negative numbers.

The situation is different when we consider $y = x^4$ in the equations $x^4 = 81$ and $x^4 = {}^-81$. Because the graph of $y = x^4$ extends into the second quadrant, $x^4 = 81$ has two solutions. We can use the symbol $\pm$ to indicate the two solutions as $x = \pm 3$, as we did with quadratic equations in Chapter 7. Because 3 is the principal fourth root of 81, we can also write $x = \pm\sqrt[4]{81} = \pm 81^{1/4}$.

The equation $x^4 = {}^-81$ has no real solution because the graph of $y = x^4$ does not go below the x-axis. Therefore it does not make sense to define $\sqrt[4]{{}^-81}$ or $({}^-81)^{1/4}$. There is no real number that gives ${}^-81$ when raised to the fourth power.

The conclusions we have drawn from Example 1 are generalized as the odd- and even-root properties.

The Odd- and Even-Root Properties

Suppose n is a positive integer and a is a real number. Consider the equation $x^n = a$.

- If n is odd, then $x = a^{1/n}$.
- If n is even, then
  - If $a > 0$, $x = \pm a^{1/n}$;
  - If $a = 0$, then $x = 0$;
  - If $a < 0$, there is no real solution to the equation.

The odd- and even-root properties may be used to find algebraic solutions to some equations involving powers.

EXAMPLE 2 Solve these equations algebraically.

a. $(x+1)^5 + 4 = 36$
b. $4x^4 = 36$
c. $x^6 = {}^-100$

Solution **a.** $(x+1)^5 + 4 = 36$

$(x+1)^5 = 32$	*Adding* $^-4$ *on both sides*
$x + 1 = 32^{1/5}$	*Using the odd-root property*
$x + 1 = (2^5)^{1/5}$	*Using a known power of* 2
$x + 1 = 2$	*Using the power-of-a-power rule:* $5 * \frac{1}{5} = 1$
$x = 1$	*Adding* $^-1$ *on both sides*

b. $4x^4 = 36$

$x^4 = 9$	*Dividing on both sides by* 4
$x = \pm 9^{1/4}$	*Using the even-root property for* $a > 0$
$x = \pm(3^2)^{1/4}$	*Using a known power of* 3
$x = \pm 3^{1/2}$	*Using the power-of-a-power rule:* $2 * \frac{1}{4} = \frac{1}{2}$
$x = \pm\sqrt{3}$	

There are thus two solutions to the equation. In simplest radical form they are $\sqrt{3}$ and $^-\sqrt{3}$. Decimal approximations to these irrational solutions are $x \approx \pm 1.732$.

c. According to the even-root property for $a < 0$, there is no real solution to this equation.

Power Functions of the Form $y = x^{1/n}$

From the odd- and even-root properties, we can see that when n is odd, $x^{1/n}$ is defined for all values of x, but when n is even, $x^{1/n}$ is not defined for negative values of x. In Figure 4 the graphs of $y = x^{1/2}$, $y = x^{1/3}$, $y = x^{1/4}$, and $y = x^{1/5}$ are shown.

From these graphs and from the odd- and even-root properties we determine the domain and range for power functions when the power is a unit fraction.

Domain and Range for $y = f(x) = x^{1/n}$

Let n be a positive integer. Then the domain and range of $y = f(x) = x^{1/n}$ depend upon whether n is odd or even.

(*continued*)

	Domain	Range
n is odd	All real numbers	All real numbers
n is even	$x \geq 0$	$y \geq 0$

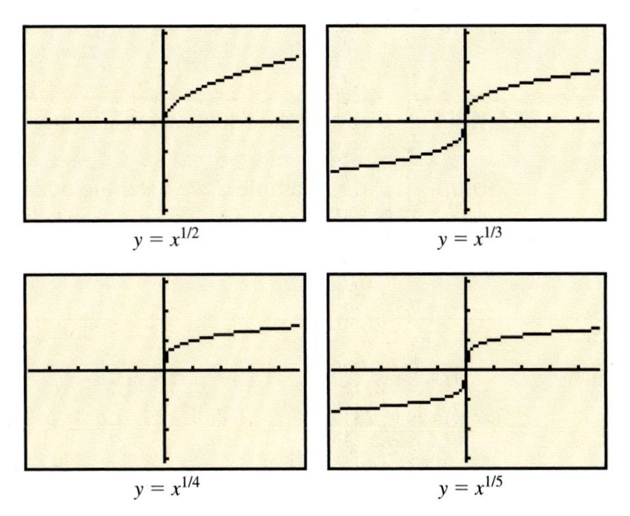

Figure 4 ZDecimal Windows: Xmin = $^-4.7$, Xmax = 4.7, Ymin = $^-3.1$, Ymax = 3.1

Example 3 provides further illustration of these ideas.

EXAMPLE 3 Solve these equations by graphing.

a. $x^{1/3} = 2$
b. $x^{1/3} = {}^-2$
c. $x^{1/2} = 1.5$
d. $x^{1/2} = {}^-1.5$

Solution

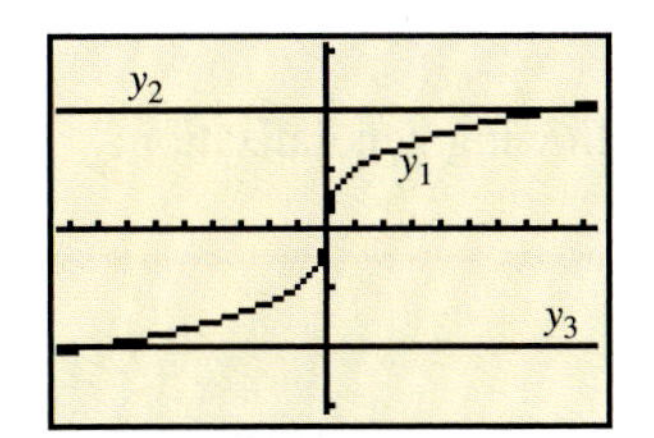

Figure 5 Xmin = $^-9.4$, Xmax = 9.4, Ymin = $^-3.1$, Ymax = 3.1

a. In Figure 5, the functions Y1 = X ∧ (1/3), Y2 = 2 and Y3 = $^-2$ have been drawn. The horizontal line Y2 = 2 intersects the graph of Y1 = X ∧ (1/3) at (8, 2). Therefore, $x^{1/3} = 2$ has one solution, $x = 8$.

b. Likewise the horizontal line Y3 = $^-2$ intersects the graph of Y1 = X ∧ (1/3) at ($^-8$, $^-2$). Therefore $x^{1/3} = {}^-2$ has one solution, $x = {}^-8$.

c. In Figure 6, the functions Y1 = X ∧ (1/2), Y2 = 1.5 and Y3 = $^-1.5$ are shown. The graphs of Y1 and Y2 intersect at (2.25, 1.5). Therefore, $x = 2.25$ is the solution to the equation $x^{1/2} = 1.5$.

d. Because the graphs of Y1 and Y3 do not intersect, the equation $x^{1/2} = {}^-1.5$ has no solution.

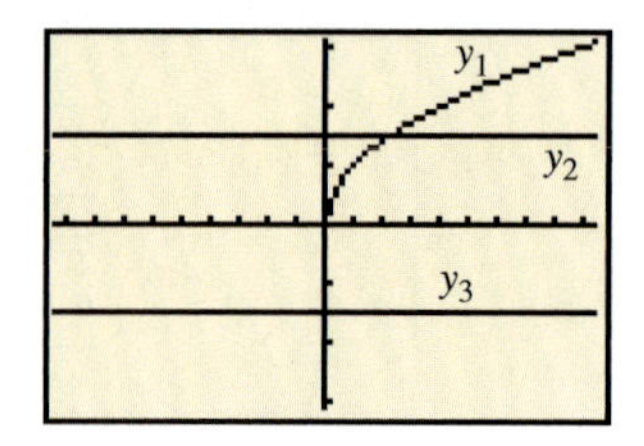

Figure 6 (Xmin = $^-9.4$, Xmax = 9.4, Ymin = $^-3.1$, Ymax = 3.1)

EXAMPLE 4 ■ Solve the equations in Example 3 by an algebraic method. Check each solution.

Solution In Example 2 we used the odd- and even-root properties to undo the raising of a number to an integer power by raising both sides of the equation to the reciprocal of that power. Let us try a similar technique with these equations.

a.	b.	c.	d.	
$x^{1/3} = 2$	$x^{1/3} = {}^-2$	$x^{1/2} = 1.5$	$x^{1/2} = {}^-1.5$	*Cubing both sides*
$(x^{1/3})^3 = 2^3$	$(x^{1/3})^3 = ({}^-2)^3$	$(x^{1/2})^2 = 1.5^2$	$(x^{1/2})^2 = ({}^-1.5)^2$	*Using the power-of-a-power rule*
$x^1 = 2^3$	$x^1 = ({}^-2)^3$	$x^1 = 1.5^2$	$x^1 = ({}^-1.5)^2$	
$x = 8$	$x = {}^-8$	$x = 2.25$	$x = 2.25$	
Check	Check	Check	Check	
$8^{1/3} \stackrel{?}{=} 2$	$({}^-8)^{1/3} \stackrel{?}{=} {}^-2$	$2.25^{1/2} \stackrel{?}{=} 1.5$	$2.25^{1/2} \stackrel{?}{=} {}^-1.5$	
$2 = 2$ True	${}^-2 = {}^-2$ True	$1.5 = 1.5$ True	$1.5 = {}^-1.5$ False	

Raising both sides to the same power seems to work for the equations in parts (a), (b), and (c). In each of these cases we get the solution that was found by graphing, and our solution checks when substituted into the original equation.

In the case of part (d), however, the method leads us to a mistake. We know from the graph that the equation has no solution; yet we appear to get $x = 2.25$. When we check this "solution," however, we end up with a false statement. The "solution" $x = 2.25$ is called an **extraneous** solution, meaning that it does not satisfy the original equation. We can see why this happens when we consider that the left side of the equation in part (d), $x^{1/2}$, is *defined to be the nonnegative square root of x, whereas the right side of the equation is a negative number.* ■ ■

Our experience in Example 4 leads us to the following generalization:

The Odd- and Even-Power Properties

Suppose n is a positive integer and a is a real number. Consider the equation $x^{1/n} = a$.

(*continued*)

- If n is odd, then $x = a^n$.
- If n is even, then
  - If $a \geq 0$, $x = a^n$;
  - If $a < 0$, there is no real solution to the equation. The "solution" a^n is an extraneous solution.

Of course, it is always a good idea to check the solutions you find to an equation to make sure that they satisfy the equation. Our experience with part (d) in Example 4 indicates that when we raise both sides of an equation to an even power, it is absolutely necessary to check for possible extraneous solutions.

EXAMPLE 5 ■ Solve the equation $\sqrt{x+1} = x - 1$.

Solution The radical on the left side of the equation may be written as a fractional power, $(x+1)^{1/2}$. We can apply the even-power property and square both sides of the equation. In applying that property, the right side, $x - 1$ serves as a. The even-power property tells us that if $a \geq 0$, we have a solution, and that if $a < 0$, we do not have a solution. However, because a contains a variable, we do not know in advance whether it will be positive or negative. So we just go ahead and solve the equation and then check the solutions.

$$\sqrt{x+1} = x - 1$$

$$(x+1)^{1/2} = x - 1 \qquad \textit{Writing the radical as a fractional power}$$

$$\left((x+1)^{1/2}\right)^2 = (x-1)^2 \qquad \textit{Squaring both sides}$$

$$(x+1) = (x-1)(x-1) \qquad \textit{Using the power-of-a-power rule}$$

$$x + 1 = x^2 - 2x + 1 \qquad \textit{Expanding the right side}$$

$$0 = x^2 - 3x \qquad \textit{Adding } {}^{-}x \textit{ and } {}^{-}1 \textit{ on both sides}$$

$$0 = x(x-3) \qquad \textit{Factoring the right side}$$

Applying the zero product property gives $x = 0$ or $x = 3$. There are two solutions to check.

Check for $x = 0$	Check for $x = 3$
$\sqrt{x+1} = x - 1$	$\sqrt{x+1} = x - 1$
$\sqrt{0+1} \stackrel{?}{=} 0 - 1$	$\sqrt{3+1} \stackrel{?}{=} 3 - 1$
$\sqrt{1} \stackrel{?}{=} {}^{-}1$	$\sqrt{4} \stackrel{?}{=} 2$
$1 = {}^{-}1$ False	$2 = 2$ True

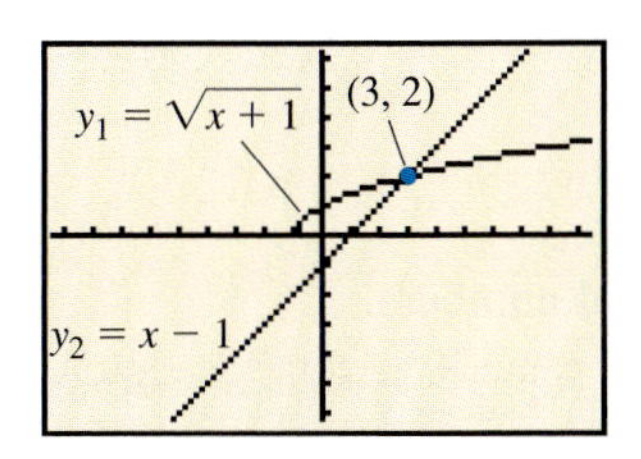

Figure 7

The solution $x = 0$ is extraneous and must be rejected. The solution $x = 3$ is a true solution and can be accepted. Figure 7 confirms the fact that the graphs of the functions represented by the two sides of the equation have just one point in common. That point, (3, 2), corresponds to the true solution, $x = 3$.

Inverse Functions

We see that when we have an equation of the form x^3 = some number, we can find a solution by taking the cube root of that number. Similarly, if we start with $x^{1/3}$ = some number, we find a solution by cubing the number. Cubing and finding the cube root are two operations that *undo* each other. The two functions $f(x) = x^3$ and $g(x) = x^{1/3}$ are called *inverses* of each other.

Definition Inverse Functions

$f(x)$ and $g(x)$ are **inverses** of each other if $f(g(x)) = x$ for every x in the domain of $g(x)$ and $g(f(x)) = x$ for every x in the domain of $f(x)$.

Think about what this definition means. Finding the composition of the two functions by performing one after the other is like doing nothing at all; one function cancels the effect of the other function.

You are familiar with operations that undo each other. For example, if you take a number such as 6 and multiply it by 3, the result is 18. If you now take 18 and divide it by 3, you are back to the original number, 6. In functional terms, suppose $f(x) = 3x$ and $g(x) = \frac{x}{3}$. Then

$$\begin{aligned} g(f(6)) &= g(3 * 6) \\ &= g(18) \\ &= \frac{18}{3} \\ &= 6 \end{aligned}$$

EXAMPLE 6 Show that these pairs of functions are inverses of each other.

a. $f(x) = x + 4$; $g(x) = x - 4$

b. $f(x) = 2x + 6$; $g(x) = \frac{1}{2}x - 3$

c. $f(x) = \frac{1}{x}$; $g(x) = \frac{1}{x}$

Solution **a.** The domain of both functions is the set of all real numbers.

$$f(g(x)) = f(x - 4) = (x - 4) + 4 = x$$

for all real numbers. Also

$$g(f(x)) = g(x+4) = (x+4) - 4 = x$$

for all real numbers.

b. The domain of both functions is the set of all real numbers.

$$f(g(x)) = f\left(\tfrac{1}{2}x - 3\right) = 2\left(\tfrac{1}{2}x - 3\right) + 6 = x - 6 + 6 = x$$

for all real numbers. Also,

$$g(f(x)) = g(2x+6) = \tfrac{1}{2}(2x+6) - 3 = x + 3 - 3 = x$$

for all real numbers.

c. The domain for both functions is the set of all real numbers x such that $x \neq 0$.

$$f(g(x)) = \frac{1}{g(x)} = \frac{1}{\frac{1}{x}} = x \text{ for all } x \neq 0$$

$$g(f(x)) = \frac{1}{f(x)} = \frac{1}{\frac{1}{x}} = x \text{ for all } x \neq 0$$

When two functions are inverses of each other, the output variable for the first function is the input variable for the second function, and vice versa. In other words, x and y swap places. The following table shows how tables for the two inverse functions $f(x) = x^3$ and $g(x) = x^{1/3}$ are related.

$f(x) = x^3$		$g(x) = x^{1/3}$	
Input (x)	Output (y)	Input (x)	Output (y)
$^-2$	$^-8$	$^-8$	$^-2$
$^-1$	$^-1$	$^-1$	$^-1$
$-\frac{1}{2}$	$-\frac{1}{8}$	$-\frac{1}{8}$	$-\frac{1}{2}$
0	0	0	0
$\frac{1}{2}$	$\frac{1}{8}$	$\frac{1}{8}$	$\frac{1}{2}$
1	1	1	1
2	8	8	2

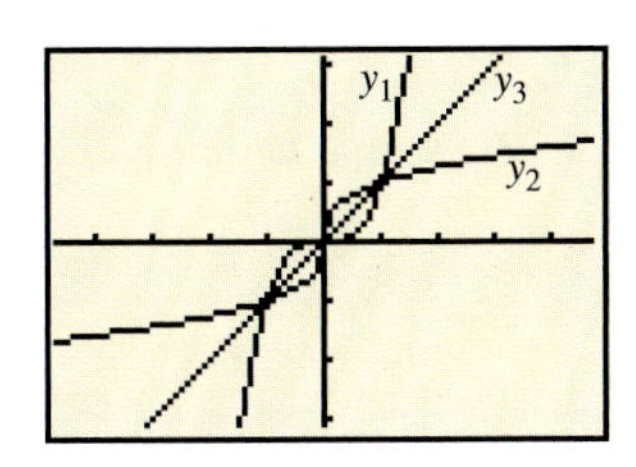

Figure 8 Y1 = X ∧ 3, Y2 = X ∧ (1/3), Y3 = X in ZoomDecimal window

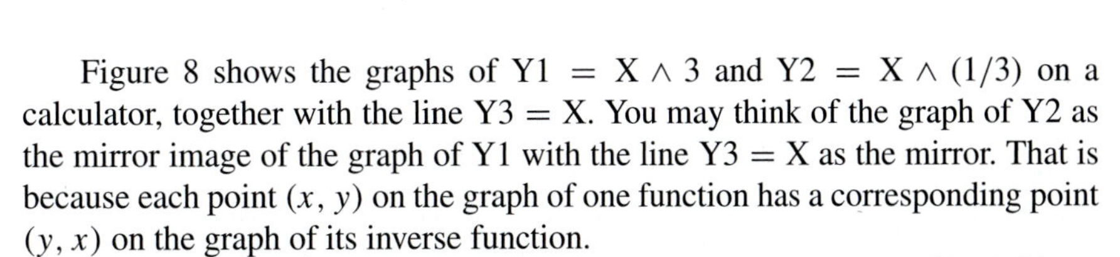

Figure 8 shows the graphs of Y1 = X ∧ 3 and Y2 = X ∧ (1/3) on a calculator, together with the line Y3 = X. You may think of the graph of Y2 as the mirror image of the graph of Y1 with the line Y3 = X as the mirror. That is because each point (x, y) on the graph of one function has a corresponding point (y, x) on the graph of its inverse function.

For example, we trace along Y1 to the point (1.3, 2.197). Theoretically, we should be able to find the point with the reversed coordinates, (2.197, 1.3), on the graph of Y2. Because of the limitations of the calculator, you can get only as close as (2.2, 1.3005914). These two points are on opposite sides of the mirror

Y3 = X and the same distance from it. This mirror-image property of $f(x) = x^3$ and $g(x) = x^{1/3}$ applies to every pair of inverse functions.

Definition Inverse Function Graph Property

Suppose $y = f(x)$ and $y = g(x)$ are inverse functions. Then the graph of $y = g(x)$ is the mirror image of the graph of $y = f(x)$ with the line $y = x$ as the mirror.

EXAMPLE 7 Explain why $f(x) = x^2$ and $g(x) = x^{1/2}$ are not considered to be inverse functions of each other.

Solution There are several ways to show this. First, according to the definition of inverse functions, $g(f(x))$ must equal x for every x in the domain of $f(x)$. Let's take a negative value of x, say, $x = {}^{-}2$. Then $g(f(x)) = \sqrt{({}^{-}2)^2} = \sqrt{4} = 2$. Because $2 \neq {}^{-}2$, the condition required by the definition is not met.

Figure 9 provides another way to look at the relationship between these two functions. If we imagine the line $y = x$ as our mirror, the right half of the parabola $y = x^2$ is the mirror image of the graph of $y = x^{1/2}$. The left half of $y = x^2$, however, has no corresponding points on the graph of $y = x^{1/2}$. Thus $y = x^{1/2}$ is not a complete inverse for the function $y = x^2$. If we flip the parabola to swap the axes, we get *more* than the graph of the square root, and if we flip the square root, we get *less* than the parabola.

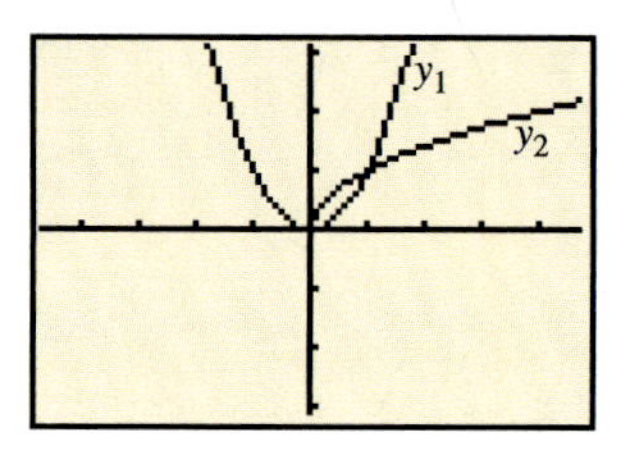

Figure 9 Y1 = X ∧ 2, Y2 = X ∧ (1/2), Y3 = X in ZDecimal Window

Exercises 9.2

1. a. What is the domain of $f(x) = x^7$?

b. What is the range of $f(x) = x^7$?

c. In what quadrants does the graph of $f(x) = x^7$ lie?

d. Which functions in Figure 1 (page 353) are similar to $f(x) = x^7$? What do these functions have in common?

2. a. What is the domain of $f(x) = x^6$?

b. What is the range of $f(x) = x^6$?

c. Compare the graphs of $f(x) = x^7$ and $f(x) = x^6$ for negative inputs. Why does $f(x) = x^7$ turn down but $f(x) = x^6$ turns up?

d. Explain what is meant by the statement that the graph of $f(x) = x^6$ can be split into two pieces that are mirror images of each other.

3. Solve each equation by graphing.

a. $3x^5 = 96$

b. $5x^3 + 2 = {}^-133$

c. $7x^4 = 7$

4. Explain, after graphing, why there are no real solutions to the equation $x^6 = {}^-17$.

5. Solve each equation algebraically.

a. $2x^4 = 50$

b. $x^8 = {}^-81$

c. $(x - 5)^2 + 6 = 55$

6. Find the domain and range of the following functions.

a. $f(x) = x^{1/3}$ **b.** $f(x) = x^{1/6}$

7. Solve each equation by graphing.

a. $x^{1/4} = 2.3$ **b.** $x^{1/4} = {}^-2.3$

c. $x^{1/5} = 1.8$ **d.** $x^{1/5} = {}^-1.8$

8. Solve each equation algebraically and check your solutions.

a. $12x^{1/4} = 24$ **b.** $12x^{1/4} = {}^-24$

c. $2x^{1/3} + 5 = 7.1$ **d.** $2x^{1/3} - 5 = {}^-7.1$

9. Solve the equation $\sqrt{10x} = 2x$ for x.

10. Consider the equation $x - 3 = \sqrt{{}^-2x + 6}$.

a. Solve the equation algebraically.

b. Does a graph of Y1 = X − 3 and Y2 = $\sqrt{({}^-2X + 6)}$ support your choice of solutions from (a)? Explain.

11. Use the definition of inverse functions (page 360) to show that each pair of functions are inverses of each other.

a. $f(x) = x + 5;\quad g(x) = x - 5$

b. $f(x) = x^7;\quad g(x) = x^{1/7}$

c. $f(x) = {}^-x;\quad g(x) = {}^-x$

12. Use a table to show that each pair of functions in Exercise 11 are inverses of each other. In the [Y=] menu, enter Y3 = Y1(Y2(X)) and Y4 = Y2(Y1(X)). To write the symbols Y1 and Y2 in the [Y=] menu, use these keystrokes: [VARS], scroll right Y-VARS (on the TI-82 start at [2nd] Y-VARS), Function [ENTER], Y1 [ENTER]. Repeat the process for the other Y. Enter each pair of inverse functions in Y1 and Y2 and make a table showing X, Y3, and Y4. Turn off Y1 and Y2 in the [Y=] menu by scrolling left until you are on the = symbol; then, [ENTER]. Go to [2ND] TABLE. What do you notice?

a. Y1(X) = X + 5; Y2(X) = X − 5

b. Y1(X) = X^7; Y2(X) = $X^{1/7}$

c. Y1(X) = ⁻X; Y2(X) = ⁻X

13. The graph shows the line Y1 = X and two linear functions that are inverses of each other. Complete the missing coordinates in the ordered pairs.

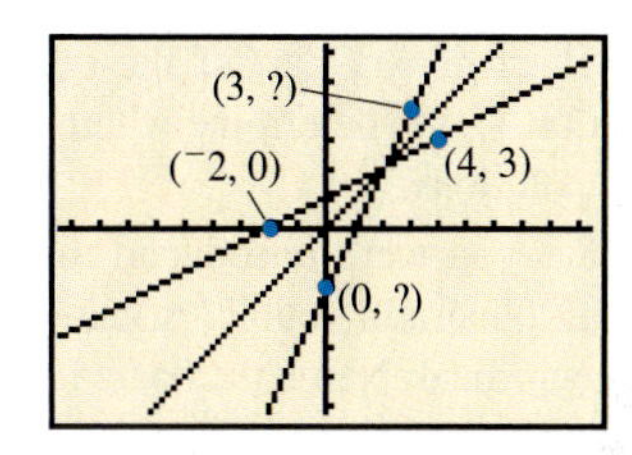

14. Use the ordered pairs on the graph to plot and label four ordered pairs for the function's inverse.

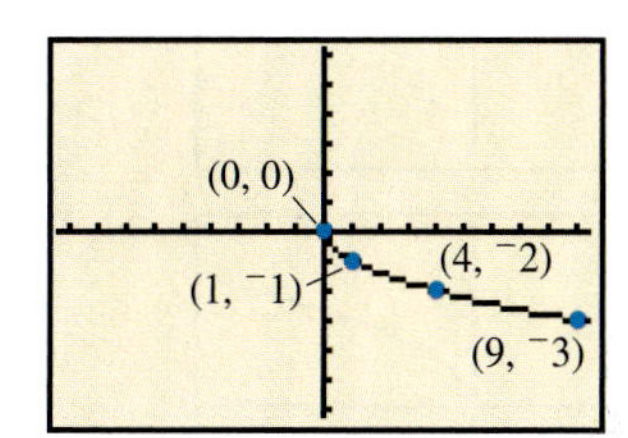

15. When one company wants to determine the number of items N it can sell at a given price P, it uses the function $N(P) = {}^-23P + 480$. When the company wants to determine the price P it can charge by selling N items, it uses the function

$$P(N) = \frac{({}^-N + 480)}{23}$$

a. Enter the functions in Y1 and Y2 and graph using the window $0 \le X \le 188$, $0 \le Y \le 124$.

b. Do the graphs of Y1 and Y2 appear to be inverses of each other? Explain.

c. Tracing along the graph of Y1, we find the points (16, 112) and (18, 66). Trace the graph of Y2; can you find the points with reversed coordinates (112, 16) and (66, 18)?

d. Are the functions $N(P)$ and $P(N)$ inverses of each other? Explain.

Skills and Review 9.2

16. Simplify $(8x^6)^{1/3}$.

17. Find the exact solutions to the equation $x^2 = 32$. Express your solutions in simple radical form.

18. The quadratic equation $x^2 - (2\sqrt{2})x + 2 = 0$ has no rational solution.

  a. Does this mean there are no real solutions?

  b. Does a graph of $y = x^2 - (2\sqrt{2})x + 2$ seem to support your conclusion?

  c. Use the discriminant part of the quadratic formula (see Section 7.3) to determine whether there are 0, 1, or 2 real solutions.

19. In Section 8.3 you were introduced to six basic functions: linear, quadratic, cubic, square root, absolute value, and reciprocal. Name the basic function that corresponds to each graph.

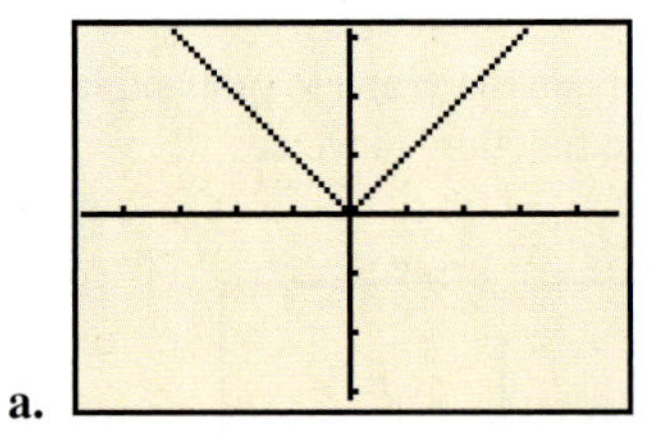

a.

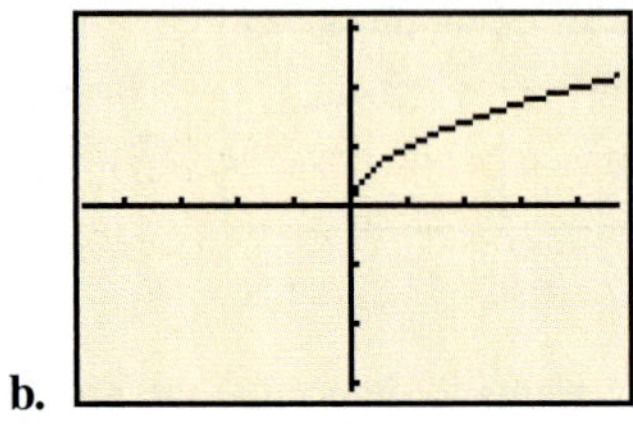

b.

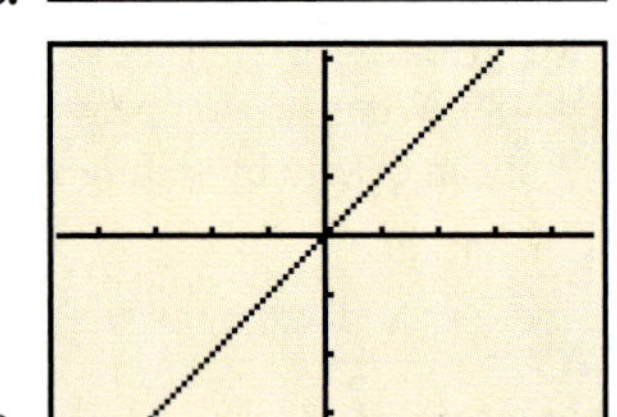

c.

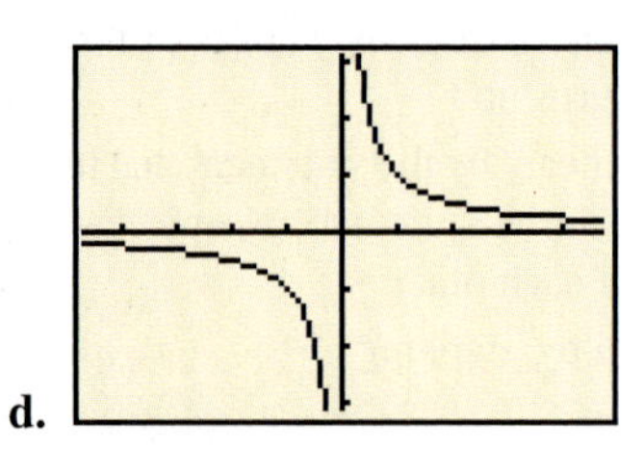

d.

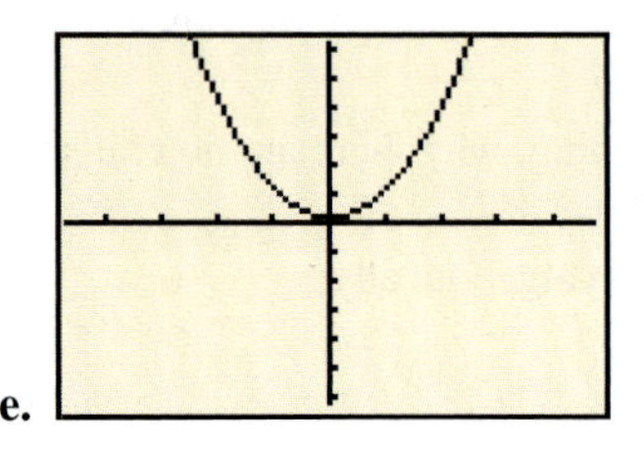

e.

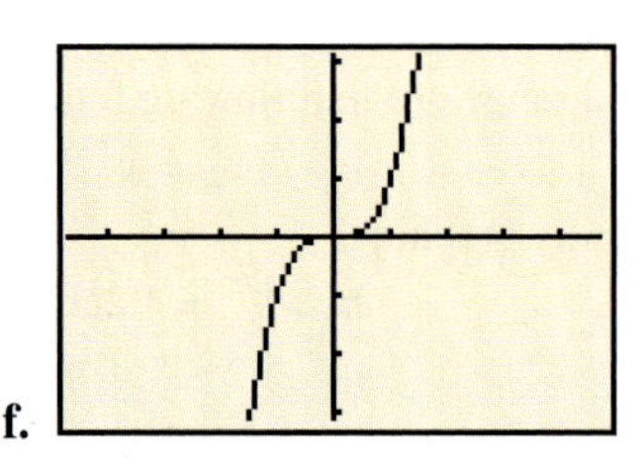

f.

20. Let $f(x) = x^2 + 2x + 5$ and $g(x) = x - 6$.

  a. Find $f(g(2))$. **b.** Find $f(g(x))$.

For Exercises 21–24, use the function $g(x) = \frac{1}{x} + 7$.

21. How does the graph of $g(x)$ relate to the graph of $f(x) = \frac{1}{x}$?

22. **a.** Evaluate $g(10)$.

  b. Evaluate $g(0)$. Why do you get an error?

23. Show the domain of $g(x)$ on a number line.

24. Show the range of $g(x)$ in set-builder notation.

25. Find the exact vertex of $y = x^2 - 4x - 12$.

9.3 Direct Variation

AFTER STUDYING THIS SECTION YOU WILL BE ABLE TO

- Identify a direct variation situation.
- Solve problems involving a direct variation model.

A Discovery in Astronomy

In 1619 the German astronomer Johannes Kepler discovered a relationship between the time it takes a planet to complete one revolution around the sun and its average distance from the sun. That relationship, which is now known as one of Kepler's three laws, is an example of a direct variation model.

Kepler began his investigation by looking at the available data. At that time there were six known planets; Uranus, Neptune, and Pluto were discovered with telescopes centuries later. Here are the data Kepler had for the four planets that are closest to the sun.

	Input Variable, x	Output Variable, y
Planet	Distance from Sun (in millions of miles)	Time for One Revolution (in Earth days)
Mercury	36	88
Venus	67	225
Earth	93	365
Mars	142	687

Kepler, of course, did not have a graphing calculator to aid him in his work. In fact, his work predated the invention of the coordinate plane. (Recall from Section 2.3 that this occurred in 1637.) But let us use our calculators to help us arrive at Kepler's discovery.

The table contains four ordered pairs. They may be entered as data under the STAT menu. Press STAT and then ENTER. Enter the distances from the sun in L1 and the times for one revolution in L2, as shown on the left in Figure 10. Then set up a scatter plot of the data using 2ND STATPLOT, as shown on the right in Figure 10.

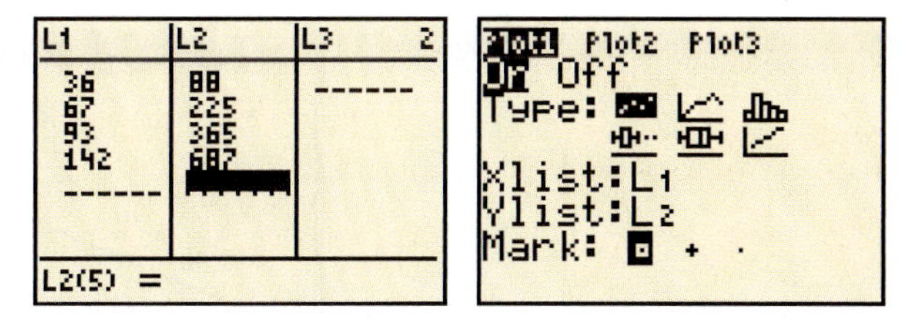

Figure 10

Set windows with Xmin = 0, Xmax = 200, Ymin = 0, and Ymax = 800. GRAPH will then give the four data points, as shown in Figure 11.

It is obvious from the table and the graph that as the distance from the sun increases, so does the time for one revolution. The question arises, What kind of function will fit these data?

We could start with the assumption that we have a linear function. Because a planet at a distance of 0 miles from the sun does not take any time to make a

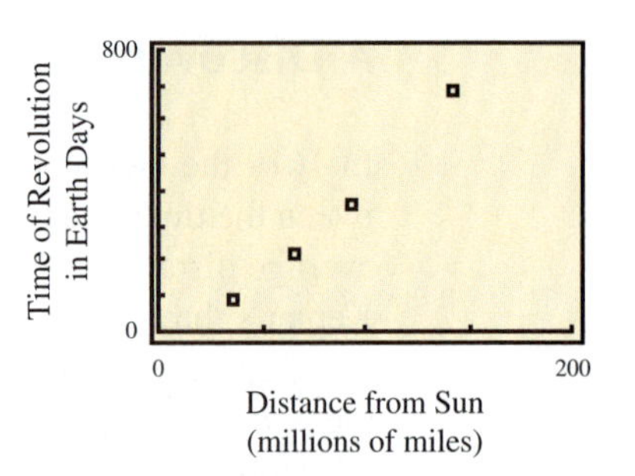

Figure 11

revolution, we can also assume that the linear function has a y-intercept of 0 and an equation of the form $y = mx$. Of course, we can tell from our graph that the four points don't seem to lie on a line, but let's try this approach anyway and see what happens.

Because we are looking only for the value of m, we need to use just one point from the table. Remember that two points determine a line, and we know the point (0, 0) is on the line. Using the data for Mercury, we have

$$y = mx$$

$$88 = m * 36$$

$$2.44 \approx m$$

So the linear function is $y = 2.44x$.

If we graph Y1 = 2.44X, we get the line shown on the left in Figure 12. It passes through the origin and the data point for Mercury, but, not surprisingly, it lies below the data points for the other planets.

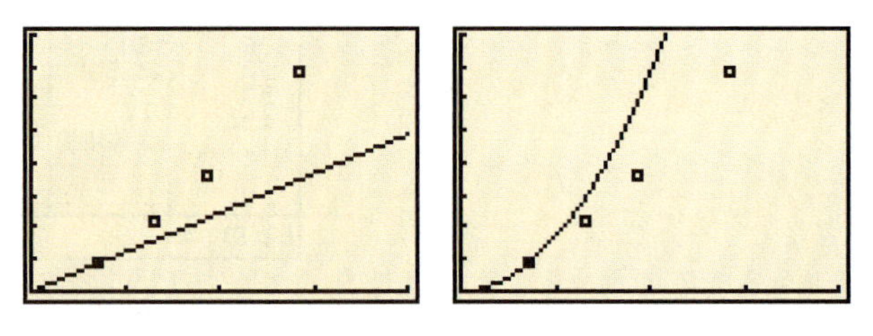

Figure 12

Because a linear function does not work very well, let us try a quadratic. From our work in Chapter 7, we know that a parabola with a vertex at the origin will have an equation of the form $y = ax^2$.

Again, because we need to find only one number, a, we need to use just one point from the table of orbital periods. Using the data for Mercury, we have

$$y = ax^2$$
$$88 = a * 36^2$$
$$88 = a * 1296$$
$$0.0679 \approx a$$

So, the quadratic function is $y = .0679x^2$.

If we graph Y2 = .0679 * X ∧ 2, we get the curve shown on the right in Figure 12. Again, it passes through the origin and the point for Mercury, but this time it lies above the data points for Venus, Earth, and Mars.

Now we may reason as follows: A function based on x to the power 1 (linear) falls short of the data points for the other planets. A function based on x to the power 2 (quadratic) gives y-values that are too large for the other planets. Let's try a power that is between 1 and 2. How about x to the power 1.5?

So, we again start with data for Mercury, this time using the equation $y = ax^{1.5}$.

$$y = ax^{1.5}$$
$$88 = a * 36^{1.5}$$
$$88 = a * 216$$
$$.407 \approx a$$

So, our new function is $y = .407x^{1.5}$.

Now let's graph Y3 = .407 * X ∧ 1.5. We get the curve shown on the left in Figure 13. It appears to pass through the data points for all four planets. We have discovered Kepler's law.

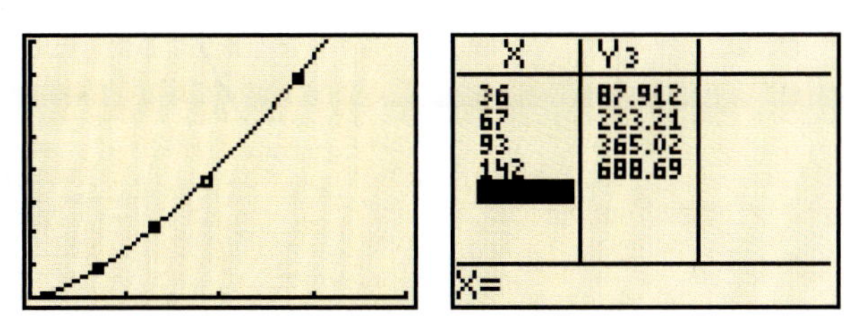

Figure 13

We can test this law using the TABLE feature of the calculator. We use 2ND TBLSET and select Ask for the independent variable. We then access 2ND TABLE and enter the four x-values from the table of periods (36, 67, 93, and 142). We get the predicted y-values shown on the right in Figure 13. Note that these come close to the actual y-values shown in the table. We seem to have a fairly good model for predicting a planet's time of revolution given its distance from the sun.

Direct Variation Models

Kepler's law is a function of a special form. The ouput variable, y (time), is equal to a constant a (.407) times the input variable, x (distance from the sun), raised to the 1.5 (or $\frac{3}{2}$) power. Another way of saying this is that the time for one revolution varies directly as the $\frac{3}{2}$ power of the distance of the planet from the sun.

The concept of *direct variation* may be described as follows.

Definition Direct Variation

If $b > 0$, then a function of the form $y = ax^b$ is called a **direct variation function**. We can say that y **varies directly** as x to the power b. The number a is called the **variation constant**.

In a real-world situation, direct variation may be a good model if the following conditions apply:

1. For positive values of x, as x increases, so does y.
2. As x gets close to zero, so does y.

The graph of a direct variation function is a line or a curve passing through the origin and sloping upward in the first quadrant (Figure 14).

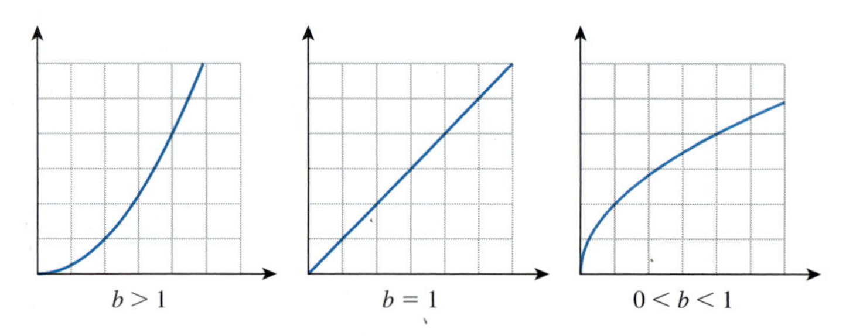

Figure 14

Comments on Direct Variation

1. The definition of direct variation is a generalization of the definition presented in Chapter 5. There, only the case of linear direct variation was considered. In other words, b was assumed to equal 1.
2. In the case where $b = 1$, the variation constant, a, is the slope of the line. Using m for slope, the formula $y = ax^b$ becomes $y = mx$.

(*continued*)

3. When $b > 1$, the graph of the direct variation function opens up. When $0 < b < 1$, the curve opens down. See Figure 14.
4. In most applications, the value of b is known. Often the first task is to find the variation constant, a. This procedure is illustrated in Examples 2 and 3.
5. The variation constant, a, is constant for all ordered pairs (x, y) in a given situation. For instance, in our example for Kepler's law, the value of a was determined by the fact that all the planets revolve around the same star, our sun. In other solar systems, the value of a may differ.

EXAMPLE 1 In 1846 the planet Neptune was discovered. Its average distance from the sun is 2,796 million miles. Use Kepler's law to find the time, in Earth years, that it takes Neptune to make one revolution around the sun.

Solution From the preceeding discussion, $y = .407x^{1.5}$ is a good model for Kepler's Law. Recall that x represents distance in millions of miles, so $x = 2796$; y represents the time of one revolution in Earth days.

Substituting for x, we have

$$y = .407 * 2796^{1.5}$$
$$\approx 60{,}173.$$

Because 365 Earth days represent one Earth year, we can find the time in Earth years by dividing by 365: $60173 \div 365 \approx 165$. It takes Neptune about 165 Earth years to make one revolution about the sun.

EXAMPLE 2 Translate each verbal statement into a general direct variation equation.

a. The amount you pay for gasoline varies directly as the number of gallons you pump.
b. The amount of power generated by a windmill varies directly as the cube of the wind speed.
c. The time it takes a pendulum to swing back and forth varies directly as the square root of its length.

Solution

a. $y = ax$, where x is the number of gallons pumped and y is the cost of the gasoline
b. $y = ax^3$, where x is the wind speed and y is the amount of power generated
c. $y = ax^{1/2}$, where x is the length of the pendulum and y is the time it takes to swing back and forth

EXAMPLE 3 Use the equations you wrote in Example 2 to find the variation constant in each situation. Then write the specific direct variation equation.

a. The amount you pay for gasoline varies directly as the number of gallons you pump. Nine gallons cost \$12.51. Find the variation constant.

b. The amount of power generated by a windmill varies directly as the cube of the wind speed. When the wind speed is 20 miles/hour, one windmill generates 1200 watts (W) of power. Find the variation constant.

c. The time it takes a pendulum to swing back and forth varies directly as the square root of its length. A pendulum that is 6 inches long takes about .785 seconds to swing back and forth. Find the variation constant.

Solution Substitute the known values of x and y into the general variation equation found in Example 2.

a. $y = ax$. When $x = 9$ gallons, $y = \$12.51$. So, $12.51 = a * 9$. Therefore, $a = 1.39$. The specific direct variation equation is $y = 1.39x$.

b. $y = ax^3$. When $x = 20$ miles/hour, $y = 1200$ W. So, $1200 = a * 20^3$. Therefore, $a = \frac{1200}{8000} = .15$. The specific direct variation equation is $y = .15x^3$.

c. $y = ax^{1/2}$. When $x = 6$ inches, $y = .785$ seconds. So, $.785 = a * 6^{1/2}$. Therefore, $a \approx .320$. The specific direct variation equation is $y = .320x^{1/2}$ or $y = .320\sqrt{x}$.

EXAMPLE 4 Use the specific direct variation equation from Example 3 to find the unknown quantity.

a. The amount you pay for gasoline varies directly as the number of gallons you pump. Nine gallons cost \$12.51. How much gasoline can you buy for \$20?

b. The amount of power generated by a windmill varies directly as the cube of the wind speed. When the wind speed is 20 miles/hour one windmill generates 1200 W of power. How much power is generated when the wind is 35 miles/hour?

c. The time it takes a pendulum to swing back and forth varies directly as the square root of its length. A pendulum that is 6 inches long takes about .785 seconds to swing back and forth. Find the length of a pendulum needed to give a period of 1 second.

Solution **a.** From Example 3, $y = 1.39x$. When $y = \$20$, we can find x by substituting:

$$20 = 1.39x$$

$$14.4 \approx x$$

You can buy about 14.4 gallons of gasoline with \$20.

b. From Example 3, $y = .15x^3$. When $x = 35$ miles/hour, we can find y by substituting:

$$y = .15(35)^3$$

$$y \approx 6431$$

The windmill generates about 6431 W of power when the speed of the wind is 35 miles/hour.

c. From Example 3, $y = .32x^{1/2}$. When $y = 1$ second, we can find x by substituting.

$$1 = .32x^{1/2}$$

$$3.125 = x^{1/2}$$

$$3.125^2 = (x^{1/2})^2$$

$$9.77 \approx x$$

The pendulum should be about 9.77 inches long in order to swing back and forth in 1 second.

Examples 2, 3, and 4 illustrate that there are three steps for solving problems involving direct variation models.

Solving Direct Variation Problems

Step 1 Identify the two variables, x and y, and write a general direct variation equation with a variation constant a. Be sure you have the correct power of x.

Step 2 Use one data point to solve for a by substituting the values for x and y. Substitute for a in the general equation to get a specific direct variation equation.

Step 3 Use the specific direct variation equation to substitute for a known value of either x or y. Then solve for the unknown value of either y or x.

The complete three-step process is shown in Example 5.

EXAMPLE 5 The surface area of a planet varies directly as the square of its radius. Earth has a radius of 3890 miles and a surface area of 190,000,000 square miles. Find the radius of the planet Mars, which has a surface area of 54,000,000 square miles.

Solution Step 1 is to write the general direct variation equation. If y represents the surface area of the planet and x is the radius, we have $y = ax^2$. Or if we prefer to use s for surface area and r for radius, we could write $s = ar^2$. Step 2 is to solve for a and find the specific direct variation equation. For Earth we know that $\boldsymbol{x = 3890}$

and $y = 190{,}000{,}000$. Substituting these values we have

$y = ax^2$	*Using the general equation from step 1*
$190{,}000{,}000 = a * 3890^2$	*Substituting for x and y*
$190{,}000{,}000 = a * 15{,}132{,}100$	
$12.56 \approx a$	*Dividing by 15,132,100 on both sides*

We now have a value for a, so we write the specific direct variation equation as

$$y = 12.56x^2$$

Step 3 is to substitute the known surface area for Mars and solve for the radius of Mars.

$y = 12.56x^2$	
$54{,}000{,}000 = 12.56x^2$	
$4{,}299{,}363 \approx x^2$	*Dividing by 12.56 on both sides*
$2073 \approx x$	*Taking the square root on both sides and keeping only the positive solution*

The radius of Mars is approximately 2073 miles.

Exercises 9.3

1. Recall that the time in days that it takes a planet to revolve around the sun can be modeled by the function $y = .407x^{1.5}$, where x is the planet's distance in millions of miles from the sun. Use this function to confirm the time in days it takes Earth to complete one revolution around the sun, using the fact that Earth is 93 million miles from the sun.

2. There are at least 31 natural satellites (moons) that orbit Saturn. The table in the next column provides data on four of them. The time in days it takes for a satellite to revolve once around Saturn is modeled by the function $y = ax^{1.5}$, where x is the distance in thousands of kilometers from Saturn. (http://seds.lpl.arizona.edu)

Natural Satellite	Distance from Saturn (thousands of kilometers)	Time for One Revolution (Earth days)
Mimas	186	.94
Dione	377	2.74
Titan	1222	15.95
Iapetus	3561	79.33

a. Find the variation constant that fits these data.

b. Enceladus has an average distance from Saturn of 238 thousand kilometers. Find the number of days it takes this satellite to complete one revolution around Saturn.

c. Find the time it takes in hours for Enceladus to complete one revolution.

3. a. Write a general function for linear direct variation.

b. Write the specific linear function that has a variation constant of $a = .5$.

c. Find y when $x = 3$.

d. Find x when $y = 20$.

4. a. Write the general direct variation function for y varying directly as the square of x.

b. Write the specific function that has a variation constant of 1.

c. Find the output when the input is 4.

d. Find the input when the output is 64.

5. a. Write the general direct variation function for $f(x)$ varying directly as the $\frac{5}{2}$ power of x.

b. Write the specific function that has a variation constant of 1.7.

c. Find $f(4)$.

d. Find x, if $f(x) = 413.10$.

In Exercises 6 and 7, use the given input and output to find the variation constant for each function.

6. $y = ax^{2/5}$, $y = 36$ when $x = 32$.

7. $y = ax^{1/2}$, $y = \frac{3}{2}$ when $x = 36$.

8. Let $y = ax^{1/4}$.

a. Find the variation constant if $y = 1.2$ when $x = 16$.

b. Find an exact value for y if $x = 6$.

c. Find x if $y = 1.8$.

9. Match each graph with the appropriate exponent b in the direct variation function $y = ax^b$.

a. $b = 1$ **b.** $0 < b < 1$ **c.** $b > 1$

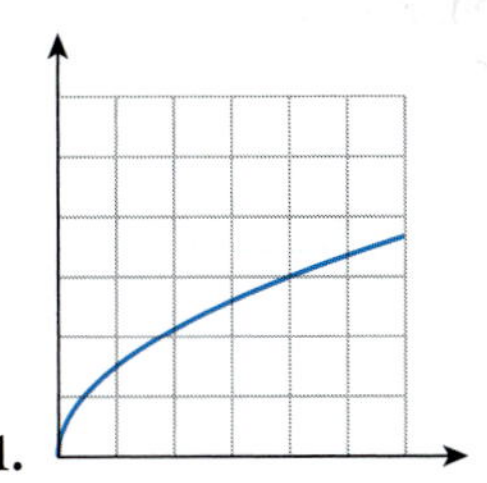
1.

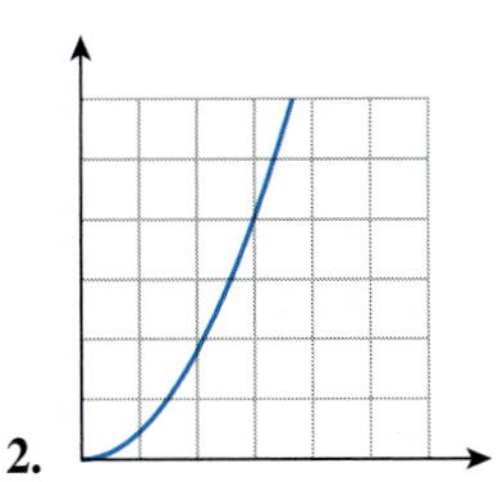
2.

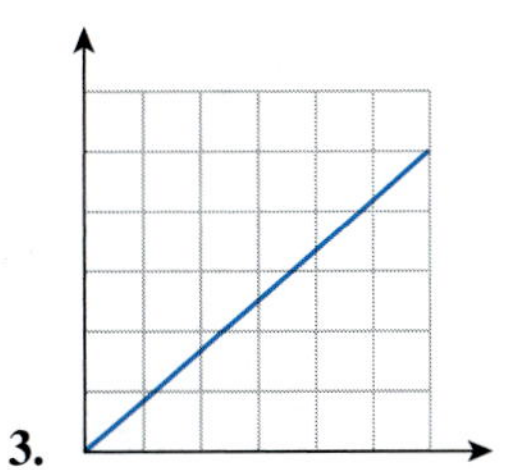
3.

Exercises 10 and 11 deal with the following table of values.

Input, x	Output, y
8	1.6
13	1.9
27	2.4
42	2.8

10. In the opening presentation of this section you were shown how to enter data into your calculator. Enter the data from the table, and adjust your WINDOW settings so that only the first quadrant shows on the graph. Be sure Xmax and Ymax slightly exceed the greatest input and output value, respectively.

a. Draw a scatter plot on your calculator. Does the plot of these data indicate a direct variation function with an exponent b of $0 < b < 1$, $b = 1$, or $b > 1$? Explain.

b. Given your choice for the exponent in part (a), which direct variation function should provide the best fit for the data: $y = ax^{1/3}$, $y = ax$, or $y = ax^3$? *Note:* Assume one of these three functions provides the best fit.

c. Use an ordered pair from the data to find the variation constant for the function you chose in part (b).

d. Write the specific direct variation function that models these data.

11. Enter the equation from Exercise 10(d) into your Y= menu and graph the equation on your calculator.

a. Does the curve appear to fit the four data points graphed in Exercise 10?

b. Go to TblSet and set Indpnt: on Ask. While in the TABLE mode, enter the inputs: $x = 8, 13, 27, 42$. Compare your function's output with the output given for the data. Does the comparison support your choice for the direct variation function? Explain.

c. Use your equation to find the output for an input of 64.

d. Use a graphing method to find the approximate input for an output of 1.8.

In Exercises 12–14, (a) identify the general direct variation function, (b) calculate the variation constant and write a specific equation for the given situation, and (c) solve the specific equation for the unknown value of one of the variables.

12. The force on a falling object varies directly as the mass of the object. When the mass is 3 kg, the force is 29.4 N. Find the mass of a falling object experiencing a force of 53.9 N.

13. The radius of a sphere varies directly as the cube root of its volume. When the volume is 33 cm^3, the radius is approximately 2 cm. Find the radius of a sphere with a volume of 48 cm^3.

14. A device for focusing light, radio waves, radar, etc., is a three-dimensional parabola called a *paraboloid of revolution*. A picture is shown here. The height of a paraboloid varies directly as the square of its radius. When the radius is 5 feet, the height is 3.125 feet. Find the radius of a paraboloid with a height of 8 feet.

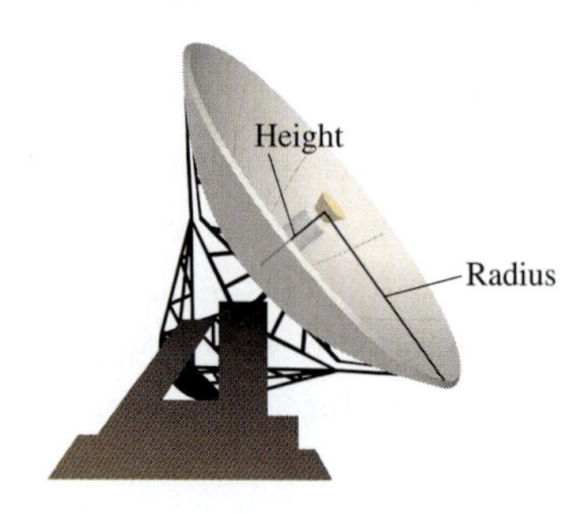

15. The direct variation function $s = \frac{1}{2}gt^2$ models the distance s, in feet, an object falls after an elapsed time of t seconds. The variation constant is $\frac{1}{2}g$, where g is a *gravitational constant*. Find the gravitational constant if $s = 100$ feet when $t = 2.5$ seconds.

Skills and Review 9.3

16. Determine whether the functions in each pair are inverses of each other.

a. $f(x) = 3x - 5$; $g(x) = \frac{x}{3} + \frac{5}{3}$

b. $f(x) = \frac{1}{x}$; $g(x) = {}^{-}\frac{1}{x}$

17. Solve the equation $\sqrt{6x - 5} = x$ algebraically. Check the solution(s).

18. Solve each equation algebraically and check the solution(s) graphically.

a. $4x^3 - 6 = 34$ **b.** $5x^{1/3} = {}^{-}30$

19. Write each expression in simple radical form.

a. $\sqrt{12}$ **b.** $18^{3/2}$ **c.** $\frac{1}{\sqrt{5}}$

20. Find exact solutions to the equation $x^2 - 2x - 5 = 0$. Express the solutions in simple radical form.

21. Simplify the expression $(c^{1/2})^{2/3}$.

22. Evaluate on a calculator $\sqrt[4]{11}$.

23. Let $f(x) = x^2$. Sketch a graph of each function.

a. Two is subtracted from the output: $f(x) - 2$.

b. The output is multiplied by ${}^{-}1$: ${}^{-}f(x)$.

c. Five is added to the input: $f(x + 5)$.

24. Given the following graph (each tick is one unit).

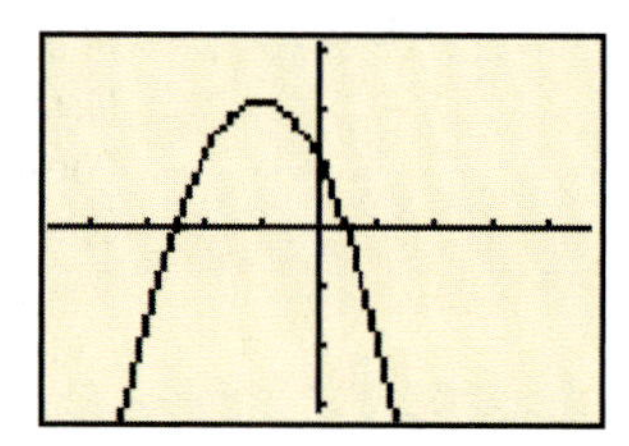

a. Estimate the y-intercept.

b. Estimate the x-intercepts.

c. Estimate the vertex.

25. Factor $x^2 - 5x - 14$.

9.4 Regression (Optional)

AFTER STUDYING THIS SECTION YOU WILL BE ABLE TO

- Fit a linear function to a set of data.
- Fit a power function to a set of data.

In this book you have seen a variety of functions that could serve as models for real-world relationships. For instance, going back to Section 2.4, the linear function $y = 3x + 20$ is a good model for the cost (y) of one month's membership in a health club, where x is the number of visits to the club. In Chapter 7 you learned that a quadratic function such as $y = {}^{-}16x^2 + 50x$ is a good model for the height (y) in feet after x seconds when a ball is tossed in the air. And in this chapter you saw how the function $y = .407x^{1.5}$ models the relationship between the distance (x), in millions of miles, from a planet to the sun and the time (y), in Earth days, it takes to complete one revolution.

Unfortunately there isn't always a function that perfectly fits every set of data. For example, consider the relationship between the economic well being of a country and the amount of money it spends on health care. The following table shows data for eight countries in the year 1998 (Sources: World Health Organization, Encyclopedia Britannica).

Country	Gross Domestic Product per Person (in U.S. dollars)	Spending on health care (in dollars per person)
Canada	20,140	1847
Greece	12,110	960
Mexico	4,440	234
Netherlands	24,410	2166
Portugal	11,030	859
Spain	14,800	1026
Sweden	26,750	2144
United States	31,910	4055

It seems that as the gross domestic product of a country increases, the amount that it spends on health care also increases. But is there an equation for that relationship? It is not obvious from the table.

EXAMPLE 1 Graph the data in the table with a calculator and describe the relationship between the two variables.

Solution Use the method shown in Section 9.3. Press STAT and select EDIT to enter data in L_1 (for gross domestic product) and L_2 (for spending on health care). Then set

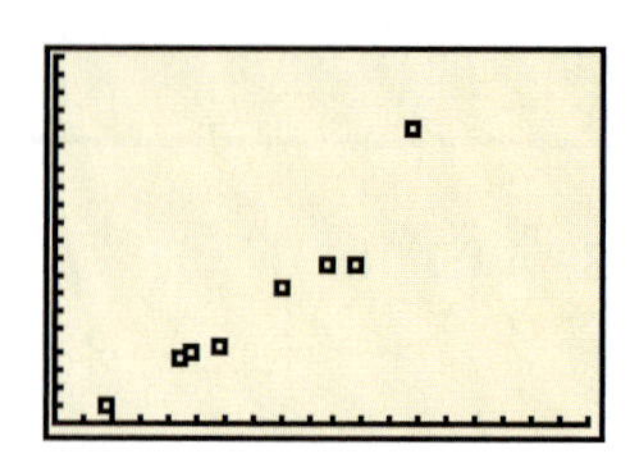

Figure 15 Health-cost data displayed in the window $0 \le X \le 47{,}000$, $0 \le Y \le 5000$

up a scatter plot of the data using 2ND STATPLOT, as shown in Figure 15. Choose a window that contains the data and graph the points to obtain the graph in the figure. Clearly the points show an upward trend; as one variable increases, so does the other. But they do not fit exactly along a straight line or a smooth curve.

Fitting a Line to Data

Looking at the data points in Figure 15, it is not difficult to imagine a line that comes close to most of the points. Take your pencil and place it on the graph to show where the line lies. When we draw such a line, we say we are **fitting** a line to the data.

Statisticians are interested in a line of best fit. There are several ways to determine which line is best. The one most commonly used is called the **least squares regression line**. (It gets that name from the fact that it minimizes the sum of the squares of the vertical distances from the data points to the line. You don't need to understand this now; you will learn more about that in a statistics course.)

Most graphing calculators are able to find an equation for the least squares regression line, as the next example demonstrates.

EXAMPLE 2 Find an equation for the least squares regression line for the data in the table on page 375.

Solution Let us assume that the data are already entered in a TI-83 calculator. Press STAT and then select CALC. Scroll down to 4, LinReg (ax + b), and press ENTER so that the command for linear regression appears on the home screen. Enter the two lists where the data are located and the line in the Y= menu where the equation is to be stored. (To access L_1 and L_2, use the 2ND key with 1 and 2. To access Y1 press VARS, scroll over to Y-VARS, and press ENTER twice.) The following should appear on the home screen.

LinReg (ax + b) L1, L2, Y1

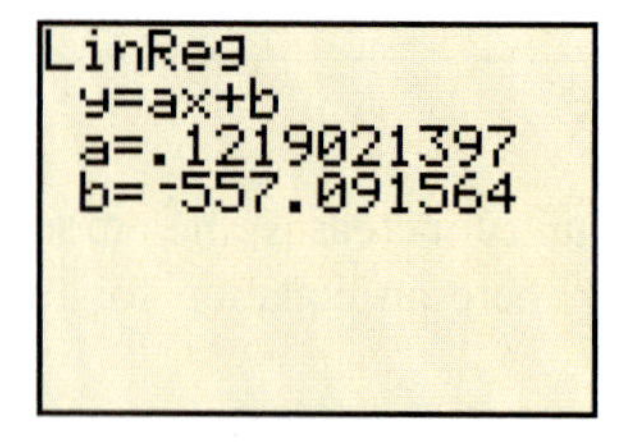

Figure 16

(For a TI-82 you enter only LinReg(ax + b) L1, L2.) After pressing ENTER, the display in Figure 16 will appear.

The regression line has a slope of approximately .122 and a y-intercept of about $^{-}557$. Press GRAPH to see the regression line displayed along with the data points, as in Figure 17. (On a TI-82 you may enter the equation in the Y= menu by hand.)

EXAMPLE 3 Suppose a country has a gross domestic product per person of $15,000. Use the regression equation from Example 2 to predict how much money it spends on health care.

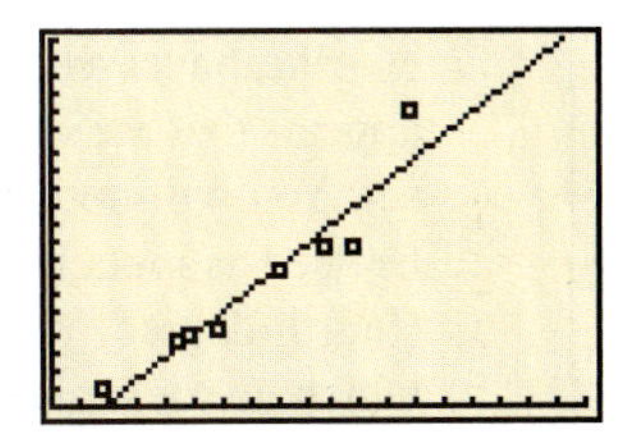

Figure 17 Regression line $y = .122x - 557$ displayed with data points

Solution You may substitute $x = 15{,}000$ into the regression equation or trace along the regression line. In either case we predict that this country will spend about \$1270 per person on health care. Because the actual data points lie above or below our regression line, we realize that \$1270 is just a good estimate; the actual amount is likely to be a little more or a little less.

Fitting a Curve to Data

Sometimes a nonlinear equation may fit the data better than a line does. In the case of the data in Figure 15, we may imagine a curve that opens up and comes close to the data points. It makes sense that if a country had zero income, they would have nothing to spend on health care, so (0, 0) should be on or near the curve. Thus direct variation may be a good model for this data set.

EXAMPLE 4 ■ Fit a direct variation model to the data set in Figure 15.

Solution Proceed as you did with linear regression, but under STAT CALC scroll down until you reach the command PwrReg (power regression). The following should appear on the home screen.

PwrReg L1, L2, Y1.

After we press ENTER, the display in Figure 18 will appear.

Figure 18

Pressing GRAPH, we find that the curve comes very close to most of the data points, as shown in Figure 19.

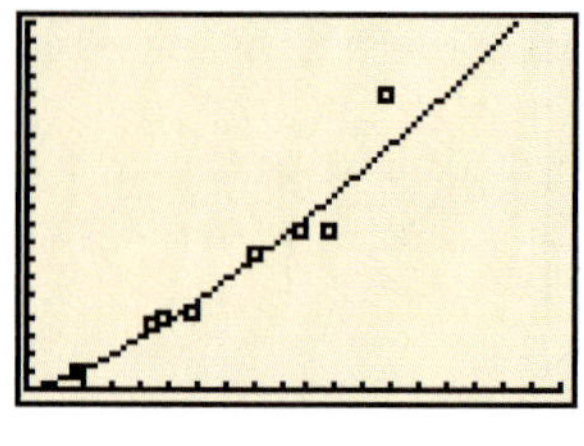

Figure 19

We have seen two different functions that model the same set of data. It appears that the second one, the power function, is a slightly better fit for this data set. You will notice that on the TI-83 you have several other functions that may be used as models to fit a set of data. These include quadratic, cubic, and quartic (fourth-degree) polynomials as well as exponential and logarithmic functions, which are described in Chapter 10. The first step in regression is to choose one of these models. The specific equation for that model is easily found with a computer or graphing calculator. Once we have found that equation, we can then use it to make predictions, as illustrated in Example 3.

Exercises 9.4

1. In Example 4 the curve comes close to most of the data points. There is one point that appears to lie significantly above the regression curve. This point is called an *outlier*. Which country is the outlier for this model?

2. Thanks to environmental-protection measures, the population of bald eagles, which once were threatened with extinction, has steadily increased. The table shows the number of pairs of bald eagles in the United States for the years 1986–1998. (*Source:* U.S. Fish and Wildlife Service.)

Calendar Year	No. of Years since 1984	Population (no. of pairs)
1986	2	1875
1988	4	2475
1990	6	3035
1992	8	3749
1994	10	4449
1996	12	5094
1998	14	5748

a. Enter the number of years since 1984 in L_1 and the population in L_2. Choose an appropriate window and draw a scatter plot. Do the data points appear to lie on a straight line?

b. Use the STAT CALC menu to find a linear equation with LinReg(ax + b) to fit these data. Draw the graph of the line. How well does the line appear to fit the data points?

c. What is the y-intercept of this line? What is its significance?

d. What is the slope of this line? What is its significance?

e. Use the equation found in (b) to predict the bald eagle population in 2006.

f. Use the equation found in (b) to predict what the population was in 1963. (The actual population in that year was 417 pairs). Explain the discrepancy.

3. This table shows the latitude and average January temperature for selected cities in the United States.

City	Latitude (°N)	Average temperature in January (°F)
Anchorage, AK	61	13
Austin, TX	30	49.1
Bismarck, ND	47	6.7
Charleston, WV	38	32.9
Grand Rapids, MI	43	22
Honolulu, HI	21	72.6
Los Angeles, CA	34	56
Madison, WI	43	15.6
Minneapolis, MN	45	11.2
Olympia, WA	47	37.2
San Francisco, CA	38	48.5
Tampa, FL	28	59.8

a. Enter the latitude in L_1 and the average January temperature in L_2. Choose an appropriate window and draw a scatter plot. Do the data points appear to lie on a straight line?

b. Use the STAT CALC menu to find a linear equation with LinReg(ax + b) to fit these data. Draw the graph of the line. How well does the line appear to fit the data points?

c. What is the y-intercept of this line? What is its significance?

d. What is the slope of this line? What is its significance?

e. What factors besides latitude might influence the average January temperature? (Hint: Look at pairs of cities with the same latitude such as Bismarck and Olympia. You may want to refer to a map.)

CHAPTER 9 ▪ KEY CONCEPTS

Unit Fraction Power For all nonnegative values of a ($a \geq 0$), $a^{1/n}$ is defined to be the nonnegative value of x that satisfies the equation $x^n = a$. When $n = 2$, we call this number the **principal square root** of a; in general this number is called the **principal nth root** of a and may be written $\sqrt[n]{a}$. The symbol $\sqrt{\ }$ is called a **radical**, n is called the **index**, and a is called the **radicand**. When the index is 2, the radical is a square root, and the index may be omitted. (Page 346)

Rational Power For all nonnegative values of a and for all positive integers m and n, $a^{m/n} = (a^m)^{1/n} = (a^{1/n})^m$. (Page 347)

Simple Radical Form An expression consisting of a single term is in simple radical form if (Page 349)

- It contains only one radical;
- The radicand is an integer;
- The radical does not appear in the denominator of a fraction;
- The radical is not raised to a power;
- Both the radicand and the index take on the smallest possible absolute values.

The Odd- and Even-Root Properties Suppose n is a positive integer and a is a real number. Consider the equation $x^n = a$. (Page 355)

- If n is odd, then $x = a^{1/n}$.
- If n is even, then
  - If $a > 0$, $x = \pm a^{1/n}$;
  - If $a = 0$, then $x = 0$;
  - If $a < 0$, there is no real solution to the equation.

The Odd- and Even-Power Properties Suppose n is a positive integer and a is a real number. Consider the equation $x^{1/n} = a$. (Page 358)

- If n is odd, then $x = a^n$.
- If n is even, then
  - If $a \geq 0$, $x = a^n$;
  - If $a < 0$, there is no real solution to the equation. The "solution" a^n is **extraneous**.

Inverse Functions $f(x)$ and $g(x)$ are inverses of each other if $f(g(x)) = x$ for every x in the domain of $g(x)$ and $g(f(x)) = x$ for every x in the domain of $f(x)$. (Page 360)

Inverse Function Graph Property Suppose $y = f(x)$ and $y = g(x)$ are inverse functions. Then the graph of $y = g(x)$ is the mirror image of the graph of $y = f(x)$ with the line $y = x$ as the mirror. (Page 362)

Direct Variation If $b > 0$, then a function of the form $y = ax^b$ is called a direct variation function. We can say that y *varies directly* as x to the power b. The number a is called the **variation constant**. (Section 9.3)

Solving Direct Variation Problems (Page 371)

Steps

Step 1 Identify the two variables, x and y, and write a general direct variation equation with a variation constant a. Be sure you have the correct power of x.

Step 2 Use one data point to solve for a by substituting the values for x and y. Substitute for a in the general equation to get a specific direct variation equation.

Step 3 Use the specific direct variation equation to substitute for a known value of either x or y. Then solve for the unknown value of either y or x.

Exploration

Recall from the chapter opener that the system of equal temperament was created for keyboard instruments to equalize the ratios of frequencies between notes.

The chart shows the 13 notes on a piano from middle C to high C and their frequencies in the traditional system of tuning used during the middle ages. Frequency is measured in cycles per second.

Note	No. of Half-Steps Above Middle C	Frequency in Traditional System	Frequency with Equal Temperament
Middle C	0	264	264
C sharp/D flat	1	278.4	279.7
D	2	297	
D sharp/E flat	3	312.8	
E	4	330	
F	5	352	
F sharp/G flat	6	371.2	
G	7	396	
G sharp/A flat	8	417.6	
A	9	440	
A sharp/B flat	10	469.4	
B	11	495	
High C	12	528	528

The equal-temperament system is based on the function $y = 264 * 2^{x/12}$, where x is the number of half-steps above middle C shown in the second column of the chart. Use this function to fill in the fourth column. (*Hint:* You may want to use the TABLE feature on your calculator.)

a. For which notes are the two frequencies exactly the same?

b. For which note is the difference between the two frequencies the greatest?

c. These intervals are supposed to have the same ratio of frequencies:

G to middle C   A to D   B flat to E flat

Find these ratios for both the traditional system and the equal-temperament system. What do you notice?

CHAPTER 9 ▪ REVIEW EXERCISES

Section 9.1 Rational Exponents and Radicals

1. Use the table feature of your calculator to show that $\sqrt[4]{x} = x^{1/4}$.

2. Evaluate each expression without a calculator.

a. $\sqrt[4]{\frac{16}{81}}$ **b.** $27^{1/3}$

3. Evaluate $\sqrt[5]{3}$ to the nearest .001.

4. Evaluate each expression. For friendly powers, find an exact answer without a calculator; otherwise use a calculator to evaluate to the nearest .01.

a. $25^{3/2}$ **b.** $81^{3/4}$ **c.** $(\frac{8}{27})^{2/3}$ **d.** $10^{5/4}$

5. Write each expression with a radical.

a. $18^{1/2}$ **b.** $49^{1/4}$

c. $3^{2/5}$ **d.** $(\frac{1}{8})^{1/2}$

6. Write each radical expression in simple radical form.

a. $\sqrt{98}$ **b.** $\sqrt{108}$ **c.** $\frac{1}{\sqrt{7}}$ **d.** $\sqrt{\frac{2}{3}}$

7. Solve these quadratic equations. Give the roots as exact numbers in simple radical form.

a. $x^2 + 1 = 55$

b. $x^2 + 4x - 14 = 0$

c. $3x^2 - 13 = 12x$

Section 9.2 Power Functions and Their Graphs

8. Solve each equation by graphing. Not all have real solutions.

a. $x^4 = 16$ **b.** $x^4 = {}^-16$

c. $x^3 = 125$ **d.** $x^3 = {}^-125$

9. Solve each equation algebraically. Not all have real solutions.

a. $2x^6 = 98$

b. $(x - 2)^3 - 5 = 22$

c. $x^4 = {}^-30$

10. Solve each equation by graphing.

a. $x^{1/2} = 3$ **b.** $x^{1/2} = {}^-3$

c. $7x^{1/5} = 11.2$ **d.** $7x^{1/5} = {}^-11.2$

11. Solve the equations in Exercise 10 by an algebraic method and check your answers for extraneous solutions.

12. a. Solve the equation $x - 4 = \sqrt{x - 2}$.

b. Are there any extraneous solutions? Explain.

13. Show that the functions $f(x) = \frac{1}{3}x - 4$ and $g(x) = 3x + 12$ are inverses of each other.

14. Explain why $f(x) = x^4$ and $g(x) = x^{1/4}$ are not inverses of each other.

Section 9.3 Direct Variation

15. In Example 1 you were given the function $y = .407x^{1.5}$, where y is the number of days it takes a planet to revolve around our sun and x is the distance, in millions of miles, from the sun. Jupiter's average distance from the sun is 487.4 million miles. Find the time, in Earth years, that it takes Jupiter to make one revolution around the sun.

For Exercises 16–18, find the variation constant, *a*.

16. $y = ax$; $y = 2$ when $x = 6$

17. $y = ax^2$; $y = 45$ when $x = 3$

18. $y = ax^{1/3}$; $y = .4$ when $x = 8$

19. Let $y = ax^{1/2}$.

a. Find the variation constant if $y = 100$ when $x = 10$.

b. Using the value of the variation constant found in part (a), find y if $x = 4$.

c. Find x if $y = 64$.

In Exercises 20 and 21, (a) identify the general direct variation function, (b) calculate the variation constant and write a specific equation for the given situation, and (c) solve the specific equation for the unknown value of one of the variables.

20. The amount a resident pays for electricity varies directly as the number of kilowatt-hours used. 651 kWh cost

\$71.61. Find how many kilowatt-hours are used if the cost is \$50.

21. The amount of power generated by a windmill varies directly as the cube of the wind speed. When the wind speed is 10 miles/hour, a windmill will generate 130 W of power. Find how much power will be generated when the wind is 25 miles/hour.

Section 9.4 Regression

22. Graph the following data with a calculator and describe the relationship between the two variables.

x	y
2	89
5	81
11	63
20	26
22	19
25	14

23. **a.** Find an equation for the least squares regression line for the data in the previous exercise.
 b. Use your regression equation to predict y when $x = 7$.

CHAPTER 9 ▪ TEST

1. Evaluate each expression without a calculator.
 a. $1000^{1/3}$ **b.** $49^{1/2}$ **c.** $\sqrt[3]{8}$
2. Evaluate each expression to the nearest .1.
 a. $\sqrt[4]{3}$ **b.** $10^{2/5}$
3. Evaluate each expression without a calculator.
 a. $100^{3/2}$ **b.** $125^{2/3}$ **c.** $(\frac{1000}{8})^{4/3}$
4. Write each expression with a radical.
 a. $54^{1/3}$ **b.** $36^{1/6}$ **c.** $4^{5/3}$
5. Write each expression in simple radical form.
 a. $\sqrt{48}$ **b.** $\sqrt{28}$ **c.** $\frac{2}{\sqrt{10}}$ **d.** $\sqrt{\frac{3}{5}}$
6. Use a graph to determine the number of solutions.
 a. $x^2 = 100$ **b.** $x^3 = {}^-27$ **c.** $x^2 = {}^-100$
7. Use algebra to solve the equations from Exercise 6.
8. Use algebra to solve each equation. Note any extraneous solutions.
 a. $(x + 4)^3 - 1 = 999$ **b.** $2x^4 = 72$
 c. $\frac{1}{8}x^{1/2} = {}^-2$ **d.** ${}^-5x^{1/3} - 4 = 1$
9. Solve the equation $\sqrt{x + 7} = x + 1$. Note any extraneous solutions.
10. Show that the functions $f(x) = 8x^3$ and $g(x) = \frac{1}{2}\sqrt[3]{x}$ are inverses of each other.
11. Let $y = ax^3$, where a is a variation constant.
 a. Find the variation constant if $y = 24$ when $x = 2$.
 b. Using the value of the variation constant found in part (a), find y if $x = 4$.
 c. Find x if $y = 3$.

In Exercises 12 and 13, (a) identify the general direct variation function, (b) calculate the variation constant and write a specific equation for the given situation, and (c) solve the specific equation for the unknown value of one of the variables.

12. The height of an equilateral triangle (three equal sides) varies directly as the length of its sides. The height of an equilateral triangle is $\sqrt{3}$ cm when the sides are 2 cm each. Find the height when the sides are $\frac{\sqrt{3}}{2}$ cm each. You may use $\sqrt{3} \approx 1.732$ and report your answer to the nearest .001 cm.
13. The distance an object drops varies directly as the square of the time since it was dropped. In 3.5 s the object drops 60.025 m. Find how long it takes the object to drop 490 m.

CHAPTER

10 Exponential and Logarithmic Functions

A Matter of Life and Death

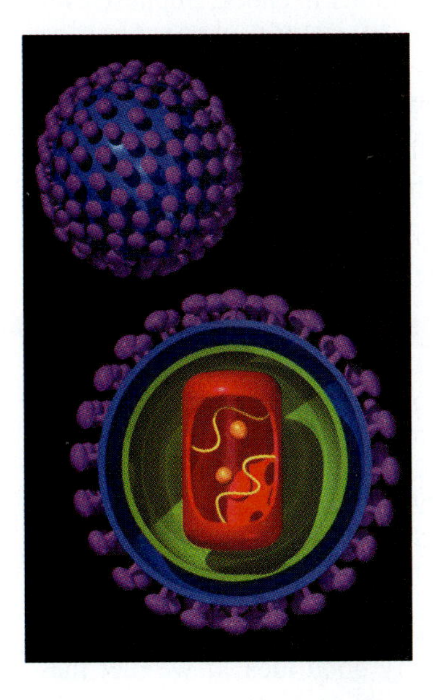

During the 1990s while progress in controlling the AIDS epidemic was made in North America, the number of people infected in Asia, Africa, and the Caribbean grew dramatically.

Public health officials in these countries are concerned about the spread of the AIDS virus and the number of people who will become infected if the disease is left unchecked. Specifically they would like to predict the number of AIDS cases in the year 2010 if the trend continues. In this chapter you will learn about a type of function used to model situations like the spread of AIDS. (See Exploration on p. 422.)

Need help? For on-line resources, visit this web site: **math.college.hmco.com/students.**

10.1 Exponential Growth

AFTER STUDYING THIS SECTION YOU WILL BE ABLE TO

- Determine when a problem involves exponential growth.
- Write an exponential function to model different growth patterns.
- Use a formula to model compound interest problems.
- Find out about how long it will take a quantity to double under exponential growth.

In Australia, the phrase *breeding like rabbits* isn't very funny. At some time in the past, European immigrants to Australia brought along a few rabbits with them, and two escaped. Unfortunately, in Australia there are no animals in the wild that eat rabbits, and so very quickly the countryside was overrun with them. You couldn't throw a rock without hitting a rabbit.

Unchecked Growth

How could this happen? Let's think about how a population of rabbits might grow if there weren't any predators to keep the rabbits in check. Suppose every pair of rabbits has 16 offspring every year (8 males and 8 females). Start with one pair of rabbits.

In the first year 16 baby rabbits are born, for a total of 18, or 9 pairs. In the second year these 9 pairs produce 144 additional babies, for a total of 162. If the process continues, we get the results shown in the table.

Year	Number of Rabbits
0	2
1	18
2	162
3	1,458
4	13,122
5	118,098
6	1,062,882

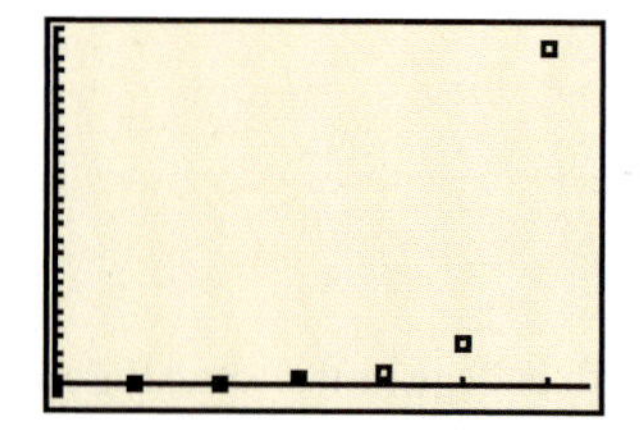

Figure 1 Dot plot of the rabbit population.

In six years, 2 rabbits turned into 1,062,882 rabbits. We are ignoring rabbits that die a natural death, but this still gives us a pretty good idea. If we plot these points, we see something very interesting (see Figure 1). Notice that we need to make Ymax 1,100,000 on the vertical axis.

The higher the points get (the more rabbits we have), the bigger the jump to the next point is. This makes sense, because the more rabbits there are, the more baby rabbits are born. Fifty rabbits means 400 new rabbits; 100 rabbits means 800 new rabbits. In fact, the number of new rabbits is always 8 times the current rabbit population, so the new rabbit population is 9 times the old population (8 times the old population plus the old population makes 9 times). Functions that work this way have a special name; they are called *exponential functions*. It's a bit hard to see what an algebraic rule for the rabbit population might be, so let's try a simpler example.

EXAMPLE 1 Yeast grow by splitting—one yeast cell becomes two, two yeast cells become four, etc. Suppose we start with one yeast cell and that yeast cells split once every second. Find a function that will tell us how many yeast cells there are after x seconds.

Solution If we make a table of the number of yeast cells, we get the following table.

Seconds	Number of Yeast Cells
0	1
1	2
2	4
3	8
4	16
5	32
6	64
7	128
8	256

If we plot these points (Figure 2), we see a pattern something like the pattern for the rabbit population (Figure 1), although the numbers are not quite so large. More yeast cells at any given point means a bigger jump to the next point. This makes sense, because every yeast cell becomes two yeast cells. Once again, the more we have, the more we get.

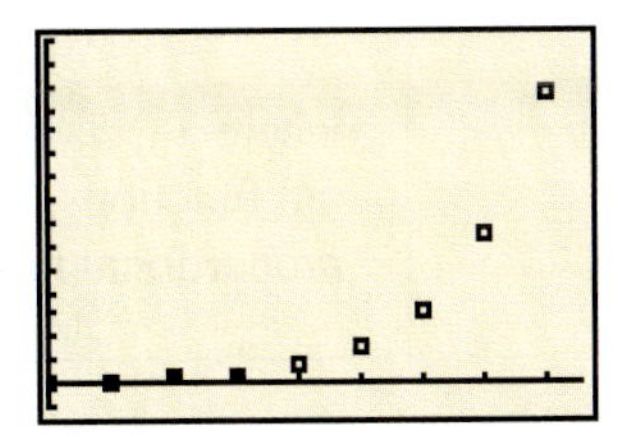

Figure 2 Dot plot of the yeast population. The shape of the graph is easier to see here than in Figure 1 because the scale on the output axis is much smaller.

Look at the number of yeast cells in each line of the table, and forget the input *seconds* for now. We see

$$1 \to 2 \to 4 \to 8 \to 16 \ldots$$

This is a nice pattern, because each number is twice the previous number. If we let x be the number of seconds and y be the number of yeast cells, we can try to write a function $y = f(x)$. We know $f(0) = 1$, because we start with 1 cell. Because that first cell doubles, we get $f(1) = 2(f(0)) = 2$. To compute $f(2)$, we again double the previous population, so $f(2) = 2(f(1)) = 2(2(f(0))) = 4$. Notice that we had to substitute back into the formula to get rid of $f(1)$, using the fact that $f(1) = 2(f(0))$. We can keep going and get $f(3) = 2(f(2)) = 2(2(2(f(0))))$ and so on. With the rabbit population we always multiplied by 9, here we always multiply by 2.

Unfortunately, this gets messy very quickly. But let's look more closely at $f(3)$.

$$f(3) = 2(2(2(f(0)))) = 2^3 f(0)$$

Multiplying by 2 a set number of times is really just multiplying by a *power* of 2, in this case the *cube*. In fact, *every* function value $f(x)$ will just be a power of 2 times the first value $f(0) = 1$, so we have

$$y = f(x) = 2^x$$

Exponential Functions

Unlike all the previous functions and expressions in this book, where we took a variable and raised it to some constant exponent (such as x^2, or $x = x^1$, or $\sqrt{x} = x^{1/2}$) here the variable is *in* the exponent.

Definition Exponential Function

An **exponential function** is any function that can be written in the form

$$f(x) = y_0 b^{kx}$$

where y_0, b, and k are real numbers, with $y_0 > 0$ and $b > 0$. Exponential functions describe situations where the *rate of change* of a variable is proportional to its value.

We need to make sure b is positive because raising negative numbers to some powers isn't possible. For example, we can't consider $(^-1)^{1/2}$, because no real number is the square root of a negative number.

This definition suggests that every exponential function may be described by three values, y_0, b, and k. In this section and the next, we learn the significance of each of these values.

Let's start with y_0. Figure 3 shows the graph of $y = y_0 2^x$ for three different values of y_0. In the case where $y_0 = 1$, we have the original function for the yeast population. Notice that the point at which the graph of an exponential function crosses the y-axis is exactly the value of y_0.

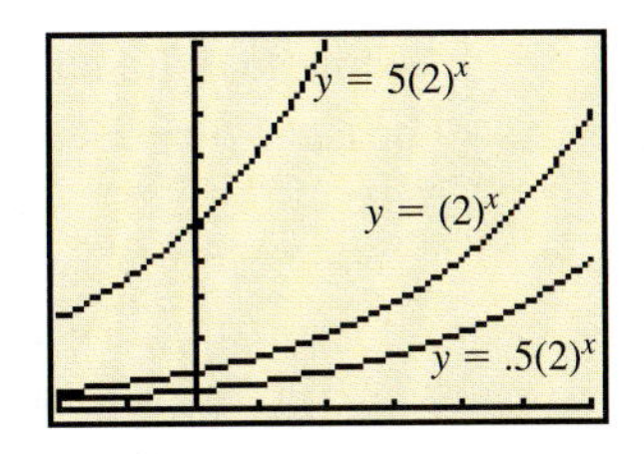

Figure 3 The graph of $y = y_0 2^x$ crosses the y-axis at the point $(0, y_0)$.

This should remind us of what we learned in Section 8.4. Large values of y_0 stretch the function away from the x-axis—in other words, they make all the outputs larger. And when $y_0 < 1$, the output values shrink and the graph looks compressed.

Initial Value of an Exponential Function

Given an exponential function $y = f(x) = y_0 b^{kx}$, the y-intercept of the graph of f is the point $(0, y_0)$. The y-coordinate of this point is the **initial value** of the function—in other words, the value for the starting time $x = 0$.

Now that we have learned something about y_0, let's turn our attention to b.

EXAMPLE 2 Every time a bank credits interest to a savings account, it bases the amount of interest on two things, the interest rate and the amount of money in the account. The more money in the account, the more interest paid. Suppose a bank gives interest each month, and the interest rate per year is 6%. The account has \$1000 in it. Find a function for the amount A in the account after x months.

Solution First, we know that $A(0) = 1000$. There are 12 months in a year, and we are getting 6% interest each year, so that is $\frac{6}{12} = \frac{1}{2}\%$ interest each month.

If we want to determine how much money we have after 1 month, we can multiply our starting amount by $100\% + .5\% = 1.005$. Notice that we add the .5% to 100% because we keep our original money and add the interest, as we did in Section 5.2. So after 1 month, we have $A(1) = 1000(1.005) = 1005$ dollars in the bank.

Each month we get another half-percent interest on whatever we have, so after 2 months we have $A(2) = A(1) * (1.005) = 1000(1.005)(1.005)$; after 3 months $A(3) = A(2) * (1.005) = 1000(1.005)(1.005)(1.005)$, etc. This looks just like the yeast example and the rabbit example, except instead of multiplying by 2 or by 5 we are multiplying by 1.005, and our function becomes

$$A(t) = 1000(1.005)^x$$

with an initial value of $A_0 = 1000$. The interest rate appears as the value of b in the definition of exponential functions on page 386. This is similar to the yeast example, where the fact that the yeast population grew by doubling was evident because $b = 2$.

When $b > 1$ in an exponential function, we have **exponential growth**.

Definition Exponential Growth Factor

In an exponential function $y = f(x) = y_0 b^{kx}$ with $b > 1$, the value of b, called the **growth factor**, changes as the growth rate changes. The faster the growth rate, the larger b is.

EXAMPLE 3 Suppose we find another bank in which to deposit the \$1000. This bank offers 8% interest, but it is paid quarterly (four times a year). What function describes the amount of money in this account after t years?

Solution This function should look a lot like the function for the amount on page 387. The interest rate every 3 months is $\frac{8\%}{4} = \frac{.08}{4} = .02$, and a first guess for the function might be

$$A(t) = 1000(1.02)^4$$

The problem is that t is the number of years, but we get interest every 3 months and not every year. How could we compute $A(1)$? This is the amount after 1 year, which means we will have gotten interest four times, and so

$$A(1) = 1000(1.02)(1.02)(1.02)(1.02) = 1000(1.02)^4$$

Every year we get interest four times, so we multiply by 1.02 four times, which is the same as multiplying by 1.02^4. Therefore,

$$A(t) = 1000(1.02^4)^t = 1000(1.02)^{4t}$$

and we can see that the number of times we get interest appears in the exponent of the exponential function.

We can now rewrite the function $A(t) = 1000(1.005)^x$ from Example 2 in terms of years t instead of months x,

$$A(t) = 1000(1.005)^{12t}$$

because in every year we get interest 12 times; that is, we multiply 1000 by 1.005 twelve times. In fact, we can write a function for *any* sort of situation where we are given a certain interest rate r every year and we get interest n times per year. This is called **compounding**, or **compound interest**, because we are getting interest on our interest.

Definition Compound Amount

Suppose an account pays r percent interest each year (expressed as a decimal, so 12% means $r = .12$), and interest is compounded n times per year. If we start with an initial amount of A_0 (A_0 is frequently written as P for *principal*), then the amount after t years is given by the exponential function

$$A(t) = A_0 \left(1 + \frac{r}{n}\right)^{nt}$$

Caution: When we enter formulas such as the formula in the definition for compound amount into our calculators, we have to be extra careful. Calculators blindly follow the order of operations discussed in Section 2.1. The first operations carried out are anything in parentheses, then powers and roots, then multiplication and division, and, finally, addition and subtraction. Exponential functions always involve a power, and the only operation that may be performed before a power is something in parentheses.

Suppose we use the keystrokes

$$(1 + r/n) \wedge n * t$$

Following the order of operations, the calculator sees the parentheses around $1 + r/n$. There are two operations inside the parentheses, $+$ and $/$, so following the rules, the calculator does the $/$ first and then does the $+$. This is just what we want it to do.

Now, the calculator looks at this expression, and doesn't see any other sets of parentheses. There are two operations left, a power ($\wedge$) and a multiplication ($*$). Because powers come before multiplications, the calculator takes the quantity inside parentheses, which it already has computed, and raises it to the power n. Only then does it do the multiplication by t.

So the calculator is *actually* computing as if we had written the expression

$$\left(1 + \frac{r}{n}\right)^{n} * t$$

which isn't right. How can we get the calculator to compute the right thing? We have to make sure that the calculator performs the power ($\wedge$) last. The only operations that come before powers are things in parentheses, so we have to make sure that every operation except for $\wedge$ is inside parentheses and enter our formula as

$$(1 + r/n) \wedge (n * t).$$

Doubling Time

Example 4 considers how long it takes to double an amount of money.

EXAMPLE 4 ■ In the bank account in Example 3, how long will it take to double the money?

Solution We could try to use the function we have for the amount of money after x years to set up an equation and solve it algebraically. Because we start with \$1000 in the account, what we are asking is when will we have \$2000, or—algebraically—what value of x will make the equation $1000(1.02)^{4x} = 2000$ true.

None of the algebraic techniques that we know seem to work. About all we can do to make the equation simpler is to divide both sides by 1000:

$$(1.02)^{4x} = 2$$

Unfortunately, the variable for which we are trying to solve is in the *exponent*, and we don't yet know how to move a term out of an exponent. Clearly, addition, subtraction, multiplication, and division won't work, and even if we take a power or a root of both sides, x will still be in the exponent.

Instead of trying to use algebra to get an *exact* solution, we may use our calculators to get an *approximate* solution. Set Y1 = (1.02) ∧ (4 ∗ X). A table will give us a good idea how Y1 grows, so use 0 for the starting value and increment the table by 1, by setting ΔTbl = 1. You will need to scroll down the table to discover that when X = 8, Y ≈ 1.8845 and that when X = 9, Y ≈ 2.0399.

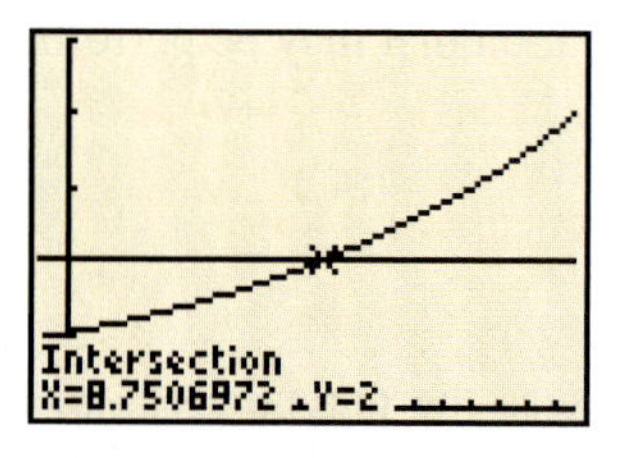

Figure 4

To solve our equation we can graph Y1 = (1.02) ∧ (4 ∗ X) and Y2 = 2. TRACE shows us that the point of intersection lies between X = 8.7 and X = 8.8. 2ND CALC intersect gives a more precise solution, X ≈ 8.7507, as shown in Figure 4. It will take about 8.75 years, that is, 8 years and 9 months, for the money to double.

Knowing the doubling time for an exponential function can be useful. For example, if you know that it will take 5 years to double your money, then you know that in 10 years you will have 4 times your initial investment, and in 15 years you will have 8 times your initial investment. There is a nice trick that can help you quickly estimate doubling times, called the **rule of 72**.

Rule of 72

If you know the yearly interest rate, then the number of years it takes to double is about 72% divided by the rate.

EXAMPLE 5 ■ What does the rule of 72 say the doubling time is for the bank account in Example 3?

Solution We divide 72% by the yearly interest rate of 8%, which gives

$$\frac{72}{8} = 9 \text{ years}$$

This is not as good an approximation as the one in Example 4, $x \approx 8.75$, which we got by using the graph, but this took much less work.

EXAMPLE 6 If a United Nations report says that the world's population is increasing at the rate of 2% per year, about how long will it take the world's population to double?

Solution We just divide:

$$\frac{72}{2} = 36$$

It should take about 36 years for the population to double.

Exercises 10.1

1. Suppose an investment's growth is modeled by the function

$$A(t) = 1000\left(1 + \frac{.06}{12}\right)^{12t}$$

   a. Find the initial value, A_0.
   b. Find the annual interest rate, r.
   c. Find the number of compoundings per year, n.
   d. On a calculator, what keystrokes are necessary to evaluate $A(4)$ correctly?

2. Consider the linear function $f(x) = 3x$ and the exponential function $g(x) = 3^x$.
   a. Complete the outputs in each table.

x	$f(x) = 3x$
0	
1	
2	
3	
4	

x	$g(x) = 3^x$
0	
1	
2	
3	
4	

   b. For each function, describe how the outputs change from the previous output.
   c. Describe the slope of each function. Is it constant? Does it change? Explain.
   d. How does each function's slope affect the shape of the function's graph?

3. a. On your calculator, graph $Y1 = 1.1^x$, $Y2 = 1.2^x$, $Y3 = 1.5^x$, $Y4 = 2^x$, and $Y5 = 3^x$, all on the same graph with a window range of $0 \le x \le 4.7$; $0 \le y \le 10$.
   b. What is the y-intercept of each function?
   c. What do the graphs show about increasing the base of the exponential function?

4. For parts (a)–(d), find the initial value for each exponential function.
   a. $f(x) = 2^x$
   b. $f(x) = .3(4)^x$
   c. $f(x) = (\frac{1}{2})(5)^x$
   d. $f(x) = 3(6)^x$
   e. Where does the initial value occur for any exponential growth function?

5. The exponential function $f(x) = y_0 b^x$ has an initial value y_0 and growth factor b. Write an exponential function whose initial value is $\frac{1}{3}$ and whose growth factor is 5.

6. In the beginning of this section we discussed the growth of a rabbit population. Look at the table of values on page 384.
   a. What is the initial number of rabbits, y_0?
   b. What is the growth factor, b?
   c. Write an exponential function, $f(x) = y_0 b^x$, that models this set of data.

7. Consider the data in this table:

x	y
0	5
1	7.5
2	11.25
3	16.875

a. How can you tell that this set of data represents an exponential function?

b. Write the exponential function, $f(x) = y_0 b^x$, that models the data.

8. Write the exponential function that models the ordered pairs given in the graph in the figure.

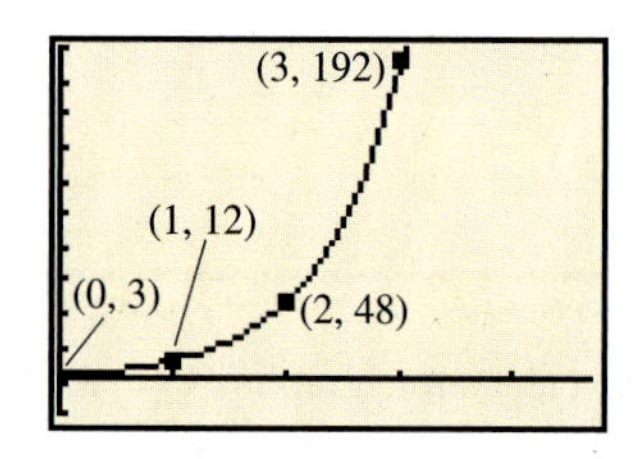

9. Use a graphing method to solve $2(1.5)^x = 15$.

10. An initial investment of \$500 earns 6% interest per year compounded monthly.

a. Use the definition of compound amount to write the amount $A(t)$ as a function of the number of years.

b. Find $A(5)$.

11. Which interest rate yields more, 5% interest compounded daily or 5.1% interest compounded yearly? Explain your choice. (*Hint:* Compare the value of \$1000 invested at each rate after one year.)

12. Suppose a \$1,500 investment earns 4.8% interest compounded quarterly.

a. Use the rule of 72 to estimate the time in years for the investment to double.

b. Approximate the doubling time using a graphing method.

13. If trends continue, the highest pay for a baseball player will increase 17% each year. Beginning in 2002 Alex Rodriguez was the highest-paid player, with a salary of \$22,000,000. We can represent the highest pay in any given year as a function of the number of years since 2002. (http://asp.usatoday.com)

a. Represent this situation numerically with a table of values for the years 2002 through 2010.

b. Represent this situation symbolically with a formula.

c. Represent this situation graphically.

14. During the 1990s the population of white-tailed deer grew 10% per year in New England. Assuming the rate of increase continues, estimate how many years it takes for the deer population to double. (http://www.gazetteonline.com)

15. The expansion of cities, or urban sprawl, is a problem in many regions. Suppose a city is growing by 4.5% each year. Estimate how long it will take for the city to double in size.

Skills and Review 10.1

16. Let $y = ax^2$ represent a direct variation function.

a. Find the variation constant if $y = 2.5$ when $x = 5$.

b. Find y if $x = 7$.

c. Find an exact value for x if $y = 60$. (*Hint:* You'll need to write your answer in simple radical form.)

17. Suppose an output varies directly as the $\frac{5}{2}$ power of its input. If the output is 12 when the input is 4, write the direct variation function.

18. Show that the functions $f(x) = 2x - 10$ and $g(x) = \frac{1}{2}x + 5$ are inverses of each other.

19. Solve for x: $2x^5 - 7 = 57$.

20. Consider the equation $\sqrt{x} = {}^-9$.

a. Find the solution(s) algebraically.

b. Solve the equation graphically. Does the graph support your answer from part (a)?

21. Solve for x: $(x + 6)^2 = 52$. Write the solutions in simple radical form.

22. Write the expression $3 * 2^{-x}$ so that a negative sign does not appear in the exponent.

23. Let $f(x) = \sqrt{x}$. Sketch a graph of each function.

a. $f(x - 1)$ **b.** $f(x) - 1$

c. ${}^-f(x)$ **d.** $\frac{1}{4}f(x)$

24. Find the domain and range of the function $f(x) = x^2 + 4$.

25. Investigators want to determine the speed of a car involved in an accident. They find that the stopping distance of the car y, measured in feet, is given by the equation $y = .03x^2 + .2x$, where x is the speed of the car in miles per hour. If skid marks measured at the scene were 260 feet, use a graph to approximate the speed of the car. Round your answer to the nearest mile per hour.

10.2 Exponential Decay

AFTER STUDYING THIS SECTION YOU WILL BE ABLE TO

- Determine when a problem involves exponential decay.
- Understand the meaning of half-life.
- Write an exponential function to model different decay patterns.
- Identify which exponential functions model growth and which model decay.

Population growth and compound interest are not the only situations where the rate of change of a function depends directly on the function's value. There are situations when we have a function for which the rate of change is proportional to its value and is *negative*.

Radioactive Decay

Radioactive materials are dangerous because they emit energy that can damage living tissue. In the process of emitting this radiation, the material changes form, as when an atom of carbon 14 decays to become a nitrogen atom. Eventually, all radioactive materials decay into a safe, stable form; however, it can take a long time for this to happen. The rate at which radioactive materials decay is described in terms of *half-life*.

Definition Half-Life

The **half-life** of a radioactive substance is the time it takes for half the substance to undergo radioactive decay. (The same term is often used for nonradioactive substances, and then it means the time it takes for half of something to disappear.)

We know that carbon 14 has a half-life of 5730 years. This means that if we start with 1 pound of carbon 14, after 5730 years we will have only $\frac{1}{2}$ pound left. This situation is like what we encountered in Section 10.1, except that instead of a quantity that is increasing, we have a quantity that is decreasing. The rate of change of the quantity is still proportional to the amount, because every 5730 years we lose half of what we had.

EXAMPLE 1 The half-life of radioactive carbon 14 is 5730 years. Explain what this means with a table, a function rule, and a graph.

Solution We can create a table in which the time increment is 5730 and the amount of carbon 14 in each row is half of what it is in the previous row.

Years	Pounds of carbon 14
0	1
5,730	$\frac{1}{2}$
11,460	$\frac{1}{4}$
17,190	$\frac{1}{8}$
22,920	$\frac{1}{16}$

This table looks a lot like the table for the yeast population (page 385), except that each entry in the right column (the output) is a power of $\frac{1}{2}$ instead of a power of 2. Unfortunately, the numbers in the left column are not as nice as they were in the table for yeast, and the function is not quite so simple as

$$A(t) = \left(\frac{1}{2}\right)^t,$$

which would give an amount that is cut in half every *year*. How can we modify this function to account for the fact that it takes a long time to cut the amount of carbon 14 in half?

We need to make sure that it takes 5730 years instead of 1 year for each new power of $\frac{1}{2}$ to occur. In other words, we want an input of 5730 to act like an input of 1 does for the function $A(t) = (\frac{1}{2})^t$. We can get this if we change the t in the exponent to

$$t * \frac{1}{5730}$$

because

$$5730 * \frac{1}{5730} = 1$$

t	$t * \frac{1}{5730}$
0	0
5,730	$\frac{5,730}{5,730} = 1$
11,460	$\frac{11,460}{5,730} = 2$
17,190	$\frac{17,190}{5,730} = 3$

Dividing by 5730 *scales* the input, making large numbers act much smaller. This is just like multiplying in the formula in the definition of compound amount on page 389 to make small numbers seem larger when interest is compounded *more* than once per year. The amount of carbon 14 after t years should be given by

$$A(t) = \left(\frac{1}{2}\right)^{t*(1/5730)}$$

and we can check this by substituting the values 0, 5730, 11,460, etc., for t. We can enter the function as Y1 = (1/2) ∧ (X ∗ 1/5730) to produce a graph on a calculator (Figure 5).

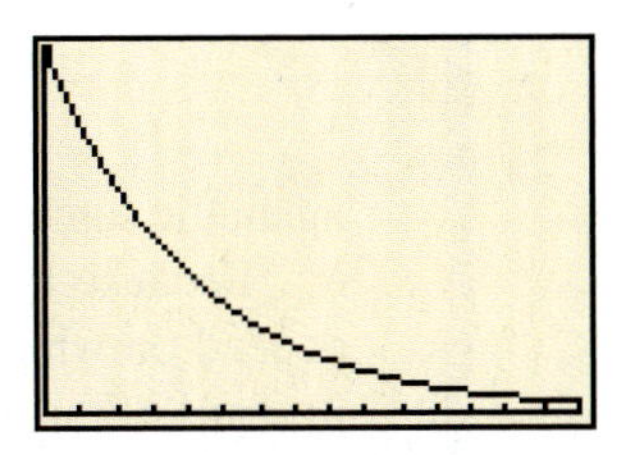

Figure 5 Graph of the amount of carbon 14 left after t years.

EXAMPLE 2 Suppose ecologists are studying the decline in size of a coral reef due to the warming of ocean waters. Each year, they measure the size of the reef in square meters (m^2), and they plot their results (Figure 6).

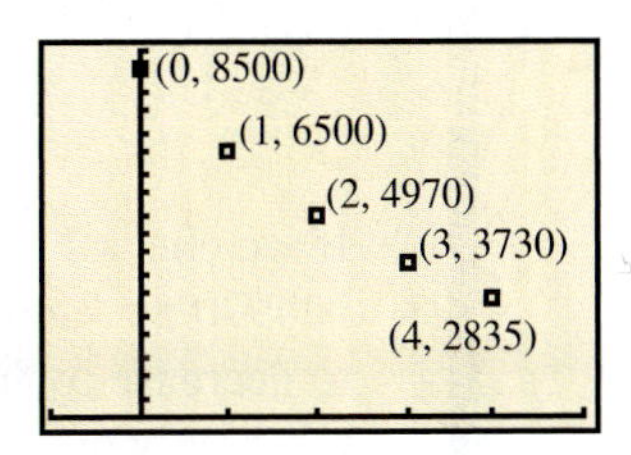

Figure 6 A dot plot of the area of the reef versus the number of years. Notice how the shape of this decline looks like the shape of Figure 5.

a. How can the scientists tell if the size of the reef is decaying exponentially?

b. If it is, what is the half-life?

Solution **a.** If a function is exponential, the time it takes for the output of the function to change by the same factor will always be the same. For example, the time it takes for any amount of carbon 14 to be cut in half is *always* 5730 years.

We can use this fact and the graph of our data to determine if the function *could* be exponential. By checking a number of points that are separated by the same amount of time, we can see if successive outputs have the same ratio. If we look at Figure 6, we can see that at time $t = 0$, the area of the reef is

8500 m^2. One year later, the area is 6500 m^2, so the part of the reef left is

$$\frac{6500}{8500} \approx .76$$

If we wait another year, the area of the reef has decreased to 4970 m^2, so during the second year, the reef decreased to

$$\frac{4970}{6500} \approx .76$$

of what it had been at the beginning of the year. Let's check one more year, say the fourth year. After 3 years, the area of the reef is 3730 m^2, and after 4 years it is 2835 meters, so the part of the reef left is

$$\frac{2835}{3730} \approx .76$$

again the same as before.

Because the part of the reef left after a year of decay doesn't seem to depend on which year we look at, the function that gives the area of the reef in terms of the number of years since measurement started could be exponential. There is no way to be certain that it is, since future measurements may show the proportion changing. However, with the data we have, it is certainly reasonable to use an exponential decay function to model the area of the reef and predict what the area will be in the future.

b. The scientists can approximate how long it takes the coral reef to lose half its area by looking at the graph. Because the initial area is 8500 m^2, they need to determine when the area is

$$\frac{8500}{2} = 4250 \text{ m}^2$$

If we check Figure 6, we see that the measurement of the area after 2 years is 4970 m^2, and after 3 years it is 3730 m^2, so the time it takes for the area to decrease by half is between 2 and 3 years. Because the measurements are taken only once a year, we can't do much better than that, although if we trace a smooth curve connecting the data points, we can try to find the point on that graph that corresponds to an area of 4250 m^2 (see Figure 7).

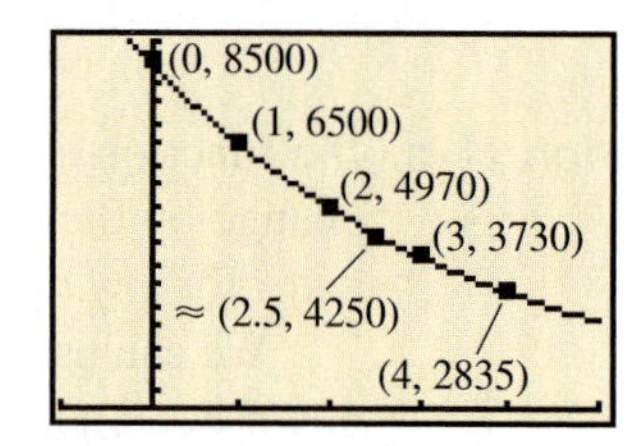

Figure 7 We can *model* the decline in the reef with a smooth curve. It looks as if the area is 4250 m^2 after about 2.5 years.

EXAMPLE 3 Let's revisit Example 5 on page 142. This time, we will try using an *exponential decay function* to model the concentration of PCBs in brown bullheads.

Solution We use an expanded table of values,

Years Since 1980, x	Concentration (mg/kg), y
4	46
5	42.5
6	39
7	36.6
8	33.9
9	31.4

We can enter the data from the table into our calculators as we did with Kepler's law on page 365. Press STAT and then ENTER. Enter the number of years since 1980 in L1 and the concentrations in L2, as shown in Figure 8. Then set up a scatter plot of the data using 2ND STATPLOT, as shown in Figure 9.

Set windows with Xmin = 0, Xmax = 30, Ymin = 0, and Ymax = 50. Pressing GRAPH then gives the six data points, as shown in Figure 10.

The points in Figure 10 certainly look as if they *might* fit an exponential decay model. We can find an *approximate* model to fit the points using the *exponential regression* feature of our calculators; this allows the calculator to find the exponential function that best fits the data in the lists L1 and L2.

Press STAT, scroll over to CALC, and then scroll down to ExpReg and press ENTER. On the home screen enter ExpReg L1, L2, Y1. After pressing ENTER your calculator will look like Figure 11. Pressing GRAPH, we find a curve that comes close to most data points, as shown in Figure 12.

Figure 8

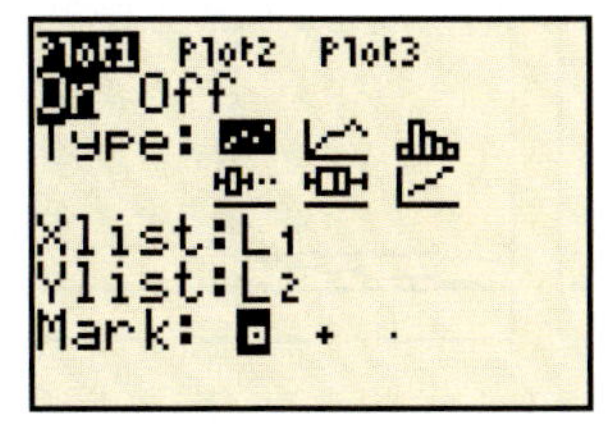

Figure 9

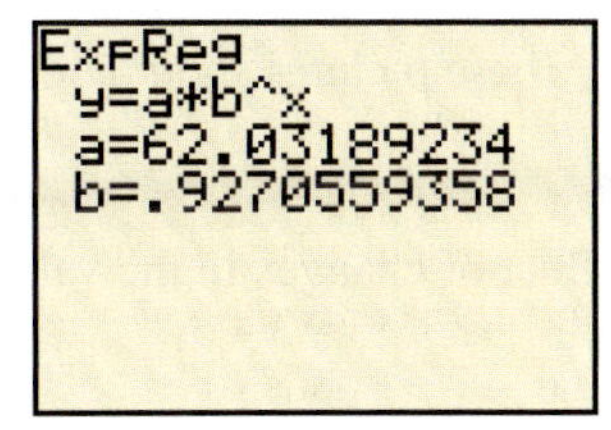

Figure 11

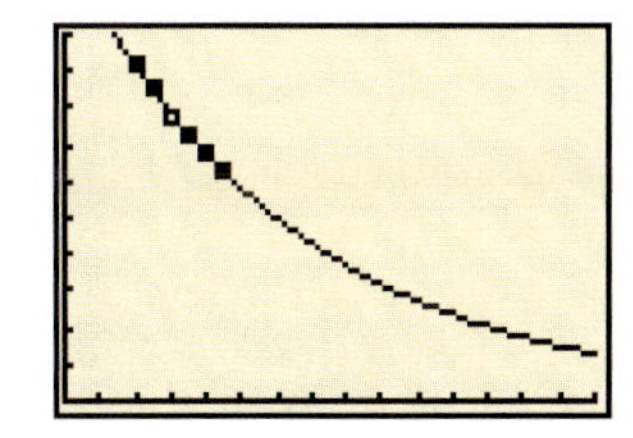

Figure 12

Figure 10

General Exponential Functions

We can now use the ideas explored in this section and the previous one to describe *any* exponential growth or decay process. If we replace t with x (because x is the standard variable on the calculator), we have

$$y = y_0 b^{kx}$$

For each specific situation that we believe we can model with an exponential growth or decay function, we just need to decide on what values to put in for y_0, b, and k.

Facts about Exponential Functions

Fact 1 For an exponential growth or decay function, the coefficient in front of the exponential part of the function, y_0, is the **initial value**—in other words, the y-value that corresponds to $x = 0$.

Fact 2 The base of the exponent, b, is the number we multiply by repeatedly to generate our new function values, and so it is the **growth** or **decay factor**—that is, the ratio between the new function value and the old one over a fixed period of time. If $b > 1$, we have exponential growth, and if $0 < b < 1$, we have exponential decay. If $b = 1$ we have the constant function $f(x) = A_0$, a special case that isn't growth or decay at all. Sometimes the growth factor b is given in the form $1 + r$ or $1 - r$, where r is the percentage rate of growth or decay in a given time period.

Fact 3 The coefficient of x in the exponent—that is, the value k—lets us adjust the **time scale** for the growth or decay, because it tells us how often per time unit we should multiply by b.

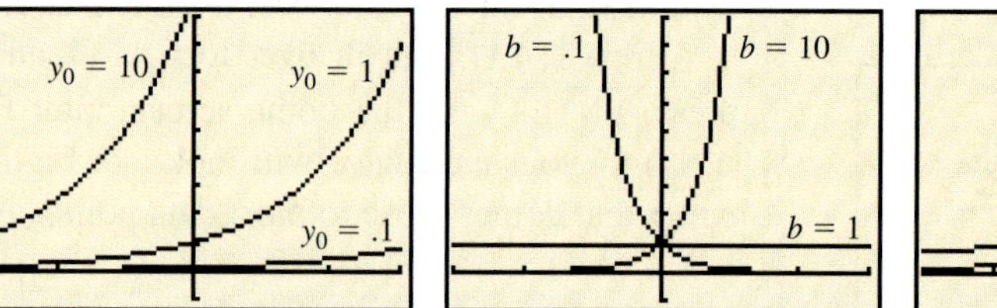

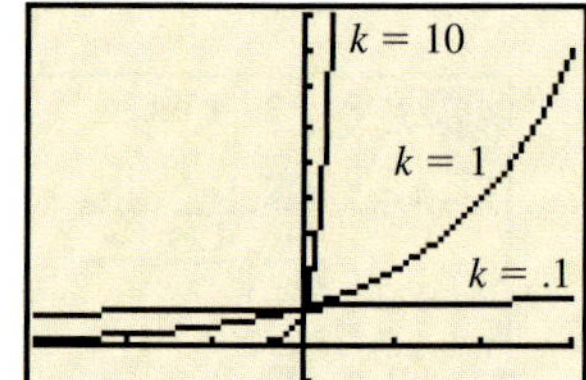

Figure 13 These pictures should help you visualize what the values of y_0, b, and k do.

The effect of changes in the values of y_0, b, and k is shown in the graphs in Figure 13.

Caution: Exponential functions have domain $\{x | ^{-}\infty < x < \infty\}$ and range $\{y | 0 < y < \infty\}$. However, in most applications, x represents time. Because we usually don't talk about negative time, we consider only the part of the graph that lies in the first quadrant.

These three facts are applied in the next example.

EXAMPLE 4 Plutonium 239 has a half-life of 24,100 years. Give a rule for the function that tells how much plutonium 239 is left after x years if the initial amount is 5 g.

Solution The initial value, y_0, is 5 g, so we know that the function is of the form

$$y = 5b^{kx}$$

Because we specify the half-life, we know the decay factor, b, is $\frac{1}{2}$, and the time scale k is $\frac{1}{24,100}$, because we want to multiply by $\frac{1}{2}$ only every 24,100 years. This gives the function

$$y = 5\left(\frac{1}{2}\right)^{x(1/24,100)}$$

We can use the rules of exponents (see Chapter 6) to rewrite exponential decay functions so that they look more like exponential growth functions. Recall that

$$B^{-x} = \frac{1}{B^x} = \left(\frac{1}{B}\right)^x$$

This means that if we have an exponential decay function b^x, with $0 < b < 1$, we can write

$$y = B^{-x}$$

with $B > 1$, where B is the reciprocal of b.

EXAMPLE 5 Write the rule for plutonium 239 in Example 4 in the form $y = y_0 B^{kx}$ with $B > 1$.

Solution To eliminate the fraction in the base, recall that

$$\frac{1}{2} = 2^{-1}$$

We can then rewrite $y = 5(\frac{1}{2})^{x(1/24,100)}$ with a 2 in the base and a negative exponent:

$$y = 5(2)^{-x(1/24,100)}$$

Notice that to change the base from $\frac{1}{2}$ to its reciprocal 2, we had to change the value of k from $\frac{1}{24,100}$ to its opposite, $-\frac{1}{24,100}$.

Exponential decay also appears in business applications. The value of an item often decreases by a certain percentage for each year you have owned it; this is called **depreciation**.

EXAMPLE 6 Suppose a particular model of car decreases in value by $\frac{1}{5}$ every year, and the cost of the car when new was $20,000. How long will it take until the car is worth half of its original value?

Solution We can use graphical methods to approximate the length of time it will take for the car to lose half its value. This is really just estimating half-life, just as in Example 1. What we need is a function for the value of the car after x years. Following the examples of growth and decay from this section and the previous one, we write

$$y = 20000\left(\frac{4}{5}\right)^x$$

because the initial value of the car is \$20,000, and each year the car will be worth $\frac{4}{5}$ of what it was worth the previous year. The decay factor $\frac{4}{5}$ is determined by the fact that the car loses $\frac{1}{5}$ of its value each year, so the value *left* after 1 year is $1 - \frac{1}{5} = \frac{4}{5}$.

Or, if we prefer a base greater than 1, we may write

$$y = 20000\left(\frac{5}{4}\right)^{-x}$$

because $\frac{5}{4}$ is the reciprocal of $\frac{4}{5}$.

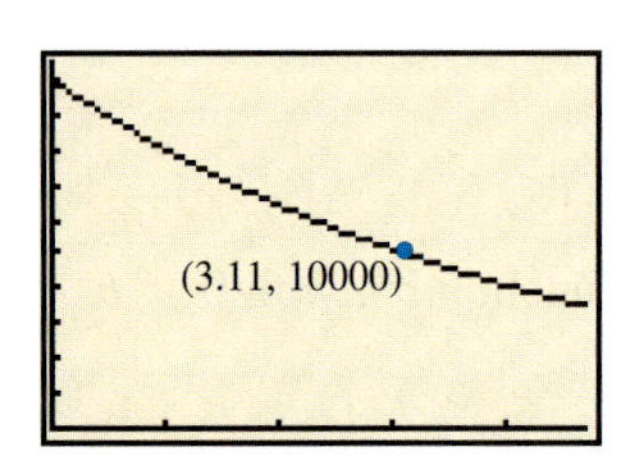

Figure 14 It looks as though the output y for the function $y = 20{,}000(\frac{4}{5})^x$ is 10,000 when the input x is about 3.11.

We can plot this function (see Figure 14) and use TRACE or Intersect to find the x-value that makes the output \$10,000, or half of the intial value; the solution is $x \approx 3.11$ years. You might not be able to get quite so close, depending on the calculator you are using and what your window settings are.

If the growth or decay rate is not a nice number like $\frac{4}{5}$, it is often easier to use the formula from the definition for compound amount on page 389. We can do this for both *positive and negative* rates.

EXAMPLE 7 The population of a rural county in Kansas is decreasing by 3% per year. About how long will it be until the population is decreased by half?

Solution Halving is just doubling in *reverse*—all we change is the sign on the percent. Instead of entering .03 into the formula in the definition of compound amount on page 389, we enter **⁻.03**. Because the rate is given per year, we have $n = 1$.

$$A(t) = A_0\left(1 + \frac{r}{n}\right)^{nt}$$

$$A(t) = A_0(1 - \mathbf{.03})^t$$

$$A(t) = A_0(.97)^t$$

Because we want to find when the population is cut in half, we have $A(t) = \frac{1}{2}A_0$. Substituting and dividing both sides of the last equation by A_0, we have $\frac{1}{2} = (.97)^t$, which we must solve for t.

We can graph Y1 = (.97) ∧ X and Y2 = .5 and find the point of intersection, which gives us an approximate value for t of 22.8 years.

In Section 10.1 we used the rule of 72 to estimate the doubling time if we know the percent increase. If we think of halving as the reverse process of doubling, the rule of 72 should work for halving times as well as doubling times.

In this case, because the rate of decrease in the population is 3% per year, we estimate the halving time as

$$\frac{72}{3} = 24 \text{ years}$$

Exercises 10.2

1. Both exponential growth and decay can be modeled by the general equation $y = y_0 b^{kx}$.
   - **a.** Use an inequality to express the values of the base b that will make this function represent *exponential growth.*
   - **b.** Use an inequality to express the values of the base b that will make this function represent *exponential decay.*
2. On your calculator, graph $Y1 = .9 \wedge X$, $Y2 = .8 \wedge X$, $Y3 = .4 \wedge X$, and $Y4 = .1 \wedge X$ all on the same axes with a window range of $0 \leq X \leq 4.7$; $0 \leq Y \leq 1.5$. What do the graphs show about decreasing the base b of an exponential function from 1 toward 0?
3. Find the initial value for each exponential decay function.
   - **a.** $f(x) = .2^x$
   - **b.** $f(x) = 5(\frac{1}{3})^x$
   - **c.** $f(x) = .4(.6)^x$
   - **d.** $f(x) = (\frac{1}{2})(\frac{1}{7})^x$
4. Compare and contrast exponential growth and decay for each of the following.
   - **a.** Initial value, or y-intercept
   - **b.** x-intercept(s)
   - **c.** Rate of change
5. Graph $y = 2(3^x)$ and $y = 2(3^{-x})$ in a four quadrant window on the same set of axes. What do you notice?
6. Rewrite $(\frac{1}{2})(4)^{-x}$ without a negative sign.
7. Describe the meaning of *half-life* for radioactive decay.
8. The half-life of sodium 24 is 15 hours. Suppose we start with 8 g of sodium 24. Show what this means using each of the following.
   - **a.** Table of values
   - **b.** Formula
   - **c.** Graph
9. The weight of a substance is decreasing with time. The data are shown in the table.

Days	Weight (kg)
0	1
13	$\frac{1}{2}$
26	$\frac{1}{4}$
39	$\frac{1}{8}$

   - **a.** How can you tell that these data fit an exponential decay function?
   - **b.** What is the initial value, y_0?
   - **c.** What is the decay factor, b?
   - **d.** What is the time scale, k?
   - **e.** Use the information collected in parts (a)–(d) to write an exponential decay function that models the data.
10. Write an exponential decay function that models the following table of values.

Months	Weight (g)
0	1800
4	600
8	200
12	66.67

11. Suppose the dollar value of a video game decreases exponentially with time and is given by the function $A(x) = 90(\frac{2}{3})^x$, where x is the number of years since purchase.
    - **a.** What is the initial value?
    - **b.** What is half of the initial value?

c. Use a table of values to find a rough estimate for the half-life.

d. Use a graphical method to find a better approximation for the half-life.

12. A serving of a popular cola soft drink has 46 mg of caffeine. Suppose the half-life for caffeine remaining in the body of a typical adult is 6 h. Write an exponential decay function that gives the amount of caffeine in the body after t hours. (http://ask.yahoo.com)

13. 25% of thorium 234 decays in 10 days. Suppose you begin with 60 g.

a. Write a function for the amount of thorium 234 that remains after t days of decay.

b. Ue a graph to approximate the number of days before the thorium 234 is decreased to half of its original value.

14. A strain of bacteria starts with 1,000,000 cells. Every 4 minutes $\frac{2}{5}$ of the cells die. Use a graph to approximate the time it takes for the number of bacteria to be half of its original cell count.

15. A $15,000 machine loses 14% of its value each year. Use the rule of 72 to estimate the number of years until the machine is worth half its original value.

Skills and Review 10.2

16. Use the formula

$$A(t) = A_0\left(1 + \frac{r}{n}\right)^{nt}$$

to find how much money is in an account after 2 years if the account starts with $1000 and the bank pays 6% annual interest compounded quarterly.

17. Use the rule of 72 in Section 10.1 to find the approximate amount of time it will take a population increasing at the rate of 7% per year to double.

18. Let $f(x) = 4(2)^{3x}$.

a. Find $f(2)$ without a calculator.

b. Use a calculator to check your answer from part (a).

19. Suppose y varies directly as the square of x.

a. Find the variation constant if $y = \frac{5}{2}$ when $x = 1$.

b. Find y when $x = 3$.

c. Find x when $y = 67.6$.

20. Consider the two functions $f(x) = 64x^3$ and $g(x) = \frac{1}{4}x^{1/3}$.

a. Use your calculator to graph the two functions on the same set of axes. Do the functions appear to be inverses of each other? Explain.

b. Use the definition of inverse functions from Section 9.2 to determine if the two functions are inverses of each other.

21. Evaluate $81^{1/4}$.

22. Use the product law of exponents to simplify $10^x * 10^{3x}$.

23. Given $f(x) = 2x - 3$ and $g(x) = x + 1$.

a. Find $f(g(^{-}5))$.

b. Find $f(g(x))$.

c. Find $g(f(x))$.

24. Determine which graphs represent functions.

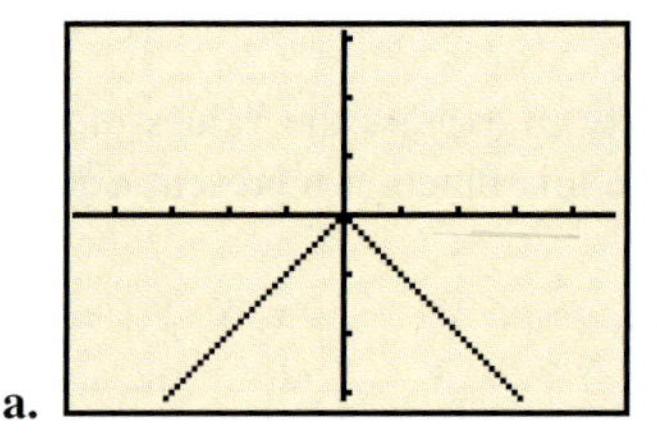

a.

b.

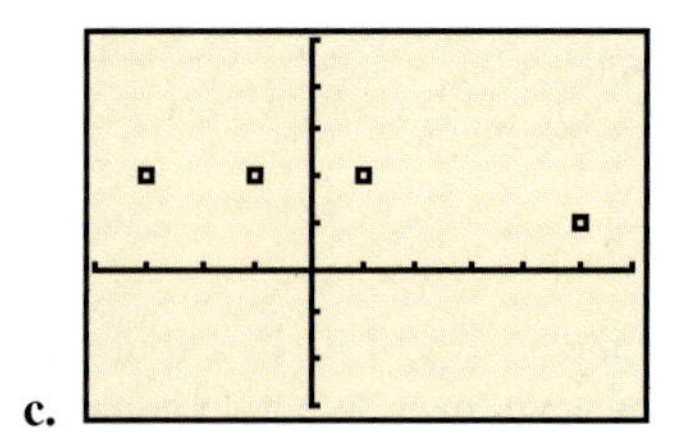

c.

d.

25. Factor $x^2 - x - 20$.

10.3 Logarithms

AFTER STUDYING THIS SECTION YOU WILL BE ABLE TO

- Solve simple exponential and logarithmic equations.
- Write exponential equations in either exponential or logarithmic form.

In the previous two sections, we studied functions of the form

$$y = y_0 b^{kx}$$

called *exponential functions*. We encountered applications of these functions to growth and decay and learned how to determine the output value, y, for any particular input value, x, using our calculators. However, if we need to find what input value gives a particular output value, our tools are limited. We can approximate a solution using a graph of the function; in special cases involving doubling or halving, we may use the rule of 72. In this section we extend our ability to work with exponents by learning about a new type of function, the **logarithm function**.

Solving Exponential Equations

Example 1 illustrates how to solve an exponential equation.

EXAMPLE 1 Suppose

$$y = (10)^x$$

What value of x will make $y = 100$?

Solution We can graph this function and use the TRACE capability of our calculators to see that the solution is $x = 2$ (Figure 15).

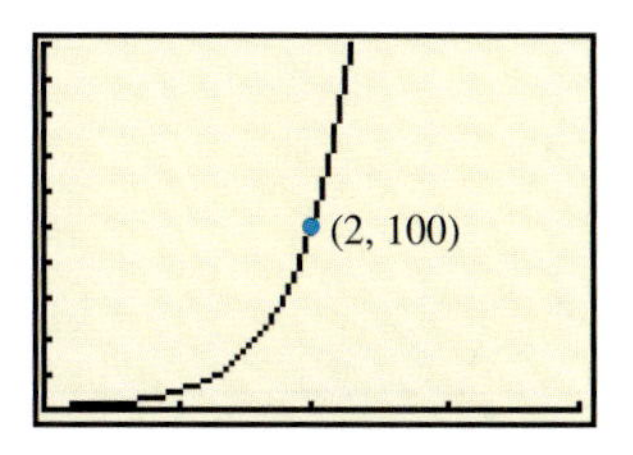

Figure 15 When $x = 2$, y appears to be 100.

Suppose we were to try to solve the problem in Example 1 algebraically. We want to know what x-value will make $y = 100$, so we substitute 100 into the function:

$$100 = 10^x$$

Now we have a problem, because the variable for which we want to solve is the *exponent*, and we have no algebraic tool to help us. We know how to solve some problems that look something like this, but only when x is in the *base* and not the exponent, such as

$$100 = x^2$$

To solve this equation, we have to answer the question, "What number can we square to get 100?" We can answer this question by doing the reverse of squaring: taking the square root of both sides to get

$$\pm\sqrt{100} = \sqrt{x^2}$$
$$\pm 10 = x$$

The square root removes the exponent from x—it *undoes* what squaring did to it.

We can do the same thing with other simple powers of x. If we know that

$$x^3 = 27$$

then we need to *undo* cubing, and we take a cube root and see that $x = 3$. We can't do anything like this with the function in Example 1. Solving the equation

$$10^x = 100$$

requires answering the question, "To what power do I raise 10 to get 100?" In this case, the problem is simple enough that we should be able to see that the answer is 2.

The Logarithm

Is there a function that is the reverse of raising 10 to the xth power, like taking a square root is the reverse of squaring? It certainly isn't one we have seen before, but it does exist.

Definition Common Logarithm

The **common logarithm** of a number x, $\log(x)$, is the number that gives x if 10 is raised to that power. We can see what this means by using two different forms for equations involving logarithms. If we remember that *logarithms are exponents*, then we get an *exponential form*, which looks like

$$10^{\log(x)} = x.$$

Alternatively, we can write

$$y = \log(x) \quad \text{if} \quad 10^y = x$$

It is often helpful to think of the common logarithm function as a question. Just as the square root function $\sqrt{x}$ is really the question, "What number, when squared, gives x?" the common logarithm $\log(x)$ is the question, "What number, when written as an exponent of 10, gives x?" If we use exponential form, we can write this question as an equation:

$$10^{?} = x$$

where $? = \log(x)$.

EXAMPLE 2 ■ Find log (1000).

Solution The value of log (1000) is the answer to the question, "What number, when written as an exponent of 10, gives 1000?" If we write

$$10^{?} = 1000$$

and remember that 1000 is the cube of 10, then we can see that the **?** should be replaced by a **3**, so $\log(1000) = 3$.

■ ■

EXAMPLE 3 ■ Find log (500).

Solution Using exponential form, with the question mark replacing the log, we get

$$10^{?} = 500$$

This causes a problem, because 500 is not an integer power of 10. We are forced to use the log key on our calculators to compute that $\log(500) \approx$ **2.69897**. We can do several things to check that this makes sense. First, note that $10^2 = 100$ and $10^3 = 1000$ and $100 < 500 < 1000$, so it seems reasonable that the number for which we are looking lies between 2 and 3. Second, we can test this result using the definition of logarithm. If $\log 500 \approx 2.69897$, then $10^{2.69897}$ should be about 500. According to a calculator, $10^{2.69897} \approx 499.999995$, which is very close.

■ ■

When one function is the reverse of another function (*undoes* the other function), we say the functions are *inverses* of each other (see Section 9.2). The pair of functions $f(x) = 10^x$ and $g(x) = \log(x)$ are a good example. In Section 9.2 you learned that $f(x) = x^3$ and $g(x) = x^{1/3}$ are also inverse functions. The functions $f(x) = x^2$ and $g(x) = \sqrt{x}$ also undo each other—but only for nonnegative values of x. So, as we observed in Example 7 on page 362, $\sqrt{x}$ is not a complete inverse for the function $f(x) = x^2$.

Equations involving powers of 10 or the common logarithm can be written in either form, exponential or logarithmic.

EXAMPLE 4 ■ Write $y = \log(200)$ in exponential form.

Solution To write an equation involving a logarithm in exponential form, we must remember two facts.

1. Anything we do on one side of an equals sign we must also do on the other side.
2. The common logarithm is the inverse of raising 10 to a power, so that to undo a log we may use 10^x.

Therefore, to write an equation with a log in another form, write both sides of the equation as exponents of 10, giving

$$10^y = 10^{\log(200)}$$

There is nothing we can do on the left-hand side of this equation, but on the right-hand side the fact that the log (200) is in the exponent of a 10 means that these two functions will undo each other, and we have

$$10^y = 200$$

■ ■

EXAMPLE 5 ■ Write $10^{3x} = \mathbf{400}$ in logarithmic form.

Solution Taking a common logarithm undoes raising 10 to a power, so we can take the log on both sides:

$$\log(10^{3x}) = \log(\mathbf{400})$$

$$\mathbf{3x} = \log(\mathbf{400})$$

■ ■

In both of the previous examples, we used the fact that the functions $f(x) = 10^x$ and $g(x) = \log(x)$ are inverses and cancel each other out. Let's put this in a nice form.

> **Logarithms and Exponentials Undo Each Other**
>
> The common logarithm function and the exponential function with base 10 are *inverses*; in other words,
>
> $$10^{\log(x)} = \log(10^x) = x$$
>
> We can think of these two functions as undoing each other.

This is exactly the same as what happens with the functions $f(x) = x^2$ and $g(x) = \sqrt{x}$, because

$$\sqrt{x^2} = (\sqrt{x})^2 = x$$

for nonnegative values of x. Just as we sometimes take square roots on both sides of an equation to solve for a variable, we may use the inverse relationship of the logarithm and exponential functions to solve equations, as illustrated by the next two examples.

EXAMPLE 6 Solve

$$5 \log(x) = 2$$

for x.

Solution Because this equation has x inside the logarithm function, we can get x outside the log term by using its inverse.

$\log(x) = \frac{2}{5}$	*Isolating the* $\log(x)$ *term by dividing on both sides of the equation by* 5
$10^{\log(x)} = 10^{2/5}$	*Raising* 10 *to the power on both sides to cancel the log*
$x = 10^{2/5} \approx 2.51$	*Approximating* $10^{2/5}$ *with a calculator*

EXAMPLE 7 Solve

$$10^{2x-1} = 100$$

for x.

Solution Because this equation has x in the exponent of 10, we need to undo the exponential function by using its inverse, the logarithm.

$\log(10^{2x-1}) = \log(100)$	*Taking the logarithm on both sides*
$2x - 1 = \log(100)$	*Using the inverse relationship* $\log(10^x) = x$
$2x - 1 = 2$	*Using the fact that* $10^2 = 100$
$2x = 3$	
$x = \frac{3}{2} = 1.5$	

We can check our answer by substituting $x = 1.5$ back into the original equation:

$$10^{2*1.5-1} = 10^{3-1} = 10^2 = 100$$

which is correct.

We can use logarithms to solve a banking problem from Section 10.1.

EXAMPLE 8 In Example 4 on page 390 we used a graph to find out how long it will take to double our money when \$1000 is invested at 8% interest compounded quarterly. Use logarithms to solve this problem algebraically.

Solution We need to solve

$$(1.02)^{4x} = 2$$

from Section 10.1. From the definition of common logarithm on page 404 we know that any number x may be written as $10^{\log(x)}$. We apply this fact to 1.02 and 2 in the equation to get

$$\left(10^{\log(1.02)}\right)^{4x} = 10^{\log(2)}$$

The expression on the left side involves taking the power of a power, so we simplify using the *power-of-a-power rule* to obtain

$$10^{\log(1.02)*(4x)} = 10^{\log(2)}$$

Now if $10^x = 10^y$, it is clear that $x = y$, because x and y must be the logarithm of the same number. So,

$$\log(1.02) * (4x) = \log(2)$$

Remembering that log (1.02) is just a number, which we are multiplying by $4x$, we can solve for x:

$$x = \frac{\log(2)}{\log(1.02) * 4}$$

$$x \approx \frac{.301030}{.00860017 * 4}$$

$$x \approx 8.7507$$

It will take about 8.75 years for the money in this account to double.

Exercises 10.3

1. Here are the graphs of the exponential function, $y = 10^x$, and the logarithm function, $y = \log(x)$.

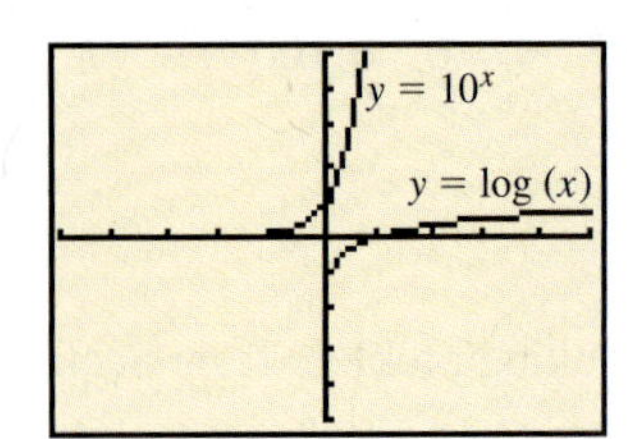

Explain why the two graphs suggest that the functions are inverses of each other. *Note:* You may wish to review Section 9.2.

2. Now compare the table of values for the same two functions.

$y = 10^x$		$y = \log(x)$	
Input (x)	Output (y)	Input (x)	Output (y)
$^-1$	$\frac{1}{10} = .1$	.1	$^-1$
0	1	1	0
1	10	10	1
2	100	100	2

Explain why the table of values also indicates that the exponential and logarithm functions are inverses of each other.

3. Rewrite each number as a power of 10 and then find the log.
   a. $\log(1000)$
   b. $\log(10)$
   c. $\log(1)$
   d. $\log\left(\frac{1}{100}\right)$
4. Notice that the logarithms are not friendly powers of ten. Evaluate each logarithm on your calculator. Round each answer to the nearest .1.
   a. $\log\left(\frac{1}{2}\right)$
   b. $\log(15)$
   c. $\log(140.7)$
5. Work this problem without a calculator: The value of $\log(3.2)$ will fall between what two consecutive whole numbers? (*Hint:* $\log(1) < \log(3.2) < \log(10)$.)
6. Use the method from Exercise 5 to estimate $\log(16)$ without a calculator.
7. Describe each step Anita used to solve the equation $6(10^{2x}) = 600$.

$$6(10)^{2x} = 600$$
$$(10)^{2x} = 100$$
$$\log(10^{2x}) = \log(10^2)$$
$$2x = 2$$
$$x = 1$$

8. Write each exponential equation in its equivalent logarithmic form.
   a. $600 = 10^x$
   b. $\frac{1}{3} = 10^{1.5x}$
   c. $.2 = 10^{3x+1}$
9. Solve the equations from Exercise 8 using your equivalent logarithmic equations.
10. Write each logarithmic equation in exponential form.
    a. $\log(x) = 2$
    b. $4\log(x) = 12$
    c. $\frac{\log(x)}{2} = \frac{1}{2}$
11. Solve the equations from Exercise 10 using your equivalent exponential equations.
12. Solve each equation in either logarithmic or exponential form.
    a. $10^{5x} = 10000$
    b. $\frac{1}{3}\log(x) = 1$
    c. $10^{\log(x)} = 5.7$
13. Solve the equation $2.7 = \log(x)$ graphically and then check your answer. Your Xmax setting in the WINDOW menu may need adjusting.
14. Solve the equation $2.7 = \log(x)$ using an algebraic method. Compare your solution with the solution from Exercise 13.
15. In Example 7 on page 400, we arrived at the equation $.5 = .97^t$. We solved this equation graphically. Now solve this equation algebraically.

Skills and Review 10.3

16. Given $\sqrt[7]{3}$.
    a. Write the expression in exponential form.
    b. Use your calculator to evaluate the expression to the nearest .001.
17. a. Write an exponential decay function for the value of an item after x years if its initial value is \$500 and the item loses one-quarter of its value each year.
    b. Use a graph to estimate how many years it will take for the item to be worth half of its original value.
18. Suppose a species of animal is decreasing by 4% per year. Use the rule of 72 to estimate the number of years before the population is decreased by half.
19. A bank account containing \$750 is earning 3.8% annual interest compounded monthly.
    a. Write an exponential growth function that models the bank account's value after t years.
    b. Use a graphing method to find the approximate number of years it takes for the account to be worth \$1500.
20. Let $f(x) = |x|$. How does the graph of $f(x+6)$ relate to the graph of $f(x)$?
21. Given the direct variation function $y = ax^2$, find the variation constant a if $y = 175$ when $x = 5$.
22. Suppose y varies directly as the square root of x and $y = 3$ when $x = 36$. Find y if $x = 12$. Write your answer in simple radical form.
23. Solve for c: $(c+3)^4 = 16$.
24. Use the quadratic formula to solve $x^2 - 12x + 9 = 0$. Write your solutions in simple radical form.
25. Solve for x: $\frac{2}{3} + \frac{4}{5} = \frac{x}{30}$.

10.4 Logarithmic Scales

After Studying This Section You Will Be Able To

- Understand that the common logarithm counts zeros.
- Use logarithmic scales such as the Richter and pH scales.

We can take advantage of the fact that *logarithms are exponents* in a variety of real-life applications. If we look at a table of values for the log function, we see something interesting.

x	$\log(x)$
0	Undefined
1	0
10	1
100	2
1000	3

The Domain and Range of the Common Logarithm

Why is log (0) undefined? Remember the definition of log from the last section (page 404). The common logarithm of a number x is the number that goes in the exponent of 10 to give x.

$$10^{\log(x)} = x$$

If $x = 0$, we have

$$10^{\log(0)} = 0$$

but no power of 10 can *ever* be zero. If the number in the exponent is a negative number with a large absolute value, then the result will be very small, but it will never be zero. For example,

$$10^{-10} = .0000000001$$

which is almost, but not quite, zero.

In fact, this means that log (x) works only for *positive* values of x, because there is no way to raise 10 to a power and get a negative number. In other words, the *domain* of the function $f(x) = \log(x)$ is the set of positive numbers. We can see this by looking at a graph of the common logarithm (Figure 16).

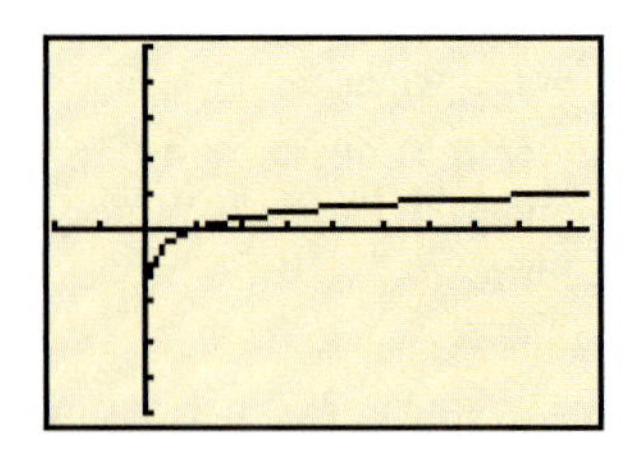

Figure 16 The function $y = \log(x)$.

Looking at Figure 16 and the following table, we start to see a pattern. If we make a table with the powers of 10 explicit, the pattern becomes evident.

x	Power	$\log(x)$
0.001	10^{-3}	$^{-}3$
0.01	10^{-2}	$^{-}2$
0.1	10^{-1}	$^{-}1$
1	10^0	0
10	10^1	1
100	10^2	2
1000	10^3	3

When the input of the logarithm function is a power of ten, the logarithm *counts the zeros*. If the log of the number is bigger than zero, it tells you how many zero digits there are in the number. If the log of the number is negative, it counts one zero digit to the left of the decimal point plus the number of zero digits to the right of the decimal point and in front of the digit 1. (Note that we must write decimals with a leading zero in this case.)

Looking again at Figure 16, we can see this behavior in the fact that the graph of $y = \log(x)$ flattens out as x gets big. As the input to the common logarithm function gets bigger and bigger, the output also gets bigger and bigger, but at a slower and slower rate: 1000 is ten times as big as 100, but $\log(1000)$ is only 1 more than $\log(100)$. In fact, every time the input increases by a factor of 10, the log increases by 1. There is, however, no limit as to how large a logarithm can become. The range of the common logarithmic function is the set of all real numbers.

Using Logarithms

Scientists and engineers find logarithms useful for describing quantities that have a wide range of values. It can be very hard to quickly scan numbers such as 1993.8, .9832333, .0000038847, or 399,928. It is much easier to compare the logarithms of these numbers rather than the numbers themselves.

One particularly good example of this phenomenon is encountered by *seismologists*, who study earthquakes. A very weak earthquake (which people wouldn't even feel) might barely move the ground at all, whereas a very strong earthquake can move the ground many feet. Dr. Charles Richter devised a scale to report the strength of earthquakes using the properties of the common logarithm function. Some recent earthquakes, with their Richter magnitude and the number of fatalities they caused, are shown in the table.

Location and Date	Richter Magnitude	Fatalities
Turkey, 1999	7.8	At least 15,000
Russia, 1995	7.5	1,989
Japan, 1995	7.2	6,430
India, 1993	6.5	Up to 22,000
Iran, 1990	7.7	50,000

Richter used a *seismograph*, a very sensitive pen that moves up and down when an earthquake hits, scratching a line on a sheet of paper. When the earthquake was over, he measured the largest wave made by the pen in millimeters and adjusted that number to account for how far away the earthquake was. He then substituted that reading into this formula (now known as the *Richter scale*):

$$M(A) = \log(A)$$

where A is the amplitude in millimeters of the largest wave recorded on the seismograph, adjusted for a distance of 100 km from the center of the quake. The output M is the **magnitude** of the earthquake.

EXAMPLE 1 What is the Richter magnitude of an earthquake with an adjusted amplitude of 945 mm?

Solution In order to compute the magnitude, we just substitute **945** for A in the formula:

$$M(945) = \log(\mathbf{945}) \approx 2.98$$

Let's look at a table of amplitudes and magnitudes.

Amplitude (mm)	Richter Magnitude
1	0
50	1.7
100	2
500	2.7
1,000	3
5,000	3.7
10,000	4

Each time the amplitude increases by a factor of 10 (say from 500 to 5000), the magnitude increases by 1 (from 2.7 to 3.7). If one earthquake is 10 times as strong as another, its magnitude is 1 greater.

EXAMPLE 2 How much stronger is a Richter 7.4 earthquake than a Richter 5.4 earthquake?

Solution Because the stronger earthquake is two units higher on the Richter scale and each unit means that the earthquake is 10 times as strong, the 7.4 quake is $10 * 10 = 100$ times as strong as the 5.4 quake.

EXAMPLE 3 If one earthquake is five times as strong as another, how do their Richter scale magnitudes compare?

Solution From the table of Richter values, we can find examples where one amplitude is five times the other. For instance, when we compare 500 mm and 100 mm, we see that the difference in magnitudes is $2.7 - 2 = .7$. Similarly, comparing 5000 mm and 1000 mm gives a difference of $3.7 - 3 = .7$. It appears that when one amplitude is five times the other, the difference in magnitudes is .7 on the Richter scale.

We can verify this by using the fact that *logarithms are exponents*. Let a represent the amplitude of the smaller earthquake and A represent the amplitude of the larger earthquake. Then because $A = 5a$, we can write

$$10^{\log(A)} = 10^{\log(5)} * 10^{\log(a)}$$

and

$$10^{\log(A)} = 10^{\log(5)+\log(a)}$$

using the product law for exponents (Section 6.1). Equating the powers—in other words, taking the log of both sides—gives

$$\log(A) = \log(5) + \log(a)$$

A calculator tells us that $\log(5) \approx .699$, so the Richter magnitudes of the two earthquakes differ by about .7.

Notice that because the Richter scale is logarithmic rather than linear, there is a wide range of numbers in the amplitude column of the table of Richter values but a small range of numbers in the magnitude column. Just imagine if every time an earthquake was reported on television the anchors said "the seismograph needle moved 13,474 mm during that one." Would you be able to tell how big an earthquake it was? Probably not, because the range of numbers they could report is so large. However, if they say it was a Richter 4.1 earthquake, you know it is a small one.

Another example of a scientific scale that uses logarithms is the pH scale used in chemistry to measure the acidity of a liquid. Some substances, with their pH, are shown in the table on the following page. Acids have a pH less than 7, and bases have a pH more than 7.

The pH scale is given by the formula

$$\text{pH}(\text{H}^+) = {}^-\log(\text{H}^+)$$

where H^+ is the concentration of hydrogen ions measured in *moles per liter* (mol/L). (If you haven't taken a chemistry course and don't know what a mole is, don't worry about the units. You can still understand the concept of pH.)

The formula for pH looks a lot like the formula for Richter magnitude on page 412, except for the minus sign. Because H^+ is never more than 1, the log of H^+ is always 0 or negative (look at the graph of the common logarithm in Figure 16). Therefore, the pH is always 0 or positive (because the opposite of a

pH	Substance	
14	**Sodium hydroxide**	**Base**
13	**Lye**	**Base**
12.4	**Lime (calcium hydroxide)**	**Base**
11	**Ammonia**	**Base**
10.5	**Milk of magnesia**	**Base**
8.3	**Baking soda**	**Base**
7.4	**Human blood**	**Base**
7	**Pure water**	**Neutral**
6.6	**Milk**	**Acid**
4.5	**Tomatoes**	**Acid**
4	**Wine and beer**	**Acid**
3	**Apples**	**Acid**
2.2	**Vinegar**	**Acid**
2	**Lemon juice**	**Acid**
1	**Battery acid**	**Acid**
0	**Hydrochloric acid**	**Acid**

negative is positive), but the scale goes the *opposite direction* from the Richter scale—the bigger H^+ is, the *lower* the pH.

EXAMPLE 4 Which substance has a lower pH and hence is more acidic, one where $H^+ = 2.1 * 10^{-5}$ mol/L or one where $H^+ = 3.5 * 10^{-7}$ mol/L?

Solution The concentrations are expressed in scientific notation (see Section 1.4), and you should be able to tell that the first H^+ is bigger than the second, because $\mathbf{2.1 * 10^{-5}} = .000021$ and $\mathbf{3.5 * 10^{-7}} = .00000035$.

The first substance has a pH given by

$$\text{pH}(\mathbf{2.1 * 10^{-5}}) = {}^{-}\log(2.1 * 10^{-5}) = 4.68$$

and the second pH is given by

$$\text{pH}(\mathbf{3.5 * 10^{-7}}) = {}^{-}\log(3.5 * 10^{-7}) = 6.46$$

so the first substance has a lower pH and is more acidic.

■ ■

In addition to measuring earthquakes and acidity, a logarithmic scale is used to measure the intensity of sound. Sounds audible to the human ear vary greatly in intensity. In fact, a sound loud enough to cause some discomfort is 10^{12} times as intense as a sound that can just be barely heard.

The sound that can just be barely heard is given the value 0 on a logarithmic scale. Every time the intensity of sound is multiplied by 10, 1 bel (B) is added on the scale. The unit *bel* is named after Alexander Graham Bell, the inventor of the telephone.

A more common unit for sound intensity is the **decibel** (dB), which is defined as one-tenth of a bel. Thus, every time the number of decibels is increased by 10, the intensity of the sound is multiplied by 10.

EXAMPLE 5 ■ How do two sounds measuring 80 dB and 50 dB compare in intensity?

Solution The difference in intensities is 30 dB, or 3 B. Because each bel represents multiplication by 10, 3 B represents a multiplication by 10^3 or 1000. The 80 dB sound is 1000 times more intense than the 50 dB sound.

EXAMPLE 6 ■ Ordinary conversation at a distance of about three feet produces sound measuring 60 dB. Noise from traffic on a busy street typically produces 75 dB. Compare the two levels of sound.

Solution The difference between the two sounds is 15 dB, or 1.5 B. The ratio of the sound intensities, therefore, is $10^{1.5} \approx 31.6$. So, the noise from the traffic is about 30 times louder than the sound of ordinary conversation.

Exercises 10.4

1. Why are log (0) and the log of a negative number undefined?

2. Find each logarithm.

   a. log(10000000) b. log(.000001)

3. Suppose each number represents an adjusted amplitude reading on a seismograph. Without a calculator, find the Richter magnitude.

   a. 1 mm

   b. 100 mm

   c. 100,000 mm

4. Suppose a seismologist records an earthquake with an adjusted amplitude of 12,843 mm and doesn't have a calculator. Estimate the Richter magnitude of the earthquake by indicating between which two consecutive whole numbers it lies.

5. Use a calculator to find the Richter magnitude for each of the given adjusted earthquake amplitudes. Round your answer to the nearest .1.

   a. 12,843 mm

   b. 240 mm

   c. 351,060,000 mm

For Exercises 6 and 7, complete the tables without a calculator. Use the given ordered pair in each table and your knowledge of logarithmic scales from this section.

6.

x	$\log(x)$
4.2	.6
42	?
420	?
4,200	?
42,000	?

7.

x	$\log(x)$
?	$^-1.2$
?	$^-.2$
?	.8
?	1.8
580	2.8

8. In the early part of 2001, four major earthquakes were reported. (http://neic.usgs.gov)

- On January 13, an earthquake measuring 7.6 on the Richter scale hit El Salvador.
- On January 26, an earthquake measuring 7.7 on the Richter scale hit India.
- On February 13, a second earthquake hit El Salvador. This one measured 6.6 on the Richter scale.
- On February 28, an earthquake measuring 6.8 hit the Seattle, Washington, area.

a. The January 13 earthquake was how many times as intense as the February 13 earthquake?

b. Compare the intensity of the India earthquake to that of the Seattle earthquake.

9. The table shows several measured earthquake amplitudes and their corresponding magnitudes on the Richter scale. Explain why it is helpful to describe the strength of an earthquake with a logarithmic scale (Richter scale) rather than an amplitude.

Amplitude, A (mm)	Richter Magnitude, $\log(A)$	Strength of Earthquake
20	1.3	Microearthquake
199,526	5.3	Moderate
1,995,262	6.3	Strong
199,526,231	8.3	Great

(Strength ratings: CVO Website, USGS/Cascades Volcano Observatory)

10. a. Find the pH for each of the hydrogen ion (H^+) concentrations. (http://www.ilpi.com)

Substance	Hydrogen Ions, H^+ (moles/L)	pH
Bleach	.0000000000001	?
Black coffee	.0000016	?
Tears	.00000004	?
Stomach acid	.01	?

b. Rank the substances according to their acidity, going from most acid to least.

11. Refer to the table in Exercise 10 to fill in the blank. The lower the pH, the ______ (more or less) acidic the substance.

12. Suppose a mixture's hydrogen ion concentration is increased by a factor of 100. By how much and in what direction will the pH change?

13. Lemon juice has a pH of 2.0 and ammonia has a pH of 11.0.

a. Which liquid has a larger concentration of hydrogen ions?

b. Given your choice from part (a), how many times larger is this hydrogen ion concentration than the other liquid's concentration?

14. A sound loud enough to cause some discomfort is 10^{12} times as intense as a sound that can just be barely heard. Recall that a sound that can just barely be heard measures 0 dB. How many decibels measure the sound that causes discomfort?

15. The sound level of light traffic measures 50 dB. The sound level of a soft whisper measures 30 dB. How many times louder is the sound of light traffic?

Skills and Review 10.4

In Exercises 16–19, algebraically solve each equation and then check your answer(s) by substitution.

16. Linear equation: $2x - 5 = 10$

17. Quadratic and square root equations:

a. $4x^2 + 7 = 43$ **b.** $4\sqrt{x} - 1 = 11$

18. Power equations:

a. $4x^5 - 2 = 126$ **b.** $5x^{1/5} + 3 = 8$

19. Exponential and logarithmic equations:

a. $3(10)^{2x} = 3000$ **b.** $4\log(x) = {}^-4$

20. Compare the steps used to solve the equations in Exercises 16–19. Describe the similarities.

21. A radioactive substance is decaying exponentially. The initial value is 3 kg and the half-life is 800 years.

a. Write an exponential decay function for the amount of substance left after x years.

b. How much substance is left after 800 years?

c. Without a calculator, find the amount of radioactive substance remaining after 2400 years.

22. A person's weekly wages vary directly as the number of hours they work. Suppose a person earns $257.50 for working 25 hours. Find the weekly wages when the person works 40 hours.

23. Show that $f(x) = x + 2$ and $g(x) = x - 2$ are inverses of each other.

24. Evaluate $(\frac{8}{125})^{2/3}$.

25. Completely factor each expression by attempting any or all of the following techniques: factoring out the greatest common factor, factoring as a difference of two squares, or factoring by splitting the middle term.

a. $2x^2 + 6x$

b. $x^2y - 16y$

c. $x^2 + 3x - 4$

10.5 The Number *e* and the Natural Logarithm (Optional)

After Studying This Section You Will Be Able To

- Write exponential functions using the natural base, *e*.
- Write exponential equations in logarithmic form using the natural logarithm, $\ln(x)$.

When we studied exponential growth and decay in Sections 10.1 and 10.2, we said that all exponential functions may be written in the form $y = y_0 b^{kx}$. We learned that the value of b is the ratio between the new amount of something and the old amount and that the value of k can be determined by seeing how often the amount changes by a factor of b. In the various applications we studied, b and k took on different values.

Scientists and mathematicians like to try to *simplify* situations like this if they can. One way to do this is to use only one base; then we have to find values only for y_0 and k. The problem is what base to choose. Ten might seem like a reasonable choice, but for a number of reasons, mathematicians have chosen to use a different number instead. This number is e (called the *natural base*), and it can be defined in several ways.

> **Definition The Natural Base *e***
>
> The **natural base** e is the ratio $A(t)/A_0$ found from the compound interest formula, assuming that the rate of interest is 100% per year and that interest is compounded continuously for 1 year. In other words, e may be approximated by the expression $(1 + \frac{1}{n})^n$ for very large values of n, indicating very small compounding periods.

Let's see what happens to $(1 + \frac{1}{n})^n$ when we make n larger and larger.

$$n = 10 \qquad \left(1 + \frac{1}{10}\right)^{10} = 2.59374$$

$$n = 100 \qquad \left(1 + \frac{1}{100}\right)^{100} = 2.70481$$

$$n = 1000 \qquad \left(1 + \frac{1}{1000}\right)^{1000} = 2.71692$$

$$n = 10{,}000 \qquad \left(1 + \frac{1}{10{,}000}\right)^{10{,}000} = 2.71815$$

It certainly looks as though the value of $(1 + \frac{1}{n})^n$ does not become too large as n gets very big; in fact, it looks as though it never even gets to 3. It can be shown that the *limit* as n gets larger and larger is a particular number, e, and that $e \approx 2.71828$.

Also, e is a number like π; it cannot be written as a fraction, so the decimal expansion goes on forever and never repeats. You can see a very good approximation to e by using the e^x key on your calculator and entering 2ND $e \wedge 1$, which gives e to nine decimal places. On many calculators, you will see 2.718281828, so it looks as if the decimal repeats, but if you could see a few more digits, you would see that it doesn't.

EXAMPLE 1 What does the exponential decay function for carbon 14 in Example 1 on page 394 look like if we write it with a base of e?

Solution We know that by using a base of $\frac{1}{2}$, the function for the amount of carbon 14 left after t years is

$$A = A(t) = \left(\frac{1}{2}\right)^{t/5730}$$

Dividing by 5730 in the exponent means that it takes 5730 years to add another power of one-half. If we want to write this using a base of e, the number in the power will have to change:

$$F(t) = e^{^{-}.0001209681t}$$

You should try substituting some values for t into both $A(t)$ and $F(t)$ to make sure you get the same outputs; for instance, $A(1000) = (\frac{1}{2})^{1000/5730} \approx .88606$, and $F(1000) = e^{^{-}.0001209681*1000} = e^{^{-}.1209681} \approx .88606$. If you plot both $A(t)$ and $F(t)$ on your calculator on one screen, you will only see one curve because the graphs are the same.

Any exponential growth or decay function may be written with a base of e. This eliminates the worry of working out what number to use as a base, although it means that the coefficient in the exponent is often more difficult to work with.

The Natural Logarithm

In order to be able to solve equations involving 10 raised to a power, we used the *common logarithm*, $\log(x)$. Recall that we can think of the function $\log(x)$ as answering the question, "What number, if written as an exponent of 10, will give x?" The functions 10^x and $\log(x)$ are *inverses*; in other words, they undo each other.

There is another logarithm that works with e^x in the same way that the *common logarithm* works with 10^x. We call this the *natural logarithm*, and write $\ln(x)$; e^x and $\ln(x)$ are *inverses*, and so we may use $\ln(x)$ to solve equations involving expressions with powers of e.

EXAMPLE 2 Find the natural logarithm of each of these numbers:

a. e^2
b. e^3
c. 10
d. e^{-1}
e. 1

Solution

a. Because $\ln(x)$ and e^x are inverse functions, $\ln(e^2) = 2$.
b. Similarly, $\ln(e^3) = 3$.
c. Because we don't know 10 as a power of e, we can't use the fact that $\ln(x)$ and e^x are inverse functions. A calculator gives us an approximate value of $\ln(10) \approx 2.303$. This tells us that $\ln(10)$ lies between 2 and 3. It also indicates that 10 must lie between e^2 and e^3. Checking these numbers on a calculator, we find that $e^2 \approx 7.390$ and $e^3 \approx 20.086$.
d. Because $\ln(x)$ and e^x are inverse functions, $\ln(e^{-1}) = {}^{-}1$. Be sure to recognize that e^{-1} itself is a positive number, because it is the reciprocal of e. In fact $e^{-1} \approx .368$.
e. Because $1 = e^0$, $\ln(1) = 0$.

■ ■

EXAMPLE 3 The apparent brightness of stars is reported using a logarithmic scale for the **magnitude** of the star. One formula for this scale is

$$M(l) = {}^{-}.92 \ln\left(\frac{l}{l_0}\right)$$

where l_0 is the amount of light given off by a star of zero magnitude and l is the amount of light given off by the star whose magnitude we wish to find. How much more light does a magnitude ${}^{-}1$ star give than a 0 magnitude star?

Solution We need to solve the equation

$$M(l) = {}^-1 = {}^-.92 \ln\left(\frac{l}{l_0}\right)$$

for l.

$$^-1 = {}^-.92 \ln\left(\frac{l}{l_0}\right)$$

$$1.087 \approx \ln\left(\frac{l}{l_0}\right) \quad \textit{Dividing on both sides by } {}^-.92$$

$$e^{1.087} = e^{\ln(l/l_0)} = \frac{l}{l_0} \quad \textit{Raising e to the power of both sides to undo the natural logarithm}$$

$$2.97 \approx \frac{1}{l_0}$$

$$2.97 * l_0 \approx l \quad \textit{Multiplying on both sides by } l_0$$

So a magnitude $^-1$ star gives off almost 3 times as much light as a magnitude 0 star (notice this scale runs *backward*—the more light a star gives off, the *lower* its magnitude).

Exercises 10.5

1. Suppose the fraction of a radioactive material left after x years is given by the function

$$f(x) = e^{^-.0157x}$$

Approximately what is the half-life of this material?

2. Find the natural logarithm of the following numbers.
   - **a.** e^7
   - **b.** $e^{^-3.5}$
   - **c.** 25
   - **d.** 100

Chapter 10 ▪ Key Concepts

Exponential Growth Exponential growth is any process where a quantity is growing so that over the same time interval, the new quantity is some constant multiple of the old quantity. Examples include population growth (if a population is growing by 10% per year, then the new year's population is 1.10 times the old population) and the amount of money in a savings account. (Page 384)

Exponential Function A function that describes a quantity growing or decaying exponentially. The input to an exponential function is usually *time*, and the output is an *amount* or *number*. Exponential functions look like

$$f(x) = y_0 b^{kx}$$

with b and y_0 both positive. Because $f(0) = y_0 b^0 = y_0$, y_0 is the y-intercept, or initial amount (page 387). b is the growth or decay factor and k determines how often per time period we multiply by b. (Page 386)

Compound Amount Compound amount is the amount of money in an account after t years, if the account pays r percent interest yearly, with interest compounded n times per year, and there is an opening balance of A_0, and given by the function

$$A(t) = A_0\left(1 + \frac{r}{n}\right)^{nt} \qquad \text{(Page 389)}$$

Rule of 72 The rule of 72 is a rule for estimating the time it will take for a quantity undergoing exponential growth to double. The doubling time is given *approximately* by dividing 72 by the yearly growth (interest) rate. So, for example, if a quantity is growing at 10% per year, it will take about $\frac{72}{10} = 7.2$ years to double. (Page 390)

Exponential Decay Exponential decay is like exponential growth, but quantities *decrease* by a constant multiple over the same time interval. Radioactive materials undergo exponential decay. (Page 393)

Half-life Half-life is the amount of time it takes for a quantity decaying exponentially to be reduced by half. The rule of 72 can be used to estimate half-life by ignoring minus signs; for example, if something is decaying by 5% per year, it's half-life is about $\frac{72}{5} = 14.4$ years. (Page 393)

Common Logarithm The common logarithm of a number is the exponent to which 10 must be raised in order to give the number. For example, the common logarithm of 100 is 2, because $10^2 = 100$. It is sometimes helpful to think of the logarithm as a question: $\log(x)$ goes where the question mark is in

$$10^{?} = x \qquad \text{(Page 404)}$$

Exponential and Logarithmic Form An equation involving logarithms can be written in two forms, logarithmic,

$$y = \log(x)$$

or exponential,

$$10^y = x$$

Notice that the logarithm (y) belongs in the *exponent* in exponential form. (Page 404)

Exponentials and Logarithms as Inverses The common logarithm and the function 10^x are *inverses* of each other, and so they can each be used to solve equations involving the other. If $10^x = y$, then if we take the common logarithm on both sides, we get

$$\log(10^x) = \log(y) \qquad \textit{Log undoes exponential}$$

$$x = \log(y)$$

If $x = \log(y)$, then we can raise 10 to the power of both sides, and we get

$$10^x = 10^{\log(y)} \qquad \textit{Exponential undoes log}$$

$$10^x = y \qquad \text{(Page 406)}$$

Logarithmic Scales A logarithmic scale is a way of scaling a quantity using logarithms. Scientists often report quantities using a logarithmic scale to make numbers easier to understand. For example, the strength of earthquakes is defined using the Richter scale (page 412) because the intensity of an earthquake as measured by a *seismograph* is usually a messy number.

Exploration

The following table of data shows the number of AIDS cases in the 21 English and Dutch speaking countries served by the Caribbean Epidemiology Centre (CAREC). (Source: Caribbean Epidemiology Centre)

Year	Number of AIDS Cases
1996	1800
1997	1900
1998	2100
1999	2350
2000	2750

An exponential function that fits these data is $y = 922.5 * 1.112^x$, where x is the number of years since 1990 and y is the number of AIDS cases.

a. Show that this function is a fairly good fit for these five data points by graphing the data and the function in the same window on a graphing calculator.

b. Estimate the time it takes for the number of AIDS cases to double.

c. Use this function to predict the number of AIDS cases in this region in the year 2010.

d. What factors might cause the prediction made in part (c) to be inaccurate?

Chapter 10 ▪ Review Exercises

Section 10.1 Exponential Growth

1. a. Use the exponential function $y = 2(4)^x$ to complete the table by hand.

x	y
0	
1	
2	
3	

b. What is the initial value, y_0?

c. By what factor b do the outputs increase?

2. Write an exponential growth function, $y = y_0 b^x$, for an initial value of 200 and a growth factor of 7.

Exercises 3–5 deal with the following situation. A $2000 investment in an individual retirement account (IRA) earns 6.8% annual interest compounded monthly.

3. a. Write an exponential function that models the amount, A, of money in this account in terms of the number of years, t.

b. What is the IRA worth after 7 years?

4. Use the rule of 72 to estimate how many years it takes for the IRA to double.

5. Use a graphing method to find a better approximation for the number of years until the IRA doubles.

6. Census data show that the population of the United States is growing by 1.2% each year. Estimate the time for the population to double. (http://quickfact.census.gov)

Section 10.2 Exponential Decay

7. Which exponential decay function (a), (b), or (c) fits the data? *Note:* You may enter the functions in your calculator and make a table to test your choice.

x	y
0	1400
3	700
6	350
9	175

a. $y = 1400(2)^{x/3}$
b. $y = 1400(\frac{1}{2})^{3x}$
c. $y = 1400(\frac{1}{2})^{x/3}$

8. Uranium 234 (U-234) has a half-life of 245,000 years. Suppose we begin with 15 g of U-234.

a. Define the term half-life.

b. How much U-234 is left after 245,000 years?

c. How much U-234 is left after 490,000 years?

d. Complete a table of values that shows the amount A of U-234 left after $t = 0$; 245,000; 490,000; and 735,000 years.

e. Write a function for the amount A of U-234 in terms of the number of years, t.

9. Write an exponential decay function for a substance that begins at 12 g and loses 8% of its value every 5 hours.

10. Suppose the snow line on a popular ski mountain is receding exponentially. The snow line can be found from the decay function $y = 1.2(.98)^t$, where the distance y from the top of the mountain (in miles) depends upon the number of years, t.

a. How far from the top of the mountain is the snow line after 15 years?

b. Approximate, using your grapher, the number of years before the snow line is .6 miles from the top of the mountain.

11. Explain why the two exponential decay functions $y = 20(6)^{-x}$ and $y = 20(\frac{1}{6})^x$ are equivalent.

12. Suppose an $8000 piece of equipment decreases in value by one-third every year.

a. Write an exponential decay function to model this situation.

b. Approximate the equipment's half-life.

13. Every exponential function of the form $y = y_0 b^{kx}$ with $y_0 > 0$ and $b > 0$ has the same domain and range. What are the domain and range for these exponential functions?

Section 10.3 Logarithms

14. Evaluate each logarithm without a calculator.

a. log (100)

b. log (10,000)

c. $\log(\frac{1}{10})$

15. The logarithms below are not of friendly powers of ten. Evaluate them on your calculator. Round your answers to the nearest .001.

a. log (.2)

b. $\log(\frac{5}{3})$

c. log (78)

16. Solve the equation $2\log(x) = 6$.

17. Given the equation $4\log(x) = 5$

a. Find an exact solution.

b. Write the solution as a decimal number rounded to the nearest .01.

18. Solve each equation.

a. $10^x = 100$

b. $2(10)^{x/5} = 20$

c. $10^{4x-1} = 1000$

19. Solve for x: $1 = 100(10)^{-x/3}$.

Section 10.4 Logarithmic Scales

20. The common logarithm function is $y = \log(x)$.

a. Sketch a graph of this function.

b. What are the domain and range of this function?

21. Use the formula magnitude $= \log(A)$ to find the Richter magnitude for each of the earthquake measurements. Round to the nearest .01.

a. 11 mm

b. 1412 mm

c. 53,764,720 mm

22. On March 3, 2002, an earthquake measuring 7.4 on the Richter scale hit Afghanistan. On April 22, 2002, an earthquake measuring 4.4 on the Richter scale hit Peru. How many more times intense was the earthquake that hit Afghanistan? (http://neic.usgs.gov)

23. Use the formula $\text{pH} = -\log(\text{H}^+)$ to find the pH of the substance with the given hydrogen ion (H^+) concentrations. (http://www.aqd.nps.gov)

a. Liquid drain cleaner: $1.0 * 10^{-14}$.

b. Oranges: $1.0 * 10^{-3}$.

c. Shrimp: $1.3 * 10^{-7}$.

24. Suppose a mixture's hydrogen ion concentration is increased by a factor of 10. By how much and in what direction will the pH change?

25. How do two sounds measuring 20 dB and 100 dB compare in intensity?

26. The World Health Organization recommends maximum sound levels for many situations. Here are two: 90 dB at a nightclub and 45 dB outdoors in a residential area at night. Compare the intensities of these two noise levels. (http://www.nonoise.org)

Section 10.5 The Number e and the Natural Logarithm

27. Suppose an initial investment of $1000 earns 6% annual interest compounded continuously. The value of the investment after t years is given by the function $A(t) = \$1000e^{.06t}$. Use a graph to approximate the number of years for the investment to double in value.

28. Find the natural logarithm of each number.

a. e^6 **b.** e^{-9} **c.** $e^{1.8}$ **d.** e **e.** 40

CHAPTER 10 ▪ TEST

1. Between the years 1990 and 2000, the population in Nevada grew exponentially. The population y any given number of years t since 1990 can be modeled by the function $y = 1{,}200{,}000(1.052)^t$.
   - **a.** What was the initial population ($t = 0$)?
   - **b.** By what percentage is the population growing each year?
   - **c.** Assuming exponential growth continues at this rate, what will the population be in the year 2010 ($t = 20$)?
   - **d.** Use the function to predict when the population in Nevada will exceed 2,400,000 people. (Exponential function derived from U.S. Census Bureau data.)

2. Recall that to compute the value A of an initial investment of A_0 dollars earning an annual interest rate r, compounded n times per year for t years, you may use the formula:

$$A = A_0\left(1 + \frac{r}{n}\right)^{nt}$$

Suppose you inherit $10,000 and invest the money in an account paying an annual interest rate of 7% compounded daily ($n = 365$).
   - **a.** Find the value of the investment after 6 years.
   - **b.** Estimate, to the nearest year, the time it takes for the investment to grow to $20,000. Explain what you did.

3. A colony of bacteria grows by 4% each minute. How many minutes is it before the colony doubles in size?

4. Radium 226 has a half-life of 1600 years. How long will it take for 50 g of radium 226 to decay to 25 g?

5. Suppose there are 8600 cases of a disease and the number of cases is decreasing by 13% each year.
   - **a.** Write an exponential decay function to model this situation.
   - **b.** How many cases will there be in 5 years?

6. A $3000 piece of hardware declines by one-quarter of its value each year.
   - **a.** Write an exponential decay function that models the value of the hardware after t years.
   - **b.** What is the value of the hardware after 2 years?
   - **c.** Approximate the number of years before the hardware is worth one-half of its initial value. Explain your work.

7. Evaluate each logarithm without a calculator and show your work.
   - **a.** $\log(10)$
   - **b.** $\log(1)$
   - **c.** $\log(10^{-3})$
   - **d.** $\log\left(\frac{1}{100}\right)$

8. Solve the logarithmic equation, $3\log(x) = 2$.
   - **a.** Find the exact solution.
   - **b.** Write your answer as a decimal number rounded to the nearest .01.

9. Solve $10^{5x} = 10{,}000$.

10. Given the formula magnitude $= \log(A)$, find the Richter magnitude of an earthquake with an adjusted amplitude (A) of 25,270 mm. Round to the nearest .1.

11. On September 25, 2002, an earthquake measuring 5.1 on the Richter scale hit Mexico. On March 31, 2002, an earthquake 100 times more intense than the Mexican earthquake hit Taiwan. What was the Richter-scale measure of the earthquake that hit Taiwan? (http://neic.usgs.gov)

12. At a distance of three feet, the sound from a vacuum cleaner measures 70 dB. The sound from a power lawn mower measures 100 dB at the same distance. How many times more intense is the sound from a lawn mower than the sound from a vacuum cleaner?

CHAPTER

11

Rational Expressions and Functions

Let It Shine

In designing or decorating a home, it is important to pay careful attention to the selection and placement of light fixtures. The intensity of the light falling on any particular surface determines our perception of the space immediately surrounding the surface. Furthermore, certain areas of a room, where tasks such as food preparation or reading take place, require more intense light than others. Interior decorators must take into account properties of light and features of various light fixtures. One of the most important principles they use in their designs is called the inverse square law. In this chapter you learn how to apply the inverse square law to determine the intensity of light at a particular distance from its source. (See Exploration on p. 470.)

Need help? For on-line resources, visit this web site: **math.college.hmco.com/students.**

11.1 Rational Expressions

After Studying This Section You Will Be Able To

- Find equivalent rational expressions.
- Add, subtract, multiply, and divide rational expressions.
- Simplify rational expressions.

In this section you will extend what you already know about fractions to the set of *rational expressions*. The definition for rational expression uses the term *polynomial*, which was introduced in Chapter 6.

Definition **Rational Expression**

A **rational expression** is a fraction in which both the numerator and denominator are polynomials. For instance,

$$\frac{x^2 + 4x - 5}{2x - 3}$$

is a rational expression: the numerator is a polynomial of degree 2, and the denominator is a polynomial of degree 1. Because constant terms are considered polynomials of degree zero, a rational number such as $\frac{4}{5}$ may also be considered to be a rational expression.

Equivalent Rational Expressions

Just as there are many different fractions that represent the same rational number, given a rational expression, there are other rational expressions that are *equivalent* to it. You may find equivalent expressions by multiplying or dividing both the numerator and the denominator by the same expression, which is equivalent to multiplying the entire expression by 1. This is demonstrated in Example 1.

In Example 1 and in most of the other examples in this section, we show three cases. The first case is most familiar because it involves ordinary fractions. The second case has expressions in which the numerator and denominator are *monomials* (polynomials with only one term). In the third case the numerator and denominator are *binomials* or multiples of binomials. Because the cases proceed from simple to complex, study how the principles applied in the first case are extended to the remaining cases.

EXAMPLE 1 For each expression, find three equivalent rational expressions.

a. $\frac{3}{5}$

b. $\frac{x^2}{y}$

c. $\frac{x+2}{x^2-9}$

Solution This is an open-ended question. There are many solutions. Here are some.

a. $\frac{3}{5} = \frac{3*4}{5*4} = \frac{12}{20}$

$\frac{3}{5} = \frac{3*10}{5*10} = \frac{30}{50}$

$\frac{3}{5} = \frac{3*{}^-2}{5*{}^-2} = \frac{{}^-6}{{}^-10}$

b. $\frac{x^2}{y} = \frac{x^2*x}{y*x} = \frac{x^3}{xy}$

$\frac{x^2}{y} = \frac{x^2*a}{y*a} = \frac{ax^2}{ay}$

$\frac{x^2}{y} = \frac{x^2*(x+1)}{y*(x+1)} = \frac{x^3+x^2}{xy+y}$

c. $\frac{x+2}{x^2-9} = \frac{(x+2)*5}{(x^2-9)*5} = \frac{5x+10}{5x^2-45}$

$\frac{x+2}{x^2-9} = \frac{(x+2)*x^3}{(x^2-9)*x^3} = \frac{x^4+2x^3}{x^5-9x^3}$

$\frac{x+2}{x^2-9} = \frac{(x+2)*(x-1)}{(x^2-9)*(x-1)} = \frac{x^2+x-2}{x^3-x^2-9x+9}$

Suppose two rational expressions are equivalent. Then when almost any values are substituted for the variables, the result will be the same for both expressions. An exception occurs when values of the variable make one of the denominators equal to zero because division by 0 is not allowed.

EXAMPLE 2 Show that

$$\frac{x+2}{x^2-9} \quad \text{and} \quad \frac{x^4+2x^3}{x^5-9x^3}$$

are equal when the following values are substituted for x: 2, $^-5$, and 100. Then show what happens when 0 and 3 are substituted for x.

Solution When $x = 2$,

$$\frac{x+2}{x^2-9} = \frac{2+2}{4-9} = \frac{4}{^-5} = {}^-\frac{4}{5}$$

$$\frac{x^4+2x^3}{x^5-9x^3} = \frac{16+2*8}{32-9*8} = \frac{32}{^-40} = {}^-\frac{4}{5}$$

The others may be computed similarly, or you may use the Table feature on a TI-83 calculator. When entering the expressions in Y1 and Y2, be sure to use grouping symbols for the numerator and the denominator:

$$Y1 = (X+2)/(X \wedge 2 - 9)$$
$$Y2 = (X \wedge 4 + 2 * X \wedge 3)/(X \wedge 5 - 9 * X \wedge 3)$$

To make a table, go to [2ND] TBLSET; for Indpnt: choose Ask. Press [2ND] TABLE and enter the values 2, $^-5$, 100, 0, and 3. Your table should look like this:

X	Y1	Y2
2	-.8	-.8
-5	-.1875	-.1875
100	.01021	.01021
0	-.2222	ERROR
3	ERROR	ERROR

X=

Notice that in the first three rows, columns Y1 and Y2 appear to have the same value. An exception occurs when $x = 0$. When $x = 0$, the first expression is defined because its denominator is $x^2 - 9 = 0 - 9 = {}^-9$. However, when $x = 0$ the second expression is undefined, because $x^5 - 9x^3 = 0 - 0 = 0$. Because division by zero is not allowed, the calculator prints ERROR in the Y2 column. In the case of $x = 3$, both expressions are undefined and ERROR appears in both the Y1 and Y2 columns.

In Example 1 we found several expressions equivalent to a given rational expression. Often, however, we need to find an equivalent rational expression with a specific denominator.

EXAMPLE 3 Find the missing numerator to make the two expressions equivalent.

a. $\frac{5}{6} = \frac{?}{12}$

b. $\frac{x}{y} = \frac{?}{xy^2}$

c. $\frac{x+1}{x-2} = \frac{?}{x^2-4}$

Solution

a. $\dfrac{5}{6} = \dfrac{?}{12}$

$\dfrac{5}{6} = \dfrac{?}{6 * 2}$ *Factoring the second denominator so that one factor is the first denominator, 6; you can see that the first denominator was multiplied by 2*

$\dfrac{5}{6} = \dfrac{5 * 2}{6 * 2}$ *To make an equivalent fraction, the first numerator must also be multiplied by 2*

The missing numerator is $5 * 2 = 10$.

b. $\dfrac{x}{y} = \dfrac{?}{xy^2}$

$\dfrac{x}{y} = \dfrac{?}{y * xy}$ *Factoring the second denominator so that one factor is the first denominator, y; you can see that the first denominator was multiplied by xy*

$\dfrac{x}{y} = \dfrac{x * xy}{y * xy}$ *To make an equivalent fraction, the first numerator must also be multiplied by xy*

The missing numerator is $x * xy = x^2y$.

c. $\dfrac{x+1}{x-2} = \dfrac{?}{x^2 - 4}$

$\dfrac{x+1}{x-2} = \dfrac{?}{(x-2)(x+2)}$ *Factoring the second denominator and noticing that one factor is the first denominator, x − 2; you can see that the first denominator was multiplied by x + 2*

$\dfrac{x+1}{x-2} = \dfrac{(x+1)(x+2)}{(x-2)(x+2)}$ *To make an equivalent fraction, the first numerator must also be multiplied by x + 2*

The missing numerator is $(x + 1)(x + 2)$, which can be expanded to $x^2 + 3x + 2$.

Combining Rational Expressions

Recall that you may add or subtract two fractions if they have a *common denominator*. Similarly, the sum or difference of two rational expressions can be expressed as a single rational expression. To add or subtract the expressions, we need to find a common denominator.

EXAMPLE 4 Express each sum as a single rational expression.

a. $\frac{1}{3} + \frac{4}{9}$

b. $\frac{5}{x} + \frac{7}{x^3}$

c. $\frac{x}{x+1} + \frac{x-2}{x^2+4x+3}$

Solution In each of these three cases, the first denominator is a factor of the second denominator. Therefore, we need only to change the first expression to an equivalent expression with that denominator. We then write both expressions with the same denominator and add the numerators.

a. $\frac{1}{3} + \frac{4}{9} = \frac{1 * \mathbf{3}}{3 * \mathbf{3}} + \frac{4}{9}$ *Multiplying both numerator and denominator of the first fraction by 3*

$= \frac{3}{9} + \frac{4}{9}$ *The fractions now have a common denominator*

$= \frac{7}{9}$ *Adding the numerators*

b. $\frac{5}{x} + \frac{7}{x^3} = \frac{5 * \mathbf{x^2}}{x * \mathbf{x^2}} + \frac{7}{x^3}$ *Multiplying both numerator and denominator of the first fraction by* x^2

$= \frac{5x^2}{x^3} + \frac{7}{x^3}$ *The fractions now have a common denominator*

$= \frac{5x^2 + 7}{x^3}$ *Adding the numerators*

c. $\frac{x}{x+1} + \frac{x-2}{x^2+4x+3} = \frac{x}{x+1} + \frac{x-2}{(x+1)(x+3)}$

Factoring the denominator of the second fraction

$= \frac{x\mathbf{(x+3)}}{(x+1)\mathbf{(x+3)}} + \frac{x-2}{(x+1)(x+3)}$

Multiplying both the numerator and denominator of the first fraction by $x + 3$

$= \frac{x\mathbf{(x+3)} + x - 2}{(x+1)\mathbf{(x+3)}}$ *Adding the numerators*

$= \frac{x^2 + 3x + x - 2}{(x+1)(x+3)}$ *Expanding* $x(x+3)$ *in the numerator*

$= \frac{x^2 + 4x - 2}{(x+1)(x+3)}$ *Combining like terms in the numerator*

The last expression can also be written with an expanded denominator as

$$\frac{x^2 + 4x - 2}{x^2 + 4x + 3}.$$

EXAMPLE 5 Express each sum as a single rational expression.

a. $\dfrac{3}{8} - \dfrac{1}{3}$

b. $\dfrac{a}{x} + \dfrac{b}{y}$

c. $\dfrac{2}{x-1} - \dfrac{3}{x+2}$

Solution In each of these cases the two denominators have no common factors other than 1. A common denominator is found by multiplying the two denominators. In fact, this gives the *least common multiple (LCM)* of the two denominators.

a. $\dfrac{3}{8} - \dfrac{1}{3} = \dfrac{3 * \mathbf{3}}{8 * \mathbf{3}} - \dfrac{1 * \mathbf{8}}{3 * \mathbf{8}}$ *Multiplying the numerator and denominator of the first fraction by 3 and multiplying the numerator and denominator of the second fraction by 8*

$= \dfrac{9}{24} - \dfrac{8}{24}$ *The fractions now have the same denominator*

$= \dfrac{1}{24}$ *Subtracting the numerators*

b. $\dfrac{a}{x} + \dfrac{b}{y} = \dfrac{a * \mathbf{y}}{x * \mathbf{y}} + \dfrac{b * \mathbf{x}}{y * \mathbf{x}}$ *Multiplying the numerator and denominator of the first fraction by y and multiplying the numerator and denominator of the second fraction by x*

$= \dfrac{ay}{xy} + \dfrac{bx}{xy}$ *The fractions now have the same denominator*

$= \dfrac{ay + bx}{xy}$ *Adding the numerators*

c. $\dfrac{2}{x-1} - \dfrac{3}{x+2} = \dfrac{2(\mathbf{x+2})}{(x-1)(\mathbf{x+2})} - \dfrac{3(\mathbf{x-1})}{(x+2)(\mathbf{x-1})}$

Multiplying the numerator and denominator of the first fraction by $x + 2$ and multiplying the numerator and denominator of the second fraction by $x - 1$

$$= \frac{2x+4}{(x-1)(x+2)} - \frac{3x-3}{(x-1)(x+2)}$$

The fractions now have the same denominator; expanding the numerators

$$= \frac{2x+4-(3x-3)}{(x-1)(x+2)}$$ *Subtracting the numerators*

$$= \frac{2x+4-3x+3}{(x-1)(x+2)}$$

Applying the distributive property and the special property of $^{-}1$ *in the numerator*

$$= \frac{^{-}x+7}{(x-1)(x+2)}$$ *Combining like terms in the numerator*

The last expression can also be written with an expanded denominator as

$$\frac{^{-}x+7}{x^2+x-2}.$$

EXAMPLE 6 Express each sum as a single rational expression.

a. $\frac{5}{12} + \frac{3}{16}$

b. $\frac{3}{x^3y} - \frac{7}{xy^2}$

c. $\frac{1}{(x+3)^2} + \frac{1}{x^2-9}$

Solution This set of problems differs from those in the two previous examples. Unlike Example 4, one denominator is not a factor of the other. Unlike Example 5, the two denominators do have common factors other than 1. In each case we can find prime factors for denominators and then find the *least common multiple (LCM)* of the denominators by taking the highest power of each prime factor.

a. $12 = 4 * 3 = 2^2 * 3^1 \quad 16 = 2^4 * 3^0$ *4 is the highest power of 2, and 1 is the highest power of 3*

$$\text{LCM} = 2^4 * 3^1 = 16 * 3 = 48$$

$$\frac{5}{12} + \frac{3}{16} = \frac{5 * \mathbf{4}}{12 * \mathbf{4}} + \frac{3 * \mathbf{3}}{16 * \mathbf{3}}$$

Multiplying the numerator and denominator of the first fraction by 4 and multiplying the numerator and denominator of the second fraction by 3

$$= \frac{20}{48} + \frac{9}{48}$$ *The fractions now have the same denominator*

$$= \frac{29}{48}$$ *Adding the numerators*

b. $x^3y = x^3y^1 \qquad xy^2 = x^1y^2$ *3 is the highest power of x and 2 is the highest power of y*

$$\text{LCM} = x^3y^2$$

$$\frac{3}{x^3y} - \frac{7}{xy^2} = \frac{3 * \mathbf{y}}{x^3y * \mathbf{y}} - \frac{7 * \mathbf{x^2}}{xy^2 * \mathbf{x^2}}$$ *Multiplying the numerator and denominator of the first fraction by y and multiplying the numerator and denominator of the second fraction by x^2*

$$= \frac{3y}{x^3y^2} - \frac{7x^2}{x^3y^2}$$ *The fractions now have the same denominator*

$$= \frac{3y - 7x^2}{x^3y^2}$$ *Subtracting the numerators*

c. $(x+3)^2 = (x+3)^2(x-3)^0$

$$x^2 - 9 = (x+3)(x-3) = (x+3)^1(x-3)^1$$

2 is the highest power of $x + 3$ and 1 is the highest power of $x - 3$

$$\text{LCM} = (x+3)^2(x-3)^1$$

$$\frac{1}{(x+3)^2} + \frac{1}{x^2-9} = \frac{1\mathbf{(x-3)}}{(x+3)^2\mathbf{(x-3)}} + \frac{1\mathbf{(x+3)}}{(x+3)(x-3)\mathbf{(x+3)}}$$

Multiplying the numerator and denominator of the first fraction by $x - 3$ and multiplying the numerator and denominator of the second fraction by $x + 3$

$$= \frac{x-3}{(x+3)^2(x-3)} + \frac{x+3}{(x+3)^2(x-3)}$$

The fractions now have the same denominator

$$= \frac{x-3+x+3}{(x+3)^2(x-3)}$$ *Adding the numerators*

$$= \frac{2x}{(x+3)^2(x-3)}$$

Combining like terms in the numerator

Simplifying Rational Expressions

In the previous examples we found equivalent expressions in order to combine two rational expressions. We can also use the principle of equivalence to *simplify* a rational expression. A rational expression is considered simplified when the numerator and denominator have no common factor other than 1.

EXAMPLE 7 Simplify these rational expressions.

a. $\frac{20}{35}$

b. $\frac{4x^3y^4}{6x^2y}$

c. $\frac{x^3 + 5x^2}{x^2 - 25}$

Solution The first step is to factor both numerator and denominator in order to identify common factors. Then we remove all identical common factors from both numerator and denominator to form a simplified rational expression.

a. $\frac{20}{35} = \frac{2^2 * \not{5}}{\not{5} * 7}$ *5 is a common factor*

$= \frac{2^2}{7}$

$= \frac{4}{7}$

The slash marks identify factors common to the numerator and denominator.

b. $\frac{4x^3y^4}{6x^2y} = \frac{\not{2} * 2 * \not{x}^2 * x * y^3 * \not{y}}{\not{2} * 3 * \not{x}^2 * \not{y}}$ *Common factors are 2, x^2, and y*

$= \frac{2xy^3}{3}$

Note: In this example you could apply the quotient law for exponents introduced in Chapter 6.

c. $\frac{x^3 + 5x^2}{x^2 - 25} = \frac{x^2\cancel{(x+5)}}{\cancel{(x+5)}(x-5)}$ *$x + 5$ is a common factor*

$= \frac{x^2}{x - 5}$

Caution: You may only remove common *factors* from the numerator and denominator. Do *not* remove terms. For example, in Example 7(c) we may not write this:

$$\frac{x^3 + 5x^2}{x^2 - 25} = \frac{\cancel{x}^2 * x + \cancel{5}x^2}{\cancel{x}^2 - \cancel{5} * 5} = \frac{x + x^2}{-5}.$$

Multiplying and Dividing Rational Expressions

When multiplying two fractions we multiply numerators to find the numerator of the product and we multiply denominators to find the denominator of the product. The same is true for rational expressions. For division, we find the reciprocal of the divisor and multiply. When all numerators and denominators are written in factored form, we simplify the answer by removing common factors.

EXAMPLE 8 Multiply or divide, as indicated, and simplify the answer.

a. $\frac{5}{18} \div \frac{25}{9}$

b. $\frac{ax^3}{by} * \frac{b^2y^3}{2a^2}$

c. $\frac{(x+4)^2}{x^2+5x} \div \frac{x^2-16}{x^2+4x-5}$

Solution

a. $\frac{5}{18} \div \frac{25}{9} = \frac{5}{18} * \frac{9}{25}$ *Changing division by $\frac{25}{9}$ to multiplication by the reciprocal, $\frac{9}{25}$*

$= \frac{\cancel{5}}{2 * \cancel{3} * \cancel{3}} * \frac{\cancel{3} * \cancel{3}}{\cancel{5} * 5}$ *Finding common factors in numerators and denominators*

$= \frac{1}{2 * 5}$ *When all prime factors have been removed from the numerator, 1 is the remaining factor*

$= \frac{1}{10}$

b. $\frac{ax^3}{by} * \frac{b^2y^3}{2a^2} = \frac{\cancel{a} * x * x * x}{\cancel{b} * \cancel{y}} * \frac{\cancel{b} * b * \cancel{y} * y * y}{2 * \cancel{a} * a}$

$= \frac{bx^3y^2}{2a}$

Note: In this example you could also apply the quotient law for exponents introduced in Chapter 6.

c. $\frac{(x+4)^2}{x^2+5x} \div \frac{x^2-16}{x^2+4x-5} = \frac{(x+4)^2}{x^2+5x} * \frac{x^2+4x-5}{x^2-16}$

Changing division by $\frac{x^2-16}{x^2+4x-5}$ to multiplication by the reciprocal, $\frac{x^2+4x-5}{x^2-16}$

$$= \frac{(x+4)\cancel{(x+4)}}{x\cancel{(x+5)}} \cdot \frac{\cancel{(x+5)}(x-1)}{\cancel{(x+4)}(x-4)}$$

Finding common factors in numerator and denominator

$$= \frac{(x+4)(x-1)}{x(x-4)}$$

Exercises 11.1

1. Define and give an example of a rational expression.
2. For each rational expression, find two equivalent rational expressions. (*Note:* There are many possible answers.)
   a. $\frac{2}{3}$ **b.** $\frac{x}{y^3}$ **c.** $\frac{x-3}{x+2}$
3. Consider the rational expression

$$\frac{x-3}{x+2}$$

   a. Explain why the expression is undefined at $x = {}^{-}2$.
   b. How should the expression be entered into the Y= menu on your calculator?
   c. How does the undefined output appear in a table on your calculator?
4. The rational expressions

$$\frac{x-3}{x+2} \quad \text{and} \quad \frac{x^2-5x+6}{x^2-4}$$

   are equivalent.
   a. Use the 2ND TABLE feature on your calculator to show that the expressions are equal for the values $x = {}^{-}3, 0, 5, 12$.
   b. Explain why the two expressions are not equal at $x = 2$.
5. Find the missing numerators to make the rational expressions equivalent.
   a. $\frac{3}{4} = \frac{?}{20}$
   b. $\frac{x}{y^2} = \frac{?}{xy^4}$
   c. $\frac{x-5}{x+3} = \frac{?}{x^2-9}$

For Exercises 6–8, express each sum or difference as a single rational expression.

6. **a.** $\frac{1}{2} + \frac{3}{10}$
   b. $\frac{4}{y} - \frac{5}{y^5}$
   c. $\frac{x-1}{x+7} + \frac{x}{x^2+8x+7}$
7. **a.** $\frac{2}{9} + \frac{1}{4}$
   b. $\frac{3}{x} - \frac{2}{y}$
   c. $\frac{1}{x+2} + \frac{4}{x+1}$
8. **a.** $\frac{1}{6} - \frac{3}{10}$
   b. $\frac{5}{x^2y} + \frac{2}{x^2y^2}$
   c. $\frac{3}{(x-2)^2} - \frac{1}{x^2-4}$
9. Simplify each rational expression.
   a. $\frac{6}{32}$
   b. $\frac{8x^2y}{10xy^3}$
   c. $\frac{x^2+3x+2}{x^2+x}$
10. Simplify the rational expression

$$\frac{2x-10}{x^3-25x}$$

11. Describe at least one method for checking that your simplified expression in Exercise 10 is equivalent to the original expression.

In Exercises 12–15, perform the indicated operation and simplify.

12. $\frac{49}{42} * \frac{6}{35}$

13. $\frac{by^2}{x^3} \div \frac{b^3y^5}{ax}$

14. $\frac{x^2-3x}{x^2-36} * \frac{(x+6)^2}{x^2}$

15. $\frac{x+8}{5x-10} \div \frac{x^2-64}{x^2-4x+4}$

Skills and Review 11.1

16. The sound from a blue whale can reach levels up to 188 dB. The sound from a jet engine measures 140 dB. How many times louder is the sound from a blue whale than the sound from a jet engine? (http://www.enchantedlearning.com)

17. Show that the basic exponential function $f(x) = 10^x$ and the basic logarithmic function $g(x) = \log(x)$ are inverses of each other.

18. Find each logarithm without using a calculator.
    a. $\log(100)$ **b.** $\log(\frac{1}{10})$ **c.** $\log(1)$

19. Use an algebraic method to solve each logarithmic equation.
    a. $\log(x) = 4$ **b.** $\log(x) = {}^{-}3$
    c. $\frac{1}{5}\log(x) = 1$

20. Use a graphing method to find an approximate solution to the logarithmic equation $\log(x) = 3.6$.

21. Excluding Alaska and Hawaii, the number of bald eagle pairs in the United States is closely modeled by the exponential function $y = 4449(1.065)^x$, where x is the number of years since 1994. (Function derived from data from the U.S. Fish and Wildlife Service.)
    a. Is the number of eagle pairs increasing or decreasing? Explain.
    b. How many eagle pairs were there in 1994?
    c. By what percent is the number of eagle pairs increasing or decreasing each year?
    d. Predict the number of eagle pairs in the year 2008.

22. Evaluate the expression $(\frac{9}{100})^{3/2}$.

23. Use your knowledge of modifying basic functions to describe how the graph of each function relates to the graph of a basic function from Section 8.3.
    a. $g(x) = |x+1|$ **b.** $g(x) = {}^{-}x^3$
    c. $g(x) = 4\sqrt{x}$ **d.** $g(x) = \frac{1}{x} + 5$

24. What is the range of the function $f(x) = x^2 - 4$?

25. Let $f(x) = x^2 - 3x - 8$. Find exact answers for each of the following.
    a. The y-intercept
    b. The x-intercepts
    c. The vertex

11.2 Rational Functions and Their Graphs

AFTER STUDYING THIS SECTION YOU WILL BE ABLE TO

- Graph a rational function.
- Identify vertical and horizontal asymptotes.
- Find the domain and range for a rational function.

Rational Functions

In this section you will learn to recognize and interpret graphs of *rational functions*.

Definition **Rational Function**

A **rational function** is a function of the form $f(x) =$ rational expression in x.

You have already seen one example of a rational function, the reciprocal function, $f(x) = \frac{1}{x}$. This function satisfies the definition of rational function because $\frac{1}{x}$ is a rational expression.

We begin our study of rational functions with an application.

EXAMPLE 1 Producto Inc. manufactures gizmos. Each week the fixed cost of production is \$500. It costs an additional \$8 to produce each gizmo. Recall from Section 2.4 that total cost = variable cost + fixed cost. If x gizmos are produced in one week, then the variable cost is $8x$ dollars for that week. Therefore, the total cost for the week is $8x + 500$ dollars. The management of Producto is interested in knowing the *average cost* of producing gizmos. By definition, average cost is the total cost for a week divided by the number of gizmos produced in that week. Average cost can be expressed as the function

$$f(x) = \frac{8x + 500}{x} = \frac{\text{total cost}}{\text{number produced}}$$

Because $f(x)$ is defined using a rational expression, it is considered a rational function.

a. Make a table showing the total cost and average cost for producing gizmos. Use these values for x: 0, 20, 40, 60, 80, 120.

b. Observe patterns in the table.

c. Explore what happens to the average cost when x takes on very large values.

Solution **a.** You may substitute values for x in the expressions for total cost and average cost. Or you may use the table feature of the calculator. Let Y1 = 8X + 500 for total cost. Let Y2 = (8X + 500)/X for average cost.

Under [2ND] TBLSET, let TblStart = 0 and ΔTbl = 20, in order to start X at 0 and increment by 20 gizmos. Select Auto for both independent and dependent variables. The result is the following table.

X (No. of gizmos)	Y1 (total cost)	Y2 (average cost)
0	500	ERROR
20	660	33
40	820	20.5
60	980	16.333
80	1140	14.25
100	1300	13
120	1460	12.167

b. Here are some patterns you may observe in the table:

1. As X increases, total cost increases at a constant rate of \$160 for every 20 gizmos, that is, \$8 per gizmo.
2. As X increases, average cost decreases, but the rate of decrease appears to slow down. (From X = 20 to X = 40, Y2 drops from 33.3 to 20.5, but from X = 100 to X = 120, it drops only from 13 to 12.167.)
3. When X = 0, the average cost is undefined because division by 0 is not allowed. That is why the first entry in the Y2 column reads ERROR.

c. Go back to [2ND] TBLSET and change the independent variable from Auto to Ask. Then substitute increasingly large values of X in the table, as shown.

X (No. of gizmos)	Y1 (total cost)	Y2 (average cost)
100	1300	13
500	4500	9
1000	8500	8.5
10000	80500	8.05
100000	800500	8.005

It appears that as X becomes very large, the value of Y2 (average cost) continues to decrease, but it always remains above \$8/gizmo.

EXAMPLE 2 Use the gizmo-production functions from Example 1. Describe each graph.

a. Total cost as a function of X.

b. Average cost as a function of X.

Solution

a. The total cost is given by Y1 = 8X + 500. You should recognize this as a linear function in slope-intercept form (Section 4.2). The slope is 8; it represents the rate of change in total cost as the number of gizmos increases, that is, \$8 per gizmo. The y-intercept is 500, which represents the cost (in dollars) of producing zero gizmos, which is the same as the fixed cost. From our study of linear functions, we know that the graph of this function is a straight line with a positive slope. (See Figure 1.)

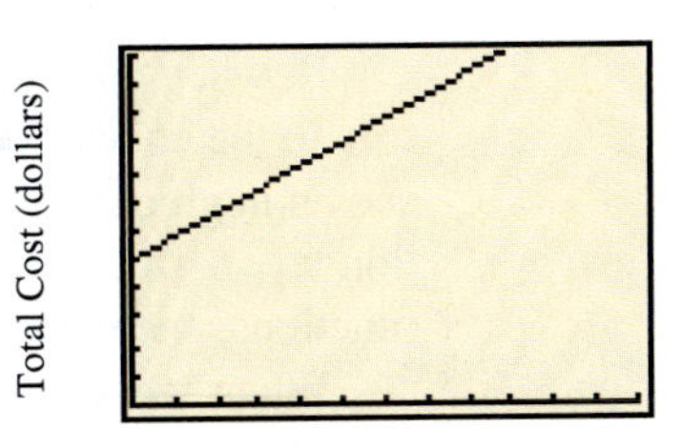

Figure 1

b. The average cost is given by Y2 = (8X + 500)/X. Because you are not familiar with rational functions, you won't be able to recognize immediately the shape of the graph. You may plot points from the table for the average cost function by hand or use the graph feature of a calculator. After some experimentation you will find an appropriate window. The graph for the window $0 \le X \le 940$, $0 \le Y \le 35$ is shown in Figure 2(a). The graph may be described as a downward-sloping curve in the first quadrant. When X is close to zero, the curve falls steeply. Then it appears to level off.

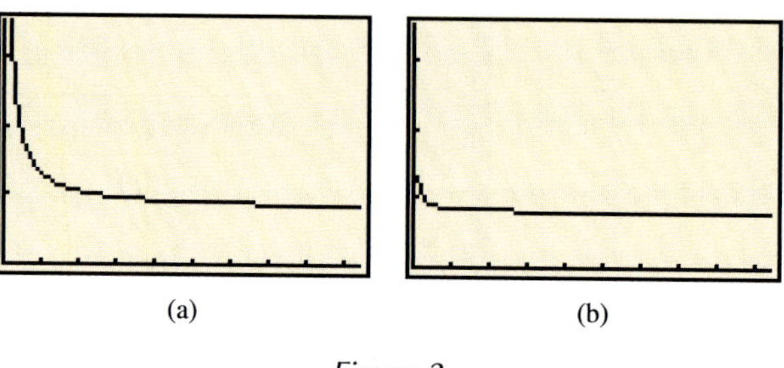
(a) (b)

Figure 2

To get a picture of how it levels off, change the scale on the x-axis. The graph in Figure 2(b) gives a picture of the function with the window $0 \le X \le 9400$, $0 \le Y \le 35$.

Notice that the right two-thirds of the graph appears to be a horizontal line. Tracing along this portion of the graph, however, reveals that as X increases, the value of Y changes. It slowly decreases and comes closer and closer to 8.0.

Hyperbolas

The graph of the average cost function, shown in Figure 2, is typical of the graphs of many rational functions. The shape of this curve has a special name. It is called a **hyperbola**. The basic reciprocal function $f(x) = \frac{1}{x}$ introduced in Section 8.3 is also a hyperbola. In fact, all the hyperbolas you will encounter in this course can be derived from the reciprocal function by stretching, reflecting, and shifting.

If, when a curve is traced, it approaches a line, the line is called an **asymptote** to the curve. Hyperbolas have two asymptotes. In Figure 4 the asymptotes, shown in green, are the y-axis (a vertical line) and the horizontal line with equation $y = 8$. In Chapter 10 you observed the graphs of exponential and logarithmic functions, each of which has only one asymptote. See Figure 3.

When we view our function in all four quadrants, we discover that a hyperbola has two **branches**, both with the same pair of asymptotes, as shown in Figure 4. In Example 2 we needed to look at only one branch because only nonnegative values of the variable make sense in the context of the application.

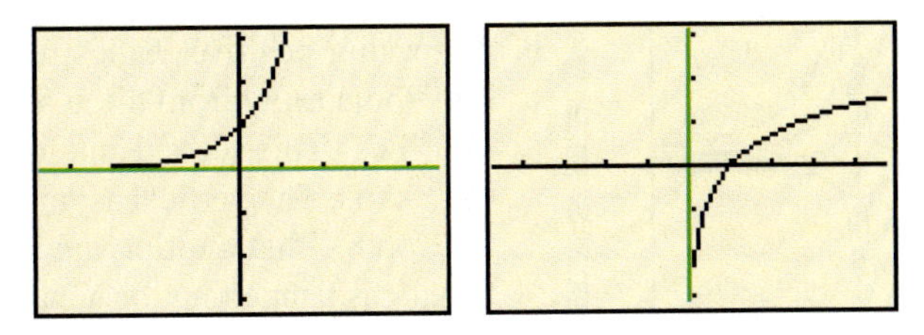

Figure 3 An exponential function has a horizontal asymptote. A logarithmic function has a vertical asymptote. Asymptotes are shown in green.

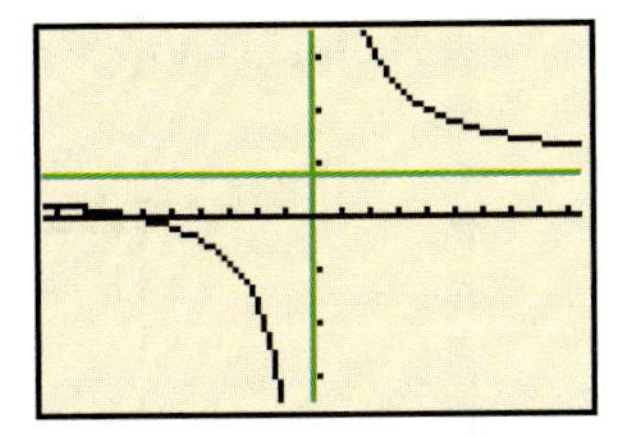

Figure 4 $Y1 = (8X + 500)/X$ in the window $^{-}94 \leq X \leq 94$, $^{-}35 \leq Y \leq 35$.

Stretching and Shifting Hyperbolas

In Section 8.4 we learned how a function may be modified by multiplying the output variable by a constant or by adding or subtracting a constant from the input or output variable. We observed the resulting changes in the graph of the function. Let us now look at how these changes affect the graph of the basic reciprocal function, $y = \frac{1}{x}$.

EXAMPLE 3

a. Graph $Y1 = 1/X$ in the friendly window $^{-}9.4 \leq X \leq 9.4$; $^{-}6.2 \leq Y \leq 6.2$. Where are the asymptotes for the graph?

b. Observe what happens when you replace the numerator, 1, with another constant. Graph these functions: $Y2 = 4/X$, $Y3 = 9/X$, $Y4 = {}^{-}1/X$. Identify the asymptotes for each graph.

c. Observe what happens when you replace the denominator with X plus or minus a constant. Graph these functions: $Y2 = 1/(X + 5)$ and $Y3 = 1/(X - 5)$. Identify the asymptotes for each graph.

d. Observe what happens when you add a constant to the expression $1/X$. Graph these functions: $Y2 = 1/X + 3$ and $Y3 = 1/X - 3$. Identify the asymptotes for each graph.

e. Observe what happens when you first add one constant in the denominator and then add another to the rational expression that is formed. Graph $Y2 = 1/(X + 5) + 3$. Identify the asymptotes.

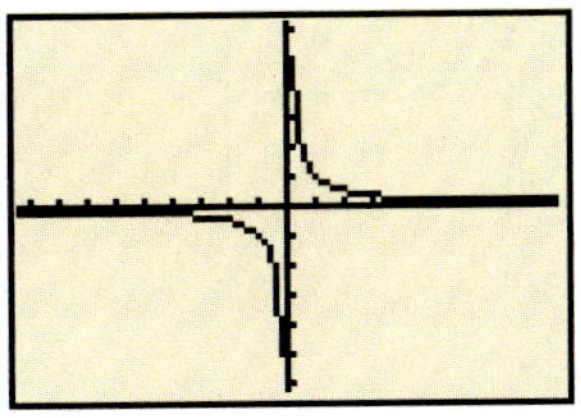

Figure 5

Solution

a. The branches of the hyperbola lie in the first and third quadrants. The horizontal asymptote is the x-axis. The vertical asymptote is the y-axis. See Figure 5.

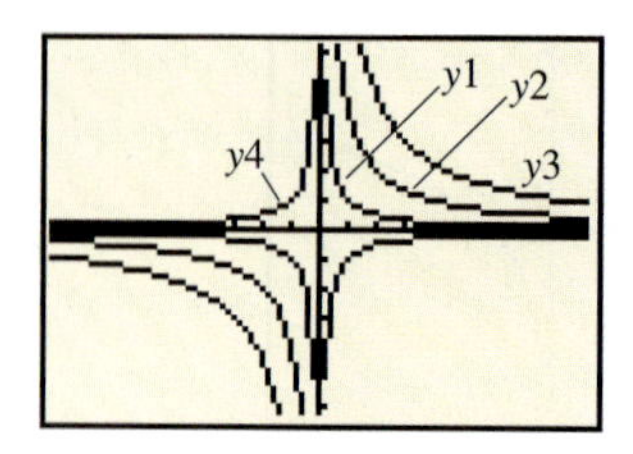

Figure 6

b. Increasing positive values in the numerator moves the two branches of the hyperbola away from the x-axis. A negative value in the numerator moves the branches to the second and fourth quadrants. The horizontal asymptote is still the x-axis. The vertical asymptote is still the y-axis. See Figure 6. Recall from Section 8.4 that multiplying a function by a constant stretches the graph. This is what is happening for Y2 and Y3. Y4 shows how multiplying a function by $^-1$ reflects the graph across the x-axis.

c. Let $f(x) = \frac{1}{x}$. Then, as we learned in Section 8.4, the graph of $y = f(x + 5) = \frac{1}{x+5}$ is the graph of $f(x)$ shifted to the left 5 units, as shown in Figure 7(a). The horizontal asymptote is still the x-axis, but the vertical asymptote is now $x = {}^-5$.

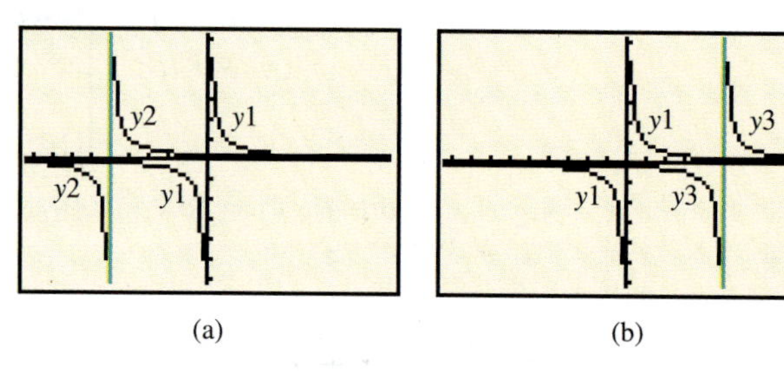

Figure 7

Similarly, the graph of $y = f(x - 5) = \frac{1}{x-5}$ is the graph of $f(x)$ shifted to the right 5 units, as shown in Figure 7(b). The horizontal asymptote is still the x-axis, but the vertical asymptote is now $x = 5$.

d. Again, the principles of Section 8.4 apply. The graph of $y = f(x) + 3 = \frac{1}{x} + 3$ is the graph of $f(x)$ shifted up 3 units, as shown in Figure 8(a). The vertical asymptote is the y-axis, and the horizontal asymptote is now $y = 3$.

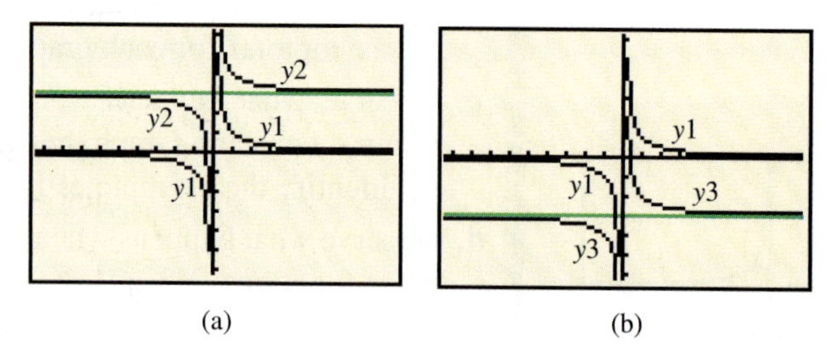

Figure 8

The graph of $y = f(x) - 3 = \frac{1}{x} - 3$ is the graph of $f(x)$ shifted down 3 units as shown in Figure 8(b). The vertical asymptote is the y-axis, and the horizontal asymptote is now $y = {}^-3$.

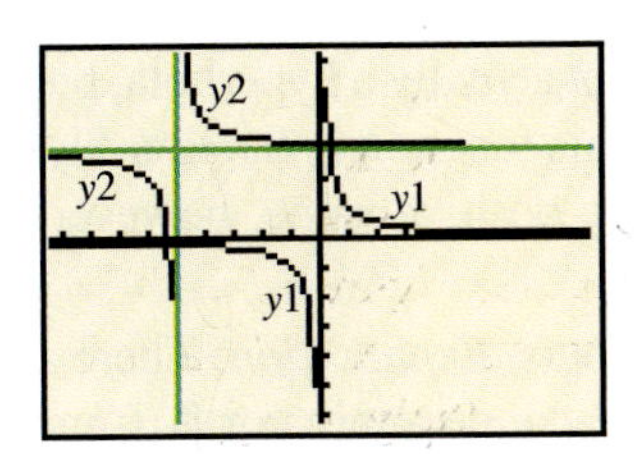

Figure 9

e. The graph of $y = f(x+5) + 3 = \frac{1}{x+5} + 3$ is the graph of $f(x)$ shifted 5 units to the left *and* 3 units up, as shown in Figure 9. The vertical asymptote is the line $x = {}^-5$, and the horizontal asymptote is the line $y = 3$.

Domain and Range of Rational Functions

The domain of a rational function is the set of all real numbers except those that give a denominator of zero. The numbers that don't belong to the domain usually correspond to the vertical asymptotes of the graph.

We can find the range of a rational function by examining its graph. In most cases that we consider, any y-value that is a horizontal asymptote is not in the range, but all other real numbers are. We must check to see if the graph is both above and below the horizontal asymptote.

EXAMPLE 4 Determine the domain and range for each of these functions.

a. $y = \frac{1}{x}$

b. $y = \frac{1}{x+5}$

c. $y = \frac{1}{x} + 3$

d. $y = \frac{10}{x^2}$

e. $y = \frac{x}{x^2 - 1}$

Solution

a. Domain: $x \neq 0$; range: $y \neq 0$. *Explanation:* The vertical asymptote is determined by the value of x that makes the denominator 0. In this case, it is $x = 0$, which is the y-axis. For all other values of x, the function is defined. As we saw in Figure 5, the x-axis ($y = 0$) appears to be a horizontal asymptote. It also appears from Figure 5 that for all other values of y, there is a point on the graph.

b. Domain: $x \neq {}^-5$; range: $y \neq 0$. *Explanation:* The denominator is 0 when $x = {}^-5$. This corresponds to the vertical asymptote for the graph shown in Figure 7(a). The horizontal asymptote is still the x-axis and again it appears that for all values of y other than 0, there is a point on the graph.

c. Domain: $x \neq 0$; range $y \neq 3$. *Explanation:* The denominator is 0 when $x = 0$, so the vertical asymptote is the y-axis and the function is defined everywhere except at 0. As we see in Figure 8(a), the horizontal asymptote is now $y = 3$, and it appears that for all values of y other than 3, there is a point on the graph.

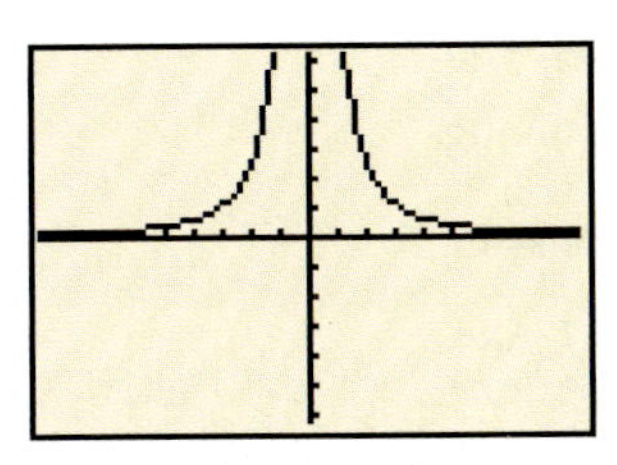

Figure 10

d. Domain $x \neq 0$; range $y > 0$. *Explanation:* Again the denominator is 0 when $x = 0$ and the function is defined for all other values of x. A graph of this function is on the left in Figure 10. Note that the curve has two branches lying

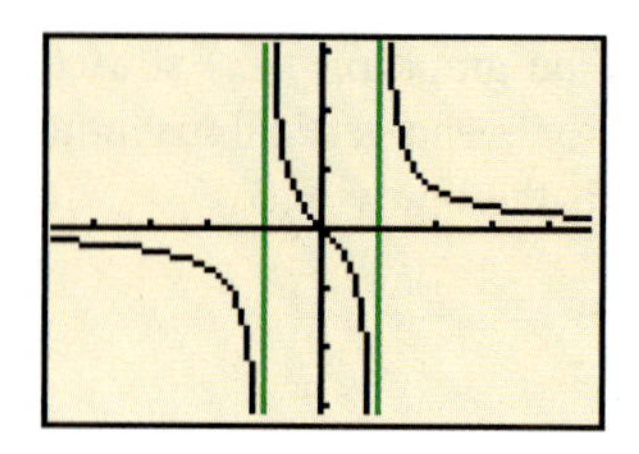

Figure 11

in the first and second quadrants. (This graph is not actually a hyperbola, but it shares certain features. You'll learn more about this function in Section 11.4.) From the graph it appears that y can take on any positive value; however, y can never be 0 or take on a negative value.

e. Domain: $x \neq {}^{-}1$ and $x \neq 1$; range: all real numbers. *Explanation:* There are two values of x, $x = {}^{-}1$ and $x = 1$, that make the denominator 0. Consequently, there are vertical asymptotes at $x = {}^{-}1$ and $x = 1$. See Figure 11. In this case there are actually three branches to the curve. The middle branch passes through the origin (0, 0), so all real numbers are included in the range even though $y = 0$ is a horizontal asymptote.

Vertical Asymptotes on a Calculator

When you graph rational functions on a calculator, strange things may occur with vertical asymptotes. For example, Figure 12 shows the graph of

$$y = \frac{2}{x - 3}$$

in a friendly window (${}^{-}9.4 \leq X \leq 9.4$; ${}^{-}6.2 \leq Y \leq 6.2$). Clearly, the hyperbola has two distinct branches. Tracing shows what happens when X gets close to 3.

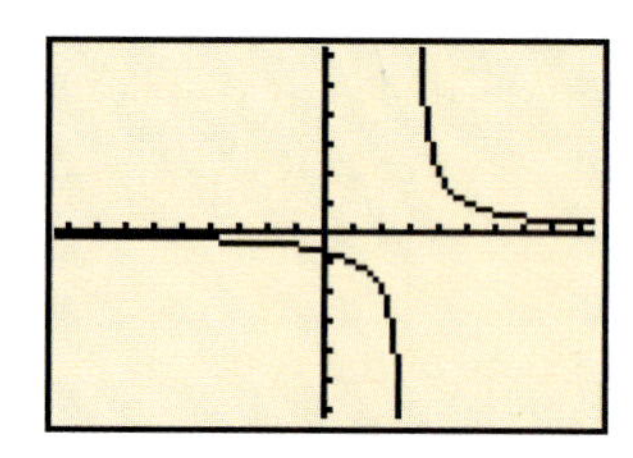

Figure 12

$X = 2.6$	$Y = {}^{-}5$
$X = 2.8$	$Y = {}^{-}10$
$X = 3$	$Y =$
$X = 3.2$	$Y = 10$
$X = 3.4$	$Y = 5$

Notice the break that occurs when $X = 3$. For that value of X, the function is undefined and the calculator is programmed to separate the two branches of the curve. Now graph the same function in the standard window (Zoom6, ${}^{-}10 \leq X \leq 10$; ${}^{-}10 \leq Y \leq 10$). This gives the graph shown in Figure 13. It now looks as if the vertical asymptote has been drawn as part of the function. But the asymptote is not part of the graph, and it really should not be shown. Tracing along the curve near the asymptote shows what has actually happened.

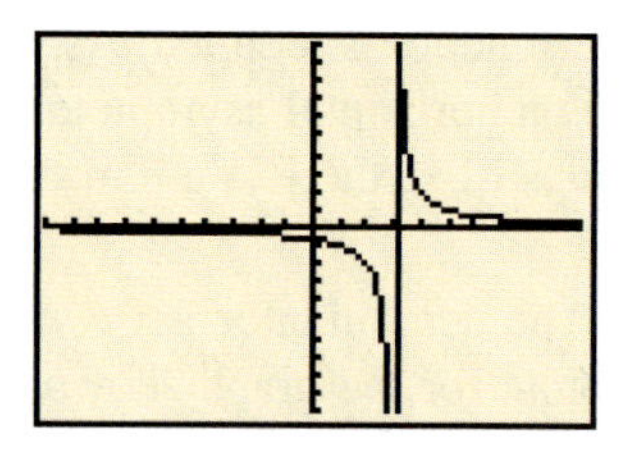

Figure 13

$X = 2.7659575$	$Y = {}^{-}8.545455$
$X = 2.9787234$	$Y = {}^{-}93.99998$
$X = 3.1914894$	$Y = 10.444445$
$X = 3.4042553$	$Y = 4.9473684$

The calculator connected the points (2.9787234, ${}^{-}93.99998$) and (3.1914894, 10.444445) in what appears to be a vertical line.

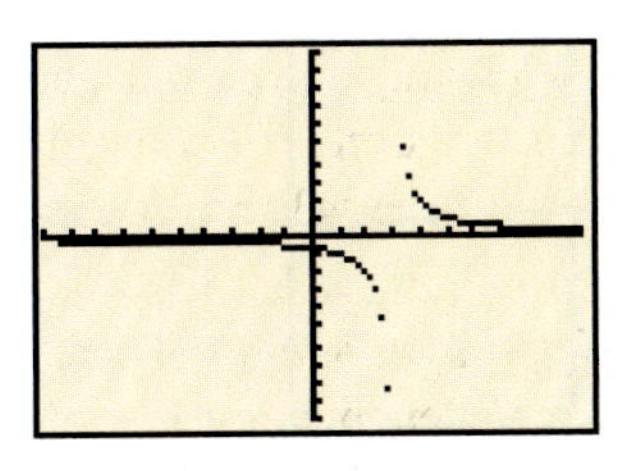

Figure 14

You can avoid this deceptive representation of the function in two ways. One is to choose a friendly window that includes as pixel values those values of x for which the function is not defined. This is not always possible when the values in question are fractional. The other way is to change the mode from *connected* to *dot*. On a TI-83 you can do this by pressing the MODE key and making the change in the fifth line of the screen. The graph in Figure 14 shows the graph with the standard window in the dot mode.

Exercises 11.2

1. Suppose the average cost for a company is given by the function

$$f(x) = \frac{7x + 400}{x},$$

where x is the number of items produced.

a. Make a table showing the average cost for producing 0, 25, 50, 75, and 100 items.

b. What patterns do you notice in your table?

c. Experiment with extremely large values of x. What happens to the average cost?

d. Explain why the output for $x = 0$ is undefined or shows an error on your calculator.

2. Again, consider the function

$$f(x) = \frac{7x + 400}{x}.$$

a. Create a graph with your calculator using the window $0 \le X \le 940$, $0 \le Y \le 35$.

b. Describe the shape of the graph.

c. Does your description support the patterns you noticed in the table from Exercise 1? Explain.

d. Does the function appear to have any x- or y-intercepts in the first quadrant?

3. Use the given table of values for the function $f(x) = \frac{1}{x-1} + 3$ to identify the horizontal and vertical asymptotes.

X	Y1
0	2
1	ERROR
10	3.1111
100	3.0101
1000	3.001
10000	3.0001

Y1=1/(X−1)+3

4. Sketch a graph of the basic reciprocal function $f(x) = \frac{1}{x}$ shifted 4 units to the right.

5. Sketch the graph of the basic reciprocal function reflected across the x-axis.

For Exercises 6–10, (a) describe how the graph of each function relates to the basic reciprocal function $f(x) = \frac{1}{x}$ and (b) identify the horizontal and vertical asymptotes.

6. $g(x) = \frac{3}{x}$

7. $g(x) = \frac{^-2}{x}$

8. $g(x) = \frac{1}{x + 4}$

9. $g(x) = \frac{1}{x} - 1$

10. $g(x) = \frac{1}{x - 2} + 6$

11. Find the domain and range of each function in Exercises 6–10.

12. Recall the average cost function from Example 1, $f(x) = \frac{8x+500}{x}$.

a. Write this function as the sum of two rational expressions. For instance, $\frac{3x+2}{x}$ can be written as $\frac{3x}{x} + \frac{2}{x}$.

b. If possible, simplify each rational expression.

c. Now use your knowledge of modifications to discuss how your graph of the new equivalent function relates to the graph of the basic rational function $g(x) = \frac{1}{x}$.

13. Given the rational function $f(x) = \frac{^-5}{x^2}$.

a. The graph of this function appears in which quadrants?

b. Find the domain and range.

14. Consider the function

$$f(x) = \frac{5x - 12}{x - 4}.$$

a. How can you find the domain and vertical asymptote?

b. How can you find the horizontal asymptote?

15. The graph of the rational function $f(x) = \frac{5}{x+4}$ is shown as viewed with the standard (Zoom 6) window.

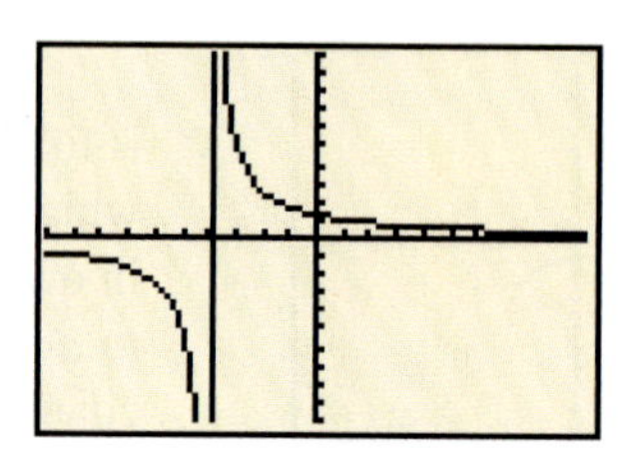

a. Should the vertical asymptote be part of the graph? Explain.

b. Explain how you can prevent the asymptote of this function from appearing on the graph.

Skills and Review 11.2

16. Simplify the rational expression

$$\frac{b - 3}{b^2 + b - 12}$$

In Exercises 17 and 18, perform the indicated operation and simplify.

17. $\dfrac{x}{x+3} - \dfrac{x+1}{x^2 - 9}$

18. $\dfrac{x^2 - x}{x^2 + 10x + 25} * \dfrac{x+5}{(x-1)^2}$

19. In the following table the outputs of the common logarithm are increasing by 1. Use your knowledge of logarithmic scales to complete the missing inputs.

x	$\log(x)$
?	$^{-}.9$
?	.1
13	1.1
?	2.1
?	3.1

20. Solve for x in the equation $250 = 2.5(10)^{6x}$.

21. A tool initially worth \$1500 decreases in value by 20% each year. Estimate the half-life of this tool using the rule of 72.

22. The amount of power generated by a windmill varies directly as the cube of the wind speed. When the wind speed is 10 miles/hour, a windmill generates 300 watts of power. Find the variation constant.

23. Let

$$f(x) = \frac{x+2}{x}.$$

Find each value without using a calculator.

a. $f(^{-}2)$ **b.** $f(0)$

c. $f(2)$ **d.** Estimate $f(200)$.

24. Factor each expression.

a. $x^2 - 14x + 49$ **b.** $^{-}2x^2 - 32x$

c. $3x^2 + 11x - 4$

25. Solve for x: $\dfrac{6}{x} = 20$.

11.3 Solving Rational Equations

AFTER STUDYING THIS SECTION YOU WILL BE ABLE TO

- Solve rational equations.
- Use rational equations to solve problems involving work rate.
- Use rational equations to solve problems involving distance, rate, and time.

Rational Equations

We begin with the definition of a *rational equation.*

> **Definition Rational Equation**
>
> An equation in which an unknown appears in one or more denominators is called a **rational equation**.

In this section we learn how to solve equations of this type and to solve problems that lead to rational equations.

EXAMPLE 1 ■ Solve for x if $\dfrac{1}{x-3} = {}^-2.$

Solution There are two ways we can approach this problem.

a. Use a graph. Consider the expression on the left side of the equation as a rational function and graph it in Y1. Graph the constant on the right side in Y2.

$$Y1 = 1/(X - 3)$$
$$Y2 = {}^-2$$

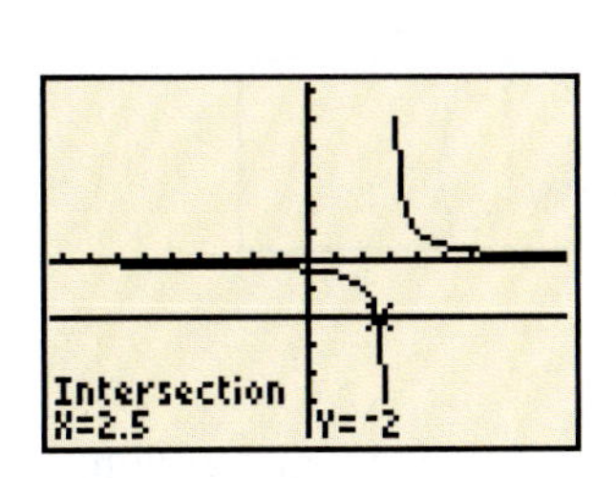

Figure 15

Graph both functions and find the point of intersection, as shown in Figure 15. Use trace or intersect to find the point of intersection as (2.5, ⁻2). It appears that there is one solution to the equation, $x = 2.5$.

b. Use algebra.

$$\frac{1}{x-3} = {}^-2$$

$$\frac{1}{x-3} * (\boldsymbol{x-3}) = {}^-2 * (\boldsymbol{x-3}) \qquad \textit{Multiplying by } x - 3 \textit{ on both sides to remove } x - 3 \textit{ from the denominator}$$

$$1 = {}^-2x + 6 \qquad \textit{Applying the distributive property on both sides}$$

$$1 - 6 = {}^-2x + 6 - 6 \qquad \textit{Adding } {}^-6 \textit{ on both sides}$$

$${}^-5 = {}^-2x$$

$$\frac{5}{2} = x \qquad \textit{Dividing by } {}^-2 \textit{ on both sides}$$

Again, it appears that there is one solution, $x = \frac{5}{2} = 2\frac{1}{2} = 2.5$. We may check the solution by substituting 2.5 for x in the original equation.

EXAMPLE 2 ■ Solve for x if $\dfrac{2}{x-3} + \dfrac{3}{x+3} = 1$.

Solution Again, there are two ways we can approach this problem.

a. Use a graph. Consider the expression on the left side of the equation as a rational function and graph it in Y1. Do the same with the right side in Y2.

$$\text{Y1} = 2/(\text{X} - 3) + 3/(\text{X} + 3)$$

$$\text{Y2} = 1$$

Graph both functions and find points of intersection, as shown in Figure 16. Use TRACE or 2ND CALC: intersect to find one point of intersection at $(^-1, 1)$ and another at $(6, 1)$. It appears that there are two solutions to the equation: $x = {}^-1$ and $x = 6$.

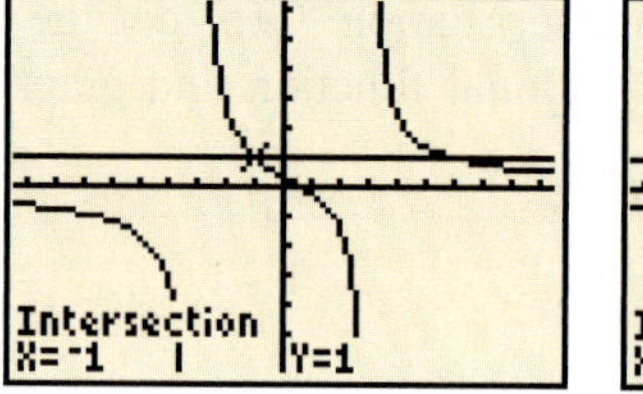

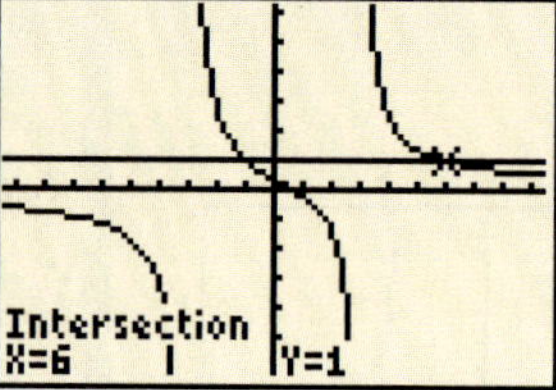

Figure 16

b. Use algebra. First express the left side as a single rational expression. The common denominator is $(x - 3)(x + 3)$, which equals $x^2 - 9$. Finding equivalent fractions gives

$$\frac{2}{x-3} + \frac{3}{x+3} = 1$$

$$\frac{2(\mathbf{x+3})}{(x-3)(\mathbf{x+3})} + \frac{3(\mathbf{x-3})}{(x+3)(\mathbf{x-3})} = 1 \qquad \textit{Finding a common denominator}$$

$$\frac{2x + 6 + 3x - 9}{(x-3)(x+3)} = 1 \qquad \textit{Writing the left side as a single fraction}$$

$$\frac{5x - 3}{x^2 - 9} = 1 \qquad \textit{Simplifying the left side}$$

$$5x - 3 = x^2 - 9 \qquad \textit{Multiplying on both sides by } x^2 - 9$$

$$0 = x^2 - 5x - 6 \qquad \textit{Collecting all terms on the right side}$$

This is a quadratic equation, which can be solved by factoring to obtain

$$0 = (x - 6)(x + 1).$$

We use the zero-product property to get $x = 6$ or $x = {}^-1$. We can find the same solutions using the quadratic formula. The result is consistent with what we found by the graphing method. Both $x = {}^-1$ and $x = 6$ check when substituted into the original equation.

Applications to Work-Rate Problems

Recall the definition of work rate from Section 1.2:

$$\text{rate} = \frac{\text{work}}{\text{time}}$$

We have used this relation to solve problems involving combined work rates. We can now solve problems in which one of the times is unknown.

EXAMPLE 3 Working together, Roger and his brother Alan can clean their family's home in $1\frac{1}{2}$ hours. Working alone, Roger can do the job in 2 hours. How long will it take Alan to do the job alone?

Solution Make a chart for each worker using the work-rate equation. Let x represent the unknown time (in hours) that it takes Alan to do the job alone.

	Work	Time	Rate
Roger	1 job	2 hours	$\frac{1}{2}$ job/hr
Alan	1 job	x hours	$\frac{1}{x}$ job/hr
together	1 job	$1\frac{1}{2}$ hours	$\frac{1}{1\frac{1}{2}} = \frac{2}{3}$ job/hr

When Roger and Alan are working together, we can *add* the rates at which they work. Working together results in a combined rate that is the sum of the individual rates.

Roger's rate + Alan's rate = combined rate

$$\frac{1}{2} + \frac{1}{x} = \frac{2}{3}$$

This equation can be solved graphically, but it will require some experimenting to get a good window. The graphical solution is left as an exercise. The equation can also be solved algebraically.

$$\frac{1}{2} + \frac{1}{x} = \frac{2}{3}$$

$$\frac{x}{2x} + \frac{2}{2x} = \frac{2}{3}$$ *2x is a common denominator on the left side*

$$\frac{x+2}{2x} = \frac{2}{3}$$ *Writing the left side as a single fraction*

$$3 * (x + 2) = 2 * 2x$$ *Multiplying by 2x and by 3 on both sides, or cross multiplying*

$$3x + 6 = 4x$$

$$6 = x$$

It would take Alan 6 hours to do the job alone. (He's lucky to have brother Roger helping him!)

Check: If Roger's rate is $\frac{1}{2}$ job/hr and Alan's rate is $\frac{1}{6}$ job/hr, then their combined rate is $\frac{1}{2} + \frac{1}{6} = \frac{3}{6} + \frac{1}{6} = \frac{4}{6} = \frac{2}{3}$ job/hr, and it will take them $\frac{3}{2} = 1\frac{1}{2}$ hours to do one job.

EXAMPLE 4 It takes 2 hours longer to drain a swimming pool (through the outtake pipe) than to fill it through the intake pipe. (See Figure 17.) When the pool is empty and both pipes are turned on, the pool is filled in $7\frac{1}{2}$ hours. How long would it take to fill an empty pool with the outtake pipe shut?

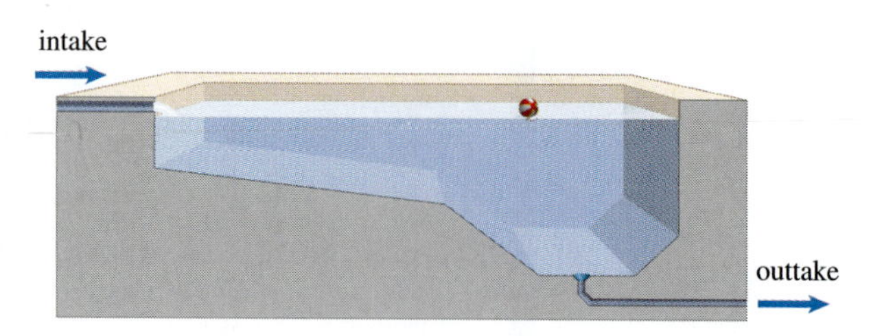

Figure 17

Solution As with Example 3, this is a work-rate problem. Here, however, the two rates (intake and outtake) work against each other instead of with each other. Again it is helpful to use a chart. Let x represent the time (in hours) for the intake pipe, working alone, to fill the pool. Then $x + 2$ represents the time (in hours) for the outtake pipe, working alone, to drain the pool. We now have enough information to complete the chart.

	Work	Time	Rate
Intake	1 pool	x hours	$\frac{1}{x}$ pools/hr
Outtake	1 pool	$x + 2$ hours	$\frac{1}{x+2}$ pools/hr
Together	1 pool	$7\frac{1}{2}$ hours	$\frac{1}{7\frac{1}{2}} = \frac{2}{15}$ pools/hr

From the chart we can find an equation using the relation

$$\text{filling} - \text{draining} = \text{combined rate}$$

$$\frac{1}{x} - \frac{1}{x+2} = \frac{2}{15}$$

To solve this equation, we need a common denominator for the left side:

$$\frac{1 * (x+2)}{x * (x+2)} - \frac{1 * x}{(x+2) * x} = \frac{2}{15}$$ *A common denominator is $x(x+2)$*

$$\frac{x + 2 - x}{x^2 + 2x} = \frac{2}{15}$$ *Writing the left side as a single fraction*

$$\frac{2}{x^2 + 2x} = \frac{2}{15}$$ *Combining like terms in the numerator of the left side*

$$2 * 15 = 2 * (x^2 + 2x)$$ *Cross multiplying*

$$30 = 2x^2 + 4x$$

$$0 = 2x^2 + 4x - 30$$ *Bringing all the terms to one side*

$$0 = 2(x^2 + 2x - 15)$$ *Factoring out 2, a common factor*

$$0 = 2(x + 5)(x - 3)$$ *Factoring $x^2 + 2x - 15$*

$$x = {}^{-}5 \quad \text{or} \quad x = 3$$

One solution, $x = {}^{-}5$ is impossible, because x represents time, which must be positive in this context. In conclusion, the intake pipe can fill the pool in 3 hours.

Check: If the intake pipe fills in 3 hours, then the outtake pipe drains the pool in $x + 2 = 5$ hours. Combined, they fill at the rate of $\frac{1}{3} - \frac{1}{5} = \frac{5}{15} - \frac{3}{15} = \frac{2}{15}$ pool/hr. At this combined rate they fill the pool in

$$\frac{1}{\frac{2}{15}} = \frac{15}{2} = 7\frac{1}{2}$$

hours. ■ ■

Applications to Distance, Rate, and Time

Rational equations can also be used to solve problems relating distance, rate, and time. One application involves motion with or against the current of a river or the wind. This situation is similar to a combined work-rate application in that two rates can work together (in which case they should be added) or against each other (in which case they should be subtracted).

If a boat travels down a river, as in Figure 18, the speed of the river current is added to the speed of the boat to give a combined rate. If a boat travels up a river, as in Figure 19, the speed of the river current is subtracted from the speed of the boat to give a combined rate.

Figure 18

Figure 19

EXAMPLE 5 A canoeist paddles **30** miles downstream in 5 hours. The river current is known to be **2** miles per hour.

a. What is the speed of the canoe in still water?

b. How long will it take the canoe to return upstream?

Solution

a. Let x represent the speed of the canoe in still water (in miles per hour). Then the combined rate for going downstream is $x + 2$ miles per hour. Using the relationship

$$\text{distance} = \text{rate} * \text{time}$$

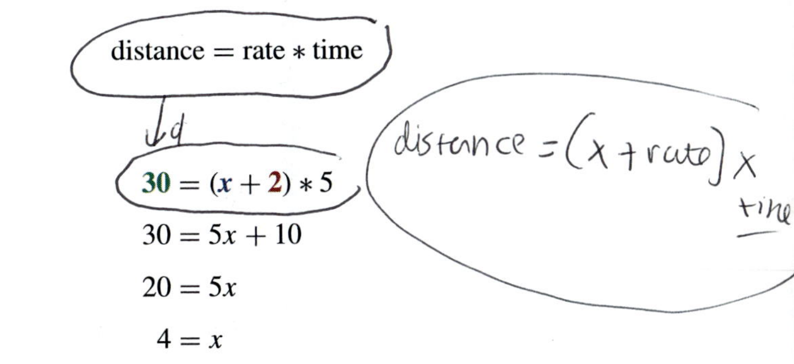

we have

$$30 = (x + 2) * 5$$
$$30 = 5x + 10$$
$$20 = 5x$$
$$4 = x$$

The canoe will travel 4 miles/hour in still water.

b. Going back upstream the canoe will travel at a combined rate of $x - 2$ miles per hour. Because x is known to be 4 miles per hour, $x - 2 = 4 - 2 = 2$ miles per hour. We solve the equation distance = rate $*$ time for time to get time = distance/rate.

Then

$$\text{time} = \frac{\text{distance}}{\text{rate}}$$

$$\text{time} = \frac{30 \text{ miles}}{2 \text{ miles/hour}} = \frac{30 \text{ miles}}{1} * \frac{1 \text{ hour}}{2 \text{ miles}} = \frac{30 \text{ hours}}{2} = 15 \text{ hours}$$

It takes the canoe 15 hours to return upstream.

Airplanes in wind behave much the same way as boats in currents, as the next example shows.

EXAMPLE 6 An airplane flies from Boston, Massachusetts, to Pittsburgh, Pennsylvania, a distance of 600 miles, and then returns to Boston. On the first part of the trip, it heads into the wind. On the return trip, it has a tail wind. The speed of the wind is known to be 20 miles per hour, and the entire trip takes $5\frac{1}{2}$ hours. How fast would the plane fly in still air?

Solution In Figure 20 the flight of the airplane and direction of the wind are shown. Let x represent the speed of the airplane in still air (in miles per hour). Then $x - 20$ miles per hour represents the plane's speed when it heads into the wind and $x + 20$ represents the speed when it has a tail wind. We make a chart using the relationship

$$\text{time} = \frac{\text{distance}}{\text{rate}}$$

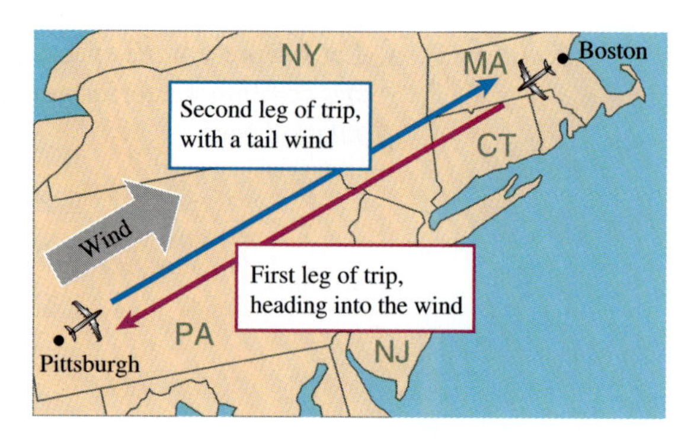

Figure 20

	Distance	Rate	Time
Boston to Pittsburgh	600 miles	$x - 20$ miles/hour	$\frac{600}{x-20}$ hours
Pittsburgh to Boston	600 miles	$x + 20$ miles/hour	$\frac{600}{x+20}$ hours
Complete trip	1200 miles	—	$5\frac{1}{2}$ hours

The time going to Pittsburgh and the time returning to Boston must add up to the total time. We get the equation

$$\frac{600}{x-20} + \frac{600}{x+20} = 5\frac{1}{2}$$

A common denominator for the left side is $(x - 20)(x + 20)$, or $x^2 - 400$.

$$\frac{600(x+20)}{(x-20)(x+20)} + \frac{600(x-20)}{(x+20)(x-20)} = \frac{11}{2}$$

$$\frac{600x + 1200 + 600x - 1200}{x^2 - 400} = \frac{11}{2}$$

$$\frac{1200x}{x^2 - 400} = \frac{11}{2}$$

$$2(1200x) = 11(x^2 - 400)$$

$$2400x = 11x^2 - 4400$$

$$0 = 11x^2 - 2400x - 4400$$

The coefficients for this equation are so large that factoring seems impractical. Use the quadratic formula with $a = 11$, $b = {}^{-}2400$, and $c = {}^{-}4400$ to obtain $x = 220$ or $x \approx {}^{-}1.8$. The negative root obviously does not apply to this problem. So we conclude the speed of the plane in still air is 220 miles per hour.

Check: Boston-to-Pittsburgh rate $= 220 - 20 = 200$ miles/hour. Time $= \frac{600}{200} =$ 3 hours. Pittsburgh-to-Boston rate $= 220 + 20 = 240$ miles/hour. Time $= \frac{600}{240} =$ 2.5 hours. The time for the complete trip is 5.5 hours.

Exercises 11.3

1. In Example 3 we wrote the rational equation

$$\frac{1}{2} + \frac{1}{x} = \frac{2}{3}$$

To solve this equation by graphing, we enter Y1 = 1/2 + 1/X and Y2 = 2/3.

   a. Choose a suitable window to show the intersection of Y1 and Y2. *Hint:* Start with Xmin = 0, Xmax = 9.4, and Ymin = 0. Adjust Ymax to show clearly the intersection of the curve with the straight line.

   b. Use [2ND] CALC: intersect to solve the equation. Compare your solution with the one from Example 3.

2. In Example 4, we wrote the rational equation

$$\frac{1}{x} - \frac{1}{x+2} = \frac{2}{15}$$

   a. What do you enter for Y1 and Y2 to solve this equation graphically?

   b. Choose a suitable window that shows the intersection of Y1 and Y2.

   c. Draw the graph as it appears on your calculator.

   d. Use [2ND] CALC: intersect to solve the equation.

3. Given the rational equation

$$\frac{1}{x-2} + 3 = 7$$

find a suitable window showing all four quadrants and solve this equation by graphing.

4. In Example 1 on page 438 the average cost of producing x gizmos was given by the rational function

$$f(x) = \frac{8x + 500}{x}$$

Suppose the average cost was \$12. Use a graph to find the number of gizmos produced. (*Hint:* Set Xmax = 188.)

5. Repeat Example 4 with the following changes: Assume it takes 3 hours longer to drain a swimming pool than it does to fill it. When the pool is empty and both pipes are turned on, the pool is filled in 6 hours.

6. Repeat Example 5 with the following changes: Assume the canoeist paddles 28 miles downstream in 4 hours. The river current is still 2 miles per hour.

In Exercises 7–10, use algebra to solve each equation for x.

7. $6 = \dfrac{20}{x-2}$

8. $\dfrac{1}{2x} + \dfrac{2}{x} = \dfrac{1}{8}$

9. $\dfrac{3}{x-1} + \dfrac{8}{x} = 3$

10. $\dfrac{400}{x-10} + \dfrac{400}{x+10} = 9$

11. Working alone, Joan can detail a car in 4 hours. Her partner takes 6 hours to detail the car alone. Working together, how long does it take Joan and her partner to do the job?

12. It takes 1 minute longer to drain a bathtub than it does to fill it. Suppose you forget to close the drain and the bathtub fills in 20 minutes. How long will it take to fill the tub with the drain closed?

13. A rowing team can travel 3000 m upstream in 50 s. The team knows they row 80 m/s in still water.

   a. What is the speed of the river current?

   b. How long does it take the team to travel downstream?

14. An airplane flies a distance of 360 miles into a head wind. On the return trip over the same route it flies with a tail wind. The entire trip takes 4 hours 12 minutes, and the wind speed is known to be 25 miles/hour. How fast will the plane fly in still air?

15. Consider the rational equation

$$\frac{7x+3}{x}=7$$

a. Graph Y1 = (7X + 3)/X and Y2 = 7. Is there a point of intersection for Y1 and Y2? Explain.

b. What happens when you try to solve the equation

$$\frac{7x+3}{x}=7$$

algebraically?

c. What conclusion can you draw from your results in parts (a) and (b)?

Skills and Review 11.3

16. Find the domain and range of the rational function

$$f(x)=\frac{1}{x-3}+2.$$

17. Consider the rational function

$$f(x)=\frac{x+1}{x}.$$

a. Graph the function.

b. Identify the asymptotes.

18. Perform the indicated operation and simplify.

$$\frac{1}{3}+\frac{1}{x}$$

19. Perform the indicated operation and simplify.

$$\frac{x}{x-7}\div\frac{3x^2}{7-x}$$

20. An earthquake with an adjusted amplitude of 250 mm measures 2.4 on the Richter scale. What is the Richter scale measure of an earthquake with an adjusted amplitude of 2500 mm?

21. Evaluate $\log(10^{-2})$.

22. Suppose the initial population of a town is 850 and is increasing by 3% each year. Use the rule of 72 to estimate the time it will take for the population to double.

23. Simplify and write without a negative exponent $(z^{3/5})^{-20}$.

24. The measures of three sides of a rectangle total 210 m. Find the dimensions of the rectangle that maximize its area.

25. Use algebra to solve each equation.

a. $x^2+5x={}^{-}6$

b. $4(10)^x=400$

c. $x=\log(10)$

d. $\frac{3}{5}x^4=\frac{48}{5}$

e. $\frac{1}{2}+\frac{2}{x}=\frac{1}{x}$

11.4 Indirect Variation

After Studying This Section You Will Be Able To

- Write a general equation for a situation involving indirect variation.
- Find a variation constant and write a specific indirect variation equation.
- Find the value of an unknown variable in an indirect variation model.

Indirect Variation Models

As a child you may have experienced the law of the lever when playing on a seesaw (or teeter-totter). In order to balance the seesaw, the heavier child must sit closer than the lighter child to the pivot point. For instance, suppose a 30-pound child sits 4 feet from the pivot point on one side of the seesaw. Then the distance the other child must sit depends upon his or her weight, as shown in this table in Figure 21.

Weight	Distance
24 pounds	5 feet
30 pounds	4 feet
40 pounds	3 feet
50 pounds	2.4 feet
60 pounds	2 feet

Figure 21

Notice that as the weight increases, the distance decreases. This is a characteristic of **indirect variation** (also called *inverse variation*), which we study in this section.

EXAMPLE 1 Graph the data from the table in Figure 21 and draw a curve through the data points.

Solution We may do this by hand or use the STAT feature of the TI-83 calculator.

a. Enter the data. Press STAT and select EDIT. Enter the weight data in L1. Enter the distance data in L2.

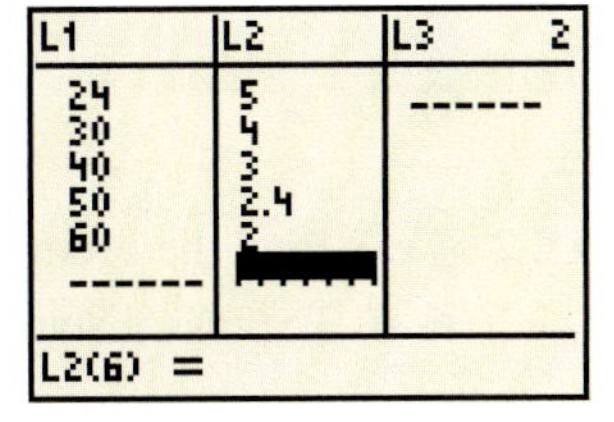

b. Now plot the data points. Press 2ND STATPLOT. Select Plot1. Type is scatter, Xlist is L_1, and Ylist is L_2. A good window for this graph is $0 \leq X \leq 94$, $0 \leq Y \leq 6.2$. Once both STATPLOT and WINDOW have been set, press GRAPH.

Five data points are displayed on the graph.

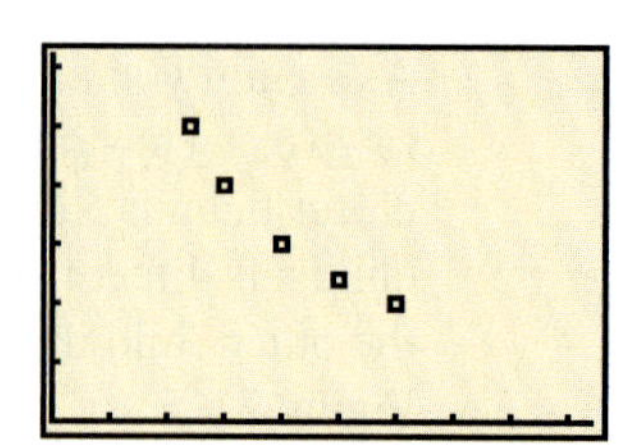

c. To draw a curve through the data points, press STAT and select CALC.

Scroll down to A: PwrReg (power regression). Press ENTER to place this command on the home screen; then type L_1, L_2, Y1. Be sure to separate with commas. (L_1 and L_2 are second functions above the 1 and 2 keys. To access Y1 Press VARS, scroll right to Y-VARS and press ENTER three times.)

The home screen display now reads

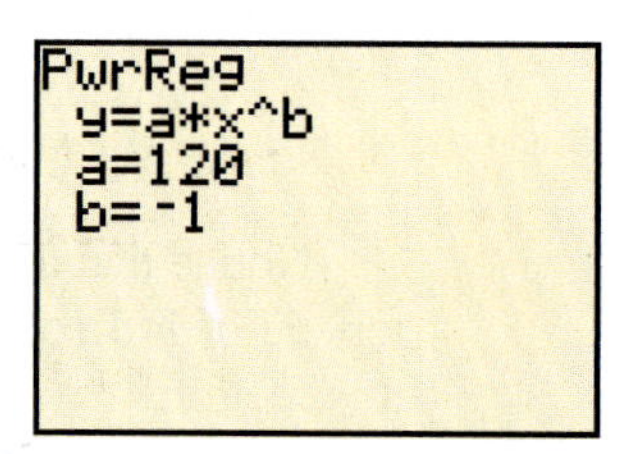

Press GRAPH and you should see the graph shown in Figure 22.

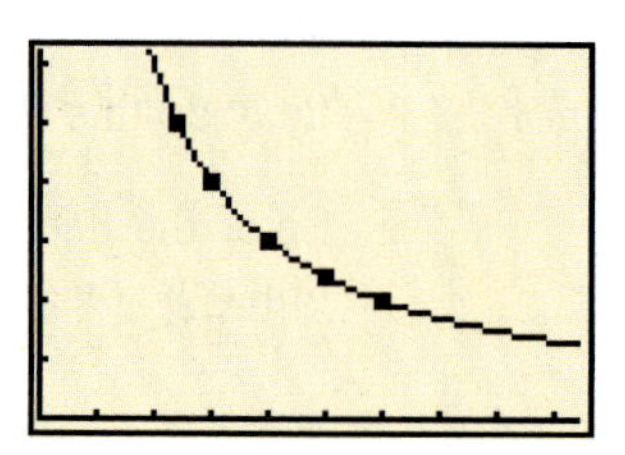

Figure 22

EXAMPLE 2 **a.** Find the product weight * distance for each row of the table in Figure 21.
b. Express distance as a function of weight.

Solution **a.**

Weight	Distance	Weight * Distance
24 pounds	5 feet	120 pound-feet
30 pounds	4 feet	120 pound-feet
40 pounds	3 feet	120 pound-feet
50 pounds	2.4 feet	120 pound-feet
60 pounds	2 feet	120 pound-feet

The product weight $*$ distance is a constant, 120. The unit for this constant, pound-feet, has significance in physics but can be ignored for our purposes.

b. Using the fact that the product is constant, we can write the equation weight $*$ distance $= 120$. Solving for distance, we have

$$\text{distance} = \frac{120}{\text{weight}}$$

The equation found in Example 2 gives distance as a rational function of weight. Using y for the dependent variable, distance, and x as the independent variable, weight, we have $y = \frac{120}{x}$. From our work in Section 11.2, we recognize that the graph of this function is a *hyperbola*. The graph drawn in Example 1 is the branch of this hyperbola in the first quadrant. The equation may be written as $y = 120x^{-1}$. In this form it may remind us of the direct variation formula $y = ax^b$. In Chapter 9 we studied cases of this formula for positive values of b ($b > 0$). In all these cases, as the variable x increases, y increases as well. Now we extend the concept of variation to situations in which $b < 0$.

In this type of variation, as one variable increases, the other decreases.

Definition Indirect Variation

If $b < 0$, then a function of the form $y = ax^b$ is called an **indirect variation function**. We say that y varies indirectly as x. The number a is called the **variation constant**. Sometimes indirect variation is called **inverse variation.**

Characteristics of Indirect Variation Models

The definition just given describes all types of indirect variation. In practice the two most common forms occur when $b = {}^{-}1$ and $b = {}^{-}2$. When $b = {}^{-}1$, we have $y = ax^{-1}$, or $y = \frac{a}{x}$. Here we say, "*y varies indirectly (or inversely) as x*." When $b = {}^{-}2$ we have $y = ax^{-2}$, or $y = \frac{a}{x^2}$. Here we say, "*y varies indirectly (or inversely) as the square of x*."

In a real-world situation, indirect variation may be a good model if the following conditions apply:

- For positive values of x, as x increases, y decreases.
- As x gets close to zero, y becomes very large, that is, *y increases without bound.*

The graph of an indirect variation function is a downward-sloping curve in the first quadrant. Both the x-axis and the y-axis are asymptotes to this curve.

Problem Solving with Indirect Variation

As you will recall from Section 9.3, solving variation problems usually requires several steps.

Solving Indirect Variation Problems

Step 1 Translate a verbal statement into a general variation equation.

Step 2 Use one pair of values to solve for the variation constant. Then write a specific variation equation with this constant.

Step 3 Use the specific equation to find an unknown value.

These three steps are illustrated in Examples 3–5.

EXAMPLE 3 Translate these verbal statements about indirect variation situations into equations.

a. A group of students wants to rent a van to attend a concert being held out of town. The cost per person for the van rental varies indirectly with the number of riders.

b. Astronauts use the fact that the weight of a person varies indirectly with the square of the person's distance from the center of the earth.

Solution

a. Representing variables with words,

$$\text{cost} = \frac{\text{constant}}{\text{number of riders}}$$

If we let x be the number of riders and y be the cost per person, then we may also write $y = \frac{a}{x}$.

b. Representing variables with words,

$$\text{weight} = \frac{\text{constant}}{\text{distance}^2}$$

If we let d be the distance from the center of the earth and w be the person's weight, then $w = \frac{a}{d^2}$.

EXAMPLE 4 Use the equations you wrote in Example 3 to find the variation constant in each situation. Then write the specific variation equation.

a. A group of students wants to rent a van to attend a concert being held out of town. The cost per person for the van rental varies indirectly with the number of riders. If 5 students make the trip, the cost will be $30 per person. Find the variation constant and the specific variation equation.

b. Astronauts use the fact that the weight of a person varies indirectly with the square of the person's distance from the center of the earth. Susan and Bob have just been selected to be astronauts. On Earth Susan weighs 125 pounds and Bob weighs 190 pounds. Find the variation constant and the specific equation for each astronaut. The radius of the earth is 4000 miles.

Solution

a. $y = \frac{a}{x}$. We know that when $x = 5$, $y = 30$. Substituting, we have $30 = \frac{a}{5}$. So $a = 150$. The specific equation is $y = \frac{150}{x}$.

b. $w = \frac{a}{d^2}$. For Susan, we know that when $d = 4000$, $w = 125$. Substituting, we have $125 = \frac{a}{4000^2}$. Solving for a, we find Susan's variation constant to be 2,000,000,000, or $2 * 10^9$. Her specific variation equation is $w = \frac{2*10^9}{d^2}$.

Similarly, for Bob $d = 4000$ and $w = 190$. Substituting, we have $190 = \frac{a}{4000^2}$. When we solve for a, we find that Bob's variation constant is 3,040,000,000, or $3.04 * 10^9$. His equation is $w = \frac{3.04*10^9}{d^2}$.

EXAMPLE 5 Use the variation equations from Example 4 to find the unknown quantities.

a. A group of students wants to rent a van to attend a concert being held out of town. The cost per person for the van rental varies indirectly with the number of riders. If 5 students make the trip, the cost is $30 per person. Find the cost per person if 8 students go to the concert.

b. Astronauts use the fact that the weight of a person varies indirectly with the square of the person's distance from the center of the earth. Susan and Bob have just been selected to be astronauts. On Earth Susan weighs 125 pounds and Bob weighs 190 pounds. They are curious to know how much they will weigh when they are 2000 miles above the surface of the earth. The radius of the earth is 4000 miles.

Solution

a. From Example 4, $y = \frac{150}{x}$. When $x = 8$, we can find y by substituting: $y = \frac{150}{8} = 18.75$. With 8 riders, they will each have to pay $18.75.

b. From Example 4, $w = \frac{a}{d^2}$. For Susan, we found $a = 2 * 10^9$, so her equation is $w = \frac{2*10^9}{d^2}$.

When the astronauts are 2000 miles above the earth, their distance from the center is 6000 miles, as shown in Figure 23. Therefore, we must substitute $d = 6000$ into Susan's equation:

$$w = \frac{2 * 10^9}{6000^2}$$

$$w \approx 55.6$$

Susan will weigh about 56 pounds when she is up there.

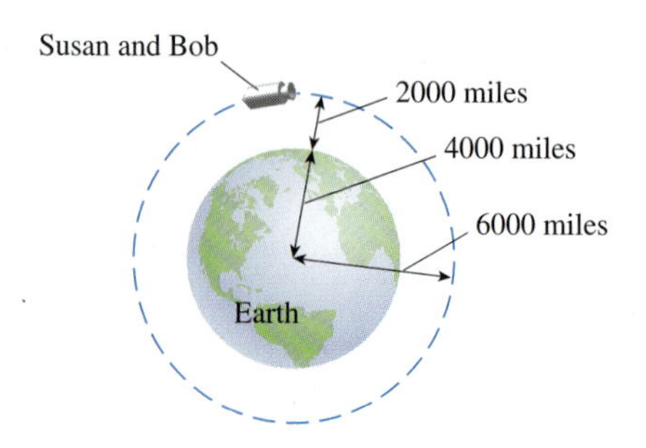

Figure 23 The astronauts are 6000 miles from the center of the earth.

For Bob, we found $a = 3.04 * 10^9$, so his equation is $w = \frac{3.04*10^9}{d^2}$. Again for Bob, $d = 6000$, so $w = \frac{3.04*10^9}{6000^2} \approx 84.4$. Bob will weigh about 84 pounds.

In Example 6, all three steps are applied in the same problem.

EXAMPLE 6 For most flying birds and insects, the number of wing beats per second varies indirectly with the length of the wing. The bee hummingbird has an average wing length of 30 mm. Its wings beat at a rate of 60 beats/s. The Puerto Rican emerald hummingbird beats its wings at a rate of 38 beats/s. About how long are its wings? At what rate does the wandering albatross, whose wingspan measures 1.5 m (1500 mm) beat its wings?

Solution Step 1 is to write the general variation equation. Representing variables with words,

$$\text{wing beat rate} = \frac{\text{constant}}{\text{wing length}}$$

If we let R be the beat rate, L be the wing length, and a be the variation constant, then $R = \frac{a}{L}$.

In Step 2 we substitute the information about the bee hummingbird into the equation. When $L = 30$, $R = 60$.

$$R = \frac{a}{L}$$

$$60 = \frac{a}{30}$$

$$1800 = a$$

The specific equation is $R = \frac{1800}{L}$.

In Step 3 we use the specific equation to solve for the unknown values of the variables. For the Puerto Rican emerald hummingbird, $R = 38$, so

$$R = \frac{1800}{L}$$

$$\mathbf{38} = \frac{1800}{L}$$

$$38L = 1800 \qquad \textit{Multiplying by L on both sides}$$

$$L = \frac{1800}{38} \approx 47.4$$

The wing length of the Puerto Rican emerald hummingbird is about 47 mm.

For the wandering albatross, $\boldsymbol{L} = \mathbf{1500}$, so

$$R = \frac{\mathbf{1800}}{\boldsymbol{L}} = \frac{1800}{\mathbf{1500}} = 1.2$$

The wings of the wandering albatross beat about 1.2 times per second.

Exercises 11.4

1. Refer to the data in Figure 21 on page 457. What characteristics of the data suggest that the weight varies indirectly with the distance?
2. Use the characteristics of indirect variation models to describe the features of the graph in Figure 22.
3. Recall the indirect variation equation from Example 2,

$$\text{distance} = \frac{120}{\text{weight}}$$

   a. What is the variation constant?

   b. Name the independent variable.

   c. Name the dependent variable.

4. The table shows the intensity of light measured in lux given off by a 50-W bulb at a given distance measured in meters. (http://faculty.millikin.edu)

Distance (m)	Intensity of Light (lux)
1	53.2
2	13.3
4	3.325
5	2.128

Follow the calculator steps from Example 1 to complete parts (a)–(c).

   a. Graph the data using the STAT feature of the TI-83 calculator. Use a window of $0 \leq X \leq 9.4$, $0 \leq Y \leq 70$.

   b. Does the graph of the data appear to represent an indirect variation model? Explain.

   c. Draw a curve through the data points using STAT CALC and scrolling down to PwrReg as shown in Example 1.

   d. Write your direct variation equation with the coefficient and exponent rounded to the nearest .1.

   e. Use your value of b to pick the correct indirect variation statement for this situation.

   1. The intensity of light varies indirectly with the distance from the light.
   2. The intensity of light varies indirectly with the square of the distance from the light.

   f. Use your indirect variation equation to predict the intensity of light at a distance of 10 m.

5. Race-car drivers must be acutely aware of how the velocity of their vehicles is related to the time it takes to complete one circuit of the racetrack. Shown in the table are average velocities (miles per hour) for various times in seconds to go around a 1-mile track.

Time	Velocity	Time * Velocity
15	240	
20	180	
24	150	
30	120	
36	100	

a. Complete the time * velocity column. What do you notice?

b. Refer to Example 2 and use your findings in part (a) to write an equation expressing velocity as a function of time.

In Exercises 6–8, use the following problem situation. In an electric circuit, resistance (in ohms), varies indirectly with current (in amps), assuming constant voltage.

6. Let I be the current and R be the resistance. Translate the verbal statement into a general variation equation.

7. In a circuit with a current of 8 amps, the resistance is 50 ohms. Find the variation constant and the specific variation equation.

8. Use the specific variation equation from Exercise 7 to find the resistance when the current is 20 amps.

In Exercises 9–11, use the following problem situation. The force of attraction between two bodies varies indirectly with the square of the distance between them.

9. Translate the verbal statement into a general variation equation.

10. Two objects 100 m apart have a force of attraction of .5 newton. Find the variation constant and the specific variation equation.

11. Use the specific variation equation from Exercise 10 to find the force of attraction between two objects that are 2 m apart.

12. y varies indirectly with the square of x.

a. Translate the verbal statement into a general variation equation.

b. If $y = \frac{1}{2}$ when $x = 6$, solve for the variation constant. Write a specific variation equation with this constant.

c. Use your specific variation equation from part (b) to find y when $x = 3$.

13. Solve each indirect variation problem.

a. y varies indirectly with the square of x, and $y = 5$ when $x = 2$. Find y when $x = 10$.

b. y varies indirectly with x, and $y = 4$ when $x = 9$. Find x when $y = 3$.

c. y varies indirectly with the cube of x, and $y = 100$ when $x = 4$. Find y when $x = 8$.

14. Consider the indirect variation situation where a function's output is halved when the input is doubled. What happens to the output when the input is increased by a factor of 4?

15. The pressure on a gas varies indirectly as the volume of the gas (assuming the temperature and quantity are fixed). When the volume of the gas is 3.5 L the pressure is 1.03 atmospheres.

a. Find the pressure of the gas if the volume is 4 L.

b. Find the volume of the gas if the pressure is 1.05 atmospheres.

Skills and Review 11.4

16. a. Perform the addition and simplify:

$$\frac{2}{x-1} + \frac{4}{x+1}$$

b. Solve for x in the equation

$$\frac{2}{x-1} + \frac{4}{x+1} = 2$$

17. How does the graph of $f(x) + 6$ relate to the graph of the basic reciprocal function $f(x) = \frac{1}{x}$?

18. Solve for x: $5 = \log(x)$.

19. Write an exponential decay function for a substance with an initial value of 200 that decreases by $\frac{1}{10}$ of its value each year.

20. The growth of a \$2000 investment earning 6% interest compounded quarterly is modeled by the compound amount function $A(t) = 2000(1 + \frac{.06}{4})^{4t}$, where t is the time in years. Use a graphing method to find the number of years for this investment to double.

21. a. Solve for x if $\sqrt{4x} = x - 3$.

b. Is $x = 1$ a true solution? Explain.

22. The expressions in each pair are equivalent. Show the work necessary to place the left expression into the simple radical form shown on the right.

a. $\sqrt{180}; 6\sqrt{5}$ **b.** $\frac{\sqrt{3}}{\sqrt{6}}; \frac{\sqrt{2}}{2}$

23. Evaluate each expression without using a calculator.

a. $27^{1/3}$ **b.** $16^{1/4}$ **c.** $\sqrt[3]{125}$

24. Let $f(x) = x^3 + 1$ and $g(x) = x^2$.

a. Find $f(g(^-4))$.

b. Find $f(g(x))$.

c. Find $g(f(x))$.

25. Consider the function $y = 60x^{-2}$.

a. Write this equation so the expression on the right side has a positive exponent.

b. Write this equation using function notation.

c. Find $f(2)$.

d. Find x when $f(x) = \frac{5}{3}$.

11.5 Division of Polynomials (Optional)

AFTER STUDYING THIS SECTION YOU WILL BE ABLE TO

- Use an area model to divide two polynomials.
- Interpret the graph of the quotient of two polynomials.

In this chapter, most of the rational functions we studied had constants in the numerator and first- or second-degree polynomials in the denominator. Some rational functions, however, may be expressed as the quotient of two polynomials both with degree greater than or equal to 1. To study these functions, it helps to know something about dividing polynomials.

EXAMPLE 1 Graph the functions Y1 = (X ∧ 2 − X − 6)/(X + 2) and Y2 = (X ∧ 2 − X − 4)/(X + 2) in the standard friendly window ($^-9.4 \le X \le 9.4$, $^-10 \le Y \le 10$) and observe carefully how the two graphs are related.

Solution The calculator screen is shown in Figure 24. We may make the following observations.

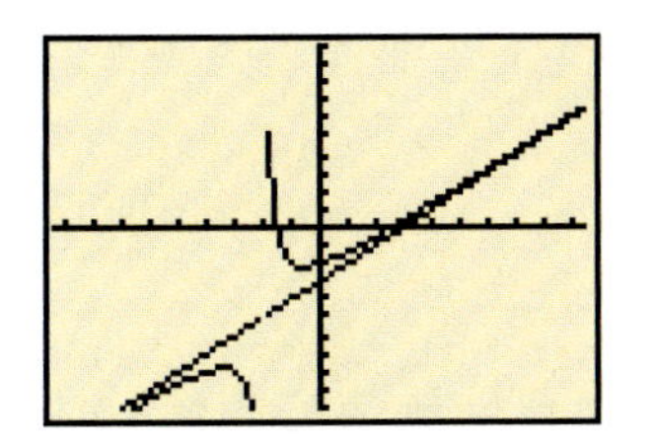

Figure 24

- The graph of Y1 appears to be a straight line. However, upon closer examination, there is a gap at $x = {}^-2$. There the function is undefined, because when $x = {}^-2$, $x + 2 = 0$.
- The graph of Y2 has two branches. There appears to be a vertical asymptote at $x = {}^-2$.
- Both branches of the Y2 graph appear to get close to the Y1 graph as x assumes large positive values and large negative values.

X	Y1	Y2
-5	-8	-8.667
-4	-7	-8
-3	-6	-8
-2	ERROR	ERROR
-1	-4	-2
0	-3	-2
1	-2	-1.333

X= -5

Figure 25

You may use the table feature to explore these properties further. Figure 25 shows a table with TblStart = $^-5$ and ΔTbl = 1. Note that for both functions, the value is ERROR for X = $^-2$, indicating that the function is undefined.

In Chapter 6 you used an area model to represent multiplication and factoring of polynomials. This model may also be used to show division. Examples 2 and 3 show how division may be used to find equivalent expressions for the rational functions Y1 and Y2 from Example 1.

EXAMPLE 2 ■ Use an area model to divide: $(x^2 - x - 6) \div (x + 2)$.

Solution The dividend (numerator) is $x^2 - x - 6$. The divisor (denominator) is $x + 2$. From our experience with factoring, we expect that the quotient will be of the form "$x+$ something" and that a two-by-two area model may be used. Start with the known factor, $x + 2$, on one side and x^2 in a corner. Then one step at a time, fill in the other cells of the two-by-two model to represent the dividend.

Step 1 Place x^2 in the upper-left corner.

x	x^2	
2		

Step 2 $x * x = x^2$, so x goes on top left. $x * 2 = 2x$, so $2x$ goes in the lower-left corner.

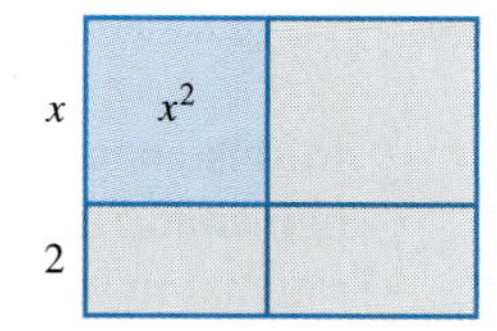

Step 3 $2x + {}^-3x = {}^-x$ (to get the middle term), so ^-3x goes in the upper-right corner.

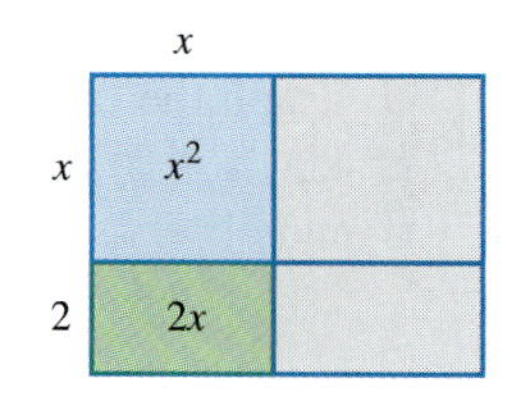

Step 4 $x * {}^-3 = {}^-3x$, so $^-3$ goes in the top-right corner. $2 * {}^-3 = {}^-6$, so $^-6$ goes in the lower-right corner.

	x	$^-3$
x	x^2	^-3x
2	$2x$	$^-6$

The last diagram shows that $(x + 2)(x - 3) = x^2 - x - 6$. The dividend appears as the product of two factors without a remainder. The pattern is

$$\frac{\text{dividend}}{\text{divisor}} = \text{quotient} \quad \text{or} \quad \frac{x^2 - x - 6}{x + 2} = x - 3.$$

EXAMPLE 3 ■ Use an area model to divide: $(x^2 - x - 4) \div (x + 2)$.

Solution ■ The dividend is $x^2 - x - 4$. The divisor is $x + 2$. Use the area model, as before.

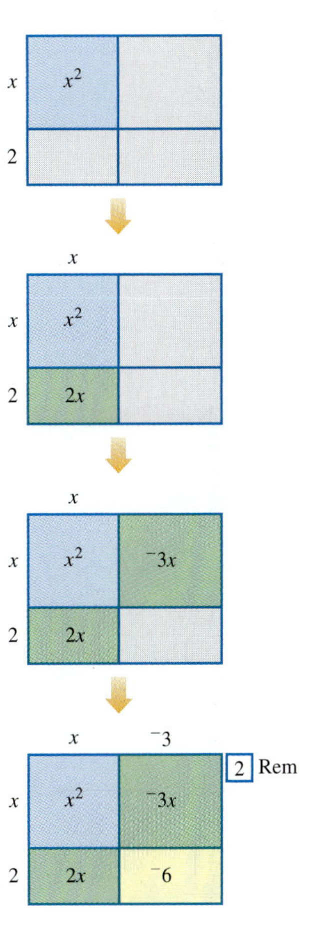

Notice that the steps here are identical to those in Example 2. However, when we get to the lower-right corner, $^{-}6$ is not the last term of the dividend. So we have a remainder, which must be 2 in order to make the last term come out to $^{-}4$.

We thus show that $(x+2)(x-3)+2 = x^2 - x - 4$. The division takes the form

$$\frac{\text{dividend}}{\text{divisor}} = \text{quotient} + \frac{\text{remainder}}{\text{divisor}}, \quad \text{or} \quad \frac{x^2 - x - 4}{x+2} = x - 3 + \frac{2}{x+2}$$

Notice that in the divided form, we see that for all values of x except $x = {}^{-}2$, the function Y2 from Example 1 is the sum of a linear function, $x - 3$, and a rational function, $\frac{2}{x+2}$. That explains why its graph gets close to the line $y = x - 3$ and also has a vertical asymptote at $x = {}^{-}2$.

EXAMPLE 4 ■ Use an area model to divide: $(x^3 - 9x^2 + 21x + 2) \div (x - 4)$.

Solution The procedure is shown in the following area model.

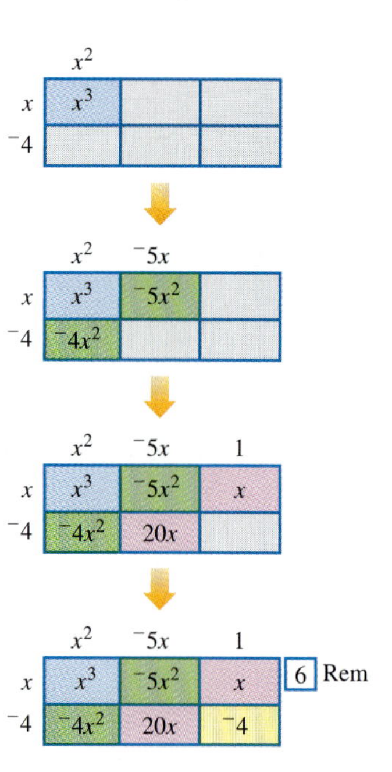

The division takes the form

$$\frac{\text{dividend}}{\text{divisor}} = \text{quotient} + \frac{\text{remainder}}{\text{divisor}}, \quad \text{or}$$

$$\frac{x^3 - 9x^2 + 21x + 2}{x - 4} = x^2 - 5x + 1 + \frac{6}{x - 4}$$

$$
\begin{array}{r}
x^2 - 5x + 1 = \text{quotient} \\
x - 4\overline{)\,x^3 - 9x^2 + 21x + 2} \\
\underline{x^3 - 4x^2 } \\
{}^-5x^2 + 21x \\
\underline{{}^-5x^2 + 20x } \\
1x + 2 \\
\underline{1x - 4} \\
6 = \text{remainder}
\end{array}
$$

Figure 26

There are other ways to write out the division of polynomials. Figure 26 shows Example 4 worked out with long division, similar to the algorithm for whole numbers. Figure 27 illustrates a technique called *synthetic division*, which might be encountered in a precalculus course.

$$
\begin{array}{r|rrrr}
4 & 1 & {}^-9 & 21 & 2 \\
 & & 4 & {}^-20 & 4 \\
\hline
 & 1 & {}^-5 & 1 & 6
\end{array}
$$

quotient $= 1x^2 - 5x + 1$ remainder $= 6$

Figure 27

Exercises 11.5

1. Find the quotient and remainder, if there is one.
   a. $(x^2 + 12x + 35) \div (x + 7)$
   b. $(x^2 + 12x + 30) \div (x + 7)$
   c. $(x^2 + 12x + 38) \div (x + 7)$
2. Find the quotient and remainder, if there is one.
   a. $(x^2 - 14x + 48) \div (x - 6)$
   b. $(x^2 - 14x + 30) \div (x - 6)$
   c. $(x^2 - 14x + 60) \div (x - 6)$
3. **a.** On your calculator, enter the three functions from Exercise 2 on the Y= screen:

   $$\text{Y1} = (\text{X} \wedge 2 - 14\text{X} + 48)/(\text{X} - 6)$$

   $$\text{Y2} = (\text{X} \wedge 2 - 14\text{X} + 30)/(\text{X} - 6)$$

   $$\text{Y3} = (\text{X} \wedge 2 - 14\text{X} + 60)/(\text{X} - 6)$$

   b. Set the standard friendly window ${}^-9.4 \leq \text{X} \leq 9.4$, ${}^-10 \leq \text{Y} \leq 10$ and graph the three functions. Observe the graphs in this window and then zoom out by a factor of 2 to the window ${}^-18.8 \leq \text{X} \leq 18.8$, ${}^-20 \leq \text{Y} \leq 20$.
   c. Describe how the three graphs in (b) are related to each other and to the quotients and remainders found in Exercise 2.
4. Find the quotients and remainders.
   a. $(x^3 + 3x^2 - 42x + 16) \div (x - 5)$
   b. $(x^3 + 3x^2 - 11x - 32) \div (x + 4)$
   c. $(2x^4 - x^3 - 7x^2 + x + 2) \div (x - 2)$
5. Study Figure 26. Explain how this long division is related to the area model shown on page 468.
6. Study Figure 27. Explain how this synthetic division is related to the area model shown on page 468.

CHAPTER 11 • KEY CONCEPTS

Rational Expression A rational expression is a fraction in which both the numerator and denominator are polynomials. (Page 426)

Equivalent Rational Expressions Two rational expressions are equivalent if one is found from the other by multiplying numerator and denominator by the same expression. (Page 426)

Combining Rational Expressions The sum or difference of two rational expressions may be expressed as a single rational expression. To add or subtract the expressions, a common denominator must be found. (Page 429)

Simplifying Rational Expressions A rational expression may be simplified by removing a common factor from both numerator and denominator. (Page 434)

Rational Function A rational function is a function of the form $f(x) =$ rational expression in x. (Page 438)

Hyperbola The graph of the function $y = f(x) = \frac{1}{x}$ is called a hyperbola. When the graph is stretched, shifted, or reflected across the x-axis, the curve remains a hyperbola. These hyperbolas have one horizontal asymptote and one vertical asymptote. (Page 440)

Asymptote If a curve approaches a line when it is traced, the line is called an asymptote to the curve. (Page 440)

Rational Equation An equation in which an unknown appears in one or more denominators is called a rational equation. (Page 447)

Applications of Rational Equations Work-rate and distance-rate-time problems are often solved using rational equations. (Pages 449, 452)

Indirect Variation If $b < 0$, then a function of the form $y = ax^b$ is called an indirect variation function. We can say that y varies indirectly as x. The number a is called the **variation constant**. Sometimes indirect variation is called **inverse variation.** (Page 459)

Characteristics of Indirect Variation In a real-world situation, indirect variation may be a good model if the following conditions apply:

1. For positive values of x, as x increases, y decreases.

2. As x gets close to zero, y becomes very large, that is, y increases without bound.

The graph of an indirect variation function is a downward-sloping curve in the first quadrant. Both the x-axis and the y-axis are asymptotes to this curve. (Page 459)

Solving Indirect Variation Problems (Page 460)

Step 1 Translate a verbal statement into a general variation equation.

Step 2 Use one pair of values to solve for the variation constant. Then write a specific variation equation with this constant.

Step 3 Use the specific equation to find an unknown value.

Exploration

The inverse square law for illumination states that illumination (measured in foot-candles) varies indirectly as the square of the distance from the source of light. The unit *foot-candle* originally referred to the intensity of light at a point 1 foot away from a standard candle. The variation constant depends upon the type of lighting source used.

a. Let I represent illumination, d represent distance, and a be a variation constant. Write a general formula for I in terms of d and a.

b. A source of light produces an illumination of 350 foot-candles at a distance of 10 feet. Find the variation constant a.

c. The same source of light is now viewed from a distance of 15 feet. Find the illumination.

d. Suppose you don't have enough light on the surface of your desk to do your homework. The inverse square law suggests two things you may do to improve this situation. What are they?

CHAPTER 11 ▪ REVIEW EXERCISES

Section 11.1 Rational Expressions

1. For each expression, find two equivalent rational expressions. *Note:* There are many possible answers.

a. $\frac{5}{6}$ **b.** $\frac{x^3}{y^2}$

c. $\frac{x-4}{x+5}$

2. Show that

$$\frac{x-4}{x+5} \quad \text{and} \quad \frac{x^2-3x-4}{x^2+6x+5}$$

are equal when substituting the following values of x: $^-4, 3, 10$. Why are the two expressions not equal when $x = {}^-1$?

3. Find the missing numerators.

a. $\frac{2}{5} = \frac{?}{15}$ **b.** $\frac{y}{x} = \frac{?}{x^2 y}$

c. $\frac{x-1}{x+3} = \frac{?}{x^2+3x}$

4. Express each sum or difference as a single rational expression.

a. $\frac{2}{x} + \frac{3}{x^4}$

b. $\frac{1}{x-2} - \frac{3x}{x^2-5x+6}$

c. $\frac{4}{y} - \frac{1}{x}$

d. $\frac{5}{x+1} + \frac{2}{x-4}$

e. $\frac{3}{yz^2} + \frac{8}{y^2z^3}$

f. $\frac{4}{x^2+2x} - \frac{1}{x^2+4x+4}$

5. Simplify each rational expression.

a. $\frac{6xy^3}{9x^3y^2}$ **b.** $\frac{x^2-36}{x^2+6x}$

c. $\frac{x^2-9}{x^2-6x+9}$ **d.** $\frac{x^2+7x}{x(x+7)^2}$

e. $\frac{x-1}{1-x}$ (*Hint:* Factor out a $^-1$ from the denominator.)

6. Perform the indicated operation and simplify.

a. $\frac{2x^3}{5y^3} \div \frac{x^4}{10y^2}$

b. $\frac{15x^2}{x^2-x-12} * \frac{x-4}{3x}$

c. $\frac{7x}{x+7} * \frac{x^2+15x+56}{2x^2+16x}$

d. $\frac{x^2}{3x-24} \div \frac{6x^3}{x^2-64}$

Section 11.2 Rational Functions

7. We wish to compare the total cost function, Y1 = 300 + 5X, to the average cost function, Y2 = (300 + 5X)/X.

a. Make a table showing the total cost and average cost for the following values of x: 0, 10, 20, 30, 40, 50. Use the table for parts (b)–(d).

b. Compare the outputs at $x = 0$. What do you notice?

c. As x increases, how do the outputs of the two functions change?

d. Compare the outputs when x takes on very large values. What do you notice?

8. Describe the graph of the average cost function from Exercise 7, Y1 = (300 + 5X)/X. Use a window of $0 \leq X \leq 94, 0 \leq Y \leq 100$.

9. Recall that the vertical asymptote of a rational function should not show on its graph. Shown, with an unfriendly window, is the graph of the function $y = \frac{3}{x+1}$. Describe two methods for preventing the vertical asymptote from appearing on the graph of this function.

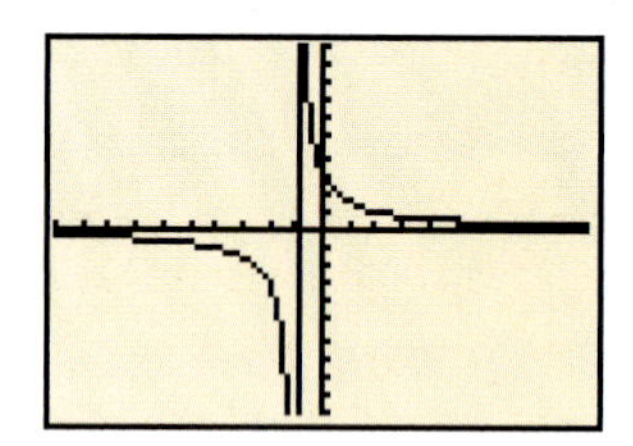

10. Use your knowledge of modifying basic functions to describe how the graph of each function relates to the graph of the basic reciprocal function $f(x) = \frac{1}{x}$.

a. $g(x) = \dfrac{1}{x+3}$ **b.** $g(x) = \dfrac{1}{x} - 2$

c. $g(x) = \dfrac{^-1}{x}$ **d.** $g(x) = \dfrac{6}{x}$

e. $g(x) = \dfrac{1}{x-2} + 1$

11. Identify the horizontal and vertical asymptotes of each function from Exercise 10.

12. Determine the domain and range of each function from Exercise 10.

Section 11.3 Solving Rational Equations

13. Use a graphing method to solve for x if

$$\frac{5}{x+2} + \frac{2}{x-2} = 3$$

14. Use algebra to find the exact solutions to the rational equation in Exercise 13.

15. Working alone it takes Lindsey 2 hours and Raul 3 hours to complete the same job. How long does it take them to complete the job if they work together? Pick the most reasonable answer without setting up an equation or performing any calculations. Explain your choice.

a. Less than 2 hours

b. More than 2 hours but less than 3 hours

c. More than 3 hours

d. 5 hours

16. Working alone, Thaddeus can prepare dinner in 45 minutes. Working together, Thaddeus and Lynn can prepare dinner in 20 minutes. Complete parts (a)–(c) to find how long it will take Lynn to prepare dinner alone.

a. Let x represent the number of minutes that it takes Lynn to prepare dinner alone. Make a chart such as the one in Example 3 on page 449.

b. Write an equation for the work rates.

c. Solve the equation and answer the question.

17. It takes 4 minutes longer to drain a Jacuzzi than it does to fill it. If the drain is left open, it takes 15 minutes to fill it. How many minutes does it take to fill the Jacuzzi when the drain is closed?

18. A person in a kayak paddles 9 km upstream in $\frac{2}{3}$ h. The river current is known to be 3 km/h. Complete parts (a)–(c) to find the speed of the kayak in still water.

a. Let x represent the speed of the kayak in still water. Then, going upstream, the speed of the kayak is reduced by the speed of the current. Write an expression for this combined rate upstream.

b. Write an equation using the combined rate upstream, the known time, and distance in the relationship time = distance/rate.

c. Solve the equation in order to find the speed of the kayak in still water.

d. How long will it take the kayak to return downstream?

19. An airline passenger is curious about the speed of the wind affecting his flight. The airplane flies, with a tailwind, from Memphis, Tennessee, to Cincinnati, Ohio, a distance of 480 miles. Then it returns, flying into the wind. The round trip takes 3 hours 6 minutes and the speed of the plane in still air is known to be 310 miles/hour. What is the wind speed?

Section 11.4 Indirect Variation

20. Suppose you have only \$10 to buy gasoline for your vehicle. Then the number of gallons of gasoline you can buy depends upon the price per gallon. When the price is \$1.00 per gallon, you can buy 10 gallons. When the price is \$1.50, you can buy $6\frac{2}{3}$ gallons. When the price is \$2.00, you can buy 5 gallons. These and additional entries are summarized in the following table.

Price per Gallon (x)	Number of Gallons (y)	Cost
1.00	10	10.00
1.50	$6\frac{2}{3}$	10.00
2.00	5	10.00
2.50	4	10.00
3.00	$3\frac{1}{3}$	10.00

a. What characteristics of the data make indirect variation a good model?

b. Find the variation constant and describe its meaning in this problem.

c. Write an indirect variation equation that fits the data.

d. Use your equation to determine the amount of gasoline that can be purchased when the price is \$1.45 per gallon.

21. Suppose y varies indirectly with the square of x and $y = 240$ when $x = 10$.

a. Find the variation constant, a.

b. Find y when $x = 20$.

22. Suppose y varies indirectly with x. Find the variation constant and then the missing value, x or y.

a. $y = 1$ when $x = 3$. Find y when $x = 6$.

b. $y = 2$ when $x = 6$. Find x when $y = .8$.

c. $y = 9$ when $x = 10$. Find y when $x = \frac{1}{2}$.

d. $y = 12.5$ when $x = 20$. Find x when $y = 2\frac{1}{3}$.

23. A group of students wants to rent a cottage in the Caribbean. The cost per person for the cottage varies indirectly with the number of renters. If 4 students share the cottage, the cost will be \$250 per person.

a. Let x represent the number of students and y represent the cost per person. Then the indirect variation equation is $y = \frac{a}{x}$. Find the variation constant a.

b. Find the cost per person if 7 students share the cottage.

24. The time it takes a jogger to complete a lap around a track varies indirectly with his speed. It takes 150 seconds to run a lap at a speed of 9 feet/second. How long does the lap take at a speed of 11 feet/second?

25. The weight of a person varies indirectly with the square of the person's distance from the center of the earth. In a recent shuttle orbit 200 miles from the earth's surface, an astronaut weighed 160 pounds. How much more does she weigh on the earth? Assume the radius of the earth is 4000 miles.

Section 11.5 Division of Polynomials

26. Find the quotient and remainder, if there is one.

a. $(x^2 - 3x - 40) \div (x + 5)$

b. $(x^2 - 3x - 36) \div (x + 5)$

c. $(x^2 - 3x - 44) \div (x + 5)$

27. Find the quotients and remainders.

a. $(2x^3 + 5x^2 - 3x + 2) \div (x + 2)$

b. $(10x^3 + x^2 + 12x + 9) \div (5x + 3)$

c. $(3x^4 + 2x^3 - 4x^2 + 7x - 5) \div (3x - 1)$

CHAPTER 11 ▪ TEST

1. Express the sum as a single rational expression:

$$\frac{5}{x+2} + \frac{1}{x+3}.$$

2. Express the difference as a single rational expression:

$$\frac{1}{x^2-4} - \frac{3}{x^2-2x-8}.$$

3. Perform the multiplication and simplify:

$$\frac{x^2-x}{x^2} * \frac{(x+1)^2}{x^2-1}.$$

4. Perform the division and simplify:

$$\frac{x+5}{6-x} \div \frac{x^2+10x+25}{(x-6)^2}.$$

5. Consider the function

$$g(x) = \frac{1}{x-4} + 5.$$

   a. Describe how the graph of $g(x)$ relates to the basic reciprocal function, $f(x) = \frac{1}{x}$.
   b. Identify the horizontal and vertical asymptotes.
   c. Find the domain and range.

6. Solve for x:

$$\frac{1}{x+3} + \frac{2}{x} = \frac{9}{4}.$$

7. Working together, Ron and Leah can send out all their party invitations in $3\frac{1}{3}$ hours. Working alone, Leah can do the job in 5 hours. How long does it take Ron to do the job alone?

8. Suppose a giant sieve for screening sand takes 1 hour more to empty than it does to fill. It takes 20 hours to fill the sieve when material is allowed to pass through. How long does it take to fill the sieve when material does not pass through?

9. A small motorboat can travel 24 miles downstream and return upstream in 6 hours. The river current is known to be 3 miles/hour. How fast does this boat move in still water?

10. Suppose y varies indirectly with x and $y = \frac{1}{20}$ when $x = 100$. Find y when $x = 30$.

11. The volume of a gas varies indirectly with the pressure upon it (assuming the temperature and quantity are fixed). Suppose the volume of a gas is 4 L when the pressure is 1.06 atmospheres. Find the volume if the pressure is 1 atmosphere.

12. In an earlier exercise, we found that the intensity of light from a bulb varied indirectly with the square of the distance to the light source. The light from a particular bulb is 3 lux at a distance of 10 m. Find the intensity of light at 5 m from the bulb.

CHAPTER

12 Inequalities

Simply the Best

George Dantzig is considered to be the father of a branch of applied mathematics called *linear programming*, which is part of a larger field known as operations research. During World War II Dantzig worked for the U.S. Air Force analyzing problems of scheduling, logistics, and supplies. Many of these problems involved systems of inequalities in many variables. Shortly after the war, he discovered a famous algorithm called the *simplex method*, which has become widely applied to a variety of problems in business and other fields. This method allows a computer to determine quickly the optimal (best) solution to a problem involving many variables. The first problem to which Dantzig applied his method was to determine a balanced diet for his family that would provide all necessary nutrients at minimal cost.

In this chapter we use systems of inequalities to model situations that can be analyzed using linear programming techniques. We then have the opportunity to apply inequalities to the problem of designing a personal fitness program. (See Exploration on p. 498)

Need help? For on-line resources, visit this web site: **math.college.hmco.com/students.**

12.1 Inequalities and the Number Line

AFTER STUDYING THIS SECTION YOU WILL BE ABLE TO

- Represent inequalities in one variable on a number line.
- Represent inequalities in one variable with set-builder notation.
- Solve linear inequalities in one variable.

There are many real-world situations that cannot be modeled mathematically using equations but instead require the use of **inequalities**. Problems with solutions that involve a *range* of values cannot be modeled using an equation.

Methods of Solving Inequalities

Examples 1–3 illustrate how to solve inequalities.

EXAMPLE 1 Let's return to Gridville in Section 2.3. Suppose Melissa lives on the corner of Street 0 (the x-axis) and Avenue 2. Melissa likes to walk only along Street 0, and she never gets farther than 4 blocks from her house. Where can Melissa go? Show the answer on a number line.

Solution Because Melissa walks only along Street 0, we have to keep track only of her x-coordinate. If she walks 4 blocks west, she will be at Avenue $^{-}2$, and if she walks 4 blocks east, she will be at Avenue 6. We can show this by shading in her possible locations on a number line (see Figure 1). We can also write this set of points as $\{x \mid {}^{-}2 \le x \text{ and } x \le 6\}$. A more compact way to show the set is $\{x \mid {}^{-}2 \le x \le 6\}$.

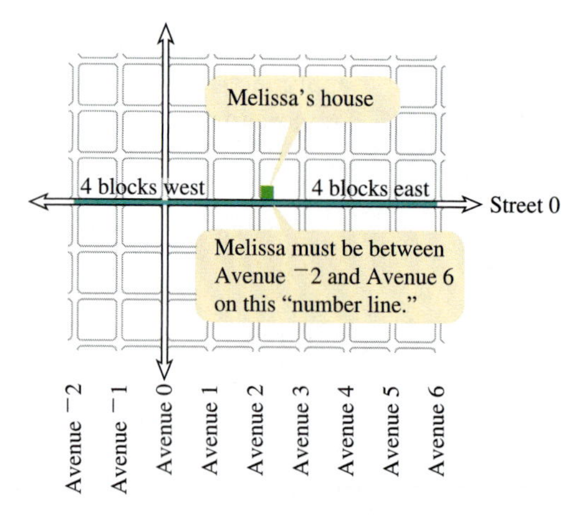

Figure 1 Melissa's position stays between $^{-}2$ and 6 on the x-axis.

EXAMPLE 2 Ms. Chang's cellular phone company charges her 10¢ per minute, and they give her a free gift if she spends more than \$20 per month. Ms. Chang always wants to get the free gift, but she can't afford to spend more than \$40 per month on phone calls. How many minutes of calls should Ms. Chang make each month? Show the answer on a number line.

Solution Because Ms. Chang always wants to get the free gift, she has to spend more than \$20. Each minute costs 10¢, so we can work out how many minutes cost \$20 = 2000¢ by division:

$$\frac{2000¢}{10¢\ /\text{minute}} = 200 \text{ minutes}$$

To get her free gift, Ms. Chang has to spend more than 200 minutes on the phone. We can also determine how many minutes cost \$40 = 4000¢ : $\frac{4000}{10} = 400$. We can then plot the allowable number of minutes on a number line (see Figure 2). In set-builder notation, this is shown as $\{x \mid 200 < x \leq 400\}$.

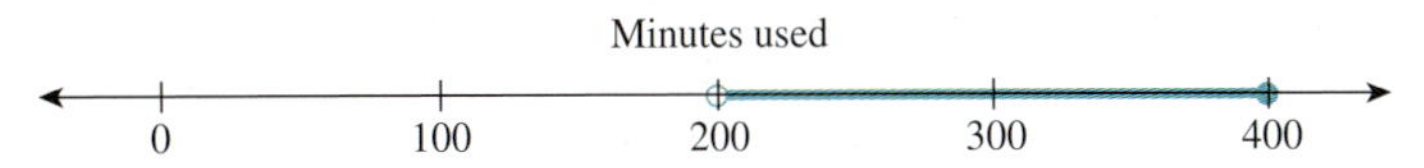

Figure 2 Ms. Chang must use more than 200 minutes and no more than 400 minutes.

EXAMPLE 3 A water main breaks at the corner of Avenue 3 and Street 2 in Gridville, flooding Avenue 3 for 6 blocks in either direction. Buses that run along Avenue 3 must stay either north or south of the flooding. Which streets can Avenue 3 buses reach? See Figure 3.

Solution Because the buses stay on Avenue 3, we need to keep track only of their y-coordinate, and we can mark the streets they can travel to on a number line. The center of the flooding is Street 2, so 6 blocks north is Street 8, and 6 blocks south is Street $^-4$. Therefore, the y-coordinate of the position of any bus on Avenue 3 must be either greater than 6 if they stay north of the flooding or less than $^-4$ if they stay south of the flooding. The possible bus positions can be marked on a y number line (see Figure 4).

We can also express the answer algebraically using set-builder notation (page 304). First, we need to write inequalities using the variable y and the information we have about the flooding. The flooding starts on Street 2 and goes 6 blocks in both directions, and the buses must stay *out* of the flooded area; in other words, they must be either north of the flood **or** south of the flood. North of the flood translates to at least 6 blocks north of street 2, so $y \geq 2 + 6 = 8$. South of the flood becomes $y \leq 2 - 6 = {}^-4$. The set of streets the buses can travel is, therefore, $\{y \mid y \leq {}^-4 \text{ or } 8 \leq y\}$.

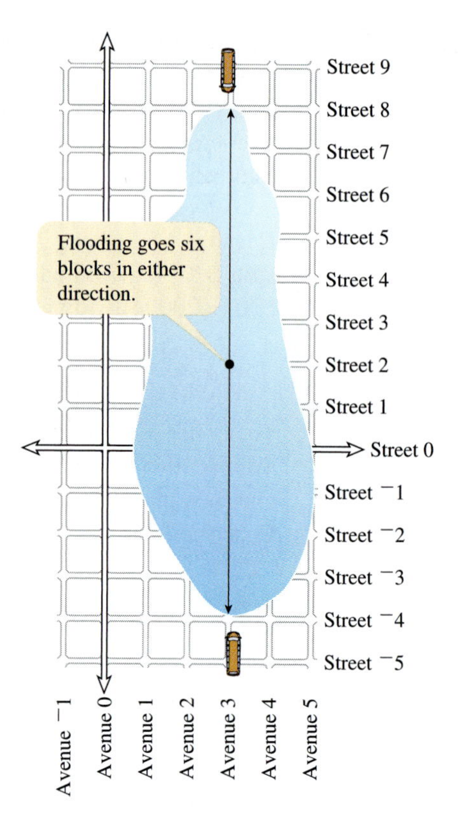

Figure 3 The flooding keeps buses away from the corner of Avenue 3 and Street 2 for 6 blocks in either direction.

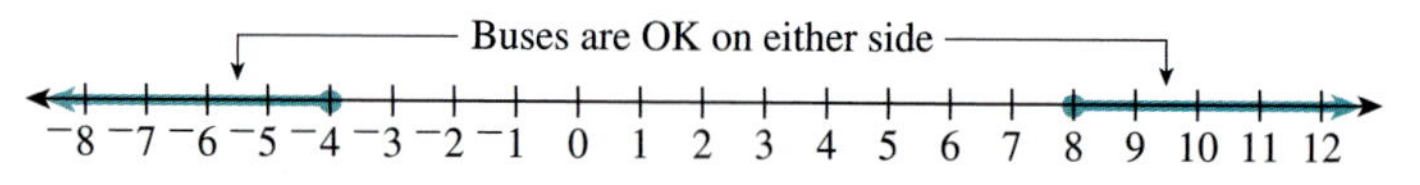

Figure 4 Buses cannot get too close to $y = 2$.

The solutions to Examples 1 and 2 are different than the solution to Example 3. In the first two examples, the possible values of the variable x-coordinate (or number of minutes) are within a given range. The variable must satisfy both conditions put on it, for example the number of minutes must be greater than 200 *and* no more than 400. On the other hand, in Example 3, the variable (y-coordinate) can satisfy either one *or* the other condition; either the bus is north of the flooding *or* it is south of it.

Inequalities that involve *and*, such as Examples 1 and 2, often have solutions that look like Figures 1 and 2, with a bounded range of values shaded. Inequalities that involve *or*, such as Example 3, often have solutions that look like Figure 4, with two (or more) regions shaded (the regions might overlap). We can specify the solution to a single inequality or to a set of inequalities using either a number line or set-builder notation.

EXAMPLE 4 Suppose the number of people who will sign up to take a writing workshop is given by the function $N = 180 - 2x$, where x is the price in dollars charged for the workshop. If the organizers want to make sure at least 30 people sign up and the room the workshop will be in will hold no more than 75 people, how much can they charge for the workshop?

Solution Here, we have a formula, $N = 180 - 2x$, that must satisfy two conditions, so we should use *algebraic* methods to try to obtain a solution. We can write

$$30 \leq 180 - 2x \quad \text{and} \quad 180 - 2x \leq 75$$

Algebraically, we can treat inequalities *almost* the same way we treat equations. If we are given an inequality, we can add the same term on both sides or multiply on both sides by the same *positive* number and still have a valid inequality. Multiplying by a negative number on both sides is a little trickier, as we soon see.

Let's try this with the two inequalities.

$$30 \leq 180 - 2x$$

$$0 \leq 150 - 2x \qquad \textit{Subtracting 30 on both sides}$$

$$2x \leq 150 \qquad \textit{Adding 2x on both sides}$$

$$x \leq 75 \qquad \textit{Multiplying on both sides by } \tfrac{1}{2}$$

Using the first inequality, we know that the charge can be no more than \$75. The second inequality gives

$$180 - 2x \leq 75$$

$$^{-}2x \leq {}^{-}105 \qquad \textit{Subtracting 180 on both sides}$$

The next step is to multiply both sides by $^{-}\frac{1}{2}$, but so far we can multiply only by a positive number. What happens if we multiply both sides of an inequality by a negative number?

We can think of an inequality as living on a number line; if x is greater than some number, then it is *to the right of it* on the number line, and if x is less than some number, then it is *to the left of it*. If we multiply by a negative number, then positive numbers become negative, and negative numbers become positive; in other words, the number line *rotates*. So, if x is to the right of a number, after the rotation it is to the left of it, and vice versa. (See Figure 5.) Therefore, if we multiply an inequality by a negative number on both sides, we need to *reverse* the inequality symbol, so $<$ becomes $>$, $\leq$ becomes $\geq$, etc.

Going back to our algebraic manipulation,

$$^{-}2x \leq {}^{-}105 \qquad \textit{Multiplying on both sides by } {}^{-}\tfrac{1}{2}$$

$$x \geq 52.5 \qquad \textit{Reversing the direction of the inequality symbol}$$

Therefore, the organizers must charge at least \$52.50 to keep the crowd small enough to fit into the room.

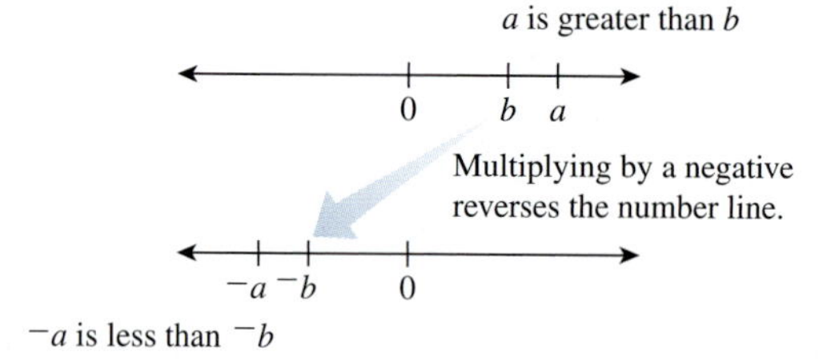

Figure 5 Multiplying an inequality by a negative reverses the relationship.

The solution to this example involves two inequalities, $x \geq 52.5$ and $x \leq 75$, and there is a set of x-values that solve the problem. Therefore, we can specify the solution set of the problem using either a number line or set-builder notation. In set-builder notation, the inequalities $x \geq 52.5$ and $x \leq 75$ become the solution set $\{x \mid 52.5 \leq x \leq 75\}$.

Manipulating Inequalities

In the previous example we discovered the following facts about manipulating inequalities.

Manipulating Inequalities

You can add the same quantity (which might be negative) to both sides of an inequality. You can multiply both sides of an inequality by a *positive* number (which might be less than 1). If you multiply both sides of an inequality by a *negative* number, you must reverse the direction of the inequality symbol.

Example 5 illustrates how to apply these ideas.

EXAMPLE 5 ■ Find the solution set if x must satisfy $2x + 3 \geq 5$ and $5x < 10$.

Solution First, we solve each inequality.

$$2x + 3 \geq 5 \quad \text{and} \quad 5x < 10$$

$$2x \geq 2 \quad \text{and} \quad x < \tfrac{10}{5}$$

$$x \geq 1 \quad \text{and} \quad x < 2$$

Because x must satisfy *both* inequalities, the solution set in set-builder notation is $\{x \mid 1 \leq x < 2\}$.

Exercises 12.1

1. In Example 1 the entire solution set was shown on a number line.
   a. Give four individual solutions from the number line.
   b. Are $^{-}2$ and 6 solutions to this problem? Explain.
2. Suppose a laser beam is shot east along Street 0 beginning at Avenue 1. *Note:* Assume a laser beam can travel an infinite distance.
   a. Use a number line to show all the points on Street 0 that this laser beam will cross.
   b. Express your answer in set-builder notation.
3. Review Example 2.
   a. Give two individual values that are solutions to Example 2.
   b. Are $x = 200$ minutes and $x = 400$ minutes solutions to this problem?
4. The number of spectators attending a college basketball game is between 3000 and 5000 people.
   a. Show the solution set on a number line.
   b. Write the solution set in set-builder notation.
5. Fill in the blank with each phrase below and choose the correct inequality symbol ($<, >, \leq, \geq$). The first four have been done for you. A number x is __4.
   a. Less than: $x < 4$
   b. Greater than: $x > 4$
   c. Less than or equal to: $x \leq 4$
   d. Greater than or equal to: $x \geq 4$
   e. At most
   f. At least
   g. No more than
   h. No less than
   i. Above
   j. Below
6. It is illegal for a motorist to travel less than 40 miles/hour or more than 65 miles/hour on many interstate highways.
   a. Show the speeds that are illegal on a number line.
   b. Write your answer from part (a) in set-builder notation.
7. Show the solution set for each condition on a number line.
   a. $^{-}5 < x$ and $x \leq 20$
   b. $x \leq {}^{-}5$ or $20 < x$
8. Write the set indicated on each number line in set-builder notation.

   a. 

   b. 

9. Review Example 4 and describe what happens to an inequality when you multiply or divide by a negative number. Demonstrate by solving the inequalities.
   a. $^{-}3x \leq 18$
   b. $-\frac{1}{2}x > {}^{-}7$

In Exercises 10–14, use algebra to solve each inequality.

10. $2x - 3 < 15$
11. $4 + x \geq 2 - (7x + 1)$
12. Write the solution set in set-builder notation if
$$\frac{x}{2} - 6 > 4 \quad \text{and} \quad \frac{x}{2} - 6 < 8$$
13. Write the solution set as a single inequality in set-builder notation if $5 \leq 20 - 3x$ and $20 - 3x < 12$.
14. Show the solution set on a number line if $4x - 12 > {}^{-}5$ or $4x - 12 \leq {}^{-}10$.
15. A rectangular region surrounded by a fence is 40 m wide (see the figure). The perimeter of the region is at least 200 m and no more than 300 m.

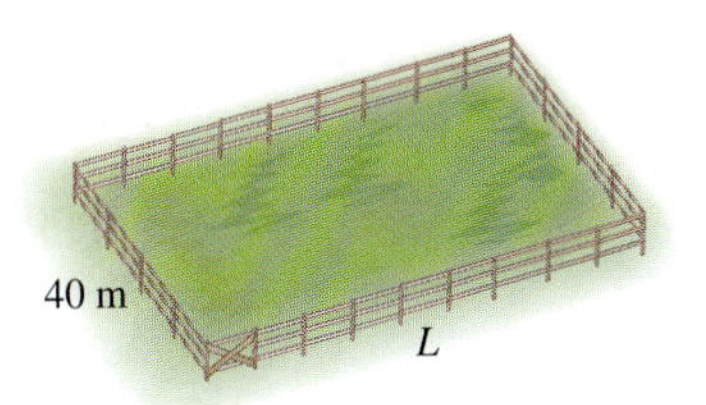

   a. Write an expression for the perimeter of the rectangle.
   b. Use your expression to write two inequalities, one with the minimum perimeter and one with the maximum perimeter.

c. Should the inequalities be connected by *and* or *or* in order to describe the possible lengths of the rectangle?

d. Solve the inequalities to find the possible lengths of the rectangle.

Skills and Review 12.1

16. y varies indirectly as the square of x and $y = 1.25$ when $x = 2$. Find y when $x = \sqrt{5}$.

17. y varies directly as the cube root of x and $y = 21$ when $x = 1000$. Find x when $y = 4.2$.

18. Simplify the rational expression

$$\frac{7-b}{b^2-49}.$$

19. Solve for x: $\frac{2}{x} - \frac{3}{x-4} = \frac{5}{2}$.

20. Show the domain and range of the function $f(x) = \frac{1}{x} + 1$ in set-builder notation.

21. Solve for y: $10^{y+3} = 100$.

22. Solve for c: $6c^{1/5} - 1 = 17$.

23. Use a graphing method to solve for x: $x^3 = 2$.

24. Let $y = (x+3)^2 - \frac{15}{7}$. Notice that this function is modified from the basic quadratic $y = x^2$. Every point on the graph of $y = x^2$ is moved 3 units left and $\frac{15}{7}$ units down. Given that the vertex for the basic quadratic function is $(0, 0)$, mentally find the vertex for this modified function.

25. The revenue R that a company generates is modeled by the equation $R = {}^{-}3.2P^2 + 64P$, where P is the price of the product in dollars.

a. Use factoring to find the price(s) at which the company generates zero revenue.

b. Recall that the x-coordinate of the vertex for a parabola is midway between the roots. Use this fact to find the vertex.

c. Review the facts about parabolas to determine if this parabola opens upward or downward.

d. Use the information from parts (a)–(c) to either sketch the graph by hand or find a suitable window for viewing the graph on your calculator.

12.2 Solving Nonlinear Inequalities by Graphing

AFTER STUDYING THIS SECTION YOU WILL BE ABLE TO

- Use a graph to solve a nonlinear inequality in one variable.
- Apply the formula for distance on a number line.

EXAMPLE 1 Suppose the supply of computers available remains relatively constant, but demand peaks at two times during the year, once at the end of the summer, when students are heading back to school, and once in December, when people are buying holiday presents (see Figure 6). When is demand greater than supply?

Solution We don't have any algebraic formula for demand; the only information we have about supply and demand is graphical, in Figure 6. We can think of this as an inequality—but one that is not linear. If we call the supply function $S(t)$ and the demand function $D(t)$, then we want to know when $D(t) > S(t)$, because this

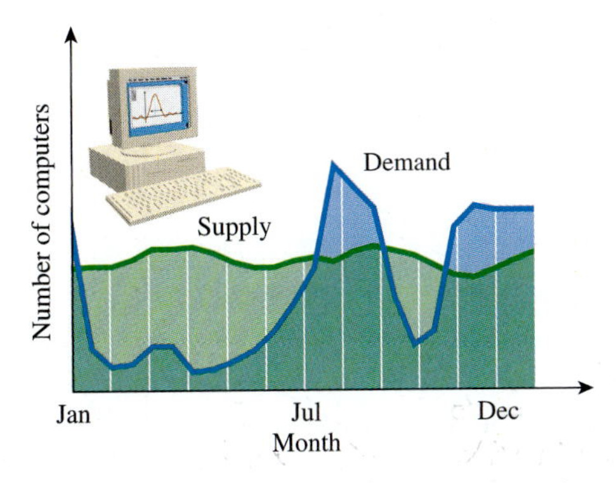

Figure 6 The demand curve is not linear. There are two times when demand is higher than supply.

will be when demand is greater than supply. Looking at Figure 6, the curve $D(t)$ is *above* the curve $S(t)$ from around the end of July until sometime near the end of the summer and then from around Thanksgiving until about the end of the year.

In general, we can solve an inequality graphically by following the basic rule that when the graph of one function is *above* the graph of another, then the values of that function are *greater than* the values of the other, and vice versa. So, if we have two functions $f(x)$ and $g(x)$ and we want to know when $f(x) > g(x)$, we can just look at a graph of both $f(x)$ and $g(x)$ and find where the graph of $f(x)$ is above $g(x)$. If we want $f(x) < g(x)$, we just look for when the graph of $f(x)$ is below $g(x)$.

EXAMPLE 2 Tashi and Tenzing are watching fireworks from their balcony, which is 50 feet high. They have the best view when the fireworks are higher than they are. The height of the fireworks is given by the formula $h(t) = {}^{-}16t^2 + 100t$, where $h(t)$ is the height of the fireworks shell in feet t seconds after it is fired. When are the shells lower than the balcony, ruining their view?

Solution If we graph the height function $h(t)$, we can compare it to the height of the balcony, which is the line $h = 50$ (see Figure 7). We need to determine when the function $h(t)$ crosses the line $h = 50$. We can enter Y1 = ⁻16X ∧ 2 + 100X and Y2 = 50 into a graphing calculator and use TRACE or 2ND CALC:intersect to get approximate solutions, $t \approx .55$ seconds and $t \approx 5.70$ seconds. (See Figure 8.) Because the shells are below the balcony before the first time and after the second, there are two time intervals in the solution: the first from the firing of the shells at $t = 0$ until $t \approx .55$, and the second from $t \approx 5.70$ until the shells hit the ground, which happens when $t = 6.25$. We can write the solution set using set-builder notation as

$$\{t \mid 0 \le t < .55 \quad \text{or} \quad 5.70 < t \le 6.25\}$$

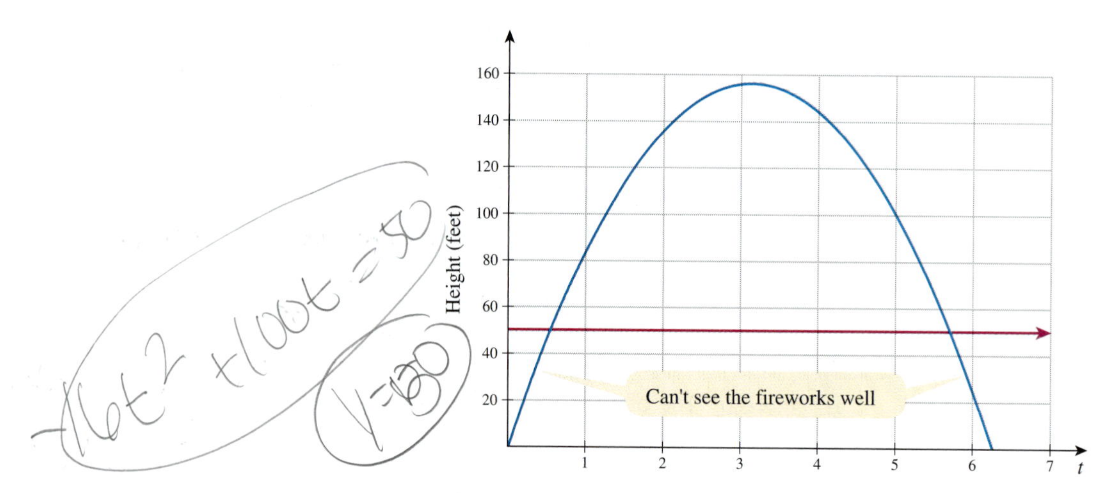

Figure 7 The height of the shell is the same of the height of the balcony at two times. Before the first time and after the second, Tashi and Tenzing can't see the shells well.

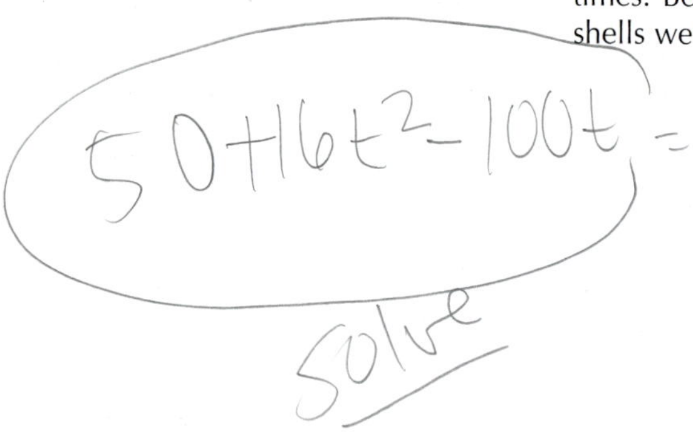

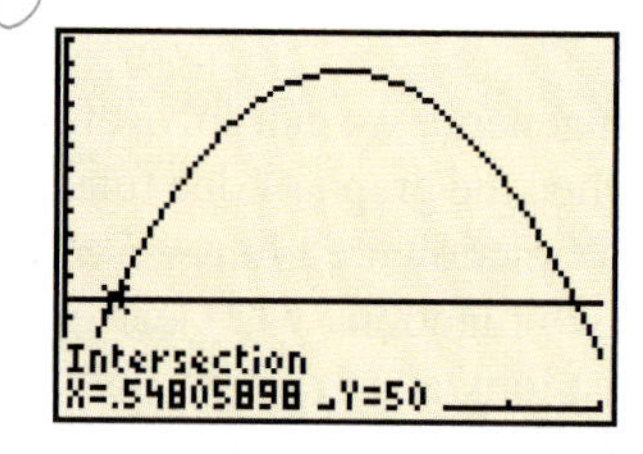

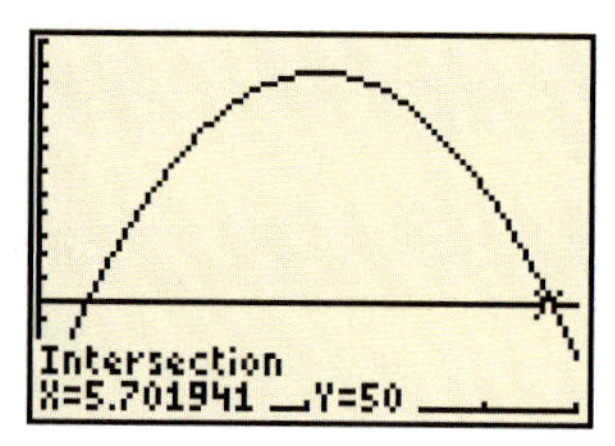

Figure 8

If, for some reason, we need exact endpoints for our solution set, we can solve a quadratic equation. In Example 2, we want to know when the height function $h(t) = {}^-16t^2 + 100t$ equals 50, so we just need to solve the quadratic equation

$$^-\mathbf{16}t^2 + \mathbf{100}t = \mathbf{50}$$

$$^-\mathbf{16}t^2 + \mathbf{100}t - \mathbf{50} = 0 \qquad \textit{Subtracting 50 on both sides}$$

The solutions given by the quadratic formula are

$$t = \frac{^-\mathbf{100}}{^-32} \pm \frac{\sqrt{\mathbf{100}^2 - 4(^-\mathbf{16})(^-\mathbf{50})}}{^-32}$$

$$= \frac{25}{8} \pm \frac{\sqrt{6800}}{^-32}$$

$$= \frac{25}{8} \pm \frac{\sqrt{400 * 17}}{^-32} \qquad \textit{Factoring 6800 to show largest perfect square}$$

$$= \frac{25}{8} \pm \frac{20\sqrt{17}}{{}^{-}32} \qquad \textit{Writing the result in simple radical form}$$

$$= \frac{25}{8} \pm \frac{{}^{-}5\sqrt{17}}{8}$$

If we evaluate these *exact solutions* to two decimal places, we get

$$\frac{25}{8} + \frac{{}^{-}5\sqrt{17}}{8} \approx .55$$

$$\frac{25}{8} - \frac{{}^{-}5\sqrt{17}}{8} \approx 5.70$$

which agrees with our graphical solution.

EXAMPLE 3 Suppose it costs the Acme widget company \$100 to produce each widget, and \$250,000 to set up the production plant. The more they charge for each widget, the fewer widgets they sell. They work out a formula for the number of widgets they will sell at a given price x in dollars,

$$N = 2450 - 2.5x$$

How much can they charge for each widget and make a profit? We assume that Acme makes exactly as many widgets as they can sell at any given price x.

Solution In this case we need to know the revenue Acme will make from selling widgets at a given price x. Revenue is given by price times number of items sold:

$$\begin{aligned} R(x) &= \text{price} * \text{number} \\ &= x(2450 - 2.5x) \\ &= {}^{-}2.5x^2 + 2450x \end{aligned}$$

The cost is given by the fixed cost plus the number sold times the cost per item,

$$C(x) = 250{,}000 + (2450 - 2.5x)100 = {}^{-}250x + 495{,}000$$

We can plot both functions on the same set of axes and try to find the x-values for which $R(x)$ is greater than $C(x)$ (Figure 9). The revenue function is *above* the cost function for a wide range of prices. Entering the functions into a graphing calculator and using either TRACE or 2ND CALC: intersect, we can find the approximate values for which the two functions are equal (when profit would be zero): $x \approx 234$ and $x \approx 846$. (See Figure 10.) So Acme will make a profit if they charge more than \$234 and less than \$846, and the solution set is

$$\{x \mid 234 < x < 846\}.$$

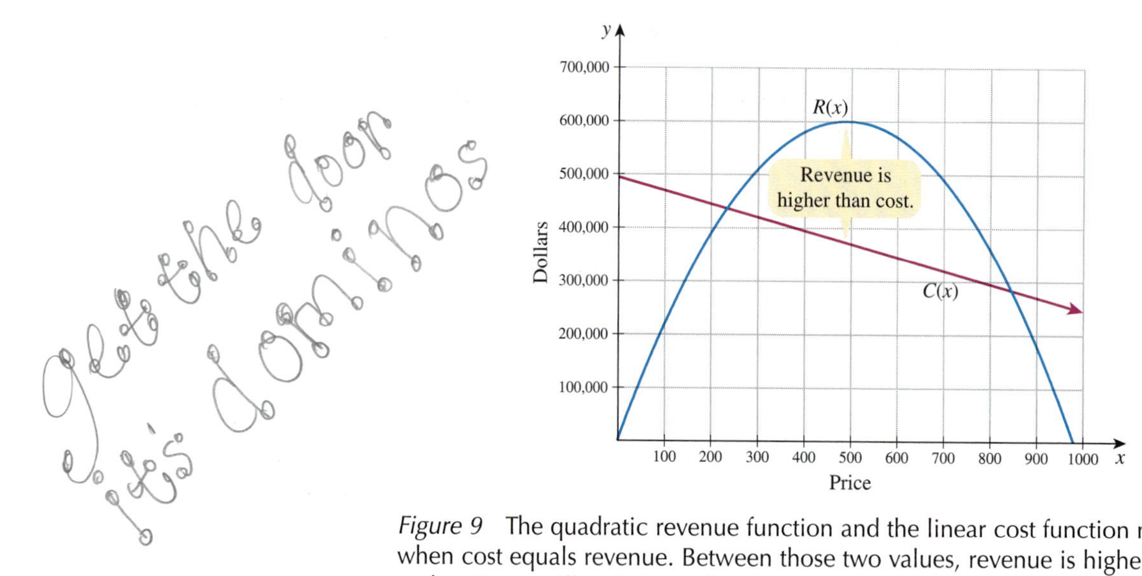

Figure 9 The quadratic revenue function and the linear cost function meet twice, when cost equals revenue. Between those two values, revenue is higher than cost, and so Acme will make a profit.

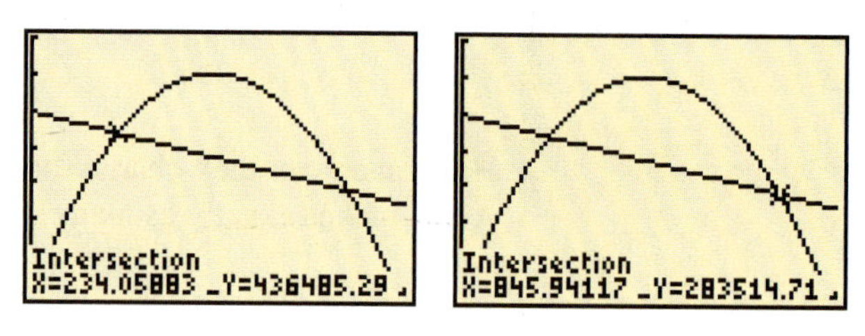

Figure 10

Distance and Absolute Value

The distance from one point to another on a number line must always be *positive*. We can measure distance using the *absolute value function*.

Distance on a Number Line

The distance from a given point a on a number line to any point x on the number line can be determined using the **absolute value function**:

$$\text{distance from } x \text{ to } a = |x - a|$$

EXAMPLE 4 A cellular phone company is trying to expand their service along a major highway, so they put a microwave tower at milepost 33. The highway is straight for 100 miles in either direction, and the signal will carry for 22 miles. If you are driving along the highway, between what mileposts will your cell phone work?

Solution Because the highway is straight for all the distance we care about, we can think of it as a number line. Using the formula for distance, if we call the position of the car (the milepost you are at) x, and the microwave tower's position $a = 33$, then your distance to the microwave tower is given by

$$D = |x - 33|$$

Because we want to know when your cell phone will work, we want $D \leq 22$, so

$$|x - 33| \leq 22$$

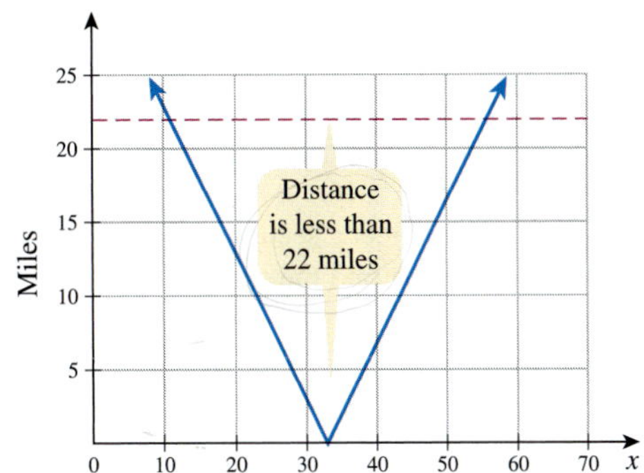
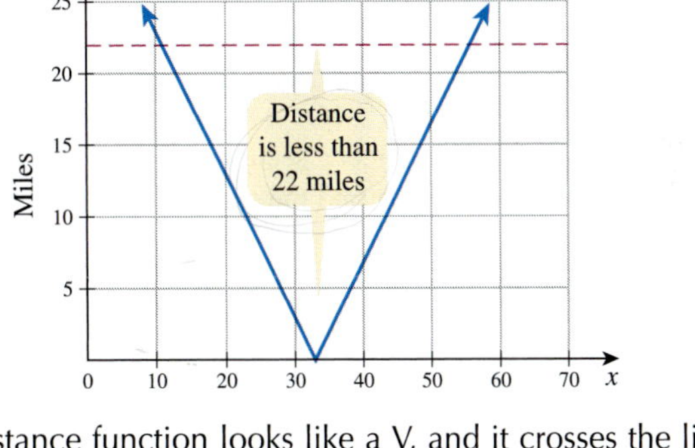

Figure 11 The distance function looks like a V, and it crosses the line $D = 22$ at two points.

We may set y equal to each side of the inequality and produce a graph (see Figure 11). It is clear from the graph that the distance is less than 22 miles *between* the points where the graphs intersect. Entering Y1 = abs (X − 33) and Y2 = 22 into a graphing calculator and using TRACE or 2ND CALC: intersect, we determine that the graphs intersect when $x = 11$ and $x = 55$. Therefore, the solution set is

$$\{x \mid 11 \leq x \leq 55\}$$

See Figure 12.

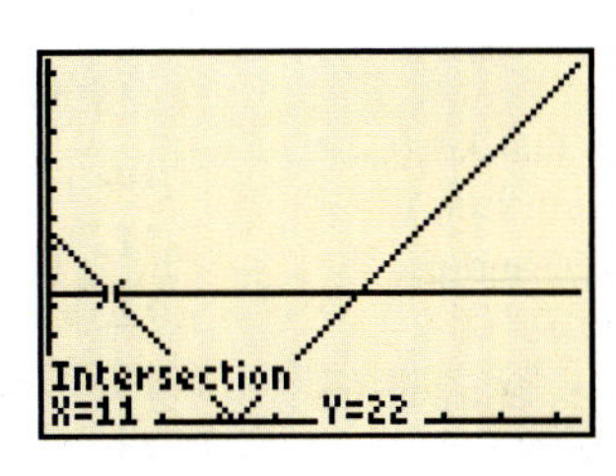

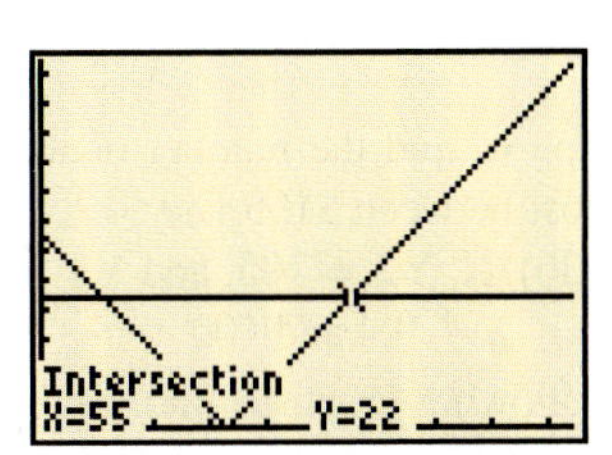

Figure 12

Exercises 12.2

1. In Example 2, the times in which Tashi and Tenzing were not able to see the fireworks were given by the solution set $\{t \mid 0 \leq t < .55 \text{ or } 5.70 < t \leq 6.25\}$.
   a. Interpret the meaning of the endpoints $t \approx .55$ second and $t \approx 5.70$ seconds.
   b. Interpret the meaning of the partial solution set
   $$\{t \mid 0 \leq t < .55\}.$$
   c. Interpret the meaning of the partial solution set
   $$\{t \mid 5.70 < t \leq 6.25\}.$$
2. An object is fired from a cannon. The height of the object is given by the formula $h(t) = {}^{-}16t^2 + 40t$, where $h(t)$ is the height of the object in feet t seconds after it is fired. Complete the following to find the times when the object is at least 16 feet high.
   a. Graph $\text{Y1} = {}^{-}16\text{X} \wedge 2 + 40\text{X}$ and $\text{Y2} = 16$ in the window $0 \leq \text{X} \leq 4.7$ and $0 \leq \text{Y} \leq 31$.
   b. When is the height of the object exactly 16 feet?
   c. When is the height of the object at least 16 feet? Write your solutions in set-builder notation.
3. Reconsider Example 3. Use a graph to find the prices at which Acme will lose money. (*Hint:* Look for when the revenue function is below the cost function.)
4. Review Example 4. On a graph, explain how you can tell that $x = 60$ miles is not a solution to this problem, and $x = 21$ miles is a solution.
5. In a poll for town council, a candidate is given 51% of the vote, plus or minus 5%. The solution of the absolute value inequality $|x - 51| > 5$ represents the percentage of the vote that is outside this candidate's 5% margin of error. Use a graph to find these percentages. (*Hint:* Use a window of $0 \leq \text{X} \leq 94$ and $0 \leq \text{Y} \leq 20$.)
6. Suppose the average cost, C, for producing x bicycles per hour is given by the function
   $$C(x) = \frac{100x + 30}{x}$$
   Use a graph to find the number of items that keep the average cost between \$105 and \$120. (*Hint:* Let $\text{Y1} = (100\text{X} + 30)/\text{X}$, $\text{Y2} = 105$, and $\text{Y3} = 120$ and use the window $0 \leq \text{X} \leq 9.4$ and $100 \leq \text{Y} \leq 125$.)
7. Use the following graph to find the values of x where the graph of $\text{Y1} = \sqrt{}(\text{X})$ is at or below the line $\text{Y2} = 2$. This is the solution to the inequality $\sqrt{x} \leq 2$.

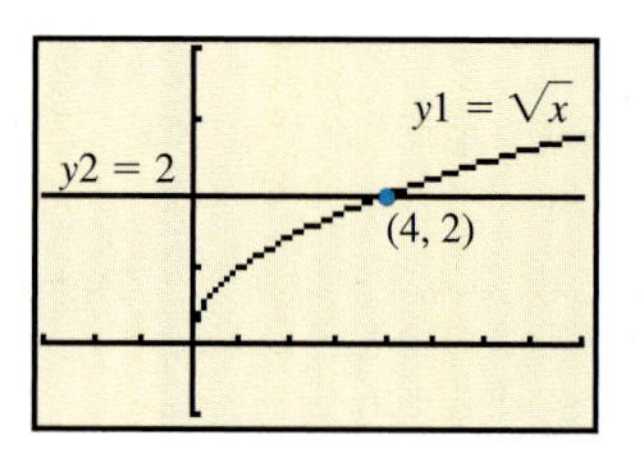

8. The following graph shows the intersections of $\text{Y1} = {}^{-}\text{X} \wedge 2 + 6\text{X}$ and $\text{Y2} = \text{X} + 4$. Find the solution set for each inequality and test the sets with the values $\text{X} = .5, 2, 5$.

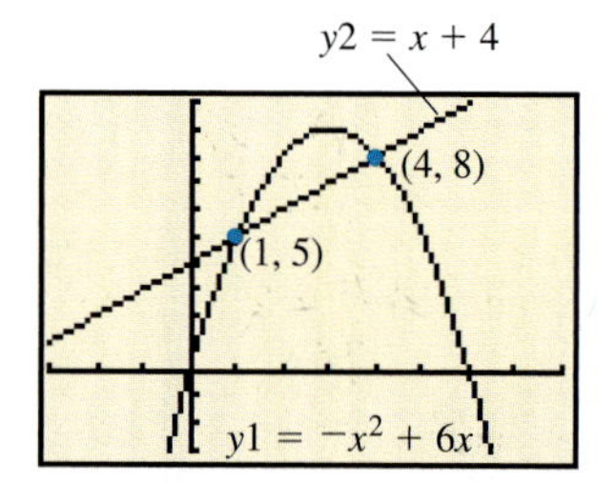

   a. ${}^{-}x^2 + 6x > x + 4$ **b.** ${}^{-}x^2 + 6x \leq x + 4$
9. The following graph shows the intersection of $\text{Y1} = \text{X} + 11$ and $\text{Y2} = {}^{-}\text{X} + 3$. Find the solution set to each inequality and test the sets with the values $x = {}^{-}5$ and $x = 0$.

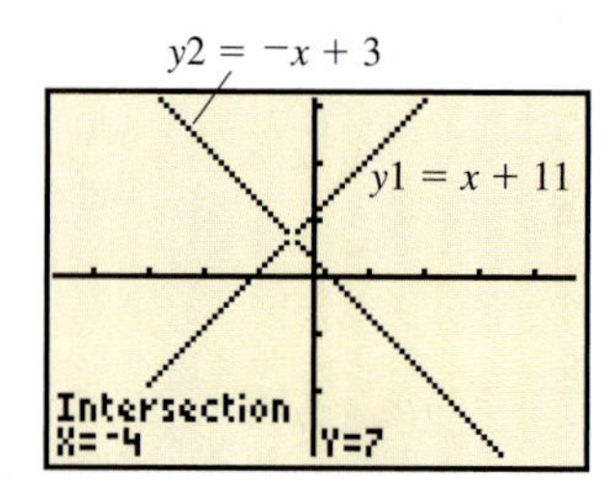

   a. $x + 11 < {}^{-}x + 3$ **b.** $x + 11 \geq {}^{-}x + 3$

In Exercises 10–15, use a graph to find the solution set. Write the solutions in set-builder notation.

10. $|x + 5| \leq 10$
11. $(x - 2)^2 > 29$
12. $x^3 + 7 \leq 143$
13. $4x^2 \geq 20$
14. $\sqrt{x} \leq 6$ or $\sqrt{x} > 14$
15. $0 < \frac{4}{x^2} < 1$

Skills and Review 12.2

16. Show your solutions on a number line if x must satisfy $^-3x > 6$ or $x + 5 > 8$.

17. Find the solution set if x must satisfy $7 \le {}^-4x + 5$ and $^-4x + 5 < 19$. Show the solutions on a number line.

18. The time, t, it takes to run around a track varies indirectly with your speed, r. It takes 1.25 minutes to run around the track when your speed is .2 mile/minute. How long will it take to run around the track when your speed is .15 mile/minute?

19. Solve for x if

$$\frac{7}{x} + \frac{5}{3x} = 13$$

20. Simplify

$$\frac{x^2 + 4x}{x^2 - 16}.$$

21. Solve for x if $(x - 9)^{1/4} = 2$.

22. Solve for x if $\sqrt{30x - 29} = x$.

23. Suppose $f(x) = x^2 + 1$ and $g(x) = 3x - 1$.

a. Find $f(g(^-2))$.

b. Find $f(g(x))$.

24. Given the following two graphs.

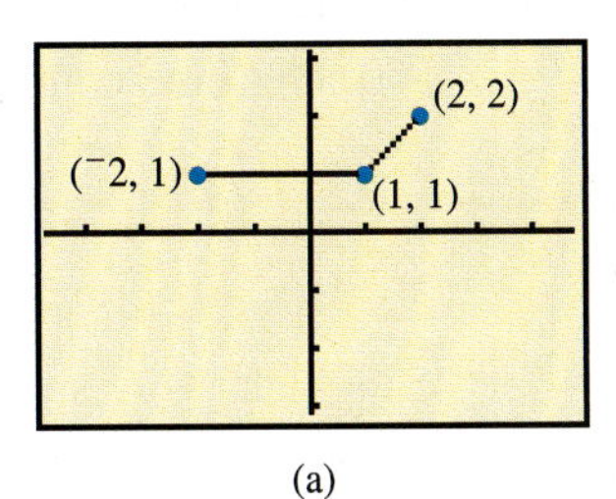

(a)

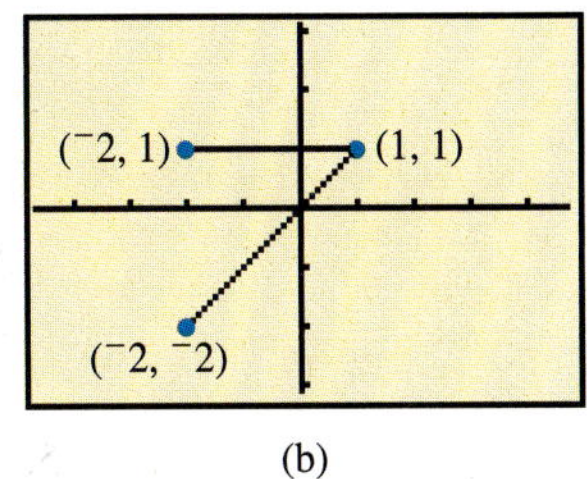

(b)

a. Which graph represents a function? Explain.

b. Show the domain and range of this function in set-builder notation.

c. Explain why the other graph cannot represent a function.

25. Use the quadratic formula to find the solutions to the equation $x^2 + 8x + 11 = 0$. Write the solutions in simple radical form.

12.3 Inequalities in Two Variables; Systems

After Studying This Section You Will Be Able To

- Graph inequalities in two variables.
- Represent a problem situation with a system of inequalities.
- Find a graphical solution to a system of inequalities in two variables.

In Sections 12.1 and 12.2 we looked at problems involving one variable, which had to satisfy conditions that could be expressed as inequalities. Many actual problems involve situations where *two* quantities must satisfy conditions that hold for *both*; what can we do then?

Systems of Inequalities

Example 1 shows how to use inequalities in two variables to solve a problem.

EXAMPLE 1 Julissa's farm has 750 plantable acres. She can't always use all these acres, but she never plants less than 500 acres. Julissa grows soybeans and wheat on her farm. How many acres of each can she plant in any year?

Solution If we read the problem carefully, we can see that there are many solutions. Julissa might plant 500 acres of soybeans and 250 of wheat (if she is lucky enough to plant all her acres one year), or she might plant 250 of each, or 600 of soybeans and no wheat, and so on. Clearly, we need two variables for the problem, one representing the number of acres of soybeans planted, which we can call x, and one for the number of acres of wheat, say, y.

There are two conditions that must be satisfied; Julissa must plant at least 500 acres but no more than 750 acres. We can translate these into two inequalites,

$$x + y \geq 500$$

$$x + y \leq 750$$

This looks very similar to the systems of equations we examined in Chapter 4. Because x and y represent acres planted, they must both be positive, so we really have two more inequalities,

$$y \geq 0$$

$$x \geq 0$$

To solve a system of equations graphically, we drew lines for each equation and then found the point that was on both lines. Because that point was a solution for the equation corresponding to each line, it solved both of them simultaneously and was a solution for the system.

We can try the same thing here. First, we must rewrite our **system of inequalities** in a form that looks something like the equation for a line, $y = mx + b$.

$$y \geq 500 - x$$

$$y \leq 750 - x$$

We cannot graph these yet, but we can graph the line associated with each inequality:

$$y = 500 - x$$

$$y = 750 - x$$

(See Figure 13.) To account for the second set of inequalities, we restrict the window on our caculators to the first quadrant, which guarantees that we see only positive x- and y-values. Basically, we are *pretending* for a minute that we actually have a system of equations.

All points on the higher line in Figure 13 satisfy the equation $y = 750 - x$, so points on or *below* the line all satisfy the inequality $y \leq 750 - x$. Points on or *above* the line $y = 500 - x$ satisfy the inequality $y \geq 500 - x$. We can indicate that points are solutions to a given inequality by shading the graph (see Figure 14). The lines associated with each inequality are called the **boundary lines**, because they form the boundary of the region of solutions.

Any point that is shaded twice satisfies both inequalities and, therefore, is a solution to this system of inequalities. In Figure 14, these points are all points

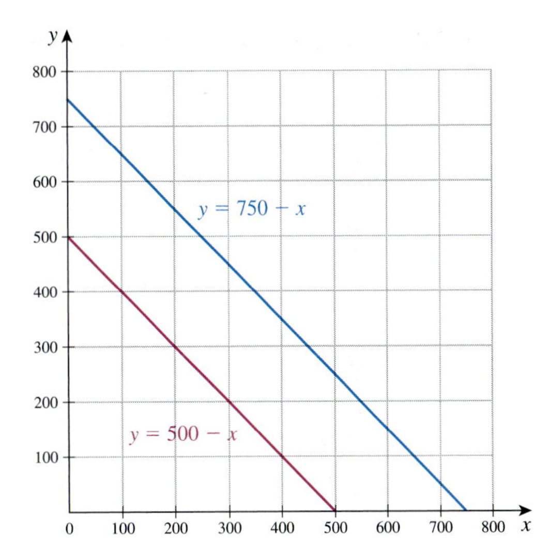

Figure 13 What points on the graph will solve both inequalities?

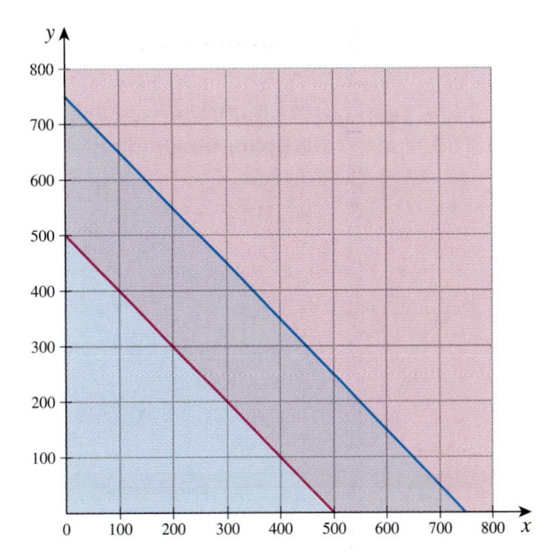

Figure 14 We have shaded below $y = 750 - x$ and above $y = 500 - x$, and points in between (or on either line) are shaded twice.

between the two lines or on either line that are in the first quadrant. For example, the point (300, 350) is a solution, because it is in the region shaded twice. This point represents Julissa planting 300 acres of soybeans and 350 acres of wheat.

You can shade the correct portion of the plane on your TI-83 by using the menu that is to the left of each equation in the Y= screen; move the cursor left until the line is blinking, and use the ENTER key to choose the correct portion to shade (see Figure 15).

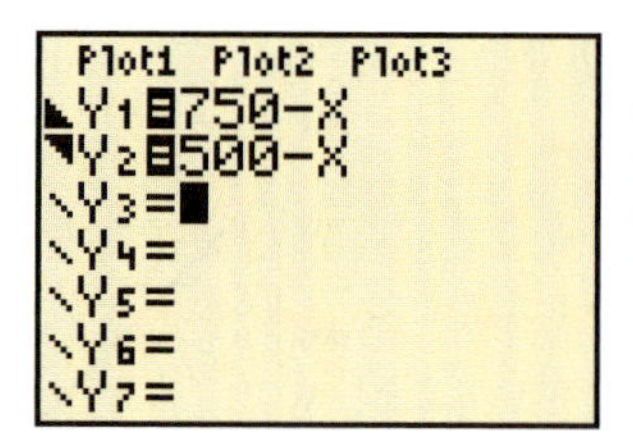

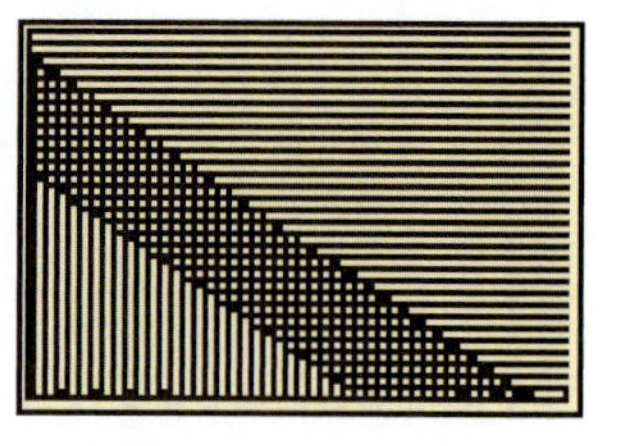

Figure 15 You can chose to shade above or below any given graph.

Being Careful About Solution Sets

It is important, as conditions are added, to interpret the information accurately.

EXAMPLE 2 Julissa realizes that the price of soybeans will go up, because people are using soy in all sorts of new products. She still needs to grow some wheat, because there is a bakery that buys from her regularly. She decides that she will always grow more soybeans than wheat but never more than twice as much. How many acres of each can she plant now?

Solution Julissa is *adding* conditions to a problem we have already solved, so we can start with the graph in Figure 14, and add to it. We have two new inequalities to satisfy,

$x > y$ *We use >, not ≥, because the problem says more.*

$x \leq 2y$

We use the related equations again to get

$$y = x$$

$$y = \frac{x}{2}$$

which we can add to Figure 14, giving Figure 16.

Notice that one of the new lines is drawn as a dashed line; this allows us to tell the difference between $>$ and $\geq$ and between $<$ and $\leq$. Unfortunately, there is no way to get a calculator to do this, so you must remember which is which. The solutions are all the points that are shaded *four* times, which can be tricky to see, as shown in Figure 16.

Many real-world problems (such as Examples 1 and 2) involving inequalities include the inequalities $x \geq 0$ and $y \geq 0$ because such problems often deal only with positive numbers. Often this is not stated explicitly, and you can think of these as *hidden inequalities*. It is not always necessary to include these in your graph, because you can guarantee that your solution will include only postive numbers if you graph only the part contained in the first quadrant.

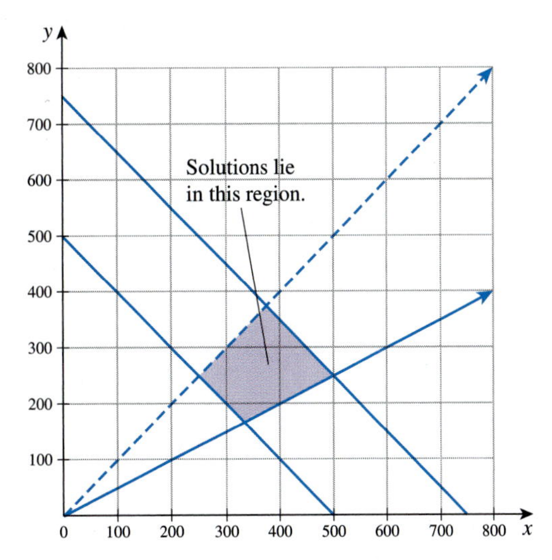

Figure 16 We add the lines $y = x$ and $y = \frac{x}{2}$. Because $y < x$, we shade below the first line. Because $y \geq \frac{x}{2}$, we shade above the second line. The line $y = x$ is drawn as a dashed line because y must be *strictly less than* x, so points on the line do not satisfy the inequality.

EXAMPLE 3 Find all pairs of numbers whose sum is less than 100 and for which the second number is at least twice as great as the first number.

Solution We call the first number x and the second number y. The numbers must satisfy the conditions

$$x + y < 100$$
$$y \geq 2x$$

Because the problem doesn't say that the numbers must be positive, we need to make sure that our graph includes both positive and negative values for x and y.

We can rearrange the inequalities in the form of lines (with y alone on the left-hand side):

$$y < 100 - x$$
$$y \geq 2x$$

The next step is to graph the *boundary lines* (see Figure 17).

The double-shaded region of solutions is a slice of the plane, because we are allowing for negative solutions.

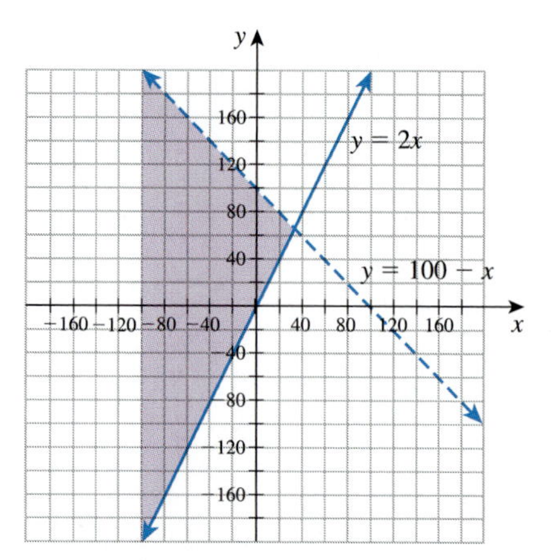

Figure 17 Graph of the boundary lines $y = 100 - x$ and $y = 2x$. We shade below the first line and above the second and draw $y = 100 - x$ as a dashed line because points on it do not satisfy $y < 100 - x$.

EXAMPLE 4 Find the solution set for the system of inequalities:

$$x + y \geq 25$$
$$y < 40$$
$$y - x \geq 10$$

Solution Here we must draw three boundary lines: $y = 25 - x$, $y = 40$ (dashed), and $y = x + 10$. Then, we shade above the first line, below the second, and above the third (Figure 18). The solutions are points that are shaded three times; the solution set is the interior of a triangle.

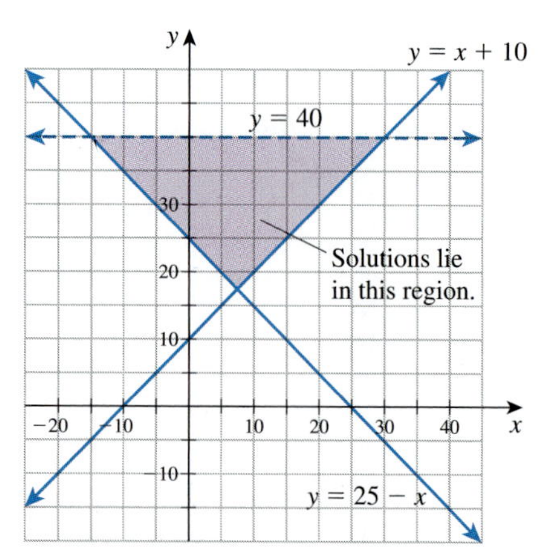

Figure 18 The solution set is below $y = 40$ and above (or on) $y = 25 - x$ and $y = x + 10$.

Exercises 12.3

1. In Example 2, the solution set to the system of inequalities was represented by the shaded region in Figure 16. Visually determine which of the ordered pairs are solutions to the system.

 a. (400, 250) **b.** (400, 100)

 c. (400, 500) **d.** (350, 300)

2. The system of inequalities in Example 2 was $y \geq 500 - x$, $y \leq 750 - x$, $y < x$, $y \geq \frac{x}{2}$. The following ordered pairs are on the boundary lines. Determine which are solutions to the system by testing them in all four inequalities.

 a. (375, 375) **b.** (500, 250)

 c. (250, 250) **d.** $(333\frac{1}{3}, 166\frac{2}{3})$

3. Review Example 3. Find two ordered pairs that *are* solutions to the system of inequalities and two that *are not.*

4. Review Example 4 and use 2ND CALC: intersect on your calculator to find the three intersection points of the boundary lines. Determine which of the intersection points are solutions to the system of inequalities.

5. Match each system of inequalities to its graph.

 a. $y > {}^-x + 4$
 $y \geq x + 1$

 b. $y < {}^-x + 4$
 $y \geq x + 1$

 c. $y \geq {}^-x + 4$
 $y < x + 1$

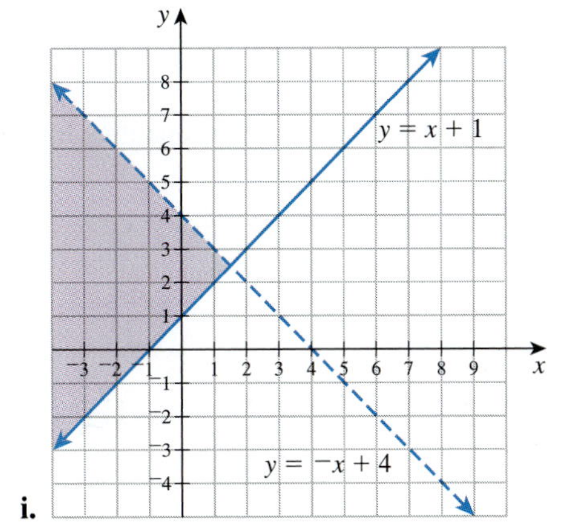

i.

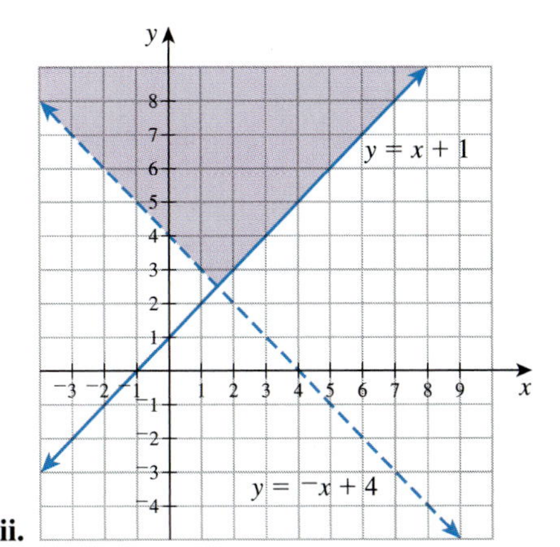

ii.

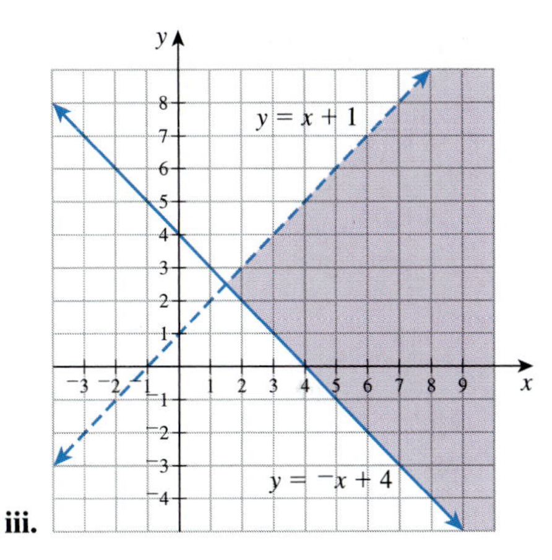

iii.

In Exercises 6–9, write a system of linear inequalities whose solution set is the shaded region of each graph.

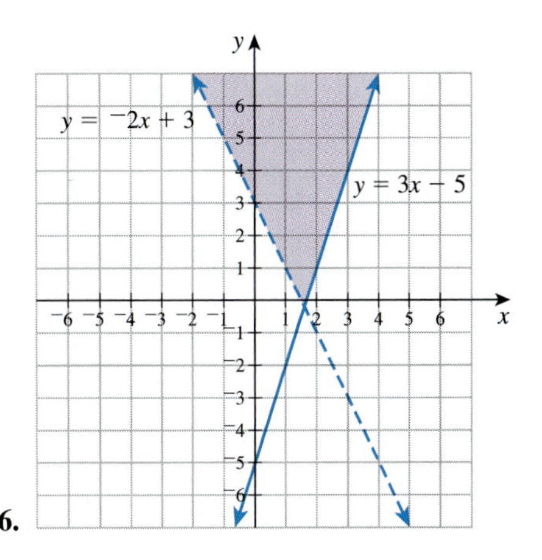

6.

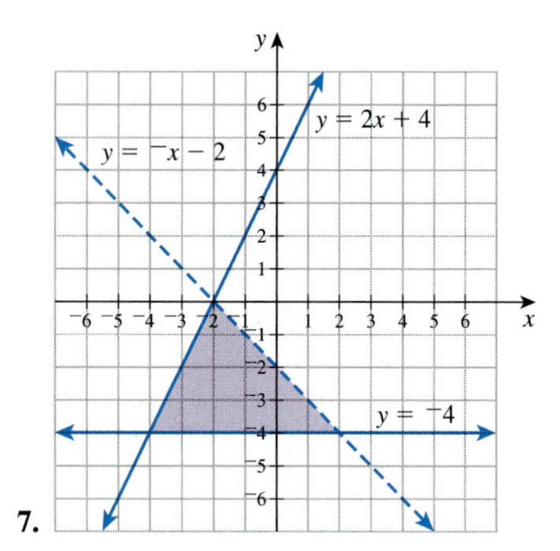

7.

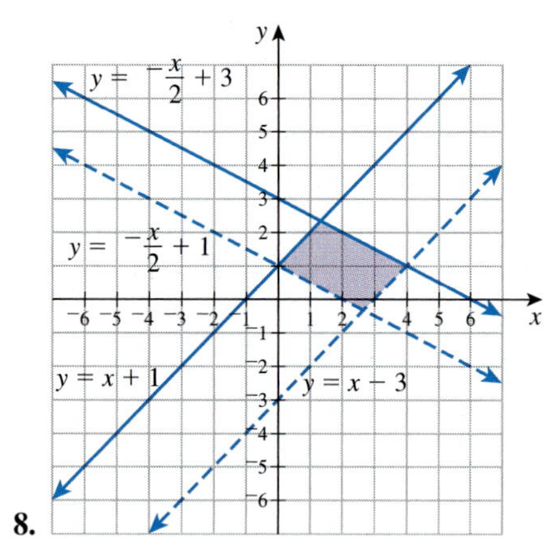

8.

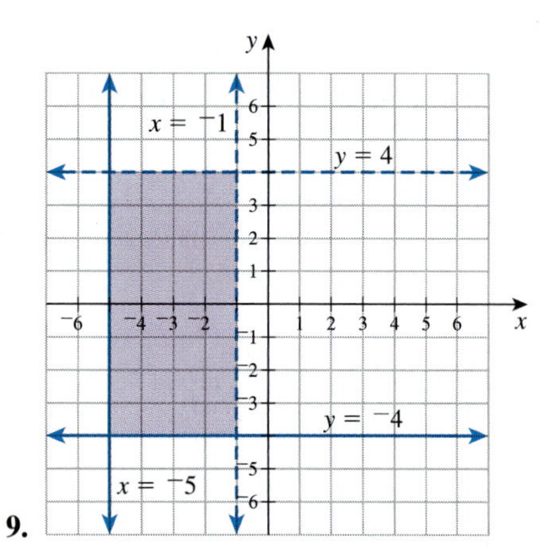

9.

In Exercises 10–13, graph the system of linear inequalities. Shade the region that represents the solution set.

10. $y \leq 2x$

$y < x - 3$

11. $x - 3y > 6$

$x + y \leq 1$

12. $y \geq 2$

$2y \leq x + 4$

$2x - y - 6 < 0$

13. $y \geq 0$

$x + y < 6$

$4x - y > 10$

14. While at the movies, a couple allows their children to buy no more than $\frac{1}{2}$ pound of peanuts and almonds. The almonds cost \$4 per pound and the peanuts cost \$1.50 per pound. The children have checked their pockets and found they have less than \$1.25 among them. We wish to represent the amount of each type of nut they can buy with a system of linear inequalities. Let x be the number of pounds of peanuts and y be the number of pounds of almonds.

a. Write two inequalities for this situation, one for the pounds of nuts and the other for the cost of buying the nuts.

b. Explain why this real-life problem requires that $x \geq 0$ and $y \geq 0$ along with the other inequalities.

c. Graph the inequalities and shade the region representing the solution set.

15. After attending to their children's needs, a couple decides to buy some M&Ms and Peppermint Patties. They decide to buy no more than 1 pound of candy. There must be at least $\frac{1}{4}$ pound of M&Ms and the Peppermint Patties must be less than twice the weight of M&Ms. Let x be the number of pounds of M&Ms and y be the pounds of Peppermint Patties.

a. Write a system of linear inequalities for this situation.

b. Graph the system and shade the region of the solution set.

Skills and Review 12.3

16. Find the solution set for $2x^3 + 5 > 50$. Show your solutions in set-builder notation.

17. Find the solution set for $|x - 4| \leq 7$. Show your solutions in set-builder notation.

18. Suppose the cost of repairing a vehicle is given by $200 + 65x$, where x is the number of hours of work. Use a number line to show the number of hours that make an estimate between \$330 and \$525.

19. Working together, two painters can complete a job in 5 hours. Working alone, it takes the second painter 2 hours more than the first. Complete parts (a)–(d) to find how long it will take the first painter to complete the job alone.

a. Set up a chart and identify any variables.

b. Write an equation for the work rate.

c. Solve the equation.

d. How long (to the nearest $\frac{1}{4}$ hour) will it take the first painter to complete the job alone?

20. Perform the subtraction and simplify

$$\frac{x}{x+5} - \frac{1}{x-5}.$$

21. Perform the division and simplify

$$\frac{x+4}{x^2+8x+15} \div \frac{x^2-16}{x+3}.$$

22. A formula for the value of an investment after t years is given by $A(t) = A_0(1 + \frac{r}{n})^{nt}$. How long will it take an initial investment of \$500 to grow to \$1000 if the investment earns 6% annual interest and is compounded monthly?

23. Suppose y varies directly as the cube of x and $y = 9$ when $x = 1$. Find x if $y = 72$.

24. Simplify the expression and write with only positive exponents.

$$\left(\frac{x^{-3/2}}{x^{1/3}}\right)^6$$

25. Consider the function $f(x) = x^2 - 5x - 14$.

a. Find the exact y-intercept.

b. Find the exact x-intercepts.

c. Find the exact vertex.

d. Sketch a graph of the function. Locate and identify the ordered pairs from parts (a)–(c).

CHAPTER 12 ▪ KEY CONCEPTS

Solving Inequalities in One Variable Graphically You can solve inequalities in one variable graphically by drawing a number line to represent the variable and then determining where on the number line the variable is allowed to be if it satisfies the inequalities. (Page 476)

Solving Inequalities in One Variable Algebraically You can solve inequalities in one variable algebraically in almost the same way you solve *equations* algebraically, by doing the same thing on both sides. You must remember that if you multiply on both sides of an inequality by a *negative* number, you reverse the direction of the inequality. (Page 480)

Solving Inequalities That Are Not Linear To solve inequalities involving expressions that are not linear, you can use a graphical method. Just remember that when one graph is above another, the function whose graph is higher will have *greater* values, and vice versa. For quadratic inequalities, you can use the quadratic formula to solve algebraically if you wish. (Page 482)

Solving Inequalities in Two Variables Inequalities in two variables can be solved by plotting the linear equations related to the inequalities and then shading the correct region of the plane. The solution sets for these inequalities are best expressed graphically. (Page 489)

Exploration

Carl belongs to a gym, where he uses a treadmill and a ski machine to work out. His trainer has told him that he burns 5 kilocalories per minute working on the treadmill and 8 kilocalories per minute working on the ski machine. Carl has at most 90 minutes per day for exercise and his goal is to burn at least 500 kilocalories per day. Because the ski machine is so popular, no one is allowed to use it for more than 30 minutes per day.

Carl's situation presents a typical linear programming problem. There are several inequalities that must be satisfied. Let x represent the time (in minutes) he spends each day on the treadmill and let y represent the time spent each day on the ski machine. Because there are only two variables, this situation is a lot less complex than most linear programming problems.

a. Write an inequality in x and y to show that Carl has at most 90 minutes per day for exercise.

b. Write an inequality in x and y to show that Carl wants to burn at least 500 kilocalories per day.

c. Write an inequality that shows that he can't use the ski machine more than 30 minutes per day.

d. Graph the three inequalities you have found in parts (a), (b), and (c). The region where they overlap is the *feasible region.*

e. The feasible region is a triangle in the coordinate plane. Find the coordinates of the vertices of the feasible region.

CHAPTER 12 ▪ REVIEW EXERCISES

Section 12.1 Inequalities and the Number Line

1. Use set-builder notation to show the values of x given by the number line.

2. Solve each inequality. Show the solutions on a number line and in set-builder notation.

a. $3x + 7 > 19$ **b.** $6x - 4 \leq 4 - 2x$

c. $^{-}5x + 2 \geq 12$ **d.** $\frac{1}{2}x - 6 > {}^{-}\frac{3}{2}x$

e. $^{-}9x > x - (2 + 4x)$ **f.** $2x - \frac{2}{3} \leq \frac{7}{3} - x$

3. Use a number line to show the values of x given by $x \leq {}^{-}2$ or $x > 5$.

4. Use set-builder notation to show the solution set if x must satisfy the given inequalities.

a. $x + 3 \leq 10$ and $x - 1 \geq 2$

b. $2x < {}^{-}6$ or $x + 5 \geq 7$

c. $\frac{1}{4}x \geq {}^{-}3$ and $6x < {}^{-}30$

d. $^{-}3x + 1 > 4$ or $5x > 20$

e. $^{-}(8 - x) < {}^{-}3$ and $^{-}4x \leq 2x + 12$

f. $7 < 2x - 1$ and $2x - 1 \leq 13$

5. Suppose a friend tells you, "It's hot outside; the temperature is *at least* 100°F." Let x represent the temperature. Choose the inequality that matches the friend's comment: $x < 100$°F, $x > 100$°F, $x \leq 100$°F, $x \geq 100$°F.

6. A vacation resort has both winter and summer activities. Business is best if the temperature is below 32°F or above 75°F. What temperatures are best for business? Show your answer in set-builder notation.

7. A basketball player for the WNBA scored 18, 20, and 15 points during the first three games of the season. Complete parts (a)–(c) to find how many points this player may score in the fourth game and still average at least 19 points per game.

a. Let x represent the points this player scores in the fourth game. Write an expression for the average number of points scored in the first four games.

b. Write an inequality showing that the average must be at least 19 points.

c. Solve the inequality and answer the question in the problem.

d. Show your solutions on a number line.

8. The number of people who attend a dance is given by the expression $125 - 5x$, where x is the cover charge (in dollars) paid at the door. The dance organizer requires at least 50 people and room capacity is no more than 100. Complete parts (a)–(b) to find what cover charges will maintain these attendance limits.

a. Write two inequalities, one for the minimum number of people and one for the maximum number.

b. Solve the inequalities and answer the question.

c. Show the solutions in set-builder notation.

Section 12.2 Solving Nonlinear Inequalities by Graphing

9. Use the graph to solve each inequality. Show your solutions on a number line.

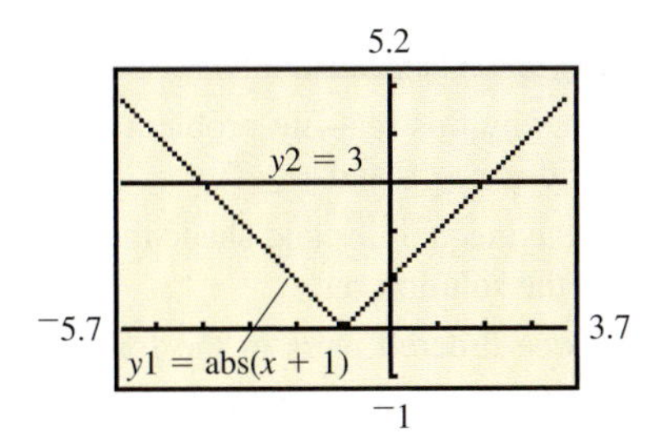

a. $|x+1| > 3$ **b.** $|x+1| < 3$

10. Use a graph to solve the inequality $^{-}4x^2 + 25x > 10$. Use the test points $x = 0, 3,$ and 6 to check your solutions. Show your solution in set-builder notation.

11. Match each inequality with its number line solutions. It may be helpful to graph the inequalities.

a. $x^2 \geq 9$ **1.** No solution

b. $x^2 \leq {}^{-}1$ **2.**

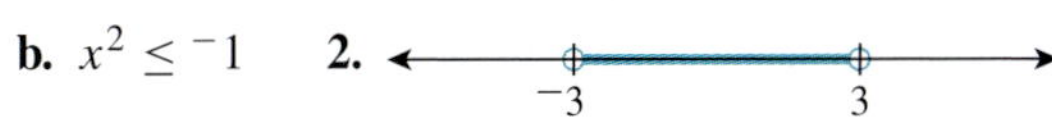

c. $x^2 < 9$ **3.** $^{-}3$ 3

12. Find each solution set and give it in set-builder notation.

a. $2x^2 - 5 < 13$ **b.** $x^2 - 8x > 0$

c. $x^2 + x \leq 12$ **d.** $|x - 3| \geq 7$

e. $x^3 < 64$ **f.** $\sqrt{x} \leq \frac{5}{2}$

13. Suppose sales from a product generate revenue given by the expression $^{-}7.3p^2 + 320p$, where p is the price of the product in dollars. Find the prices of the product that generate revenue greater than $3000.

14. At a packaging plant grapefruits are removed if their diameter differs from 5 inches by more than 2 inches. Let d be the diameter of the grapefruit. Solve the inequality $|d - 5| > 2$ to find the sizes of grapefruit that are removed.

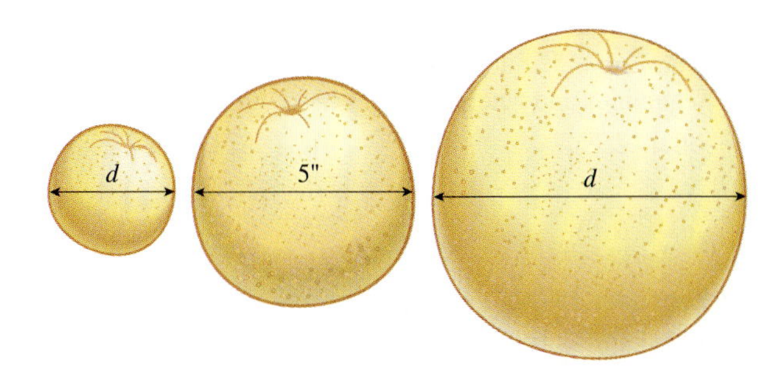

15. A couple owns a house with a square lot. They are told that their lot is at least 40,000 square feet and no more than 50,625 square feet. What are the possible lot dimensions? Show your solutions in set-builder notation.

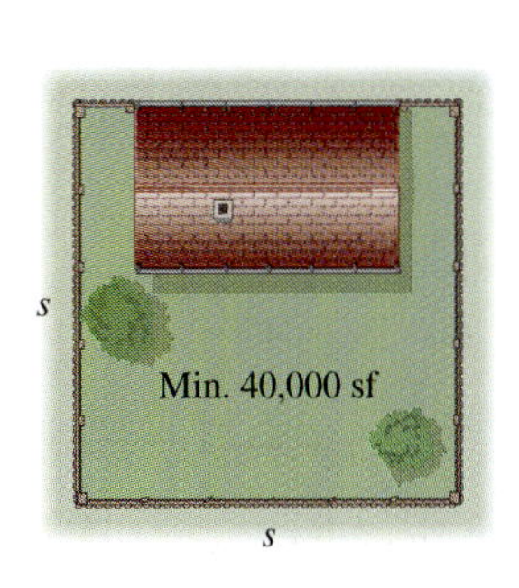

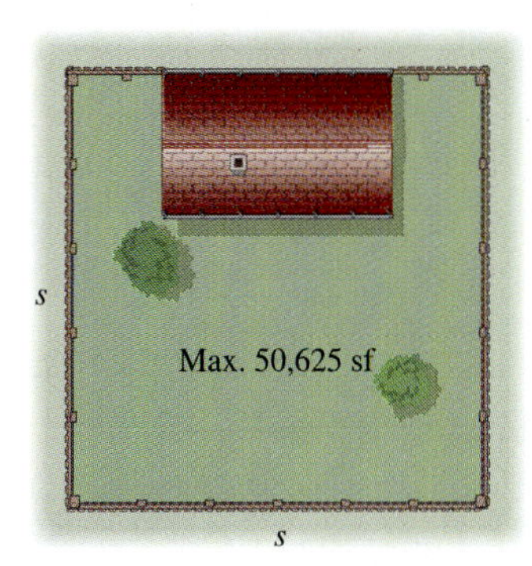

Section 12.3 Inequalities in Two Variables; Systems

16. In the following graph, the shaded region is the solution set of the system of linear inequalities $y > {}^{-}x + 3$ and $y \leq 2x$. Use the graph to determine if each ordered pair is a solution to the system.

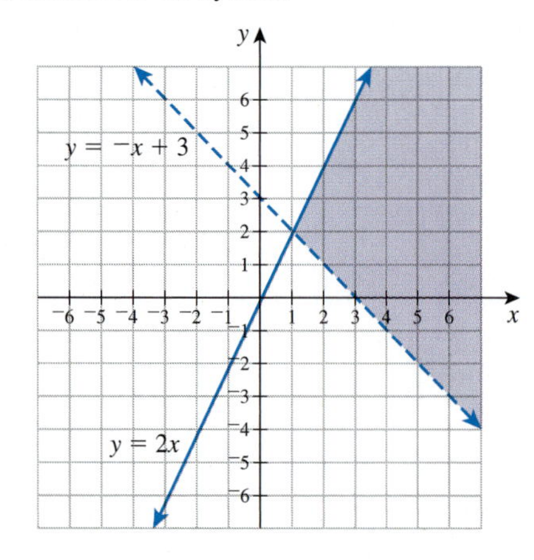

a. $(7, 5)$ **b.** $(1, {}^{-}5)$

c. $(9, {}^{-}2)$ **d.** $({}^{-}6, 3)$

17. Given the system of linear inequalities $y \leq 3$ and $y > x + 5$. Substitute each ordered pair into the inequalities and determine if it is a solution to the system.

a. $({}^{-}2, 3)$ **b.** $({}^{-}5, 1)$

c. $({}^{-}5, 4)$ **d.** $({}^{-}5, 3)$

e. $({}^{-}5, 0)$

In Exercises 18–21, graph the system of linear inequalities and shade the region that represents their solution set.

18. $y > 3 - \frac{1}{2}x$

$y \geq x - 1$

19. $y \geq 3 - \frac{1}{2}x$

$y < x - 1$

20. $x - y \leq 5$

$y \leq x + 1$

$x \geq 0$

21. $x + 4y \leq 12$

$2x + y < 6$

$x \geq 0$

$y \geq 0$

22. The sum of two numbers is more than 100 and less than 150. Also, the larger number is at least three times the smaller.

a. Let x represent the smaller number and y represent the larger number. Write a system of inequalities in two variables to represent this problem.

b. Graph the inequalities and shade the region representing their solution set.

c. Give two pairs of numbers that satisfy the system and two that don't.

23. Suppose an enclosure around a rectangular space must meet the following guidelines. The perimeter must be more than 60 feet and no more than 100 feet and the length must be at most $2\frac{1}{2}$ times the width.

a. Write a system of inequalities in two variables to represent this situation.

b. Graph the inequalities and shade the solution set.

c. Determine if a width of 6 feet and a length of 30 feet is a solution to this problem.

24. Suppose you need to buy pants and shirts and can spend no more than $200. You find a sale where each pair of pants costs $30 and each shirt costs $20. You need at least twice as many shirts as pants.

a. Let y represent the number of shirts and x represent the number of pants. Write an inequality that shows the total cost for the pants and shirts is no more than $200.

b. Write an inequality that shows there are at least twice as many shirts as pants.

c. Explain why this real-life problem also requires the inequalities $x \geq 0$ and $y \geq 0$.

d. Graph the inequalities and shade the region that represents the solution set.

e. Give three ordered pairs in the form (pants, shirts) that are solutions to this system of equations.

CHAPTER 12 ▪ TEST

1. Find the solution set if x must satisfy

$$3x + 8 < 32$$

and

$$3x + 8 \geq {}^{-}1.$$

Give your answer in set-builder notation.

2. Find the solution set if x must satisfy ${}^{-}2x + 1 \geq 7$ or $x - 5 > 0$. Give your answer in set-builder notation.

3. A soccer team keeps statistics on the number of goals scored against them. In their first four games, opposing teams scored 3, 5, 4, and 6 goals. How many goals could this team allow in the fifth game and still average less than 4 opposing goals per game?

4. A carpenter charges a flat rate of \$200 plus \$30 per hour for work. A customer receives an estimate between \$1400 and \$1700 to build a closet. How many hours does the carpenter plan to spend on the job?

In Exercises 5–7, use a graph to find each solution set and show the solutions in set-builder notation.

5. $x^3 > 4$
6. $5x^2 - 13x < 8$
7. $|2x + 5| > 9$
8. The height of an object, in meters above the ground, is given by the expression ${}^{-}4.9t^2 + 28t + 10$, where t is the time in seconds since the object was released. Use a graph to determine when this object is at least 40 m above the ground.

In Exercises 9–10, graph the system of linear inequalities and shade the region that represents the solution set.

9. $$\begin{aligned} y &> 2x - 7 \\ x - y &\leq 4 \end{aligned}$$

10. $$\begin{aligned} x &\geq 0 \\ y &\geq 0 \\ y &\leq 5 \\ 3x + 1 &> y \end{aligned}$$

11. The sum of two positive numbers is at most 250 and the larger number is at least 4 times the smaller number.
    - **a.** Let x represent the smaller number and y represent the larger number. Write a system of inequalities in two variables to represent this problem.
    - **b.** Graph the system of linear inequalities and shade the solution set.
    - **c.** Determine if the ordered pair (10, 240) is a solution.

12. A machine shop makes two types of parts, bearings and washers. In order to make a profit the company must ship at least 5000 parts per day. In order to maintain quality standards, however, the company can ship at most 10,000 parts per day. Graph a system of linear inequalities and shade the solution set.

13. Two people plan to share the driving on a road trip. They decide the trip should take less than 8 hours and the distance be at least 360 miles. One person averages 40 miles per hour and the other averages 60 miles per hour.
    - **a.** Let x represent the slower driver's number of hours behind the wheel and y represent the faster driver's number of hours. Write a system of inequalities in two variables to represent this problem.
    - **b.** Graph the inequalities and shade the region that represents the solution set.
    - **c.** Determine whether driving times of 4.5 hours for the slower driver and 3 hours for the faster driver represents a solution to this problem.

APPENDIX

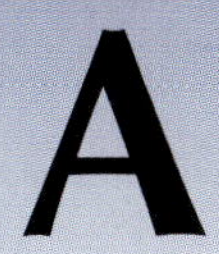

Transition to Intermediate Algebra

Use this guide as a reference to review material from Chapters 1–6 in preparation for Chapters 7–12.

A. Evaluating Expressions

Topic	Main Ideas	Introduced	Forward Links
Signed numbers	Know the rules for addition, subtraction, multiplication and division of signed numbers	Sec. 1.1	Throughout book
Exponents	Positive Integer Exponents represent repeated multiplication	Sec. 1.3	Chap. 9
Zero and negative exponents	$a^0 = 1$ $a^{-n} = \frac{1}{a^n}$	Sec. 1.3	Chap. 9
Scientific notation	$a * 10n$ where $1 \leq a < 10$ and n is an integer. Notice how scientific notation appears of calculators with the letter E	Sec. 1.3	Throughout book
Square roots	Use mental math to find square roots of perfect squares between 1 and 100 and to estimate square roots of other positive numbers less than 100. If a positive integer is not a perfect square, then its square root is an irrational number.	Sec. 1.3	Chap. 7
Properties of numbers	Addition and multiplication are commutative: $a + b = b + a$ and $a * b = b * a$: Addition and multiplication are associative: $(a + b) + c = a + (b + c)$ and $(a * b) * c = a * (b * c)$	Sec.1.4 and 2.1	Throughout book
Distributive property	$a * (b + c) = (a * b) + (a * c)$ Demonstrate with area model Expand and Factor	Sec. 2.1	Chap. 7 Chap. 11
Order of operations (very important!)	1. Do operations inside grouping symbols first ( ) 2. Evaluate exponents and roots from left to right $\wedge\sqrt{}$ 3. Perform multiplication and division from left to right $*$ / 4. Perform negation, addition, and subtraction from left to right $^{-}$ $+$ $-$	Sec. 2.1	Throughout book

B. Equivalent Expressions

Topic	Main Ideas	Introduced	Forward Links
Equivalent expressions	Two expressions are equivalent if when any values are substituted for the variable (or variables), the results are the same, e.g., $2(x+1)$ is equivalent to $2x+2$.	Sec. 2.2	Throughout book
Like terms	An application of the distributive property is combining like terms: e.g., $3x + 9x = 12x$.	Sec. 2.2	Throughout book
Procedure for simplifying an expression	1. Use the special properties of 1 and $^{-}1$ if necessary 2. Remove parentheses using the distributive property 3. Group like terms using the commutative property 4. Combine like terms	Sec. 3.2	Throughout book
Laws of exponents	$a^m * a^n = a^{m+n}$ $(a^m)^n = a^{mn}$ $(ab)^m = a^m * b^m$ $\frac{a^m}{b^m} = a^{m-n}$ $\frac{a}{b^m} = \frac{a^m}{b^m}$	Sec.6.1	Chap. 9
Negative exponent in denominator	$\frac{1}{a^{-n}} = a^n$	Sec. 6.1	Chap. 9 Chap. 10

C. Solving Equations

Topic	Main ideas	Introduced	Forward Links
Equation solving (very important)	BASIC FACT: If we are given an equation, we can add the same term on both sides or multiply by the same number on both sides and still have a valid equation. Think of subtraction as adding an opposite. Think of division as multiplying by a reciprocal.	Sec. 3.1	Throughout book
Procedure for solving linear equations	1. If necessary, simplify expression on left side. 2. If necessary, simplify expression on right side. 3. If necessary, add same terms on both sides until you have a variable term on one side and a constant term on the other side. 4. If necessary, multiply both sides by the reciprocal of the coefficient of the variable.	Sec. 3.2	Throughout book
Cross-multiplying	If $\frac{a}{b} = \frac{c}{d}$ then $ad = bc$	Sec. 5.1	Chap. 11
Substitution method for solving systems	1. Put one equation in slope intercept form $y = mx + b$ 2. Substitute $mx + b$ for y in the second equation. 3. Solve for x. 4. Substitute for x in first equation to find y.	Sec. 4.4	Chap. 12
Elimination method for solving systems	1. Put both equations in standard form $Ax + By = C$ 2. If necessary multiply one or both equations by the same number on both sides to get x or y terms to be opposites. 3. Add the two equations to get an equation in one variable. Solve for that variable. 4. Substitute back into one of the original equations and solve for the other variable.	Sec. 4.4	Chap. 12
Zero product property	Suppose $a * b = 0$. Then either $a = 0$ or $b = 0$.	Sec. 6.2 Sec. 6.4	Chap. 7

D. Functions and Their Representations

Topic	Main Ideas	Introduced	Forward Links
Functions	Dependent (output) variable is often y. Independent (input) variable is often x. The value of the output variable depends upon the value of the input variable.	Sec. 2.4	Chap. 8
Representing functions	The Rule of Four states that functions may be represented by: 1. description of the situation in words 2. a formula 3. a table of values 4. a graph	Sec. 2.4	Chap. 7 Chap. 8
Functions on calculators	Use Y=, WINDOW, and GRAPH to draw a graph. Use Y=, TblSet, and TABLE to make a table.	Sec. 2.4	Chap. 7
Friendly windows	For TI-82 and TI-83 calculators, PIXELWIDTH = (Xmax − Xmin)/94 PIXELHEIGHT = (Ymax − Ymin)/62 Friendly Windows give nice values for PIXELWIDTH	Sec. 2.4	Chap. 7

E. Linear Equations in Two Variables

Topic	Main Ideas	Introduced	Forward Links
Slope of a line	Slope $= \frac{\text{change in output}}{\text{change in input}}$ Slope $= \frac{\text{rise}}{\text{run}}$ Slope represents rate of change	Sec. 4.1	Sec. 7.1
Linear equations	Straight lines have constant slope	Sec. 4.1	Sec. 7.1
Slope-intercept form for a linear equation	$y = mx + b$ y is the dependent variable x is the independent variable m is the slope (rate of change) b is the y-intercept (initial value)	Sec. 4.2	Chap. 8
Graph a linear equation with a calculator	If necessary, solve for y in terms of x.	Sec. 4.2	Throughout book
System of two linear equations in two variables	Solution is the intersection point (if it exists).	Sec. 4.3	Chap. 12
Inconsistent system	Lines are parallel. There is no solution.	Sec. 4.3	Chap. 12

F. Expanding and Factoring with an Area Model

Topic	Main Ideas	Introduced	Forward Links
Mulitiplying monomial by polynomial	Use the distributive property Illustrate with area model	Sec. 6.2	Chap. 7
Factor out a monomial	Use the distributive property (in reverse) Illustrate with area model	Sec. 6.2	Chap. 7
Finding the greatest common factor (GCF)	1. List all variables and prime factors of coefficients. 2. When a variable or prime factor does not appear include that factor to the zero power. 3. Take the smallest power of each to form the GCF.	Sec. 6.2	Sec. 11.1
Zero product property	Suppose $a * b = 0$, then either $a = 0$ or $b = 0$ Some quadratic equations may be solved using this property	Sec. 6.2 Sec. 6.4	Sec. 7.2
Multiplying polynomials	Use area model	Sec. 6.3	Chap. 7
Factor by grouping	Use the diagonal test. A polynomial with four terms may be factored by grouping if the product of one pair of terms is equal to the product of the other two terms.	Sec. 6.3	Chap. 7
Squaring a binomial	$(a + b)^2 = a^2 + 2ab + b^2$	Sec. 6.3	Sec. 7.2
Factor quadratic trinomials	One method is to split the middle term and use an area model to factor by grouping.	Sec. 6.4	Sec. 7.2 Sec. 11.1

G. Problem Solving

Topic	Main Ideas	Introduced	Forward Links
Seven steps to problem solving (very important)	1. Understand 2. Visualize 3. Assign variable(s) 4. Write equation(s) 5. Solve equation(s) 6. Answer the question 7. Check the answer.	Sec. 3.3	Throughout book
Perimeter and area	Perimeter measures the distance around the outside of the figure and is measured in linear units such as feet. Area measures the surface enclosed by a plane figure and is measured in square units such as square feet.	Sec. 1.4	Chap. 7
Geometry formulas	Area of rectangle, parallelogram, triangle, trapezoid Circumference and area of circles	Sec. 1.4	Chap. 7
Distance-Rate-Time	$\frac{\text{Distance}}{\text{Time}} = \text{Rate}$ miles divided by hours = miles per hour, etc. Distance = Rate ∗ Time (miles = miles per hour ∗ hours, etc.)	Sec. 1.2	Sec. 11.3
Work rate problems	$\frac{\text{Work}}{\text{time}} = \text{work rate}$ Work rate is often expressed in jobs per hour, etc.	Sec. 1.2	Sec. 11.3
Percent change	$\text{percent change} = \frac{\text{change}}{\text{original}}$ New amount = original ± change (refer to number line model)	Sec. 5.2	Chap. 10
Linear combinations	Typically a system of equations with $x + y = \text{something}$ $\text{something} * x + \text{something} * y = \text{something}$ Applications to solutions, mixtures, interest problems	Sec. 5.3	Sec. 12.3
Simple interest	Interest = Principal ∗ rate ∗ time	Sec. 5.3	Sec. 10.1

APPENDIX

B Friendly Windows for Graphing Calculators

Friendly Windows for TI-82 and TI-83 Calculators

PIXEL WIDTH	FOUR-QUADRANT WINDOWS			FIRST QUADRANT WINDOWS	
	Xmin	Xmax	Shortcut for square windows	Xmin	Xmax
.1	$^{-}4.7$	4.7	ZOOM 4 ZDecimal	0	9.4
.2	$^{-}9.4$	9.4	ZOOM 4 ZDecimal	0	18.8
			ZOOM 3[b] ZoomOut ENTER		
.5	$^{-}23.5$	23.5	ZOOM 8 ZInteger[a]	0	47
			ZOOM 2[b] Zoom In ENTER		
1	$^{-}47$	47	ZOOM 8 ZInteger[a]	0	94
2	$^{-}94$	94	ZOOM 8 ZInteger[a]	0	188
			ZOOM 3[b] Zoom In ENTER		
5	$^{-}235$	235		0	470
10	$^{-}470$	470		0	940

Suggested values for Xscl: Xscl = 10* PIXELWIDTH

[a]ZInteger will always give a screen with PIXELWIDTH = 1 and PIXELHEIGHT = 1. The center of the screen depends upon the last location of the cursor. If necessary, first activate ZOOM 6 ZStandard before pressing ZOOM 8 ZInteger.

[b]To use ZOOM 2 and ZOOM 3, set ZoomFactors = 2

Feature	TI-82/83	TI-85/86	TI-89
Set Xmin, Xmax, Ymin, Ymax	WINDOW	GRAPH F2	diamond F2
Create special windows	ZOOM	GRAPH F3	diamond F2 F2
Display graph	GRAPH	GRAPH F5	diamond F3
Enter function	Y =	GRAPH F1	diamond F1
Set up table	2nd TblSet	TABLE F2 (86 only)	diamond F4
Display table	2nd TABLE	TABLE F1 (86 only)	diamond F5
Trace points on a graph	TRACE	GRAPH F4	diamond F3 F3
Find point of intersection	2nd CALC 5:Intersect	GRAPH MORE F1	diamond F3 F5 5:Intersect
Find turning points	2nd CALC 3:Min and 4:Max	GRAPH MORE F1	diamond F3 F5 3:Min and 4:Max

Friendly Windows for TI-85 and TI-86 Calculators

PIXEL WIDTH	FOUR-QUADRANT WINDOWS			FIRST QUADRANT WINDOWS	
	Xmin	Xmax	Shortcut for square windows	Xmin	Xmax
.1	−6.3	6.3	ZOOM 4 ZDecimal	0	12.6
.2	−12.6	12.6	ZOOM 4 ZDecimal	0	25.2
			ZOOM 3[b] ZoomOut ENTER		
.5	−31.5	31.5	ZOOM 8 ZInteger[a]	0	63
			ZOOM 2[b] Zoom In ENTER		
1	−63	63	ZOOM 8 ZInteger[a]	0	126
2	−126	126	ZOOM 8 ZInteger[a]	0	252
			ZOOM 3[b] Zoom In ENTER		
5	−315	315		0	630
10	−630	630		0	1260

Friendly Windows for TI-89 Calculators

PIXEL WIDTH	FOUR-QUADRANT WINDOWS			FIRST QUADRANT WINDOWS	
	Xmin	Xmax	Shortcut for square windows	Xmin	Xmax
.1	−7.9	7.9	ZOOM 4 ZDecimal	0	15.8
.2	−15.8	15.8	ZOOM 4 ZDecimal	0	31.6
			ZOOM 3[b] ZoomOut ENTER		
.5	−39.5	39.5	ZOOM 8 ZInteger[a]	0	79
			ZOOM 2[b] Zoom In ENTER		
1	−79	79	ZOOM 8 ZInteger[a]	0	158
2	−158	158	ZOOM 8 ZInteger[a]	0	316
			ZOOM 3[b] Zoom In ENTER		
5	−395	395		0	790
10	−790	790		0	1580

Suggested values for Xscl: Xscl = 10* PIXELWIDTH

[a]ZInteger will always give a screen with PIXELWIDTH = 1 and PIXELHEIGHT = 1. The center of the screen depends upon the last location of the cursor. If necessary, first activate ZOOM 6 ZStandard before pressing ZOOM 8 ZInteger.

[b]To use ZOOM 2 and ZOOM 3, set ZoomFactors = 2

Answers to Exercises

Chapter 1 Geometry and Numbers

1.1 Exercises

1. a. $^{+}6 + ^{-}4 = ^{+}2$ **b.** $^{+}5 + ^{-}9 = ^{-}4$

3. a. $(^{-})(^{-})(^{-})(^{-})(^{-})(^{-})$
$(+)(+)(+)(+)(+)(+)(+)$
$^{+}1$

b. $(^{-})(^{-})(^{-})(^{-})(^{-})(^{-})(^{-})(^{-})$
$(+)(+)$
$^{-}6$

c. $(^{-})(^{-})(^{-})(^{-})(^{-})(^{-})$
$(^{-})(^{-})$
$^{-}8$

d. $(+)(+)(+)(+)(+)$
$(+)(+)(+)(+)$
$^{+}9$

e. $(^{-})(^{-})(^{-})(^{-})(^{-})(^{-})$
$(+)(+)(+)(+)(+)(+)$
0

5. a. $^{-}14°C$ **b.** $^{-}6 + ^{-}8 = ^{-}14$

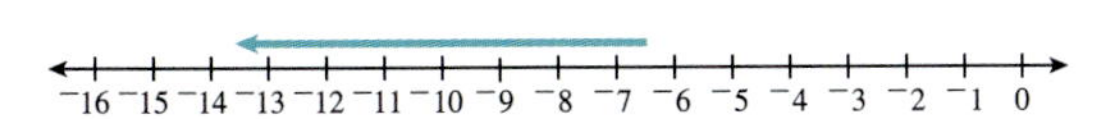

7. a. $^{+}\$275$

b. $^{+}100 + ^{-}250 + ^{+}475 + ^{-}50 = ^{+}275$

9. a. $^{+}5 + ^{-}13 = ^{-}8$

b. $^{-}37 + ^{-}42 = ^{-}79$

c. $^{+}11.8 + ^{-}5.7 = ^{+}6.1$

d. $^{-}\frac{2}{5} + ^{+}\frac{1}{10} = ^{-}\frac{3}{10}$

11. An odd number of negatives results in a negative number. An even number of negatives results in a positive number.

13. a. Starting position: 5; final position: $^{-}3$; change: $^{-}8$

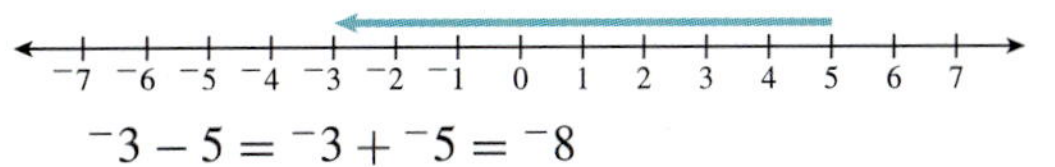

$^{-}3 - 5 = ^{-}3 + ^{-}5 = ^{-}8$

b. Starting position: $^{-}3$; final position: 5; change: 8

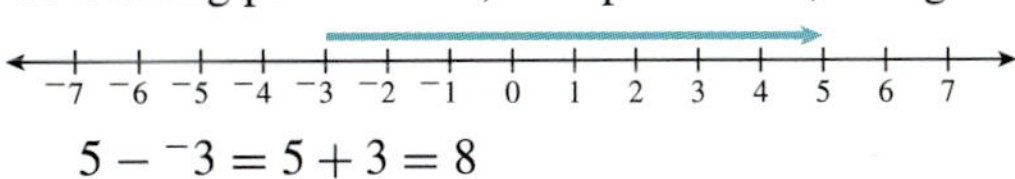

$5 - ^{-}3 = 5 + 3 = 8$

c. Starting position: 6; final position: 2; change: $^{-}4$

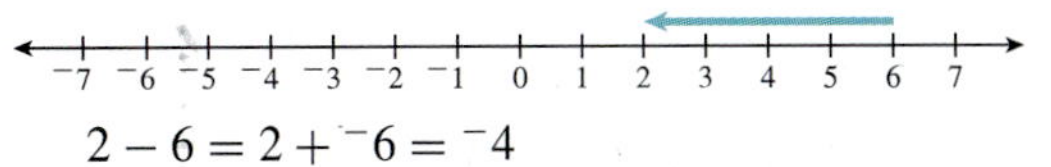

$2 - 6 = 2 + ^{-}6 = ^{-}4$

d. Starting position: $^{-}3$; final position: 4; change: 7

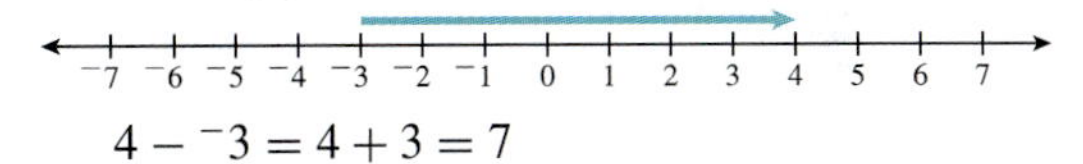

$4 - ^{-}3 = 4 + 3 = 7$

15. a. $^{+}6 \div ^{+}2 = ^{+}3$. The temperature rose $^{+}3$ degrees per hour. The numbers $^{-}6$ and $^{+}2$ are the same sign so the quotient is positive.

b. $^{-}6 \div ^{+}2 = ^{-}3$. The temperature rose $^{-}3$ degrees per hour. The numbers $^{-}6$ and $^{+}2$ are opposite signs so the quotient is negative.

c. $^{-}6 \div ^{-}2 = ^{+}3$. The temperature rose $^{+}3$ degrees per hour. The numbers $^{-}6$ and $^{-}2$ are the same sign so the quotient is positive.

d. $^{+}6 \div ^{-}2 = ^{-}3$. The temperature rose $^{-}3$ degrees per hour. The numbers $^{+}6$ and $^{-}2$ are opposite signs so the quotient is negative.

1.1 Skills and Review

17. **a.** 1

b. Do not add denominators.

c. 1

d. Multiplication is repeated addition. Adding five one-fifths is the same as multiplying one-fifth by 5.

19. **a.** 10 m/s

b. $\frac{1}{2} * 4$ miles $= 2$ miles

c. $68 + 6 = 74$

d. $81 - 9 = 72$

21. **a.** $\frac{^-21}{10}$ **b.** 2 **c.** $^-.9$

23. 5400 seconds

25. .1 is greater. The digit 1 in .1 is in the tenths place, whereas the digit 7 in .07 is in the hundredths place.

1.2 Exercises

1. **a.**

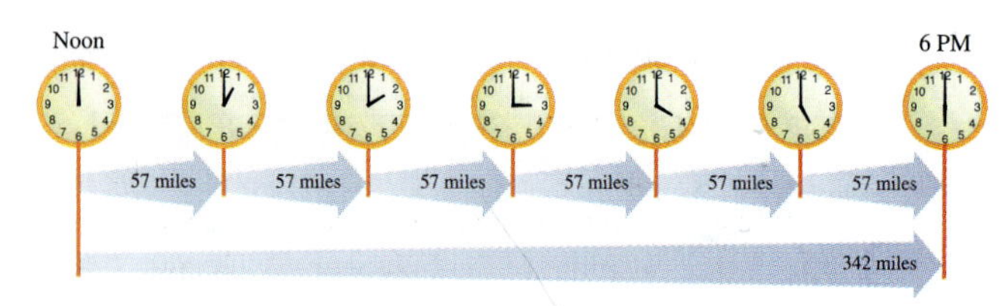

b. 57 miles/hour

3. You will travel 3 times as far in 1 hour as you will in $\frac{1}{3}$ of an hour.

5. **a.** 125 apples/hour

b. $\frac{1}{4}$ house/day

c. $\frac{1}{3}$ bike/hour

7. **a.** $= \frac{3}{4}$ **b.** $= \frac{2}{5}$ **c.** $= \frac{7}{10}$

d. $\approx \frac{1}{3}$ **e.** $= \frac{5}{8}$ **f.** $\approx \frac{2}{3}$

9. Error division by 0. Division by 0 is undefined.

11. f b e a d c 0 0.5 1.0

13. **a.** $\frac{13}{20}$ **b.** $\frac{31}{25}$ **c.** $\frac{23}{40}$

d. $\frac{2}{5}$ **e.** $\frac{201}{250}$

15. **a.** $.3 = \frac{3}{10}$

b. $.33 = \frac{33}{100}$

c. $.333 = \frac{333}{1000}$

d. $.3333 = .3333$

e. $.3333333333 = .3333333333$

f. $.333333333333 = \frac{1}{3}$ (*Caution:* This is an approximation by the calculator.)

g. No. $\frac{1}{3} = .\overline{3}$. There must be an infinite number of 3s for the decimal to equal $\frac{1}{3}$.

1.2 Skills and Review

17. If the signs are the same, the product is positive. If the signs are opposite, the product is negative.

19. **a.** $\frac{^-2}{5}$ **b.** 0 **c.** $\frac{^-1}{6}$

21. **a.** Starting position: 6; change: $^-8$; final position: $^-2$

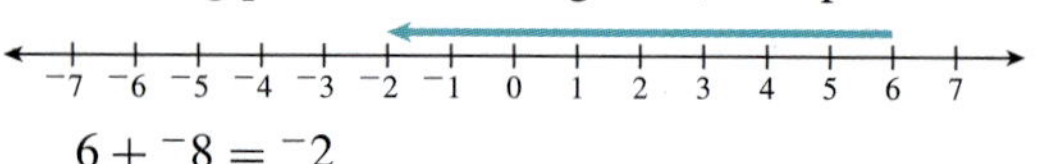

$6 + {^-8} = {^-2}$

b. Starting position: $^-5$; change: 9; final position: 4

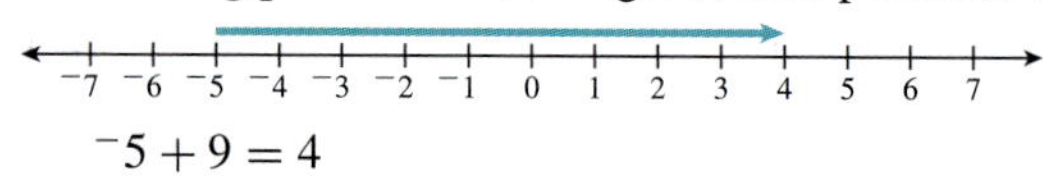

${^-5} + 9 = 4$

c. Starting position: 4; final position: $^-2$; change: $^-6$

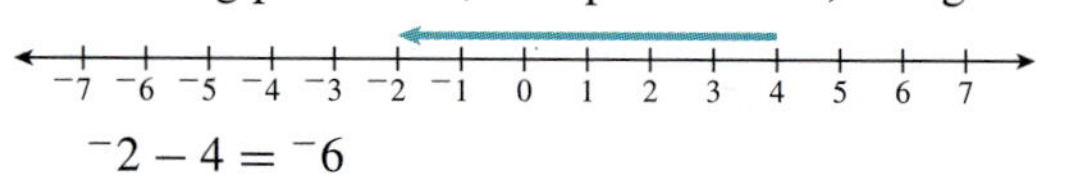

${^-2} - 4 = {^-6}$

d. Starting position: $^-1$; final position: 3; change: 4

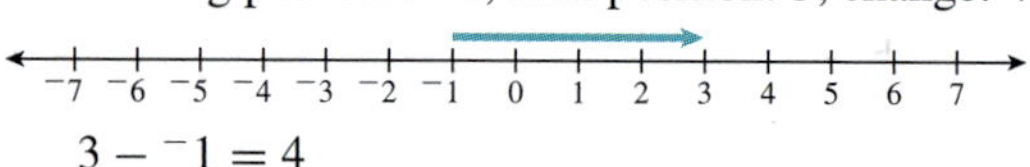

$3 - {^-1} = 4$

23. $\frac{4 \text{ yards}}{1} * \frac{3 \text{ feet}}{1 \text{ yard}} * \frac{12 \text{ inches}}{1 \text{ foot}} = 144$ inches. Write the original quantity as a fraction. Use the facts given in the hint to set up ratios. The ratios should be set up so that when you multiply, all the units cancel except the units to which you are converting.

25. $\frac{1}{2}$. Convert to fractions with common denominators.

$$0, \frac{1}{2}, \frac{2}{3}, 1 \qquad \frac{0}{6}, \frac{3}{6}, \frac{4}{6}, \frac{6}{6}$$

$\frac{4}{6}$ is closer to $\frac{3}{6}$ than $\frac{6}{6}$, so $\frac{1}{2}$ is the answer.

1.3 Exercises

1. **a.** $3 + 3 + 3 + 3 = 12$

b. $3 * 3 * 3 * 3 = 81$

3. **a.** 16 **b.** 1

c. $\frac{1}{16}$ **d.** 9

e. $^-27$

5. **a.** 1

b. Any nonzero real number raised to the zeroth power equals 1.

7. a. The numbers in c, d, and e are different but are so close to 0 that they appear to lie in the same spot on the number line.

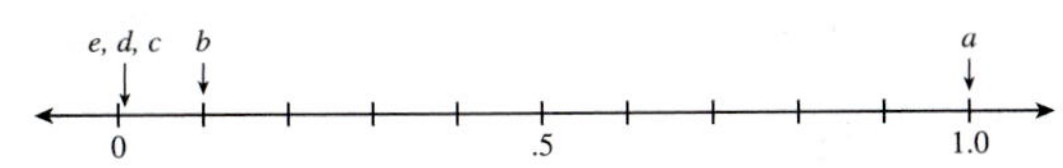

b. 0

c. No. The numbers will get closer to 0 but never become negative. Conclusion: Negative powers do not make negative numbers.

9. a. $1.089 * 10^3$ feet/second

b. $4.84 * 10^8$ miles

c. $2.0 * 10^{-9}$ mile

11. a. $9.59 * 10^{-5} = .0000959$°F/day

b. $1.377 * 10^{10} = 13{,}770{,}000{,}000$ g/day

13. c

15. a. 3.1 **b.** 6.5 **c.** 8.9

1.3 Skills and Review

17. a. $\frac{1}{4} = .25$ hour **b.** 0 hour

c. $\frac{1}{2} = .5$ hour **d.** 1 hour

19. a. 19.304 cm

b. $\frac{7.6 \text{ inches}}{1} * \frac{2.54 \text{ cm}}{1 \text{ inch}} = 19.304$ cm

21. 2 cars/day

23. a. 21.0 m

b. 21 feet/minute

c. 927.6 cm

25. 543 feet

1.4 Exercises

1. Perimeter is the distance around the outside of the figure. Area is the space inside the figure.

3. Answers may vary.

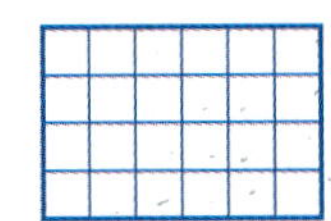

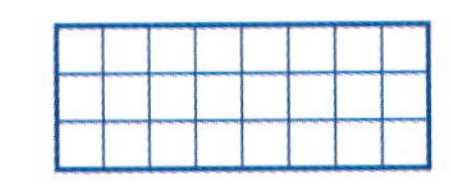

5. a. 14 cm

b.

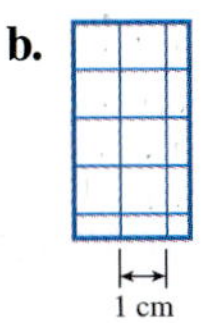

Area = 11.25 cm^2, answers may vary slightly.

c. Area = 11.25 cm^2

d. The areas are the same. (Estimates within 2 cm^2 are good.)

7. a. $\frac{1}{8}$

b. $\frac{1}{8}$ mile^2; the answers from (a) and (b) are the same.

c. $P = 1\frac{1}{2}$ miles = 1.5 miles

d. $A = 12{,}800$ rods^2, $P = 480$ rods

e. 80 acres

f. 80 acres

9. a. $A = 25.2$ square feet

b. $P = 28$ feet

11. a. $A \approx 24 \text{ cm}^2$ **b.** $A = 24 \text{ cm}^2$

13. Circumference is the perimeter of a circle and Carl was running around the perimeter of the circle.

15. $C = 2 * \pi * r \approx 2 * (3.14) * (8.5) = 53.38$ m

$C = \pi * d \approx 3.14 * (17) = 53.38$ m

1.4 Skills and Review

17. a. Reorder the multiplication: $5 * 7 = 7 * 5 = 35$.

b. Most of the facts in the multiplication table are duplicates if you use the commutative property.

19. a. 6 **b.** 10 **c.** 100 **d.** .9 **e.** $\frac{3}{5}$

21. 7000

23. $\frac{1}{25} = .04$

25. e

Chapter 1 Exploration

a. $^{-}335.58$

b. 9589.67

c. Use the formula start + change = final.

$9925.25 + {}^{-}335.58 = 9589.67$

Final DJIA — $^{-}335.58$ Change in DJIA — Starting DJIA

9589.67 — 9925.25

Chapter 1 Review Exercises

1. a. (+)(+)(+)

(+)(+)(+)(+)

$^{+}3 + {}^{+}4 = {}^{+}7$

b. (+)(+)

(⁻)(⁻)(⁻)(⁻)(⁻)

$^{+}2 + {}^{-}5 = {}^{-}3$

c. $(^-)$

$(^+)(^+)(^+)$

$^-1 + {}^+3 = {}^+2$

d. $(^-)(^-)(^-)(^-)$

$(^-)(^-)$

$^-4 + {}^-2 = {}^-6$

2. a. The numbers have opposite signs. Subtract the absolute value of the numbers $(71.2 - 2.5 = 68.7)$ and take the sign of the number with the larger absolute value $(^+71.2)$. The answer is 68.7.

b. The numbers have the same sign. Add the absolute value of the numbers $(15.1 + 56.76 = 71.86)$ and keep the sign. The answer is $^-71.86$.

3. a. $^+6 - {}^+5 = {}^+6 + {}^-5 = {}^+1$

b. $^+18 - {}^-7 = {}^+18 + {}^+7 = {}^+25$

c. $\frac{^-3}{8} - \frac{^+1}{4} = \frac{^-3}{8} + \frac{^-1}{4} = \frac{^-5}{8}$

d. $^-12.1 - {}^-4.3 = {}^-12.1 + {}^+4.3 = {}^-7.8$

4. 45 feet

5. $^-120$ feet

6. a. 6 **b.** $^-200$ **c.** $\frac{^-2}{3}$ **d.** 8

7. a. 2 **b.** $^-51$ **c.** $\frac{^-1}{15}$ **d.** 3

8. a. 70 km/h **b.** 12 km/h **c.** 8 km/h

9. a. 8 feet^2/minute **b.** $\frac{1}{3}$ job/hour **c.** $\frac{1}{4}$ job/hour **d.** $\frac{7}{12}$ job/hour

10. a. 1.8 **b.** .567 **c.** .057

11. a. 180% **b.** 56.7% **c.** 5.7%

12.

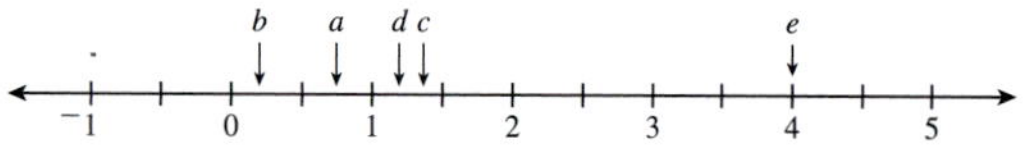

13. a. 64 **b.** 81 **c.** 8 **d.** $\frac{1}{32}$ **e.** 1

14. a. $^-3$ **b.** 9 **c.** $^-27$ **d.** 81

15. a. $\frac{1}{16}$ **b.** $\frac{1}{15}$ **c.** 1

16. a. $3.48 * 10^8$ **b.** $7.4 * 10^{^-4}$ **c.** $5.772 * 10^3$

17. a. $1.39128 * 10^{10} = 13{,}912{,}800{,}000$

b. $5.0 * 10^{^-4} = .0005$

18. a. 3 **b.** 8 **c.** 6

19. a. 2.8 **b.** 6.1 **c.** 8.5

20. $s \approx 6.9$ inches

21. $P = 20$ cm

$A = 21\ \text{cm}^2$

22. $P = 20$ cm

$A = 21\ \text{cm}^2$

23. $P = 28$ cm

$A = 23\ \text{cm}^2$

24. $3 + 27 + 18 + 2 = 50$

25. $2 * 5 * 16 = 160$

26. $P = 34$ inches, $A = 55\ \text{inches}^2$

27. $C \approx 43.96$ feet

Chapter 1 Test

1. $^-3$

2. $^-10$

3. 25°C

4. a. $^-4$

b. $^-6$

c. $\frac{1}{4} = .25$

5. $\frac{8}{3} \approx 2.67$ ounces/hour

6. a. $.40 = 40\%$ **b.** $.75 = 75\%$ **c.** $.30 = 30\%$ **d.** $\approx .333 = 33.3\%$ **e.** $2.25 = 225\%$

7. 2^0, 145%, $\frac{3}{2}$, 1.512, $1\frac{5}{8}$

8. a. 8 **b.** $^-8$ **c.** $\frac{1}{8}$ **d.** 1 **e.** 1

9. a. $1.15 * 10^5$ **b.** $7.4 * 10^{^-6}$

10. a. $1.664 * 10^{^-6}\ \text{mm}^2$ **b.** $.000001664\ \text{mm}^2$

11. 7.2

12. $P = 22$ cm, $A = 24\ \text{cm}^2$

13. $14 + 16 + 33 + 7 + 2 + 8 = 80$

14. ≈ 1758.4 mm

15. 11.2 feet

Chapter 2 Algebraic Expressions

2.1 Exercises

1. a.

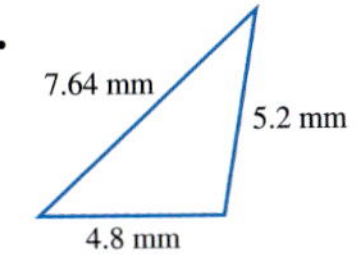

b. 7.64 mm + (5.2 mm + 4.8 mm)

c. 17.64 mm

3. $220 * 891 = 220 * (198 + 297 + 396)$
$= 220 * 198 + 220 * 297 + 220 * 396$
$= 196{,}020$ feet^2

5. a. $4 * 512 = 4 * (500 + 12) = 4 * 500 + 4 * 12 = 2048$

b. 4433

7. Expansion

9. Division before subtraction; 1

11. a. $\frac{21}{4} = 5.25$

b. $(3 + 24 - 6)/(6 - 2)$

13. Answers may vary.

15. a. $a + b - 8 - c$ **b.** $^-y - 6 + z$ **c.** $3 - s - t$

2.1 Skills and Review

17. 31.4 inches

19. a. 81 **b.** $\frac{1}{8}$ **c.** 1

d. 1 **e.** $\frac{1}{81}$

21. a. $\frac{1}{2^{10}}$

b. $9.765625 * 10^{-4}$

c. .0009765625

23. a. 50 feet^2/hour

b. $\frac{5}{6}$ feet^2/minute

c. $\frac{1}{9}$ wall/hour

25. 19. One morning the temperature was $^-3$°F. By afternoon the temperature had risen to 16°F. What was the change in temperature?

2.2 Exercises

1. a. 1 **b.** 4 **c.** 4

d. 9.42 **e.** 15 **f.** 5

3. a. 0 **b.** 14 **c.** 20 cm^2

d. 48 mm^2 **e.** 10 inches **f.** 75.36 feet^2

5. a. $\frac{7}{3} = 2\frac{1}{3}$ cups

b. $S * \frac{1}{3} = \frac{1}{3}S$ cups

c. $F = \frac{1}{3}S$ or $F = \frac{S}{3}$

7. $12 * 1 - 12 * 1 \neq 1$

9. a. $C = .06 * M + .25 * M$ or $C = .06M + .25M$

b. $C = .31 * M$ or $C = .31M$

c. 31 cents

d. \$3.10

11. a.

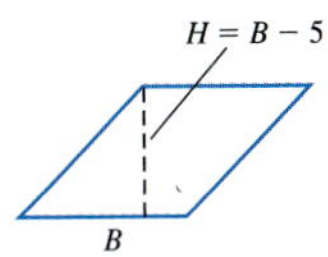

b. Let B represent the length of the base; then the height is $B - 5$.

c. $B * (B - 5) = B^2 - 5 * B$

d. 300 inches^2

13. 30 feet

15. a. 125 miles **b.** Highway

2.2 Skills and Review

17. $^-23$

19. 3

21. a. $2 * 5 - 3$ **b.** 7 inches

c. 35 inches^2 **d.** 24 inches

23. a. 64 **b.** 1 **c.** 1

d. $\frac{1}{25} = .04$ **e.** 16

25. a. 3 **b.** $^-9$

c. $^-2$ **d.** 11

2.3 The Coordinate Plane

1.

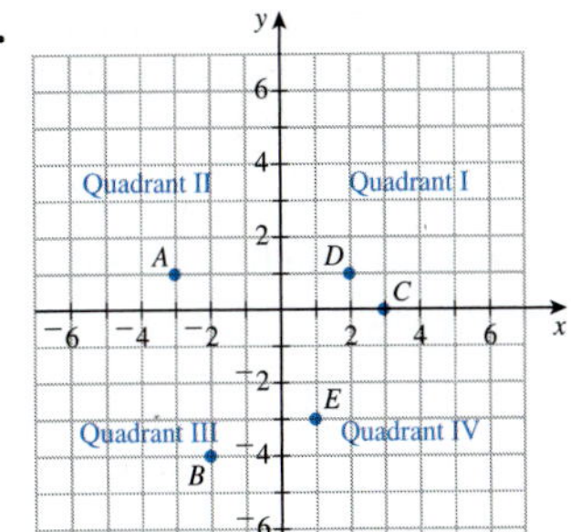

3. $(^-5, 4)$

5. a. Answers may vary; $(^-5, ^-3)$, $(^-2, ^-3)$, $(0, ^-3)$, $(4, ^-3)$.

b. The y-coordinate is always $^-3$.

c. The x-coordinate varies.

7. a. x **b.** increase

9. Run = 1, rise = 9; run = $^-1$, rise = $^-9$

11. a.

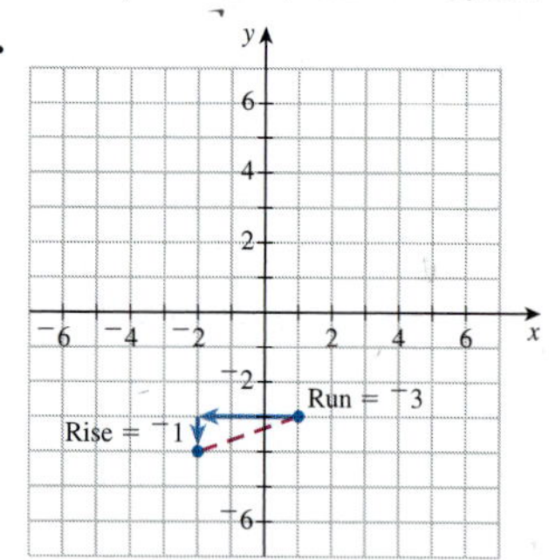

b. run = $^-3$, rise = $^-1$

c. $\sqrt{10} \approx 3.16$

13. $\sqrt{17} \approx 4.12$

15. 9 square units

2.3 Skills and Review

17. a. $12y$ **b.** $5x$

c. $3x + 3$ **d.** $^-2x^2 + x$

19. $\frac{3}{2} = 1.5$

21. 400

23. a. 32 **b.** 9

c. $^-9$ **d.** $\frac{1}{25} = .04$

e. $^-1$

25. a. $^-3$ **b.** 3

c. 13 **d.** $^-13$

2.4 Exercises

1. The variable cost is the part of the total cost that depends upon the value of the independent variable. The fixed cost is the part of the total cost that remains constant and is not affected by the value of the independent variable.

3. b. $c = 42t + 75$ **c.** Yes **d.** $159.00

5. a. No, you can't have a negative number of boxes.

b.

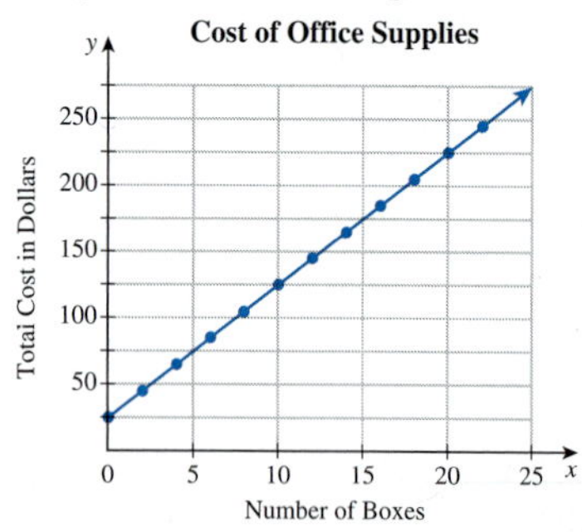

c. yes

d. 14-box shipment costs $165.00.

7. a. Answers may vary

$$\text{Xmin} = {}^-23.5, \text{Xmax} = 23.5, \text{Xscl} = 5,$$

$$\text{Ymin} = {}^-50, \text{Ymax} = 50, \text{Yscl} = 10$$

b. A straight line passing through quadrants I, II, and III

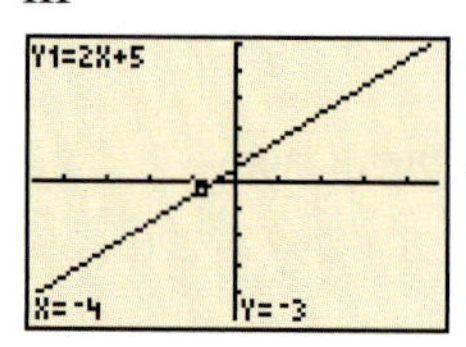

c. Y = $^-3$

9. a. $C = \frac{5}{9} * (F - 32)$

b.

°Fahrenheit	°Celsius
10	$^-12.2$
30	$^-1.1$
50	10
70	21.1
90	32.2

c. **Converting Fahrenheit to Celsius Temperatures**

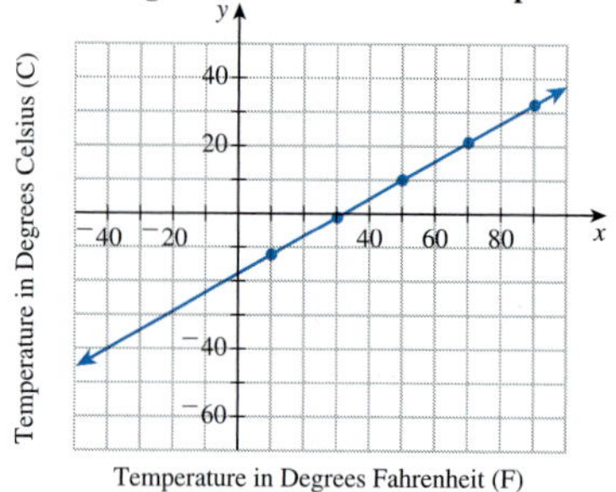

11. a. Answers may vary. Let i represent the interest (in dollars) and p represent the principal.

b. $i = .06 * p$

c.

Investment ($)	Interest Earned ($)
500	30
1000	60
1500	90
2000	120
2500	150

d.

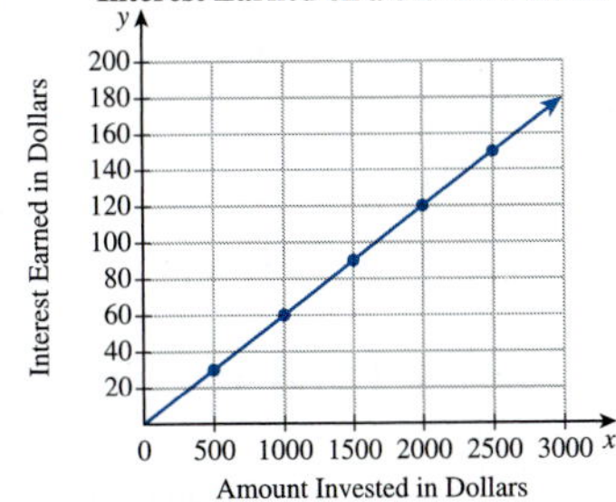

13. a. Let V represent the value of the product and t represent the age of the product (in years).

Formula: $V = 10{,}000 - 1500t$

Table of values:

Age of Product (years)	Value of Product ($)
0	10,000
1	8500
2	7000
3	5500
4	4000
5	2500
6	1000

Graph:

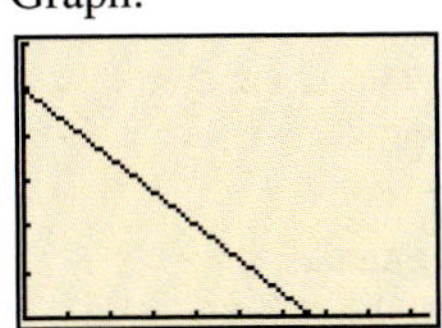

b. 0 to about 6.67 years

c. $0 to $10,000

d. Pick friendly values for Xmin and Xmax that include the range of years. Ymin and Ymax should include the range of product values.

15. a. A four-quadrant window shows both positive and negative values for x and y. A first-quadrant window shows only nonnegative values for x and y.

b. Cost of office supplies: first quadrant;
$y = 2x + 5$: four quadrant;
Fahrenheit and Celsius temperatures: four quadrant;
Paint required to cover walls: first quadrant

2.4 Skills and Review

17. $(^-2, 3)$

19. a. $4x + 2.8$ **b.** 18 inches

21. a. The commutative property states that for any numbers a and b, $a + b = b + a$. Because $50 - 4x = 50 + {}^-4x$, we have $50 - 4x = {}^-4x + 50$.

b. Number substitutions may vary.
When $x = 1$, both expressions equal 46.
When $x = 2$, both expressions equal 42.
When $x = 7$, both expressions equal 22.

23. 6.4

25. a. 3 miles/hour **b.** 6 miles/hour
c. 9 miles/hour **d.** 15 miles/hour

Chapter 2 Exploration

a. $x_2 - x_1 = {}^-10{,}315$
$y_2 - y_1 = {}^-5659$
$z_2 - z_1 = 4978$

b. About 12,775 miles

c. Yes. The closest the automobile could be from the satellite is 11,000 miles. So we expect the distance will exceed that number.

Chapter 2 Review Exercises

1. a. $(6 + 124) + 58 = 188$
b. $189 + (43 + 17) = 249$

2. No; $(30 - 10) - 5 = 20 - 5 = 15$;
$30 - (10 - 5) = 30 - 5 = 25$

3. a. $7 * (8 * 5) = 280$ **b.** $\left(\frac{1}{3} * 3\right) * 467 = 467$

4. No; $(16 \div 4) \div 2 = 4 \div 2 = 2$;
$16 \div (4 \div 2) = 16 \div 2 = 8$

5. a. $3 * 40 + 3 * 7 = 141$
b. $4 * 90 - 4 * 2 = 352$
c. $6 * 8 + 6 * 10 = 108$
d. $2 * x + 2 * y = 2x + 2y$
e. $9 * a - 9 * b = 9a - 9b$
f. $8 * x - 8 * y = 8x - 8y$

6. a. $4 * (52 + 8) = 240$ **b.** $3 * (9 + 81) = 270$
c. $5 * (x + y)$ **d.** $7 * (a - b)$

7. a. 9 **b.** 19 **c.** $^-1$

8. $(5 + 11) \div (6 - 8) = {}^-8$

9. a. $^-4$ **b.** 0 **c.** 1

10. a. Special property of 1, c
b. Special property of $^-1$, c

11. Answers may vary.

12. Answers may vary.

13. Answers may vary.

14. a. $C \approx 12.56$ inches **b.** $\frac{1}{2}$ ft/second

c. $A \approx 50.24$ square feet **d.** 6

e. 10

15. a. yes

x	$5 * x - 2$	$6 * x - (x + 2)$
4	18	18
3	13	13
$^{-}2$	$^{-}12$	$^{-}12$

b. No, substitution of $^{-}2$ gives different outputs.

x	$\sqrt{x^2 + 2 * x + 1}$	$x + 1$
4	5	5
3	4	4
$^{-}2$	1	$^{-}1$

c. yes

x	$(x - 1) * (x + 1)$	$x^2 - 1$
4	15	15
3	8	8
$^{-}2$	3	3

16. $\text{Area} = \frac{1}{2} * (b_1 + b_2) * h$

$\text{Area} = \left(\frac{1}{2} * b_1 + \frac{1}{2} * b_2\right) * h$

$\text{Area} = \frac{1}{2} * b_1 * h + \frac{1}{2} * b_2 * h$

17. a. $15y$ **b.** $3x$ **c.** $3a^2 - 12a$

18. a. $\dfrac{11}{a}$ **b.** $\dfrac{12}{a} - \dfrac{5}{9}$

c. $\dfrac{2 + x}{y}$ **d.** $x^2 + x$

19. a. $x + 5$ **b.** $x + 4$ **c.** $x - 4$

d. $x - 2$ **e.** $2x$ **f.** $\frac{1}{3}x$

g. $7x$ **h.** $\frac{x}{6}$ **i.** $\frac{x}{3}$

j. $3x + 4$ **k.** $3x - 4$

20. a. $10x + 6$ **b.** 156 yards

21. 18.03 cm

22.

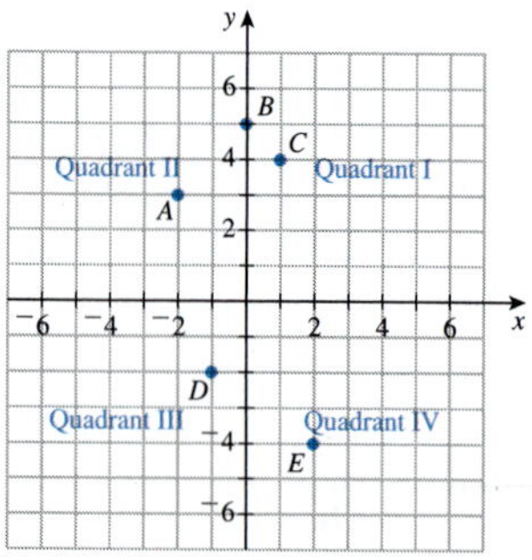

23. Only the y-coordinate varies on a vertical line.

24. They are all 3.

25. a. Run = $^{-}4$, rise = 6

b. Run = $^{-}4$, rise = $^{-}9$

26. $(4, {^{-}2})$

27. $(1, {^{-}3})$

28. Xmin = $^{-}4.7$, Xmax = 4.7, Xscl = 1, Ymin = $^{-}3.1$, Ymax = 3.1, Yscl = 1

29. The corners of the screen are located at (10, 10), ($^{-}10$, 10), ($^{-}10$, $^{-}10$), and (10, $^{-}10$).

30. 5

31. a. 2nd $\sqrt{}$((⁻6 − 2) ∧ 2 + (7 − ⁻1) ∧ 2)

b. 11.31

32. $300

33. a. $y = .25x + 100$

b. x (the number of travel miles)

c. y (the cost of the trip)

34.

x = number of miles traveled	y = cost of trip (in $)
0	100
100	125
200	150
300	175
400	200

35.

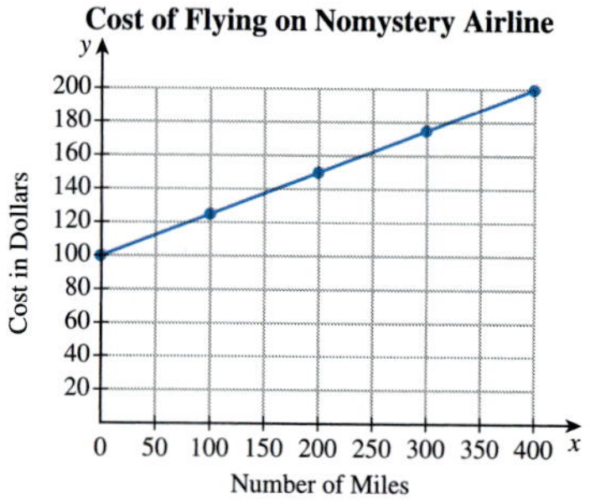

36. a. Y1 = .25X + 100

b. TblStart = 0, Δ Tbl = 100

c. Xmin = 0, Xmax = 470, Xscl = 50, Ymin = 0, Ymax = 250, Yscl = 25

Chapter 2 Test

1. a. $(^{-}8 + {^{-}2}) + 51 = 41$

b. $9 * (5 * 6) = 270$

c. $9 + (^{-}73 + 73) = 9$

2. **a.** $5 * 3 - 5 * 8 = {}^-25$ **b.** $6 * (a + b)$
c. $7 * a + 7 * b$ **d.** $2 * (x - y)$

3. **a.** 6 **b.** 81 **c.** 4

4. **a.** 0 **b.** 0
c. $\frac{1}{5}$ **d.** $^-1$

5. **a.** $y = 21$ **b.** $A = 20 \text{ cm}^2$

6. Yes

x	$x + 4 * x$	$x * (1 + 4)$
5	25	25
2	10	10
$^-3$	$^-15$	$^-15$

7. $A = P + P * r * t$

8. **a.** $6x$ **b.** $\frac{3}{x} + \frac{1}{8}$
c. $a - 5$ **d.** $\frac{11}{a}$

9. **a.** $4w + 270$ **b.** 1170 feet

10. 340 feet

11.

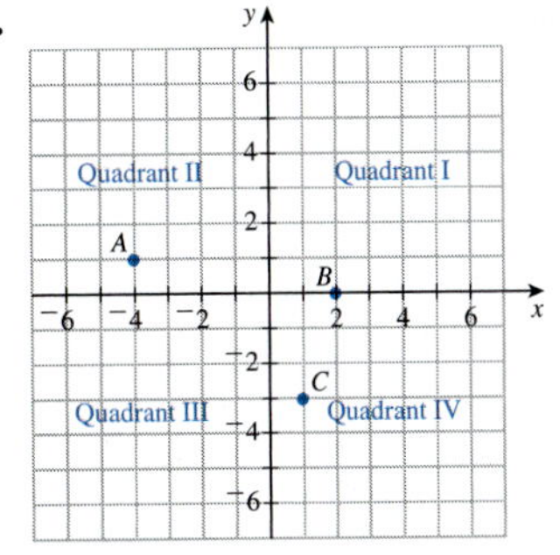

12. **a.** run $= {}^-1$, rise $= 2$
b. run $= 4$, rise $= 1$

13. 10

14. Let C represent the monthly charge in dollars and t represent the number of minutes of calls; $C = .07t + 4.95$.

15.

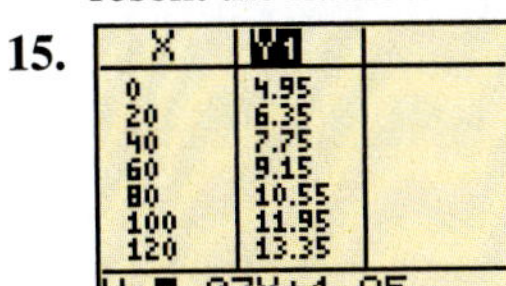

16.

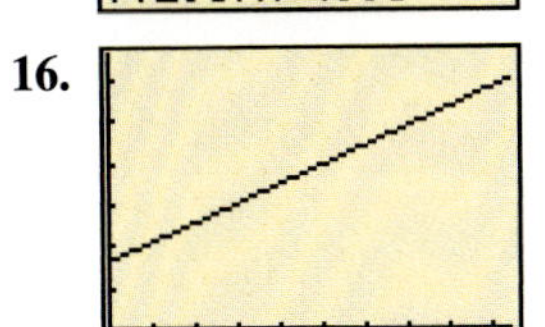

Chapter 3 Linear Equations

3.1 Exercises

1. $t = 2.5$ hours

3. $x = 7$

5. $b = -\frac{34}{9}$

7. $x = 2$

9.

$3x + 20 + {}^-20 = 6x + 2 + {}^-20$	*Adding $^-20$ on both sides*
$3x = 6x + {}^-18$	*Combining like terms*
$3x + {}^-6x = 6x + {}^-18 + {}^-6x$	*Adding ^-6x on both sides*
$^-3x = {}^-18$	*Combining like terms*
$\frac{^-3x}{^-3} = \frac{^-18}{^-3}$	*Dividing by $^-3$ on both sides*
$x = 6$	*Solution*

The health club costs the same for 6 visits per month.

11. $x = {}^-24$

13. $x = {}^-\frac{1}{4} = {}^-.25$

15. **a.** provider one: $12 + 2h$; provider two: $15 + 1.5h$
b. $12 + 2h = 15 + 1.5h$
c. $h = 6$
d. Both Internet providers charge the same amount for 6 hours of use.

3.1 Skills and Review

17. **a.** $Y1 = 7 + 5X$
b.

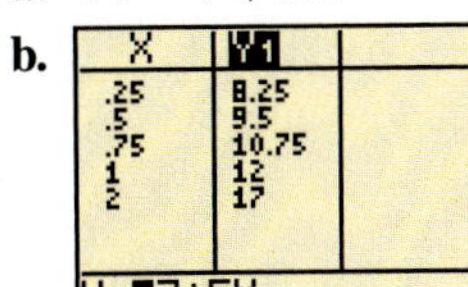

c.

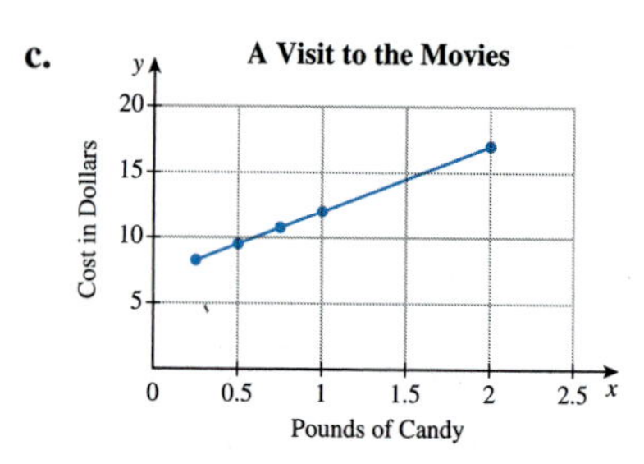

19. Run = $^-3$, rise = 8

21. $^-9$

23. $^-4$

25. b. $^-25$

3.2 Exercises

1. Distributive property: $5(x+3) = 5*x + 5*3 = 5x + 15$

3. $\frac{3}{5}x - \frac{1}{10}x + 2 - 6 = \frac{1}{2}x - 4$

5. $8x + 23$

7. a. $^-c^2 + 11c - 5$

b. $^-4x + 5$

c. 9 inches^2 + 5 inches

d. $5x - 5$

e. $\frac{8}{3}x - 14$

9. a. $x = \frac{19}{4} = 4.75$ **b.** $x = {}^-7$

c. $x = {}^-1.2$ **d.** $x = \frac{3}{4} = .75$

11. a. $40t, 55t$

b. $40t + 55t = 57$

c. $t = \frac{3}{5}$ hour or 36 minutes

13. a.

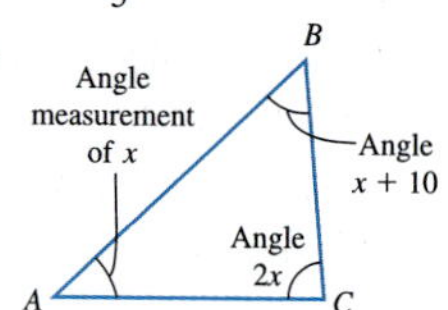

b. $x + x + 10 + 2x = 180$

c. $x = 42.5°$

15. The measure of angle A is 47.5°, B is 100°, and C is 32.5°.

3.2 Skills and Review

17. $2*1 - 5 = {}^-10*1 + 7$

$^-3 = {}^-3$ *OK*

19. Yes, $\frac{1}{3}x + \frac{8}{5} = \frac{5x}{15} + \frac{24}{15} = \frac{5x+24}{15}$

21. $^-10$°C

23. $^-13$; $^-8 + ((^-1) \wedge 19 + 6)/(9 - 10)$

25. $\frac{^-16 \text{ feet}}{\text{sec}^2}(4 \text{ sec}^2) + \frac{50 \text{ feet}}{\text{sec}}(2 \text{ sec}) + 150 \text{ feet}$

Evaluating exponents

$\frac{^-64 \text{ feet} * \text{sec}^2}{\text{sec}^2} + \frac{100 \text{ feet} * \text{sec}}{\text{sec}} + 150 \text{ feet}$

Performing multiplication

$^-64$ feet + 100 feet + 150 feet

Canceling units of measurement

186 feet

Performing addition and subtraction

3.3 The Problem Solving Process

1. (1) *Understand* the problem.

(2) *Visualize* the problem.

(3) *Assign* variable(s).

(4) *Write* equation(s).

(5) *Solve* equation(s).

(6) *Answer* the question.

(7) *Check* your answer.

3. (1) Find the dimensions of a rectangle with perimeter 900 feet and length twice the width.

(2) See Figure 9 on page 108.

(3) Let x = the width (in feet); then $2x$ = the length.

(4) $900 = x + 2x + x + 2x$

(5) $150 = x$

(6) The width is 150 feet and the length is 300 feet.

(7) 150 feet + 300 feet + 150 feet + 300 feet = 900 feet, the answer checks.

5. No, we need to answer the question in the problem and check the answer. Although $t = 3$ hours is the solution to the equation, the answer to the question is that Jacob catches Ruth at noon.

7. (4) $10x - 5(11 - x) = 8$

(5) $x = \frac{21}{5} = 4.2$

(6) The smaller number is 4.2 and the larger number is 6.8.

(7) $4.2 + 6.8 = 11$ and $10 * 4.2 - 5(11 - 4.2) = 8$. The answer checks.

9. (3) Let w = the width of the rectangle, then $4w$ = the length.

(4) $100 = w + 4w + w + 4w$

(5) $10 = w$

(6) The width is 10 meters and the length is 40 meters.

(7) 10 m + 40 m + 10 m + 40 m = 100 m. The answer checks.

11. 58 minutes

13. a. (2)

	Rate (miles/hour)	Time (hours)	Distance (miles)
Train A	$R+5$	2	$2(R+5)$
Train B	R	2	$2R$
Total	N/A	N/A	370 mi

b. **(3)** Let R = the speed of train B (in miles/hour), then $R+5$ = the speed of train A.

(4) $2(R+5)+2R=370$

(5) $R=90$

(6) The average rate of train B is 90 miles/hour and the rate of train A is 95 miles/hour.

(7) Train B travels 2 hours at a rate of 90 miles/hour, for a distance of 180 miles. Train A travels 2 hours at a rate of 95 miles/hour, for a distance of 190 miles. The sum of the distances is 370 miles, so the answer checks.

15. Kyle bought 20 tickets, Missy bought 32 tickets.

3.3 Skills and Review

17. $6x-30$

19. $x=4$

21. $y={}^{-}6$

23. ${}^{-}4$

25. run = 5, rise = 3

3.4 Exercises

1. a. Multiply each term by 2 to find the next term: 16, 32, 64.

b. Add 10 to each term to find the next term: 53, 63, 73.

c. Subtract 2 from each term to find the next term: 33, 31, 29.

d. Starting at 50, successive terms are found by adding 1, subtracting 2, adding 3, subtracting 4, adding 5, etc.: 47, 54, 46.

3. a. $a_1=17, d=3$

b. $a_1=85, d={}^{-}5$

c. $a_1=15, d=2.5$

d. $a_1=1, d={}^{-}3$

5. 15 posts

7. Yes, 76th term

Chapter 3 Exploration

a. $6H_2O$

b. $6CO_2$

c. $6O_2$

d. $6H_2O+6CO_2=6O_2+1C_6H_{12}O_6$

e.

	Atoms on Left Side	Atoms on Right Side
Carbon (C)	6	6
Hydrogen (H)	12	12
Oxygen (O)	18	18

f. Answers may vary.

Chapter 3 Review Exercises

1. a. $x=1$ **b.** $x=\frac{{}^{-}4}{5}={}^{-}.8$

c. $x={}^{-}6$ **d.** $x=1$

e. $x=\frac{{}^{-}3}{2}={}^{-}1.5$ **f.** $x=\frac{21}{2}=10.5$

g. $x=9$ **h.** $x=21$

2. $2*9-3=9+6$

$15=15$ *OK, the solution $x=9$ checks.*

3. a.

Time (hours)	Distance (miles)
1	50
2	100
3	150
4	200
t	$50t$

b. Distance = rate $*$ time; the rate is 50 miles/hour and the time is t. Therefore, distance = $50t$.

c. The solution represents the number of hours that it takes to travel a distance of 120 miles at a rate of 50 miles/hour.

d. $t=\frac{12}{5}=2.4$ hours

4. a. $373=C+273$

b. $C=100$; water boils at 100°C.

5. $28=3S+1$

$S=9$; a row made of 28 toothpicks has 9 squares.

6. a. $2.50+1.50d$

b. $4+1d=2.50+1.50d$

c. $d=3$; the fees of each club are the same when exactly 3 CDs are purchased.

7. a. $2x-3$ **b.** $5x-2$

c. $7x-6$ **d.** $x^2+6x-24$

e. $3x^2 + 2y$ **f.** $3x + \frac{1}{2}$

g. $\frac{1}{6}x$ **h.** $x + 6$

8. The special property of $^-1$, *c* on page 53 lets us change the subtraction in front of the left parenthesis to addition if we insert a $^-1$; $4x$.

9. a.

$$\begin{array}{rll} 3x + 8 &= 32 & \textit{LS} \\ +^-8 & \;\; +^-8 & \\ 3x &= 24 & \textit{A} \\ \frac{3x}{3} &= \frac{24}{3} & \textit{M} \\ x &= 8 & \textit{LS, RS} \end{array}$$

b.

$$\begin{array}{rll} 5x - 10 + 9 &= 7x + 21 & \textit{LS, RS} \\ 5x - 1 &= 7x + 21 & \textit{LS} \\ +^-5x & \;\; +^-5x & \\ ^-1 &= 2x + 21 & \textit{A} \\ +^-21 & \;\; +^-21 & \\ ^-22 &= 2x & \textit{A} \\ \frac{^-22}{2} &= \frac{2x}{2} & \textit{M} \\ ^-11 &= x & \textit{LS, RS} \end{array}$$

10. a. $x = 5$ **b.** $x = 1$

c. $x = 35$ **d.** $x = 40$

e. $x = \frac{1}{4} = .25$ **f.** $x = {^-9}$

g. $x = \frac{^-5}{13}$ **h.** $x = \frac{3}{2} = 1.5$

11. $L = 6$; the rectangle has a length of 6 inches.

12. $x = \frac{3}{4}$

13. a. Less than 1 hour; in 1 hour the combined distance of the trains is 100 miles.

b. $45t$

c. $45t + 55t = 36$

d. $t = \frac{9}{25} = .36$; 2:22 P.M.

14. a. Angle *B* is $x + 20$; angle *C* is $x + 70$

b. $x + x + 20 + x + 70 = 180$

c. $x = 30$

d. The measure of angle *A* is 30°, *B* is 50°, and *C* is 100°.

15. a. $90 + x + x = 180$

b. $x = 45$

c. Both unknown angles measure 45°.

16. a.

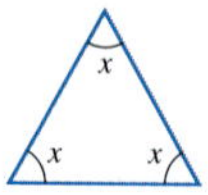

b. $x + x + x = 180$; $x = 60$; each angle measures 60°.

17. a. Find the total distance the pilot flies.

b.

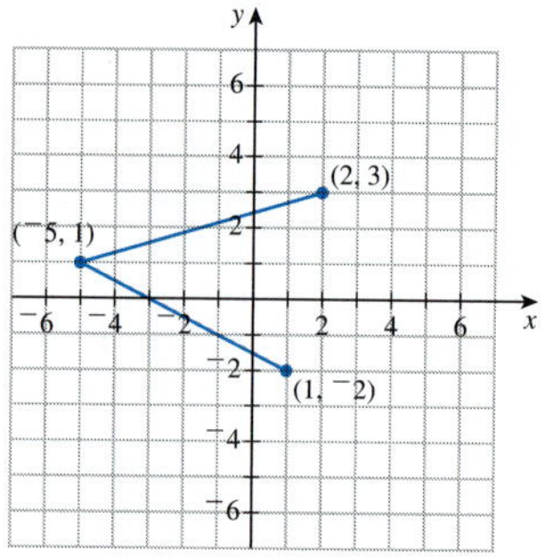

c. Let d_1 = the distance (in blocks) from (2, 3) to ($^-5$, 1).

Let d_2 = the distance from ($^-5$, 1) to (1, $^-2$).

d. $d_1 = \sqrt{(^-7)^2 + (^-2)^2}$

$d_2 = \sqrt{6^2 + (^-3)^2}$

e. $d_1 \approx 7.3$

$d_2 \approx 6.7$

f. The pilot flew approximately 14 blocks.

g. The answer checks with an approximate count of the blocks in the picture.

18. (1) Find the dimensions of a tennis court with perimeter 228 feet and a length that is 6 feet more than twice the width.

(2)

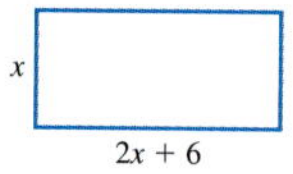

(3) Let x = width, then $2x + 6$ = length.

(4) $228 = x + 2x + 6 + x + 2x + 6$

(5) $36 = x$

(6) The width is 36 feet and the length is 78 feet.

(7) 36 feet + 78 feet + 36 feet + 78 feet = 228 feet, so the answer checks.

19. About 478 minutes

20. a. 12.9 seconds

b. 113.52 m

c. Marion Jones

21. a. Arithmetic: 22, 26, 30

b. 405, 1215, 3645

c. 0, 17, 0

d. Arithmetic: 11, 9, 7

22. a. $a_{10} = 64$

b. $d = 1$

c. $a_1 = 8$

Chapter 3 Test

1. $x = {}^{-}2$

2. $x = \frac{-5}{3}$

3. $6\left(\frac{3}{2}\right) - 1 \stackrel{?}{=} 2\left(\frac{3}{2}\right) + 5$

$8 = 8$ *OK!*

Therefore, $x = \frac{3}{2}$ is a solution to the equation.

4. $2x + 9 - 5x = 4(x - 3)$

$x = 3$

5. $C = \frac{290}{9} \approx 32.2°$

6. **a.** $20 - \frac{x}{13}$

b. Approximately 19.6 gallons

c. 260 miles

7. **a.** Let t = the number of hours it took for the boats to meet.

b. $15.5t + 1.5t = 58$

c. $t = \frac{58}{17} \approx 3.41$; the boats met at about 3:50 A.M.

8. **c.** Angle A measures 36° and angles B and C measure 72°.

9. **a.** 1.625 hours or 1 hour 37 minutes 30 seconds

b. Yes. Lance runs about 8 miles/hour. In 15 minutes, or one-quarter of an hour, Lance gets about a 2-mile head start on Doug. Doug is faster than Lance by a little more than 1 mile/hour. Therefore, in less than 2 hours Doug will have made up Lance's 2-mile head start.

10. The length of each leg is 3.2 cm and the remaining side is 13.6 cm.

Chapter 4 Systems of Linear Equations

4.1 Exercises

1.

	Variable Cost	Fixed Cost	Rate of Change
a.	.25 * (number of minutes)	$1.25	$.25 per minute
b.	.23 * (number of ounces)	$.14	$.23 per ounce
c.	.25 * (number of miles)	$8.	$.25 per mile

3. **a.** Answers may vary; (0, 1000), (100, 1002)

b. .02

c. Ordered pairs may vary; (200, 1004), (1000, 1020). Slope = .02

d. The slopes from (b) and (c) are the same. Although the elevation changes, the slope of the hill is always .02.

5.

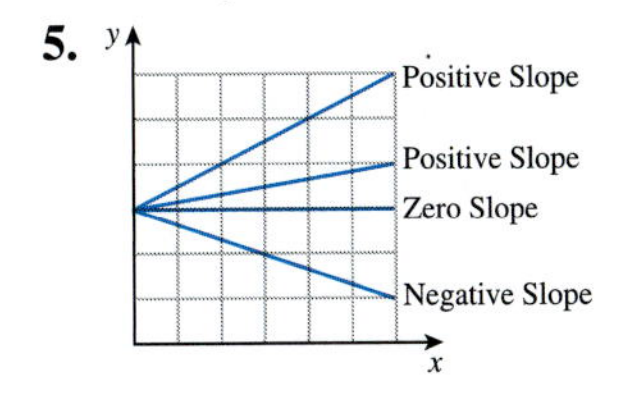

7. **a.** (5, 7) **b.** $(3, {}^{-}6)$ **c.** (2, 9) **d.** (1, 4), (3, 5)

9. **a.** ${}^{-}1$ **b.** $\frac{-11}{5}$ **c.** $\frac{3}{4}$ **d.** 0

11. **a.** $1000

b. No. In the first year the investment increased $100. In the second year it increased $110.

c. No, the rate of change is not constant.

d. Although both investments have an initial value of $1000 and a first year growth of $100, this investment is better because it increases by a larger amount every year thereafter.

13. **a.** 1.09

b. 1.09 > 1.00; no, the company did not tell the truth.

15. **a.**

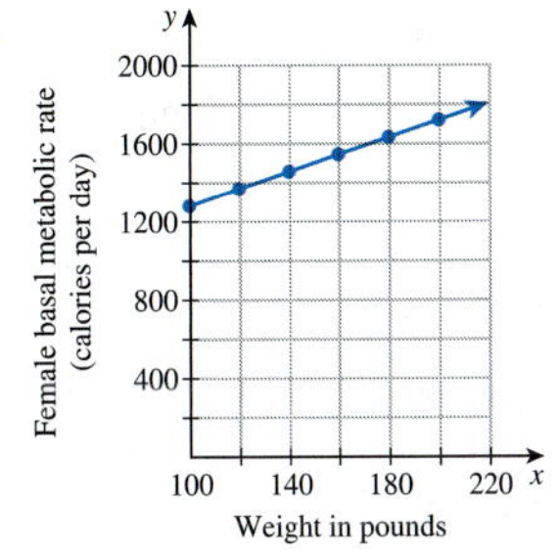

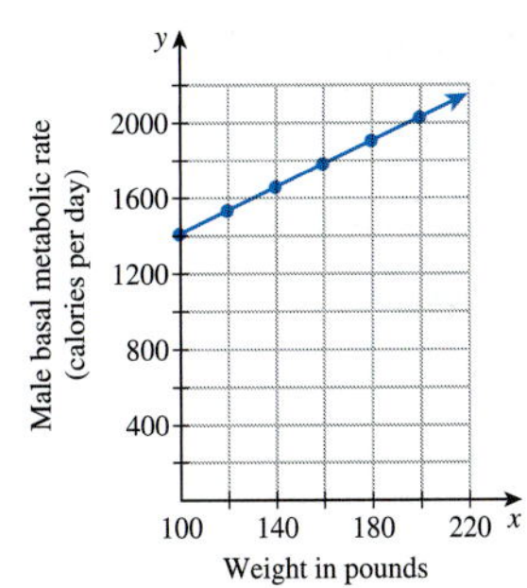

b. Female BMR: 4.4, Male BMR: 6.2

c. A female needs 4.4 more calories per day for every 1-pound increase in weight. A male needs 6.2

more calories per day for every 1-pound increase in weight.

4.1 Skills and Review

17. **(1)** Find the number of minutes at which the two phone plans cost the same.

(2) Example 2 of Section 3.2 is a similar problem.

(3) Let t = the number of calling minutes.

(4) $.07t + 22.95 = .03t + 28.45$

(5) $t = 137.5$

(6) At 137.5 minutes the two cellular phone plans cost the same.

(7) Plan A: At 7 cents per minute, 137.5 minutes costs \$9.63. Add the fixed cost of \$22.95 for a total of \$32.58.

Plan B: At 3 cents per minute, 137.5 minutes costs \$4.13. Add the fixed cost of \$28.45 for a total of \$32.58. The answer checks.

19. $^{-}4 * 5 + 23 = 2 * 5 - 7$

$3 = 3$ *OK. $x = 5$ is a solution*

21. $r = \frac{11}{12}$ cm $\approx .92$ cm

23. $11c^2 - 6d + 16$

25. 4 inches/second

4.2 Exercises

1. $^{-}8x + y = {}^{-}10$ or $8x - y = 10$

$y = {}^{-}x + 7$

$y = 3x - \frac{7}{2}$

3. **a.** Slope-intercept form **b.** $m = {}^{-}3$

c. y-intercept = (0, 5) **d.** x-intercept = $(\frac{5}{3}, 0)$

5. **a.** x-intercept = (2, 0), y-intercept = (0, $^{-}4$)

b.

c. $m = 2$

d. $y = 2x - 4$

7. **a.** $y = \frac{3}{5}x + 4$ **b.** $y = 2x - 1$

c. $y = 7$ **d.** $y = {}^{-}\frac{1}{2}x$

9. **a.** $y = {}^{-}x + 4$ **b.** $y = x + 3$

c. $y = {}^{-}\frac{1}{2}x - 3$ **d.** $y = 4$

11. **a.** $y = 3500$ feet

b. y will decrease by 22.

c. $y = {}^{-}22x + 3500$

13. **a.**

Hour of Day	Elapsed Hours since 2 P.M.	Temperature (°F)
10:00 P.M.	8	75
11:00 P.M.	9	73.5
12:00 A.M.	10	72
1:00 A.M.	11	70.5
2:00 A.M.	12	69
3:00 A.M.	13	67.5
4:00 A.M.	14	66

From the table, the low temperature is 66°F.

b. $y = {}^{-}1.5x + 87$ for elapsed times between 0 and 14 hours.

c. Windows may vary: Xmin = 0, Xmax = 18.8, Xscl = 2, Ymin = 65, Ymax = 90, Yscl = 4

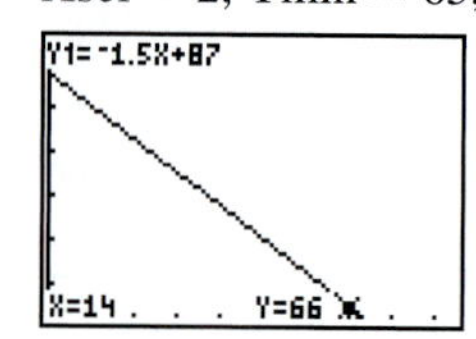

From the trace of the graph, the low temperature is 66°F.

d. The exact low temperature is 66°F.

15. **a.** $\frac{1}{3}x$

b. $\frac{1}{2}y$

c. $\frac{1}{3}x + \frac{1}{2}y = 6$

d. $y = {}^{-}\frac{2}{3}x + 12$

e. Friendly windows may vary: Xmin = 0, Xmax = 18.8, Xscl = 2, Ymin = 0, Ymax = 15, Yscl = 2, Y1 = ($^{-}$2/3)X + 12

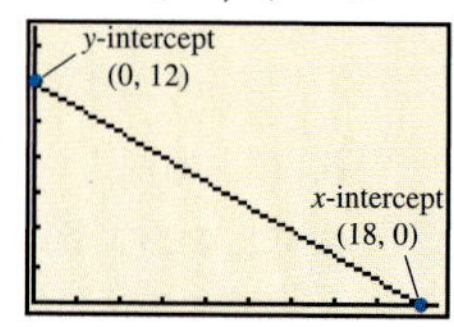

f. The x-intercept is (18, 0). It takes Paul 18 hours to complete 6 jobs alone. The y-intercept is (0, 12). It takes Joan 12 hours to complete 6 jobs alone.

g. The slope is $^-\frac{2}{3}$. For each 3-hour increase in Paul's hours, Joan's hours decrease by 2, or for each 3 hour decrease in Paul's hours, Joan's hours increase by 2.

4.2 Skills and Review

17. a. Run = 4, rise = 5

b. $\frac{5}{4}$

c. Distance = $\sqrt{41} \approx 6.4$

19. $x = 33$

21. a. Diagonal $\approx$ 10.8 m

b. Diagonal = $\sqrt{117}$ m

23. $4 * (83 - 3) = 320$

25. 0

4.3 Exercises

1. Intersection: $(^-1, 1)$

$y = {}^-x$	$y = x + 2$
$1 \stackrel{?}{=} {}^-({}^-1)$	$1 \stackrel{?}{=} ({}^-1) + 2$
$1 = 1$ *OK*	$1 = 1$ *OK*

The solution $(^-1, 1)$ checks.

3. a. Answers may vary; statement 1: (1, 9), (2, 8), (3, 7); statement 2: (10, 5), (9, 4), (8, 3).

b. and **c.**

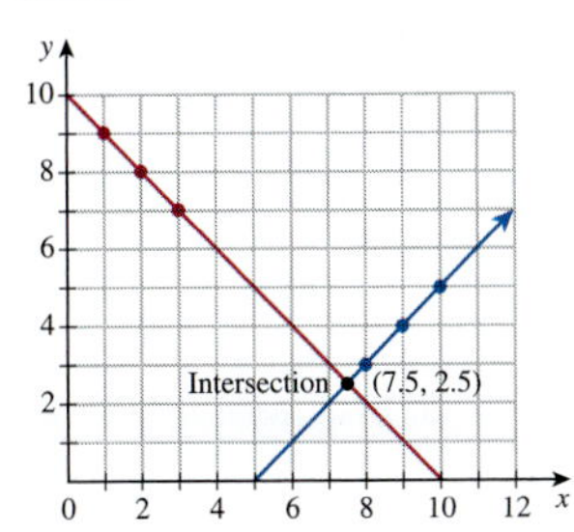

d. The intersection must satisfy both statements; the pair of numbers must sum to 10 and have a difference of 5.

e. (7.5, 2.5); see above graph.

5. $(2, ^-1)$

7. No solution (inconsistent system). These are two different lines that have the same slope of 1. Therefore, the lines are parallel and the system has no solution.

9. (.25, 0)

11. (1, 7)

13. a.

Miles Driven	Pinnacle's Cost	Competitor's Cost
50	\$20.50	\$22.50
60	\$23.00	\$23.00
70	\$25.50	\$23.50

b. 60 miles

c. Intersection point

15. a. (500, 85)

b. At 500 miles, both companies charge \$85.

c. Udrive is cheaper for rentals of more than 500 miles. Each additional mile at Udrive is 10 cents, whereas for Mile Rent-a-Car it is 15 cents.

4.3 Skills and Review

17. a. Look at the table and find Pinnacle's cost when 0 miles are driven.

b. The rate of change is constant, so pick any two points to find the slope.

$$m = \frac{10.50 - 8.00}{10 - 0} = .25,$$

so the rate of change is \$.25 per mile.

19. a. $m = {}^-3$, y-intercept = (0, 2)

b. $m = \frac{3}{2}$, y-intercept = $(0, ^-3)$

21. $^-750$ vehicles/hour

23. $W = 3$ cm

25. a. $^-5x + 10$ **b.** 25

c. 25 **d.** yes

4.4 Exercises

1. (c) and (d) are complete solutions.

3. a. We need both an x- and y-value to be a solution of a system of two equations.

b. The solution is $(^-1, 3)$.

5. $(^-2, \frac{^-4}{15})$

7. a. Multiplying the second equation by 3 leads to a ^-6y term when the equations are added. We want to eliminate the y-term in the resulting equation.

b.

$4x - 3y = 22$

$^-9x + 3y = {}^-42$ *Multiplying on both sides of the second equation by* $^-3$

$^-5x = {}^-20$ *Adding the equations*

$x = 4$

$3 * 4 - y = 14$ *Substituting 4 for x into $3x - y = 14$*

$y = {}^{-}2$ *Solving for y*

Solution: $(4, {}^{-}2)$

9. $(7, 2)$

11. a. Elimination. The equations are in standard form. Multiply one of the equations by $^{-}1$ and add the equations to eliminate the y-term.

b. $\left(0, -\frac{4}{3}\right)$

13. $({}^{-}10, 7)$

Check the solution in both equations:

$${}^{-}10 + 2 * 7 \stackrel{?}{=} 4 \qquad {}^{-}({}^{-}10) \stackrel{?}{=} 7 + 3$$

$$4 = 4 \quad OK \qquad 10 = 10 \quad OK$$

The solution $({}^{-}10, 7)$ checks.

15. No solution (inconsistent system); the graphs form parallel lines.

4.4 Skills and Review

17. Approximately $(1.18, 5.55)$

19. a. x-intercept $= (3, 0)$, y-intercept $= (0, {}^{-}2)$

b.

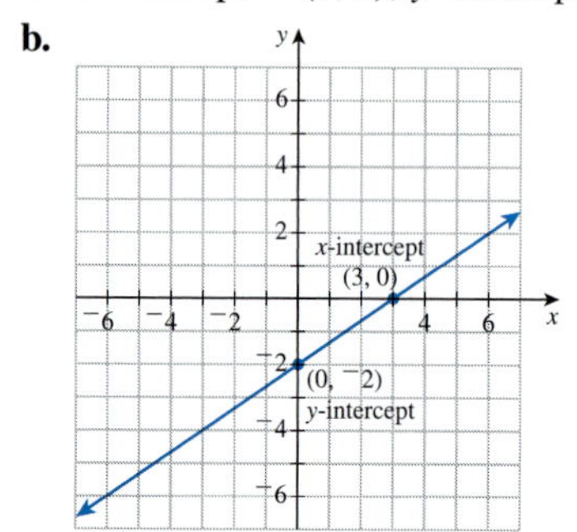

21. $^{-}5$; the larger the absolute value of the slope, the steeper the road.

23. a. $\sqrt{52}$ **b.** about 7.2

25. a. 2 **b.** 1 **c.** $\frac{1}{2}$

d. $^{-}4$ **e.** .15

4.5 Exercises

1. $m_1 = m_2 = 2$

3. The lines are parallel, $m_1 = m_2 = 1$.

5. a. None of these

b. Parallel

c. Perpendicular

d. Same

7. Yes. Although the lines appear to be parallel, the slopes are different ($m_1 = 1, m_2 = .99$). Lines with unequal slopes intersect.

Chapter 4 Exploration

a.

Friendly window $(0 \le X \le 9.4, 0 \le Y \le 2000)$

b. 199

c. 199 per year

d. 2682

e. $^{-}502$; the function yields a negative number, which does not make sense in the context of the problem.

Chapter 4 Review Exercises

1. a. $m = 4$ **b.** $m = \frac{-8}{3}$ **c.** $m = {}^{-}1$

d. $m = 0$ **e.** $m = \frac{-1}{2}$

2. a. 0 inches

b. 2 inches/hour

c.

Time (hours)	Amount of Snow on Ground (inches)
0	0
1	2
2	4
3	6

d.

$m = 2$; the slope and rate of change (2 inches/hour) are the same.

3. a. $-\frac{100}{3}$ supplements per dollar

b. For every \$3 decrease in price, Vitomax can sell 100 more supplements. Apply this rate of change to the first ordered pair. \$20 − \$3 = \$17, and 5000 + 100 = 5100, giving a second ordered pair of (17, 5100).

c. (14, 5200)

d. $m = -\frac{100}{3}$

4. $m = \frac{3}{2}$

5. a ⇔ 4, b ⇔ 3, c ⇔ 1, d ⇔ 2

6. a.

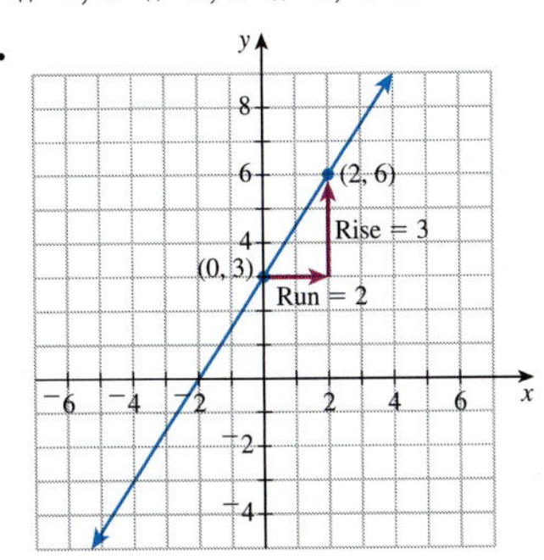

7. a. 5 inches of snow

b. 2 inches/hour

c. $y = 2x + 5$

8. a. $y = {}^{-}300x + 1800$

9. a. $y = 6x + 2$ **b.** $y = \frac{1}{3}x + 4$

c. $y = 2x + 1$ **d.** $y = {}^{-}5x + \frac{1}{2}$

e. $y = x - 3$ **f.** $y = {}^{-}x + 5$

10. a. $m = 4$, y-intercept = (0, 3)

b. $m = {}^{-}1$, y-intercept = (0, ${}^{-}1$)

c. $m = 3$, y-intercept = (0, ${}^{-}5$)

d. $m = 1$, y-intercept = (0, ${}^{-}3$)

e. $m = {}^{-}\frac{2}{3}$, y-intercept = (0, 4)

f. $m = {}^{-}1$, y-intercept = (0, 2)

11. $y = \frac{3}{4}x - 3$

12. a. $y = 5x + 11$ **b.** $y = x + 9$

c. $y = {}^{-}3x + 7$ **d.** $y = \frac{1}{2}x - 4$

13. a. $y = 2x + 2$ **b.** $y = 4x - 17$

c. $y = \frac{-7}{2}x + 13$ **d.** $y = -\frac{1}{3}x + 4$

14. a. $\frac{4}{1}$

b. $4y$

c. $3x + 4y = 12$

d. $\frac{9}{4}$ or $2\frac{1}{4}$ months

e. x-intercept = (4, 0), y-intercept = (0, 3). To complete the album, the first musician must write songs for 4 months when the second musician writes for 0 months. To complete the album, the second musician must write songs for 3 months when the first musician writes for 0 months.

f.

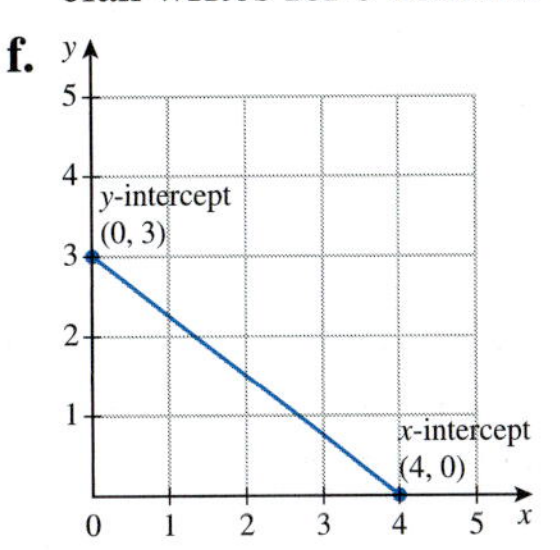

15. a. x-intercept: (5, 0), y-intercept: (0, 5)

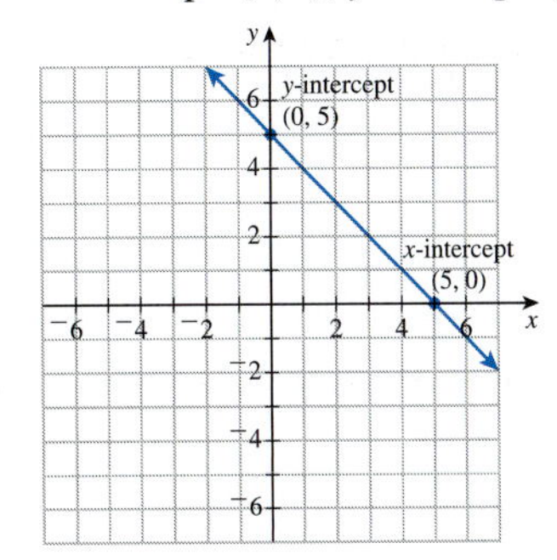

b. (4, 0) (0, ${}^{-}2$)

c. (2, 0) (0, 3)

d. (20, 0) (0, ${}^{-}2$)

e. (2, 0) (0, ${}^{-}12$)

f. $(\frac{9}{4}, 0)$ $(0, \frac{9}{5})$

16. a.

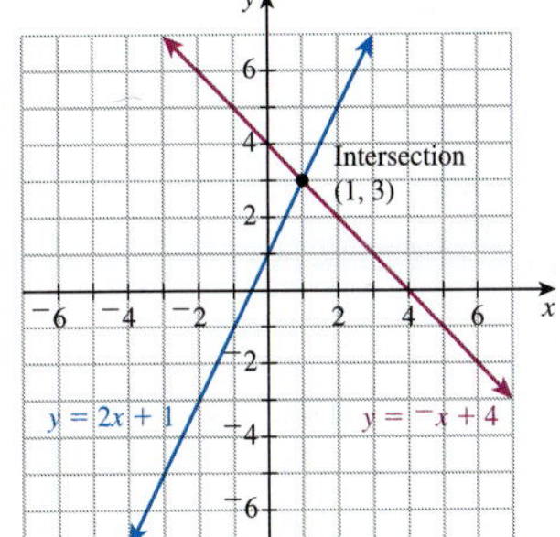

b. Intersection point

c. (1, 3)

17. (1, 4)

18. (${}^{-}2$, ${}^{-}3$)

19. No solution (inconsistent system)

20. Approximately (1.02, ${}^{-}7.49$)

21. Approximately (3.67, 1.33)

22. The numbers are 53.75 and 24.45.

23. The newspapers charge the same amount for 10 lines of advertising.

24. $\left(\frac{11}{3}, \frac{4}{3}\right)$
25. $(^-2, ^-9)$
26. $\left(\frac{1}{2}, \frac{3}{2}\right)$
27. $(5, ^-2)$
28. $(5, ^-1)$
29. $(^-1, 2)$
30. $(0, ^-6)$
31. $\left(^-4, \frac{3}{4}\right)$
32. **a.** $x + y = 180$
  b. $y = 2x - 30$
  c. smaller angle: 70°; larger angle: 110°
33. The cat will catch the mouse in $1\frac{1}{2}$ seconds.
34. **a.** parallel
  b. perpendicular
  c. same
  d. none of these
35. perpendicular

Chapter 4 Test

1. $m = ^-\frac{1}{4}$

2. **a.**

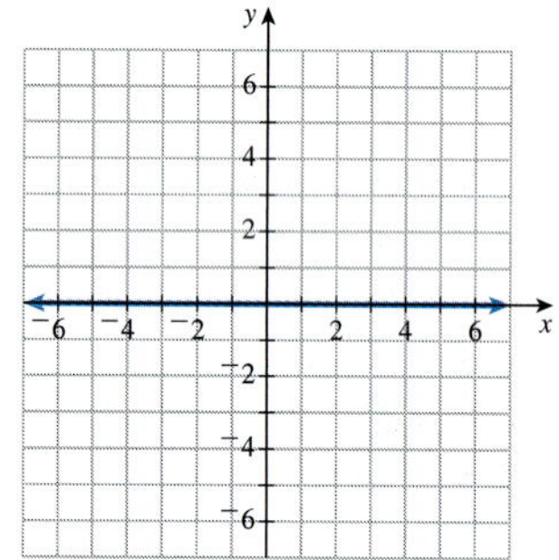

b.

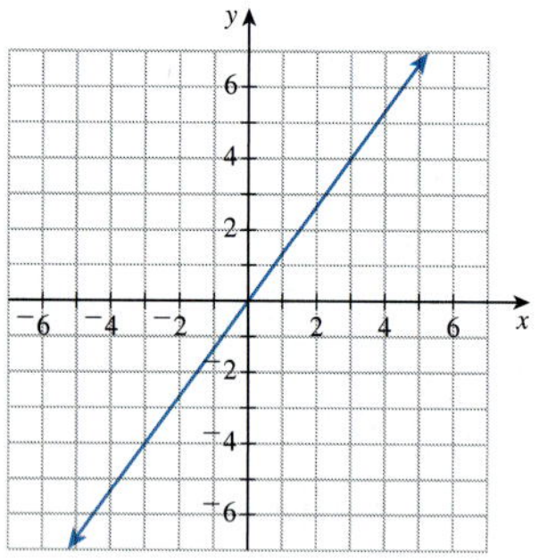

c.

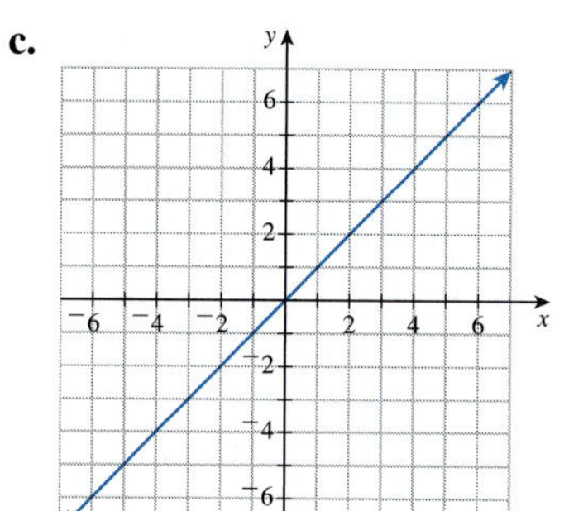

d.

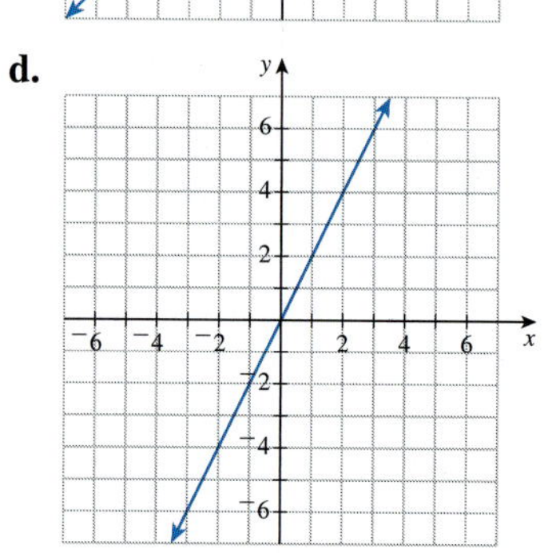

3. **a.** $\frac{5}{2}$ inches/year

b.

Time (years)	Height (inches)
0	250
2	255
4	260

4. $y = \frac{5}{2}x + 250$
5. **a.** $m = ^-\frac{2}{3}$
  b. y-intercept $= (0, 2)$
6. $y = ^-\frac{3}{4}x + 2$
7. $y = 5x - 7$
8. x-intercept: $(5, 0)$, y-intercept: $(0, ^-4)$

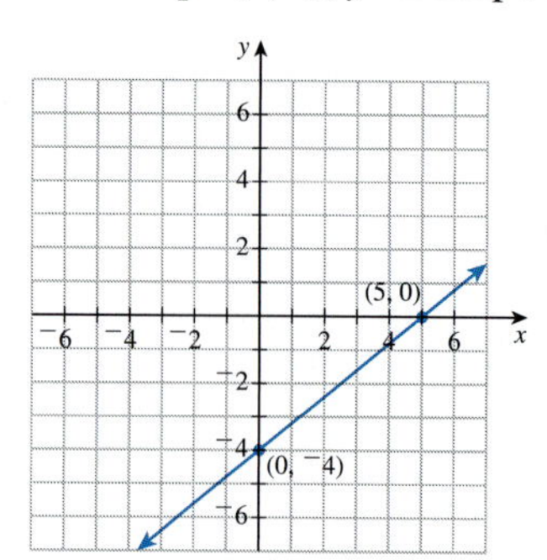

9. Both companies charge the same amount when their trucks are driven 100 miles.
10. Approximately $(.67, ^-.14)$

11. $\left(\frac{2}{3}, {}^-\frac{1}{7}\right)$

12. $({}^-6, 1)$

13. Smaller angle: 52.5°; larger angle: 127.5°

14. The lines are parallel.

Chapter 5 More Applications of Linear Equations

5.1 Exercises

1. Two of several proportions produce the same cross product.

$$\frac{2}{50} = \frac{6}{150} \qquad \frac{6}{2} = \frac{150}{50}$$
$$2 * 150 = 50 * 6 \qquad 6 * 50 = 2 * 150$$
$$300 = 300 \qquad 300 = 300$$

3. $x = \frac{76}{3} \approx 25.33$

5. Length of AD $= \frac{21}{2}$ cm $= 10.5$ cm

7. D

9. Two of several proportions: $\frac{x}{84} = \frac{6}{7}$ and $\frac{84}{x} = \frac{7}{6}$

11. $\frac{10}{7} \approx 1.4$ minutes

13. $\frac{725}{3}$ minutes ≈ 242 minutes

15. 85 penguins

5.1 Skills and Review

17. Smaller angle: 34°; larger angle: 146°

19. a. $m = \frac{4}{3}$ **b.** $y = \frac{4}{3}x - \frac{17}{3}$

21. $x = {}^-\frac{5}{13} \approx {}^-.38$

23. 36π inches^2

25. 77 mm^2

5.2 Exercises

1. a.

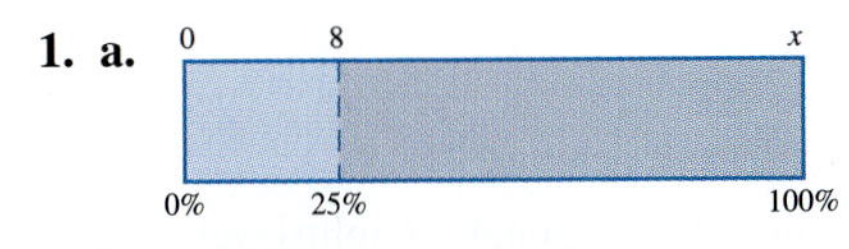

b. $x = 32$

c. $x = 32$

3. About 48.1%

5.

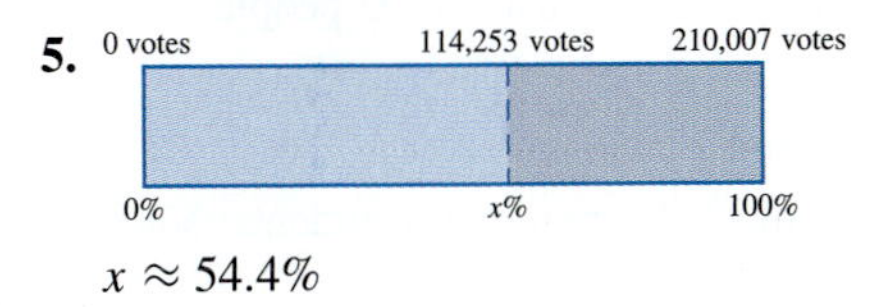

$x \approx 54.4\%$

7.

0 g x g 250 g
NaCl Water (H_2O)
0% 9% 100%

Sodium chloride: 22.5 g; water: 227.5 g

9.

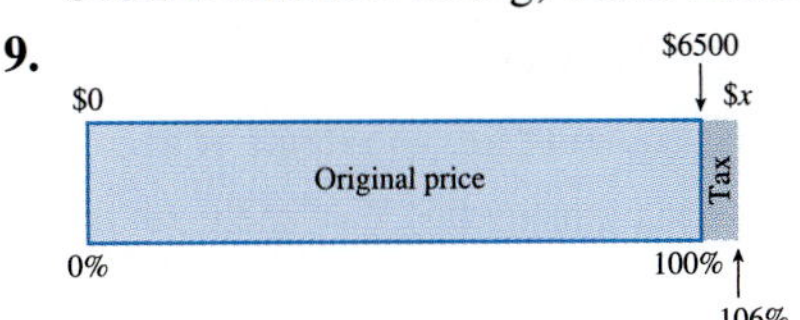

$6890

11.

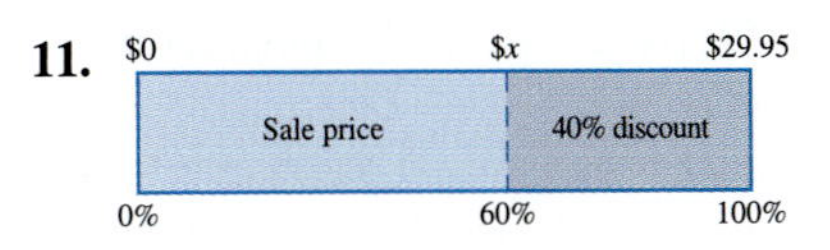

$x = \$17.97$

13. a. About 40%

b. About 13.8%

c. Compare each state's change in population with its 1990 population. Although California had a larger change in population, the change relative to the 1990 population was less.

15. a. approximately .124 gallons

b. approximately 15.8 ounces

5.2 Skills and Review

17. a. $\frac{4}{3} = \frac{x}{10}$ or $\frac{x}{4} = \frac{10}{3}$

b. $x = \frac{40}{3} \approx 13.33$

19. approximately (1.91, .55)

21. The equation is in slope-intercept form, $y = mx + b$. Therefore, the slope is the coefficient of x; $m = 3$.

23.

Days Rented	Charge ($)
1	$3 + 1 * 1 = 4$
2	$3 + 1 * 2 = 5$
4	$3 + 1 * 4 = 7$

25. 21

5.3 Exercises

1. **a.** $2.80 per pound
   b. Peanuts
   c. Peanuts, because $2.80 is closer to the $2 peanut price than it is to the $5 cashew price.

3. **a.**

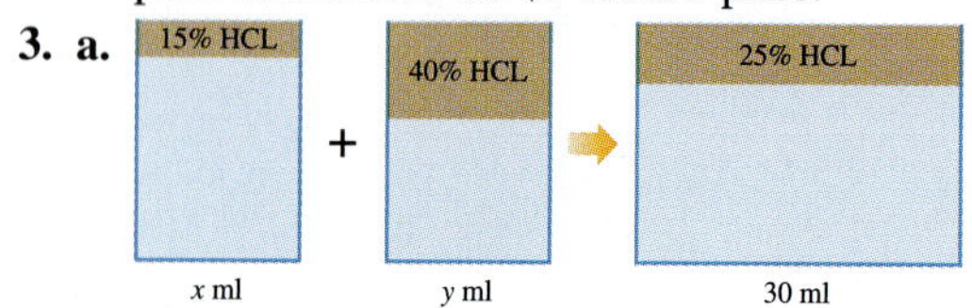

   b. $x + y = 30$
   c. $.15x + .40y = .25(30)$
   d. (1) Solve each equation for y in terms of x.
   (2) Enter the equations in the [Y =] menu on the calculator.
   (3) Find a window that includes the intersection of the two lines.
   (4) Use [TRACE] or [2nd] CALC 5: intersect to find the intersection.

5. Distance equation: $4x$ is the distance Jason travels, $10y$ is the distance Nathan travels, and 16 is their combined distance.
   Time equation: x is the number of hours Jason travels, y is the number of hours Nathan travels, and $\frac{1}{2}$ hour is Nathan's later start time.

7. **a.** Let x = the amount of chocolate raspberry, in pounds. Let y = the amount of French vanilla, in pounds.

   b.

	Chocolate Raspberry	French Vanilla	Mixture
Rate ($/pound)	4	6	
Weight (pounds)	x	y	3
Cost ($)	$4x$	$6y$	13

   c. $x + y = 3$ *weight equation*
   $4x + 6y = 13$ *money equation*
   d. $x = \frac{5}{2} = 2\frac{1}{2} = 2.5,\ y = \frac{1}{2} = .5$
   e. Chocolate raspberry: $2\frac{1}{2}$ pounds, French vanilla: $\frac{1}{2}$ pound

9. **a.**

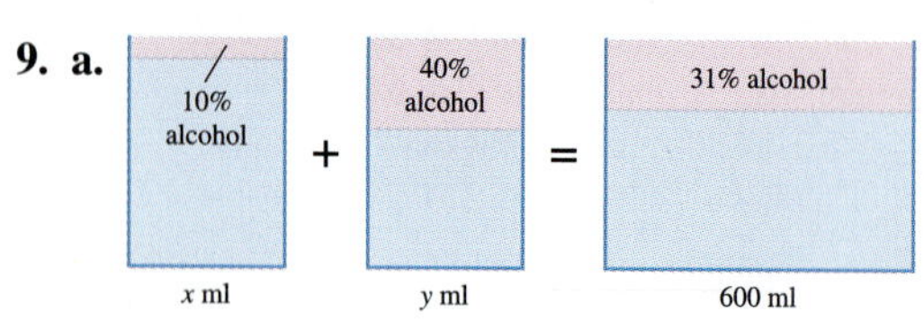

   b. Let x = the amount (in mL) of 10% solution.
   Let y = the amount (in mL) of 40% solution.
   c. $x + y = 600$ *solution equation*
   $.10x + .40y = .31(600)$ *alcohol equation*
   d. $x = 180,\ y = 420$
   e. 10% solution: 180 mL, 40% solution: 420 mL
   f. Quick check: There should be quite a bit more of the 40% solution in the mixture because 31% is much closer to 40% than it is to 10%. *Thorough check:* The numbers should meet the conditions of the problem.

11. **a.** Total driving time 6 hr 30 min
   Total distance
   Wichita — 446 mi — St. Louis
   $61x$ miles $72y$ miles
   b. Let x = the time in hours Thelma drives.
   Let y = the time in hours Louise drives.
   $x + y = 6.5$ *time equation*
   $61x + 72y = 446$ *distance equation*
   c. $x = 2,\ y = 4.5$
   d. Thelma: 2 hours; Louise: 4.5 hours

13. Ordinary seats: 520; special seats: 80

15. 7.5% source: $4100, 11.25% source: $1500

5.3 Skills and Review

17. Hydrochloric acid: 18 g; water: 282 g

19. $x \approx {}^{-}10.97$

21. x-intercept = (4, 0), y-intercept = (0, $^{-}9$)

23. $x = \frac{3}{11} \approx .27$

25. 10 inches

5.4 Exercises

1. 12 quarters, 8 dimes, 4 pennies

3. 5 small colonials, 3 big colonials, 3 split-levels

Chapter 5 Exploration

a. $\frac{119}{500} = \frac{x}{18500}$; $x = 4403$; estimate: 4403 people
b. 23.8%
c. Between 20.1% and 27.5%
d. Minimum students: 3718; maximum students: 5088

Chapter 5 Review Exercises

1. **a.** $x = 45$ **b.** $x = \frac{27}{2} = 13.5$

c. $x = 15$ **d.** $x = \frac{11}{5} = 2.2$

2. Two possible proportions:

$$\frac{6 \text{ gallons}}{9 \text{ dollars}} = \frac{20 \text{ gallons}}{c \text{ dollars}} \quad \text{or} \quad \frac{c \text{ dollars}}{9 \text{ dollars}} = \frac{20 \text{ gallons}}{6 \text{ gallons}}$$

3. 68°F

4. one gallon: \$1.20; 150 gallons: \$180

5. 21.25 feet

6. 10.8 gallons

7. **a.** The time it takes for two people to complete a job is less than the fastest person working alone.

b. $1\frac{1}{2} = 1.5$ hours

8. Estimate; 103 deer

9. **a.** Part = percent $*$ whole

b. whole $= \frac{\text{part}}{\text{percent}}$

10.

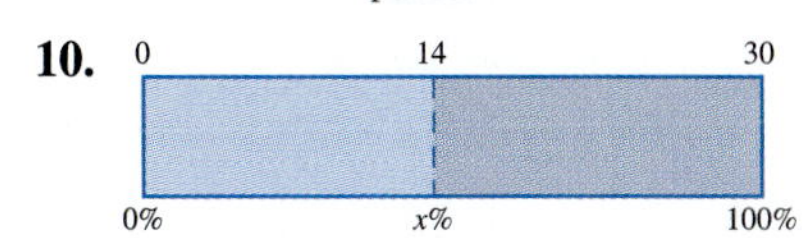

11. estimate: 50%

12. About $46\frac{2}{3}\%$

13. **a.**

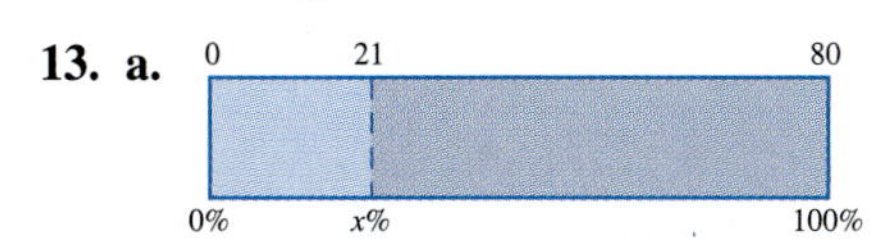

b. estimate: 25%

c. 26.25%

14. 10.8

15. 200

16. 30%

17. 3 g

18. 25,000 mg

19. **a.** The *solute* is the substance that is dissolved in another.

b. The *solvent* is the substance in which another is dissolved.

c. The *solution* consists of the solute and the solvent mixed together.

20. 9%

21. \$33.36

22. 27.38%

23. approximately 10,684 people

24. **a.** Let x = the amount of peanut butter fudge, in pounds.

Let y = the amount of vanilla fudge, in pounds.

b. $x + y = 2$ *weight equation*

$5x + 7y = 13$ *money equation*

c. $\text{Y} = {}^{-}\text{X} + 2$

$\text{Y2} = ({}^{-}5/7)\text{X} + 13/7$

d. Peanut butter fudge: .5 pounds, vanilla fudge: 1.5 pounds

25. 10% solution: 80 mL, 30% solution: 120 mL

26. Approximately \$112.90 was deposited in the checking account.

Approximately \$587.10 was deposited in the savings account.

27. $x = {}^{-}22.33$, $y = 4.83$, and $z = {}^{-}9.5$

28. 6 pennies, 24 nickels, and 1 quarter

Chapter 5 Test

1. $x = \frac{80}{3} \approx 26.67$

2. about \$233.10

3. $\frac{24}{7} \approx 3.43$ hours $\approx$ 3 hours 26 minutes

4. approximately 38 polar bears

5.

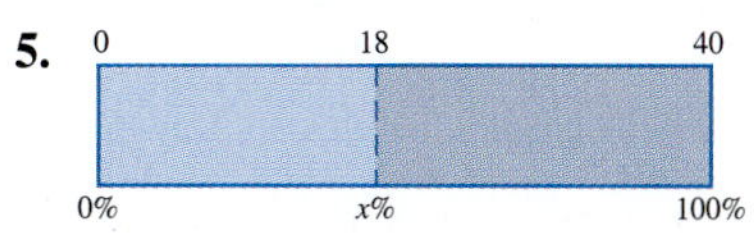

6.

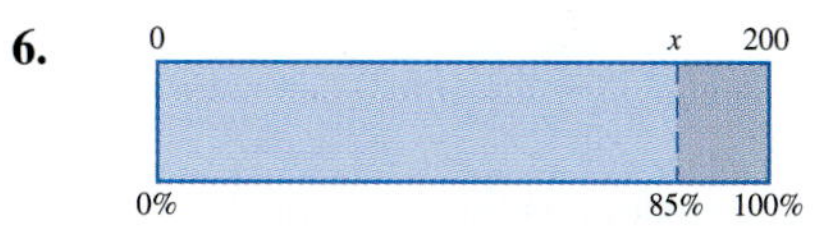

$x = 45\%$

7.

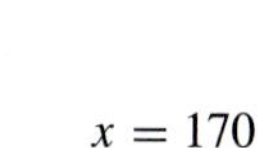

$x = 170$

$x \approx 57.14$

8. 80%

9. 12.5%

10. The 15% solution contains more sulfuric acid.

11. Almonds: $1\frac{1}{3}$ pounds; peanuts: $2\frac{2}{3}$ pounds

12. AAA bonds: \$3000; CD: \$2000

13. 1.75 L

Chapter 6 Exponents and Factoring

6.1 Exercises

1. a. x^7 **b.** y^{12} **c.** x^6y^{12}

3. a. $12a^9$; product law $a = a, m = 4, n = 5$

b. b^{18}; power-of-a-power law $a = b, m = 3, n = 6$

c. $a^{14}b^{28}$; power-of-a-product law and power-of-a-power law $a = a^2, b = b^4, m = 7$

d. $64b^3$; power-of-a-product law $a = 4$, $b = b$, and $m = 3$

5. a. k^4 **b.** $\frac{1}{k^4}$ **c.** $\frac{1}{j^{15}}$ **d.** $\frac{j^{20}}{k^8}$

7. a. y^4 **b.** $\frac{1}{y^6}$ **c.** $\frac{y^9}{z^9}$

9.
$$\begin{aligned}\frac{(2x^3y^4)^3}{x^{-2}y^{12}} &= \frac{8x^9y^{12}}{x^{-2}y^{12}}\\ &= 8x^{11}y^0\\ &= 8x^{11}(1)\\ &= 8x^{11}\end{aligned}$$

11. xy^3

13. $\frac{4x^2}{y}$

15. $^-2$

6.1 Skills and Review

17. a. $x + y = 6$ weight equation: x is the weight of the plain chocolate, y the weight of the deluxe chocolate, and 6 is the weight of the mixture, in pounds.

$3x + 5y = 27$ money equation: $3x$ is the cost of the plain chocolate, $5y$ is the cost of the deluxe chocolate, and 27 is the cost of the mixture, in dollars.

b. $x = 1.5$ and $y = 4.5$. The mixture contains 1.5 pounds of plain chocolate and 4.5 pounds of deluxe chocolate.

19. $x = 3.6$

21. $x = {}^-6$

23. The height is 4 cm.

25. 53

6.2 Exercises

1.

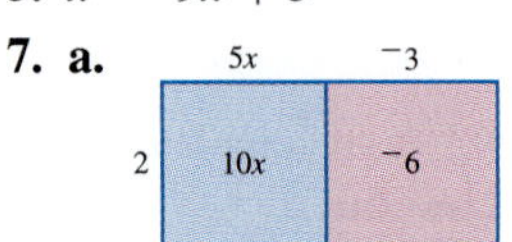

3. $40z^2 - 24z$

5. $x^2 - 9x + 5$

7. a.

b. GCF $= 18$; $18(3y + 4)$

c. GCF $= 7x$; $7x(2x^2 + y - 6x)$

d. $7x(2x^2 + y - 6x) = 14x^3 + 7xy - 42x^2$

9. a. $6x^3(3x^2 + 1)$ **b.** $3xy^3(7x - 5y)$

11. $2\pi r(h + r)$

13. a. $(x + 3)$ **b.** $(x + 3)(4y + 9)$

15. a. $x = 0$ or $x = {}^-4$

b. $x = 0$ or $x = 11$

6.2 Skills and Review

17. $\frac{x^8}{9}$

19. .5 L

21. The width of the enlargement is 9 inches and the length is 15 inches.

23. a. $m = \frac{^-5}{6}$

b. x-intercept: $(14, 0)$, y-intercept: $\left(0, \frac{35}{3}\right) = \left(0, 11\frac{2}{3}\right)$

25. 13

6.3 Exercises

1. a. $(x + 6)(3x^2 + 1) = 3x^3 + 18x^2 + x + 6$

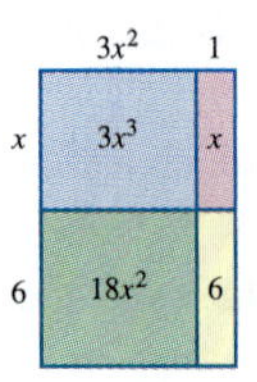

b. $3x^3 + 18x^2 + x + 6$

c. Answers may vary. Let $x = 3$, then $(x + 6)(3x^2 + 1) = 252$ and $3x^3 + 18x^2 + x + 6 = 252$.

3. $4x^2 + 19x - 63$

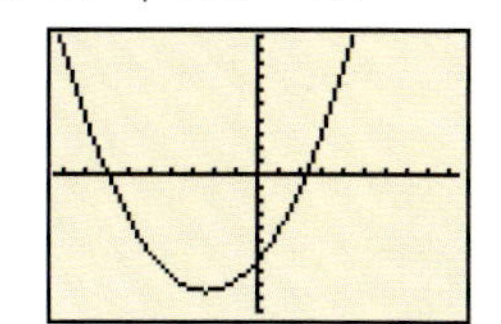

The graph of Y1 is the same as the graph of Y2, so the expansion is correct.

5. $x^2 + 8xy + 16y^2$

7. Expansion is the reverse of factoring.

9. $2y^2 + 7x + 2xy + 7y = (y + x)(2y + 7)$

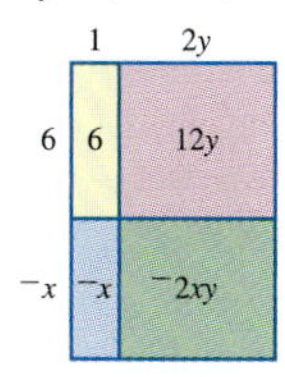

11. a

13. The expression can be factored because the diagonal test for grouping works,

$$6x^3 * 5 = {}^{-}2x^2 * {}^{-}15x$$

$$6x^3 - 2x^2 - 15x + 5 = (2x^2 - 5)(3x - 1)$$

15. a. $\frac{1}{2}ab$

b. $2ab$

c. $2ab + c^2$

d. $a^2 + 2ab + b^2$

e. $2ab + c^2 = a^2 + 2ab + b^2$

f. Subtract $2ab$ from both sides.

6.3 Skills and Review

17. $6x^3 - 12x^2 + 66x$

19. Randy should invest \$3125 in the 8.2% annuity and \$1875 in the 5% CD.

21. 257,600

23. $x = \frac{13}{8} = 1\frac{5}{8} = 1.625$

25. $\frac{15}{4} = 3\frac{3}{4} = 3.75$

6.4 Exercises

1. a. $x^2 + x - 6$; quadratic trinomial

b. $x^3 - x + 6$

c. $x^2 + 3$

d. $4x^2 - x + 5$; quadratic trinomial

3.

	$2x$	1
x	$2x^2$	$1x$
3	$6x$	3

$2x^2 + 7x + 3 = (2x + 1)(x + 3)$

5. a. $(x^2)(6) = ({}^{-}3x)({}^{-}2x) = 6x^2$

b.

	x	$^{-}3$
x	x^2	$^{-}3x$
$^{-}2$	$^{-}2x$	6

$x^2 - 5x + 6 = (x - 3)(x - 2)$

c. George made his mistake in the grouping of the second step: $^{-}2x + 6 \neq {}^{-}(2x + 6)$. Use the rule for subtraction to write the four terms as a sum.

$$\begin{aligned} x^2 - 5x + 6 &= x^2 - 3x - 2x + 6 \\ &= x^2 + {}^{-}3x + {}^{-}2x + 6 \\ &= (x^2 + {}^{-}3x) + ({}^{-}2x + 6) \\ &= x(x + {}^{-}3) + {}^{-}2(x + {}^{-}3) \\ &= (x + {}^{-}3)(x + {}^{-}2) \\ &= (x - 3)(x - 2) \end{aligned}$$

7. $(3x + 2)(2x - 5)$

9. Prime: No factor pairs of $40x^2$ add to $^{-}3x$:

$$\begin{aligned} {}^{-}1x + {}^{-}40x &= {}^{-}41x \quad \textit{No} \\ {}^{-}2x + {}^{-}20x &= {}^{-}22x \quad \textit{No} \\ {}^{-}4x + {}^{-}10x &= {}^{-}14x \quad \textit{No} \\ {}^{-}5x + {}^{-}8x &= {}^{-}13x \quad \textit{No} \end{aligned}$$

11. Diagonal product $= (2x^6)({}^{-}28) = {}^{-}56x^6$

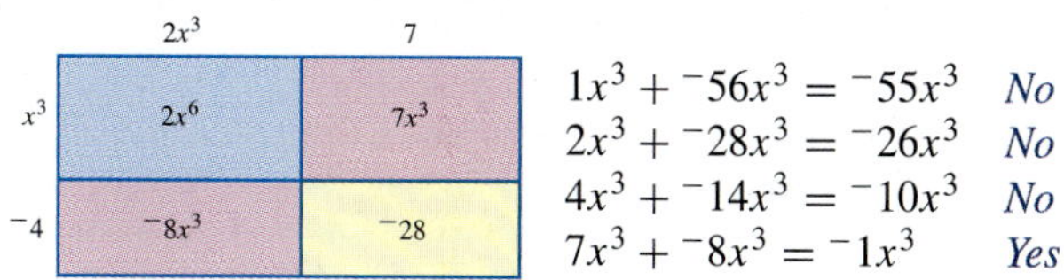

$1x^3 + {}^{-}56x^3 = {}^{-}55x^3$ *No*
$2x^3 + {}^{-}28x^3 = {}^{-}26x^3$ *No*
$4x^3 + {}^{-}14x^3 = {}^{-}10x^3$ *No*
$7x^3 + {}^{-}8x^3 = {}^{-}1x^3$ *Yes*

$2x^6 - x^3 - 28 = (2x^3 + 7)(x^3 - 4)$

13. a. $x = 4$ or $x = 8$

b. $x = {}^{-}2$ or $x = 5$

c. $x = {}^{-}3$ or $x = -\frac{1}{2}$

15. **a.** $x^2 + x - 2 = (x+2)(x-1)$

$$(^-2)^2 + {}^-2 - 2 \stackrel{?}{=} (^-2+2)(^-2-1)$$
$$0 = 0 \quad \textit{OK}$$

$$1^2 + 1 - 2 \stackrel{?}{=} (1+2)(1-1)$$
$$0 = 0 \quad \textit{OK}$$

b. 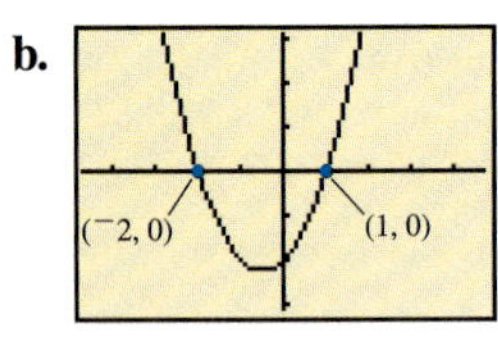

x-axis

c. $x = 6$ or $x = {}^-3$

d. It's easy to recognize the values of x that give an output of 0 in a factored polynomial.

6.4 Skills and Review

17. 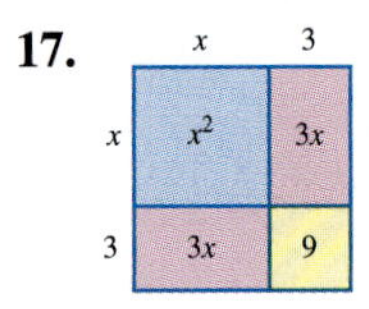

$(x+3)^2 = x^2 + 6x + 9$

19. $x = 0$ or $x = 3$

21. 5:04 P.M.

23. Method 1: Place both equations in *slope-intercept* form and compare slopes.

$$y = {}^-2x + 12$$
$$y = {}^-2x + 10$$

The equations have the same slope but have different y-intercepts, so the lines are parallel and the system has no solution.

Method 2: Solve algebraically. Substitute the second equation into the first equation.

$$x + \tfrac{1}{2}(^-2x + 10) = 6$$
$$x - x + 5 = 6$$
$$5 = 6 \quad \textit{But } 5 \neq 6\textit{, so there is no solution to this system of equations.}$$

25. **a.** $x = \frac{22}{3}$ **b.** $x \approx 7.333$

6.5 Exercises

1. **a.** $(y+5)(y^2 - 5y + 25)$
b. $(3 - 2x)(9 + 6x + 4x^2)$
c. $(x-6)(x^2 + 6x + 36)$
d. $(10z + 1)(100z^2 - 10z + 1)$
e. $(x^2 - y)(x^4 + x^2y + y^2)$

3. **a.** Prime
b. $(x+9)(x^2 - 9x + 81)$
c. Prime
d. $(x-3)(x^2+3x+9)(x+3)(x^2-3x+9)$
e. $(x^2+8)(x^4 - 8x^2 + 64)$

Chapter 6 Exploration

a. Step 1 $x = 64$
Step 2 $y = 18$
Step 3 $82 * 46 = x^2 - y^2 = 4096 - 324 = 3772$

b. By Step 1, $x = \frac{m+n}{2}$
By Step 2, $y = \frac{m-n}{2}$
Then

$$x + y = \frac{m+n}{2} + \frac{m-n}{2} = m$$

and

$$x - y = \frac{m+n}{2} - \frac{m-n}{2} = n.$$

Therefore, $m * n = (x+y)(x-y) = x^2 - y^2$. This is Step 3 of the algorithm.

c. When both numbers are odd or both are even, the sum is divisible by 2 and the difference is divisible by 2. This is necessary because the table has squares only of whole numbers.

d. We may write 77 as the sum of 76 and 1, then multiply by 44. This gives a product that can be found with the algorithm multiplying using squares on page 209.

$$77 * 44 = (76 + 1) * 44$$
$$= 76 * 44 + 44 \quad \textit{Using the distributive property}$$

Chapter 6 Review Exercises

1. **a.** x^9 **b.** y^8 **c.** x^5y^5

2. **a.** $64x^6y^{18}$ **b.** $28y^6$ **c.** z^8

3. volume $= 9.261 * 10^{12}$ inches^3

4. a. $\frac{1}{x^3}$ **b.** x^3

c. $\frac{1}{x^9}$ **d.** $\frac{x^{12}}{y^6}$

5. $1.13 * 10^{-3}$ feet $= .0013$ feet

6. a. x^4 **b.** $\frac{1}{x^4}$ **c.** $\frac{x^6}{y^6}$

7. $a \Leftrightarrow 3$ Quotient law

$b \Leftrightarrow 5$ Power-of-a-power law

$c \Leftrightarrow 2$ Quotient law and definition of negative integer exponents

$d \Leftrightarrow 1$ Product law

$e \Leftrightarrow 4$ Negative exponent in the denominator

8. a. $\frac{x^3}{y^3}$ **b.** $\frac{64x^6}{y^3}$

c. $\frac{y^4}{x^{16}}$ **d.** $\frac{27y^6}{8x^{12}}$

9.

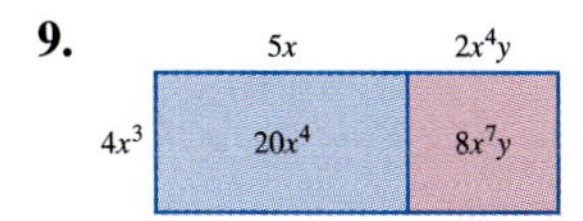

$4x^3(5x + 2x^4y) = 20x^4 + 8x^7y$

10. a. $30x^2 - 5x$ **b.** $8x^2z + 14xyz - 6xz$

c. $^{-}x^2 + 8x - 2$

11. a. xy **b.** $3x^2$ **c.** 15

12.

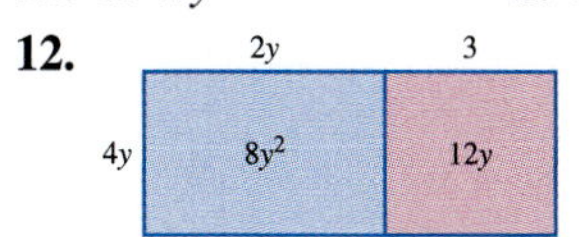

$8y^2 + 12y = 4y(2y + 3)$

13. $5x^2y(z^3 - 3xy^3 + 2x^3y)$

14. a. $^{-}2x(4x - 5)$ **b.** $^{-}3(3x + 4y)$

15. a. $z(4x - 3)$

b. $(y + 7)(4x - 3)$

c. (a) and (b) are basically the same. Let z replace $(y + 7)$ in part (b).

16. a. $x = 0$ or $x = {}^{-}5$ **b.** $x = 0$ or $x = 8$

17. a.

	x	2
x	x^2	$2x$
4	$4x$	8

$(x + 4)(x + 2) = x^2 + 6x + 8$

b.

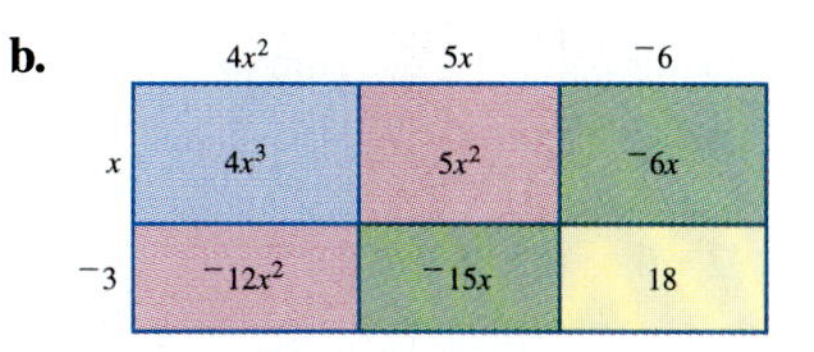

$(x - 3)(4x^2 + 5x - 6) = 4x^3 - 7x^2 - 21x + 18$

18. a.

	x	$^{-}9$
y	xy	$^{-}9y$
2	$2x$	$^{-}18$

$2x - 9y + xy - 18 = (y + 2)(x - 9)$

b.

	$3z$	1
$^{-}4y$	$^{-}12yz$	$^{-}4y$
1	$3z$	1

$^{-}4y + 3z - 12yz + 1 = ({}^{-}4y + 1)(3z + 1)$

19. a. Factorable: $(4x^3)({}^{-}5) = (10x^2)({}^{-}2x) = {}^{-}20x^3$; $(2x + 5)(2x^2 - 1)$

b. Not factorable because we can't pair terms to make the product of one pair equal the product of the other pair.

c. Factorable: $(3x)(8y) = (6xy)(4) = 24xy$; $(2y + 1)(3x + 4)$

20. a. $x^2 + 2xy + y^2$ **b.** $x^2 + 4xy + 4y^2$

c. $25x^2 - 30xy + 9y^2$

21. a. $24x^2$

b. $1x$ and $24x$, $2x$ and $12x$, $3x$ and $8x$, $4x$ and $6x$

c. $3x + 8x = 11x$

d. $x^2 + 3x + 8x + 24$

e.

	x	8
x	x^2	$8x$
3	$3x$	24

$(x + 3)(x + 8)$

22. a. $x^2 + 2x - 7x - 14 = (x + 2)(x - 7)$

b. $5x^2 - 1x + 10x - 2 = (5x - 1)(x + 2)$

c. $3x^2 + 1x - 12x - 4 = (3x + 1)(x - 4)$

d. $7x^2 - 2x + 28x - 8 = (7x - 2)(x + 4)$

23. a. Prime, no factor pairs of $5x^2$ sum to $^{-}2x$.

$^{-}1 * {}^{-}5$ $\quad$ $^{-}1x + {}^{-}5x = {}^{-}6x$ $\quad$ *No*

b. Prime, no factor pairs of $12x^2$ sum to $1x$.

$1 * 12$	$1x + 12x = 13x$	*No*
$2 * 6$	$2x + 6x = 8x$	*No*
$3 * 4$	$3x + 4x = 7x$	*No*

24. a. $(x + 3y)(x + 7y)$ **b.** $(2x^2 + 5)(x^2 - 3)$

25. a. $(x - 5)(x + 5)$ **b.** $(3 - y)(3 + y)$

c. $(2x - 9y)(2x + 9y)$

26. a. $2(9x^2 - 18x + 8) = 2(9x^2 - 6x - 12x + 8)$

$2(3x(3x - 2) - 4(3x - 2)) = 2(3x - 2)(3x - 4)$

b. $y(xy - x + 2y - 2) = y(x(y - 1) + 2(y - 1))$

$= y(y - 1)(x + 2)$

c. $3y(x^2 - 100) = 3y(x - 10)(x + 10)$

27. a. $x = {}^-5$ or $x = {}^-8$ **b.** $x = {}^-3$ or $x = 1$

c. $x = \frac{1}{3}$ or $x = 6$

28. a. $(x + 4)(x^2 - 4x + 16)$

b. $(1 - y)(1 + y + y^2)$

c. $(2x + 10y)(4x^2 - 20xy + 100y^2)$

d. $(y + 3z^2)(y^2 - 3yz^2 + 9z^4)$

29. a. $6(1 - 2z)(1 + 2z + 4z^2)$

b. $2(5x + y)(25x^2 - 5xy + y^2)$

c. $(2 - y)(4 + 2y + y^2)(2 + y)(4 - 2y + y^2)$

Chapter 6 Test

1. $\frac{x^2y^6}{z^8}$

2. $\frac{3x^3}{10y^3}$

3. $\frac{1}{27x^9}$

4. $1.728 * 10^{12}$ signals

5. a. $3a^2 + 12a$

b. $6x^3y^2 + 12x^2y^3 - 30xy^4$

6. ${}^-3bc^2(2abc + 1 + 4a^3b^3c)$

7. $x = 0$ or $x = 5$

8. a. $2x^2 - 9x - 35$

b. $3x^3 - 13x^2 + 10x - 24$

9. a. $(1 + 2y)(2x + 3)$ **b.** $(3x - 1)(2x^2 - 3)$

10. $4x^2 - 28xy + 49y^2$

11. a. $(4x + 1)(x - 6)$ **b.** $(x - 5y)(x + 7y)$

12. $(8x - y)(8x + y)$

13. a. $2y(x - 5y)(x + 8y)$ **b.** $3(2 - x)(2 + x)$

14. $x = \frac{1}{3}$ or $x = \frac{5}{2}$

Chapter 7 Solving Quadratic Equations

7.1 Exercises

1. a. See Figure 2. Note that some data is rounded.

b. No

c. No, linear models have constant slope. The graph of these data has changing slope.

3. 8 inches

5. a. $x = \pm 3$

b. $s = \pm 5$

c. $s = \pm\sqrt{26}$

$s \approx \pm 5.1$. (a) and (b) are exact, but (c) is approximate; 9 and 25 are perfect squares, but 26 is not.

7. $d = \pm\sqrt{\frac{1000}{3}}$; exact

$d \approx \pm 18.26$; approximate, $\frac{1000}{3}$ is not a perfect square.

9. $x \approx \pm 4.08$

11.

$2x - 3 = 0$	$2x^2 - 3 = 0$
$2x = 3$	$2x^2 = 3$
$x = \frac{3}{2}$	$x^2 = \frac{3}{2}$
	$x = \pm\sqrt{\frac{3}{2}}$

Solving quadratic equations of the form $ax^2 + c = 0$ uses the same process as solving linear equations, with the extra step of taking a square root on both sides of the equation.

13. a. $x = {}^-1$ or $x = {}^-3$ **b.** $x = 1$ or $x = 2$

c. $y = 2$ or $y = {}^-7$ **d.** $x = 2$ or $x = 9$

e. $x = {}^-2$ or $x = {}^-3$

15. a. $x = \pm\frac{7}{2}$

b. $x = \frac{7}{2}$ or $x = \frac{{}^-7}{2}$

c. The answers are the same.

7.1 Skills and Review

17. $x = 8$ or $x = {}^-8$

19. $4x^2 - 12x + 9$

21. a. $x + y = 212$

b. $x - y = 83$

c. Graphical solutions are approximate: 147.5 and 64.5.

23. $z = \frac{15}{8} = 1.875$

$\frac{9}{8} = \frac{9}{8}$; *OK, the answer checks*

25. 4

7.2 Exercises

1. y-intercept, x-intercepts (if any), and the vertex.

3. a. y-intercept **b.** y-intercept

c. y-intercept **d.** $x = 0$

5. a. $x \approx .7$ or $x \approx 4.3$ **b.** $(0, 3)$

c. $(2.5, {}^-3.25)$ **d.** Minimum

7. a. 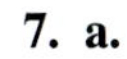

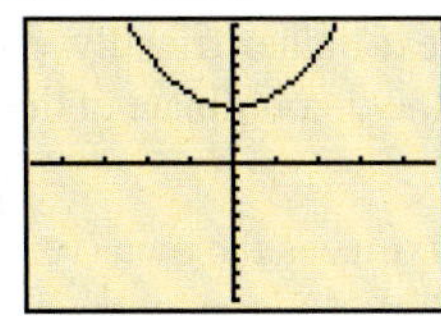

No x-intercept

b. 

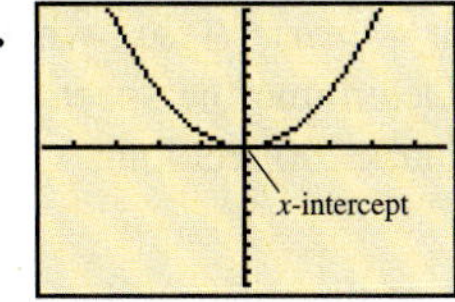

One x-intercept

c.

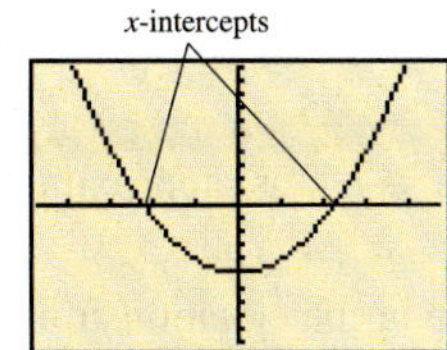

Two x-intercepts

9. a. ${}^-16(x^2 - 2x - 3)$

${}^-16(x - 3)(x + 1)$

b. $x = 3$ or $x = {}^-1$

c. Windows may vary: Xmin $= {}^-4.7$, Xmax $= 4.7$

d. Ymin $= {}^-3.1$, Ymax $= 70$

11. a. Between 5 and 6 seconds

b. Between 2 and 3 seconds

13. a. No

b. The water will reach an approximate height of 104 feet.

15. The football will remain airborne approximately 4.6 seconds.

7.2 Skills and Review

17. a. $y = {}^-20$

b. $x = \frac{3}{2}$ or $x = {}^-5$

19. $x = 4$ or $x = {}^-7$

21. $4y^2 - 9$

23. $\dfrac{y^2}{4x^2}$

25. a. Area $= 93.5$ inches^2

b. Perimeter $= 39$ inches

7.3 Exercises

1. a. Greatest common factor (GCF) and zero-product property (ZPP); the expression on the left side of the equation factors so that the ZPP can be used.

b. Factor and ZPP; the expression on the left side of the equation factors into the product of two binomials allowing the use of ZPP.

c. Graphing; find approximate solutions with the graphing calculator.

d. Square root principle; the equation is of the form $ax^2 + c = 0$.

e. Quadratic formula; the expression on left side of the equation does not factor with integer coefficients and we need exact solutions.

3. a. $a = 2, b = 12, c = {}^-5$

b. $x = {}^-3 \pm \frac{\sqrt{184}}{4}$ *This square root simplifies (we learn more about simplifying radicals in Chapter 9).*

$x = {}^-3 + \frac{\sqrt{184}}{4}$ or $x = {}^-3 - \frac{\sqrt{184}}{4}$

$x \approx .39$ or $x \approx {}^-6.39$

5. $x = {}^-1$

7. a. $x = {}^-\frac{b}{2a}$ **b.** $x = {}^-\frac{3}{2} = {}^-1.5$

c. $y = {}^-\frac{29}{2} = {}^-14.5$ **d.** x-intercepts

9. a., b.

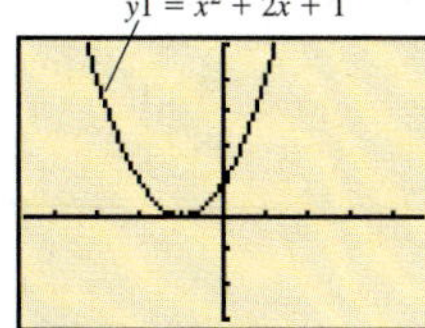

$b^2 - 4ac = 0$

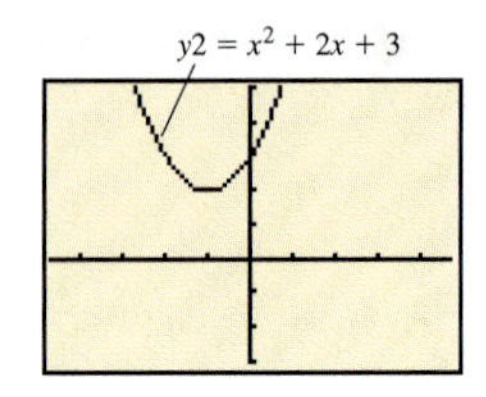

$b^2 - 4ac = {}^-8$

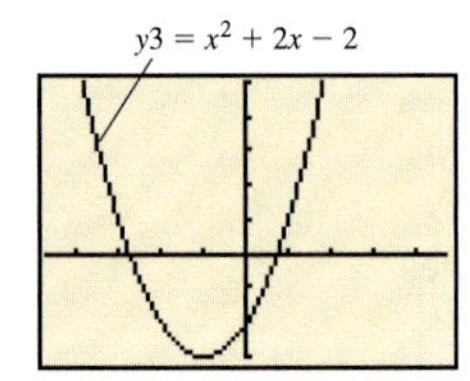

$b^2 - 4ac = 12$

c. The quadratic equation $0 = ax^2 + bx + c$ has

1. No real root when the discriminant is negative;
2. One real root when the discriminant is zero;
3. Two real roots when the discriminant is positive.

11. a.

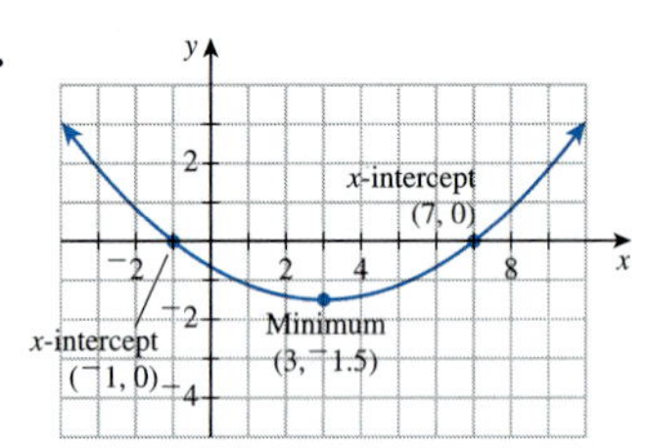

b. (7, 0)

13. No solution

15. a. $x = \frac{1}{5}$ or $x = {}^-1$

b. $x = \frac{1}{5}$ or $x = {}^-1$

c. $x = .2 = \frac{1}{5}$ or $x = {}^-1$

7.3 Skills and Review

17. $x \approx .22$ or $x \approx 3.33$

19. a. Cheese $= 1.3d^2$

b.

Diameter (inches)	Cheese (grams)
6	47
10	130
14	255

c.

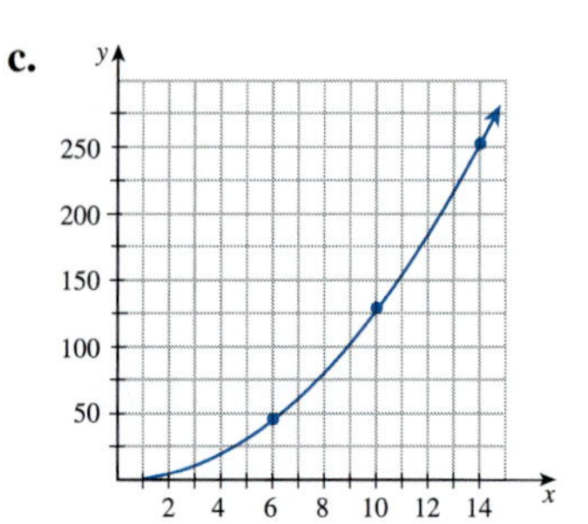

21. $\frac{4y^2}{x^{10}}$

23. $b_1 = \frac{7}{2}$ cm $= 3.5$ cm

25. a. $\frac{73}{100}$

b. 73%

7.4 Exercises

1. In a real-life maximum problem, such as Example 1, Xmax can be found by doubling the x-coordinate of the vertex and choosing the next-higher friendly value. Set Ymax slightly larger than the y-coordinate of the vertex.

3. x-intercepts

5. We want the width that maximizes the area (vertex), not the width that makes zero area (x-intercepts).

7. Step 1 The stopping distance of a car is given by the quadratic equation $y = ax^2 + bx$, where $a = .04$ and $b = .2$. Determine the speed of a car that leaves a skid mark 200 feet long.

Step 2 See Figure 22.

Step 3 Let x = the speed of the car in miles/hour and y = the stopping distance in feet.

Step 4 $200 = .04x^2 + .2x$

Step 5 $x \approx 68.25$ or $x \approx {}^-73.25$

Step 6 The speed of the car was approximately 68 miles/hour.

Step 7 Substitute 68 for x in the equation from step 4. The answer checks.

9. The speed of the car was approximately 58.79 miles/hour.

11. a. The spaceship is 1,056,000 feet from the sun.

b. The probe would hit the sun in approximately 48.6 seconds.

c. The probe would melt in approximately 34.3 seconds.

13. a.

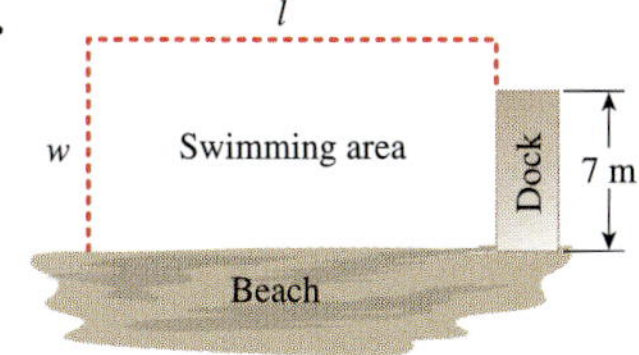

b. $l + w + w - 7 = 30$

c. $l = 37 - 2w$

d. $A = lw$, substituting the expression for l gives $A = {}^{-}2w^2 + 37w$.

e. The area is maximized with a width of 9.25 m and a length of 18.5 m.

f. The maximum swimming area is 171.125 m^2.

15. a. Prices of \$0 or approximately \$27.59 generate zero revenue.

b. Revenue is maximized at a price of approximately \$13.79.

c. Prices of approximately \$9.57 or \$18.02 generate \$500 of revenue.

7.4 Skills and Review

17. $b^2 - 4ac = {}^{-}79$; ${}^{-}79 < 0$, no x-intercept.

19. $x \approx {}^{-}12.43$ or $x \approx 6.43$

21. $\dfrac{y^3}{x^9}$

23. $x = {}^{-}3$

25.

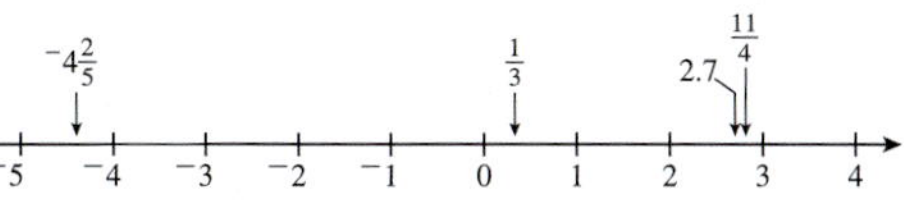

7.5 Exercises

1. a. $(x - 6)(x^2 - 9) = 0$

b. $(x - 6)(x - 3)(x + 3) = 0$

c. $x = 6$ or $x = 3$ or $x = {}^{-}3$

3. The calculator displays 6.

Chapter 7 Exploration

a.

X	Y1
20	100
15	300
10	500
5	700

Y1 = -40X+900

b. $y = {}^{-}40x^2 + 900x$

c. $x = 11.25$; a price of \$11.25 maximizes revenue.

d. The charity will make \$3562.50 profit.

Chapter 7 Review Exercises

1. The slope of a linear equation remains constant. The slope of a quadratic equation changes.

2. a. Linear **b.** Quadratic **c.** Other

3. $x \approx {}^{-}.83$ or $x = 1.2$

4. $x = \pm\sqrt{8}$

$x \approx \pm 2.83$

5. $x = \pm\sqrt{\frac{3}{7}}$

$x \approx \pm .65$

6. $x = 0$ or $x = {}^{-}\frac{4}{3}$

7. $x = 5$ or $x = {}^{-}\frac{3}{2}$

8. $x = 0$ or $x = {}^{-}2$

9. a. The pizza is approximately 13.1 inches.

b. The pizza is exactly $\sqrt{\frac{180}{1.05}} = \sqrt{\frac{1200}{7}}$ inches.

10. a. A parabola given by the equation $y = ax^2 + bx + c$ opens upward when $a > 0$ and opens downward when $a < 0$.

b. The turning point is the vertex. The x-coordinate is $\frac{{}^{-}b}{2a}$ and the y-coordinate is found by evaluating the quadratic when $x = \frac{{}^{-}b}{2a}$.

c. If the discriminant, $b^2 - 4ac$, is

1. Positive, there are two x-intercepts;

2. Zero, there is one x-intercept;

3. Negative, there is no x-intercept.

d. The y-intercept is the point where the graph intersects the y-axis. This point is $(0, c)$.

11. y-intercept: $(0, {}^{-}4)$

x-intercepts: $(1, 0)$, $({}^{-}1.33, 0)$

Vertex: minimum at $({}^{-}.17, {}^{-}4.08)$

12. y-intercept: $(0, 0)$

x-intercept: $(0, 0)$

Vertex: maximum at $(0, 0)$

13. y-intercept: $(0, 15)$

x-intercepts: none

Vertex: minimum at $({}^{-}2.5, 2.5)$

14. a.

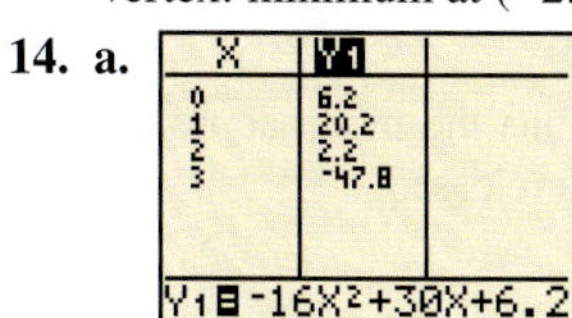

X	Y1
0	6.2
1	20.2
2	2.2
3	-47.8

Y1 = -16X²+30X+6.2

b. Windows may vary: Xmin = 0, Xmax = 4.7, Ymin = 0, Ymax = 25

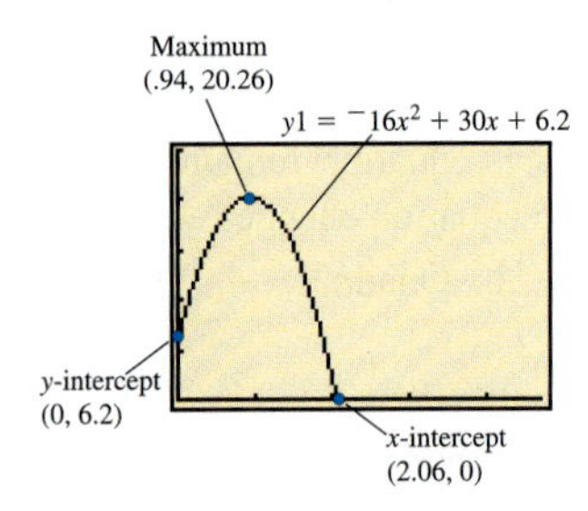

c. The ball hits the ground in approximately 2.06 seconds.

d. Approximate maximum at (.94, 20.26)

e. (0, 6.2); this is the height of the ball at release.

15. The larger x-intercept is used to set Xmax and the y-coordinate of the vertex is used to set Ymax. Xmin and Ymin are usually set at zero.

16. a. The x-intercepts are the points where the graph intersects the x-axis. These occur when $y = 0$.

b. $(1, 0), \left(^-\frac{4}{3}, 0\right)$

17. $x = {}^-6 \pm \sqrt{38}$

18. $x = {}^-\frac{1}{6} \pm \frac{\sqrt{97}}{6}$

19. $x = \frac{1}{6}$ or $x = \frac{^-1}{2}$

20. a. (3, 14)

b. x-intercepts: $\left(3 + \frac{\sqrt{56}}{2}, 0\right), \left(3 - \frac{\sqrt{56}}{2}, 0\right)$

21. a.

150 cm
s
s

b. Let s = the length of a side of the square in centimeters; $s^2 + s^2 = 150^2$.

c. $s = \pm\sqrt{11250}$

d. Each side of the square is $\sqrt{11250} \approx 106.1$ cm.

22. a. The width is approximately 1.67 m and the length is 6 m.

b. The width is $\frac{5}{3} = 1\frac{2}{3}$ m and the length is 6 m.

23. a. Vertex: maximum $\approx (6.1, 34.49)$

b. Approximately 6.1 seconds after release, the ball is 34.49 m above the surface of the moon.

c. x-intercept $\approx (12.58, 0)$

d. The astronaut's throw hits the moon approximately 12.58 seconds after it is released.

24. The ball is 18 m high at exactly

$$\frac{10}{1.64} + \frac{\sqrt{54.08}}{1.64} \text{ seconds}$$

or

$$\frac{10}{1.64} - \frac{\sqrt{54.08}}{1.64} \text{ seconds}$$

after release.

25. a.

72 in.
w
w + 16

b. Let w = the width of the table top in inches; then $w + 16$ is the length. Equation: $w^2 + (w + 16)^2 = 72^2$

c. $w = {}^-8 + \frac{\sqrt{10112}}{2} \approx 42.3$

d. The table top is approximately 42.3 inches wide by 58.3 inches long.

26. The area of the space is maximized with a width of 51 feet and a length of 102 feet.

27. The vehicle was traveling exactly

$$\frac{^-.22}{.09} + \frac{\sqrt{31.5484}}{.09}$$

miles/hour or approximately 60 miles/hour.

28. a. $1 + 4i$

b. $9 - 4i$

c. $^-41 + 23i$

d. 34

29. a. $x = \pm 2i$

b. $x = 3 \pm 4i$

c. $x = {}^-\frac{5}{2} + \frac{\sqrt{11}}{2}i$

d. $x = {}^-3 \pm \frac{\sqrt{72}}{2}i$

Chapter 7 Test

1. $x = \pm 2$

2. $x = 0$ or $x = {}^-3$

3. $x = 2$ or $x = 6$

4. $x = \frac{1}{4}$ or $x = {}^-\frac{5}{2}$

5. $x = 1$ or $x = {}^-5$

6. $x \approx .98$ or $x \approx {}^-8.18$

7. a. The parabola opens downward so the vertex is a maximum.

b. $\left(^-\frac{1}{6}, \frac{109}{12}\right)$

c. $b^2 - 4ac = 109$; $109 > 0$, two x-intercepts

d. $(1.57, 0)$, $(^-1.91, 0)$

e. $(0, 9)$

8. a. y-intercept: $(0, 400)$; x-intercepts: $(10, 0)$, $(40, 0)$; the vertex might vary $\approx (25, ^-225)$

b. Windows may vary: Xmin $= 0$, Xmax $= 47$, Ymin $= ^-300$, Ymax $= 500$

9. a. The water attains a maximum height of approximately 12.98 m.

b. 1.5 m

c. The water hits the ground at exactly

$$\frac{15}{9.8} + \frac{\sqrt{254.4}}{9.8} \text{ seconds}$$

or approximately 3.16 seconds after it is shot.

10. The TV has a height of 3 feet and a width of 4 feet.

11. a. A price of \$13.46 (rounded to the nearest penny) generates zero revenue.

b. A price of \$6.73 (rounded to the nearest penny) generates maximum revenue.

12. The driver was not speeding. The driver's approximate speed was 59.3 miles/hour.

Chapter 8 What Is a Function

8.1 Exercises

1. a.

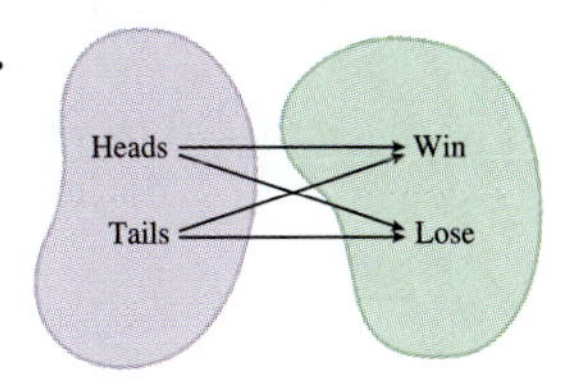

b. A function requires that each input be associated with one and only one output. This is not a function; each input is associated with two outputs, win and lose.

3. a. Function; domain: {lab, husky, poodle}; range: {dog}

b. Not a function

5. a. 0 seconds and 2.5 seconds

b. Domain: $\{x \mid 0 \le x \le 2.5\}$

Domain
0 2.5

c. 0 feet and 100 feet

d. Range: $\{x \mid 0 \le y \le 100\}$

Range
0 100

e. In a real-life problem, the limits of the domain are used to set Xmin and Xmax and the limits of the range are used to set Ymin and Ymax.

f. Quadrant I; the domain and range are both nonnegative.

7. $f(0) = 0$

$f(2) = 12$

$f(3) = 27$

9. a. $f(2) = ^-3$ **b.** $g(^-4) = 28$

c. $h(3) = ^-9.1$

11. $f(^-2.15) = ^-16.641$

13. a. Domain: $\{x \mid ^-\infty < x < \infty\}$; range: $\{y \mid y \le 2\}$

b. Domain: $\{x \mid ^-\infty < x < \infty\}$; range: $\{y \mid y \ge ^-2\}$

c. Domain: $\{x \mid ^-\infty < x < \infty\}$; range: $\{y \mid y = 5\}$

15. a. Range: $\{y \mid y \ge 0\}$; the outputs will be nonnegative so the graph will appear only in Quadrants I and II.

b. The absolute value of any real number results in an output greater than or equal to zero.

8.1 Skills and Review

17. The width is 45 feet and the length is 90 feet.

19. a. $x = ^-4$ **b.** $x = ^-4$ **c.** $x = ^-4$

21. $x = \pm\sqrt{35}$

$x \approx 5.92$

23. a. x^2 **b.** $\frac{1}{x^2}$

25. $k = \frac{3}{2}$

8.2 Exercises

1. Not a function

3. Function

5. Not a function

7. a.

x	y
$^{-}4$	$^{-}2$
$^{-}1$	$^{-}1$
0	0
$^{-}1$	1
$^{-}4$	2

b.

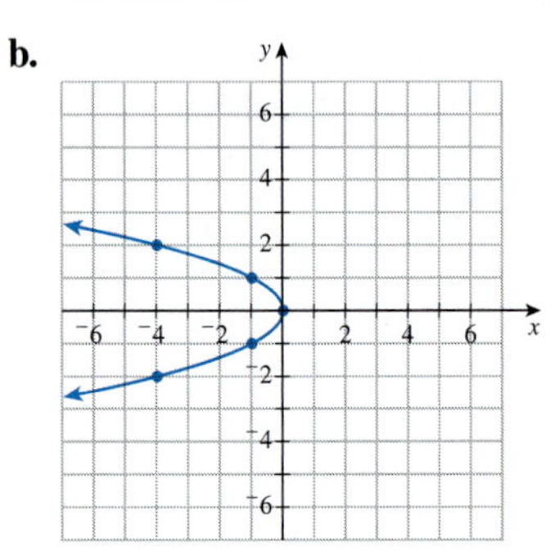

c. Not a function; the graph fails the vertical-line test.

9. a. Yes; there is one and only one output associated with each input.

b. (1996, 52), (1997, 53), (1998, 56), (1999, 56), (2000, 56), (2001, 55)

c. Domain: {1996, 1997, 1998, 1999, 2000, 2001}; range: {52, 53, 55, 56}

11. a. 3 **b.** 1 **c.** 2

13.

Domain

4

Domain: $\{x|4 \leq x < \infty\}$

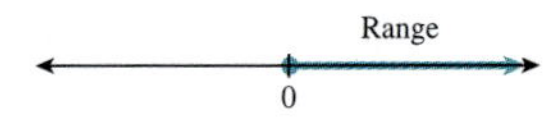

Range: $\{y|0 \leq y < \infty\}$

15.

Domain

$^{-}6$

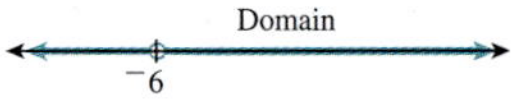

Domain: $\{x|x \neq {}^{-}6\}$

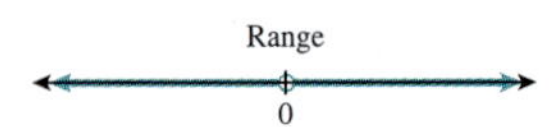

Range: $\{y|y \neq 0\}$

8.2 Skills and Review

17. Domain: {31, 32, 33, 34, 35}; range: {Packers, Broncos, Rams, Ravens}

19. Step 1 Find the price(s) of a product that generate zero revenue.

Step 2 A graph of the revenue function, $R = {}^{-}5p^2 + 90p$

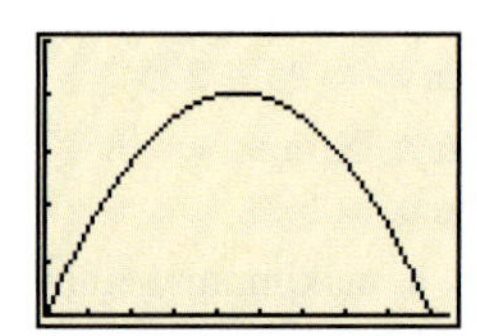

Step 3 Let p = price of the product in dollars and R = revenue in dollars.

Step 4 $0 = {}^{-}5p^2 + 90p$

Step 5 $p = 0$ or $p = 18$

Step 6 Prices of \$0 or \$18 generate zero revenue.

Step 7 Substitute 0 and then 18 into the equation from step 4. The answers check.

21. a. $x = 2$ **b.** $x = 3$ or $x = {}^{-}4$

23. $\dfrac{25x^8}{y^2}$

25. $^{-}23a^2 + 48a - 16$

8.3 Exercises

1.

Basic Function	Function	Graph
Linear	$f(x) = x$	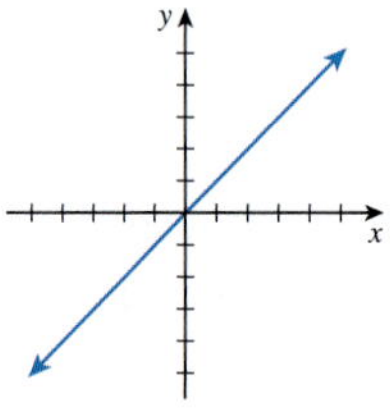
Quadratic	$f(x) = x^2$	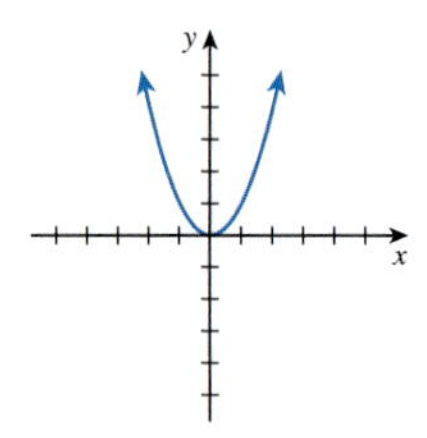
Cubic	$f(x) = x^3$	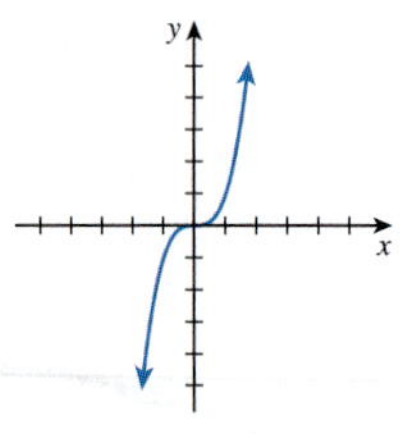

1. *(continued)*

Basic Function	Function	Graph
Square root	$f(x) = \sqrt{x}$	
Absolute value	$f(x) = \lvert x \rvert$	
Reciprocal	$f(x) = \frac{1}{x}$	

3. a. The square of any number is nonnegative.

b. Vertex = $(0, 0)$

c. The graph is symmetric about the y-axis.

5. a. We can't take the square root of a negative number in the real number system.

b. The domain excludes negative values: domain: $\{x \mid 0 \leq x\}$.

7. Both graphs are symmetric about the y-axis.

9. The graph has two branches with the same shape, one in the first quadrant and the other in the third.

11. a. $(3.01, {}^{-}4.21)$

b. $({}^{-}.34, 3.32)$

13. a. $91.25

b. $c(g(x)) = 50 + .0825x$

15. a. $f(g(3)) = 0$

b. $f(g(x)) = x^2 - 4x + 3$

c. $g(f({}^{-}1)) = 2$

d. $g(f(x)) = x^2 - 2x - 1$

8.3 Skills and Review

17. Domain: $\{x \mid {}^{-}\infty < x < \infty\}$; range: $\{y \mid {}^{-}3 \leq y\}$

19. $x = 0$ or $x = 6$

21. a. $a > 0$, opens upward

b. y-intercept: $(0, {}^{-}28)$

c. x-intercepts: $({}^{-}4, 0), (7, 0)$

d. Vertex: minimum $\left(\frac{3}{2}, \frac{{}^{-}121}{4}\right)$

e.

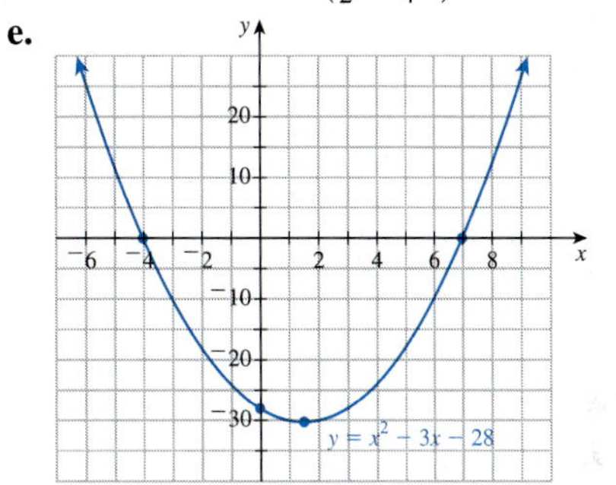

23. a. $3x^3 + 7x^2 - 5x$

b. $2x^2 + 13x - 24$

25. $x = \frac{{}^{-}5}{2}$

8.4 Exercises

1. Stretching a function stretches the graph away from the x-axis. Compressing a function compresses the graph toward the x-axis.

3. a. $c > 1$

b. $0 < c < 1$

c. $c = 1$

5. The original function is reflected (flipped) across the x-axis.

7. The graph of $f(x) = 0$ is a line on the x-axis. Reflecting $f(x)$ across the x-axis does not change the graph.

9. The table shows that values of Y2 correspond to values of Y1 one row earlier. Typical of this movement is the x-intercept. The table shows that the x-intercept occurs when $x = 0$ for Y1 and when $x = 1$ for Y2.

11. a. 2 **b.** 3 **c.** 4 **d.** 1

13. a. $g(x) = \lvert x - 2 \rvert$

b. $g(x) = \sqrt{x + 2}$

c. $g(x) = x^3 - 1$

d. $g(x) = (x + 1)^2 + 1$

15. Vertex: $(9.5, 52)$

8.4 Skills and Review

17.

Basic Function	Function	Graph	Domain and Range
Linear	$f(x) = x$	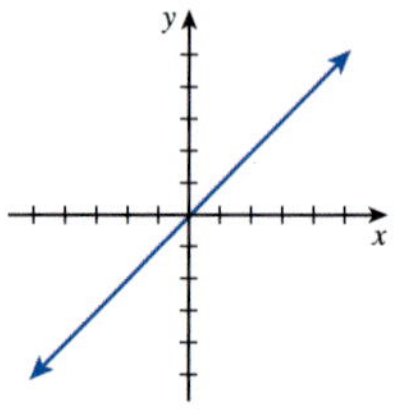	Domain: $\{x \mid {}^{-}\infty < x < \infty\}$ Range: $\{y \mid {}^{-}\infty < y < \infty\}$
Quadratic	$f(x) = x^2$	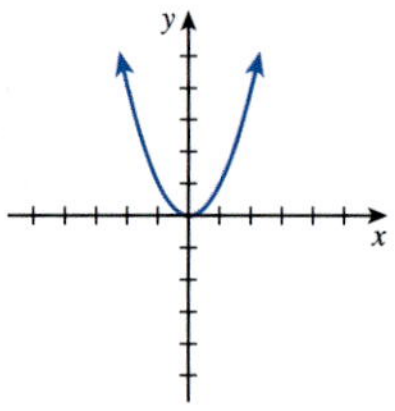	Domain: $\{x \mid {}^{-}\infty < x < \infty\}$ Range: $\{y \mid 0 \leq y\}$
Cubic	$f(x) = x^3$	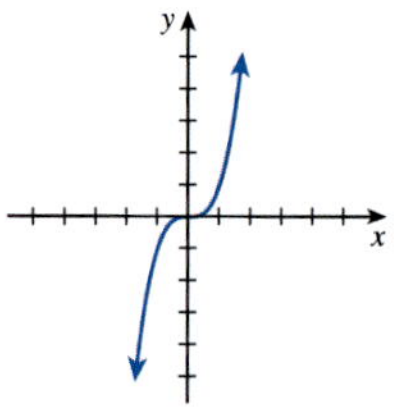	Domain: $\{x \mid {}^{-}\infty < x < \infty\}$ Range: $\{y \mid {}^{-}\infty < y < \infty\}$
Square root	$f(x) = \sqrt{x}$	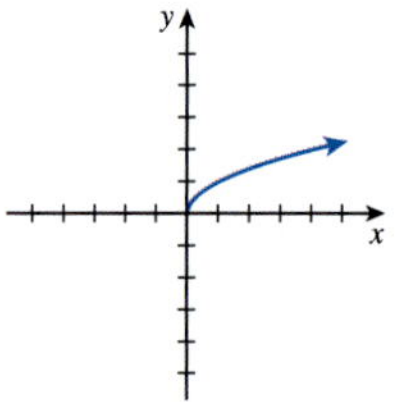	Domain: $\{x \mid 0 \leq x\}$ Range: $\{y \mid 0 \leq y\}$
Absolute value	$f(x) = \lvert x \rvert$	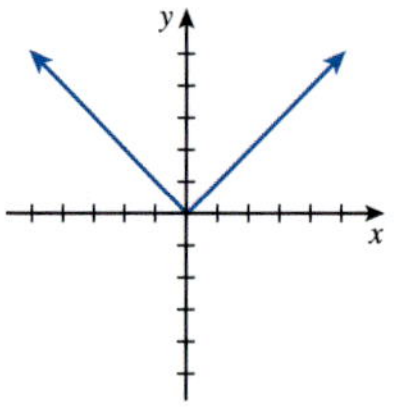	Domain: $\{x \mid {}^{-}\infty < x < \infty\}$ Range: $\{y \mid 0 \leq y\}$
Reciprocal	$f(x) = \frac{1}{x}$	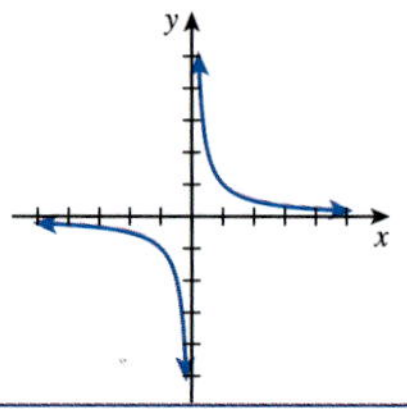	Domain: $\{x \mid x \neq 0\}$ Range: $\{y \mid y \neq 0\}$

19. a. Yes, the graph of the function passes the vertical-line test.

b. $\left(\frac{5}{2}, \frac{11}{4}\right) = (2.5, 2.75)$

c. Range: $\{y|2.75 \le y\}$

21.

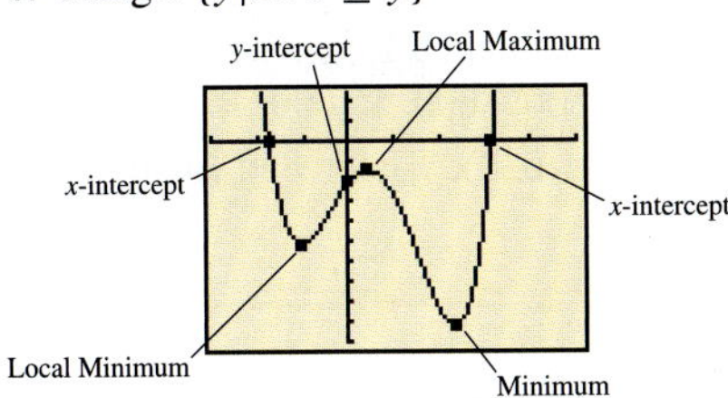

23. $24b(3a^2 - 2b^2)$

25. About $(^-.08, ^-1.85)$

Chapter 8 Exploration

a. The browser links to the page for athletics.

b. The browser remains on the home page.

c. Domain: $\{(x, y)|110 \le x \le 170 \text{ and } 30 \le y \le 90\}$

d. Range: set of 18 links.

Chapter 8 Review Exercises

1. A function is a rule that associates to an allowable input value one and only one output value.

2. The set of all allowable inputs to a function is called the *domain* of the function.

The allowable inputs are those that correspond to an output.

The set of possible outputs of a function is the *range* of the function.

3. Yes. Each input has associated with it a unique output value.

a. Domain: $\{\alpha, \beta, \gamma, \varepsilon\}$

b. Range: $\{1, 2, 3\}$

4. No. The input x has two outputs, 3 and 8, associated with it.

5. a. Domain 0 35

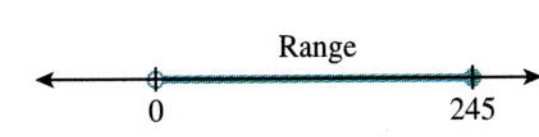

6. a. $f(x) = x^2 - 3$ **b.** $f(^-2) = 1$

c. $f(1.46) = {}^-.8684$

7. a. $f(3) = 4$ **b.** $g(2) = 2$

c. $h(^-1) = {}^-35$ **d.** $p(4) = 3$

e. $q(^-2) = {}^-28$ **f.** $r(1) = {}^-3$

8. Function

9. Not a function

10. Function

11. Domain: {A, B, C, D}; range: {4, 6, 8, 10}

12. Domain: $\{^-2, ^-1, 3, 4\}$; range: $\{^-2, ^-1, 1, 2\}$

13. a. Domain

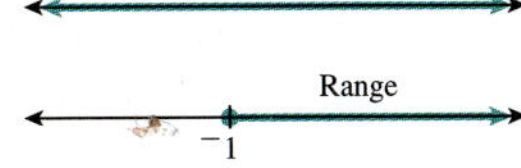

b. Domain: $\{x|^-\infty < x < \infty\}$; range: $\{y|^-1 \le y\}$

14. a. Domain: $\{x|2 \le x\}$; range: $\{y|0 \le y\}$

b. Domain: $\{x|^-2 \le x\}$; range: $\{y|0 \le y\}$

c. Domain: $\{x|x \ne {}^-2\}$; range: $\{y|y \ne 0\}$

d. Domain: $\{x|x \ne 1\}$; range: $\{y|y \ne 0\}$

15. a. Linear: I, III;
quadratic: I, II;
cubic: I, III;
square root: I;
absolute value: I, II;
reciprocal: I, III

b. Reciprocal

16. Local maximum: $(^-2, 4)$; local minimum: $(1, ^-1)$

17. Local Maximum (2, 1.3) — (0, 0) Local Minimum

Local minimum: (0, 0);
Local maximum: (2, 1.3) (answers for the y-coordinate may vary slightly)

18. a. It costs \$90 to print 4500 pages.

b. It costs $30 * \left(\frac{s}{1500}\right) = \frac{s}{50} = .02s$ dollars to print s pages.

c. $c(p(s)) = .02s$. Yes. We can find the cost in two steps by using the functions $p(s)$ and then $c(p)$, or we can use the composite function $c(p(s))$ to find the cost in one step.

19. a. $f(g(2)) = 2$ **b.** $f(g(3)) = 7$

c. $f(g(x)) = x^2 - 2$ **d.** $g(f(0)) = 0$

e. $g(f(^-2)) = {}^-4$ **f.** $g(f(x)) = x^2 + 4x$

20. a. $k(j(^-1)) = {}^-4$ **b.** $k(j(x)) = \frac{1}{x} - 3$

c. $j(k(2)) = {}^-1$ **d.** $j(k(x)) = \frac{1}{x-3}$

21. a. 4 **b.** 2 **c.** 1 **d.** 5 **e.** 3

22. a. The graph of $f(x)$ moves left 3 units.

b. The graph of $f(x)$ moves down 3 units.

c. The graph of $f(x)$ is reflected (flipped) across the x-axis.

d. The graph of $f(x)$ is compressed toward the x-axis.

23. The basic square root function moves left 4 units and moves down 2 units.

24. The graph of $f(x)$ is stretched away from the x-axis, reflected across the x-axis, and moved right 2 units.

25. The graph of $f(x)$ is moved down 5 units and moved left 1 unit.

Chapter 8 Test

1. The graph passes the vertical-line test.

2. a. $g(x) = x^3 - 3x^2$

b. $g(2) = {}^-4$

c. $g({}^-3.27) = {}^-67.044483$

3. Domain

${}^-1$

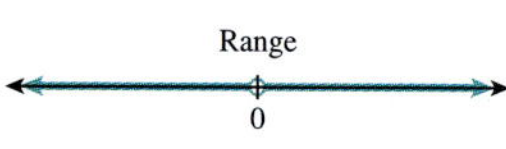

4. a.

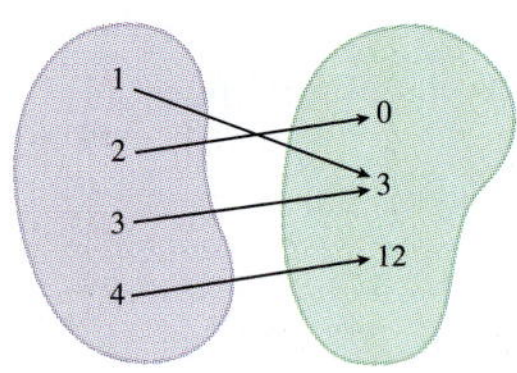

Yes, each input is associated with a single output.

b.

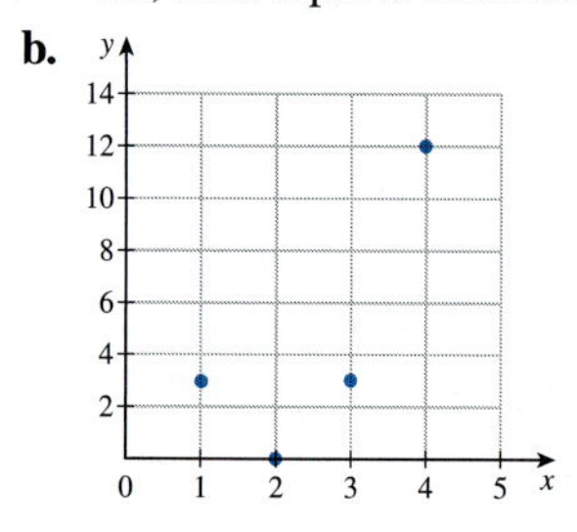

Yes, the graph passes the vertical-line test.

5. No, the graph fails the vertical-line test.

6. a. Yes, there is a single output associated with each input.

b. Domain: {DSL, cable, 56 Kbs, 33.6 Kbs, 28.8 Kbs}; range: {1000, 2600, 8600, 15,000, 15,500}

7. a. One x-intercept: (0, 0), y-intercept: (0, 0)

b. No

8.

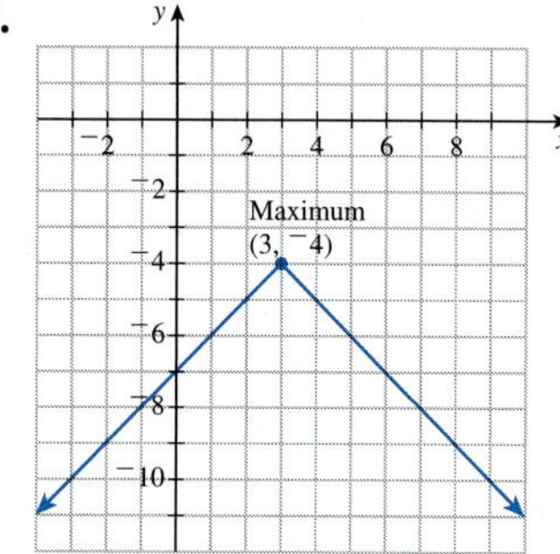

9. a. $c(t(b)) = 45b$

b. 78.75 cups of coffee

10. a. $f(g(4)) = \frac{1}{3}$

b. $f(g(x)) = \frac{1}{x-1}$

c. $g(f({}^-3)) = {}^-\frac{4}{3}$

d. $g(f(x)) = \frac{1}{x} - 1$

11. The graph of $f(x)$ is stretched away from the x-axis and moved right 3 units.

12. a. The graph of $f(x)$ moves right 2 units.

b. The graph of $f(x)$ moves up 2 units.

c. The graph of $f(x)$ is reflected across the x-axis.

d. The graph of $f(x)$ is stretched away from the x-axis.

Chapter 9 Powers and Roots

9.1 Exercises

1. $16^{-1} = \frac{1}{16}$; $16^0 = 1$; $16^1 = 16$

a. $1 < 16^{1/2} < 16$

b. $\sqrt{16}$

c. ${}^-4$ does not lie between 1 and 16; see part (a).

d. No, the solutions of $x^2 = 16$ are $x = 4$ and $x = {}^-4$, whereas the solution of $x = \sqrt{16}$ is only $x = 4$.

3. a. $\left(\sqrt{7}\right)^5$ or $\sqrt{7^5}$; in simple radical form, $49\sqrt{7}$

b. $\sqrt[3]{13}$

c. $5^{3/2}$

d. $11^{1/5}$

5. $\sqrt{(^-5)^2} = 5$; $\left(\sqrt{^-5}\right)^2$ is undefined.

The two expressions differ in the placement of the parentheses. In the first, we square $^-5$ before taking the square root. In the second, we must attempt the square root of $^-5$ before squaring. Negative numbers, however, do not have real number square roots.

$\sqrt[3]{(^-5)^3} = {}^-5$; $\left(\sqrt[3]{^-5}\right)^3 = {}^-5$

These two expressions result in a cube root of a negative number. The cube root of a negative number is a negative number. For example, $\sqrt[3]{^-8} = {}^-2$ because $(^-2)(^-2)(^-2) = {}^-8$. Section 9.2 provides further information on odd and even roots.

7. a. 2 **b.** 6 **c.** 4 **d.** 2

9. Each side of the cube is exactly $\sqrt[3]{20}$ cm or approximately 2.7 cm.

11. a. 9 **b.** 8 **c.** .00032
d. $\frac{27}{125}$ **e.** 129.64

13. a. $\frac{\sqrt{5}}{5}$ **b.** $\frac{3\sqrt{2}}{2}$ **c.** 2
d. $\frac{\sqrt{70}}{5}$ **e.** $\frac{\sqrt{21}}{7}$ **f.** $\frac{\sqrt{30}}{6}$

15. a. $49\sqrt{7}$
b. $7^{5/2} = 49\sqrt{7} \approx 129.64$

9.1 Skills and Review

17. a. 0
b. 3 is not in the domain of $g(x)$; $g(3) = \sqrt{3-4} = \sqrt{^-1}$, and negative numbers do not have real square roots.

19. 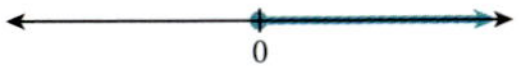

21. $c = {}^-3$ or $c = {}^-5$

23. a. $(x-4)(x-6)$
b. $^-4(x-3)(x+3)$
c. $5x(2x-7)$

25. $x = 6$

9.2 Exercises

1. a. Domain: $\{x \mid {}^-\infty < x < \infty\}$
b. Range: $\{y \mid {}^-\infty < y < \infty\}$
c. I and III
d. $f(x) = x^3$, $f(x) = x^5$. n is odd, the domain and range are the set of real numbers, and the graphs appear in quadrants I and III.

3. a. $x = 2$
b. $x = {}^-3$
c. $x = {}^-1$ or $x = 1$

5. a. $x = \pm\sqrt[4]{25} = \pm\sqrt{5}$
b. No real solution; you can't take the even root of a negative number.
c. $x = {}^-2$ or $x = 12$

7. a. $x = 27.9841$
b. No real solution; the graph of Y1 = X ∧ (1/4) does not go below the x-axis. Therefore, the graph of Y1 does not intersect the graph of Y2 = $^-2.3$.
c. $x = 18.89568$
d. $x = {}^-18.89568$

9. $x = 0$ or $x = 2.5$

11. a. $f(g(x)) = x$ and $g(f(x)) = x$
b. $f(g(x)) = x$ and $g(f(x)) = x$
c. $f(g(x)) = x$ and $g(f(x)) = x$

13. (3, 4) and (0, $^-2$)

15. a. 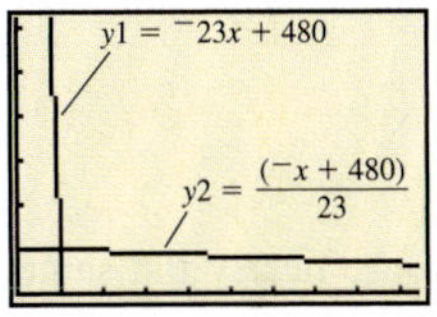

b. Yes; the graph of Y2 appears to be the mirror image of the graph of Y1, with the line $y = x$ as the mirror.
c. Yes
d. Yes; when two functions are inverses of each other, each point (x, y) on the graph of one function has a corresponding point (y, x) on the graph of its inverse function.

9.2 Skills and Review

17. $x = \pm 4\sqrt{2}$

19. a. Basic absolute value function
b. Basic square root function
c. Basic linear function
d. Basic reciprocal function
e. Basic quadratic function
f. Basic cubic function

21. The graph of $f(x)$ is shifted up 7 units.

23. 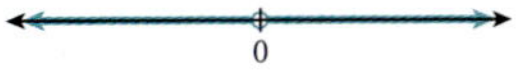

25. The vertex is a minimum at (2, $^-16$).

9.3 Exercises

1. $y = .407(93)^{1.5}$; $y \approx 365$ days

3. a. $y = ax$ **b.** $y = .5x$
c. $y = 1.5$ **d.** $x = 40$

5. a. $f(x) = ax^{5/2}$ **b.** $f(x) = 1.7x^{5/2}$

c. $f(4) = 54.4$ **d.** $x = 9$

7. $a = .25$

9. a. 3; the graph looks like the basic linear function, $y = x$.

b. 1; the graph is similar to the graph of the basic square root function, $y = \sqrt{x} = x^{1/2}$.

c. 2; the graph is similar to the graph of the basic quadratic function, $y = x^2$.

11. a. Yes

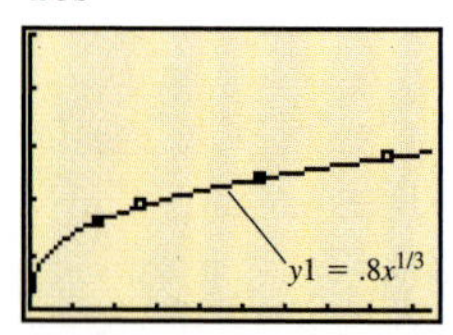

X	Y1
8	1.6
13	1.8811
27	2.4
42	2.7808

Y1=.8X^(1/3)

b. Yes. Both table outputs are nearly the same, so our choice of functions is a good one.

c. $y = 3.2$

d. $x \approx 11.4$

13. a. $y = ax^{1/3}$

b. $a \approx .624$, $y = .624x^{1/3}$

c. $y \approx 2.26$; the radius of the sphere is about 2.3 cm.

15. $g = 32$; the gravitational constant is 32 feet/second^2.

9.3 Skills and Review

17. $x = 1$ or $x = 5$

19. a. $2\sqrt{3}$

b. $54\sqrt{2}$

c. $\frac{\sqrt{5}}{5}$

21. $c^{1/3}$

23. a.

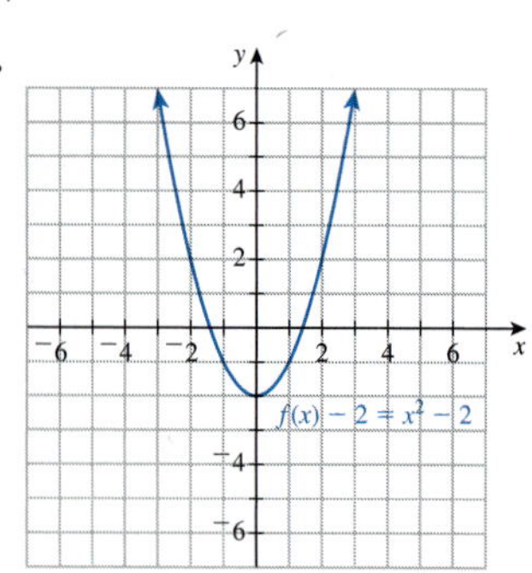

b.

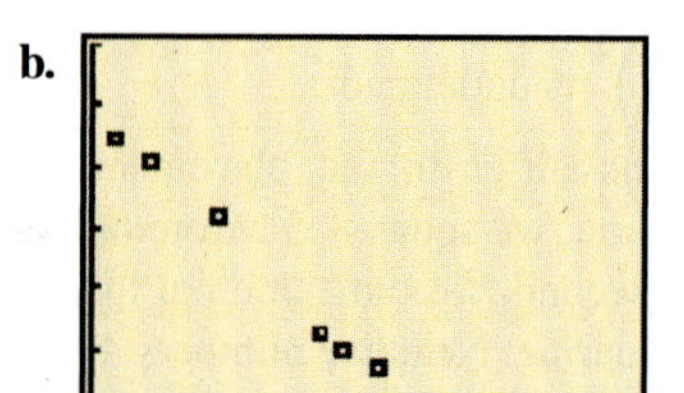

c.

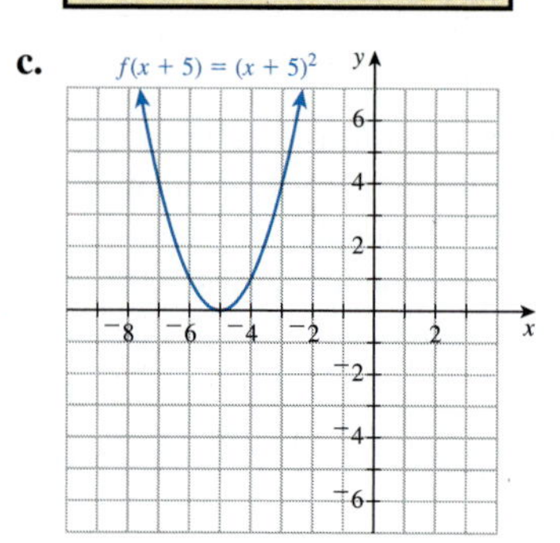

25. $(x + 2)(x - 7)$

9.4 Exercises

1. United States

3. a.

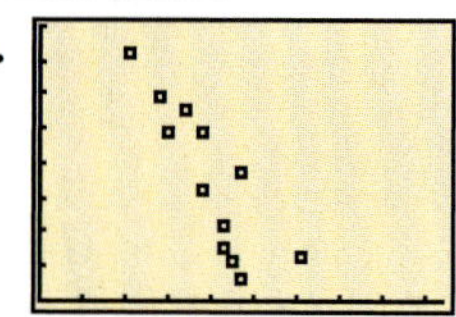

Window: $0 \le X \le 94$, $0 \le Y \le 80$; no.

b.

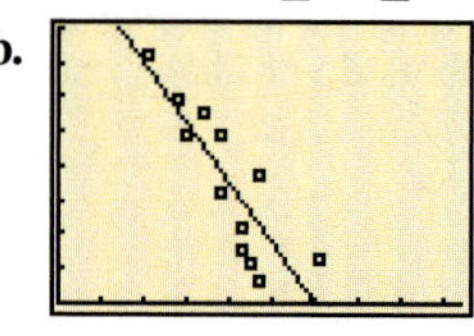

$y \approx {}^{-}1.758x + 105$; the line does not fit the points exactly but is close to most of them.

c. (0, 105); 105°F is the predicted average January temperature at 0°N (the equator).

d. ${}^{-}1.758$ is the rate of change for the temperature. The average January temperature drops about 1.76°F for each increase of 1° in latitude.

e. Other geographic factors, including proximity to large bodies of water and elevation, affect the temperature. For example, Olympia and San Francisco are both located on the Pacific coast and are much warmer than the corresponding inland cities, Bismarck and Charleston.

Chapter 9 Exploration

Note	Number of Half-Steps above Middle C	Frequency in Traditional System	Frequency with Equal Temperament
Middle C	0	264	264
C sharp/D flat	1	278.4	279.7
D	2	297	296.3
D sharp/E flat	3	312.8	314.0
E	4	330	332.6
F	5	352	352.4
F sharp/G flat	6	371.2	373.4
G	7	396	395.6
G sharp/A flat	8	417.6	419.1
A	9	440	444.0
A sharp/B flat	10	469.4	470.4
B	11	495	498.4
High C	12	528	528

a. Middle C and high C

b. Note A

c.

Ratio of Frequencies	Traditional	Equal Temperament
G to middle C	$\frac{396}{264} = 1.50$	$\frac{395.6}{264} \approx 1.498$
A to D	$\frac{440}{297} \approx 1.48$	$\frac{444.0}{296.3} \approx 1.498$
B flat to E flat	$\frac{469.4}{312.8} \approx 1.50$	$\frac{470.4}{314.0} \approx 1.498$

The ratios of frequencies for the traditional system are close to each other, but the ratios of frequencies for the equal temperament are much closer. The ratios of the equal temperament are within .001 of each other.

Chapter 9 Review Exercises

1.

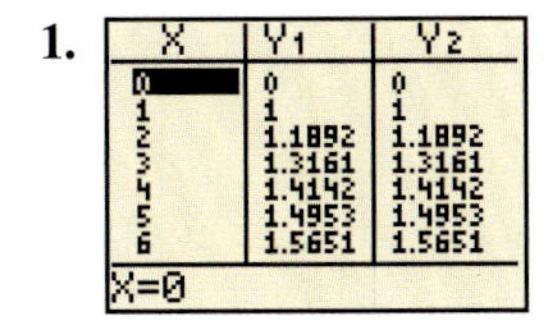

X	Y1	Y2
0	0	0
1	1	1
2	1.1892	1.1892
3	1.3161	1.3161
4	1.4142	1.4142
5	1.4953	1.4953
6	1.5651	1.5651

X=0

Y1 = 4 $^{x}\sqrt{}$(X), Y2 = X ∧ (1/4); Y1 and Y2 are the same, so the expressions are equal.

2. a. $\frac{2}{3}$ **b.** 3

3. 1.246

4. a. 125 **b.** 27

c. $\frac{4}{9}$ **d.** 17.78

5. a. $\sqrt{18}$ or $3\sqrt{2}$ **b.** $\sqrt[4]{49}$ or $\sqrt{7}$

c. $\sqrt[5]{3}^2 = \sqrt[5]{9}$ **d.** $\frac{1}{\sqrt{8}}$ or $\frac{\sqrt{2}}{4}$

6. a. $7\sqrt{2}$ **b.** $6\sqrt{3}$ **c.** $\frac{\sqrt{7}}{7}$ **d.** $\frac{\sqrt{6}}{3}$

7. a. $x = \pm 3\sqrt{6}$ **b.** $x = {}^{-}2 \pm 3\sqrt{2}$

c. $x = 2 \pm \frac{5\sqrt{3}}{3}$

8. a. $x = \pm 2$

b. No real solution; the graph of Y1 = X ∧ 4 does not go below the x-axis.

c. $x = 5$

d. $x = {}^{-}5$

9. a. $x = \pm\sqrt[6]{49} \pm \sqrt[3]{7}$.

b. $x = 5$

c. No real solution; you can't take the even root of a negative number.

10. a. $x = 9$

b. No real solution; the graph of Y1 = X ∧ (1/2) does not go below the x-axis.

c. $x = 10.48576$

d. $x = {}^{-}10.48576$

11. a. $x = 9$

b. No real solution; $x = 9$ is an extraneous solution because it doesn't satisfy the original equation.

c. $x = 10.48576$

d. $x = {}^{-}10.48576$

12. a. $x = 3$ or $x = 6$

b. $x = 3$ is an extraneous solution because it doesn't satisfy the original equation. A graph of Y1 = X − 4 and Y2 = $\sqrt{}$(X − 2) shows a single intersection of X = 6. Therefore, $x = 3$ is not a solution.

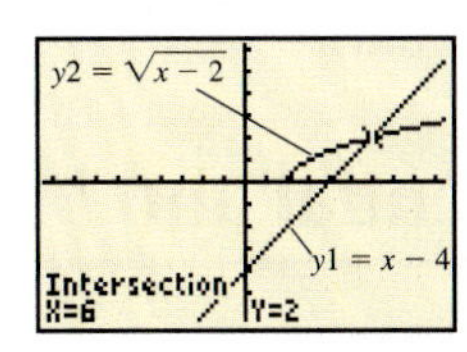

13. $f(g(x)) = x$ and $g(f(x)) = x$

14. The left half of the graph of Y1 = X ∧ 4 has no corresponding points on the graph of Y2 = X ∧ (1/4).

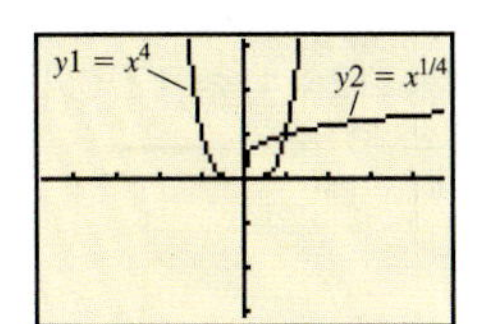

15. Jupiter completes one revolution around the sun in about 12 years.

16. $a = \frac{1}{3}$

17. $a = 5$

18. $a = .2$

19. a. $a = \frac{100}{\sqrt{10}} = 10\sqrt{10} \approx 31.62$
 b. $y \approx 63.25$
 c. $x = 4.096$

20. a. $y = ax$
 b. $a = .11$, $y = .11x$
 c. $x \approx 455$; the resident uses about 455 kWh.

21. a. $y = ax^3$
 b. $a = .13$, $y = .13x^3$
 c. $y = 2031.25$; the windmill generates 2031.25 W of power.

22. 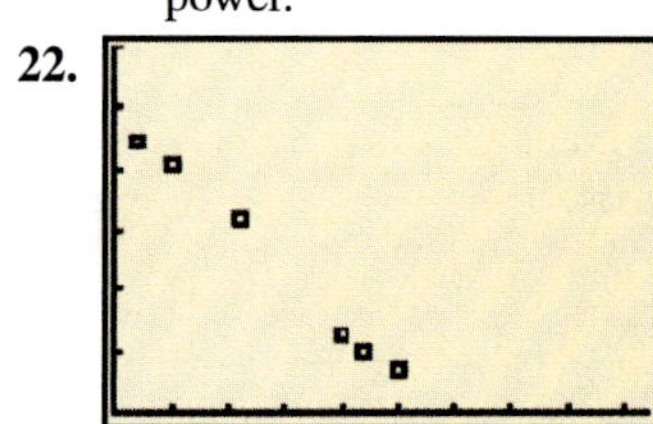

The points show a downward trend; as x increases, y decreases. The points fit fairly closely along a straight line.

23. a. $y \approx {}^{-}3.46x + 97.65$
 b. $y \approx 73.45$

Chapter 9 Test

1. a. 10 b. 7 c. 2

2. a. 1.3 b. 2.5

3. a. 1000 b. 25 c. 625

4. a. $\sqrt[3]{54}$ b. $\sqrt[6]{36}$
 c. $(\sqrt[3]{4})^5 = \sqrt[3]{4^5}$

5. a. $4\sqrt{3}$ b. $2\sqrt{7}$ c. $\frac{\sqrt{10}}{5}$ d. $\frac{\sqrt{15}}{5}$

6. a. Two solutions; the graph of Y1 = X ∧ 2 intersects the graph of Y2 = 100 twice.
 b. One solution; the graph of Y1 = X ∧ 3 intersects the graph of Y2 = ⁻27 once.
 c. No real solution; the graph of Y1 = X ∧ 2 doesn't intersect the graph of Y2 = ⁻100.

7. a. $x = \pm 10$
 b. $x = {}^{-}3$
 c. No real solution; you can't take the square root of a negative number.

8. a. $x = 6$
 b. $x = \pm\sqrt[4]{36} = \pm\sqrt{6}$
 c. No real solution; $x = 256$ is an extraneous solution because it doesn't satisfy the original equation.
 d. $x = {}^{-}1$

9. $x = 2$; $x = {}^{-}3$ is an extraneous solution because it doesn't satisfy the original equation.

10. $f(g(x)) = x$ and $g(f(x)) = x$

11. a. $a = 3$ b. $y = 192$ c. $x = 1$

12. a. $y = ax$
 b. $a = \frac{\sqrt{3}}{2} \approx .866$, $y = \frac{\sqrt{3}}{2}x$ or $y = .866x$
 c. $y = .750$; the height of the equilateral triangle is .750 cm.

13. a. $y = ax^2$
 b. $a = 4.9$, $y = 4.9x^2$
 c. $x = 10$; in 10 s the object drops 490 m.

Chapter 10 Exponential and Logarithmic Functions

10.1 Exercises

1. a. $A_0 = 1000$
 b. $r = 6\%$
 c. $n = 12$
 d. $A(4) = 1000(1 + .06/12) \wedge (12 * 4) \approx 1270.49$

3. a. 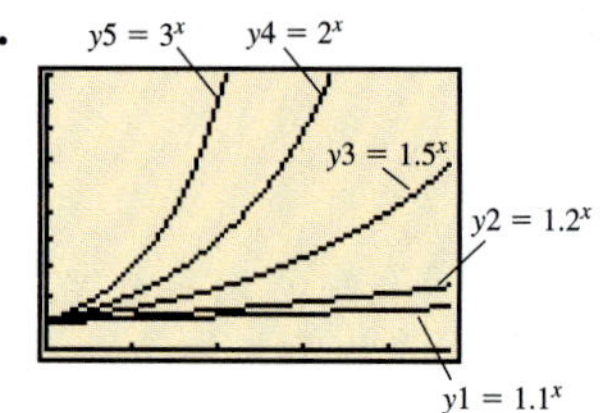

b. y-intercept: (0, 1)

c. The greater the base, b, the steeper the graph.

5. $f(x) = \frac{1}{3}(5)^x$

7. a. The outputs increase by the same factor, in this case 1.5. This is the growth factor, b.

b. $f(x) = 5(1.5)^x$

9. $x \approx 4.97$

11. Daily compounding at 5% yields more. Compare $1000 invested for 1 year. After 1 year the 5% investment appreciates to $1051.27, whereas the 5.1% investment appreciates to $1051.00.

13. a.

Year	Number of Years Since 2002	Highest Baseball Salary
2002	0	22,000,000
2003	1	25,740,000
2004	2	30,115,800
2005	3	35,235,486
2006	4	41,225,519
2007	5	48,233,857
2008	6	56,433,612
2009	7	66,027,327
2010	8	77,251,972

b. $A(t) = 22000000(1.17)^t$

c.

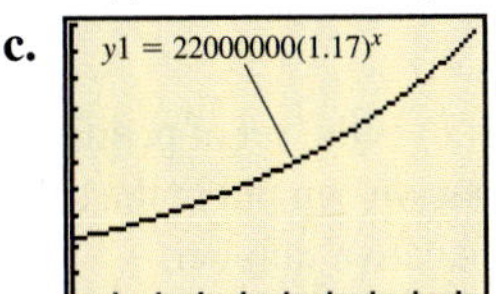

15. About 16 years

10.1 Skills and Review

17. $y = \frac{3}{8}x^{5/2}$

19. $x = 2$

21. $x = {}^{-}6 \pm \sqrt{52}$

$x = {}^{-}6 \pm 2\sqrt{13}$ *(Simple radical form)*

23. a.

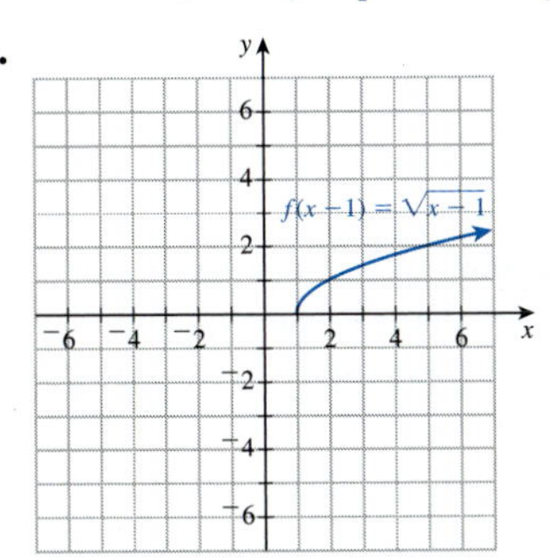

b.

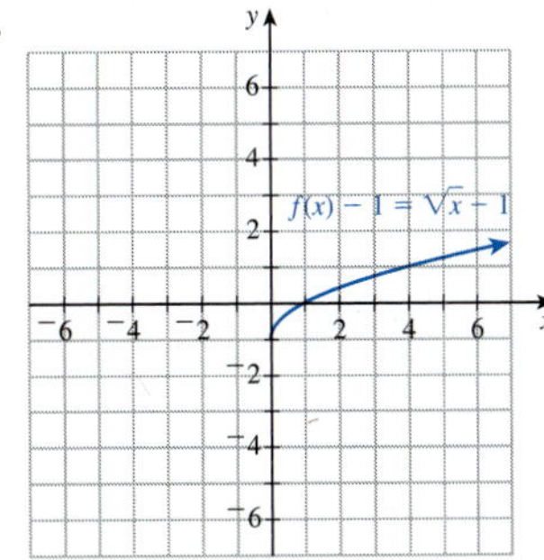

c.

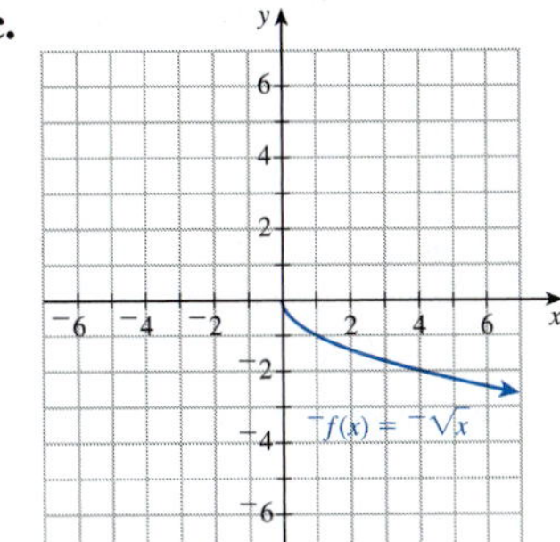

d.

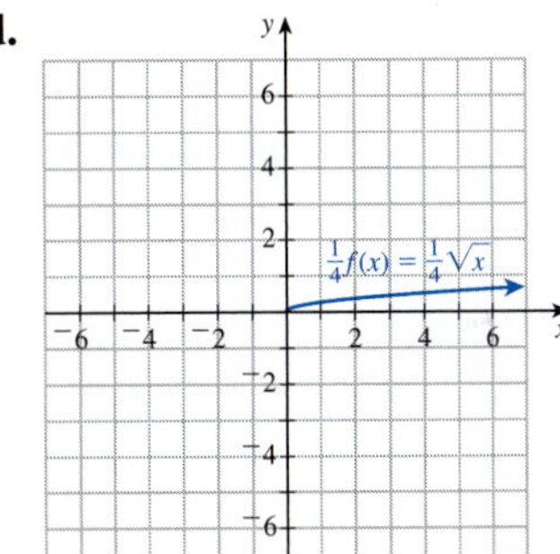

25. The car was traveling 90 miles/hour.

10.2 Exercises

1. a. $b > 1$

b. $0 < b < 1$

3. a. 1 **b.** 5 **c.** .4 **d.** $\frac{1}{2}$

5. The graph of $y = 2(3)^{-x}$ is the mirror image of the graph of $y = 2(3)^x$ with the y-axis as the mirror.

7. Half-life gives a quick method to compare rates of decay. It is the amount of time it takes for a substance to decay to one-half of its initial amount.

9. a. For equal time periods the output decreases by the same factor. In this case, after every 13 days the weight is halved.

b. $y_0 = 1$

c. $b = \frac{1}{2}$

d. $k = \frac{1}{13}$

e. $y = \left(\frac{1}{2}\right)^{(1/13)t} = \left(\frac{1}{2}\right)^{t/13}$

11. **a.** $y_0 = \$90$

b. \$45

c.

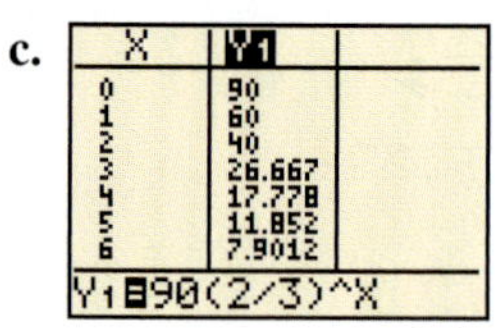

X	Y1
0	90
1	60
2	40
3	26.667
4	17.778
5	11.852
6	7.9012

Y1=90(2/3)^X

From the table of values, an estimate for the half-life of the video game is between 1 and 2 years.

d.

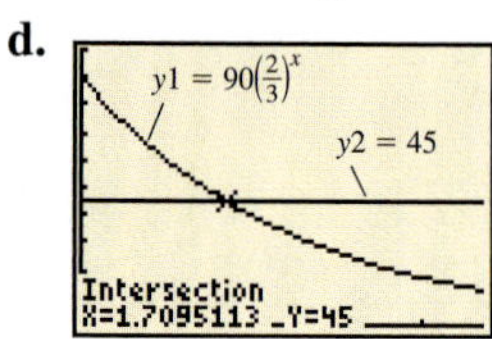

From the graph, a better approximation for the half-life is 1.7 years.

13. **a.** $A(t) = 60\left(\frac{3}{4}\right)^{t/10}$

b. In approximately 24 days there is half of the initial amount of thorium 234.

15. The machine is worth half of its initial value in about 5 years.

10.2 Skills and Review

17. The population doubles in about 10.3 years.

19. **a.** $a = \frac{5}{2}$ **b.** $y = \frac{45}{2} = 22.5$ **c.** $x = 5.2$

21. 3

23. **a.** $f(g(^{-}5)) = {}^{-}11$ **b.** $f(g(x)) = 2x - 1$

c. $g(f(x)) = 2x - 2$

25. $(x-5)(x+4)$

10.3 Exercises

1. The graph of $y = \log(x)$ appears to be the mirror image of the graph of $y = 10^x$ with the line $y = x$ as the mirror.

3. **a.** $10^3 = 1000 \rightarrow \log(1000) = 3$

b. $10^1 = 10 \rightarrow \log(10) = 1$

c. $10^0 = 1 \rightarrow \log(1) = 0$

d. $10^{^{-}2} = \frac{1}{100} \rightarrow \log\left(\frac{1}{100}\right) = {}^{-}2$

5. $\log(3.2)$ is between 0 and 1.

7. $6(10)^{2x} = 600$

$(10)^{2x} = 100$ *Dividing by 6 on both sides*

$\log(10^{2x}) = \log(10^2)$ *Taking the logarithm on both sides and writing 100 as 10^2*

$2x = 2$ *Common log undoes raising 10 to a power*

$x = 1$ *Dividing by 2 on both sides*

9. **a.** $x \approx 2.77815$ **b.** $x \approx {}^{-}.31808$

c. $x \approx {}^{-}.56632$

11. **a.** $x = 100$ **b.** $x = 1000$ **c.** $x = 10$

13. $x \approx 501.187$ *Check:* $\log(501.187) \approx 2.7$

15. $t = \frac{\log(.5)}{\log(.97)} \approx 22.76$

10.3 Skills and Review

17. **a.** $y = 500\left(\frac{3}{4}\right)^x$

b. The item is worth half of its initial value in approximately 2.4 years.

19. **a.** $A(t) = 750\left(1 + \frac{.038}{12}\right)^{12t}$

b. The account is worth \$1500 in approximately 18.3 years.

21. $a = 7$

23. $c = {}^{-}1$ or $c = {}^{-}5$

25. $x = 44$

10.4 Exercises

1. The domain of $f(x) = \log(x)$ is the set of positive numbers. There is no number that we can put in the exponent of 10 that will give 0 or a negative number.

3. **a.** $M(1) = 0$ **b.** $M(100) = 2$

c. $M(100{,}000) = 5$

5. **a.** $M(12{,}843) \approx 4.1$

b. $M(240) \approx 2.4$

c. $M(351{,}060{,}000) \approx 8.5$

7.

x	$\log x$
.058	$^{-}1.2$
.58	$^{-}.2$
5.8	.8
58	1.8
580	2.8

9. It is hard to distinguish between amplitudes that vary so much. On the other hand, Richter magnitudes usually vary between 1 and 9, making it easier to see differences in earthquake intensity.

11. The lower the pH, the more acidic the substance.

13. a. Lemon juice
b. 10^9 or 1,000,000,000
15. 10^2 or 100

10.4 Skills and Review

17. a. $x = \pm 3$ b. $x = 9$
19. a. $x = \frac{3}{2} = 1.5$ b. $x = 10^{-1} = \frac{1}{10} = .1$
21. a. $A(t) = 3\left(\frac{1}{2}\right)^{t/800}$ b. $\frac{3}{2} = 1.5$ kg
c. $\frac{3}{8} = .375$ kg
23. $f(g(x)) = x$ and $g(f(x)) = x$
25. a. $2x(x+3)$ b. $y(x-4)(x+4)$
c. $(x+4)(x-1)$

10.5 Exercises

1. The half-life for this material is approximately 44.15 years.

Chapter 10 Exploration

a.

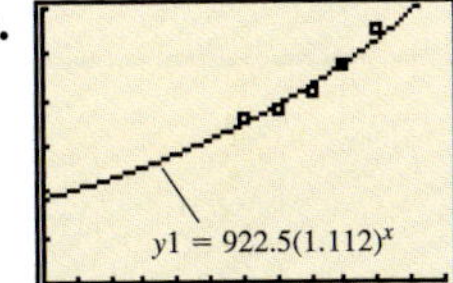

b. If the current rate continues, the number of AIDS cases will double in about 6.4 years.
c. Estimate 7710 AIDS cases in this region in the year 2010.
d. Answers may vary.

Chapter 10 Review Exercises

1. a.

x	y
0	2
1	8
2	32
3	128

b. $y_0 = 2$
c. $b = 4$
2. $y = 200(7)^x$
3. a. $A(t) = 2000\left(1 + \frac{.068}{12}\right)^{12t}$ b. $A(7) \approx \$3214.92$
4. Between 10 and 11 years
5. 10.2 years
6. Rule of 72: 60 years
7. c
8. a. Half-life is the amount of time it takes for half of a substance to decay.
b. 7.5 g
c. 3.75 g
d.

Time (years)	Amount U-234 (g)
0	15
245,000	7.5
490,000	3.75
735,000	1.875

e. $A(t) = 15\left(\frac{1}{2}\right)^{t/245{,}000}$
9. $y = 12(.92)^{t/5}$
10. a. about .9 miles b. 34.3 years
11. Using the definition of negative integer exponents from Section 6.1, we have $6^{-x} = \frac{1}{6^x}$; because 1 to any power is 1, we may write $\left(\frac{1}{6}\right)^x$. Therefore, $20(6)^{-x} = 20\left(\frac{1}{6}\right)^x$.
12. a. $y = 8000\left(\frac{2}{3}\right)^t$ b. 1.7 years
13. Domain: $\{x \mid -\infty < x < \infty\}$; range: $\{y \mid 0 < y < \infty\}$
14. a. 2 b. 4 c. -1
15. a. $-.699$ b. .222 c. 1.892
16. $x = 1000$
17. a. $10^{5/4} = 10^{1.25}$ b. 17.78
18. a. $x = 2$ b. $x = 5$ c. $x = 1$
19. $x = 6$
20. a.

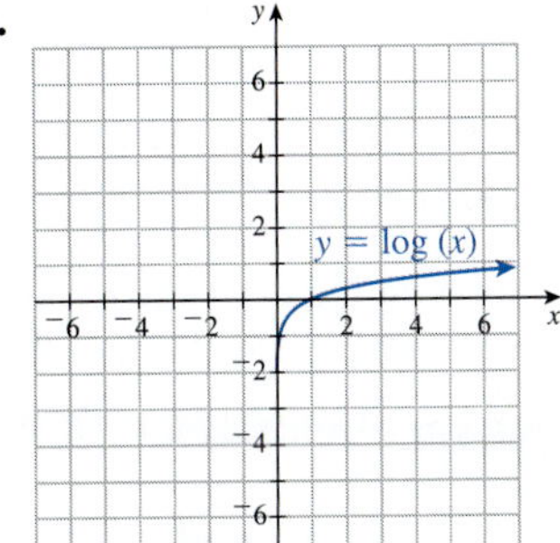

b. Domain: $\{x \mid 0 < x < \infty\}$; range: $\{y \mid -\infty < y < \infty\}$
21. a. 1.04 b. 3.15 c. 7.73
22. 1000
23. a. 14 b. 3 c. 6.9
24. The pH will decrease by 1.
25. The 100-dB sound is 10^8, or 100,000,000, times more intense than the 20-dB sound.
26. The recommended maximum sound at a nightclub is about 30,000 ($10^{4.5} \approx 31{,}623$) times more intense than the maximum sound outdoors in a residential area at nighttime.
27. The investment doubles in approximately 11.55 years.

28. a. 6

b. $^-9$

c. 1.8

d. 1

e. $\ln(40) \approx 3.69$

Chapter 10 Test

1. a. $y_0 = 1{,}200{,}000$

b. 5.2%

c. 3,307,471

d. Using the rule of 72 gives 13.8 years. A closer approximation by graphing is 13.7 years.

2. a. $A(6) \approx \$15{,}219.00$

b. Using the rule of 72 or graphing, we find the investment doubles in about 10 years.

3. Rule of 72: 18 minutes

4. 1600 years

5. a. $A(t) = 8600(.87)^t$

b. about 4286

6. a. $A(t) = 3000\left(\frac{3}{4}\right)^t$

b. $A(2) = \$1687.50$

c. The piece of hardware is worth half its initial value in approximately 2.4 years.

7. a. 1 **b.** 0 **c.** $^-3$ **d.** $^-2$

8. a. $10^{2/3}$ **b.** 4.64

9. $x = \frac{4}{5}$

10. 4.4

11. 7.1

12. The lawn mower sound is 1000 times more intense than the vacuum cleaner sound.

Chapter 11 Rational Expressions

11.1 Exercises

1. A rational expression is a fraction in which both the numerator and denominator are polynomials. Answers may vary:

$$\frac{x+3}{x^2-2x+5}.$$

3. a. Substituting $^-2$ for x in the expression will give a denominator of 0, and division by 0 is undefined.

b. Y1 = (X − 3)/(X + 2)

c. ERROR

5. a. 15 **b.** x^2y^2

c. $x^2 - 8x + 15$

7. a. $\frac{17}{36}$ **b.** $\frac{3y-2x}{xy}$

c. $\frac{5x+9}{(x+2)(x+1)}$

9. a. $\frac{3}{16}$ **b.** $\frac{4x}{5y^2}$ **c.** $\frac{x+2}{x}$

11. Method 1: Compare table outputs. Enter the expressions in the Y= menu. The table outputs for Y1 and Y2 should be equal except at X = 5.

Method 2: Compare graphs. The graph of Y1 should be the same as the graph of Y2.

13. $\frac{a}{b^2x^2y^3}$

15. $\frac{x-2}{5(x-8)}$

11.1 Skills and Review

17. $f(g(x)) = 10^{\log(x)} = x$ and $g(f(x)) = \log(10^x) = x$

19. a. 10,000 **b.** .001 **c.** 100,000

21. a. Increasing; because $b > 1 (b = 1.065)$, we have growth.

b. The initial value is $y_0 = 4449$ eagle pairs.

c. The eagle pairs are increasing by 6.5% each year.

d. 10,744 eagle pairs

23. a. The graph of the basic absolute value function $f(x) = |x|$ is shifted to the left 1 unit.

b. The graph of the basic cubic function $f(x) = x^3$ is reflected across the x-axis.

c. The graph of the basic square root function $f(x) = \sqrt{x}$ is stretched away from the x-axis.

d. The graph of the basic reciprocal function $f(x) = \frac{1}{x}$ is shifted up 5 units.

25. a. y-intercept: $(0, {}^-8)$

b. x-intercepts:

$$\left(\frac{3}{2} + \frac{\sqrt{41}}{2}, 0\right), \quad \left(\frac{3}{2} - \frac{\sqrt{41}}{2}, 0\right)$$

c. Vertex (minimum):

$$\left(\frac{3}{2}, \frac{^-41}{4}\right)$$

11.2 Exercises

1. a.

X	Y1
0	ERROR
25	23
50	15
75	12.333
100	11

Y1=(7X+400)/X

b. As the number of items x increases, the average cost decreases, but the rate of decrease slows.

c. As x gets very large, the average cost approaches \$7 per item.

d. Substituting 0 for x in the average cost function causes division by 0, which is undefined.

3. Horizontal asymptote: $y = 3$; vertical asymptote: $x = 1$.

5.

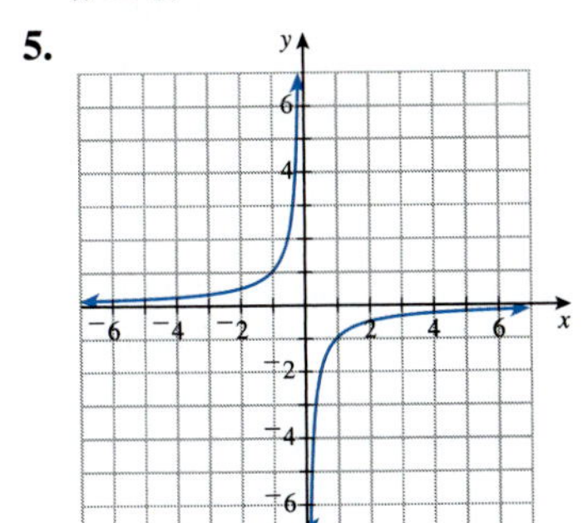

7. The branches of the graph of $f(x) = \frac{1}{x}$ are reflected across the x-axis and stretched away from the x-axis. The branches now appear in the second and fourth quadrants. Horizontal asymptote: x-axis; vertical asymptote: y-axis.

9. The graph of $f(x)$ is shifted down 1 unit. Horizontal asymptote: $y = {}^-1$; vertical asymptote: y-axis.

11. a. Domain: $\{x|x \neq 0\}$; range: $\{y|y \neq 0\}$

b. Domain: $\{x|x \neq 0\}$; range: $\{y|y \neq 0\}$

c. Domain: $\{x|x \neq {}^-4\}$; range: $\{y|y \neq 0\}$

d. Domain: $\{x|x \neq 0\}$; range: $\{y|y \neq {}^-1\}$

e. Domain: $\{x|x \neq 2\}$; range: $\{y|y \neq 6\}$

13. a. Third and fourth quadrants

b. Domain: $\{x|x \neq 0\}$; range $\{y|y < 0\}$

15. a. No, the function is undefined at $x = {}^-4$.

b. Either graph with a MODE of Dot or graph with a friendly window such as $^-9.4 \leq X \leq 9.4$, $^-6.2 \leq Y \leq 6.2$.

11.2 Skills and Review

17. $\dfrac{x^2 - 4x - 1}{(x-3)(x+3)}$

19.

x	$\log(x)$
.13	$^-.9$
1.3	.1
13	1.1
130	2.1
1300	3.1

21. 3.6 years

23. a. $f({}^-2) = 0$

b. $f(0)$ is undefined due to division by 0.

c. $f(2) = 2$

d. $f(200) \approx 1$

25. $\frac{3}{10} = .3$

11.3 Exercises

1. a. Answers may vary; Ymax = 1.

b. $x = 6$; the graphing solution is the same as the algebraic solution.

3. Windows may vary; Xmin = $^-9.4$, Xmax = 9.4, Ymin = $^-10$, Ymax = 10; $x = 2.25$

5. The empty pool fills in 3 hours with the outtake pipe shut.

7. $x = \frac{16}{3}$

9. $x = \frac{2}{3}$ or $x = 4$

11. Working together, Joan and her partner detail a car in 2.4 hours, or 2 hours 24 minutes.

13. a. The speed of the river current is 20 m/s.

b. The rowing team travels downstream in 30 s.

15. a. No, the graph of Y1 has a horizontal asymptote of $y = 7$.

b. You get the impossible solution $3 = 0$.

c. There is no solution.

11.3 Skills and Review

17. a.

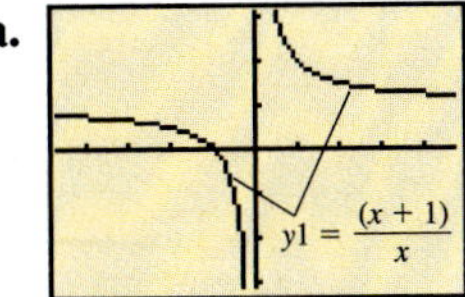

b. Vertical asymptote: $x = 0$; horizontal asymptote: $y = 1$

19. $-\frac{1}{3x}$

21. $^{-}2$

23. $\frac{1}{z^{12}}$

25. **a.** $x = {}^{-}3$ or $x = {}^{-}2$ **b.** $x = 2$
c. $x = 1$ **d.** $x = \pm 2$
e. $x = {}^{-}2$

11.4 Exercises

1. As the weight increases, the distance decreases. The distance decreases quickly at first, then the rate of decrease slows.

3. **a.** 120 pound-feet
b. Weight
c. Distance

5. **a.** Time * Velocity is a constant of 3600.
b. Velocity $= \frac{3600}{\text{time}}$

7. $a = 400$, $R = \frac{400}{I}$

9. $F = \frac{a}{d^2}$

11. $F = 1250$ newtons

13. **a.** $y = \frac{1}{5}$ **b.** $x = 12$ **c.** $y = 12.5$

15. **a.** Pressure = .90125 atmosphere
b. Volume ≈ 3.43 L

11.4 Skills and Review

17. $f(x) + 6 = \frac{1}{x} + 6$; the basic reciprocal function is shifted up 6 units.

19. $y = 200\left(\frac{9}{10}\right)^x$

21. **a.** $x = 9$; we reject $x = 1$ as extraneous because it doesn't satisfy the original equation.
b. No, substitution or graphing show that this solution is false.

23. **a.** 3 **b.** 2 **c.** 5

25. **a.** $y = \frac{60}{x^2}$
b. $f(x) = \frac{60}{x^2}$
c. $f(2) = 15$
d. $x = \pm 6$

11.5 Exercises

1. **a.** $x + 5$
b. $x + 5 - \frac{5}{x+7}$
c. $x + 5 + \frac{3}{x+7}$

3. **a.** Y1 = (X ∧ 2 – 14X + 48)/(X – 6)
Y2 = (X ∧ 2 – 14X + 30)/(X – 6)
Y3 = (X ∧ 2 – 14X + 60)/(X – 6)

b.

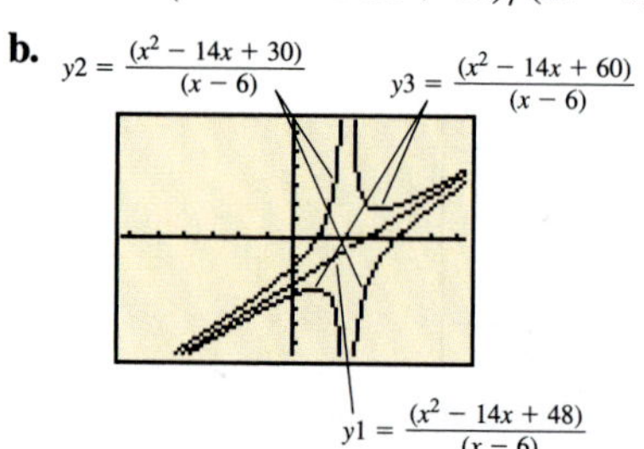

c. The graph of Y1 is a straight line with a gap at $x = 6$. The graph Y2 has two branches. There appears to be a vertical asymptote at $x = 6$, and both branches of Y2 appear to get close to the Y1 graph as x assumes large positive or negative values.

The graph of Y3 is similar to the graph of Y2 except the branches appear in the upper right and lower left.

Note that in Exercise 2 the quotient is the same in each case and only the remainders differ.

When the remainder is 0, the graph of Y1 is a graph of the linear quotient $y = x - 8$.

When the remainder is not 0 and for large positive or negative values of x, the graphs of Y2 and Y3 approach the line $y = x - 8$. As x approaches 6, the graphs of Y2 and Y3 are stretched away from the line $y = x - 8$ and approach the vertical asymptote at $x = 6$.

5. In both methods the first term of the quotient is found by finding the ? in $x * ? = x^3$.

In both methods we multiply the first term in the quotient by each term in the divisor.

In both methods we subtract the preceding result from the dividend.

The process is repeated to find the remaining terms of the quotient.

Chapter 11 Exploration

a. $I = \frac{a}{d^2}$

b. $a = 35{,}000$

c. $I \approx 155.6$ foot-candles

d. Either move closer to the light source or improve the light source.

Chapter 11 Review Exercises

1. Answers may vary.

a. $\frac{10}{12}, \frac{5x}{6x}$ **b.** $\frac{3x^4}{3xy^2}, \frac{x^3y^9}{y^{11}}$

c. $\frac{xz-4z}{xz+5z}, \frac{x^2-3x-4}{x^2+6x+5}$

2. In your calculator enter the first expression as Y1 and the second expression as Y2.

X	Y1	Y2
$^{-}4$	$^{-}8$	$^{-}8$
3	$^{-}.125$	$^{-}.125$
10	.4	.4
$^{-}1$	$^{-}1.25$	ERROR

X=

$Y1 = (X-4)/(X+5)$, $Y2 = (X^2-3X-4)/(X^2+6X+5)$

When $x = {}^{-}1$, the output of the first expression is $^{-}1.25$. The output of the second expression is undefined because of division by 0.

3. a. 6 **b.** xy^2 **c.** $x^2 - x$

4. a. $\frac{2x^3+3}{x^4}$ **b.** $\frac{{}^{-}2x-3}{(x-2)(x-3)}$

c. $\frac{4x-y}{xy}$ **d.** $\frac{7x-18}{(x+1)(x-4)}$

e. $\frac{3yz+8}{y^2z^3}$ **f.** $\frac{3x+8}{x(x+2)^2}$

5. a. $\frac{2y}{3x^2}$ **b.** $\frac{x-6}{x}$

c. $\frac{x+3}{x-3}$ **d.** $\frac{1}{x+7}$

e. $^{-}1$

6. a. $\frac{4}{xy}$ **b.** $\frac{5x}{x+3}$

c. $\frac{7}{2}$ **d.** $\frac{x+8}{18x}$

7. a.

X	Y1	Y2
0	300	ERROR
10	350	35
20	400	20
30	450	15
40	500	12.5
50	550	11

X=

$Y1 = 300 + 5X$, $Y2 = (300 + 5X)/X$

b. The total cost is $300, but the average cost is undefined because of division by 0.

c. The total cost increases at a constant rate. The average cost decreases, but the rate of decrease slows.

d.

X	Y1	Y2
100	800	8
1000	5300	5.3
10000	50300	5.03
100000	500300	5.003

X=

$Y1 = 300 + 5X$, $Y2 = (300 + 5X)/X$

The total cost continues to increase at a constant rate. The average cost approaches $5 per item.

8.

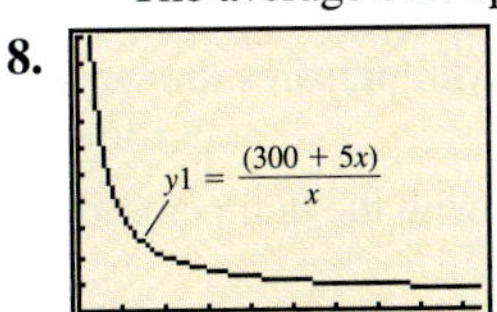

It is a downward-sloping curve in the first quadrant. When x is close to 0, the curve falls steeply. Then it appears to level off.

9. Either graph with a MODE of Dot or graph with a friendly window such as $^{-}9.4 \le X \le 9.4$, $^{-}6.2 \le Y \le 6.2$.

10. a. The graph of $f(x) = \frac{1}{x}$ is shifted left 3 units.

b. The graph of $f(x)$ is shifted down 2 units.

c. The graph of $f(x)$ is reflected across the x-axis.

d. The graph of $f(x)$ is stretched away from the x-axis.

e. The graph of $f(x)$ is shifted right 2 units and shifted up 1 unit.

11. a. Horizontal asymptote: x-axis; vertical asymptote: $x = {}^{-}3$

b. Horizontal asymptote: $y = {}^{-}2$; vertical asymptote: y-axis

c. Horizontal asymptote: x-axis; vertical asymptote: y-axis

d. Horizontal asymptote: x-axis; vertical asymptote: y-axis

e. Horizontal asymptote: $y = 1$; vertical asymptote: $x = 2$

12. a. Domain: $\{x \mid x \neq {}^{-}3\}$; range: $\{y \mid y \neq 0\}$

b. Domain: $\{x \mid x \neq 0\}$; range: $\{y \mid y \neq {}^{-}2\}$

c. Domain: $\{x \mid x \neq 0\}$; range: $\{y \mid y \neq 0\}$

d. Domain: $\{x \mid x \neq 0\}$; range: $\{y \mid y \neq 0\}$

e. Domain: $\{x \mid x \neq 2\}$; range: $\{y \mid y \neq 1\}$

13. $x \approx {}^{-}.67$ or $x = 3$

14. $x = \frac{{}^{-}2}{3}$ or $x = 3$

15. a

16. a.

	Work	Time	Rate
Thaddeus	1 dinner	45 min	$\frac{1}{45}\,\frac{\text{dinner}}{\text{min}}$
Lynn	1 dinner	x min	$\frac{1}{x}\,\frac{\text{dinner}}{\text{min}}$
Together	1 dinner	20 min	$\frac{1}{20}\,\frac{\text{dinner}}{\text{min}}$

b. $\frac{1}{45} + \frac{1}{x} = \frac{1}{20}$

c. $x = 36$; working alone, Lynn prepares dinner in 36 minutes.

17. The Jacuzzi fills in 6 minutes with the drain closed.

18. a. $x - 3$

b. $\frac{2}{3} = \frac{9}{x-3}$

c. $x = \frac{33}{2} = 16.5$; the speed of the kayak in still water is 16.5 km/h.

d. The kayak returns downstream in $\frac{6}{13} \approx .46$ hours, or about 28 minutes.

19. The wind speed is 10 miles/hour.

20. a. As the price per gallon increases, the number of gallons decreases. The number of gallons decreases quickly at first; then the decrease slows.

b. $a = 10$. The variation constant is the $10 fixed amount you have to spend on the gas. It is a product, price per gallon $(x) *$ number of gallons (y).

c. Gallons $= \frac{10}{\text{Price}}$ or $y = \frac{10}{x}$

d. You can purchase approximately 6.9 gallons of gasoline when the price is $1.45 per gallon.

21. a. $a = 24,000$ b. $y = 60$

22. a. $a = 3, y = .5$ b. $a = 12, x = 15$ c. $a = 90, y = 180$ d. $a = 250, x \approx 107.14$

23. a. $a = 1000$

b. The cost is approximately $142.86 per person when 7 students share the cottage.

24. The jogger completes the lap in approximately 122.7 seconds when his speed is 11 feet per second.

25. The astronaut weighs 16.4 pounds more on the earth than she does in an orbit 200 miles above the earth.

26. a. $x - 8$

b. $x - 8 + \frac{4}{x+5}$

c. $x - 8 - \frac{4}{x+5}$

27. a. $2x^2 + x - 5 + \frac{12}{x+2}$

b. $2x^2 - x + 3$

c. $x^3 + x^2 - x + 2 - \frac{3}{3x-1}$

Chapter 11 Test

1. $\dfrac{6x + 17}{(x + 2)(x + 3)}$

2. $\dfrac{^-2x + 2}{(x - 2)(x + 2)(x - 4)}$

3. $\dfrac{x + 1}{x}$

4. $\dfrac{^-x + 6}{x + 5}$

5. a. The graph of $f(x)$ is shifted right 4 units and up 5 units.

b. Horizontal asymptote: $y = 5$; vertical asymptote: $x = 4$

c. Domain: $\{x | x \neq 4\}$; range: $\{y | y \neq 5\}$

6. $x = \frac{^-8}{3}$ or $x = 1$

7. Working alone, Ron completes the job in 10 hours.

8. The sieve fills in 4 hours when material is not allowed to pass through.

9. The speed of the motorboat is 9 miles/hour in still water.

10. $y = \frac{1}{6}$

11. The volume is 4.24 L.

12. The intensity of the light is 12 lux at a distance of 5 m.

Chapter 12 Inequalities

12.1 Exercises

1. a. Answers may vary; $x = {^-1}, 2.5, 3, 5\frac{1}{4}$.

b. Yes. The phrase *never gets farther* than 4 blocks from her house means we include the endpoints of $^-2$ and 6.

3. a. Answers may vary; 250 min and 300 min

b. $x = 200$ min is not a solution and $x = 400$ min is a solution.

5. a. $x < 4$ b. $x > 4$ c. $x \leq 4$ d. $x \geq 4$ e. $x \leq 4$ f. $x \geq 4$ g. $x \leq 4$ h. $x \geq 4$ i. $x > 4$ j. $x < 4$

7. a. [number line: $^-5$, 20]

b. [number line: $^-5$, 20]

9. Multiplying or dividing an inequality by a negative number reverses the inequality symbol.

a. $x \geq {}^-6$

b. $x < 14$

11. $x \geq \frac{^-3}{8}$

13. $\{x \mid \frac{8}{3} < x \leq 5\}$

$\frac{1}{2}$ $\frac{7}{4}$

15. a. $80 + 2L$

b. $200 \leq 80 + 2L$ and $80 + 2L \leq 300$

c. and

d. $60 \leq L$ and $L \leq 110$. The length of the rectangle must be at least 60 m and no more than 110 m.

12.1 Skills and Review

17. $x = 8$

19. $x = 2$ or $x = \frac{8}{5}$

21. $y = {}^-1$

23. $x \approx 1.26$

25. a. $P = \$0$ or $P = \$20$

b. Vertex: (10, 320)

c. Opens downward

d.

Vertex (10, 320)

$R = {}^-3.2p^2 + 64p$

(0, 0) (20, 0)

12.2 Exercises

1. a. At .55 second the fireworks have just come into good view. At 5.70 seconds the fireworks have just gone out of good view.

b. In the interval between 0 and .55 second, the fireworks are still below the level of the balcony.

c. In the interval between 5.70 and 6.25 seconds, the fireworks have fallen below the level of the balcony.

3.

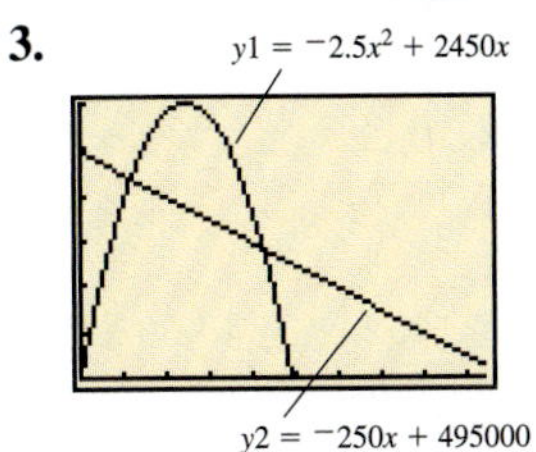

Acme will lose money when their widget is priced less than \$234 or more than \$846. In set-builder notation, $\{x \mid 0 \leq x < 234 \text{ or } 846 < x\}$.

5.

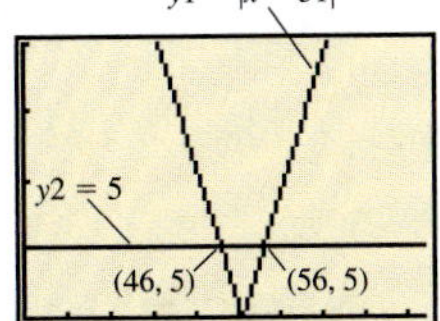

Percentages less than 46% or more than 56% are outside this candidate's margin of error. In set-builder notation, $\{x \mid 0 \leq x < 46 \text{ or } 56 < x \leq 100\}$.

7. $\{x \mid 0 \leq x \leq 4\}$

9. a. $\{x \mid x < {}^-4\}$; substitution of $x = {}^-5$, true; $x = 0$, false.

b. $\{x \mid x \geq {}^-4\}$; substitution of $x = 0$, true; $x = {}^-5$, false.

11. $\{x \mid x < {}^-3.39 \text{ or } 7.39 < x\}$

13. $\{x \mid x \leq {}^-2.24 \text{ or } 2.24 \leq x\}$

15. $\{x \mid x < {}^-2 \text{ or } 2 < x\}$

12.2 Skills and Review

17. $-\frac{7}{2}$ $-\frac{1}{2}$

19. $x = \frac{2}{3} \approx .67$

21. $x = 25$

23. a. $f(g({}^-2)) = 50$

b. $f(g(x)) = 9x^2 - 6x + 2$

25. $x = {}^-4 + \sqrt{5}$ or $x = {}^-4 - \sqrt{5}$

12.3 Exercises

1. a, d

3. Answers may vary; two solutions: (0, 99), (20, 60), two nonsolutions: (0, 100), (49, 50).

5. a. ii **b.** i **c.** iii

7. $y < {}^-x - 2$; $y \leq 2x + 4$; $y \geq {}^-4$

9. $y < 4$; $y \geq {}^-4$; $x < {}^-1$; $x \geq {}^-5$

11.

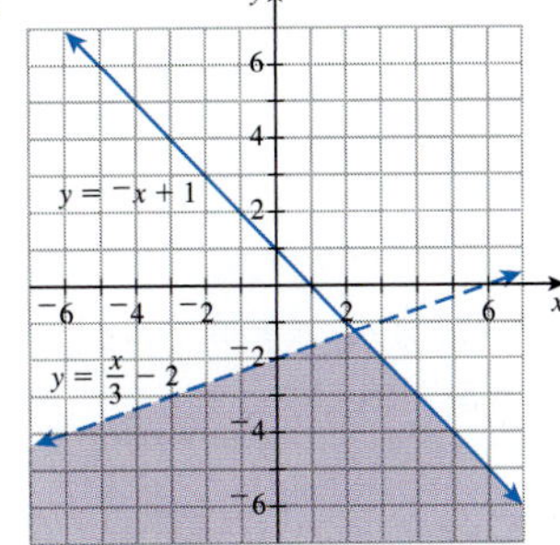

13.

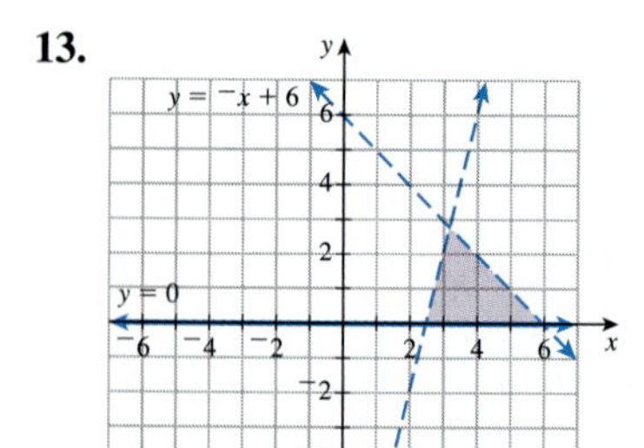

15. a. $x + y \leq 1$; $x \geq \frac{1}{4}$; $y < 2x$; $y \geq 0$

b.

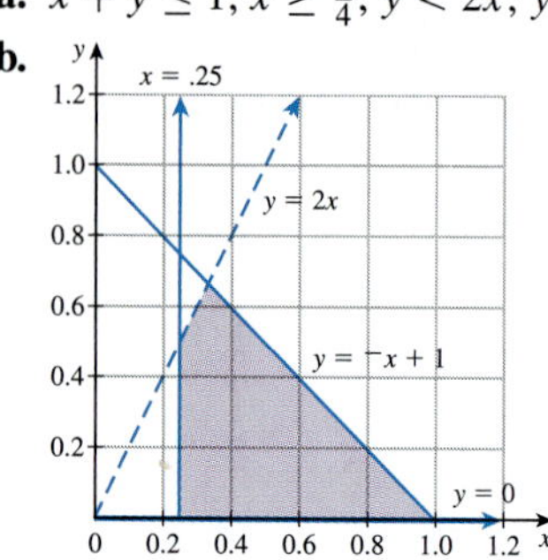

12.3 Skills and Review

17. $\{x \mid {}^{-}3 \leq x \leq 11\}$

19. a. Let x = the number of hours it takes the first painter to complete the job alone.

	Work	Time	Rate
First painter	1 job	x hours	$\frac{1}{x}$ job/hour
Second painter	1 job	$x + 2$ hours	$\frac{1}{x+2}$ job/hour
Together	1 job	5 hours	$\frac{1}{5}$ job/hour

b. $\frac{1}{x} + \frac{1}{x+2} = \frac{1}{5}$

c. $x \approx 9.1$ or $x \approx {}^{-}1.1$

d. It takes the first painter 9 hours to complete the job alone.

21. $\dfrac{1}{(x+5)(x-4)}$

23. $x = 2$

25. a. y-intercept: $(0, {}^{-}14)$

b. x-intercepts: $({}^{-}2, 0)$, $(7, 0)$

c. Vertex: $\left(\frac{5}{2}, \frac{{}^{-}81}{4}\right) = (2.5, {}^{-}20.25)$

d.

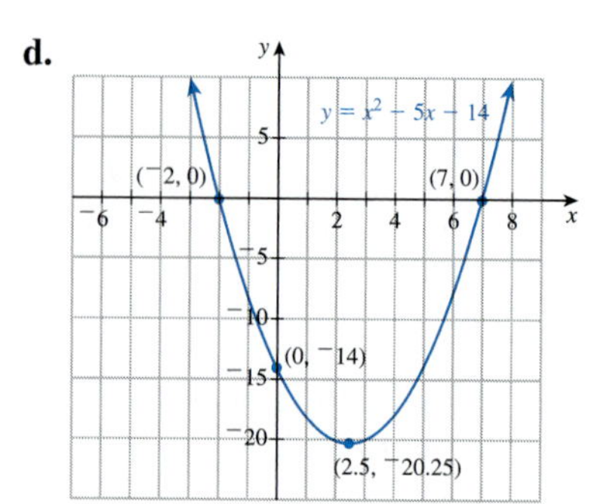

Chapter 12 Exploration

a. $x + y \leq 90$

b. $5x + 8y \geq 500$

c. $y \leq 30$

d.

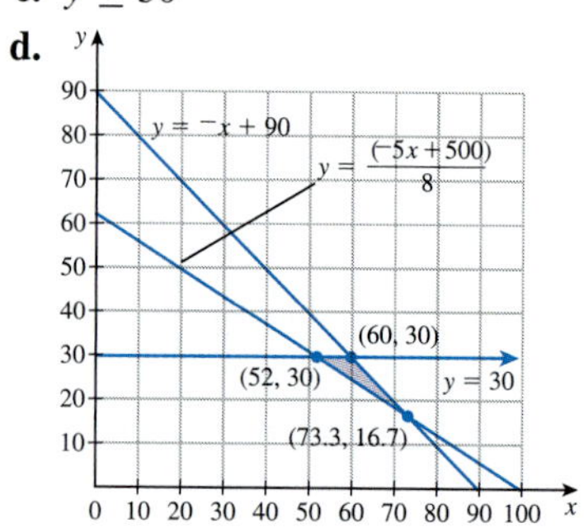

e. $(52, 30)$, $(60, 30)$, $\left(73\frac{1}{3}, 16\frac{2}{3}\right)$

Chapter 12 Review Exercises

1. $\{x \mid x \geq {}^{-}3\}$

2. a. $\{x \mid x > 4\}$

b. $\{x \mid x \leq 1\}$

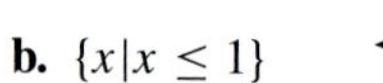

c. $\{x \mid x \leq {}^{-}2\}$

d. $\{x \mid x > 3\}$

e. $\left\{x \mid x < \frac{1}{3}\right\}$

f. $\{x \mid x \leq 1\}$

3. ${}^{-}2$ ... 5

4. a. $\{x \mid 3 \leq x \leq 7\}$

b. $\{x \mid x < {}^{-}3 \text{ or } 2 \leq x\}$

c. $\{x \mid {}^{-}12 \leq x < {}^{-}5\}$

d. $\{x < {}^{-}1 \text{ or } x > 4\}$

e. $\{x \mid {}^{-}2 \leq x < 5\}$

f. $\{x \mid 4 < x \leq 7\}$

5. $x \geq 100°\text{F}$

6. $\{x|x < 32 \text{ or } 75 < x\}$

7. a. $\frac{18+20+15+x}{4} = \frac{53+x}{4}$

b. $19 \leq \frac{53+x}{4}$

c. $23 \leq x$; the player must score at least 23 points in the fourth game to average at least 19 points per game.

d. 23

8. a. $50 \leq 125 - 5x$ and $125 - 5x \leq 100$

b. $5 \leq x$ and $x \leq 15$; the cover charge must be at least \$5 and no more than \$15.

c. $\{x|5 \leq x \leq 15\}$

9. a. $^-4$ 2

b. $^-4$ 2

10. $\{x|.43 < x < 5.82\}$; substitution of $x = 3$, true; $x = 0$, false; $x = 6$, false.

11. a. 3 **b.** 1 **c.** 2

12. a. $\{x|^-3 < x < 3\}$ **b.** $\{x|x < 0 \text{ or } 8 < x\}$

c. $\{x|^-4 \leq x \leq 3\}$ **d.** $\{x|x \leq {}^-4 \text{ or } 10 \leq x\}$

e. $\{x|x < 4\}$ **f.** $\left\{x|0 \leq x \leq \frac{25}{4}\right\}$

13. Any price between \$13.59 and \$30.25 generates revenue above \$3000.

14. Remove grapefruits less than 3 inches in diameter or greater than 7 inches in diameter.

15. $\{s|200 \leq s \leq 225\}$; the sides of the square lot are at least 200 feet and no more than 225 feet.

16. Only a and c are solutions.

17. Only b and d are solutions.

18.

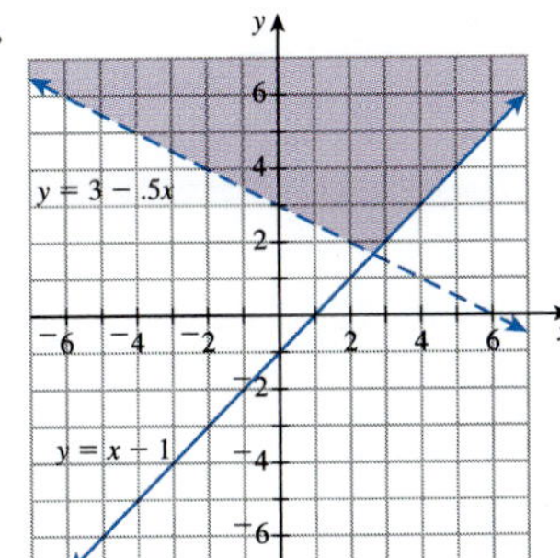

19.

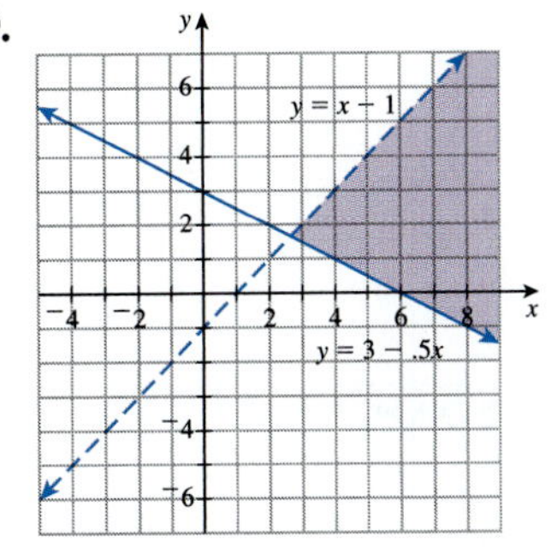

20.

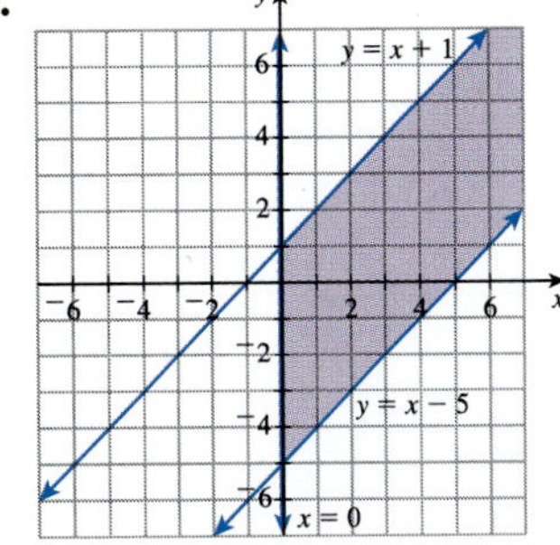

21.

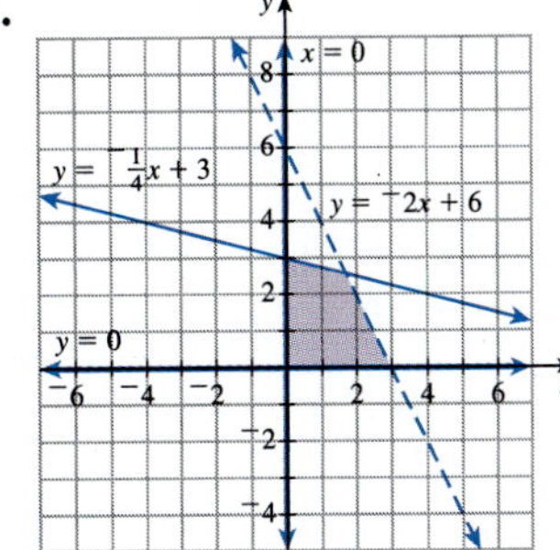

22. a. $x + y > 100$; $x + y < 150$; $y \geq 3x$

b.

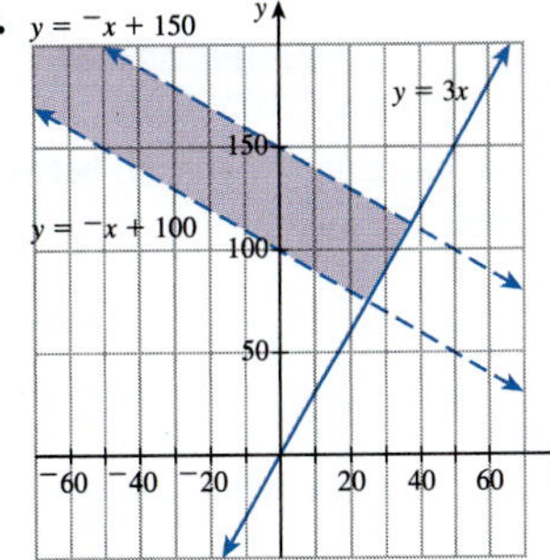

c. Answers may vary; (25, 100) and ($^-40$, 150) are solutions. (75, 50) and (50, 0) are not solutions.

23. a. $2x + 2y > 60$; $2x + 2y \leq 100$; $y \leq 2.5x$; $y \geq 0$; $x \geq 0$

b.

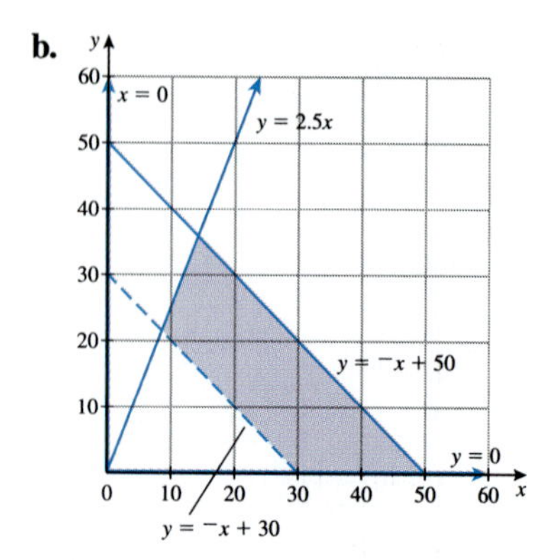

c. (6, 30) is not a solution.

24. **a.** $30x + 20y \le 200$

b. $y \ge 2x$

c. There can't be a negative number of pants or shirts.

d.

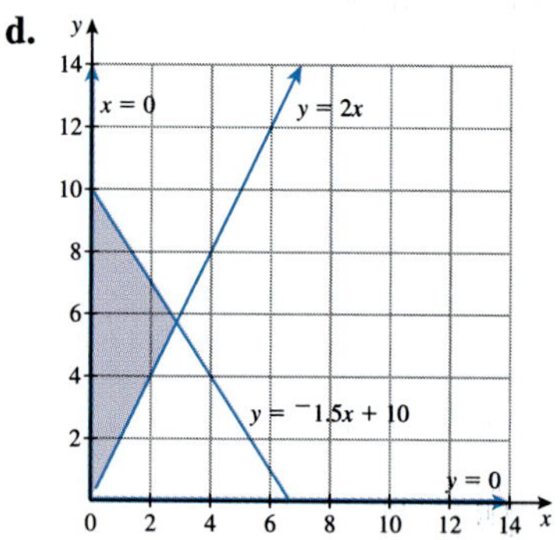

e. (0, 10), (1, 8), (2, 7)

Chapter 12 Test

1. ⁻3 8 $\{x \mid {}^{-}3 \le x < 8\}$

2. $\{x \le {}^{-}3 \text{ or } 5 < x\}$

3. $\{x < 2\}$; the opposing team may score 0 or 1 goal.

4. $40 \le x$ and $x \le 50$; the carpenter will need between 40 and 50 hours to complete the job.

5. $\{x \mid x > 1.59\}$

6. $\{x \mid {}^{-}.51 \le x \le 3.11\}$

7. $\{x \mid x < {}^{-}7 \text{ or } 2 < x\}$

8. The object is at least 40 m high anytime between 1.43 s and 4.29 s after release.

9.

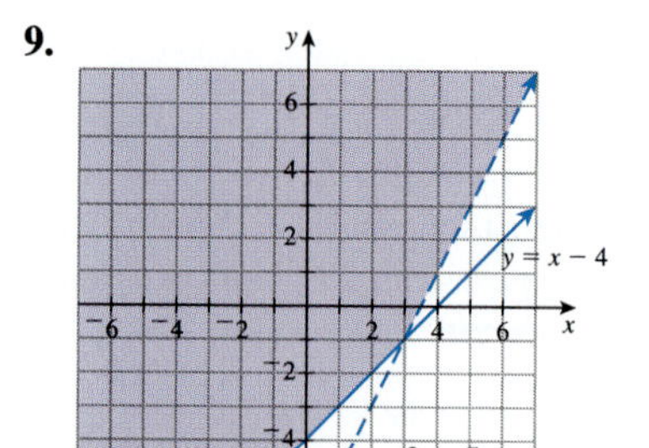

10.

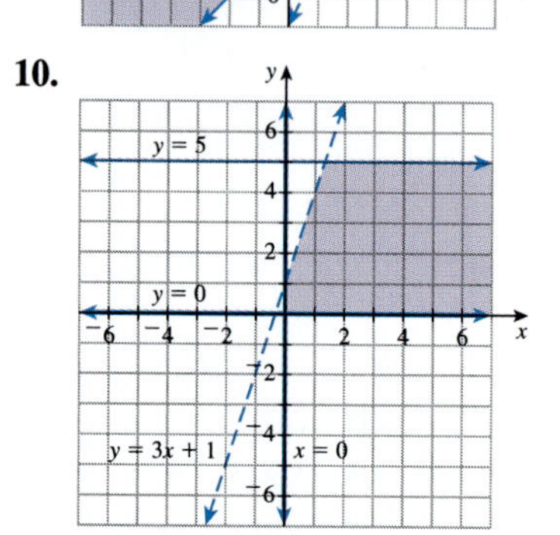

11. **a.** $x + y \le 250$; $y \ge 4x$; $x > 0$; $y > 0$

b.

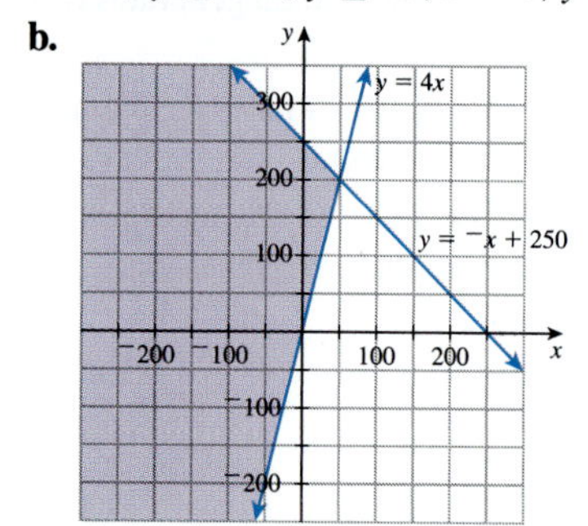

c. (10, 240) is a solution.

12.

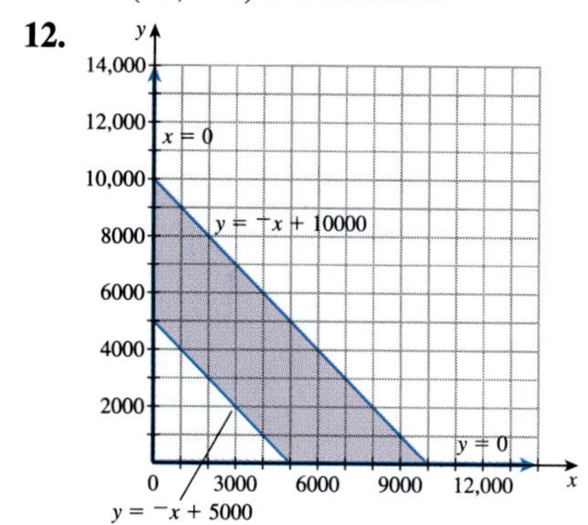

13. a. $x + y < 8$; $40x + 60y \geq 360$; $x \geq 0$; $y \geq 0$

b.

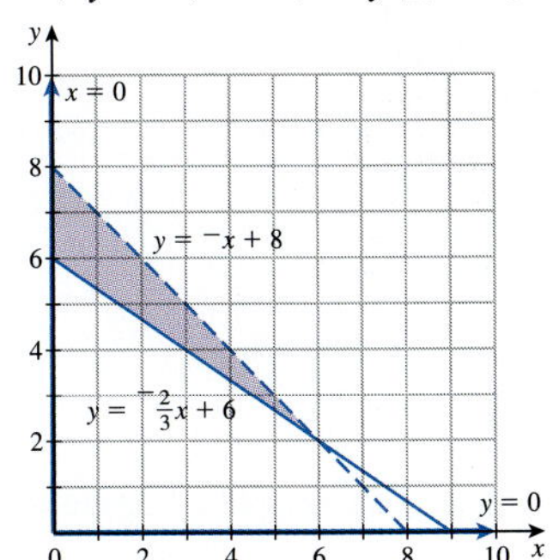

c. (4.5, 3) is a solution.

Index

Library of Functions

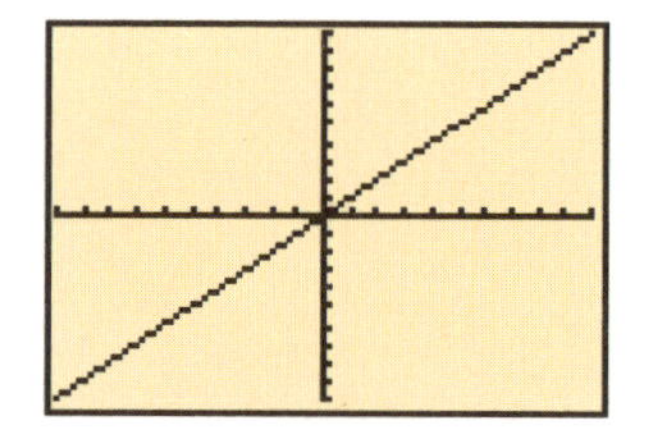

Basic Linear ($y = x$)

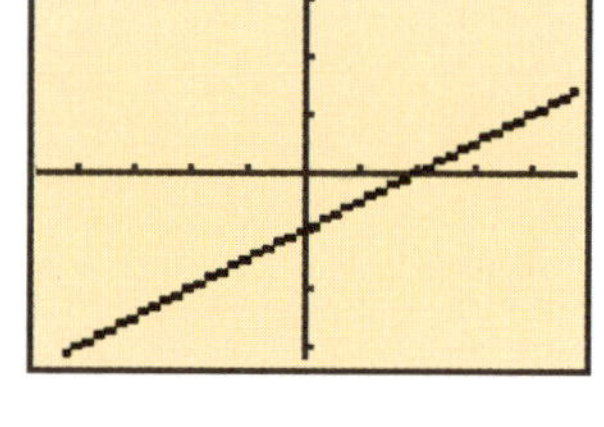

Linear ($y = mx + b$)

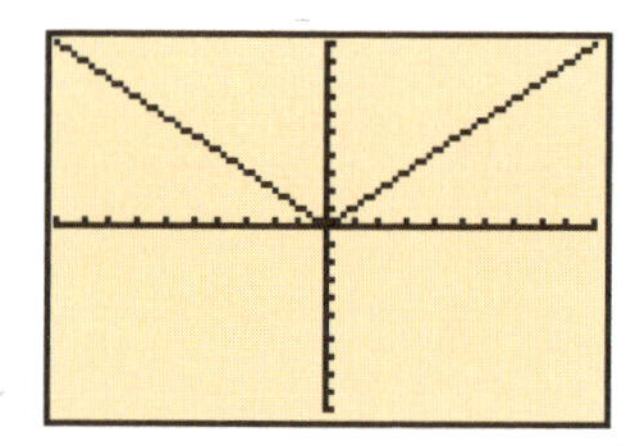

Absolute Value ($y = |x|$)

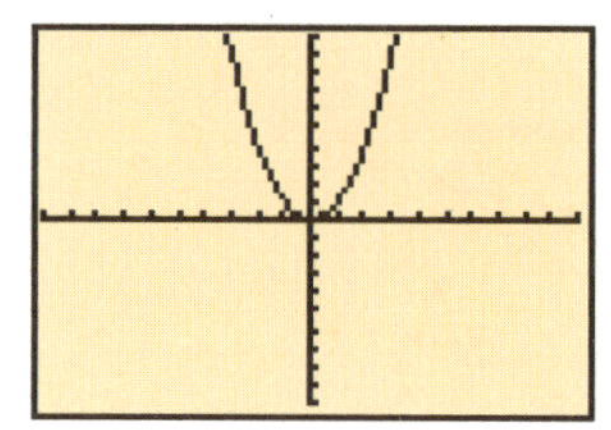

Basic Quadratic ($y = x^2$)

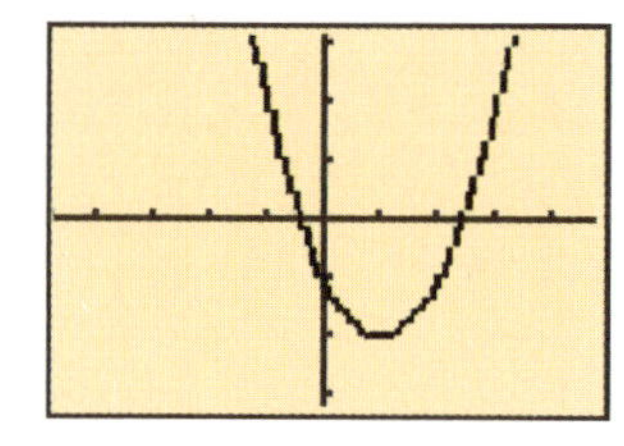

Quadratic ($y = ax^2 + bx + c$)

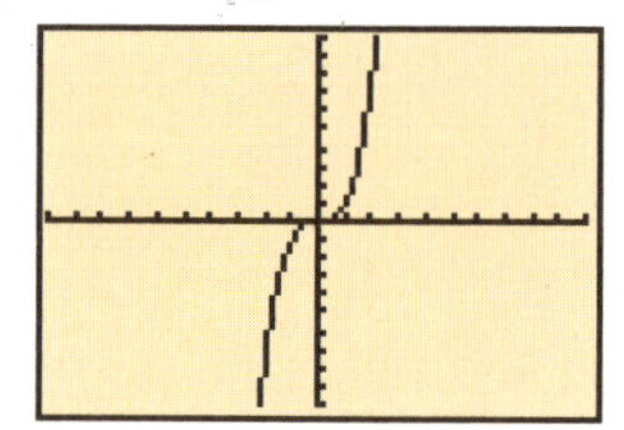

Basic Cubic ($y = x^3$)

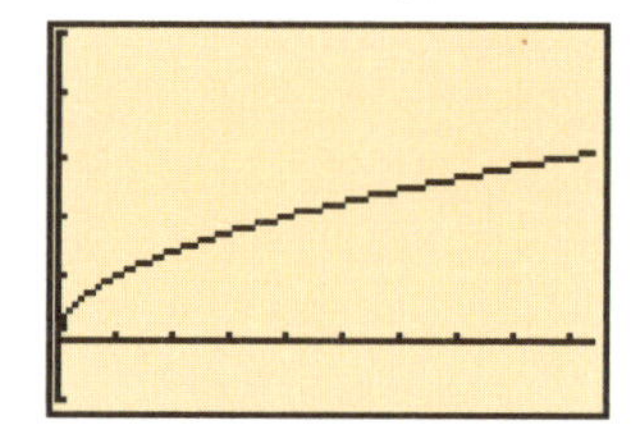

Square Root ($y = x^{1/2}$)

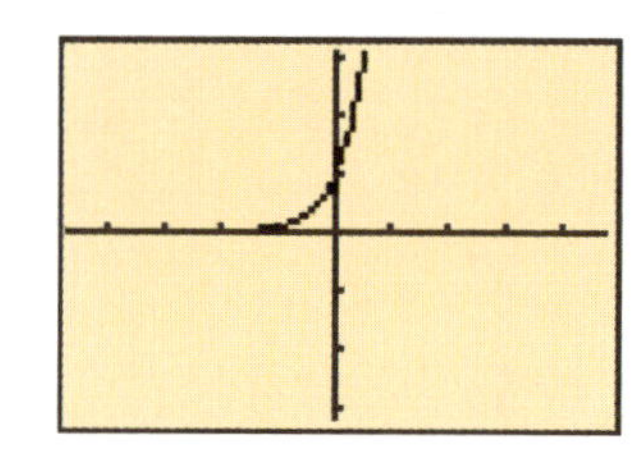

Exponential ($y = 10^x$)

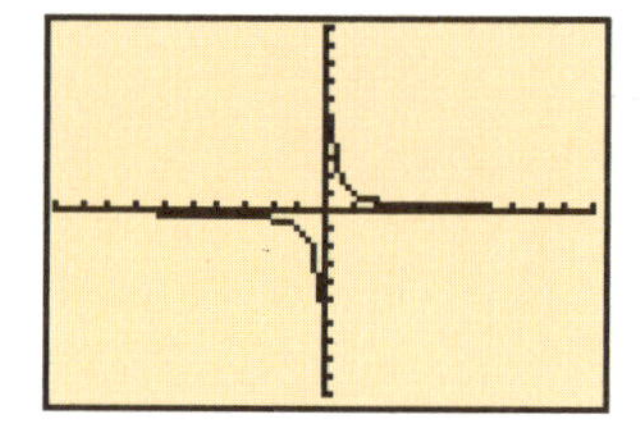

Basic Reciprocal ($y = 1/x$)

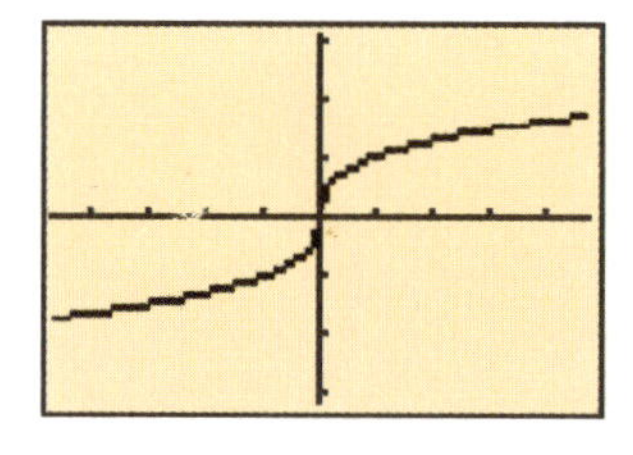

Cube Root ($y = x^{1/3}$)

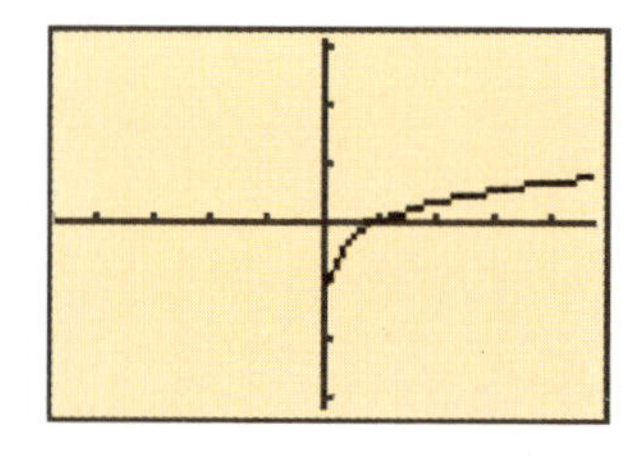

Logarithmic ($y = \log x$)

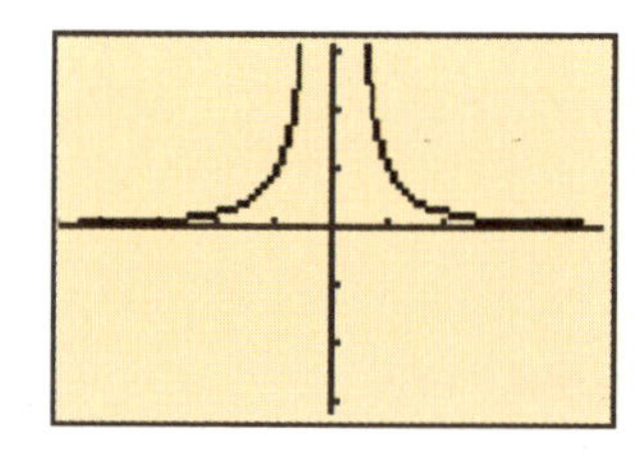

Inverse Square ($y = 1/x^2$)